W0049752

Operator Theory
Advances and Applications
Vol. 95

Editor:
I. Gohberg

Editorial Office:
School of Mathematical
Sciences
Tel Aviv University
Ramat Aviv, Israel

Editorial Board:
J. Arazy (Haifa)
A. Atzmon (Tel Aviv)
J. A. Ball (Blackburg)
A. Ben-Artzi (Tel Aviv)
H. Bercovici (Bloomington)
A. Böttcher (Chemnitz)
L. de Branges (West Lafayette)
K. Clancey (Athens, USA)
L. A. Coburn (Buffalo)
K. R. Davidson (Waterloo, Ontario)
R. G. Douglas (Stony Brook)
H. Dym (Rehovot)
A. Dynin (Columbus)
P. A. Fillmore (Halifax)
C. Foias (Bloomington)
P. A. Fuhrmann (Beer Sheva)
S. Goldberg (College Park)
B. Gramsch (Mainz)
G. Heinig (Chemnitz)
J. A. Helton (La Jolla)
M.A. Kaashoek (Amsterdam)
T. Kailath (Stanford)
H.G. Kaper (Argonne)
S.T. Kuroda (Tokyo)
P. Lancaster (Calgary)
L.E. Lerer (Haifa)
E. Meister (Darmstadt)
B. Mityagin (Columbus)
V. V. Peller (Manhattan, Kansas)
J. D. Pincus (Stony Brook)
M. Rosenblum (Charlottesville)
J. Rovnyak (Charlottesville)
D. E. Sarason (Berkeley)
H. Upmeier (Marburg)
S. M. Verduyn-Lunel (Amsterdam)
D. Voiculescu (Berkeley)
H. Widom (Santa Cruz)
D. Xia (Nashville)
D. Yafaev (Rennes)

Honorary and Advisory
Editorial Board:
P. R. Halmos (Santa Clara)
T. Kato (Berkeley)
P. D. Lax (New York)
M. S. Livsic (Beer Sheva)
R. Phillips (Stanford)
B. Sz.-Nagy (Szeged)

Topics in Interpolation Theory

Edited by

H. Dym
B. Fritzsche
V. Katsnelson
B. Kirstein

Springer Basel AG

Editors:

H. Dym
V. Katsnelson
Department of Theoretical Mathematics
The Weizmann Institute of Science
Rehovot 76100
Israel

B. Fritzsche
B. Kirstein
Mathematisches Institut
Universität Leipzig
04109 Leipzig
Germany

1991 Mathematics Subject Classification 47A57, 30E05, 42A82

Library of Congress Cataloging-in-Publication Data
Topics in interpolation theory / edited by H. Dym ... [et al.].
p. cm. – (Operator theory, advances and applications : vol. 95)
Includes bibliographical references (p. –) and index.
ISBN 978-3-0348-9838-6 ISBN 978-3-0348-8944-5 (eBook)
DOI 10.1007/978-3-0348-8944-5
1. Interpolation. I. Dym, H. (Harry), 1938– . II. Series: Operator theory. advances and applications : v. 95.
QA281.T66 1997
511´.42–dc21 97–8157
CIP

Deutsche Bibliothek Cataloging-in-Publication Data
Topics in interpolation theory / ed. by H. Dym ... – Basel ; Boston ; Berlin : Birkhäuser, 1997
(Operator theory ; Vol. 95)
ISBN 978-3-0348-9838-6

NE: Dym, Harry [Hrsg.]; GT

This work is subject to copyright. All rights are reserved, whether the whole or part of the material is concerned, specifically the rights of translation, reprinting, re-use of illustrations, recitation, broadcasting, reproduction on microfilms or in other ways, and storage in data banks. For any kind of use permission of the copyright owner must be obtained.

© 1997 Springer Basel AG
Originally published by Birkhäuser Verlag, Basel, Switzerland in 1997
Softcover reprint of the hardcover 1st edition 1997
Printed on acid-free paper produced from chlorine-free pulp.∞
Cover design: Heinz Hiltbrunner, Basel

ISBN 978-3-0348-9838-6
9 8 7 6 5 4 3 2 1

Table of contents

Vladimir Petrovich Potapov

Welcoming remarks

It is a great pleasure to welcome you all to our meeting 'Recent Developments in Schur Analysis, A Workshop in Honour of the 80th Birthday of V.P. Potapov'. Many of you have crossed thousands of miles to be with us and we appreciate it very much. We have participants from Northern America, Asia and Europe who have gathered here. We are happy that many mathematicians from Eastern Europe, mainly from the former Soviet Union, can now participate. It was not easy to organize this meeting. Much work, imagination and patience was necessary. However, finally, and a little bit unexpectedly, we succeeded in finding a sufficient number of sources to support our conference. In this connection we thank the following institutions for contributing to the support of our workshop:

The Saxonian Ministry of Science and Arts.
Deutscher Stifterverband.
German Society of the Friends of the Weizmann Institute.
The Rector of The University of Leipzig.
The Center for Higher Studies of The University of Leipzig.

The conference enjoyed strong support from all levels of Leipzig University. This is the fifth Schur analysis conference which is held at Leipzig University so that one could say there is now some sense of tradition. Our first Leipzig workshop in Schur analysis took place in the period October 16–20, 1989. This workshop was aimed at building bridges between mathematicians from the Soviet Union and the Western world. From the very beginning the Leipzig Schur analysis group was mainly influenced by the approaches worked out in Odessa and Kharkov. For this reason, we felt a need to propagate and popularize the ideas of our Ukrainian teachers by bringing them together with a circle of interested people from Western Europe and Northern America. All these aims had to be realized under socialist conditions or, more precisely, under the usual system of restrictions which were typical for science in the German Democratic Republic which was completely dominated by the Communist Party. Since the intersection of the Leipzig Schur analysis group with the Communist Party was empty it was not easy for us to organize a scientific conference at Leipzig University. However, we were able to find some reasonable and helpful members in the leadership of our Department of Mathematics at that time who were willing to support our efforts and were powerful enough to carry such an activity through. Finally, we were lucky that all that we hoped for could be arranged. From the present point of view this period in October 1989 turned out to be the beginning of the drastic changes which led to the reunification of

Germany. Although the political situation has completely changed, the main aim of the Schur analysis workshop in Leipzig remains the same, namely bringing together mathematicians from the former Soviet Union and Eastern Europe with the Western mathematical community.

This workshop, which was planned and prepared in close connection with the Department of Theoretical Mathematics of the Weizmann Institute, represented by Professors Viktor Emanuelovich Katsnelson and Harry Dym, is dedicated to the memory of an outstanding mathematician and strong personality: Vladimir Petrovich Potapov. He was born on January 24, 1914. This means that he would have been 80 years old this year. This was the motivation for dedicating this workshop to Vladimir Petrovich. He worked on the border between complex analysis and operator theory; he made important contributions to both areas, and influenced their subsequent development. Potapov did not begin his university studies in mathematics. First he studied three years at the Odessa Conservatory. There he made friends with the later world famous pianist Emil Gilels and became acquainted with Svjatoslav Richter and Mstislav Rostropovic. In 1934 he gave up his studies in music and took the entrance examinations to the Faculty of Physics and Mathematics of Odessa University. He passed these examinations brilliantly and was accepted as a student. It is a pleasure to have with us today Professor Michael Samoilovich Livšic who entered the university one year before Vladimir Petrovich. The two, having common scientific interests, became close friends. Those who influenced Vladimir Petrovich and Michael Samoilovich most during their university years were Mark Grigor'evich Krein and Boris Yakovlevich Levin. At that time the now famous Odessa school of functional analysis sprung up around Mark Grigor'evich Krein. Besides Livšic and Potapov, the younger participants included Naimark, Glazman, Rutman, Milman, Artemenko, Brodskii, Orlov and others. In 1939 Vladimir Petrovich became a Ph.D. student of B.Ya. Levin. He studied the divisors of almost periodical polynomials. In 1945 he defended his first thesis. His supervisor, B.Ya. Levin always rated it higher than Vladimir Petrovich's subsequent fundamental works. In 1944 Michael Samoilovich defended his second thesis and was working out the theory of non-unitary and nonselfadjoint operators which was generally recognized to be of fundamental importance by mathematicians all over the world.

The concept of the characteristic matrix function was the basic corner stone of the theory. Under the influence of Michael Samoilovich, the analysis of matrix functions became the focal point of Potapov's research. In the fifties he worked out the factorization theory of J-contractive matrix functions. In 1954 Vladimir Petrovich defended his second thesis. In 1955 it was published in "Trudy Moskovskogo Matematicheskogo Obshchestva". This work laid down the fundamentals of J-theory. Let me quote some sentences from the contribution of I.E. Ovcharenko to the Potapov volume which was recently published in the OT series [OT72]. He writes: M.G. Krein often regretted that, when M.S. Livšic moved to Kharkov the cooperation between the "operator expert" M.S. Livšic and the "analysis expert"

V.P. Potapov was broken. "Livšic and Potapov are a tremendous power when they are together", Krein said.

Potapov's theory gave a new approach to solving problems in the theory of electrical networks and in the theory of interpolation. During his last years together with his pupils he worked out the so-called fundamental matrix inequality method to treat matrix versions of classical interpolation and moment problems of the Nevanlinna-Pick and Hamburger type. Potapov's excellent qualities were illustrated most sharply in personal contacts with his students. He not only appreciated cleverness and ability in others highly, but he also felt a persistent need to help them. He was a teacher, an advisor, and a friend to each of his students. Even when he left Odessa he kept himself informed about the scientific and private life of his students. He was proud of their scientific achievements. The V.P. Potapov Memorial Session in the Afternoon Session on Wednesday will bring us more detailed reminiscences about Vladimir Petrovich Potapov.

Finally, I would like to say a few words about the organization of the conference itself. It was first planned as a little workshop for a small circle of invited people working mainly in interpolation theory. The original main aim was to survey the actual state of affairs in matrix and operator interpolation. While working on the organization, the list of topics expanded and the plan grew by itself and crossed the borders of the initial plan, even when we restricted it as much as possible. Moreover, a lot of mathematicians, mainly from the former Soviet Union, heard about our conference and asked for invitations. We endeavoured to find the optimal variant. We tried to avoid parallel sessions and in this we succeeded, but we have to pay for it with a very heavy schedule.

Again, let me welcome you all to the conference and wish you all a fruitful, pleasant and enjoyable stay in Leipzig.

Bernd Kirstein

Editorial introduction

About one half of the papers in this volume are based on lectures which were presented at a conference at Leipzig University in August 1994, which was dedicated to Vladimir Petrovich Potapov. He would have been eighty years old. These have been supplemented by:

(1) Historical material, based on reminiscences of former colleagues, students and associates of V.P. Potapov.
(2) Translations of a number of important papers (which serve to clarify the Potapov approach to problems of interpolation and extension, as well as a number of related problems and methods) and are relatively unknown in the West.
(3) Two expository papers, which have been especially written for this volume.

For purposes of discussion, it is convenient to group the technical papers in this volume into six categories. We will now run through them lightly, first listing the major theme, then in parentheses the authors of the relevant papers, followed by discussion. Some supplementary references are listed at the end; OT72 which appears frequently in this volume, refers to Volume 72 in the series Operator Theory: Advances and Applications. It was dedicated to V.P. Potapov.

1. Multiplicative decompositions
(Yu.P. Ginzburg; M.S. Livšic, I.V. Mikhailova; V.I. Smirnov).

This is a good place to begin because M.S. Livšic's investigations in the theory of nonselfadjoint operators with finite dimensional imaginary part generated a need for a factorization theorem for matrix valued functions which are contractive with respect to a signature matrix J in some appropriately chosen region in the complex plane (usually the disc or the upper half-plane) analogous to the Smirnov factorization for scalar functions of bounded Nevanlinna characteristic. This was an important stimulation for Potapov's research. The development of the theory of nonselfadjoint operators is described in the paper of Livšic; see also the papers [LP] and [P], which are listed at the end of this survey, for more information. The paper by Smirnov, which is neither well known nor readily available in the West, seems to be the first study of inner-outer factorization for scalar functions.

The paper by Ginzburg (which extends his paper with Svechuk in OT72) investigates the problem of recovering an operator valued function from the values of its modulus on the boundary, which is an important problem in factorization theory. This paper also contains many useful references. The paper of Mikhailova establishes the uniqueness of the multiplicative decomposition of a J-inner matrix valued function in the special case when all the factors are of the form $\exp(i\lambda A)$, where J and A are 2×2 matrices such that $JA \geq 0$.

2. Fundamental matrix inequalities (V.K. Dubovoj; Yu.M. Dyukarev; L.B. Golinskii and I.V. Mikhailova; V.E. Katsnelson; I.V. Kovalishina).

The Potapov approach to interpolation is to construct a FMI (Fundamental Matrix Inequality) for each problem in such a way that the set of solutions of the FMI coincides with the set of solutions of the original interpolation problem. This method is effective for a large class of problems which include the bitangential Nevanlinna-Pick problem, moment problems and problems of integral representation for assorted classes of functions and matrix valued functions.

The papers referred to above give some idea of the wide scope of this approach. In particular, Dyukarev considers an abstract moment problem in the Stieltjes class. This leads to two FMI's which are related by a coupling identity, which insures that they are compatible.

I.V. Kovalishina uses the Potapov method to investigate a problem of M.G. Krein on the existence of extensions of continuous positive definite kernels of a special form, which are specified in terms of the values of a scalar function that is given on a finite interval.

Dubovoj considers the description of the set of solutions to a matrix version of an interpolation problem of the type considered by Schur, wherein the first $n+1$ matrix coefficients in the power series expansion of a matrix valued function which is both analytic and contractive in the unit disc are specified. In this paper, the underlying Pick matrix is nonnegative but singular.

The paper by Golinskii and Mikhailova uses the Potapov approach to investigate the existence of a Parseval formula for a generalized Fourier transform which is associated with a family of reproducing kernel Hilbert spaces of entire functions. Such spaces were originally introduced by L. de Branges (in a far-reaching generalization of the Paley-Wiener spaces which are associated with classical Fourier analysis). Some matrix versions of this theory were considered by de Branges in [dB3]; see also [DI] and [S] for additional information, references and applications.

In order to extract information from the FMI it is often necessary to transform it into a more revealing form. In most of the literature on the subject, this is done in an ad hoc and mysterious fashion. The reasoning behind this transformation is explained in great detail on a pair of concrete examples in the expository paper by Katsnelson.

3. Canonical systems of differential equations (M.G. Krein; I.V. Mikhailova and V.P. Potapov; L.A. Sakhnovich).

An important special case of the Potapov multiplicative decomposition theory for J-contractive matrix valued functions is the special case of entire J-inner matrix valued functions. In this case, the multiplicative representation is equivalent to identifying the given (entire J-inner) matrix valued function as the monodromy matrix of a canonical system of differential equations.

Some basic facts about matrix valued entire functions and canonical systems are established in the paper of M.G. Krein. In particular, a formula for the density of the eigenvalues of a canonical "boundary value problem" in terms of the coefficients of the corresponding canonical system is formulated. This formula was discovered independently by L. de Branges (Theorem X in [dB1]; see also Theorem 39 in [dB2]). A far-reaching generalization is presented in Chapters 6 and 7 of the book [GK] by Gohberg and Krein (see especially formula (4.2) in Chapter 7). Krein also formulated a strategy for recovering the coefficient of the canonical system by studying an associated extension problem for an appropriately chosen positive definite kernel. In general, the coefficient of the canonical system is not unique, even under natural normalizing conditions. An important exception for $m = 2$ was solved by L. de Branges [dB2]. A related extension problem is considered in the paper by I.V. Mikhailova and V.P. Potapov who studied the maximum quadrant on which a kernel $k(x, y) = f(x-y)$ which is defined in terms of a given continuous Hermitian function $f(x)$ (subject to some conditions) is positive definite, and gave a formula for the resolvent matrix. This theme was elaborated upon in a paper by Kovalishina and Potapov [KP].

4. The abstract interpolation problem (V.E. Katsnelson, A.Ya. Kheifets and P.M. Yuditskii; H. Dym and B. Freydin).

The AIP is basically a method for establishing the existence and representation of solutions to a given interpolation problem from the data, when the Fundamental Identity is satisfied. The strategy is to construct an isometric operator from the data in a natural way. The set of all solutions to the interpolation problem is then identified with the set of characteristic functions of all unitary extensions of the original isometry. The paper of Katsnelson, Kheifets and Yuditskii explains this strategy and sketches a couple of examples. More information may be found in the list of supplementary references which have been added to the translation for this volume. The papers of Dym and Freydin use the general method of the Abstract Interpolation Problem to solve a bitangential interpolation problem in the general setting of upper triangular operators.

5. Spaces with an indefinite metric (N.I. Akhiezer; A. Dijksma and H. Langer; E. Russakovskii).

The paper of Akhiezer discusses interpolation in a space of functions which are holomorphic in $\mathbb{D}$, except for at most κ poles, and contractive on the boundary of $\mathbb{D}$. A similar class of problems was considered a little earlier by Takagi [Ta1], [Ta2]. Both of these papers use the methods of classical analysis. A generalization of these problems was reformulated as a problem of approximation in the theory of compact Hankel operators by Adamjan, Arov and Krein [AAK]. The general method of Potapov is also applicable to this circle of problems; see e.g., Golinskii [G].

The paper of Dijksma and Langer presents an operator theoretical approach for studying interpolation problems in spaces of meromorphic functions with at most κ poles by using operators in spaces with at most κ negative squares. Their paper contains a description of the set of all solutions. The paper of Russakovskii studies matrix valued boundary value problems with boundary conditions which depend linearly on the spectral parameter. This is reduced to an abstract spectral problem in a Pontryagin space with a finite number of negative squares.

6. Other directions (D. Alpay and V. Bolotnikov; L.B. Golinskii; V.E. Katsnelson and B. Kirstein; S. Kupin and P.M. Yuditskii).

The paper of Alpay and Bolotnikov discusses a class of interpolation problems in reproducing kernel Hilbert modules and gives a number of examples. Golinskii studies the boundary behavior of a Schur function in terms of its Schur parameters. Kupin and Yuditskii study interpolation problems in multiconnected domains, using the theory of Riemann surfaces of Carleson-Widom type. Finally, the paper of Katsnelson and Kirstein contains a systematic expository presentation of the theory of matrix valued functions of Smirnov class, which plays such a central role in this sort of analysis.

References

[AAK] V.M. Adamjan, D.Z. Arov and M.G. Krein, *Analytic properties of Schmidt pairs for a Hankel operator and the generalized Schur-Takagi problem* (Russian), Matem Sbornik **86**:1 (1971), 34–75. English transl: Math. USSR Sbornik **15**:1 (1971), 31–75.

[BaC] M. Bakonyi and T. Constantinescu, *Schur's Algorithm and Several Applications*, Pitman Res. Notes Math. Ser., Vol. **61**, Longman, Essex, 1992.

[dB1] L. de Branges, *Hilbert spaces of entire functions, II*, Trans. Amer. Math. Soc. **99** (1961), 118–152.

[dB2] L. de Branges, *Hilbert Spaces of Entire Functions*, Prentice Hall, Englewood Cliffs, NJ, 1968.

[dB3] L. de Branges, *The expansion theorem for Hilbert spaces of entire functions*, in: Entire Functions and Related Parts Analysis, Proc. Sympos. Pure Math. Vol. **11**, Amer. Math. Soc., Providence, RI, 1968, pp. 79–148.

[DI] H. Dym and A. Iacob, *Positive definite extensions, canonical equations and inverse problems*, in: Topics in Operator Theory Systems and Networks (H. Dym and I. Gohberg, eds.), **OT12**, Oper. Theory Adv. Appl., Birkhäuser, Basel, 1984, pp. 141–240.

[G] L.B. Golinskii, *A generalization of the Nevanlinna-Pick matrix problem*, Izv. Akad. Nauk Arm. SSR Math. **18** (1983), 187–205 (Russian). English transl. in: Soviet Journal of Contemporary Mathematical Analysis (Allerton Press) **18**:3 (1983), 22–39.

[GK] I. Gohberg and M.G. Krein, *Theory and Applications of Volterra Operators in Hilbert Space* (Russian) Nauka, Moscow, 1967. English transl.: Amer. Math. Soc., Providence, RI, 1970.

[KP] I.V. Kovalishina and V.P. Potapov, *The triad method in the theory of extension of Hermitian operators* (Russian), Izv. Akad. Nauk Armyan. SSR, Ser. Mat. **24** (1989), No. 3, 269–292. English transl.: Soviet J. Contemp. Math. Anal. **24** (1989), No. 3, 61–84.

[LP] M.S. Livšic and V.P. Potapov, *A theorem on the multiplication of characteristic matrix-function* (Russian), Dokl. Akad. Nauk SSSR **72** (1950), No. 4, 625–628. Private Translation, T. Ando, Collected Papers of V.P. Potapov, Sapporo, Japan, 1982.

[P] V.P. Potapov, *The multiplicative structure of J-contractive matrix functions* (Russian), Trudy Moskov. Math. Obšč. **4** (1955), 125–236. English transl.: Amer. Math. Soc. Transl. (2) **15** (1960), 131–243.

[S] A.L. Sakhnovich, *Spectral functions of a canonical system of order 2n*, Math. USSR Sbornik **71** (1992), No. 2., 355–369.

[Ta1] T. Takagi, *On an algebraic problem related to an analytic theorem of Carathéodory and Fejér and on an allied theorem of Landau.* Japanese J. Math. **1** (1924), 83–93. Reprinted in: T. Takagi, Collected Papers, Springer, Tokyo, 1990, 226–243.

[Ta2] T. Takagi, *Remarks on an algebraic problem.* Japanese J. Math. **2** (1925), 13–17. Reprinted in: T. Takagi, Collected Papers, Springer, Tokyo, 1990, 240–243.

To obtain broader perspectives on the subject matter of this volume, the following partial list of supplementary reading may be helpful.

[A] N.I. Akhiezer, *The Classical Moment Problem*, Oliver and Boyd, Edinburgh, 1965.

[AK] N.I. Akhiezer and M.G. Krein, *Some Questions in Moment Theory*, Translations of Mathematical Monographs **21**, Amer. Math. Soc., Providence, RI, 1962.

[BGR] J.A. Ball, I. Gohberg and L. Rodman, *Interpolation of Rational Matrix Functions*, Birkhäuser-Verlag, Basel, 1990.

[C] T. Constantinescu, *Schur Parameters, Factorization and Dilation Problems*, Birkhäuser-Verlag, Basel, 1996.

[D1] H. Dym, *J Contractive Matrix Functions, Reproducing Kernel Hilbert Spaces and Interpolation*, CBMS Reg. Conf. Ser. in Math. Vol. **71**, Amer. Math. Soc., Providence, RI, 1989.

[D2] H. Dym, *Book Review*, Bull. Amer. Math. Soc. **31** (1994), 1–16.

[FF] C. Foias and A. Frazho, *The Commutant Lifting Approach to Interpolation Problems*, Birkhäuser–Verlag, Basel, 1990.

[H] J.W. Helton, *Operator Theory, Analytic Functions, Matrices and Electrical Engineering*, CBMS Regional Conference, Ser. in Math., Vol. **68**, Amer. Math. Soc., Providence, RI, 1987.

[HSP] G. Herglotz, I. Schur, G. Pick, R. Nevanlinna and H. Weyl, *Ausgewählte Arbeiten zu den Ursprüngen der Schur-Analysis* (B. Kirstein and B. Fritzsche, eds.), Teubner, Leipzig, 1991.

[K] V.E. Katsnelson, *Methods of J-theory in Continuous Interpolation Problems of Analysis*, Private Translation, T. Ando, Sapporo, 1995.

[KN] M.G. Krein and A.A. Nudelman, *The Markov Moment Problem and Extremal Problems*, Transl. Math. Monographs, Vol. **50**, Amer. Math. Soc., Providence, RI, 1977.

[KP] Seven Papers (by I.V. Kovalishina and V.P. Potapov) Translated from Russian; Amer. Math. Soc. Transl. (2), Vol. **138**, 1988.

[OT72] *Matrix and Operator Valued Functions* (The Vladimir Petrovich Potapov Memorial Volume), Oper. Theory. Adv. Appl. **OT72**, (I. Gohberg and L.A. Sakhnovich, eds.), Birkhäuser, Basel, 1994

[RR] M. Rosenblum and J. Rovnyak, *Hardy Classes and Operator Theory*, Oxford Univ. Press, London, 1986.

[SNF] B. Sz.-Nagy and C. Foias, *Harmonic Analysis of Operators on Hilbert Space*, North-Holland, Amsterdam.

December 1996

H. Dym, B. Fritzsche
V. Katsnelson, B. Kirstein

Acknowledgement. We wish to express our appreciation and gratitude to Mrs. Ruby Musrie whose expert and dedicated efforts made this volume possible. She worked many long and hard days, amidst all the pressures of being Department Secretary, to prepare this elegant manuscript.

Operator Theory
Advances and Applications, Vol. 95
© 1997 Birkhäuser Verlag Basel/Switzerland

Vladimir Petrovich Potapov as remembered by colleagues, friends and former students

Moshe Livšic

Ben Gurion University of the Negev, Beer-Sheva, Israel

Fate has decreed that it falls upon me to talk to you here in this very special place about a very special person: Vladimir Petrovich Potapov. I first met Vova Potapov (as he was then known, while I was Mischa Livšic) when he entered Odessa University in 1934. He was then twenty years old and had transferred to the University after leaving the Odessa Conservatory of Music; I was a second-year student in the Faculty of Mathematics and Mechanics.

It was not an easy time for him because he was relatively old to be entering the university and to be changing in such a drastic way his major field of study, from music to mathematics. He probably considered himself as a failure in music and therefore there was no choice but to start everything from scratch.

I want to say a few words about what Odessa State University was like then, especially its mathematics department. Of course we all know what happened after the October revolution. After the civil war the party attitude towards higher education as far as the universities were concerned, was roughly as follows: To use the older professors without trusting them completely until they managed to educate the younger generation, i.e., until a new generation of Soviet professors would grow up and be ready to take their place. Of course this was a very painful process. A number of older professors, some of whom were quite old at the time, were often persecuted and sometimes imprisoned. Sometimes they would get back to the universities from the prisons, and sometimes they would not.

Let me tell you in a few words about some of the basic courses that were given to us in the mathematics department. The courses that made a deep impression on Vladimir Petrovich Potapov were Mathematical Analysis and the Theory Functions of a Real Variable. The course on the theory of functions of a real variable was given by a very very good mathematician: Mstislav Nicolaevitch Bobynin.

The course by Bobynin was given at a very advanced level with a very high level of precision. Bobynin also required precision from his students. He was very meticulous, which was very important for the theory of functions of a real variable. He was an extremely demanding examiner, but he was very fair. The popular phrase was "Bobynin is guilty for the death of many" – which rhymes in Russian but not in English.

As I have mentioned already, life at the university was a very complicated and multi-coloured picture. The theory of functions of a real variable came only in the second or the third year. In the first year when we had just entered and started our class of mathematical analysis in the big auditorium, we were confronted with a new lecturer, a professor named Maximoff, whom nobody knew at Odessa. He landed from somewhere, nobody knew from what or from where, and was assigned the task of teaching mathematical analysis.

Maximoff lived in a student dormitory not far from the mathematics department on a street that is named now and was named in the past: "the street of Peter the Great". But at the time it was named "the street of the Communist International". The Faculty of Mathematics and Mechanics was housed in an old building which was very imposing. You entered through huge doors; it had high ceilings, and opposite the entrance, there was a huge well-framed old clock made by the famous Odessa master Timchenko. More than once, when the bell rang at the university to indicate the start of the lecture in mathematical analysis, the students gathered, but there was no Professor Maximoff. So a student delegation was sent to the dormitory to pick him up. Maximoff had been drinking vodka all night long and so he simply could not physically wake up until he was picked up by the students and actually brought to the classroom.

Professor Maximoff's lectures were quite unusual and almost totally impossible to understand. I still remember how he finally managed to prove (God knows how) that the derivative of e^x is equal to e^x. But this did not make sense to us because we thought, quite naively, that the derivative of a function is something different from the function itself. It was hard to grasp that the derivative of this particular function was the same as the function.

We posed the question to him and he started to think it over. After some time he told us, "Well, that means the the derivative is constant". But by that time we already knew the graph of the exponent. So some students went to the board and started to argue with Maximoff, and the whole huge auditorium was just divided into small groups of students who were fighting and arguing about what does it mean that the derivative (which, as we knew, had to be equal to the slope of the tangent) is constant.

Then finally one of the students drew the graph of the function e^x and suddenly everybody was amazed. Maximoff was also very amazed but then he said, "Well okay, so in the drawing you see that the slope of the tangent is changing, but I have proved to you that it is not, so you shouldn't believe the drawing, you should believe what I have proved to you".

Eventually there was a kind of small student rebellion. We went with our complaints about this fashion of teaching to the Dean. But there was really very little that the administration could do because there was no way to get rid of Professor Maximoff. So we tried to survive for a semester with Maximoff and then eventually, when the semester ended, Maximoff disappeared the same way that he appeared, into nowhere.

After Maximoff left, the course of mathematical analysis was given by a very good older and competent professor: Kryzhanovskii. Maximoff was just one of those wayfarers that were arriving on the waves of time and making a good career of it.

Kryzhanovskii was an old man at the time. As far as I can recollect, he had a grey white beard. But he used a very poor teaching technique with us and in fact with everybody. At the very beginning when he had just met the students for the first time, he told us that he had studied in Germany with the great Hilbert and that he had failed the course. He was extremely proud of the fact that he had studied with Hilbert and that he had failed the course. Actually he had tried several times and he failed all those exams with Hilbert. He was very proud of this and when he was saying it we could see the pride in his eyes. Nevertheless, he knew his subject extremely well. When he lectured on mathematical analysis he would become very enthusiastic and get very easily carried away. Then, as a consequence, he would talk very very fast and it was absolutely impossible to stop him. In particular, he could answer no questions because he was just going too fast. I still remember very well how Kryzhanovskii was telling (us first year students) about rational and irrational numbers with great enthusiasm. Of course the subject was quite difficult for beginning mathematical students. He was lecturing on the Dedekind cut and the whole theory of real numbers. But he was rushing forward with such speed that he had no time to say rational number and irrational number; he would say "rat" and "irrat". The students were also very afraid of him because during the lectures he would actually go between the chairs in the auditorium and as he was talking very fast and very enthusiastically, he would spit on everybody around. I remember to this very day how Kryzhanovskii was writing on the board. He covered all the board with chalk, writing with extreme speed. We didn't understand anything. Then he had no place on the board and he had no patience to erase it, so he disappeared. It turns out that he was on the other side of the board where there was also a blackboard. We didn't see him and couldn't see what he was writing. But we could hear the sound of the chalk on the other side of the blackboard. Fortunately the bell rang and that particular lecture was over.

Beyond these anecdotal cases, I still remember the many excellent courses of lecture that we had. I already mentioned the Bobynin lectures. We were also given some lectures by Mark Grigor'evich Krein on the theory of integral equations. These lectures were very modern for that time. He introduced the notion of a Hilbert space, of a scalar product, and other ideas from functional analysis, which were very new. He would also give a lot of examples and lots of interesting applications; things that you just couldn't find in the literature. I also remember an excellent course of lectures by Krein on mathematical physics. Those courses of Krein had a very strong influence on me. Several years after Potapov and I heard these lectures, a young professor arrived at the university: Boris Yakovlevich Levin. He came from Rostov on Don and was a student of a very well-known mathematician, Mordekhai Boltovskoi. Boris Yakovlevich Levin gave a very good comprehensive course on the fundamentals of complex analysis and after that he gave a special course on theory of entire functions. In the course on the theory of entire functions, a lot of attention was paid to theorems on the growth of entire functions in different directions and the influence of this on the distribution of zeroes. The courses of Mark Krein and the courses of Boris Levin made a deep impression on Vladimir Petrovich Potapov. He became very fascinated by function theory; he became a good friend of Boris Yakovlevich Levin very quickly and would visit him often.

Vova Potapov had a very keen sense of humour but his jokes were not always innocent and harmless. We had male and female students and of course the male students considered themselves superior to the girls studying mathematics. The word "zubrit" in Russian means to learn something by heart without understanding any word of it. The female students were called "zubrilka" which means somebody who takes very nice notes in class, and learns the material by heart without even trying to understand what it means. But then the time for exams came and it turned out that the male students who were so very proud of their deep understanding of the subject had not been taking notes in the classrooms. Vova Potapov never took notes nor did I. So there was no choice; Vova Potapov would have to go in a humble way to borrow notes from some of the women students. There were several women students with whom we were friends. Frida Naimark was studying with us. She was the sister of the well-known Mark Naimark who was actually for a long time an assistant lecturer. In particular he was assigned as an assistant lecturer to my group. Another woman student was Sarah Barskaya. Their classroom notes were just ideal; you know everything was written down in a very neat and very orderly fashion. After very humble demands on his part, they would give their notes to Vova Potapov. But then he invented a method of taking revenge on them. At some places in the notes he would put question marks, even if there was no need for a question mark. Then when he would give their notes back to them and they would go through them on the night before the exam, they would suddenly see a question mark which was put there by Vova Potapov himself. That forced them of course to immediately run out and look for Vova Potapov to ask him what does this question mark mean. That was a great pleasure for him, a great revenge, because this way he would be able to explain to them that you take those nice notes but you really don't understand anything, you have to look for me so that I can explain to you what my question mark means.

I also remember how at some point the administration decided that mathematical students should study methods of teaching. So they assembled all the students, not only the mathematicians but also students from physics and chemistry. There was a huge crowd and eventually a woman professor came in and started to lecture. Well it was not exactly plain to anybody what the methods of teaching should be. So she started to talk about something and there was a lot of noise and nobody listened to her. Vova Potapov made no noise. During the lectures he would take the sulphur off a whole box of matches with his small knife and fill the case of a fountain pen with the sulphur. Then at the moment that the bell should have been ringing, a rocket would be launched.

This collection of anecdotes might give you some idea about this period. There were many other stories but it is impossible to tell all of them.

I will jump now to the period after World War II when Odessa University would go back to Odessa after it was evacuated during the war years. There were mathematicians and other scientists who also started coming back to Odessa after being scattered; some were fighting in the army, some were just scattered around the country. Potapov, Krein and Levin came back to Odessa. I also found myself in Odessa a little later. Before the war there was a functional analysis seminar in Odessa led by Krein. It was probably the first functional analysis seminar in the Soviet Union. Krein already had a school before the war, and among the participants of the seminar, there were David Milman and Vitold Shmulyan. Shmulyan was drafted into the army and often wrote to us from the front line about his newest results that he was obtaining there. I still remember getting letters from him. Unfortunately Shmulyan did not come back to Odessa; he was killed in the war.

When I came back to Odessa after the war, things had changed. There was a certain feeling of elation. Times were very hard but there was the victory over the Nazis. People were expecting new beginnings and new developments to emerge. But the so-called national (ethnic) politics of the party had changed. It had probably changed already during the war, but it became visible immediately after the war. Before the war, nationality was not an important factor. The important factor was your social origin. For instance, in order to be admitted into the university, you had to have had experience as a worker. If your parents were workers, that was okay too. If not, then you had to work in a factory for at least two years and present proof of that in order to be accepted into the university. This happened to my close friend Israel Glazman (who is probably known to most people here from the fundamental textbook by Akhiezer and Glazman: Theory of Linear Operators in Hilbert Space). We were friends from the age of eight when we were both still in school. His father was a simple accountant. Israel Glazman had an extraordinary gift for music. He had perfect pitch. He went to study in the music school of the famous Odessa music teacher Stolyarskii, who had produced some of the greatest performers in the Soviet Union. For instance, David Oistrakh was a student of Stolyarskii at the time and actually Stolyarskii ranked Glazman above Oistrakh. When Israel Glazman and I decided to enroll in the university, Glazman faced a problem because he had no working experience. I did not have this problem because my father was teaching at the Institute of Communication and there was a clause that made an exception for the children of people who were actually teaching at an institute of higher education. This was also the reason that Vova Potapov could enroll in the university without having any work experience because his father was a very well-known professor of philology.

Now it should be noted that those were the workers' rules and none of us doubted them. It was clear for us that the working class had won the revolution and now should be given the first opportunity. Isya Glazman told me that "of course I'm going to work somewhere to get some work experience". He went to work as a simple worker at the factory of the January Uprising. Sometime later an accident took place at the factory and he broke his hand and couldn't do anything with it for several months. This was the end of his career as a violinist.

All that was characteristic of the period before the war. After the war, the party's attentions switched from the class problem to the national (ethnic) problem. Probably that was Hitler's greatest victory. In the Soviet Union this was called the national question, and of course the main specialist on the national question was Joseph Stalin himself. He was a great "specialist" in many questions, but, in particular, he was the leading authority on this particular question. The national problem, or ethnic problem, was especially acute in Odessa. (By the way, Odessa is 200 years old this year. That is why it is a young city compared to Leipzig.) At the time that Odessa was founded by the Empress Catherine the Second (Catherine the Great), it was a new town on the border of the empire and not heavily populated, so you could settle there freely. There were no limitations whatsoever so the population was very mixed. There were Jews, Ukrainians, Russians of course, a few Greeks and Germans. There were two German villages near Odessa, Lustdorf and Klein Liebental. My parents and I lived in a communal apartment, i.e., an apartment that was shared by several families, as was usual at the time. One of our neighbors was a German woman. I still remember her name: Paulina Philipovna; she had a daughter, Olya. Greeks for the most part would settle near the seashore; their main occupation was fishing. Odessa had a very large Jewish population, almost one-third of the total population of the city was Jewish. I think that something like 150,000 Jews were living in Odessa before the war. All the Germans were deported immediately when the war broke out. For some reason the Greeks were also deported, nobody knows what their particular guilt was.

When we came back to Odessa after the war, we discovered that the party's politics had changed and that a new national policy had come into being. It was decided by the powers above that only a very small number of Jews, or even better, no Jews at all, should remain at the university. When Mark Grigor'evich Krein came back to the university, it was impossible of course to say directly that he was not welcome because in the Soviet Union it was not the custom to tell anybody anything straight; it was not the custom to tell them the truth. At the time there was a very famous poem of Mayakovskii which was known to everybody who was growing up in Russia and to any communist world-wide which said that, when we say Lenin we mean the party, when we say the party we mean Lenin; among the people this was ironically paraphrased: That's the way we all speak, we say one thing when we mean the other. So Mark Grigor'evich Krein was gladly welcomed back as a great scientist coming back to Odessa. But at the same time he was warned that he would not have the right to staff his own department or to choose his own students. This would be decided by other people. He was welcomed back but he would not have a say on who was going to work with him. Mark Grigor'evich was not one of those men whom you could tell what to do and what not to do, whom to hire and whom not to hire. He simply rejected the position. But then you see the university administration was not "guilty". He was "guilty"; he was offered a job and refused it! Thus he did not return to the university and the school which he formed, that was already so very well known, found itself without a home. And his students, both Jewish and non-Jewish, who really wanted to do mathematics and understood that one could no longer do it at the university, had to seek positions elsewhere. They had to find positions at all kinds of technical institutes around the city that were secondary as far as mathematics goes. Subsequently Odessa University simply ceased to exist as a mathematical center and I don't think it ever recovered. It still does not exist as far as mathematics is concerned.

Sometime later Vova Potapov started to work at the Odessa Pedagogical Institute, which was an institute for the education of teachers. He started to teach in the departments of physics and mathematics. The rector of the Pedagogical Institute at that time was Vlasenko. He was Ukrainian; of course he implemented the party line. But he was a clever man and he understood that while implementing the party line he should also somehow encourage research to go on. Besides, he also had a great respect for scientists. Then Vladimir Petrovich slowly started to hire talented mathematicians. He had very good relations with Vlasenko and was able to do it. For instance, he helped to hire one of our former university friends, Orlov. I don't remember exactly when Potapov met Michael Samoilovich Brodskii. Brodskii was not a regular student at the university, he was attending evening classes there. He was a geometer by specialization. Potapov met him, liked him and eventually he hired him for the Pedagogical Institute. So there were two of our friends, Orlov and Brodskii. Some time later he also invited me to work at the Pedagogical Institute at a half-time position. At the time I was the head of the department at the Hydro-Meteorological Institute. At the Pedagogical Institute, I gave the course in theoretical mechanics. One of the successful students from the Pedagogical Institute, Lev Aronovich Sakhnovich, is now sitting here. He is a living example of a person who would not have been a mathematician and probably would not have been sitting here among us if it would not have been for Potapov. Sakhnovich was an excellent student but he had a problem. It was not his nationality; nationality was never mentioned in the Soviet Union. The problem was item No. 5. You see, in the questionnaire you would first have to fill in your birthdate, secondly the names of parents, etc. and then item five would be nationality. But it was not nice to talk about nationality so it was just called item No. 5. Sakhnovich had troubles with item No. 5, but Potapov liked him very much and wanted to take him as a graduate student to work with me. He had to go through tremendous troubles to arrange this. I don't want to recollect what those troubles were; in fact, I cannot; Potapov probably never told me all he had to go through. But eventually he managed to get permission for Sakhnovich, a Jew, to enter the graduate program at the Pedagogical Institute. That was in 1953 just before Stalin's death. That was the time of the fight against "cosmopolitans" with the wrong item No. 5 and the prosecution of the "criminal physicians", etc.

Another case that I remember very well although I don't remember the exact order, was a graduate student by the name of Mukminoff. He was a Tartar. He had lost his sight in the war but somehow he managed; his wife was probably helping him by reading to him and he managed to do the calculations in his head. However, he was a Tartar and that was also bad for some obscure reasons. There had been some "problems" with the Tartars in Crimea who were suspected of collaborating with the Turks, or the Germans, or God knows with whom, before and during the war. Now he was not from Crimea, he was from another group of Tartars that were living on the Volga. Nevertheless, he was a Tartar and that was bad. This again led to a very heavy fight on Potapov's part, perhaps even a heavier fight than in the case of Sakhnovich. Eventually, he managed to get him into the graduate program. And then Potapov just continued "misbehaving". Ginzburg came to Odessa from Chelabinsk, a town in the Ural Mountains. His degree was actually in engineering not in mathematics, but Potapov decided to take him in the graduate program, this time as his own Ph.D. student (Mukminoff and Sakhnovich were my students). Once again this was a problem because Ginzburg was a very bad-sounding name. Around this time also my collaboration with Potapov started. Potapov persuaded Brodskii to join us, and we formed a small research group, Vladimir Petrovich, Brodskii and myself. Our collaboration lasted for about ten years.

I have many more reminiscences about Vladimir Petrovich, but unfortunately the time is running late. So perhaps we shall leave them for another occasion.

Damir Zyamovich Arov

South-Ukrainian Pedagogical University, Odessa, Ukraine

I have already written about the scientific influence on me of Mark Grigor'evich Krein and Vladimir Petrovich Potapov in the Potapov memorial volume in the OT series [OT72]. So my talk is going to be about something else.

First of all I would like to thank the two Bernds [Bernd Kirstein and Bernd Fritzsche] and Leipzig University for organizing this unique and very special conference. I understand my great responsibility. There are people in this audience who have known Vladimir Petrovich Potapov very well and can tell much more than I can. I have been thinking about why this Potapov memorial conference was organized, I think it is due in equal manner both to Potapov's achievements as a mathematician and to Potapov the person. I'm not planning to make a saint out of Vladimir Petrovich Potapov. He liked to drink and had a number of romantic relationships. The fact is that Vladimir Petrovich loved life very much. He was an extremely emotional man. He loved music, he loved his daughter [Tatyana Vladimorvna Potapova] very much, he loved the women that he was involved with, he loved his students and his colleagues and he loved his son [Andrei Vladimorovich

Kolyadincev]. He was a very unique man in those respects. He was a highly non-standard person in our standardized society.

There were very non-standard things both in his genetic print and in his upbringing. He went to school only in the tenth grade (the last grade of the Soviet Union) in order to get the final certificate. He was brought up in the home of his father who was, as you have heard from the previous speaker, a very well-known professor of Philology who is up to this day remembered very much in Odessa, in fact remembered more than his son, the mathematician. Potapov's father was very liberal with his son's upbringing. He was given quite a lot of liberty. If he wanted to play small street gambling games with the kids in the street, his father would give him the money. Also, since he did not go to school, he was not part of the Pioneers, which was the teenage communist organization. Also, he was not a member of the Comsomal, which was a communist youth organization. He did become a party member but that was later. To the best of my knowledge, he joined the party after Stalin's death.

> [M.S. Livšic remarks: *I want to add a few words concerning Potapov joining the communist party. On the day of Stalin's death, Vladimir Potapov went to the party bureau at the institute and filled in his application. But he was told, don't hurry, we won't take your application for the moment, and he was given his application form back. Vladimir Potapov came back home, got Stalin's portrait from somewhere and hung it on the wall in the most conspicuous place so that everybody would see it first and foremost. His close friend Sergey Orlov came to visit and suddenly saw Stalin's portrait on the wall. It was hanging quite close to the ceiling, so he took a chair, climbed up and tore the portrait from the wall. As far as I know, this ritual was performed daily. Vladimir would hang it up again and Sergey would tear it down.*]

I became Potapov's graduate student in 1959; it was not by chance. But I have already described this in my article in OT72 and I do not wish to repeat this now.

Most of my talk today is not going to about the influence on me of Potapov the mathematician, but on the influence that Potapov, as a man, has had on me.

After finishing Odessa State University in 1957, I had to go to a village to work as a school teacher, for 2 years. As a point of general information, for the non-Russians in the audience, people finishing the university were sent to work someplace. You couldn't just graduate from the university and then go on continuously as you might like. I had some contacts with Krein, but of course, he couldn't help me in any way because of column five. He couldn't help me with work, nor could he take me on as an assistant, or as a PhD student.

Fortunately, as my two year term was nearing completion, Vladimir Petrovich contacted me at Babrov's suggestion. I went for an interview with him and he invited me to become his Ph.D. student. This was after the period about which Michael Samoilovich Livšic was talking. We are talking about 1959 and of course

Potapov was "misbehaving" again, after all those "bad names" he was taking on yet another problematic student. He had to do a lot of possible and impossible things in order to take me. I know about some of them, there are many of them that I never knew about.

I know that he had some very serious talks with the rector, Vlasenko. He went to talk to the Chairman of the Examination Committee on the History of the Party and to the Chairman of the Examination Committee on the English language; everybody here knows about my knowledge of foreign languages. In particular he had to explain to everybody the meaning of my name. I was born in 1934. My parents were ardent communists, and named me Damir which means "giving world revolution". I might add that there are very few people with such names. Potapov used my first name to advantage, probably also having some serious drinks with the administration in order to advance the matter. In any case, he succeeded in getting me into the Ph.D. program. At that time I was interested in applications of information theory to ergodic theory. Because of this, Brodskii who was the department head at the time, suggested that it might be possible to arrange for Mark Grigor'evich Krein to be my supervisor. I declined because Vladimir Petrovich Potapov had done so much for me.

As I told you, I am going to tell you about Vladimir Petrovich in my own way, which might be different from the way other people see him. He became my supervisor and I became his Ph.D. student. However, I was given complete freedom, which was quite unusual. I went on with my own scientific interests and my own research. Also, I not only became his Ph.D. student, but for a certain period almost a part of his family. I would spend whole days in his house. If I didn't go to the library, I would go there in the morning and leave in the evening. I would eat and drink there.

At first it was on Lastochking Street, where he lived with his first wife Asia Evseevna, who was a piano teacher, and their daughter Tatyana. Later on, it was on Matrosova Street, where he lived with his second wife Natalyia Vasilyevna, who was a mathematician, their son Andrei and her daughter Nina from an earlier marriage.

I would tell Potapov about my work, telling him about the state-of-the-art and about things that I was working on. He would seem interested and would listen to me carefully. He was very helpful in at least two ways: First, he suggested that I should actually calculate the entropies of some particular cases. The computation was very hard. I had a certain feeling of uncertainty in my character. It seemed to me that I would be unable to carry this out, and his support was extremely important for me. And then, in a certain case, when I was looking for a certain representation and didn't know where to go, he actually suggested to look at a theorem of Bohr, and this was the right thing. At the time I thought that this was just a coincidence that came by chance. However, I learned later, that the suggestion was a natural consequence of Potapov's previous interest in almost periodic functions. In any event, I had complete freedom. I would go to Mark

Grigor'evich Krein's seminar once a week and, scientifically speaking, at that time I was much more under Krein's influence than under Potapov's influence.

During this period (the time of my PhD dissertation), Vladimir Petrovich. gave me complete freedom to pursue my own interests. I have just told you how helpful he was to me. I regret to say that I did not express my gratitude to him in any of my papers and he never even hinted to me that I should have done so.

Vladimir Petrovich was a spiritual father to me at those times. As I have told you, I was brought up in a staunchly communist family. I believed everything that I was told. I would have made a lot of mistakes in my life if I had not met Vladimir Petrovich. Vladimir Petrovich wrote me a letter on the first day when I entered the Phd program. In a very clear handwriting, he wrote out a list of the ten commandments from the bible and told me to study them. Nowadays religion is very fashionable in our society. You can see almost all the former communist leaders going to the Church and praying to God, the same way that they once prayed to Marx. But at that time in our antireligious society of those days this was very surprising and significant for me.

At the beginning of my talk I described Potapov's unusual life style. On the other hand you have just heard from Professor Livšic's talk about the situation just after the war when Krein and his school were forced out of the University. Potapov, who was a totally unknown scientist of the time, wrote an article in the communist paper, "The Bolshevicks' Banner", describing in very angry terms, the situation in the mathematics and physics departments in Odessa University. This did not lead to any positive results, but he did try something, he had the courage to enter the battle. Later, I learned from different sources, that he was actually offered the post of department head at Odessa University but that he did not accept the position because it did not include the freedom to hire the members of the department himself.

He joined the party, but he joined the party not in order to pursue his own career, he would have done this anyway, but in order to help other people. He went to work at the Pedagogical Institute in 1948. Shortly thereafter he became Chairman of the Department of Mathematics; he became a Dean there in 1952. As you have also heard, he gathered the best mathematical force of the city, the Krein School. The Mathematical Department at Odessa Pedagodical Institute was absolutely a unique phenomena for a Mathematical Department at a Pedagogical Institute in the Soviet Union. Such well known mathematicians as B.Ya Levin, M.S. Livšic, M.S. Brodskii, M.A. Rutman, S.A. Orlov and D.L. Kucher worked there. He was the Dean there for four years, from 1952–1956.

Finally, I would like to say a few words about Potapov's attitude towards operator theory. It may have been negative in in later years, but it was not true for his early years. He started from operator theory, his problems came from operator theory, he had common works with Professor Livšic that were grounded in operator theory. In our private conversations, Potapov never criticized operator theory.

I did hear him express a negative attitude towards operator theory in his seminar, and also heard about this from other people. But I think this negative attitude on his part was also partly due to the local situation. You have to remember that he was working in Odessa, where young talented mathematicians would automatically fall under the very strong operator theory influence of Mark Grigor'evich Krein. Also, although he was saying that function theoretical problems should be solved by function theoretical means, his own work did not always follow this advice. His approach was that scalar function theory problems should be solved by J theory, but this strategy transforms a scalar problem into a matrix function theory problem of increased dimension.

Moreover, when Potapov started to look at the Krein problems on the extension of continuous positive definite functions where algebraic methods really couldn't be brought to work, he really started to use the methods of functional analysis. Instead of saying let's look at the rigged-Hilbert space, he would say let's look at good functions and also at "not so bad" functions.

One can talk about many good deeds connected with Vladimir Petrovich's name. Personally, when I remember our communications, I don't think that I could have given him anything significant, but he gave me quite a lot.

Lev Aronovich Sakhnovich

A.S. Popov Electrotechnical Institute of Communication, Odessa, Ukraine

Michael Samoilovich has already told you how hard it was to get me into the Ph.D. program. I will tell you in only a few words about the role which Vladimir Petrovich played in my becoming a scientist. When I was a second year student, he had just completed his doctoral dissertation for the second doctoral degree, but he had not defended it yet. He gave me a copy of his doctoral dissertation with large margins for my comments. I was not particularly timid at the time, so I just started writing comments. If you remember, the first chapter of his work contains many theorems from linear algebra. I proved them in a different way. "Are you sure that you are spending your energy in the best way?" – Vladimir Petrovich asked me. He took me to the Institute library and gave me a paper by M.G. Krein and asked me to prove the theorems in this paper without using the fixed point theorem. This was very easy for me to do since I did not know the fixed point theorem at that time. Indeed, three weeks later I obtained the corresponding results without using the fixed point theorem. In the Krein paper there are two classes of operators, contractive operators and plus operators. I proved that one class can be obtained from the other by multiplying by a constant. Vladimir Petrovich liked to repeat this story; it amused him a great deal. When I was in my fourth year I was handed over in a very solemn manner by V.P. Potapov to M.S. Livšic. From that time on and all through my studies in the Ph.D. program, I worked under Michael Samoilovich. Nevertheless, Vladimir Petrovich continued to follow my work quite closely.

I want to also tell you about another person who played an important role in my being accepted into the PhD program. Michael Samoilovich has mentioned the Rector, Vlasenko. He was not only a clever man, there were quite a number of clever men around, but to be both a clever man and a decent man was quite an exception among the administrators at that time. Vladimir Petrovich characterized Vlasenko in a very interesting way. He used to say that Vlasenko is like a rusty weather vane. His body is ready to follow the direction of the wind, but since he is rusted, he can't. After I was accepted into the PhD program, Vladimir Petrovich took me aside and told me that something totally unbelievable has happened. "You have been accepted", he said. After I had graduated, the problem of where I would work arose. Vladimir Petrovich suggested the following combination to the Rector at Institute of Marine Engineers. They should engage me as a laboratory assistant and he would agree then to work there half time as a professor or head of the department. He also told them that if they gave me the job of a laboratory assistant, then they would have a professor in five years. But even under these conditions the administration of the Institute refused to engage me because I was Jewish. Under such circumstances, Vladimir Petrovich saw only one way out, to dismiss somebody from his department and to hire me in his place. In four years I did become a professor. So Vladimir Petrovich was quite prophetic as far as that went.

The last thing I want to tell about is one trait of Vladimir Petrovich's scientific approach that is very well known to everybody. Namely, his very negative attitude towards operator theory. "Operator theory is a parasite, a virus, on the pure body of function theory", he used to say. When he was in a better mood, he would express himself in a somewhat less extravagant way, and I think that was his real attitude. His point of view was that problems of function theory should be solved by the methods of function theory. One can agree with this point of view or not. However, I think that the question is: Is it a productive point of view or isn't it. Laplace's deterministic attitude was a fruitful attitude and the nondeterministic approach of Niels Bohr was a fruitful attitude as well. Therefore, I think that for Potapov this restricted approach was a fruitful one. Thanks to this approach he developed J-theory and solved a number of very important problems in their proper framework, "using proper methods", as he would say. Also operator theory did not suffer; it continued to develop thanks in no small part to Potapov's results. It also contributed a great deal to our understanding of Potapov's results and also it contributed to function theory.

In conclusion, I would like to say that I still feel the void that has been left by Potapov's death.

Adolf Abramovich Nudelman

Odessa Civil Engineering Institute, Odessa, Ukraine

Everybody around had already connected us together and was just waiting for a formal announcement on our part. Finally the day arrived when we announced to our parents that we are planning to get married. Exactly at this moment Vladimir

Petrovich appeared. He was a close friend of my wife's father. When the interesting news had been announced to Potapov he then in his characteristic slowness placed his hand into his pocket and took out a packet of cigarettes with a picture of a man riding a horse on it and slowly blessed us with this packet of cigarettes. We recalled this incident ten days ago when we were celebrating the 40th anniversary of our marriage.

Vladimir Kirillovich Dubovoj

Kharkov State University, Kharkov, Ukraine

As a mathematician, the major influence on my education was of course Professor Livšic. I asked Michael Samoilovich to be my advisor at the beginning of my fourth year at Kharkov University. So I am talking here about my two last years in the University and another three years of Graduate School. One can judge his influence on me by the following fact. Michael Samoilovich fixed a day, I remember it was Tuesday and every week at one-thirty I would have to come to tell him about my mathematical advances during the past week, This went on for five years. I first met Vladimir Petrovich when he was already the famous Vladimir Petrovich Potapov. He was giving a course of lectures at Kharkov University that was organized by Boris Yakovlevich Levin. Vladimir Petrovich moved to Kharkov in May 1976. He died there in December 1980.

The Kharkov period was much shorter than the Odessa period. But the trace that Vladimir Petrovich left in Kharkov is, I think, quite comparable to the trace he left in Odessa. It is of course hard to measure such things, but just look at the number of Kharkov people at this conference.

One can explain this, I think, by the fact that ideas of Potapov grew easily on the Kharkov mathematical soil, because that soil had already been cultivated by Michael Samoilovich for many years by that time. It was further strengthened by the fact that Vladimir Petrovich came to Kharkov shortly after Michael Samoilovich left Kharkov. Michael Samoilovich had been leading a seminar in Kharkov, which at that time was one of the leading centers in classical and functional analysis in the Soviet Union. At this seminar you would see mathematicians from all the major mathematical centers in the Soviet Union. Then, in this void that was formed by Michael Samoilovich's departure, Potapov arrived. At least all the ideas that Vladimir Petrovich was talking about were very familiar to us.

To simplify things a bit, Michael Samoilovich was doing things in the language of operator theory, while Vladimir Petrovich, as has already been told here, was insisting upon pure function theory. But those different methods were reflecting the same mathematical essence. My period of active contact with Vladimir Petrovich was from 1978–1980. In 1978 I had a sabbatical and I had asked Vladimir Petrovich to let me use this half a year to learn better analytical methods at his mathematical department at the Institute for Low Temperature Physics. The following proposition was put to me: Either I go totally free or if I want to study those methods, then I have to accept the following routine. I must come to work at 9:00, work until half past three and then at half past three I must come to his office and show him what I had done during the day and discuss it with him. I agreed to this proposition and was very glad of it.

I had four months of very intensive mathematical interaction with Vladimir Petrovich. I don't know, but perhaps Vladimir Petrovich was already feeling that he didn't have much time left. He was paying enormous attention, to everybody who wanted to enter the subject. To help me orient myself in this area, I asked Vladimir to suggest some problems, not necessarily unsolved problems, but problems in which I could orient myself and try to understand what was going on. That is how the problem which I talked about yesterday came to be. It required not only function theoretic methods, but also the operator theoretic methods that I started with Michael Samoilovich.

It should be emphasized that Vladimir Petrovich's home was always open for us in Kharkov. We often gathered at his house. Somehow it always worked out that we would gather around the table at 10:00 in the evening. We would leave late at night, between 2:00–3:00 a.m. I would like to advance an objection to Victor. He said that when Vladimir Petrovich was drunk... Vladimir Petrovich was never drunk. Sometimes he was in the state when he had drank, but he was never drunk. I would also like to point out the attention that had been given to all of us at this time by Irina Vasilyevna Kovalishina. One has to remember that all those results at this time were usually published only as short notes in different issues of Doklady. We would mainly learn things from Vladimir Petrovich and from Irina Kovalishina. In general, this was a very open house for guests and a place that is very warmly remembered by all of us.

Two more items: One could always talk to Vladimir Petrovich as to an equal. He didn't require much in his life. Once he needed to visit a doctor and I was accompanying him. He had warned me beforehand that if I would come with a cab, then he would not enter it. I understood that this is the way that things were going to be.

Another story: Once when discussing a certain point, we were walking together down the corridors of the Institute. Boris Yakovlevich joined us and cited at great length a passage from the Bible. When Boris Yakovlevich was just going to leave, Vladimir Petrovich cited the passage which followed it. Unfortunately, I don't remember which passage it was. Several days later I expressed my amazement at this episode to Vladimir Petrovich and he said, "Oh Veloda, you have grown up on the history of the communist party".

Finally I want to thank the two Bernds and the Leipzig mathematicians, for this conference, and not only for this conference. I am spending quite a lot of time here near them and I see the special attention that they are paying to us – in these difficult times – so, I just want to express my appreciation and to say, Thank You.

Victor Emanuelovich Katsnelson

The Weizmann Institute of Science, Rehovot, Israel

V.P. Potapov visited Kharkov for one semester while on sabbatical in 1975. He moved to Kharkov permanently in 1976, because the Institute for Low Temperature Physics and Engineering was able to offer him an apartment. At the Institute, Potapov was in charge of a small group in the Mathematics Division which included L.B. Golinskii, I.V. Mikhailova and two other young mathematicians.

I knew the name Potapov from his early work on almost periodic functions. I became interested in almost periodic functions when I was a third year student. There was a paper by Jessen and Tornhare in Acta Mathematica from the war years that I could not find in the University library. Boris Yakovlevich Levin told me that he had a translation of this paper by Potapov. It was in a very fat notebook written with a very tiny calligraphic script. I borrowed this notebook from Levin and I am still sorry that I returned it to him. I don't know where it is now.

That's where I made my first encounter with Potapov, as a translator and a scribe. I also saw him once at the defense of a PhD thesis at Kharkov, but that was just a two-hour encounter. I really didn't pay much attention. Then in 1975 I became interested in analytic matrix functions, and his work was pointed out to me. I got his phone number and called him in Odessa. I told him that I would like to have a scientific conversation with him and asked whether I could come. This was the winter of 1975 and I asked him where I could stay in Odessa. He told me that I could sleep on the terrace of his house. I didn't know what a terrace was, but I did know that it was supposed to be cold in Odessa in January, so I brought a very fat, warm winter sleeping bag with me. It turned out that a terrace in Odessa was just a normal room in the house. I can tell you where it was, it was the apartment of his first and only official wife. They weren't on very good personal terms at that time. The apartment was quite interesting; it had one room that was very, very large, about 50 square meters. On the ceiling there was written something in a script that I didn't understand, as well as in Arabic and Greek. He told me that it meant peace upon the house. There was a huge portrait of Brahms upon the wall. I didn't know who Brahms was, so I asked him. There was also a big polished box with a radio receiver inside. It was not one of those produced commercially.

I didn't know that Vladimir Petrovich was interested in electric circuits. It turned out that this was a radio that he had constructed himself, starting in the late 1950's. He had actually designed and built the receiver himself. It was a highly nontrivial design for those times. We had some scientific discussions and then I realized that with a scientist of this stature it doesn't make any sense to speak about concrete scientific problems. It is much more productive to talk about general scientific philosophy.

Then in 1975 Vladimir Petrovich came to give a course of lectures at Kharkov State University. I don't know who organized it; it couldn't have been Boris

Yakovlevich Levin because at the time Boris Yakovlevich Levin was hardly admitted into the university himself. He left the University and went to the Low Temperature Institute in 1971 or 1972; such behavior was looked upon negatively by the university administration. When the scientific council for a thesis defense was being formed, Boris Yakovlevich was not included even when the thesis was on the theory of functions of a complex variable. He became very angry and swore that he would never get involved with the university or any kind of activity which was affiliated with it. So, he came just for the function theory seminar, whose actual leader he was even though the official leader was I.V. Ostrovskii. He definitely couldn't have done anything for Potapov in the university; he was not the kind of person who had official influence within the university establishment.

When Potapov joined the Institute for Low Temperature Physics and Engineering in 1976, he quickly understood the atmosphere in the Mathematics Division of the Institute. It should be remarked that at that time the intersection of the Mathematics Division with the Communist Party was almost empty. Potapov understood that this was not healthy and that it was important to create a layer of party members who were decent people, because otherwise such a layer might form itself out of nondecent people. He enriched the party group in the Mathematics Department by adding E.Y. Khruslov, G.P. Chistyakov and several good physicists whose names I don't recall.

In July 1976 Potapov actually moved to to Kharkov. He came by train. He walked out of the train with his suitcase. He had no furniture with him. He came with very interesting company. In general, Russian cows prefer to eat grass, but if there is no grass some other solution has to be found. The agricultural director of one of the regions that the train was passing through, sent a group of people to Siberia in order to bring some branches from trees to feed the cows. So Potapov having travelled with such interesting company walked out of the train looking for other company, than the company of branches. He stayed in my apartment for the first two weeks. The first thing he managed to do was to get into bad terms with my wife. The best way to get my wife angry is to say that women are unable to really understand anything. He claimed that women are good for only for three things: Kinder, Kichen and Kirchen, and perhaps also voluminous computations. Several weeks later he rented an apartment for approximately one year. In Kharkov this apartment was well known to a number of mathematicians.

As far as I know, Potapov didn't give particularly many lectures at the University or elsewhere. He did organize a seminar together with Kovalishina and was very active in making remarks on other people's lectures. He preferred to educate and develop young people in personal conversations, both in his home and at the Institute. Often these talks were on non-mathematical subjects. For instance when Peter Yuditskii asked him what one should do in order to achieve success in science, the answer was very constructive, you should lose some weight and then you should report the results. Peter followed this advice. After some time he was able to report that he was down to 96 kilos, Potapov told him you're on the right path. Now we are able to understand one of the reasons for Peter's success.

In August 1976, shortly after he moved to Kharkov, he participated in a Summer School which was organized by A.V. Strauss. The main focus of the school was spectral theory. I remember a number of brilliant talks by M. S. Birman (who is an outstanding lecturer). V.P. Potapov, F.S. Rofe-Bekelov (a former student of I.M. Glazman) and I were the only participants from Kharkov. Potapov and I gave a series of lectures on analytic matrix function theory. However, there was only polite interest. The next year, we gave another series of lectures on the same theme in the Voronezh Winter School (January–February, 1977). A number of members of this audience were exposed to this circle of ideas in a systematic way for the first time at the Voronezh Winter School. The fact that they, and so many others, are here today, is testimony to V.P. Potapov's vision.

Operator Theory
Advances and Applications, Vol. 95
© 1997 Birkhäuser Verlag Basel/Switzerland

On a minimum problem in function theory and the number of roots of an algebraic equation inside the unit disc[1]

N.I. Akhiezer

Abstract. The problem of finding the least value of $\max\{|f(z)| : |z| = 1\}$ is considered in some classes K_p of functions which are analytic in the closed unit disc apart from at most p interior poles and satisfy an interpolation constraint. It is shown that if $f(z) \in K_0$ and its first $n+1$ Taylor coefficients are prescribed, then the solution of this extremal problem is essentially unique and has constant modulus on the unit circle. The structure of this function is studied. [Abstract added by editors.]

0 Introduction

Let

$$z_1, z_2, \ldots, z_{n+1} \tag{1}$$

be $n + 1$ points which lie inside the unit disc of the complex plane and consider the set of all functions $f(z)$ which are regular[2] inside the unit disc and which take on the prescribed values $w_1, \ldots, w_{n+1}$ at the points (1):

$$f(z_1) = w_1, f(z_2) = w_2, \ldots, f(z_{n+1}) = w_{n+1}. \tag{2}$$

What is the least value of the upper bound of the modulus of a function from this set and for which functions is this least value is attained?

This problem belongs to the circle of ideas associated with Carathéodory–Fejér. There is a close relationship to the problem concerning bounded functions which was proposed and solved by G. Pick in 1915, [Pi1], [Pi2]. (See also the interesting investigations of R. Nevanlinna [Nev1], [Nev2]). By means of passage to the limit, a more general problem, in which not only the values of the function f but also the values of its derivatives are prescribed, is obtained.

1) Submitted by the academician S.N. Bernstein. Translated from: Izvestiya Akademii Nauk SSSR. Otdelenie Matematicheskikh i Estestvennykh Nauk, no. 9 (1930), 1169–1189.

2) Apparently regular means holomorphic – editorial note.

A special and extremely important case of this general problem is the original Carathéodory–Fejér case in which the conditions (2) are replaced by the conditions

$$f(0) = c_0, f'(0) = 1! \cdot c_1, f''(0) = 2! \cdot c_2, \dots, f^{(n)}(0) = n! \cdot c_n. \tag{3}$$

This problem is connected with the coefficient problem for bounded power series. A new and original solution of the latter problem was presented by I. Schur in two remarkable memoirs [S1], [S2].

Among many very interesting results, the second of the Schur memoirs contains a theorem on the coefficients of those algebraic equations all if whose roots are less than one in modulus. Later this theorem was generalized and proved purely algebraically and independently of the Landau–Schur algorithm in a paper of A. Cohn [Co]. Conversely, using results of A. Cohn, one can obtain a solution of the Carathéodory–Fejér problem. L. Fejér [Fe] (see also [Eg]) showed that in the simplest case:

$$0 < c_0 \le c_1 \le c_2 \le \dots \le c_n ,$$

it is possible to use Kakeya's results instead of A. Cohn's theorems.

In the present paper an elementary solution of the Carathéodory–Fejér problem is given in the general case for arbitrary coefficients $c_0, c_1, \dots, c_n$. A solution of the Pick problem (i.e., when the conditions are given by (2)) is given as well.

Moreover, these problems are generalized in a new direction. As a simple consequence of the considered extremal problems, one of the main theorems by A. Cohn is derived.

1 An auxiliary result

Definition. A function $f(z)$ belongs to the class K_p if the following two conditions are satisfied:

1) f satisfies condition (2) or another condition which can be derived from (2) by means of passage to the limit (for example (3)). In our further considerations we fix this condition (and refer to it as **condition (B)**).

2) $f(z)$ is regular in the closed unit disc $|z| \le 1$ except for some poles situated inside the open unit disc $|z| < 1$. The total number of such poles (counted with their multiplicities) does not exceed p.

From this definition it follows that

$$K_p \subseteq K_{p+1} \subseteq K_{p+2} \subseteq \cdots .$$

We remark that the difference of two functions satisfying the condition (**B**) has at least $n+1$ zeros (counted with their multiplicities) inside the open unit disc.

Theorem 1. *Assume that there exists a rational function of the form*

$$R(z) = \lambda\, \frac{a_0 + a_1 z + \ldots + a_m z^m}{\overline{a_m} + \overline{a_{m-1}} z + \ldots \overline{a_0} z^m}, \quad (m \leq n, \lambda \text{ is positive}) \tag{4}$$

which has exactly p poles (counted with their multiplicities) in the domain $|z| < 1$. Then for every function $f(z)$ belonging to the class K_{p+n-m} the following inequality holds true

$$\max_{|z|=1} |f(z)| \geq \lambda. \tag{5}$$

Proof. Let $f(z)$ be a function which belongs to the class K_{p+n-m}. Let us consider the difference

$$\varphi(z) = f(z) - R(z) \tag{6}$$

and let us denote by σ and τ, respectively, the total number of poles and zeros of the function $\varphi(z)$ in the domain $|z| < 1$. Clearly

$$0 \leq \sigma \leq 2p + n - m \tag{7}$$

and

$$\tau \geq n + 1. \tag{8}$$

We will distinguish two mutually exclusive. cases.

Case 1. The function $\varphi(z)$ does not vanish on the circle $|z| = 1$. In other words there exists a $\delta > 0$ such that the inequality

$$|\varphi(z)| \geq \delta$$

holds for $|z| = 1$.

Since the function $R(z)$ has exactly p poles in the domain $|z| < 1$ and its modulus takes on a constant value (which is equal to λ) on the circle $|z| = 1$, the total number of its zeros in the domain $|z| < 1$ does not exceed $m - p$. Thus it follows that as z runs over the circle $|z| = 1$ in the positive direction just once, the argument of the function

$$\frac{\varphi(z)}{R(z)} \tag{9}$$

increases by at least $2\pi(\tau - \sigma + p - m + p)$ radians. In view of (7) and (8), the last expression is not less then 2π.

Therefore there exists at least one point ζ on the circle $|z| = 1$ such that the function (9) takes on a real positive value at this point. Hence, for this point ζ we obtain

$$|f(\zeta)| = |R(\zeta) + \varphi(\zeta)| = |R(\zeta)| \left\{1 + \frac{\varphi(\zeta)}{R(\zeta)}\right\} \geq \lambda + \delta.$$

The claim is proved.

Case 2. There exist points on the circle $|z| = 1$ such that

$$\varphi(z) = 0.$$

Since the considered functions are regular on the circle $|z| = 1$, there exists a $\rho \quad (0 < \rho < 1)$ such that:

1) The functions $\varphi(z)$ and $R(z)$ are regular and do not vanish in the domain $\rho \leq |z| < 1$.

2) The function $\varphi(z)$ has at least $n + 1$ zeros [3] in the domain $|z| < \rho$.

Choosing r to satisfy the inequality

$$\rho < r < 1$$

and reasoning as in Case 1, we conclude that there exist at least one point ζ_r on the circle $|z| = r$ such that

$$\frac{\varphi(\zeta_r)}{R(\zeta_r)} > 0$$

holds and hence that

$$|f(\zeta_r)| > |R(\zeta_r)|.$$

Letting r tend to 1 we get a point $\zeta \quad (|\zeta| = 1)$ such that

$$|f(\zeta)| \geq |R(\zeta)| = \lambda.$$

The theorem is proved. □

The established result allows us to solve the following extremal problem which in a certain sense is a generalization of the problems stated earlier:

Problem A. *Let p be a fixed number satisfying the condition*

$$0 \leq p \leq n.$$

What is the least possible value of $\max\limits_{|z|=1} |f(z)|$ *when* f *is runs over the whole class* K_p?

To solve this problem we will first investigate whether or not the class K_p $(p = 0, 1, \ldots, n)$ contains a rational function of the form (4). The solution of the problem A will be obtained as an immediate consequence of this investigation. If the problem A is solved, then the resolution of the Carathéodory and Pick problems mentioned in the introduction presents no difficulties.

---

3) For this it is enough that ρ exceeds the greatest value of $|z_j| \quad (j = 1, 2, \ldots, n + 1)$.

2 The Carathéodory-Fejér problem

We start with the resolution of the problem A under the assumption that the condition (**B**) is of the form (3) (i.e., the Carathéodory–Fejér conditions). Without less of generality we assume that $c_0 \neq 0$. For this goal we will study all rational functions of the form

$$F(z) = \lambda \, \frac{a_0 + a_1 z + \ldots + a_n z^n}{\overline{a_n} + \overline{a_{n-1}} z + \ldots + \overline{a_0} z^n}, \tag{10}$$

where λ is positive and the coefficients are specified by the equations

$$F(0) = c_0, F'(0) = 1! \cdot c_1, \ldots, F^{(n)}(0) = n! \cdot c_n. \tag{11}$$

Since $F(0)$ is finite, the function $F(z)$ is regular in a neighbourhood of the point $z = 0$ and it is representable in the form

$$F(z) = c_0 + c_1 z + \ldots + c_n z^n + \gamma_{n+1} z^{n+1} + \ldots. \tag{12}$$

Comparing (10) and (12) we obtain the following system of equations

$$\left\{ \begin{array}{lcl} \lambda a_0 & = & c_0 \overline{a_n} \\ \lambda a_1 & = & c_1 \overline{a_n} + c_0 \overline{a_{n-1}} \\ \lambda a_2 & = & c_2 \overline{a_n} + c_1 \overline{a_{n-1}} + c_0 \overline{a_{n-2}} \\ \ldots & \ldots & \ldots\ldots\ldots\ldots\ldots\ldots\ldots\ldots \\ \lambda a_n & = & c_n \overline{a_n} + c_{n-1} \overline{a_{n-1}} + \ldots + c_0 \overline{a_0} \end{array} \right. , \tag{13}$$

where $a_0, \ldots, a_n$ are the coefficients of the rational function (10) which satisfies (11) and λ is the same positive value as in (1). From this it follows immediately that the positive value λ has to satisfy the following equation

$$D(\lambda) = \begin{vmatrix} \lambda & 0 & 0 & \cdots & 0 & c_0 & c_1 & c_2 & \cdots & c_n \\ 0 & \lambda & 0 & \cdots & 0 & 0 & c_0 & c_1 & \cdots & c_{n-1} \\ 0 & 0 & \lambda & \cdots & 0 & 0 & 0 & c_0 & \cdots & c_{n-2} \\ \cdots & \cdots & \cdots & \cdots & \cdots & \cdots & \cdots & \cdots & \cdots & \cdots \\ 0 & 0 & 0 & \cdots & \lambda & 0 & 0 & 0 & \cdots & c_0 \\ \overline{c_0} & 0 & 0 & \cdots & 0 & \lambda & 0 & 0 & \cdots & 0 \\ \overline{c_1} & \overline{c_0} & 0 & \cdots & 0 & 0 & \lambda & 0 & \cdots & 0 \\ \cdots & \cdots & \cdots & \cdots & \cdots & \cdots & \cdots & \cdots & \cdots & \cdots \\ \overline{c_n} & \overline{c_{n-1}} & \overline{c_{n-2}} & \cdots & \overline{c_0} & 0 & 0 & 0 & \cdots & \lambda \end{vmatrix} = 0. \tag{14}$$

It is clear that to each positive root of this equation there corresponds a solution of the system (13) with respect to the variables $a_0, a_1, \ldots, a_n$ and hence that there exist some rational functions of the form (10) which satisfy the conditions (11). Conversely, if the equation (14) has no positive roots then there exist no fraction of the form (10) satisfying the condition (11). However, from the form of the equation (14) it follows that all its roots are real. Since $D(-\lambda) = D(\lambda)$, there exist positive roots of the equation (14) and their total number (counted with their multiplicities) is equal to $n+1$. Thus, we have to investigate to which classes $K_p \quad (p=0,1,\ldots,n)$ the rational fractions corresponding to the different roots of the equation (14) belong.

Lemma. *Assume that λ^* is a ν-multiple root of the equation* (14). *Then there exists a rational function of the form* (10)–(11) *which after cancellation can be represented in the form*[4]

$$\lambda^* \frac{b_0 + b_1 z + \ldots + b_{n-\nu+1} z^{n-\nu+1}}{\overline{b_{n-\nu+1}} + \overline{b_{n-\nu}} z + \ldots + \overline{b_0} z^{n-\nu+1}}. \tag{15}$$

Proof. To justify the lemma, we show that from the assumption that *the root λ^* has multiplicity ν* it follows that it is possible to choose the coefficients b_0, b_1, ..., $b_{n-\nu+1}$ in (15) in such a way that (15) is fulfilled. But this is equivalent to the solvability of the system

$$\left\{\begin{array}{rcl} \lambda^* b_0 & = & c_0 \overline{b_{n-\nu+1}} \\ \lambda^* b_1 & = & c_1 \overline{b_{n-\nu+1}} + c_0 \overline{b_{n-\nu}} \\ \cdots & \cdots & \cdots\cdots\cdots\cdots\cdots\cdots \\ \lambda^* b_{n-\nu+1} & = & c_{n-\nu+1} \overline{b_{n-\nu+1}} + \ldots + c_0 \overline{b_0} \\ 0 & = & c_{n-\nu+2} \overline{b_{n-\nu+1}} + \ldots + c_1 \overline{b_0} \\ \cdots & \cdots & \cdots\cdots\cdots\cdots\cdots\cdots \\ 0 & = & c_n \overline{b_{n-\nu+1}} + \ldots + c_{\nu-1} \overline{b_0} \end{array}\right. ,$$

with respect to the variables $b_0, b_1, \ldots, b_{n-\nu+1}$. Thus the lemma will be proved if we show that the equations

$$\begin{vmatrix} \lambda^* & 0 & \cdots & 0 & c_0 & c_1 & \cdots & c_{n-\nu+1} \\ 0 & \lambda^* & \cdots & 0 & 0 & c_0 & \cdots & c_{n-\nu} \\ \cdots & \cdots & \cdots & \cdots & \cdots & \cdots & \cdots & \cdots \\ 0 & 0 & \cdots & \lambda^* & 0 & 0 & \cdots & c_0 \\ \overline{c_0} & 0 & \cdots & 0 & \lambda^* & 0 & \cdots & 0 \\ \overline{c_1} & \overline{c_0} & \cdots & 0 & 0 & \lambda^* & \cdots & 0 \\ \cdots & \cdots & \cdots & \cdots & \cdots & \cdots & \cdots & \cdots \\ \overline{c_{n-\nu+1}} & \overline{c_{n-\nu}} & \cdots & \overline{c_0} & 0 & 0 & \cdots & \lambda^* \end{vmatrix} = 0$$

and

$$\begin{vmatrix} 0 & 0 & \cdots & 0 & c_{k+1} & c_{k+2} & \cdots & c_{k+n-\nu+2} \\ 0 & \lambda^* & \cdots & 0 & 0 & c_0 & \cdots & c_{n-\nu} \\ \cdots & \cdots & \cdots & \cdots & \cdots & \cdots & \cdots & \cdots \\ 0 & 0 & \cdots & \lambda^* & 0 & 0 & \cdots & c_0 \\ \overline{c_0} & 0 & \cdots & 0 & \lambda^* & 0 & \cdots & 0 \\ \overline{c_1} & \overline{c_0} & \cdots & 0 & 0 & \lambda^* & \cdots & 0 \\ \cdots & \cdots & \cdots & \cdots & \cdots & \cdots & \cdots & \cdots \\ \overline{c_{n-\nu+1}} & \overline{c_{n-\nu}} & \cdots & \overline{c_0} & 0 & 0 & \cdots & \lambda^* \end{vmatrix} = 0,$$

have the same solution λ^* for $k \in \{0,1,\ldots,\nu-2\}$.

4) Later (see Theorem 2) it will be shown that $b_0 \neq 0$ – *editorial note.*

The left hand side of the first equation is (up to a nonzero constant factor) the main minor of order $2n+2-\nu+1$ which is obtained from the determinant $D(\lambda^*)$ by crossing out the first $\nu-1$ rows and the first $\nu-1$ columns. Analogously, to obtain the left hand side of the second equation, we have to cross out the first $\nu-1$ columns and rows with the indices $1,2,\ldots,\nu-k-2,\nu-k,\nu-k+1,\ldots,\nu$ from the determinant $D(\lambda^*)$. Since the multiplicity of the root λ^* is equal to ν, the rank of the determinant $D(\lambda^*)$ is equal to $2n+2-\nu$. The lemma is proved. □

Theorem 2. *Assume that the positive roots of equation (14)*

$$\lambda_0 > \lambda_1 > \lambda_2 > \ldots > \lambda_q > 0$$

have multiplicities

$$\nu_0, \nu_1, \nu_2, \ldots, \nu_q$$

respectively, where

$$\nu_0 + \nu_1 + \nu_2 + \ldots + \nu_q = n+1. \tag{16}$$

Then there exists a series of fractions satisfying the conditions (3), *which (after complete cancellation) have the form*[5]

$$R_j(z) = \lambda_j \, \frac{b_0 + b_1 z + \ldots + b_{n-\nu_j+1} z^{n-\nu_j+1}}{\overline{b_{n-\nu_j+1}} + \overline{b_{n-\nu_j}} z + \ldots + \overline{b_0} z^{n-\nu_j+1}} \qquad (j = 1,2,\ldots,q). \tag{17}$$

The fraction $R_j(z)$ has exactly

$$\nu_0 + \nu_1 + \nu_2 + \ldots + \nu_{j-1} \tag{18}$$

poles in the domain $|z| < 1$. In particular, $R_0(z)$ is regular in the domain $|z| \le 1$.

Proof. From the Lemma we know that for $\lambda^* = \lambda_j$ there exists a fraction of the form (17) satisfying (11). Thus, we have to show that this fraction is not cancellable and that the total number of its poles in the domain $|z|<1$ is given by the expression (18).

First we will prove that $R_0(z)$ is regular in the domain $|z|<1$. If, to the contrary, $R_0(z)$ has $\alpha > 0$ poles in the domain $|z|<1$, then the function $R_1(z)$ has not less than $\nu_0 + \alpha$ poles in this domain (otherwise, according to Theorem 1 we would have $\lambda_1 \ge \lambda_0$). By the same reasoning, $R_2(z)$ has not less than $\nu_0 + \nu_1 + \alpha$ poles in the domain $|z|<1$, and so on. Finally, the function $R_q(z)$ has not less than

$$\nu_0 + \nu_1 + \ldots + \nu_{q-1} + \alpha = N$$

poles in the domain $|z|<1$. However,

$$N = n + 1 + \alpha - \nu_q$$

5) *For simplicity we write $b_0, b_1, \ldots$ instead of $b_0^{(j)}, b_1^{(j)}, \ldots$.*

and the degree of the function $R_q(z)$ is equal to $n+1-\nu_q$, which is less than N. This contradiction proves that $\alpha = 0$, i.e., *the function* $R_0(z)$ *is regular in the domain* $|z| < 1$. Let us turn to the function $R_1(z)$. It has not less than ν_0 poles in the domain $|z| < 1$ (otherwise, it would belong to the class $K_{\nu_0 - 1}$, and by Theorem 1 we would have $\lambda_1 \geq \lambda_0$. On the other hand, the function $R_1(z)$ can not have more than ν_0 poles, otherwise, repeating the above reasoning, we would obtain that $R_2(z)$ has more than $\nu_0 + \nu_1$ poles, and so on. As we showed, this is impossible. Thus, *the function* $R_1(z)$ *has exactly* ν_0 *poles in the domain* $|z| < 1$. Analogously, we show that *the function* $R_2(z)$ *has exactly* $\nu_0 + \nu_1$ *poles*, and so on.

It remains to prove that the fractions (17) are not cancellable. If we assume that some fraction R_j is cancellable and is such that the degrees of its numerator and denominator do not exceed the number $n - \nu_j = m$, then the total number of poles of the fraction R_j is equal to

$$\nu_0 + \nu_1 + \ldots + \nu_{j-1} = p.$$

Since the fraction R_{j+1} belongs to the class $K_{\nu_0+\nu_1+\ldots+\nu_j}$ (i.e., to the class K_{p+n-m}), it would follow from Theorem 1 that $\lambda_{j+1} \geq \lambda_j$. However, this is impossible. Thus, the theorem is proved. □

In view of Theorem 1, we can now obtain the solution of problem A in the case of the Carathéodory–Fejér conditions. The obtained result can be represented in the following form:

Theorem 3. *Let*

$$\lambda_0 > \lambda_1 > \ldots > \lambda_q > 0$$

be all the positive roots of equation (14) *and let*

$$\nu_0, \nu_1, \ldots, \nu_q \qquad (\nu_0 + \nu_1 + \ldots + \nu_q = n+1)$$

be their multiplicities. Let $f(z)$ *be a function which is analytic in the domain* $|z| \leq 1$, *regular on the circle* $|z| = 1$, *having not more than* p $(0 \leq p \leq n)$ *poles in the domain* $|z| < 1$ *and satisfying the conditions* (3), *then*

$$\max_{|z|=1} |f(z)| \geq \lambda_j \, ,$$

where the index j *is determined by the inequalities*

$$\nu_0 + \nu_1 + \ldots + \nu_{j-1} \leq p \leq \nu_0 + \nu_1 + \ldots + \nu_{j-1} + \nu_j - 1$$

(if $p \leq \nu_0 - 1$ *then* $j = 0$*).*

Now we turn to the Carathéodory–Fejér problem in its traditional formulation (i.e., in the formulation given in the introduction). The following theorem gives the solution of this problem:

Theorem 4. *Let the function*

$$f(z) = c_0 + c_1 z + \ldots + c_n z^n + \gamma_{n+1} z^{n+1} + \ldots \tag{19}$$

be regular[6] *in the domain $|z| < 1$. Then the least upper bound G of its modulus in the domain $|z| < 1$ satisfies the inequality*

$$G \geq \lambda \tag{20}$$

where λ is the maximal root of the equation (14).

Equality holds in (20) if and only if

$$f(z) = \lambda \, \frac{a_0 + a_1 z + \ldots + a_m z^m}{\overline{a_m} + \overline{a_{m-1}} z + \ldots + \overline{a_0} z^m} = R(z) \quad (m \leq n),$$

where $R(z)$ is the rational fraction corresponding to the root λ (regular in the domain $|z| \leq 1$).

Proof. Let us consider the product

$$\varphi(z) = f(z) \left(\overline{a_m} + \overline{a_{m-1}} z + \ldots \overline{a_0} z^m\right).$$

From the relations

$$\left\{\begin{array}{rcl} \lambda a_0 & = & c_0 \overline{a_m} \\ \lambda a_1 & = & c_1 \overline{a_m} + c_0 \overline{a_{m-1}} \\ \ldots & \ldots & \ldots\ldots\ldots\ldots\ldots\ldots \\ \lambda a_m & = & c_m \overline{a_m} + c_{m-1} \overline{a_{m-1}} + \ldots + c_0 \overline{a_0} \\ 0 & = & c_{m+1} \overline{a_m} + c_m \overline{a_{m-1}} + \ldots + c_1 \overline{a_0} \\ \ldots & \ldots & \ldots\ldots\ldots\ldots\ldots\ldots \\ 0 & = & c_n \overline{a_m} + c_{n-1} \overline{a_{m-1}} + \ldots + c_{n-m} \overline{a_0} \end{array}\right. ,$$

we obtain

$$\varphi(z) = \lambda \left(a_0 + a_1 z + \ldots + a_m z^m + b_{n+1} z^{n+1} + \ldots\right).$$

Let $0 \leq \rho < 1$. Then the equality

$$\lambda^2 \sum_{k=0}^{m} |a_k|^2 \rho^{2k} + \lambda^2 \sum_{k=n+1}^{\infty} |b_k|^2 \rho^{2k} = \frac{1}{2\pi} \int_0^{2\pi} \left| f\left(\rho e^{i\theta}\right) \left\{\overline{a_m} + \ldots + \overline{a_0} \rho^m e^{mi\theta}\right\}\right|^2 \, d\theta$$

is valid. From this it follows that

$$\lambda^2 \sum_{k=0}^{m} |a_k|^2 \rho^{2k} + \lambda^2 \sum_{k=n+1}^{\infty} |b_k|^2 \rho^{2k} \leq G^2 \sum_{k=0}^{m} |a_k|^2 \rho^{2(m-k)}.$$

---

6) *We do not make any assumptions concerning the behavior of the function $f(z)$ on the circle $|z| = 1$.*

Hence, the series $\sum\limits_{k=n+1}^{\infty} |b_k|^2$ converges, and

$$\lambda^2 \sum_{k=0}^{m} |a_k|^2 + \lambda^2 \sum_{k=n+1}^{\infty} |b_k|^2 \le G^2 \sum_{k=0}^{m} |a_k|^2.$$

This formula shows [7] that the least possible value of G is λ, and this value is attained only by the conditions

$$b_{n+1} = b_{n+2} = \ldots = 0,$$

which are equivalent to the identity

$$f(z)\left(\overline{a_m} + \overline{a_{m-1}}z + \ldots + \overline{a_0}z^m\right) = \lambda\left(a_0 + a_1 z + \ldots + a_m z^m\right).$$

Thus, the theorem is proved. □

3 A remark on equation (14)

The equation (14) can be represented in the form

$$\begin{vmatrix} \lambda E & C \\ \overline{C}' & \lambda E \end{vmatrix} = 0,$$

where

$$C = \begin{Bmatrix} c_0 & c_1 & \cdots & c_n \\ 0 & c_0 & \cdots & c_{n-1} \\ \cdots & \cdots & \cdots & \cdots \\ 0 & 0 & \cdots & c_0 \end{Bmatrix},$$

is an upper triangular matrix form of order $n+1$, $\overline{C}'$ is the complex conjugate matrix of the transposed matrix of C and E is the unit matrix of order $n+1$. In view of the I. Schur theorem [8]

$$\begin{vmatrix} \lambda E & C \\ \overline{C}' & \lambda E \end{vmatrix} = |\lambda^2 E - \overline{C}' C|.$$

---

7) The strategy which we use here, was first used by I. Schur in the above mentioned work. See also [Sz].

8) See the above cited paper by I. Schur. The following result is called the I. Schur theorem:
Let A, B, C and D be matrices of the same order and let the matrices A and C be permutable. Then

$$\begin{vmatrix} A & B \\ C & D \end{vmatrix} = |AD - CB|.$$

The proof is based on the identity

$$\begin{pmatrix} A & B \\ C & D \end{pmatrix} = \begin{pmatrix} A & 0 \\ C & E \end{pmatrix} \cdot \begin{pmatrix} E & A^{-1}B \\ 0 & D - CA^{-1}B \end{pmatrix}.$$

See also the above cited paper by A. Cohn [Co, p.130].

Thus, the equation (14) is representable in the form

$$|\lambda^2 E - \overline{C}' C| = 0. \tag{14_1}$$

Let us remark that the left hand side of this equation is the determinant of the Hermitian form

$$H = \lambda^2 \sum_{k=1}^{n+1} x_k \overline{x}_k - \sum_{k=1}^{n+1} |c_0 x_k + c_1 x_{k+1} + \ldots + c_{n+1-k} x_{n+1}|^2. \tag{21}$$

4 An application to the problem on the number of roots of an algebraic equation in the disc

Let the equation

$$a_0 + a_1 z + a_2 z^2 + \ldots + a_n z^n = 0 \quad (a_0 \neq 0, a_n \neq 0). \tag{22}$$

be given. We consider the function

$$F(z) = \frac{a_0 + a_1 z + \ldots + a_n z^n}{\overline{a_n} + \overline{a_{n-1}} z + \ldots + \overline{a_0} z^n} \,. \tag{23}$$

This function is regular in a neighbourhood of zero and hence, it is representable in the form

$$F(z) = c_0 + c_1 z + \ldots + c_n z^n + \ldots.$$

The coefficients c_i are determined from the equations

$$\left\{ \begin{array}{lcl} a_0 & = & c_0 \overline{a_n} \\ a_1 & = & c_0 \overline{a_{n-1}} + c_1 \overline{a_n} \\ \cdots & \cdots & \cdots\cdots\cdots\cdots\cdots\cdots\cdots\cdots \\ a_n & = & c_0 \overline{a_0} + c_1 \overline{a_1} + \ldots + c_n \overline{a_n} \end{array} \right. \,. \tag{24}$$

From here it follows that the number $\lambda = 1$ is a root of equation (14). Denoting the positive roots of this equation by

$$\lambda_0 \geq \lambda_1 \geq \lambda_2 \geq \ldots \geq \lambda_{n-1} \geq \lambda_n > 0$$

(the root $\lambda = 1$ is among these roots), we observe that squares of these roots are characteristic values of the Hermitian form

$$\sum_{k=1}^{n+1} |c_0 x_k + c_1 x_{k+1} + \ldots + c_{n+1-k} x_{n+1}|^2 \,.$$

Let us assume that $1 = \lambda_{n-k}$ and moreover that

$$\lambda_{n-k+1} < 1 < \lambda_{n-k-1}.$$

In other words, we assume that the rank of the form

$$H = \sum_{i=1}^{n+1} |x_i|^2 - \sum_{i=1}^{n+1} \left|c_0 x_i + c_1 x_{i+1} + \ldots + c_{n+1-i} x_{n+1}\right|^2 \tag{25}$$

is equal to n and that its signature is equal to $2k - n$. Since $\lambda = 1$ is a simple root of equation (14), every solution of the system (24) with respect to a_i has the form

$$a_k^* = \sigma\, a_k \qquad (k = 0, 1, \ldots, n),$$

where σ is an arbitrary complex parameter. However, according to Theorem 2, among the rational fractions corresponding to the root $\lambda = 1$, there exists a rational fraction of degree n which is uncancellable and has exactly $n-k$ zeros in the domain $|z| < 1$. Hence, the fraction (23) is uncancellable as well and has exactly k zeros in the domain $|z| < 1$.

Thus, *if the rank of the form (25) is equal to n and if its signature is equal to $2n - k$, then equation (22) has exactly k zeros in the domain $|z| < 1$ and exactly $n - k$ zeros in the domain $|z| > 1$. The polynomials*

$$A(z) = a_0 + a_1 z + \ldots + a_n z^n$$

and (26)

$$A^*(z) = \overline{a_n} + \overline{a_{n-1}} z + \ldots + \overline{a_0} z^n$$

do not have common zeros. (In other words the equation (22) does not have a pair of roots which is mirror-symmetric with respect to the circle $|z| = 1$.)

To obtain the Schur–Cohn theorem from here, we have to apply an invertible linear transformation to the form (25). This transformation should be chosen in such a way that we can obtain a form depending on the variables a_k (instead of the variables c_k). For this goal we introduce

$$x_{n+1} = \overline{a_n} t_{n+1}, \quad x_k = t_k + \overline{a_{k-1}} t_{n+1} \quad (k = 1, 2, \ldots, n). \tag{27}$$

Using these relations and the equalities (24), we obtain

$$c_0 x_i + c_1 x_{i+1} + \ldots + c_{n+1-i} x_{n+1} = c_0 t_i + c_1 t_{i+1} + \ldots + c_{n-i} t_n + a_{n+1-i} t_{n+1}.$$

Observing that

$$\begin{aligned}\sum_{i=1}^{n} \overline{a_{n+1-i}} \left(c_0 t_i + c_1 t_{i+1} + \ldots + c_{n-i} t_n\right) &= \sum_{k=1}^{n} t_k \left(\overline{a_{n+1-k}} c_0 + \ldots + \overline{a_n} c_{k-1}\right) \\ &= \sum_{i=1}^{n} a_{i-1} t_i,\end{aligned}$$

we see without difficulties that the nonsingular transformation (27) brings the form (25) to the form

$$H = \sum_{i=1}^{n} |t_i|^2 - \sum_{i=1}^{n} \left|c_0 t_i + c_1 t_{i+1} + \ldots + c_{n-i} t_n\right|^2 . \tag{28}$$

Now we apply the supplementary transformation

$$t_i = \overline{a_n} z_i + \overline{a_{n-1}} z_{i+1} + \ldots + \overline{a_i} z_n \tag{29}$$

to the form (28). In view of (24) we find that

$$c_0 t_i + c_1 t_{i+1} + \cdots c_{n-i} t_n = a_0 z_i + a_1 z_{i+1} + \cdots + a_{n-i} z_n.$$

Thus, (28) takes the form

$$H = \sum_{i=1}^{n} \left|\overline{a_n} z_i + \overline{a_{n-1}} z_{i+1} + \ldots + \overline{a_i} z_n\right|^2 - \sum_{i=1}^{n} \left|a_0 z_i + a_1 z_{i+1} + \ldots + a_{n-i} z_n\right|^2 . \tag{30}$$

The last form is expressed directly in terms of the coefficients a_j. Thus, we obtain the following theorem:

Theorem 5. [A. Cohn][9] *If a Hermitian form* (30) *has rank n and signature $2n-k$, then equation* (22) *has k zeros in the domain $|z| < 1$ and $n-k$ zeros in the domain $|z| > 1$. Under these conditions the polynomials* (26) *do not have common zeros.*

5 The G. Pick problem

Let us turn to Problem A supposing that the conditions **(B)** have the form (2) (i.e., the G. Pick conditions). Proceeding just as in Section 2, we will construct rational functions of degree n with constant modulus on the unit circle which satisfy the conditions **(B)**. Without loss of generality we assume that $w_k \neq 0 \quad (k = 1, 2, \ldots, n+1)$. Now it is more convenient to seek such a fraction not in the form (10) but in the form

$$F(z) = \frac{\displaystyle\sum_{k=1}^{n+1} \frac{A_k w_k}{z - z_k}}{\displaystyle\sum_{k=1}^{n+1} \frac{A_k}{z - z_k}}. \tag{31}$$

The expression (31) satisfies the conditions (2) for every A_k. It remains to require that the modulus of the function $F(z)$ is constant on the circle $|z| = 1$. This requirement is equivalent to the equalities

$$F\left(\frac{1}{\overline{z_j}}\right) = \frac{\lambda^2}{\overline{w_j}} \qquad (j = 1, 2, \ldots, n+1).$$

Denoting this constant value of $|F(z)|$ by λ we obtain the system of equations

$$\sum_{k=1}^{n+1} A_k \, \frac{w_k \overline{w_j} - \lambda^2}{z_k \overline{z_j} - 1} = 0 \qquad (j = 1, 2, \ldots, n+1). \tag{32}$$

9) I. Schur obtained this theorem in the special case $k = n$.

Thus, the equation from which we determine the value of λ has the form

$$\begin{vmatrix} \dfrac{w_1\overline{w_1}-\lambda^2}{1-z_1\overline{z_1}} & \dfrac{w_1\overline{w_2}-\lambda^2}{1-z_1\overline{z_2}} & \dots & \dfrac{w_1\overline{w_{n+1}}-\lambda^2}{1-z_1\overline{z_{n+1}}} \\ \dfrac{w_2\overline{w_1}-\lambda^2}{1-z_2\overline{z_1}} & \dfrac{w_2\overline{w_2}-\lambda^2}{1-z_2\overline{z_2}} & \dots & \dfrac{w_2\overline{w_{n+1}}-\lambda^2}{1-z_2\overline{z_{n+1}}} \\ \dots & \dots & \dots & \dots \\ \dfrac{w_{n+1}\overline{w_1}-\lambda^2}{1-z_{n+1}\overline{z_1}} & \dfrac{w_{n+1}\overline{w_2}-\lambda^2}{1-z_{n+1}\overline{z_2}} & \dots & \dfrac{w_{n+1}\overline{w_{n+1}}-\lambda^2}{1-z_{n+1}\overline{z_{n+1}}} \end{vmatrix} = 0. \tag{33}$$

Denoting

$$a_{ik} = \frac{w_i\overline{w_k}}{1-z_i\overline{z_k}}, \qquad b_{ik} = \frac{1}{1-z_i\overline{z_k}}$$

and observing that

$$a_{ik} = \overline{a_{ki}}, \qquad b_{ik} = \overline{b_{ki}}$$

and that the Hermitian forms

$$\sum a_{ik}x_i\overline{x_k} = \sum b_{ik}y_i\overline{y_k} = \frac{1}{2\pi}\int_0^{2\pi}\left|\sum_{\nu=1}^{n+1}\frac{y_\nu}{e^{i\theta}-z_\nu}\right|^2 d\theta,$$

(where $y_i = w_i x_i$) are positive, we conclude that all the values of λ^2 which satisfy equation (33) are positive. Thus, we obtain $n+1$ positive values for λ. In view of the nature of equation (33), we see that if a number λ is a multiple root of equation (33), then the rank of the corresponding matrix drops by the multiplicity of the root. Thus, repeating *mutatis mutandis* the reasoning used in Section 2, we obtain the following lemma:

If λ^* is a root of multiplicity ν of the equation (33), *then among the rational fractions* (31) *corresponding to the value λ^* there exists a fraction which (after reductions) can be represented in the form*

$$\frac{\displaystyle\sum_{k=1}^{n-\nu+2}\frac{B_k w_k}{z-z_k}}{\displaystyle\sum_{k=1}^{n-\nu+2}\frac{B_k}{z-z_k}} = \lambda^*\,\frac{b_0+b_1z+\ldots+b_{n-\nu+1}z^{n-\nu+1}}{\overline{b_{n-\nu+1}}+\overline{b_{n-\nu}}z+\ldots+\overline{b_0}z^{n-\nu+1}}.$$

Theorems which are analogous to Theorem 2 and Theorem 3 can be established in a similar way. Omitting such considerations, we will turn to the proof of a result which is analogous to Theorem 4:

Theorem 6. *Let $F(z)$ be a function which is regular* [10] *in the domain $|z| < 1$ and satisfies the conditions (2). Then the least upper bound of its modulus in the domain $|z| < 1$ satisfies the inequality*

$$G \geq \lambda, \tag{34}$$

where λ is the maximal root of equation (33). The equality in (34) is valid if and only if

$$F(z) = R(z) = \frac{\sum_{k=1}^{n+1} \frac{A_k w_k}{z - z_k}}{\sum_{k=1}^{n+1} \frac{A_k}{z - z_k}},$$

where $R(z)$ is the rational fraction which corresponds to the root λ and which is regular in the domain $|z| < 1$.

Proof. Let us consider the product

$$\varphi(z) = F(z) \sum_{k=1}^{n+1} \frac{A_k}{z - z_k} = \sum_{k=1}^{n+1} \frac{B_k}{z - z_k} + \psi(z), \tag{35}$$

where

$$\psi(z) = c_0 + c_1 z + c_2 z^2 + \dots$$

is a regular function in the domain $|z| < 1$.

Let

$$\rho := \max\{|z_1|, |z_2|, \dots, |z_{n+1}|\}.$$

Let r be an arbitrary number which satisfies the inequality $\rho < r < 1$ and let $z = re^{i\theta}$. In view of (35), we obtain on the one hand

$$\frac{1}{2\pi} \int_0^{2\pi} |\varphi(z)|^2 \, d\theta \leq G^2 \frac{1}{2\pi} \int_0^{2\pi} \left| \sum_{k=1}^{n+1} \frac{A_k}{z - z_k} \right|^2 d\theta = G^2 \sum_{k,j} \frac{A_k \overline{A_j}}{r^2 - z_k \overline{z_j}}$$

and on the other hand

$$\begin{aligned} \frac{1}{2\pi} \int_0^{2\pi} |\varphi(z)|^2 \, d\theta &= \sum_{k,j} \frac{B_k \overline{B_j}}{r^2 - z_k \overline{z_j}} + \frac{1}{2\pi} \int_0^{2\pi} |\psi(z)|^2 \, d\theta + \frac{1}{\pi} \Re \int_0^{2\pi} \psi(z) \sum_{k=1}^{n+1} \frac{\overline{B_k}}{\overline{z} - \overline{z_k}} \, d\theta \\ &= \sum_{k,j} \frac{A_k \overline{A_j} w_k \overline{w_j}}{r^2 - z_k \overline{z_j}} + \sum_{i=0}^{\infty} |c_i|^2 r^{2i}, \end{aligned}$$

---

10) *We do not make any assumptions concerning the behavior of the function $F(z)$ on the circle $|z| = 1$.*

since $B_k = A_k w_k$ in view of (2), and

$$\int_0^{2\pi} \psi(z) \frac{\overline{B_k}}{\overline{z} - \overline{z_k}} \, d\theta = \int_0^{2\pi} \psi(z) \frac{\overline{B_k} z}{r^2 - z\overline{z_k}} \, d\theta = i \oint_{|z|=r} \psi(z) \frac{\overline{B_k} \, dz}{z\overline{z_k} - r^2} = 0.$$

Thus,

$$G^2 \sum_{k,j} \frac{A_k \overline{A_j}}{r^2 - z_k \overline{z_j}} \geq \sum_{k,j} \frac{A_k \overline{A_j} w_k \overline{w_j}}{r^2 - z_k \overline{z_j}} + \sum_{i=0}^{\infty} |c_i|^2 r^{2i}.$$

Hence, the series

$$\sum_{i=0}^{\infty} |c_i|^2$$

converges and (in view of (32))

$$G^2 \sum_{k,j} \frac{A_k \overline{A_j}}{1 - z_k \overline{z_j}} \geq \lambda^2 \sum_{k,j} \frac{A_k \overline{A_j}}{1 - z_k \overline{z_j}} + \sum_{i=0}^{\infty} |c_i|^2.$$

This formula shows that λ is the least possible value for G and that this value is attained only if

$$c_0 = c_1 = c_2 = \ldots = 0,$$

i.e., if $\psi \equiv 0$. This is the same as

$$F(z) \sum_{k=1}^{n+1} \frac{A_k}{z - z_k} = \sum_{k=1}^{n+1} \frac{A_k w_k}{z - z_k}.$$

The theorem is proved. □

References

[CaFe] Carathéodory, C.; Fejér, L.: *Über den Zusammenhang der Extremen von harmonischen Funktionen mit ihren Koeffizienten und über den Picard-Landauschen Satz*, Rend. Circ. Mat. Palermo **32** (1911), 218–239 .

[Co] Cohn, A.: *Über die Anzahl der Wurzeln einer algebraischen Gleichung in einem Kreis*, Math. Zeitschrift **14** (1922), 110–148.

[Eg] Egerváry, E.: *Über gewisse Extremumprobleme der Funktionentheorie*, Math. Ann. **99** (1928), 542–561.

[Fe] Fejér, L.: *Über gewisse Minimalprobleme der Funktionentheorie*, Math. Ann. **97**, (1926), 104–123. Reprinted in: Fejér, L.: *Gesammelte Arbeiten. Band 2* (Herausgegeben und mit Kommentaren versehen von Pál Turán). Akadémiai Kiado. Budapest 1970, pp. 182–202.

[FrKi] *Ausgewählte Arbeiten zu den Ursprüngen der Schur-Analysis. (Series: Teubner-Archiv zur Mathematik:* 016*).* (B. Fritzsche and B. Kirstein – editors.) B.G. Teubner Verlagsgesellschaft, Stuttgart·Leipzig 1991.

[Nev1] Nevanlinna, R.: *Asymptotische Entwicklungen beschränkter Funktionen und das Stieltjessche Momentenproblem*, Ann. Acad. Sci. Fenn. **A 18** (1922), no. 5, 1–53

[Nev2] Nevanlinna, R.: *Über beschränkte analytische Funktionen*, Ann. Acad. Sci. Fenn. **A 32** (1929), no. 7, 1–75 Reprinted in: [FrKi], pp. 97–171.

[Pi1] Pick, G.: *Über die Beschränkungen analytischer Funktionen, welche durch vorgegebene Funktionswerte bewirkt werden*, Math. Ann. **77** (1916), 7–23. Reprinted in: [FrKi], pp. 77-93.

[Pi2] Pick, G.: *Über beschränkte Funktionen mit vorgegebenen Wertzuordnungen*, Ann. Acad. Sci. Fenn. **A 15** (1920), no. 3, 1–17.

[S1] Schur, I.: *Über Potenzreihen, die im Innern des Einheitskreises beschränkt sind. I.*, J. reine und angewandte Math. **147**(1917), 205–232. Reprinted in: [FrKi], pp. 22–49. English translation: On power series which are bounded in the interior of the unit circle.I, [Sch], pp. 31–59.

[S2] Schur, I.: *Über Potenzreihen, die im Innern des Einheitskreises beschränkt sind. II.*, J. reine und angewandte Math. **148** (1918), 122–145 Reprinted in: [FrKi], pp. 50–73. English translation: On power series which are bounded in the interior of the unit circle.II, [Sch], pp. 61–88.

[Sch] *I. Schur Methods in Operator Theory and Signal Processing* (Operator Theory: Advances and Applications, vol. **18**) (I.Gohberg, ed.). Birkhäuser Verlag, Basel·Boston·Stuttgart 1986.

[Sz] Szász, O.: *Ungleichungen für die Koeffizienten einer Potenzreihe*, Math. Zeitschrift **1** (1918), 163–184

The editors have added [FrKi], [Sch] and a number of collected works to the original list of references.

AMS Subject Classification: 41A20, 41A50, 41A57, 47B35.

Operator Theory
Advances and Applications, Vol. 95
© 1997 Birkhäuser Verlag Basel/Switzerland

On tangential interpolation in reproducing kernel Hilbert modules and applications

D. Alpay and V. Bolotnikov

Abstract. Nevanlinna-Pick and Carathéodory-Fejér interpolation problems are considered in reproducing kernel Hilbert modules. Applications are given to interpolation in Hardy-Sobolev, Dirichlet and Bergman spaces. The case of de Branges-Rovnyak spaces is also considered.

1 Introduction

In this paper we continue our investigations [3], [4], [5], [6] of interpolation problems in the space $\mathbf{H}_2^{p\times q}$ of $\mathbb{C}^{p\times q}$-valued functions with entries in the Hardy space $\mathbf{H}_2$ of the unit disk $\mathbb{D}$. In [4] we studied a two-sided residue interpolation problem under the constraint

$$\frac{1}{2\pi}\int_0^{2\pi} \mathrm{Tr}\,\left(H(e^{it})^*H(e^{it})\right) dt \leq k$$

and a left-sided residue problem under the constraint

$$[H,\ H] \overset{\text{def}}{=} \frac{1}{2\pi}\int_0^{2\pi} H(e^{it})^*H(e^{it})dt \leq I_q. \tag{1.1}$$

As a corollary, we solved left-sided interpolation problems of the Nevanlinna-Pick and Carathéodory-Fejér type under the restriction (1.1) and in particular, the problem

$$a_iH(z_i) = c_i \quad (i = 1,\dots,n) \tag{1.2}$$

where the interpolation data consists of n points $z_1,\dots,z_n$ in $\mathbb{D}$, and of matrices $a_i \in \mathbb{C}^{r_i\times p}$, $c_i \in \mathbb{C}^{r_i\times q}$ $(i = 1,\dots,n)$, with $[H,\ H] \leq I_q$. Condition (1.1) is equivalent (see [6]) to the nonnegativity of the kernel

$$K_H(z,\omega) = \frac{I_p}{1-z\bar{\omega}} - H(z)H(\omega)^* \tag{1.3}$$

in $\mathbb{D}$ (the definition of a nonnegative kernel is recalled in Section 2). The difference between this kernel and the more classical kernel

$$\frac{I_p - H(z)H(\omega)^*}{1-z\bar{\omega}} \tag{1.4}$$

associated with Schur functions should be noted. It will be elaborated on in Section 11. Let us mention at this stage that the nonnegativity of (1.3) is equivalent to the fact that the operator of multiplication by H is a contraction from $\mathbb{C}^q$ into $\mathbf{H}_2^p$, while the nonnegativity of the kernel (1.4) is equivalent to the fact that multiplication by H is a contraction from $\mathbf{H}_2^q$ into $\mathbf{H}_2^p$. See Theorem 11.1 for a result which covers both situations. See also [6] for a discussion of this point.

From the positivity of the kernel K_H in $\mathbb{D}$ we see that a necessary condition for the problem (1.2) to be solvable is that the $n \times n$ block matrix with ℓj block

$$a_\ell K_H(z_\ell, z_j)a_j^* = \frac{a_\ell a_j^*}{1 - z_\ell \bar{z}_j} - c_\ell c_j^*$$

should be nonnegative. It suffices to multiply the $n \times n$ block matrix with ℓj block entry $K_H(\omega_\ell, \omega_j)$ on the left by the block diagonal matrix diag $(a_1, a_2, \ldots, a_n)$ and on the right by its adjoint. The condition is in fact sufficient (see [24], [33, p.345] for the scalar case) and a description of the set of all solutions is given in [4] and [6]. Thus, the interpolation conditions (1.2) are well adapted to the constraint (1.1).

The situation is not so simple if one keeps the constraint (1.1) but replaces the left-sided interpolation conditions (1.2) by right-sided conditions

$$H(z_i)b_i = d_i \quad (i = 1, \ldots, n) \tag{1.5}$$

where now $b_i \in \mathbb{C}^{q \times r_i}$ and $d_i \in \mathbb{C}^{p \times r_i}$. This exhibits an important difference with the related problem of interpolation of Schur functions (i.e. when (1.1) is replaced by $H(z)^*H(z) \leq I_q \quad (z \in \mathbb{D})$). If H is a $\mathbb{C}^{p \times q}$-valued Schur function, then $H^\#$ defined by $H^\#(z) = H(\bar{z})^*$ is a $\mathbb{C}^{q \times p}$-valued Schur function and one can convert right-sided interpolation (1.4) into a left-sided interpolation problem (by taking adjoints) without leaving the context of Schur functions. This is not the case of $\mathbf{H}_2^{p \times q}$ functions subject to (1.1). If H satisfies (1.1) then $H^\#$ does not, in general, satisfy

$$\frac{1}{2\pi} \int_0^{2\pi} H^\#(e^{it})^* H^\#(e^{it}) dt \leq I_p.$$

We study these problems in the framework of reproducing kernel Hilbert modules, and we study tangential interpolation problems in such modules.

The outline of the paper is as follows. In Section 2 we review various definitions and properties of reproducing kernel Hilbert modules. We also give precise statements of the interpolation problems to be solved. In Sections 3 and 4 we study left-tangential Nevanlinna-Pick and Carathéodory-Fejér interpolation problems in such modules while in Section 5 and 6 we focus on right-tangential interpolation. In Sections 7 to 11 we study examples and various instances of the left-tangential interpolation problem considered in Section 3. In Section 7, we consider the case of the Hardy spaces and in Section 8 the case of Hardy-Sobolev spaces. The case of

Dirichlet spaces is treated in Section 9 and the case of Bergman spaces is considered in Section 10. Finally, in Section 11 we consider the problem of interpolation in de Branges-Rovnyak spaces.

We note that most of the analysis presented here can also be considered in a nonstationary framework (see [9], [27] for the definitions). This will be presented elsewhere.

The notation is quite standard: $\mathbb{C}^{p\times q}$ denotes the space of $p \times q$ matrices with complex entries and $\mathbb{C}^p$ is short for $\mathbb{C}^{p\times 1}$. The adjoint of a matrix A is denoted by A^*.

2 Reproducing kernel Hilbert modules

Let Ω be a set and let $\mathcal{H}$ be a Hilbert space of $\mathbb{C}^{p\times q}$-valued functions defined on Ω with inner product $\langle\ ,\ \rangle$. We suppose that if $F \in \mathcal{H}$ and $A \in \mathbb{C}^{q\times q}$ then $FA \in \mathcal{H}$ (i.e. $\mathcal{H}$ is a right $\mathbb{C}^{q\times q}$-module) and that we are given a $\mathbb{C}^{q\times q}$-valued hermitian form $[\ ,\ ] : \mathcal{H}\times\mathcal{H} \to \mathbb{C}^{q\times q}$, which is continuous with respect to $\langle\ ,\ \rangle$. Then, $\mathcal{H}$ is called a reproducing kernel Hilbert module if there exists a $\mathbb{C}^{p\times p}$-valued function $K(z,\omega)$ defined on $\Omega\times\Omega$ and such that for every choice of $\omega \in \Omega$ and $B \in \mathbb{C}^{p\times q}$ the function $K_\omega B : z \to K(z,\omega)B$ belongs to $\mathcal{H}$ and

$$[F,\ K_\omega B] = B^*F(\omega) \quad \forall\ F \in \mathcal{H}. \tag{2.1}$$

The function $K(z,\omega)$ is nonnegative, in the sense that all the finite block matrices of the form $(K(\omega_i,\omega_j))_{i,j=1}^r$, where $r = 1,2,\ldots$ and $\omega_1,\ldots,\omega_r$ vary in Ω, are nonnegative. This property will be denoted by $K(z,\omega) \succeq 0$. The function $K(z,\omega)$ is furthermore uniquely defined (as is easily verified), and is called the reproducing kernel of $\mathcal{H}$. When $p = q = 1$, the space $\mathcal{H}$ is a reproducing kernel Hilbert space, as defined by Aronszajn. We refer to [12], [18], [39] and [41] for the theory of reproducing kernel Hilbert spaces and to [30] and [36] for the theory of reproducing kernel Hilbert modules. If K is a $\mathbb{C}^{p\times p}$-valued function nonnegative on a set Ω, we denote by $\mathcal{H}^{p\times q}(K)$ the Hilbert module of $\mathbb{C}^{p\times q}$-valued functions with reproducing kernel K.

The one-to-one correspondence of Aronszajn-Moore between nonnegative functions and reproducing kernel Hilbert spaces extends to a one-to-one correspondence between $\mathbb{C}^{p\times p}$-valued nonnegative functions on Ω and Hilbert modules of $\mathbb{C}^{p\times q}$-valued functions defined on Ω.

Theorem 2.1. *Let $K(z,\omega)$ be a $\mathbb{C}^{p\times p}$-valued function which is nonnegative on a set Ω (in the sense described above). Then $\mathcal{H}^{p\times q}(K)$ is uniquely defined. It is the closure of elements of the form*

$$F(z) = \sum_{i=1}^{n} K(z,\omega_i)D_i \tag{2.2}$$

where $\omega_i \in \Omega$ and $D_i \in \mathbb{C}^{p\times q}$, in the inner product

$$\langle F, F\rangle_{\mathcal{H}^{p\times q}(K)} = \sum_{i,j=i}^{n} \operatorname{Tr} D_i^* K(\omega_i, \omega_j) D_j. \tag{2.3}$$

For F of the form (2.2), let

$$[F, F] = \sum_{i,j=i}^{n} D_i^* K(\omega_i, \omega_j) D_j. \tag{2.4}$$

Then $[\ ,\ ]$ is continuous with respect to the inner product (2.3) *and satisfies* (2.1).

The proof of this theorem is the same as for the classical case $p = q = 1$ (see [12]). The continuity of the form (2.4) is obtained using the Cauchy-Schwartz inequality.

As an example, let us take $K(z, \omega)$ of the form $K(z, \omega) = k(z, \omega) I_p$ where $k(z, \omega)$ is scalar and nonnegative on Ω. Then, $\mathcal{H}^{p\times q}(K)$ is the space of $p\times q$ matrix-valued functions $M = (m_{ij})$ with entries in the reproducing kernel Hilbert space $\mathcal{H}(k)$. Furthermore, the forms (2.3) and (2.4) are now equal to

$$\langle M, M\rangle = \sum_{i,k} \langle m_{ik},\ m_{ik}\rangle_{\mathcal{H}(k)} \quad \text{and} \quad [M, M]_{ij} = \sum_{k} \langle m_{ik},\ m_{jk}\rangle_{\mathcal{H}(k)}$$

respectively.

Example 2.1. Let

$$K(z, \omega) = \sum_{n\in S} z^n \bar{\omega}^n M_n, \tag{2.5}$$

where $S \subset \mathbb{N}$ and where the $M_n \in \mathbb{C}^{p\times p}$ are strictly positive matrices such that, if Card $S = \infty$,

$$\limsup \|M_n\|^{\frac{1}{n}} \le 1.$$

Then, $K(z, \omega)$ is nonnegative in $\mathbb{D}$, and the associated reproducing kernel Hilbert module consists of the functions of the form

$$F(z) = \sum_{n\in S} z^n D_n,$$

where $D_n \in \mathbb{C}^{p\times q}$ are such that

$$\sum_{n\in S} \operatorname{Tr} D_n^* M_n^{-1} D_n < \infty.$$

Moreover, the form (2.4) is equal to

$$[F, F] = \sum_{n\in S} D_n^* M_n^{-1} D_n.$$

When $p = q = 1$ and $M_n \stackrel{\text{def}}{=} m_n = \frac{1}{n}$, $n > 0$, $K(z,\omega) = -\ln(1 - z\bar{\omega})$, and the associated space is the Dirichlet space. When $m_n = \frac{1}{n^2+1}$, one obtains the Hardy-Sobolev space, while the choice $m_n = n$ leads to the Bergman space. Matrix analogues of these spaces are given by taking matrices M_n such that

$$k_1 m_n I_p \leq M_n \leq k_2 m_n I_p,$$

where k_1 and k_2 are strictly positive constants.

Now we introduce the problems NPL_Υ, NPR_Υ and CFL_Υ. In the following, Υ denotes a strictly positive $q \times q$ matrix. For $H \in \mathcal{H}^{p\times q}(K)$ we consider the constraint

$$[H, H] \leq \Upsilon. \tag{2.6}$$

Problem $\mathbf{NPL}_\Upsilon$. The left-sided Nevanlinna-Pick interpolation problem: give necessary and sufficient conditions for a function $H \in \mathcal{H}^{p\times q}(K)$ which satisfies (2.6) and (1.2) to exist, and describe the set of all such functions.

Problem $\mathbf{NPR}_\Upsilon$. The right-sided Nevanlinna-Pick interpolation problem: give necessary and sufficient conditions for a function $H \in \mathcal{H}^{p\times q}(K)$ which satisfies (2.6) and (1.4) to exist, and describe the set of all such functions.

When $\Omega \subset \mathbb{C}$ and $K(z,\omega)$ is analytic in z and $\bar{\omega}$, the functions $z \to \frac{\partial^j}{\partial\bar{\omega}^j}K(z,\omega)D$ belong to $\mathcal{H}^{p\times q}(K)$ for $D \in \mathbb{C}^{p\times q}$; the proof of this fact for Hilbert modules is the same as in the case of Hilbert spaces; see e.g. [7] for the latter.

Problem $\mathbf{CFL}_\Upsilon$. The left-sided Carathéodory-Fejér problem): give necessary and sufficient conditions for a function $H \in \mathcal{H}^{p\times q}(K)$ which satisfies (2.6) and the conditions

$$\sum_{j=0}^{k} a_{k-j}\frac{H^{(j)}(z_0)}{j!} = c_k \quad (k = 0, \ldots, n) \tag{2.7}$$

where $z_0 \in \Omega$ and the matrices a_j, c_j are preassigned (and of appropriate dimensions) to exist, and describe the set of all such functions.

We will denote by NPL, NPR and CFL the Nevanlinna-Pick and Carathéodory-Fejér problems without the matrix constraint (2.6).

In Lemma 2.2 we express condition (2.6) in terms of nonnegative kernels. We first need a preliminary lemma.

Lemma 2.1. *Let H, $G \in \mathcal{H}^{p\times q}(K)$ and let A, $B \in \mathbb{C}^{p\times q}$. Then*

$$[HA,\ GB]_{\mathcal{H}^{p\times q}(K)} = B^*[H,\ G]_{\mathcal{H}^{p\times q}(K)}A.$$

This proposition can be easily checked for H and G of the form (2.2) and extended to all of $\mathcal{H}^{p\times q}(K)$ by continuity.

Lemma 2.2. *Let $\Upsilon \in \mathbb{C}^{q\times q}$ be strictly positive matrix and let H be in $\mathcal{H}^{p\times q}(K)$. Then H satisfies (2.6) if and only if*

$$K(z,\omega) - H(z)\Upsilon^{-1}H(\omega)^* \succeq 0 \quad (z,\omega \in \Omega). \tag{2.8}$$

Proof. Without loss of generality, we can assume that $\Upsilon = I_q$. Then, (2.6) holds if and only if the operator M_H of multiplication by H is a contraction from $\mathbb{C}^{p\times q}$ into $\mathcal{H}^{p\times q}(K)$, i.e. if and only if the operator $I - M_H M_H^*$ is nonnegative. Taking into account the fact that, for $A \in \mathbb{C}^{p\times q}$,

$$M_H^* K(z,\omega)A = H(\omega)^* A$$

we obtain the desired equivalence with (2.8). □

In the scalar case, the above result reduces to a standard result (see e.g. [39]).

Lemma 2.3. *Let $\Upsilon \in \mathbb{C}^{q\times q}$ be positive semidefinite, let* rank $\Upsilon = \nu$ *and let*

$$\Upsilon = U^* \begin{pmatrix} \hat{\Upsilon} & 0 \\ 0 & 0_{q-\nu} \end{pmatrix} U \quad (U^*U = UU^* = I_q, \ \hat{\Upsilon} > 0). \tag{2.9}$$

Then the function $H \in \mathcal{H}^{p\times q}(K)$ satisfies (2.6) if and only if it admits a representation

$$H(z) = \begin{pmatrix} \hat{H}(z) & 0 \end{pmatrix} U \tag{2.10}$$

where $\hat{H} \in \mathcal{H}^{p\times \nu}(K)$ and $[\hat{H},\hat{H}]_{\mathcal{H}^{p\times\nu}(K)} \leq \hat{\Upsilon}$.

Proof. The inequality (2.6) is equivalent to the family of inequalities $[H,H] \leq \Upsilon + \varepsilon I_q$ ($\varepsilon \to 0$) which by Lemma 2.1 and (2.8) reduces to

$$K(z,\omega) - H(z)U^* \begin{pmatrix} (\hat{\Upsilon}+\varepsilon I)^{-1} & 0 \\ 0 & \frac{1}{\varepsilon}I \end{pmatrix} UH(\omega)^* \succeq 0.$$

Since the last kernel is nonnegative in Ω, the function $H(z)U^*$ is of the form

$$H(z)U^* = \begin{pmatrix} \hat{H}(z) & 0 \end{pmatrix}$$

and the kernel $K(z,\omega) - \hat{H}(z)\hat{\Upsilon}^{-1}\hat{H}(\omega)^*$ is nonnegative on Ω. Thus, by Lemma 2.2, $[\hat{H},\hat{H}]_{\mathcal{H}^{p\times\nu}(K)} \leq \hat{\Upsilon}$. □

3 The left-sided Nevanlinna-Pick problem

In this section we consider the problem NPL_{Υ} formulated above. We begin with the matrix

$$\mathbb{P} = \left(a_i K(z_i, z_j) a_j^*\right)_{i,j=1}^{n} \tag{3.1}$$

which is nonnegative. We can assume without loss of generality that the $r \times r$ left upper minor $\tilde{\mathbb{P}}$ of $\mathbb{P}$ is strictly positive and that its rank is equal to rank $\mathbb{P}$:

$$\mathbb{P} = \begin{pmatrix} \tilde{\mathbb{P}} & T \\ T^* & S \end{pmatrix}, \quad \tilde{\mathbb{P}} = \left(a_i K(z_i, z_j) a_j^*\right)_{i,j=1}^{r}, \quad \operatorname{rank} \tilde{\mathbb{P}} = \operatorname{rank} \mathbb{P}. \tag{3.2}$$

Indeed, by maybe reindexing the interpolation data, we can assume that the a_i are all line vectors; the matrix $\mathbb{P}$ is then the Gramm matrix of the space $\mathcal{M}$ spanned by the functions $K(z, \omega_i) a_i^*$, in the metric of $\mathcal{H}^{p\times 1}(K)$. Take a maximal subspace $\mathcal{M}_1$ of these functions which forms a basis of $\mathcal{M}$. The corresponding Pick matrix $\tilde{\mathbb{P}}$ is strictly positive. Its size is the rank of $\mathbb{P}$. Indeed, the Schur complement of $\tilde{\mathbb{P}}$ in $\mathbb{P}$ is the Gramm matrix of the orthogonal complement of $\mathcal{M}_1$ in $\mathcal{M}$ and is equal thus to the zero matrix. Thus,

$$S = T^* \tilde{\mathbb{P}}^{-1} T \tag{3.3}$$

which in view of (3.1) and (3.2) can be rewritten as

$$a_i K(z_i, z_j) a_j^* = a_i \left(K(z_i, z_1) a_1^*, \ldots, K(z_i, z_r) a_r^*\right) \tilde{\mathbb{P}}^{-1} \begin{pmatrix} a_1 K(z_1, z_j) \\ \vdots \\ a_r K(z_r, z_j) \end{pmatrix} a_j^*$$
$$(i, j = r+1, \ldots, n). \tag{3.4}$$

Lemma 3.1. *Let $H_{\min}$ be the function defined by*

$$H_{\min}(z) = G(z) \tilde{\mathbb{P}}^{-1} C_r \tag{3.5}$$

where

$$G(z) = \left(K(z, z_1) a_1^*, \ldots, K(z, z_r) a_r^*\right), \qquad C_r = \begin{pmatrix} c_1 \\ \vdots \\ c_r \end{pmatrix}. \tag{3.6}$$

Then $H_{\min}$ belongs to $\mathcal{H}^{p\times q}(K)$, satisfies the interpolation conditions

$$a_i H_{\min}(z_i) = c_i \quad (i = 1, \ldots, r) \tag{3.7}$$

and

$$[H_{\min}, H_{\min}] = C_r^* \tilde{\mathbb{P}}^{-1} C_r. \tag{3.8}$$

Proof. It follows from (3.5), (3.6) that $H_{\min} \in \mathcal{H}^{p\times q}(K)$ and moreover, the matrix $a_iG(z_i)$ coincides with the i-th block row of the matrix $\tilde{\mathbb{P}}$. Therefore,

$$a_iH_{\min}(z_i) = a_iG(z_i)\tilde{\mathbb{P}}^{-1}C_r = c_i \quad (i = 1, \ldots, r).$$

Furthermore, let $\tilde{\mathbb{P}}^{-1} = (p_{ij})$. Then

$$H_{\min}(z) = \sum_{i,j=1}^{r} K(z, z_i)a_i^*p_{ij}c_j \tag{3.9}$$

and therefore,

$$\begin{aligned}
[H_{\min}, H_{\min}] &= [\sum_{i,j=1}^{r} K(\cdot, z_i)a_i^*p_{ij}c_j, \sum_{i,j=1}^{r} K(\cdot, z_i)a_i^*p_{ij}c_j] \\
&= \sum_{t,j=1}^{r} c_t^*\left(\sum_{i,\ell=1}^{r} p_{t\ell}a_\ell K(z_\ell, z_i)a_i^*p_{ij}\right)c_j \\
&= \sum_{t,j=1}^{r} c_t^*p_{tj}c_j = C_r^*\tilde{\mathbb{P}}^{-1}C_r.
\end{aligned}$$

□

It is easy to see that every function $H \in \mathcal{H}^{p\times q}(K)$ which satisfies (3.7) is of the form

$$H(z) = H_{\min}(z) + \Psi(z) \tag{3.10}$$

where Ψ is a function in $\mathcal{H}^{p\times q}(K)$ such that

$$a_i\Psi(z_i) = 0 \quad (i = 1, \ldots, r). \tag{3.11}$$

Let $\mathcal{M}$ be the finite dimensional subspace of $\mathcal{H}^{p\times q}(K)$ defined as

$$\begin{aligned}
\mathcal{M} &= \text{span } \{K(z, z_i)a_i^*b_i, \;\; b_i \in \mathbb{C}^{r_i\times q}, \; i = 1, \ldots, r\} \\
&= \{H : \;\; H(z) = G(z)B, \quad B \in \mathbb{C}^{r\times q}\}
\end{aligned} \tag{3.12}$$

with $r = \sum r_i$. It follows from (3.5) (or (3.9)) that $H_{\min} \in \mathcal{M}$. Note that $\mathcal{M}$ is a reproducing kernel Hilbert module with reproducing kernel $G(z)\tilde{\mathbb{P}}^{-1}G(\omega)^*$. Therefore, its orthogonal complement $\mathcal{M}^\perp$ in $\mathcal{H}^{p\times q}(K)$ is the reproducing kernel Hilbert module with reproducing kernel

$$\tilde{K}(z, \omega) = K(z, \omega) - G(z)\tilde{\mathbb{P}}^{-1}G(\omega)^*,$$

i.e. $\mathcal{M}^\perp = \mathcal{H}^{p\times q}(\tilde{K})$. Since

$$[\Psi, K(\cdot, z_i)a_i^*b] = b^*a_i \; \Psi(z_i) \quad (\forall b \in \mathbb{C}^{r_i\times q}),$$

the function Ψ satisfies (3.11) if and only if it belongs to $\mathcal{H}^{p\times q}(\tilde{K})$.

In particular, $[\Psi, H_{\min}] = 0$. Thus, the representation (3.10) is orthogonal and on account of (3.8),

$$[H,\ H] = [\Psi,\ \Psi] + [H_{\min},\ H_{\min}] = [\Psi,\ \Psi] + C_r^* \tilde{\mathbb{P}}^{-1} C_r.$$

Therefore, a function H of the form (3.10) satisfies the constraint (2.6) if and only if

$$[\Psi, \Psi] \leq \Upsilon - C_r^* \tilde{\mathbb{P}}^{-1} C_r. \tag{3.13}$$

Since Ψ belongs to $\mathcal{H}^{p\times q}(\tilde{K})$, the kernel

$$K(z, \omega) - G(z)\tilde{\mathbb{P}}^{-1}G(\omega)^* - \mu\Psi(z)\Psi(\omega)^*$$

is nonnegative for sufficiently small $\mu > 0$. In particular, in view of (3.4),

$$\begin{aligned} 0 \ &\leq a_i K(z_i, z_i) a_i^* - a_i G(z_i)\tilde{\mathbb{P}}^{-1}G(z_i)^* a_i^* - \mu a_i \Psi(z_i)\Psi(z_i)^* a_i^* \\ &= -\mu a_i \Psi(z_i)\Psi(z_i)^* a_i^* \qquad (i = r+1, \ldots, n) \end{aligned}$$

and therefore, $a_i\Psi(z_i) = 0$ for $i = 1, \ldots, n$. So, the function H of the form (3.10) satisfies the last $n-r$ interpolation conditions (1.2) if and only if $a_i H_{\min}(z_i) = c_i$ for $i = r+1, \ldots, n$ or equivalently, if and only if

$$\begin{pmatrix} c_{r+1} \\ \vdots \\ c_n \end{pmatrix} = T^* \tilde{\mathbb{P}}^{-1} \begin{pmatrix} c_1 \\ \vdots \\ c_r \end{pmatrix} \tag{3.14}$$

where T is defined in (3.2).

Theorem 3.1. *The problem NPL$_\Upsilon$ has a solution if and only if*

$$\begin{pmatrix} \mathbb{P} & C_n \\ C_n^* & \Upsilon \end{pmatrix} \geq 0. \tag{3.15}$$

Proof. It follows from the above considerations that the problem NPL$_\Upsilon$ is solvable if and only if (3.14) is in force and $\Upsilon \geq C_r^* \tilde{\mathbb{P}}^{-1} C_r$. These conditions are equivalent to (3.15). To see this, we introduce the matrix $\Phi = \begin{pmatrix} I & 0 \\ -T^*\tilde{\mathbb{P}}^{-1} & I \end{pmatrix}$. Note that in view of (3.2), (3.3) and (3.14),

$$\Phi \begin{pmatrix} \tilde{\mathbb{P}} & T \\ T^* & S \end{pmatrix} \Phi^* = \begin{pmatrix} \tilde{\mathbb{P}} & 0 \\ 0 & 0 \end{pmatrix}, \qquad \Phi\, C_n = \begin{pmatrix} C_r \\ 0 \end{pmatrix}. \tag{3.16}$$

The signature of the matrix from the left side of (3.15) is the same as the one obtained by multiplying it on the left by the matrix $\begin{pmatrix} \Phi & 0 \\ 0 & I \end{pmatrix}$ and on the right by its adjoint. In view of (3.16), the resulting matrix is equal to

$$\begin{pmatrix} \tilde{\mathbb{P}} & 0 & C_r \\ 0 & 0 & 0 \\ C_r & 0 & \Upsilon \end{pmatrix}$$

which ends the proof. □

As will be illustrated in Sections 7–11, the real problem is to get a nice description of the space $\mathcal{M}^{\perp}$. In the case of $\mathbf{H}_2^{p\times q}$, the Beurling-Lax theorem ([34], [38]) insures that $\mathcal{M}^{\perp} = \Theta\mathbf{H}_2^{p\times q}$, where Θ is a $p \times p$ Blaschke product (see [4] and Section 7). In general, such a representation for $\mathcal{M}^{\perp}$ is not available.

Theorem 3.2. *Under assumption* (3.15), *all solutions to the problem NPL$_\Upsilon$ are described by* (3.10) *where $H_{\min}$ is the function defined by* (3.5) *and Ψ is an arbitrary function from $\mathcal{H}^{p\times q}(\tilde{K})$ satisfying* (3.13). *The problem has a unique solution if and only if or* $\Upsilon = C_r^*\tilde{\mathbb{P}}^{-1}C_r$ or $K(z,\omega) = G(z)\tilde{\mathbb{P}}^{-1}G(\omega)^*$.

Using the last two theorems one can describe all solutions to the problem NPL (i.e. when the constraint (2.6) is removed).

Theorem 3.3. *The problem NPL has a solution if and only if* (3.15) *holds for some* $\Upsilon \geq 0$, *i.e. if and only if* $\mathbb{P} \geq 0$ *and* $P_{Ker\mathbb{P}}C_n = 0$. *The set of all solutions is parametrized by* (3.10) *when* Ψ *varies in the space* $\mathcal{H}^{p\times q}(\tilde{K})$. *The problem has a unique solution if and only if* $K(z,\omega) = G(z)\tilde{\mathbb{P}}^{-1}G(\omega)^*$.

We note that the analysis in this section is close in spirit to [25, chapter 7].

4 The left-sided Carathéodory-Fejér problem

In this section we consider the problems CFL and CFL$_\Upsilon$ with interpolation conditions (2.7). Using the polynomials

$$\mathbf{a}(z) = \sum_{i=1}^{n} a_i(z-z_0)^i, \qquad \mathbf{c}(z) = \sum_{i=1}^{n} c_i(z-z_0)^i \tag{4.1}$$

one can rewrite (2.7) as

$$\frac{d^i}{dz^i}\left(\mathbf{a}(z)H(z) - \mathbf{c}(z)\right)_{z=z_0} = 0 \qquad (i = 0,\ldots,n). \tag{4.2}$$

Let $\mathbb{P}$ be the matrix with block entries

$$\mathbb{P}_{ij} = \frac{1}{i!j!}\frac{\partial^{i+j}}{\partial z^i\partial\bar{\omega}^j}\left(\mathbf{a}(z)K(z,\omega)\mathbf{a}(\omega)^*\right)_{z=\omega=z_0} \qquad (i,j = 0,\ldots,n) \tag{4.3}$$

and assume that $\mathbb{P} > 0$. Let $G(z)$ be the function given by

$$G(z) = \left(K(z,z_0)a_0^*,\ \frac{\partial}{\partial\bar{\omega}}\left(K(z,\omega)\mathbf{a}(\omega)^*\right)_{\omega=z_0},\ldots,\frac{1}{n!}\frac{\partial^n}{\partial\bar{\omega}^n}\left(K(z,\omega)\mathbf{a}(\omega)^*\right)_{\omega=z_0}\right). \tag{4.4}$$

It follows immediately from (4.3), (4.4) that the matrix $\frac{d^i}{dz^i}\left(\mathbf{a}(z)G(z)\right)_{z=z_0}$ coincides with the i-th block row of the matrix $\mathbb{P}$. Therefore, the function

$$H_{\min}(z) = G(z)\mathbb{P}^{-1}C, \qquad C = \begin{pmatrix} c_0 \\ \vdots \\ c_n \end{pmatrix} \tag{4.5}$$

is subject to the constraints

$$\frac{d^i}{dz^i}\left(\mathbf{a}(z)H_{\min}(z)-\mathbf{c}(z)\right)_{z=z_0}=0 \qquad (i=0,\ldots,n)$$

and hence, satisfies the interpolation conditions (2.6).
Since the functions $\frac{\partial^i}{\partial\bar{\omega}^i}K(z,\omega)B$ belong to $\mathcal{H}^{p\times q}(K)$ for $i=1,\ldots$ and $B\in\mathbb{C}^{p\times q}$, we have $H_{\min}\in\mathcal{H}^{p\times q}(K)$. Using Lemma 2.1 it can be checked by a direct computation that

$$[H_{\min},H_{\min}]=C^*\mathbb{P}^{-1}C. \tag{4.6}$$

Every function $H\in\mathcal{H}^{p\times q}(K)$ satisfying (2.6) is of the form

$$H(z)=H_{\min}(z)+\Psi(z) \tag{4.7}$$

for some function $\Psi\in\mathcal{H}^{p\times q}(K)$ such that

$$\frac{d^i}{dz^i}\left(\mathbf{a}(z)\Psi(z)\right)_{z=z_0}=0 \qquad (i=0,\ldots,n). \tag{4.8}$$

Let $\mathcal{M}$ be the finite dimensional subspace of $\mathcal{H}^{p\times q}(K)$ defined as

$$\begin{aligned}&=\operatorname{span}\left\{\frac{\partial^i}{\partial\bar{\omega}^i}\left(K(z,\omega)\mathbf{a}(\omega)^*\right)_{\omega=z_0}b_i,\ \ b_i\in\mathbb{C}^{r\times q},\ i=0,\ldots,n\right\}\\ &=\left\{H:\ \ H(z)=G(z)B,\quad B\in\mathbb{C}^{r(n+1)\times q}\right\}.\end{aligned} \tag{4.9}$$

$\mathcal{M}$ is the reproducing kernel Hilbert module with reproducing kernel $G(z)\mathbb{P}^{-1}G(\omega)^*$ and hence, its orthogonal complement in $\mathcal{H}^{p\times q}(K)$ is a reproducing kernel Hilbert module $\mathcal{M}^{\perp}=\mathcal{H}^{p\times q}(\tilde{K})$ with reproducing kernel

$$\tilde{K}(z,\omega)=K(z,\omega)-G(z)\mathbb{P}^{-1}G(\omega)^*.$$

Since $\forall b\in\mathbb{C}^{r_i\times q}$,

$$\left[\Psi,\ \frac{\partial^i}{\partial\bar{\omega}^i}\left(K(z,\omega)\mathbf{a}(\omega)^*\right)_{\omega=z_0}b\right]=b^*\frac{d^i}{dz^i}\left(\mathbf{a}(z)\Psi(z)\right)_{z=z_0}, \tag{4.10}$$

the function Ψ satisfies (4.8) if and only if it belongs to $\mathcal{H}^{p\times q}(\tilde{K})$. In particular, $[\Psi,\ H_{\min}]=0$, the representation (4.7) is orthogonal and in view of (4.6),

$$[H,H]=[\Psi,\Psi]+C^*\mathbb{P}^{-1}C.$$

As in the previous section, this leads to the following result:

Theorem 4.1. *Let the matrix $\mathbb{P}$ given by (4.2) be strictly positive. Then:*

(i) The problem CFL always has solutions. The problem CFL$_\Upsilon$ is solvable if and only if

$$\Upsilon\geq C^*\mathbb{P}^{-1}C.$$

(ii) All solutions to the problems CFL and CFL$_\Upsilon$ are parametrized by formula (4.7) where $H_{\min}$ is the function defined by (4.5) and Ψ is a function from $\mathcal{H}^{p\times q}(\tilde{K})$: arbitrary (for CFL) or such that $[\Psi,\Psi]\leq\Upsilon-C^*\mathbb{P}^{-1}C$ (for CFL$_\Upsilon$).

We conclude this section with the following analogue of Theorem 3.1.

Theorem 4.2. *The problem CFL_Υ has a solution if and only if*

$$\begin{pmatrix} \mathbb{P} & C_n \\ C_n^* & \Upsilon \end{pmatrix} \geq 0. \tag{4.11}$$

where $\mathbb{P}$ and C_n are the matrices given by (4.3) and (4.5) respectively.

Proof. Let us consider the Hilbert module

$$\mathbf{H}(K) = \mathcal{H}^{p\times q}(K) \oplus \cdots \oplus \mathcal{H}^{p\times q}(K)$$

which is a direct sum of $n+2$ copies of $\mathcal{H}^{p\times q}(K)$, with the $\mathbb{C}^{(n+2)p\times(n+2)p}$-valued form

$$\left[\begin{pmatrix} H_0 \\ \vdots \\ H_{n+1} \end{pmatrix}, \begin{pmatrix} H_0 \\ \vdots \\ H_{n+1} \end{pmatrix}\right]_{\mathbf{H}(K)} = \left([H_i,\ H_j]_{\mathcal{H}^{p\times q}(K)}\right)_{i,j=0}^{n+1}, \qquad H_i \in \mathcal{H}^{p\times q}(K). \tag{4.12}$$

Let us take $H_j(z) = \frac{\partial^j}{\partial\bar{\omega}^j}\left(K(z,\omega)\mathbf{a}(\omega)^*\right)_{\omega=z_0} b_j$ $(b_j \in \mathbb{C}^{r\times q},\ j=0,\dots,n)$ and let $H_{n+1}(z) = H(z)$ be a solution to the problem CFL_Υ. Then in view of (4.1), (4.2), (4.10) and (2.7),

$$[H_i,\ H_j]_{\mathcal{H}^{p\times q}(K)} = b_j^*\mathbb{P}_{ij}b_i, \qquad [H,\ H_j]_{\mathcal{H}^{p\times q}(K)} = b_j^*\ \frac{d^i}{dz^i}\left(\mathbf{a}(z)H(z)\right)_{z=z_0} = b_j^*c_j$$

which upon being substituted into (4.12), lead to

$$0 \leq \begin{pmatrix} b_0^* & & & \\ & \ddots & & \\ & & b_n^* & \\ & & & I_q \end{pmatrix} \begin{pmatrix} \mathbb{P} & C \\ C^* & [H,H] \end{pmatrix} \begin{pmatrix} b_0 & & & \\ & \ddots & & \\ & & b_n & \\ & & & I_q \end{pmatrix}.$$

Since the b_i are arbitrary, this last inequality implies

$$0 \leq \begin{pmatrix} \mathbb{P} & C \\ C^* & [H,H] \end{pmatrix} \leq \begin{pmatrix} \mathbb{P} & C \\ C^* & \Upsilon \end{pmatrix}.$$

Conversely, let us suppose that (4.11) holds. We consider the kernel

$$k(z,\omega) = (1 + z\bar{\omega} + \cdots + z^n\bar{\omega}^n)I_q.$$

For arbitrary $\varepsilon > 0$ the kernel $K+\varepsilon k$ is nonnegative and the corresponding matrix

$$\begin{aligned}\mathbb{P}_\varepsilon &= \left(\tfrac{1}{i!j!}\ \tfrac{\partial^{i+j}}{\partial z^i\partial\bar{\omega}^j}\ \left(\mathbf{a}(z)\left\{K(z,\omega) + \varepsilon k(z,\omega)\right\}\mathbf{a}(\omega)^*\right)_{z=\omega=z_0}\right) \\ &= \mathbb{P} + \varepsilon \begin{pmatrix} a_0 & 0 & \dots & 0 \\ a_1 & \ddots & \ddots & \vdots \\ \vdots & \ddots & \ddots & 0 \\ a_n & \dots & a_1 & a_0 \end{pmatrix} \begin{pmatrix} a_0 & 0 & \dots & 0 \\ a_1 & \ddots & \ddots & \vdots \\ \vdots & \ddots & \ddots & 0 \\ a_n & \dots & a_1 & a_0 \end{pmatrix}^*\end{aligned}$$

is strictly positive. Let $\mathcal{H}_\varepsilon^{p\times q}(K+\varepsilon k)$ be the reproducing kernel Hilbert module with reproducing kernel $K+\varepsilon k$. By Theorem 4.2, there exists a function $H_\varepsilon \in \mathcal{H}_\varepsilon^{p\times q}(K+\varepsilon k)$ which satisfies the interpolation conditions (4.2) and is such that $[H_\varepsilon, H_\varepsilon]_{\mathcal{H}_\varepsilon^{p\times q}(K+\varepsilon k)} \le \Upsilon$, or equivalently,

$$K(z,\omega) + \varepsilon k(z,\omega) - H_\varepsilon(z)\Upsilon^{-1}H_\varepsilon(\omega)^* \succeq 0 \qquad (z,\ \omega \in \Omega).$$

Thus, for $z \in \Omega$,

$$H_\epsilon(z)\Upsilon H_\epsilon(z)^* \le K(z,z) + \epsilon k(z,z),$$

and the family H_ϵ is uniformly bounded on compact subsets of Ω. Hence, there exists a limit function $H(z)$, which is analytic in Ω, is readily seen to satisfy the interpolation conditions (4.2) and is subject to

$$K(z,\omega) - H(z)\Upsilon^{-1}H(\omega)^* \succeq 0.$$

By Lemma 2.2, $H \in \mathcal{H}^{p\times q}(K)$ and $[H,H] \le \Upsilon$. The case of degenerate Υ can be easily reduced to the considered one using Lemma 2.3. □

5 Right-sided interpolation

We begin with the simplest right-sided problem: *given matrices* $b \in \mathbb{C}^{q\times s}$, $d_i \in \mathbb{C}^{p\times s}$ *and points* $z_i \in \Omega$, *find necessary and sufficient conditions for* $H \in \mathcal{H}^{p\times q}(K)$ *to exist such that* (2.6) *holds and*

$$H(z_i)b = d_i \qquad (i = 1, \ldots, n). \tag{5.1}$$

Without loss of generality we can assume Υ to be strictly positive and b to be a full rank matrix. Then, $b^*\Upsilon b$ is nonsingular.

Lemma 5.1. *There exists a function* $H \in \mathcal{H}^{p\times q}(K)$ *satisfying* (2.6) *and* (5.1) *if and only if*

$$\mathbb{P} \ge D(b^*\Upsilon b)^{-1}D^* \tag{5.2}$$

where $\mathbb{P}$ *and* D *are matrices defined by*

$$\mathbb{P} = \left(K(z_i, z_j)\right)_{i,j=1}^n, \qquad D = \begin{pmatrix} d_1 \\ \vdots \\ d_n \end{pmatrix}. \tag{5.3}$$

Proof. Let us assume that H exists. By Lemma 2.2, the kernel

$$K(z,\omega)-H(z)\Upsilon^{-1}H(\omega)^*$$

is nonnegative in Ω. Writing the corresponding matrix for the n points $z_1,\ldots,z_n$ and using (5.3) we get the following inequality

$$\mathbb{P}\geq\begin{pmatrix}H(z_1)\\ \vdots\\ H(z_n)\end{pmatrix}\Upsilon^{-1}\left(H(z_1)^*,\ldots,H(z_n)^*\right)$$

which in turn is equivalent to

$$\begin{pmatrix}\mathbb{P} & \begin{pmatrix}H(z_1)\\ \vdots\\ H(z_n)\end{pmatrix}\\ * & \Upsilon\end{pmatrix}\geq 0. \tag{5.4}$$

Multiplying the last inequality by the matrix $\begin{pmatrix}I & 0\\ 0 & b\end{pmatrix}$ on the right, by its adjoint on the left and taking into account (5.1), (5.3) we obtain

$$\begin{pmatrix}\mathbb{P} & D\\ D^* & b^*\Upsilon b\end{pmatrix}\geq 0 \tag{5.5}$$

which is equivalent to (5.2). Let us suppose now that (5.2) (or equivalently, (5.5)) is in force. Using Schur complements (see [25, formulas (0.3) and (0.4) p.3]), one sees that (5.5) is equivalent to

$$\begin{pmatrix}\mathbb{P} & D(b^*\Upsilon b)^{-1}b^*\Upsilon\\ \Upsilon b(b^*\Upsilon b)^{-1}D^* & \Upsilon\end{pmatrix}\geq 0 \tag{5.6}$$

which by Theorem 3.1 means that there exists a function $H\in\mathcal{H}^{p\times q}(K)$ such that
$[H,H]\leq\Upsilon$ and

$$\begin{pmatrix}H(z_1)\\ \vdots\\ H(z_n)\end{pmatrix}=D(b^*\Upsilon b)^{-1}b^*\Upsilon.$$

This function evidently satisfies (5.1). □

To describe all functions satisfying (2.6) and (5.1) we introduce the subspace

$$\begin{aligned}\mathcal{M}\ &=\text{span}\,\{K(z,z_i)M_i,\ \ M_i\in\mathbb{C}^{p\times q},\ i=1,\ldots,n\}\\ &=\{H:\ \ H(z)=G(z)A,\quad M\in\mathbb{C}^{pn\times q}\},\end{aligned}$$

where

$$G(z) = (K(z, z_1), \dots, K(z, z_n)). \tag{5.7}$$

As above, $\mathcal{M}$ is a finite dimensional reproducing kernel Hilbert module with reproducing kernel $G(z)\mathbb{P}^{-1}G(\omega)^*$ and its orthogonal complement $\mathcal{M}^\perp$ is the reproducing kernel Hilbert module $\mathcal{H}^{p\times q}(\tilde{K})$ with reproducing kernel $\tilde{K}(z,\omega) = K(z,\omega) - G(z)\mathbb{P}^{-1}G(\omega)^*$.

Lemma 5.2. *The function H satisfies (2.6) and (5.1) if and only if it admits a representation of the form*

$$H(z) = G(z)\mathbb{P}^{-1}F_M + \Psi(z) \tag{5.8}$$

where $\mathbb{P}$ and G are defined by (5.3) and (5.7) respectively,

$$F_M = D(b^*\Upsilon b)^{-1}b^*\Upsilon + M\left(I_q - b(b^*\Upsilon b)^{-1}b^*\Upsilon\right), \tag{5.9}$$

$M \in \mathbb{C}^{pn\times q}$ is an arbitrary matrix such that

$$M\left\{\Upsilon^{-1} - b(b^*\Upsilon b)^{-1}b^*\right\}M^* \leq \mathbb{P} - D(b^*\Upsilon b)^{-1}D^* \tag{5.10}$$

and Ψ is an arbitrary function from $\mathcal{H}^{p\times q}(\tilde{K})$ such that

$$[\Psi, \Psi]_{\mathcal{H}(\tilde{K})} \leq \Upsilon - F_M^*\mathbb{P}^{-1}F_M. \tag{5.11}$$

Proof. Let H satisfy (2.6) and (5.1). By Theorem 3.2, H admits a representation (5.8) for some $\Psi \in \mathcal{H}^{p\times q}(\tilde{K})$ and

$$F_M = \begin{pmatrix} H(z_1) \\ \vdots \\ H(z_n) \end{pmatrix}.$$

In view of (5.1), F_M is subject to $F_M\, b = D$ and hence, it admits a representation (5.9) for some $M \in \mathbb{C}^{pn\times q}$. Since the representation (5.8) is orthogonal,

$$F_M^*\mathbb{P}^{-1}F_M = \left[G\mathbb{P}^{-1}F_M,\, G\mathbb{P}^{-1}F_M\right] \leq [H, H] \leq \Upsilon. \tag{5.12}$$

Substituting (5.9) into (5.12) we obtain (5.10). Using again the orthogonality of the representation (5.8) and taking into account (5.12), one can conclude that H of the form (5.8) satisfies (2.6) if and only if the parameter Ψ is subject to (5.11).

The converse is clear: for every $\Psi \in \mathcal{H}^{p\times q}(\tilde{K})$ we have $\Psi(z_i) = 0$ and on account of (5.9), $F_M\, b = D$. Therefore, the function H satisfies conditions (2.6), (5.1) due to (5.10), (5.11). □

Similarly one can solve the following Carathéodory-Féjer problem: *given matrices* $b \in \mathbb{C}^{q\times s}$, $d_i \in \mathbb{C}^{p\times s}$ *and a point* $z_0 \in \Omega$, *to describe all functions* $H \in \mathcal{H}^{p\times q}(K)$ *satisfying* (2.6) *and such that*

$$\sum_{j=0}^{k} \frac{H^{(j)}(z_0)}{j!}\, b = d_k \qquad (j = 0, \ldots, n). \tag{5.13}$$

Lemma 5.3. *There exists a function* $H \in \mathcal{H}^{p\times q}(K)$ *satisfying* (2.6) *and* (5.13) *if and only if*

$$\mathbb{P} \geq D(b^*\Upsilon b)^{-1}D^*$$

where

$$\mathbb{P} = \left(\left(\frac{\partial^{i+j}}{\partial z^i \partial \bar{\omega}^j}\, K(z,\omega)\right)_{z=\omega=z_0}\right)_{i,j=0}^{n}, \qquad D = \begin{pmatrix} d_1 \\ \vdots \\ d_n \end{pmatrix}. \tag{5.14}$$

Lemma 5.4. *The function* H *satisfies the conditions* (2.6) *and* (5.13) *if and only if it admits a representation* (5.8) *for*

$$G(z) = \left(K(z,z_0),\ \frac{\partial}{\partial \bar{\omega}} K(z,\omega)|_{\omega=z_0}, \ldots, \frac{1}{n!}\frac{\partial^n}{\partial \bar{\omega}^n} K(z,\omega)|_{\omega=z_0}\right),$$

F_M *being defined by* (5.9), (5.10) *and arbitrary function* Ψ *from the reproducing kernel Hilbert module* $\mathcal{H}^{p\times q}(\tilde{K})$ *with the reproducing kernel* $\tilde{K}(z,\omega) = K(z,\omega) - G(z)\mathbb{P}^{-1}G(\omega)^*$ *and such that* (5.11) *is in force.*

6 On the recursive solution and the solvability criterion

We now turn to the solution of the problem NPR_Υ. The corresponding recursion is not so explicit as for the left-sided interpolation. This is not surprising, since the reproducing property of the form [,] corresponds just to left interpolation conditions.

We will assume that $K(z,z) > 0$ for all $z \in \Omega$. Let $H \in \mathcal{H}^{p\times q}(K)$ satisfy (2.6) and the first interpolation condition $H(z_1)b_1 = d_1$. Let $\mathcal{H}^{p\times q}(\tilde{K})$ be the reproducing kernel Hilbert module with the reproducing kernel $\tilde{K}(z,\omega) = K(z,\omega) - K(z,z_1)K(z_1,z_1)^{-1}K(z_1,\omega)$. By Lemma 5.2, H is of the form

$$H(z) = K(z,z_1)K(z_1,z_1)^{-1}F_{M_1} + \Psi(z) \tag{6.1}$$

where

$$F_{M_1} = d_1(b_1^*\Upsilon b_1)^{-1}b_1^*\Upsilon + M_1\left(I_q - b_1(b_1^*\Upsilon b_1)^{-1}b_1^*\Upsilon\right), \tag{6.2}$$

$M_1 \in \mathbb{C}^{p\times q}$ is a matrix such that

$$M_1\left\{\Upsilon^{-1} - b_1(b_1^*\Upsilon b_1)^{-1}b_1^*\right\}M_1^* \leq K(z_1,z_1) - d_1(b_1^*\Upsilon b_1)^{-1}d_1^* \tag{6.3}$$

and Ψ is an arbitrary function from $\mathcal{H}^{p\times q}(\tilde{K})$ such that

$$[\Psi,\Psi]_{\mathcal{H}(\tilde{K})} \leq \Upsilon - F_{M_1}^* K(z_1,z_1)^{-1}F_{M_1} \stackrel{\text{def}}{=} \Upsilon_1. \tag{6.4}$$

We want to solve, if possible, the supplementary constraint $H(z_2)b_2 = d_2$. Thus, with F_{M_1} as in (6.2) we have to solve

$$\Psi(z_2)b_2 = d_2 - K(z_2,z_1)K(z_1,z_1)^{-1}F_{M_1}b_2 \stackrel{\text{def}}{=} d_2'. \tag{6.5}$$

Assuming $\tilde{K}(z_2,z_2) > 0$, all such Ψ are of the form

$$\Psi(z) = \tilde{K}(z,z_2)\tilde{K}(z_2,z_2)^{-1}F_{M_2} + \Psi_1(z)$$

where

$$F_{M_2} = d_2'(b_2^*\Upsilon_1 b_2)^{-1}b_2^*\Upsilon_1 + M_2\left(I_q - b_2(b_2^*\Upsilon_1 b_2)^{-1}b_2^*\Upsilon_1\right),$$

$$M_2\left\{\Upsilon_1^{-1} - b_2(b_2^*\Upsilon_1 b_2)^{-1}b_2^*\right\}M_2^* \leq \tilde{K}(z_2,z_2) - d_2'(b_2^*\Upsilon_1 b_2)^{-1}d_2'^*$$

and Ψ_1 is is an arbitrary function from the reproducing kernel Hilbert module with reproducing kernel $\tilde{K}(z,\omega) - \tilde{K}(z,z_2)\tilde{K}(z_2,z_2)^{-1}\tilde{K}(z_2,\omega)$ and such that

$$[\Psi,\Psi] \leq \Upsilon - F_{M_1}^* K(z_1,z_1)^{-1}F_{M_1} - \Upsilon_1 - F_{M_2}^*\tilde{K}(z_2,z_2)^{-1}F_{M_2} \stackrel{\text{def}}{=} \Upsilon_2$$

and so, the process can be iterated. Using the solvability criterion for the full rank problem (which is a particular case of Theorem 3.1) we obtain

Theorem 6.1. *The problem NPR_Υ has a solution if and only if there exist matrices $M_1,\ldots,M_n$ such that the block matrix with ij entry*

$$K(z_i,z_j) - T_i\Upsilon^{-1}T_j^*$$

is nonnegative, where

$$T_i = d_i(b_i^*\Upsilon b_i)^{-1}b_i^*\Upsilon + M_i\left(I_q - b_i(b_i^*\Upsilon b_i)^{-1}b_i^*\Upsilon\right).$$

As already mentioned, the next five sections are devoted to various instances of the left tangential Nevanlinna-Pick interpolation problem.

7 The case of Hardy spaces

In the case where $K(z,\omega) = \frac{I_p}{1-z\bar{\omega}}$, the space $\mathcal{H}^{p\times q}(K)$ is the space $\mathbf{H}_2^{p\times q}$ of $p\times q$ matrices with entries in the Hardy space of the disk $\mathbf{H}_2$ with the form (1.1). Both the Nevanlinna-Pick and the Carathéodory-Fejér problem are particular cases of the following problem: *Given* $H_0 \in \mathbf{H}_2^{p\times q}$ *and* Θ *a* $p\times p$ *inner function, describe the set*

$$\{F \in \mathbf{H}_2^{p\times q} |\ \ F = H_0 + \Theta H,\ \ H \in \mathbf{H}_2^{p\times q}, \text{and}\ \ [F,F] \leq \Upsilon\}.$$

For similar problems in the framework of Schur functions, we refer to the works of D. Arov [13], J. Ball and J.W. Helton [15] and A. Kheifets [31]. In the present case, the solution is much simpler than the problems considered in these works, because of the Hilbert space structure. Let P denote the orthogonal projection from $\mathbf{H}_2^{p\times q}$ onto $\mathbf{H}_2^{p\times q} \ominus \Theta\mathbf{H}_2^{p\times q}$. If $\mathbf{L}_2^{p\times q}$ denotes the space of $p\times q$ matrices with entries in the Lebesgue space $\mathbf{L}_2$ of the unit circle (with matrix-valued form (1.1) and associated inner product), and if $\mathbf{p}$ denotes the orthogonal projection from $\mathbf{L}_2^{p\times q}$ onto $\mathbf{H}_2^{p\times q}$, it is easy to check that

$$P = \Theta\mathbf{p}\Theta^*|_{\mathbf{H}_2^{p\times q}}.$$

We decompose $H_0 = PH_0 + (I-P)H_0$. Then, with $U = \mathbf{p}\Theta^* H_0 + H$,

$$F = H_0 + \Theta H = (I-P)H_0 + \Theta U$$

and

$$[F,F] = [(I-P)H_0, (I-P)H_0] + [U,U]. \tag{7.1}$$

The condition $[F,F] \leq \Upsilon$ holds if and only if F can be written as $F = (I-P)H_0 + \Theta U$, where $U \in \mathbf{H}_2^{p\times q}$ satisfies (with $\mathbf{q} = I - \mathbf{p}$)

$$[U,U] \leq \Upsilon - [\mathbf{q}\Theta^* H_0, \mathbf{q}\Theta^* H_0] := \Upsilon_1. \tag{7.2}$$

Note that (7.2) is the generalization of (3.13) and of the inequality appearing in Theorem 4.1 to the present setting.

8 An example: Interpolation in Hardy-Sobolev spaces

In this section we look at the special case of Hardy-Sobolev spaces; we were made aware of this case by Laurent Baratchart, and Theorem 8.1 was developed with him. We wish to thank Laurent Baratchart for allowing us to use this material. For boundary interpolation problems in Hardy-Sobolev spaces, we refer to [17], [16].

The Hardy-Sobolev space $\mathbf{HS}_2$ of functions in the disk $\mathbb{D}$ consists of the elements of $\mathbf{H}_2$ whose derivative is still in $\mathbf{H}_2$, with norm $\|f\|^2_{\mathbf{HS}_2} = \|f\|^2_{\mathbf{H}_2} +$

$\|f'\|^2_{\mathbf{H}_2}$. The space $\mathbf{HS}_2$ is a reproducing kernel Hilbert space, with reproducing kernel

$$K(z,\omega)=\sum_0^\infty \frac{z^n\bar{\omega}^n}{1+n^2}. \tag{8.1}$$

More generally, if Γ is a strictly positive element of $\mathbb{C}^{2q\times 2q}$, one can define $\mathbf{HS}_2^{p\times q}(\Gamma)$ as the space of elements $H\in\mathbf{H}_2^{p\times q}$ with norm

$$\mathrm{Tr}\int_0^{2\pi}\left(H(e^{it})^*\ H'(e^{it})^*\right)\Gamma\begin{pmatrix} H(e^{it})\\ H'(e^{it})\end{pmatrix}dt.$$

It is a reproducing kernel Hilbert space with reproducing kernel

$$\hat{K}(z,\omega)=K(z,\omega)\Gamma^{-1},$$

and a reproducing kernel Hilbert module when endowed with the matrix-valued form

$$[H,H]_{\mathbf{HS}_2^{p\times q}}=\int_0^{2\pi}\left(H(e^{it})^*\ H'(e^{it})^*\right)\Gamma\begin{pmatrix} H(e^{it})\\ H'(e^{it})\end{pmatrix}dt.$$

Of course generalizations of these with derivatives of higher order are possible. They correspond to $M_n=\frac{1}{p(n)}$ in (2.6) where $p(n)$ is a polynomial with integral coefficients. To fix ideas, we focus here on the case where $\Gamma=I_{2p}$ and where only the first order derivative appears.

We specialize the analysis of Section 3 to the present case, and study the orthogonal complement of the space $\mathcal{M}$ defined in (3.12). Note first that for the reproducing kernel (8.1) of the space $\mathbf{HS}_2$ and the full rank matrices $a_1,\ldots,a_n$, the Gramm matrix $\mathbb{P}$ defined via (3.1) is strictly positive (since the vector-valued functions defined by the columns of the functions $z\mapsto K(z,z_i)a_i^*$ are linearly independent). From this point of view, every problem NPL_Υ in the space $\mathbf{HS}_2^{p\times q}$ is nondegenerate. This allows to construct the minimal norm solution

$$H_{\min}(z)=(K(z,z_1)a_1^*,\ldots,K(z,z_n)a_n^*)\,\mathbb{P}^{-1}\begin{pmatrix} c_1\\ \vdots\\ c_n\end{pmatrix}. \tag{8.2}$$

Theorem 8.1. *Let $z_1,\ldots,z_n$ be distinct points in $\mathbb{D}$, and let $a_1,\ldots,a_n$ be full rank matrices in $\mathbb{C}^{r_1\times p},\ldots,\mathbb{C}^{r_n\times p}$. Let $\mathcal{N}$ be the subspace of $\mathbf{HS}_2^{p\times q}$ defined as*

$$\mathcal{N}=\{f\in\mathbf{HS}_2^{p\times q}:\ \ a_if(z_i)=0,\ \ i=1,\ldots,n\}$$

and let Θ be the $\mathbb{C}^{p\times p}$-valued function given by

$$\Theta(z)=I_p-(1-z)\left(\frac{a_1^*}{1-z\bar{z}_1},\ldots,\frac{a_n^*}{1-z\bar{z}_n}\right)\mathbb{K}_n^{-1}\begin{pmatrix}\frac{a_1}{1-z_1}\\ \vdots\\ \frac{a_n}{1-z_n}\end{pmatrix} \tag{8.3}$$

where

$$\mathbb{K}_n = \left(\frac{a_i a_j^*}{1 - z_i \bar{z}_j}\right)_{i,j=1}^n . \tag{8.4}$$

Then,

$$\mathcal{N} = \Theta \mathbf{HS}_2^{p\times q} = \left\{ f \in \mathbf{HS}_2^{p\times q} : \ f(z) = \Theta(z) u(z); \ \ u \in \mathbf{HS}_2^{p\times q} \right\}.$$

Proof. By the Beurling-Lax theorem, the space of functions in $\mathbf{H}_2^{p\times q}$ which satisfy the conditions $a_i f(z_i) = 0$ is of the form $\Theta \mathbf{H}_2^{p\times q}$, where Θ is an inner function, of the form (8.3) up to a right unitary factor (see e.g. [4]). The derivative of Θ is in $\mathbf{H}_2^{p\times p}$ since it is a rational function with poles outside the closed unit disk. Let $f = \Theta u \in \mathcal{N}$, with $u \in \mathbf{H}_2^{p\times q}$. From $f' = \Theta' u + \Theta u'$, we see that $\Theta u'$ belongs to $\mathbf{H}_2^{p\times q}$; this in turns implies that u' itself belongs to $\mathbf{H}_2^{p\times q}$. Let $r_0 = \sup |z_i|$. Then $r_0 < 1$ and for $r \in (r_0, 1)$ we have for some $M > 0$

$$\|\Theta(re^{it})^{-1}\| \leq \frac{M}{(r - r_0)^n}$$

(as can be seen from the general form of a finite Blaschke product), and so

$$\begin{aligned} \int_0^{2\pi} \|u'(re^{it})\|^2 dt &= \int_0^{2\pi} \|\Theta(re^{it})^{-1} \Theta(re^{it}) u'(re^{it})\|^2 dt \\ &\leq \frac{M}{(r - r_0)^n} \int_0^{2\pi} \|\Theta(r(e^{it}) u'(re^{it})\|^2 dt \end{aligned}$$

and hence $u' \in \mathbf{H}_2^{p\times q}$, and $\mathcal{N} \subset \Theta \mathbf{HS}_2^{p\times q}$.

Conversely, if $f \in \Theta \mathbf{HS}_2^{p\times q}$, it belongs to $\mathbf{HS}_2^{p\times q}$ as is seen from $f' = \Theta u' + \Theta' u$ and hence $\mathcal{N} \supset \Theta \mathbf{HS}_2^{p\times q}$. □

The description of all solutions H to the problem NPL_Υ in the space $\mathbf{HS}_2^{p\times q}$ is given by the formula

$$H(z) = H_{\min}(z) + \Theta(z) u(z) \tag{8.5}$$

where $H_{\min}$ and Θ are given by (8.2) and (8.3) respectively and the free parameter u varies in $\mathbf{HS}_2^{p\times q}$ and satisfies

$$[\Theta u, \ \Theta u]_{\mathbf{HS}_2^{p\times q}} \leq \Upsilon - \left(\ c_1^*, \ldots, c_n^* \ \right) \mathbb{P}^{-1} \begin{pmatrix} c_1 \\ \vdots \\ c_n \end{pmatrix}. \tag{8.6}$$

The representation (8.5) is orthogonal in the $\mathbf{HS}_2^{p\times q}$-metric but the operator of multiplication by Θ is not in general an isometry in $\mathbf{HS}_2^{p\times q}$. Because of this, the restriction (8.6) on the parameter is not so explicit.

Another approach to interpolation in Hardy-Sobolev spaces is possible using the following:

Remark 8.1. *An element H is in $\mathbf{HS}_2^{p\times q}$ and satisfies $[H, H]_{\mathbf{HS}_2^{p\times q}} \leq \Upsilon$ if and only if the function*

$$F = \begin{pmatrix} H \\ H' \end{pmatrix} \tag{8.7}$$

is in $\mathbf{H}_2^{2p\times q}$ and satisfies

$$[F,\ F] \leq \Upsilon. \tag{8.8}$$

Thus the problem NPL$_\Upsilon$ for $H \in \mathbf{HS}_2^{p\times q}$ can be reduced to the following problem in the Hardy class $\mathbf{H}_2^{2p\times q}$: *find all functions $F \in \mathbf{H}_2^{2p\times q}$ of the form* (8.7) *which satisfy* (8.8) *and such that*

$$(a_i,\ 0)F(z_i) = c_i \qquad (i = 1, \ldots, n). \tag{8.9}$$

Using the description of all solutions to the interpolation problem in $\mathbf{H}_2^{2p\times q}$ (see [4]) and taking into account the special structure of the interpolation data in (8.9) we conclude that $F \in \mathbf{H}_2^{2p\times q}$ satisfies the interpolation conditions (8.9) if and only if it can be written as

$$F(z) = \begin{pmatrix} \tilde{H}_{\min}(z) \\ 0 \end{pmatrix} + \begin{pmatrix} \Theta(z) & 0 \\ 0 & I_p \end{pmatrix} U(z)$$

where $\tilde{H}_{\min}$ is the function defined as

$$\tilde{H}_{\min}(z) = \left(\frac{a_1^*}{1 - z\bar{z}_1}, \ldots, \frac{a_n^*}{1 - z\bar{z}_n}\right) \mathbb{K}_n^{-1} \begin{pmatrix} c_1 \\ \vdots \\ c_n \end{pmatrix}, \tag{8.10}$$

where $\mathbb{K}_n$ and Θ are given by (8.3), (8.4) and where $U = \begin{pmatrix} u_1 \\ u_2 \end{pmatrix} \in \mathbf{H}_2^{2p\times q}$.

Not all parameters U are permissible, since F must be of the form (8.7). Writing F and U in matrix form one gets the relationship

$$H(z) = \tilde{H}_{\min}(z) + \Theta(z)u_1(z) \tag{8.11}$$

and $u_2(z) = H'(z)$. Using the arguments from the proof of Theorem 8.1, one can check that the function H of the form (8.11) belongs to $\mathbf{HS}_2^{p\times q}$ if and only if $u_1 \in \mathbf{HS}_2^{p\times q}$. Unfortunately, the representation (8.11) is orthogonal only in the $\mathbf{H}_2^{p\times q}$-metric but not in the $\mathbf{HS}_2^{p\times q}$-metric. This is why the restriction (8.8) does not lead to any explicit condition on the parameter u_1. Nevertheless, when the "norm" constraint (2.6) is removed, formulas (8.5) and (8.11) parametrize the same set of solutions of the interpolation problem NPL (defined before Theorem 3.3) when the parameter varies in $\mathbf{HS}_2^{p\times q}$. To verify this fact directly, note that $H_{\min} - \tilde{H}_{\min} \in \Theta\mathbf{HS}_2^{p\times q}$ (where $H_{\min}$ and $\tilde{H}_{\min}$ are defined by (8.2) and (8.10) respectively.

One can also relate the problem NPL_Υ in the space $\mathbf{HS}_2^{p\times q}$ to some interpolation problem in the Schur class. This is discussed at the end of Section 11. Finally, we note that interpolation in Sobolev spaces (in a somewhat different framework) is considered in [1].

9 Interpolation in Dirichlet spaces

Theorem 8.1 suggests the following question: *Let $\mathcal{H}$ be a reproducing kernel Hilbert module included in $\mathbf{H}_2^{p\times q}$ and let (in the notation of Section 8)*

$$\mathcal{N} = \{f \in \mathcal{H} : \ a_i f(z_i) = 0, \ i = 1, \ldots, n\}.$$

When is it true that $\mathcal{N} = \Theta\mathcal{H}$ (where Θ is as in Section 8)? More generally, if $\mathcal{H}$ is a reproducing kernel Hilbert module of $\mathbb{C}^{p\times q}$-valued functions analytic in $\mathbb{D}$, when does there exists a $\mathbb{C}^{p\times p}$-valued rational function P such that $\mathcal{N} = P\mathcal{H}$?

We do not have general answers to these questions. In Section 11, we discuss the case of de Branges spaces. For these spaces, equality $\mathcal{N} = \Theta\mathcal{H}$ does not necessarily hold. In Section 10 we consider the case of Bergman spaces and also give a partial answer to the second question. In the present section we consider the case of Dirichlet spaces. These are defined as follows: the scalar Dirichlet space $\mathcal{D}$ consists of all functions f analytic in $\mathbb{D}$ for which $f(0) = 0$ and

$$\int\int_{\mathbb{D}} |f'(z)|^2 dxdy < \infty. \tag{9.1}$$

Alternatively, the function $f(z) = \sum_1^\infty a_n z^n \in \mathcal{D}$ if and only if

$$\sum_1^\infty n|a_n|^2 < \infty. \tag{9.2}$$

The space $\mathcal{D}$ endowed with the norm defined by (9.1) or (9.2) is a reproducing kernel Hilbert space with reproducing kernel $-\ln(1 - z\bar{\omega})$. It is included in the Hardy space $\mathbf{H}_2$.

Theorem 9.1. *Let $z_1, \ldots, z_n$ be distinct points in $\mathbb{D}$. Let $\mathcal{N}$ be the subspace of $\mathcal{D}$ defined by*

$$\mathcal{N} = \{f \in \mathcal{N} : \ f(z_i) = 0, \ i = 1, \ldots, n\}$$

and let $\Theta(z) = \prod_1^n \frac{z - z_i}{1 - z\bar{z}_i}$ Then,

$$\mathcal{N} = \Theta\mathcal{D} = \{f \in \mathcal{D} : \ f(z) = \Theta(z)u(z); \ u \in \mathcal{D}\}.$$

Proof. Let $f \in \mathcal{N}$. Since $\mathcal{D} \subset \mathbf{H}_2$, we can write $f = \Theta u$, where Θ is as in the statement of the theorem and $u \in \mathbf{H}_2$. It follows from $f' = \Theta u' + \Theta' u$ that $\int\int_{\mathbb{D}} |\Theta u'|^2(z)dxdy < \infty$. This in turns implies that $\int\int_{\mathbb{D}} |u'|^2(z)dxdy < \infty$. To see this it suffices to write $\mathbb{D} = (\mathbb{D} \setminus K) \cup K$, where K is a compact subset of $\mathbb{D}$ which contains the zeros of Θ. On $\mathbb{D} \setminus K$, the function $1/\Theta$ is uniformly bounded in modulus by some constant C, and we have

$$\begin{aligned}
\int\int_{\mathbb{D}} |u'|^2(z)dxdy &= \int\int_{\mathbb{D}\setminus K} |u'|^2(z)dxdy + \int\int_K |u'|^2(z)dxdy \\
&= \int\int_{\mathbb{D}\setminus K} |\Theta(z)^{-1}|^2 |\Theta u'|^2(z)dxdy + \int\int_K |u'|^2(z)dxdy \\
&\leq C\int\int_{\mathbb{D}} |\Theta u'|^2(z)dxdy + \int\int_K |u'|^2(z)dxdy < \infty.
\end{aligned}$$

Hence $\mathcal{N} \subset \Theta\mathcal{D}$. Conversely, let $f = \Theta u \in \Theta\mathcal{D}$. Since Θ and Θ' are bounded in $\mathbb{D}$, it follows from $f' = \Theta' u + \Theta u'$ that $f \in \mathcal{D}$, and hence $f \in \mathcal{N}$. □

The previous theorem was given in the scalar case for simplicity of exposition; matrix valued analogues of $\mathcal{D}$ can be built from the functions

$$K(z,\omega) = \sum_1^\infty z^n \bar{\omega}^n \frac{M_n}{n}$$

where the M_n are strictly positive matrices uniformly bounded from above and below. For these, Theorem 9.1 is still true (with conditions of the form $a_i f(z_i) = 0$ and with Θ given by (8.3)).

10 Interpolation in Bergman spaces

The scalar Bergman space $\mathcal{B}$ is the reproducing kernel Hilbert space with reproducing kernel $\frac{1}{(1-z\bar{\omega})^2}$, with $z, \omega \in \mathbb{D}$. It consists of the functions $f(z) = \sum_0^\infty a_n z^n$ such that $\int\int_{\mathbb{D}} |f(z)|^2 dxdy < \infty$, or equivalently, $\sum_0^\infty \frac{|a_n|^2}{n+1} < \infty$. It is not included in $\mathbf{H}_2$, but the following theorem holds:

Theorem 10.1. *Let $z_1, \ldots, z_n \in \mathbb{D}$. Then, the set of functions $f \in \mathcal{B}$ such that $f(z_i) = 0, \quad i = 1, \ldots, n$ is equal to $P\mathcal{B}$, where $P(z) = \prod_1^n (z - z_i)$.*

The proof is similar to the proof of Theorems 8.1 and 10.1 and will be omitted. Instead of P, one also can take the Blaschke product based on the z_i. The preceding result is a particular case of:

Theorem 10.2. *Let $\mathcal{H}$ be a reproducing kernel Hilbert space of $\mathbb{C}^{p\times q}$-valued functions analytic in $\mathbb{D}$ with the following properties:*

(i) if $\omega \in \mathbb{D}$ and if $f \in \mathcal{H}$ vanishes at ω, then $\frac{f(z)}{z-\omega} \in \mathcal{H}$.

(ii) If $f \in \mathcal{H}$, then the function $z \mapsto zf(z)$ is also in $\mathcal{H}$.

Let $z_1, \ldots, z_n \in \mathbb{D}$ and let $\mathcal{N}$ be the subspace of $\mathcal{H}$ defined as

$$\mathcal{N} = \{f \in \mathcal{H} : \ f(z_i) = 0, \ i = 1, \ldots, n\}.$$

Let $P(z) = \prod_i (z - z_i)$. Then, $\mathcal{N} = P\mathcal{H}$.

Proof. Let $f \in \mathcal{H}$ and $f(z_i) = 0$ for $i = 1, \ldots n$. Then, we can write $f(z) = p(z)g(z)$ where g is analytic in $\mathbb{D}$. By (i), $g \in \mathcal{H}$. Hence, $\mathcal{N} \subset \mathcal{H}$. The converse inclusion follows from (ii). □

The de Branges spaces which will be considered in Section 11 do not have property (ii). Conditions (i) and (ii) hold for kernels in Example 2.2 when S contains all the integers from a certain rank and the matrices M_n satisfy inequalities of the type $k_1 M_n \leq M_{n+1} \leq k_2 M_n$ for n large enough, where k_1 and k_2 are strictly positive constant. The theorem may be applied for instance to the kernels $\frac{1}{(1-z\bar{\omega})^\ell}$, with $\ell \in \mathbb{N}$, and to their matrix-valued analogues. We do not have a tangential version of the preceding theorem (i.e. when the conditions $f(z_i) = 0$ are replaced by conditions of the form $a_i f(z_i) = 0$).

11 Some remarks on interpolation in de Branges-Rovnyak spaces

To introduce the de Branges-Rovnyak spaces, let us first recall that the Schur class $\mathcal{S}^{p\times r}$ consists of all $\mathbb{C}^{p\times r}$-valued functions s which are analytic and take contractive values in the unit disk: $s(z)s(z)^* \leq I_p, \ \ \forall z \in \mathbb{D}$. For every Schur function s the kernel

$$K_s(z, \omega) = \frac{I_p - s(z)s(\omega)^*}{1 - z\bar{\omega}} \tag{11.1}$$

is nonnegative in $\mathbb{D}$ and generates a reproducing kernel Hilbert space with reproducing kernel K_s. Such spaces were first introduced and studied by L. de Branges and J. Rovnyak in [20, Appendix] [21] and are denoted by $\mathcal{H}(s)$. We refer to [19] and [25] for further information on these and related spaces. When $p = r$ and s is inner, we have $\mathcal{H}(s) = \mathbf{H}_2^p \ominus s\mathbf{H}_2^p$. In general, the space $\mathcal{H}(s)$ is only contractively included in $\mathbf{H}_2^p$. In particular, elements of $\mathcal{H}(s)$ have entries of bounded type in $\mathbb{D}$.

We discuss the following problem NP(s, σ): *given Schur functions $s \in \mathcal{S}^{p\times r}$ and $\sigma \in \mathcal{S}^{q\times r}$ find all $\mathbb{C}^{p\times q}$-valued contractive multipliers H from $\mathcal{H}(\sigma)$ into $\mathcal{H}(s)$ (i.e. all analytic functions H such that the operator $M_H : \ \mathcal{H}(\sigma) \to \mathcal{H}(s)$ of multiplication by H is a bounded operator of norm less than* 1*) which satisfy the interpolation conditions* (1.2).

The case where s and σ vanish identically corresponds to the interpolation in Schur classes; the case where s vanishes identically and $\sigma(z) = zI_p$ corresponds to interpolation in Hardy spaces with matrix-norm constraint (see [6]); the case $s \equiv 0$ and $\sigma(z) = z^N I_p$ is studied in [5]. It corresponds to interpolation in Hardy spaces with constraints on the trigonometric moments of $H(e^{it})^* H(e^{it})$.

The set of all multipliers H from $\mathcal{H}(\sigma)$ into $\mathcal{H}(s)$ can be characterized in terms of nonnegative kernels as well as in terms of Schur functions. To simplify the presentation we first consider the case where σ is strictly contractive, i.e. $\|\sigma(z)\| < 1$ for all $z \in \mathbb{D}$.

Theorem 11.1. *Let σ be a strictly contractive Schur function. The following statements are equivalent:*

(i) H is a multiplier from $\mathcal{H}(\sigma)$ into $\mathcal{H}(s)$.

(ii) The kernel

$$K_{s,\sigma}(z,\omega) = \frac{I_p - s(z)s(\omega)^*}{1 - z\bar{\omega}} - H(z)\frac{I_q - \sigma(z)\sigma(\omega)^*}{1 - z\bar{\omega}}H(\omega)^* \tag{11.2}$$

is nonnegative in $\mathbb{D}$.

(iii) There exists a Schur function $\Sigma = \begin{pmatrix} \Sigma_{11} & \Sigma_{12} \\ \Sigma_{21} & \Sigma_{22} \end{pmatrix} \in \mathcal{S}^{(p+r)\times(r+q)}$ such that

$$H(z) = \Sigma_{12}(z)\left(I_q - \sigma(z)\Sigma_{22}(z)\right)^{-1} \tag{11.3}$$

and

$$s(z) = \Sigma_{11}(z) + \Sigma_{12}(z)\left(I - \sigma(z)\Sigma_{22}(z)\right)^{-1}\sigma(z)\Sigma_{21}(z). \tag{11.4}$$

Proof. The equivalence $(i) \Leftrightarrow (ii)$ was proved in [6] for arbitrary pairs of reproducing kernel Hilbert spaces. The equivalence $(ii) \Leftrightarrow (iii)$ can be easily obtained with help of the following result of R. Leech and M. Rosenblum: if $A(z)$ and $B(z)$ are respectively $\mathbb{C}^{p\times q}$- and $\mathbb{C}^{p\times r}$-valued analytic functions of bounded type in $\mathbb{D}$, the kernel

$$K_{A,B}(z,\omega) = \frac{A(z)A(\omega)^* - B(z)B(\omega)^*}{1 - z\bar{\omega}} \tag{11.5}$$

is nonnegative in $\mathbb{D}$ if and only if $B(z) = A(z)\Sigma(z)$ for some Schur function $\Sigma \in \mathcal{S}^{q\times r}$ and for all $z \in \mathbb{D}$ (see [37], and [8] for this and links with the inverse scattering problem; see also [32]).

The kernel (11.2) is of the form (11.5) with

$$A(z) = \begin{pmatrix} I_p & H(z)\sigma(z) \end{pmatrix}, \qquad B(z) = \begin{pmatrix} s(z) & H(z) \end{pmatrix}.$$

Take $\omega_0 \in \mathbb{D}$, and consider the function $z \mapsto G(z) = H(z)\frac{I_q - \sigma(z)\sigma(\omega_0)^*}{1 - z\bar{\omega}_0}$. Since H is a multiplier, the columns of G are in $\mathcal{H}(s)$, and in particular of bounded type. Furthermore, $\det(I_q - \sigma(z)\sigma(\omega_0)^*) \not\equiv 0$ since σ is strictly contractive. It follows that H is of bounded type. Hence the functions A and B are of bounded type, and by the theorem mentioned above, there exists a Schur function $\Sigma = \begin{pmatrix} \Sigma_{11} & \Sigma_{12} \\ \Sigma_{21} & \Sigma_{22} \end{pmatrix} \in \mathcal{S}^{(p+r)\times(r+q)}$ such that

$$(I_p \;\; H(z)\sigma(z)) \;\Sigma(z) = (s(z) \;\; H(z)) \tag{11.6}$$

or equivalently,

$$\Sigma_{12} + H\sigma\Sigma_{22} = H, \qquad \Sigma_{11} + H\sigma\Sigma_{21} = s.$$

These last two equalities are in turn equivalent to (11.3) and (11.4). $\square$

For different approaches to multipliers in de Branges-Rovnyak spaces we refer to [22] and [35]. We note that (11.4) suggests links with the inverse scattering problem (see [11], [25, chapter 8], [28]). To explain these connections, we take both s and σ to be $\mathbb{C}^{p\times q}$-valued. An element $s \in \mathcal{S}^{p\times q}$ can be viewed as the transfer function of a dissipative time-invariant causal system with q inputs and p outputs (see [26]). The inverse scattering problem associated to s consists in finding all representations of the form (11.4) with Σ and σ in the adequate Schur classes. In terms of systems, it corresponds to obtaining the system with transfer function s as the connection of a system with $(p+q)$ inputs and $(p+q)$ outputs with transfer function Σ to a system with transfer function σ (see [28, chapter 3], [29]). Assume that furthermore $\det \Sigma_{21} \not\equiv 0$. Equation (11.4) can be rewritten as the linear fractional transformation

$$s = (\theta_{11}\sigma + \theta_{12})(\theta_{21}\sigma + \theta_{22})^{-1}$$

where

$$\theta_{11} = \Sigma_{12} - \Sigma_{11}\Sigma_{21}^{-1}\Sigma_{22}, \qquad \theta_{12} = \Sigma_{11}\Sigma_{21}^{-1}, \qquad \theta_{21} = -\Sigma_{21}^{-1}\Sigma_{22}, \qquad \theta_{22} = \Sigma_{21}^{-1}.$$

The matrix-function Θ with block entries θ_{ij} is the Potapov-Ginzburg transform of Σ, and is J-contractive in $\mathbb{D}$, with

$$J = \begin{pmatrix} I_p & 0 \\ 0 & -I_q \end{pmatrix}$$

(see [14], [23] and [25] for more information and references on the Potapov-Ginzburg transform). A method to find the pairs (Θ, σ) such that s admits the above linear fractional representation was developed in [11] in terms of reproducing kernel spaces.

The previous theorem considered the case where the Schur function σ is strictly contractive; the general case is treated as follows. An arbitrary Schur function can always be written as

$$\sigma(z) = U \begin{pmatrix} \sigma_0 & 0 \\ 0 & I_\nu \end{pmatrix} V$$

where U and V are unitary and $\sigma_0 \in \mathcal{S}^{(p-\nu)\times(r-\nu)}$ is strictly contractive (see [25, Lemma 0.13 p.9]). We write $HU = (h_0, h_1)$ where h_0 is $\mathbb{C}^{p\times(q-\nu)}$-valued and h_1 is $\mathbb{C}^{p\times\nu}$-valued. The kernel (11.2) can be rewritten as

$$K_{s,\sigma}(z,\omega) = \frac{I_p - s(z)s(\omega)^*}{1 - z\bar{\omega}} - h_0(z)\frac{I_q - \sigma_0(z)\sigma_0(\omega)^*}{1 - z\bar{\omega}}h_0(\omega)^* ,$$

which brings us back to the previous case. We can thus characterize h_0 using Theorem 11.1 while h_1 is itself an arbitrary analytic function.

Let σ be strictly contractive. Using (11.4) and (11.6) we reduce the problem NP(s,σ) to the following problem in the Schur class: *find all functions* $\Sigma \in \mathcal{S}^{(p+r)\times(r+q)}$ *which satisfy* (11.4) *and the interpolation conditions*

$$(a_i \ c_i\sigma(z_i))\,\Sigma(z_i) = (a_i s(z_i) \ c_i) \qquad (i = 1, \cdots, n). \tag{11.7}$$

Similarly the Carathéodory-Fejér type interpolation conditions (2.7) on the multiplier H are equivalent to the conditions

$$\sum_{j=0}^{k}\left(a_{k-j} \quad \sum_{i=0}^{j} c_{j-i}\frac{\sigma^{(i)}(z_0)}{i!}\right)\frac{\Sigma^{(j)}(z_0)}{j!} = \left(\sum_{j=0}^{k} a_{k-j}\frac{s^{(j)}(z_0)}{j!} \quad c_k\right) \tag{11.8}$$

for the function $\Sigma \in \mathcal{S}^{(p+r)\times(r+q)}$ satisfying the "scattering" condition (11.4).

Therefore, the set of all solutions to the corresponding interpolation problem is characterized by the intersection of two sets: the set of solutions Σ to the interpolation problem (11.7) (or (11.8)), and the set of solutions to the representation problem (11.4). This intersection may be empty.

The nonnegativity of the kernel (11.2) leads to the necessary condition

$$\left(a_i\frac{I_p - s(z_i)s(z_j)^*}{1 - z_i\bar{z}_j}a_j^* - c_i\frac{I_q - \sigma(z_i)\sigma(z_j)^*}{1 - z_i\bar{z}_j}c_j^*\right)_{i,j=1}^{n} \geq 0 \tag{11.9}$$

for the problem NP(s,σ) to be solvable. This condition in general is not sufficient as can be seen from the following example: let $s(z) = zI_p$, $\sigma(z) = 0_{q\times q}$; then $\mathcal{H}(s) = \mathbb{C}^p$, $\mathcal{H}(\sigma) = \mathbf{H}_2^{q\times 1}$ and there are no nontrivial multipliers between these spaces despite the existence of interpolation data z_i, a_i, c_i for which the matrix (11.9) is nonnegative.

Nevertheless, for the case $\sigma(z) = zI_q$ and $s \in \mathcal{S}^{p\times r}$ it follows from Theorem 3.2 that the corresponding condition (11.6) is sufficient for the problem NP$(s,\ zI_q)$ to be solvable. The nonnegativity of the kernel

$$K_H(z,\omega) = \frac{I_p - s(z)s(\omega)^*}{1 - z\bar{\omega}} - H(z)H(\omega)^* \tag{11.10}$$

in $\mathbb{D}$ means that the function H belongs to $\mathcal{H}(s)$ and $[H,\ H]_{\mathcal{H}(s)} \leq I_q$. Substituting $\sigma = zI_q$ in Theorem 11.1 one obtains the characterization of such H in terms of

Schur functions $\Sigma \in \mathcal{S}^{(p+r)\times(r+q)}$ (see [6] and Theorem 11.2 below). The solution to the interpolation problem (11.7) can be given in terms of a linear fractional transformation, where the parameter varies in $\mathcal{S}^{(p+r)\times(r+q)}$ when the problem is nondegenerate, and in a smaller Schur class when the problem is degenerate (we refer to [2] for the latter). The condition (11.4) (with $\sigma(z) = z$) gives a constraint on the parameter.

The study of the problem $\mathrm{NP}(s, \sigma)$ for $\sigma(z) = z$ can also be done using the results of Section 3, but the space $\mathcal{M}^{\perp}$ cannot, in general, be identified as $P\mathcal{H}(s)$ for some function P analytic in $\mathbb{D}$. As an example, let ω be some fixed point in the open unit disk and take $s(z) = z^3$ and $\mathcal{M}^{\perp} = \{f \in \mathcal{H}(s);\ f(\omega) = 0\}$. The space $\mathcal{H}(s)$ is spanned by the functions $1, z, z^2$ and has no nonconstant multipliers, and therefore, no representation of the form $\mathcal{M}^{\perp} = P\,\mathcal{H}(s)$ is possible.

The above mentioned result of Leech and Rosenblum has numerous applications. For instance, it can be used to characterize Schur functions s such that $s \in \mathcal{H}(s)$ (see [10]). It is also used to prove directly the following result (see [6]) which is a particular case of Theorem 11.1 (with $s(z) = 0_{p\times q}$ and $\sigma(z) = zI_q$) and allows to relate the interpolation problem in the Hardy-Sobolev space $\mathbf{HS}_2^{p\times q}$ to a problem in the Schur class.

Theorem 11.2. *Let $H \in \mathbf{H}_2^{p\times q}$. Then, $[H, H] \le I_q$ (where $[\,,\,]$ is defined in (1.1)) if and only if H can be written as*

$$H(z) = s_1(z)(I_q - zs_2(z))^{-1} \tag{11.11}$$

where s_1 and s_2 are respectively $\mathbb{C}^{p\times q}$- and $\mathbb{C}^{q\times q}$-valued and where $s = \begin{pmatrix} s_1 \\ s_2 \end{pmatrix}$ is analytic and contractive in $\mathbb{D}$ (i.e. $s(z)^*s(z) \le I_q$ for all $z \in \mathbb{D}$).

In the scalar case, $p = q = 1$, this theorem was proved by D. Sarason (see [40]). Combining Theorem 11.2 and Remark 8.1 one establishes the link between the Hardy-Sobolev and the Schur classes. Indeed, if H is in $\mathbf{HS}_2^{p\times q}$ and $[H, H]_{\mathbf{HS}_2^{p\times q}} \le I_q$, the function F of the form (8.7) can be written by Theorem 11.2, as

$$F(z) = \begin{pmatrix} s_1(z) \\ s_2(z) \end{pmatrix} (I - zs_0(z))^{-1} \quad \text{for some} \quad s = \begin{pmatrix} s_0 \\ s_1 \\ s_2 \end{pmatrix} \in \mathcal{S}^{(2p+q)\times q}. \tag{11.12}$$

From the equations

$$H(z) = s_1(z)(I - zs_0(z))^{-1}, \qquad H'(z) = s_2(z)(I - zs_0(z))^{-1},$$

we see that the s_i are related by the relationship

$$s_2(z) = s_1'(z) - s_1(z)(I - zs_0(z))^{-1}(s_0(z) + zs_0'(z)). \tag{11.13}$$

This leads to the following result:

Theorem 11.3. *Let H be a function analytic in $\mathbb{D}$; then, the following are equivalent:*

(i) $H \in \mathbf{HS}_2^{p\times q}$ and $[H,H]_{\mathbf{HS}_2^{p\times q}} \leq I_q$.

(ii) The kernel

$$K(z,\omega) = \frac{I_{2p}}{1-z\bar{\omega}} - \begin{pmatrix} H(z) \\ H'(z) \end{pmatrix} \begin{pmatrix} H(\omega)^* & H'(\omega)^* \end{pmatrix}$$

is nonnegative in $\mathbb{D}$.

(iii) There exist analytic functions s_0 and s_1 such that the function s from (11.12) *(with s_2 as in* (11.13)*) is a Schur function, and*

$$H(z) = s_1(z)(I - zs_0(z))^{-1}.$$

For $s \in \mathcal{S}^{p\times q}$ the de Branges-Rovnyak space $\mathcal{H}(s)$ is the state space for a coisometric realization of s (see [25]). This realization can be used to get realization theorems for functions in the Hardy and Hardy-Sobolev spaces. This will be presented elsewhere.

References

[1] J. Agler. Nevanlinna-Pick interpolation on Sobolev spaces. *Proc. Amer. Math. Soc.*, 108:341–351, 1986.

[2] D. Alpay and V. Bolotnikov. Two sided Nevanlinna-Pick interpolation for a class of matrix-valued functions. *Zeitschrift für Analysis und ihre Anwendungen*, 12:211–238, 1993.

[3] D. Alpay and V. Bolotnikov. The Carathéodory-Fejér H_2 interpolation problem. *Zeitschrift für Analysis und ihrer Anwendungen*, 13:583–597, 1994.

[4] D. Alpay and V. Bolotnikov. Two sided interpolation for matrix functions with entries in the Hardy space. *Lin. Alg. Appl.*, 223/224:31–56, 1995.

[5] D. Alpay, V. Bolotnikov, and Ph. Loubaton. On tangential H_2 interpolation with second order norm constraints. *Integral Equations Operator Theory*, 24:156–178, 1996.

[6] D. Alpay, V. Bolotnikov, and Y. Peretz. On the tangential interpolation problem for H_2 functions. *Transactions of the American Mathematical Society*, 347:675–686, 1995.

[7] D. Alpay, P. Bruinsma, A. Dijksma, and H.S.V. de Snoo. Interpolation problems, extensions of symmetric operators and reproducing kernel spaces II. *Integral Equations Operator Theory*, 14:465–500, 1991.

[8] D. Alpay, P. Dewilde, and H. Dym. On the existence and construction of solutions to the partial lossless inverse scattering problem with applications to estimation theory. *IEEE Transactions on Information theory*, 35:1184–1205, 1989.

[9] D. Alpay, P. Dewilde, and H. Dym. *Lossless inverse scattering and reproducing kernels for upper triangular operators*, volume 47 of *Operator Theory: Advances and Applications*, pages 61–133. Birkhäuser Verlag, Basel, 1990.

[10] D. Alpay, A. Dijksma, H. de Snoo, and J. Rovnyak. Schur functions, operator colligations, and reproducing kernel Pontryagin spaces. Preprint, 1995.

[11] D. Alpay and H. Dym. Hilbert spaces of analytic functions, inverse scattering and operator models, I. *Integral Equations Operator Theory*, 7:589–641, 1984.

[12] N. Aronszajn. Theory of reproducing kernels. *Transactions of the American Mathematical Society*, 68:337–404, 1950.

[13] D. Arov. The generalized bitangential Carathéodory-Nevanlinna-Pick problem and (j, J_0)-inner matrix-functions. *Russian Acad. Sci. Izv. Math.*, 42:1–26, 1994.

[14] T.Ya. Azizov and I.S. Iohvidov. *Foundations of the theory of linear operators in spaces with indefinite metric.* Nauka, Moscow, 1986. (Russian). English translation: *Linear operators in spaces with an indefinite metric.* John Wiley, New-York, 1989.

[15] J. Ball and J.W. Helton. A Beurling-Lax theorem for the Lie group $U(m, n)$ which contains most classical interpolation theory. *Journal of Operator Theory*, 8:107–142, 1983.

[16] L. Baratchart. Interpolation and Fourier coefficients in the Hardy space H_2. In M. Kaashoek, J.H. van Schuppen, and A.C.M. Ran, editors, *Proceedings of the international symposium MTNS 89*, volume 3 of *Progress in systems and control theory*, pages 387–394. Birkhäuser Verlag, Basel, 1990.

[17] L. Baratchart and M. Zerner. On the recovery of functions from pointwise boundary values in a Hardy-Sobolev space of the disk. *Journal of Computational and Applied Mathematics*, 46:255–269, 1993.

[18] S. Bergman and M. Schiffer. *Kernel functions and elliptic differential equations in mathematical physics.* Academic Press, 1953.

[19] L. de Branges. *Espaces hilbertiens de fonctions entières.* Masson, Paris, 1972.

[20] L. de Branges and J. Rovnyak. Canonical models in quantum scattering theory. In C. Wilcox, editor, *Perturbation theory and its applications in quantum mechanics.* Holt, Rinehart and Winston, New-York, 1966.

[21] L. de Branges and J. Rovnyak. *Square summable power series.* Holt, Rinehart and Winston, New-York, 1966.

[22] B. Davis and J. McMarthy. Multipliers in de Branges spaces. *Michigan Math. J.*, 38:225–240, 1991.

[23] M. Dritschel and J. Rovnyak. *Extensions theorems for contractions on Kreĭn spaces*, volume 47 of *Operator theory: Advances and Applications*, pages 221–305. Birkhäuser Verlag, Basel, 1990.

[24] P.L. Duren and D.L. Williams. Interpolation problems in function spaces. *Journal of Functional Analysis*, 9:75–86, 1972.

[25] H. Dym. *J contractive matrix functions, reproducing kernel spaces and interpolation*, volume 71 of *CBMS Lecture Notes.* Amer. Math. Soc., Rhodes Island, 1989.

[26] P.A. Fuhrmann. *Linear systems and operators in Hilbert space.* McGraw-Hill international book company, 1981.

[27] I. Gohberg, editor. *Time variant systems and interpolation*, volume 56 of *Operator theory: Advances and Applications.* Birkhäuser Verlag, Basel, 1992.

[28] J. Helton. *Operator theory, analytic functions, matrices and electrical engineering*, volume 68 of *CBMS Lecture Notes.* Amer. Math. Soc., Rhodes Island, 1987.

[29] J.W. Helton and J. Ball. The cascade decomposition of a given system vs the linear fractional decompositions of its transfer function. *Integral Equations and Operator Theory*, 5:314–385, 1982.

[30] S. Itoh. Reproducing kernels in modules over C^*-algebras and their applications. *Bull. Kyushu Inst. Tech.*, 37:1–20, 1990.

[31] A. Kheifets. Generalized bitangential Schur-Nevanlinna-Pick problem, related Parseval equality and scattering operator. deposited in VINITI, 11.05.1989. No. 3108 B89 Dep., 1–60, 1989, russian.

[32] A. Kheifets. Parseval equality in abstract interpolation problem and coupling of open systems. *Jour. Sov. Math.*, 49:1307–1310, 1990.

[33] M.G. Kreĭn and A.A. Nudelman. *The Markov moment problem and extremal problems*, volume 50 of *Translations of mathematical monographs.* American Mathematical Society, Providence, Rhode Island, 1977.

[34] P.D. Lax and R.S. Phillips. *Scattering theory.* Academic Press, New-York, 1967.

[35] B. Lotto and D. Sarason. Multipliers in de Branges-Rovnyak spaces. *Indiana University Mathematical Journal*, 42:907–920, 1993.

[36] W. Paschke. Inner product spaces over B^*-algebras. *Transactions of the American Mathematical Society*, 1982:443–468, 1973.

[37] M. Rosenblum and J. Rovnyak. *Hardy classes and operator theory.* Oxford University Press, 1985.

[38] W. Rudin. *Real and complex analysis.* Mc Graw Hill, 1982.

[39] S. Saitoh. *Theory of reproducing kernels and its applications*, volume 189. Longman scientific and technical, 1988.

[40] D. Sarason. *Exposed points in* H^1, *I, volume 41 of Operator Theory: Advances and Applications*, pages 485–496. Birkhäuser Verlag, Basel, 1989.

[41] L. Schwartz. Sous espaces hilbertiens d'espaces vectoriels topologiques et noyaux associés (noyaux reproduisants). *J. d'Analyse Math.*, 13:115–256, 1964.

Daniel Alpay
Department of Mathematics
Ben-Gurion University of the Negev
Beer-Sheva, 84105, Israel
Email: dany@math.bgu.ac.il

Vladimir Bolotnikov
Department of Theoretical Mathematics
The Weizmann Institute of Science
Rehovot, 76100, Israel
E-mail: vladi@wisdom.weizmann.ac.il

AMS Subject Classification: 46E22, 47A57, 42A82.

Operator Theory
Advances and Applications, Vol. 95
© 1997 Birkhäuser Verlag Basel/Switzerland

Notes on a Nevanlinna-Pick interpolation problem for generalized Nevanlinna functions

Aad Dijksma and Heinz Langer

Abstract. We consider a scalar Nevanlinna-Pick interpolation problem with finitely many data and assume that the Pick matrix $\mathbb{P}$ is invertible and has κ negative eigenvalues. We look for solutions of this problem in the class of meromorphic functions whose Nevanlinna kernel has κ negative squares. The set of these solutions can be written as a fractional linear transformation of a parameter in the class of Nevanlinna functions, much as in the case $\kappa = 0$. But now not the whole Nevanlinna class can be used as a parameter set. Our results are obtained through the characterization of the selfadjoint extensions of a symmetric operator in a Pontryagin space with both defect numbers equal to 1 in terms of a so called u-resolvent matrix.

1 Introduction

We consider the following **Nevanlinna-Pick interpolation problem**: Given n distinct points $z_1, z_2, \ldots, z_n$ in the upper half plane $\mathbb{C}^+$ and n points $w_1, w_2, \ldots, w_n$ in $\mathbb{C}$. Assume that the hermitian $n \times n$ matrix

$$\mathbb{P} = (p_{jk})_{j,k=1}^{n} \quad \text{with} \quad p_{jk} = \frac{\bar{w}_j - w_k}{\bar{z}_j - z_k} \tag{1}$$

is invertible and has κ negative eigenvalues, counting multiplicities. Find all functions N in the generalized Nevanlinna class $\mathbf{N}_\kappa$ which are holomorphic in z_j and satisfy

$$N(z_j) = w_j, \quad j = 1, 2, \ldots, n.$$

The matrix $\mathbb{P}$ is called the **Pick** matrix associated with the data $z_1, z_2, \ldots, z_n$ and $w_1, w_2, \ldots, w_n$. A scalar function N belongs to the **generalized Nevanlinna class** $\mathbf{N}_\kappa$ if N is meromorphic on $\mathbb{C} \setminus \mathbb{R}$, $N(\bar{z}) = \overline{N(z)}$, and the **Nevanlinna kernel**

$$K_N(z, w) = \frac{N(z) - \overline{N(w)}}{z - \bar{w}}$$

has κ negative squares on $\mathbb{C}^+$. The kernel condition means that for every choice of the number $m \in \mathbb{N}$ and of the m points $\lambda_1, \lambda_2, \ldots, \lambda_m \in \mathbb{C}^+$ the $m \times m$ hermitian matrix $(K_N(\lambda_j, \lambda_k))_{j,k=1}^{m}$ has at most κ and for at least one such choice it has exactly κ negative eigenvalues, counting multiplicities. The number of negative/positive squares of any kernel $K(z,w)$ satisfying $\overline{K(z,w)} = K(w,z)$ is defined similarly. For $N \in \mathbf{N}_\kappa$ we denote by $\mathcal{D}_N$ the domain of local holomorphy of N in $\mathbb{C}$.

A function in $\mathbf{N}_\kappa$ admits a representation in terms of a selfadjoint linear relation in a Pontryagin space. Recall that a **(closed) linear relation** A in a Pontryagin space $\mathcal{P}$ is a (closed) linear subset A of the direct sum space $\mathcal{P}^2 = \mathcal{P} \oplus \mathcal{P}$. The adjoint of A is the closed linear relation

$$A^* = \{\{u,v\} \in \mathcal{P}^2 : \langle v, f\rangle_{\mathcal{P}} - \langle u, g\rangle_{\mathcal{P}} = 0 \text{ for all } \{f,g\} \in A\}.$$

A is called **symmetric** if $A \subseteq A^*$ and **selfadjoint** if equality prevails. The graph of an operator is a linear relation; a linear relation A is (the graph of) an operator if and only if its multivalued part $A(0) = \{g \in \mathcal{P} : \{0,g\} \in A\}$ is the trivial subspace of $\mathcal{P}$. The resolvent set $\rho(A)$, the spectrum $\sigma(A)$, and the point spectrum $\sigma_p(A)$ of a closed linear relation A are defined as if A were an operator. The point infinity is a regular point of a closed linear relation A if A is a bounded everywhere defined operator, and it is an eigenvalue of A if $A(0)$ is nontrivial.

The function N belongs to $\mathbf{N}_\kappa$ for some integer $\kappa \geq 0$ if and only if it has a representation of the form

$$N(z) = \overline{N(\mu)} + (z - \bar{\mu})\langle (I + (z-\mu)(A-z)^{-1})u, u\rangle_{\mathcal{P}},$$

where $A = A^*$ is a selfadjoint linear relation with nonempty resolvent set $\rho(A)$ in a Pontryagin space $\mathcal{P}$, μ is a point in $\rho(A)$, and u (which depends upon μ) belongs $\mathcal{P}$. The representation can be chosen u-**minimal** which means that

$$\overline{\operatorname{span}}\,\{(I + (z-\mu)(A-z)^{-1})u : z \in \rho(A)\} = \mathcal{P}.$$

In that case N belongs to the class $\mathbf{N}_\kappa$ with $\kappa = \operatorname{ind}_- \mathcal{P}$, $\mathcal{D}_N = \rho(A)$, and the representation is unique up to an isomorphism; for proofs, see, for example, [KL2], [DLS2], and [DLS3]. The representation implies that N has at most 2κ poles in $\mathbb{C} \setminus \mathbb{R}$ which are symmetric with respect to the real axis; these poles correspond to the nonreal eigenvalues of the selfadjoint linear relation A. It is well known that the functions in the class $\mathbf{N}_0$ are locally holomorphic on $\mathbb{C} \setminus \mathbb{R}$. This class is the class of **Nevanlinna functions**. Alternatively, it can be characterized by: The function N belongs to $\mathbf{N}_0$ if and only if N is holomorphic on $\mathbb{C}^+$, $N(\bar{z}) = \overline{N(z)}$, $z \in \mathbb{C}^+$, and N maps $\mathbb{C}^+$ into $\mathbb{C}^+ \cup \mathbb{R}$. Observe that if a function in $\mathbf{N}_0$ is constant, then this constant is real. The Maximum Modulus Principle implies that if $N \in \mathbf{N}_0$ attains a real value r, say, at a point of $\mathbb{C}^+$, then $N(z) \equiv r$.

For $\kappa = 0$, V.P. Potapov gave an explicit formula for all solutions of the above problem in terms of a fractional linear transformation. His method was based on the Schwarz Lemma and the so called Fundamental Matrix Inequality; see, for example, [KP]. To describe this formula we set

$$Z = \operatorname{diag}\,(z_1, z_2, \ldots, z_n), \quad \hat{J} = \begin{pmatrix} 0 & -1 \\ 1 & 0 \end{pmatrix},$$

and define the 2×2 matrix function $\Theta(z) = \begin{pmatrix} \theta_{11}(z) & \theta_{12}(z) \\ \theta_{21}(z) & \theta_{22}(z) \end{pmatrix}$ by

$$\Theta(z) = I - z \begin{pmatrix} w_1 & \cdots & w_n \\ 1 & \cdots & 1 \end{pmatrix} (Z - z)^{-1} \mathbb{P}^{-1} Z^{-*} \begin{pmatrix} w_1 & \cdots & w_n \\ 1 & \cdots & 1 \end{pmatrix}^* \hat{J}. \tag{2}$$

Then, for the case $\kappa = 0$, all solutions of the Nevanlinna-Pick problem are given by

$$N(z) = \frac{\theta_{11}(z)T(z) + \theta_{12}(z)}{\theta_{21}(z)T(z) + \theta_{22}(z)} =: \Theta_T(z), \tag{3}$$

where T runs through the class $\mathbf{N}_0$ of all Nevanlinna functions or is equal to ∞ in which case $N(z) = \theta_{11}(z)/\theta_{21}(z)$. We write $\widetilde{\mathbf{N}}_0$ for the set $\mathbf{N}_0 \cup \{\infty\}$. If in (3) we use for $\theta_{jk}(z)$, $j, k = 1, 2$, the entries obtained from (2), they have poles at the interpolation points $z_1, z_2, \ldots, z_n$. Obviously, in (3) we can get rid of these poles by multiplying all these entries by the polynomial $(z_1 - z)(z_2 - z) \cdots (z_n - z)$. Another way to get rid of these poles is as follows. As $\overline{N(\bar{z})} = N(z)$ and $\overline{T(\bar{z})} = T(z)$, instead of the $\theta_{jk}(z)$ in (3) we can use the functions

$$\hat{\theta}_{jk}(z) = \overline{\theta_{jk}(\bar{z})},$$

which have their poles in the points $\bar{z}_1, \bar{z}_2, \ldots, \bar{z}_n$. Evidently, the corresponding 2×2 matrix function $\hat{\Theta}(z)$ is given by

$$\hat{\Theta}(z) = I - z \begin{pmatrix} \bar{w}_1 & \cdots & \bar{w}_n \\ 1 & \cdots & 1 \end{pmatrix} (Z^* - z)^{-1} \mathbb{P}^{-t} Z^{-1} \begin{pmatrix} \bar{w}_1 & \cdots & \bar{w}_n \\ 1 & \cdots & 1 \end{pmatrix}^* \hat{J},$$

where ^{t} stands for transpose.

In this note we derive these formulas from the u-resolvent matrix theory developed in [KL3] when it is adapted to the case of a nondensely defined symmetric operator. This has been done already in the case of two sided interpolation problems for Nevanlinna pairs (with $\kappa = 0$) in [Br1], [Br2] and in [ABDS]. Moreover, we show that the use of the u-resolvent matrix theory gives more information about the parametrization formula (3) for arbitrary κ than contained in the literature; see for example, [Ba], [G], [DGK], and [N1]. Namely, it is known that the solutions of the interpolation problem are still of the form $N(z) = \Theta_T(z)$, but, in general, not every parameter T from $\widetilde{\mathbf{N}}_0$ gives a solution of the interpolation problem. We show that the excluded parameters T only yield **partial solutions**, that is, solve the Nevanlinna-Pick interpolation problem in some of the points z_j but not in all points, and we characterize these points z_j and parameters T for which the relation $\Theta_T(z_j) = w_j$ does not hold.

To be more precise we show that the interpolation points z_j can be divided according to the following properties:

I. For all $T \in \widetilde{\mathbf{N}}_0$, $\Theta_T(z)$ is analytic near z_j and $\Theta_T(z_j) = w_j$.

II. For all but one $T \in \widetilde{\mathbf{N}}_0$, $\Theta_T(z)$ is analytic near z_j and $\Theta_T(z_j) = w_j$; the excluded parameter T is identically equal to a real number.

III. There is a number $\alpha \in \mathbb{C}^+$ such that $\Theta_T(z)$ is analytic near z_j and $\Theta_T(z_j) = w_j$ for all $T \in \widetilde{\mathbf{N}}_0$ with $T(z_j) \neq \alpha$. The equation $T(z_j) = \alpha$, which characterizes the excluded parameters, holds for infinitely many T and they are nonconstant.

The three cases I, II, and III correspond to a partition of the complex plane into three sets: $\mathbb{C} = \Delta_+ \cup \Delta_0 \cup \Delta_-$, that is, case I, II or III, occurs if and only if z_j belongs to Δ_+, Δ_0 or Δ_-, respectively. The latter property can be read of from the inverse of the Pick matrix. Moreover, we show that if T is one of the excluded parameters then the function $\Theta_T(z)$ belongs to a class $\mathbf{N}_{\kappa'}$ for some κ' with $0 \leq \kappa' < \kappa$.

The interpolation problem for arbitrary κ has been studied in [Ba], with a degenerate Pick matrix in [W1], [W2]; for $\kappa = 0$ and with a degenerate Pick matrix in [Br2] and in [D, Chapter 7]. A detailed account of various types of interpolation problems including the Nevanlinna-Pick interpolation problem with $\kappa = 0$ can be found in the monograph [BGR]. Our results are closely related to the Nevanlinna-Pick interpolation problem with $\kappa = 0$ in the case where some of the interpolation points lie on the real axis; see, for example, [N2].

In this note we consider the scalar case for simplicity. The interpolation problem for matrix functions or generalized Nevanlinna pairs will be considered elsewhere.

We assume familiarity with operator theory in Pontryagin spaces; see, for example, the monographs [AI], [Bo] and [IKL]. A study of selfadjoint linear relations in Pontryagin spaces can be found in, for example, [AI] and [DS]. In this note a Pontryagin space $\mathcal{P}$ is a Kreĭn space in which at least one (and hence each) maximal uniformly negative subspace is finite dimensional; all such subspaces have the same dimension and this number is called the (negative) **index** of the Pontryagin space $\mathcal{P}$ and is denoted by $\operatorname{ind}_- \mathcal{P}$.

2 The model

Suppose we are given the data $z_1, z_2, \ldots, z_n$, $w_1, w_2, \ldots, w_n$ as in the Introduction. We denote by $\mathcal{P}$ the linear space $\mathbb{C}^n$ of column n-vectors equipped with the indefinite inner product defined by means of the Pick matrix (1)

$$\langle x, y \rangle_{\mathcal{P}} = y^* \mathbb{P} x, \quad x, y \in \mathbb{C}^n.$$

Then $\mathcal{P}$ is a Pontryagin space with $\operatorname{ind}_- \mathcal{P} = \kappa$. We denote by e_j the n-vector with entries $(e_j)_k = \delta_{jk}, j, k = 1, 2, \ldots, n$, where δ is the Kronecker delta. Then $\langle e_j, e_k \rangle_{\mathcal{P}} = p_{kj}$. In $\mathcal{P}$ we define the operator S by:

$$Sx = Zx, \quad x \in \operatorname{dom} S = \{x \in \mathbb{C}^n : e^* x = 0\},$$

where $Z = \operatorname{diag}(z_1, z_2, \ldots, z_n)$ and e is the n-vector $e = (1\,1 \cdots 1)^{\mathrm{t}}$, so that $e^* x$ stands for the sum of the coefficients of x. We denote by $\varphi(z)$, $z \in \mathbb{C}$, the n-vector:

$$\varphi(z) = \begin{cases} \mathbb{P}^{-1}(Z^* - z)^{-1} e, & z \neq \bar{z}_1, \bar{z}_2, \ldots, \bar{z}_n, \\ \mathbb{P}^{-1} e_j, & z = \bar{z}_j,\ j = 1, 2, \ldots, n. \end{cases}$$

In the sequel we write

$$\mathbb{P}^{-1} = \left(\pi_{jk}\right)_{j,k=1}^{n}.$$

Then the n-vector function $\hat{\varphi}$ whose j-th component in $z \in \mathbb{C}$ is given by

$$(\hat{\varphi}(z))_j = \sum_k (\prod_{l \neq k} (\bar{z}_l - z))\pi_{jk}$$

is entire and for each $z \in \mathbb{C}$, $\varphi(z) = \hat{\lambda}(z)\hat{\varphi}(z)$ for some nonzero complex number $\hat{\lambda}(z)$.

The proof of the following result is straightforward and omitted.

Lemma 2.1. (a) *S is a symmetric operator:*

$$\langle Sx, y\rangle_{\mathcal{P}} = \langle x, Sy\rangle_{\mathcal{P}}, \quad x, y \in \operatorname{dom} S,$$

and $\operatorname{codim}\operatorname{dom} S = 1$.
(b) *For any fundamental symmetry J in $\mathcal{P}$, the symmetric operator JS in the Hilbert space $(\mathcal{P}, \langle J\cdot,\cdot\rangle_{\mathcal{P}})$ has both defect numbers equal to* 1.
(c) *S has no eigenvalues: $\sigma_p(S) = \emptyset$.*
(d) *For all $z \in \mathbb{C}$,* $\operatorname{codim}\operatorname{ran}(S - z) = 1$ *and* $\operatorname{ran}(S - \bar{z})^{\langle\perp\rangle_{\mathcal{P}}} = \operatorname{span}\{\varphi(z)\}$.
(e) *S is simple, that is,* $\cap_{z\in\mathbb{C}}\operatorname{ran}(S - z) = \{0\}$.

The pair $S, \mathcal{P}$ is called the **model** associated with the data of the Nevanlinna-Pick interpolation problem.

We recall some general definitions and results but apply them directly to our model. A selfadjoint linear relation $\tilde{A}$ in a Pontryagin space $\tilde{\mathcal{P}}$ is called a selfadjoint extension of S in $\mathcal{P}$ if the space $\mathcal{P}$ is a closed subspace of $\tilde{\mathcal{P}}$ and $S \subset \tilde{A}$. It is called **regular** if the **exit space** $\tilde{\mathcal{P}} \ominus \mathcal{P}$ is a Hilbert space or, equivalently, $\operatorname{ind}_- \tilde{\mathcal{P}} = \operatorname{ind}_- \mathcal{P}$; and it is called **canonical** if $\tilde{\mathcal{P}} = \mathcal{P}$. We denote by $P_{\mathcal{P}}$ the orthogonal projection in the space $\tilde{\mathcal{P}}$ onto $\mathcal{P}$. We will be interested in minimal selfadjoint extensions of S. By definition, a selfadjoint extension $\tilde{A}$ in $\tilde{\mathcal{P}}$ of S is called **minimal** if the resolvent set $\rho(\tilde{A})$ is nonempty and for some (and hence every) $\mu \in \rho(\tilde{A})$,

$$\overline{\operatorname{span}}\{(I + (z - \mu)(\tilde{A} - z)^{-1})\mathcal{P} : z \in \rho(\tilde{A})\} = \tilde{\mathcal{P}}.$$

Since S is symmetric, it admits regular minimal selfadjoint extensions. Lemma 2.1.(b) implies that S has canonical selfadjoint extensions also. They are densely defined, except for

$$\tilde{A}_\infty = S \dotplus \operatorname{span}\{\{0, \mathbb{P}^{-1}e\}\}, \quad \text{direct sum in } \mathcal{P} \oplus \mathcal{P},$$

which is a selfadjoint linear relation. All canonical selfadjoint extensions have a nonempty resolvent set. This is a well known property of densely defined selfadjoint operators in a Pontryagin space; that $\rho(\tilde{A}_\infty) \neq \emptyset$ follows, on account of [DS, Proposition 4.4 and the Corollary to Theorem 4.6], from the fact that $\sigma_p(\tilde{A}_\infty)$ is a finite set. Thus all canonical selfadjoint extensions of S are regular and minimal.

For later reference we record the following lemma. We leave the proof, which is straight forward, to the reader.

Lemma 2.2. (a) *The spectrum of* $\tilde{A}_\infty$ *is given by*

$$\sigma(\tilde{A}_\infty) = \sigma_p(\tilde{A}_\infty) = \{z \in \mathbb{C} : \sum_{j,k}(\prod_{l \neq j}(\bar{z}_l - z))\pi_{jk} = 0\} \cup \{\infty\}.$$

The defining relation on the righthand side is equivalent to the equation $\varphi(\bar{z})^* e = 0, z \in \mathbb{C}$, *and so in particular the interpolation points cannot all belong to* $\sigma(\tilde{A}_\infty)$.
(b) *For* $x \in \mathcal{P}$ *and* $z \in \rho(\tilde{A}_\infty) \setminus \{z_1, z_2, \dots, z_n\}$,

$$(\tilde{A}_\infty - z)^{-1}x = (Z - z)^{-1}\left(x - \frac{e^*(Z-z)^{-1}x}{e^*(Z-z)^{-1}\mathbb{P}^{-1}e}\mathbb{P}^{-1}e\right)$$

and the term in brackets belongs to $\operatorname{ran}(S - z) = \{\varphi(\bar{z})\}^{\langle\perp\rangle_{\mathcal{P}}}$.

Lemma 2.1.(d) and (e) imply that the $\varphi(z)$'s span $\mathcal{P}$, and this implies the following simple result.

Lemma 2.3. *The number of positive and negative squares of the kernel* K:

$$K(z,w) = \langle\varphi(z), \varphi(w)\rangle_{\mathcal{P}}, \quad z, w \neq \bar{z}_1, \bar{z}_2, \dots, \bar{z}_n,$$

is equal to the number of positive and negative eigenvalues of $\mathbb{P}$, *that is,* $n - \kappa$ *and* κ, *respectively.*

In the sequel we set $\delta(z) = \langle\varphi(z), \varphi(z)\rangle_{\mathcal{P}}$ or in full:

$$\delta(z) = \begin{cases} \sum_{j,k=1}^{n} \pi_{jk}(z_j - \bar{z})^{-1}(\bar{z}_k - z)^{-1}, & z \neq \bar{z}_1, \bar{z}_2, \dots, \bar{z}_n, \\ \pi_{jj}, & z = \bar{z}_j,\ j = 1, 2, \dots, n. \end{cases}$$

Then $\mathbb{C}$ can be partitioned into three disjoint parts:

$$\mathbb{C} = \Delta_+ \cup \Delta_0 \cup \Delta_-, \tag{1}$$

where

$$\Delta_+ = \{z \in \mathbb{C} : \delta(z) > 0\}, \quad \Delta_0 = \{z \in \mathbb{C} : \delta(z) = 0\}, \quad \Delta_- = \{z \in \mathbb{C} : \delta(z) < 0\}.$$

For the first statement of the following lemma, compare the first statement of [KL1, Theorem 3.3]; although there the symmetric operator is densely defined the proofs of the two statements are similar.

Lemma 2.4. (a) Δ_0 *has no interior points.*
(b) *If* $0 < \kappa < n$, *then* $\Delta_+ \neq \emptyset$ *and* $\Delta_- \neq \emptyset$. *Moreover,* $\Delta_+ = \mathbb{C}$ ($\Delta_- = \mathbb{C}$) *if and only if* $\kappa = 0$ ($\kappa = n$, *respectively*).
(c) Δ_+, Δ_0 *and* Δ_- *are symmetric with respect to the real axis.*
(d) *The interpolation point* z_j *belongs to* Δ_+, Δ_0 *or* Δ_-, *respectively, if and only if* $\pi_{jj} > 0$, $\pi_{jj} = 0$ *or* $\pi_{jj} < 0$, *respectively,* $j = 1, 2, \dots, n$.

Proof. To prove (a) we make use of a lemma of Shmul'yan in [S]: If $f(z)$ and $g(z)$ are holomorphic functions on a region G in $\mathbb{C}$ with values in a Hilbert space $\mathcal{H}$ such that for all $z \in G$, $\langle f(z), g(z)\rangle_{\mathcal{H}} \equiv \gamma$, a constant, then for all $z, w \in G$, $\langle f(z), g(w)\rangle_{\mathcal{H}} \equiv \gamma$. Assume that

$$\delta(z) = e^*(Z - \bar{z})^{-1}\mathbb{P}^{-1}(Z - \bar{z})^* e \equiv 0$$

on a small disc $G \subset \mathbb{C} \setminus \{\bar{z}_1, \bar{z}_2, \cdots, \bar{z}_n\}$. Set $f(z) = \mathbb{P}^{-1}(Z - \bar{z})^* e$ and $g(z) = (Z^* - z)^{-1} e$. Then, according to the lemma with $\mathcal{H} = \mathbb{C}^n$, for all $z, w \in G$,

$$\langle f(z), g(w)\rangle_{\mathbb{C}^n} \equiv 0.$$

We multiply both sides by $(\bar{z}_j - z)\overline{(\bar{z}_k - w)}$, and let $z \to \bar{z}_j$ and $w \to \bar{z}_k$. Then we get that $\pi_{jk} = e_j^* \mathbb{P}^{-1} e_k = 0$, and hence $\mathbb{P}^{-1} = 0$. This contradiction implies that Δ_0 has no interior points.

The statements in (b) follow from Lemma 2.3..

We prove (c) by considering the Cayley transform $V_z = (S - z)(S - \bar{z})^{-1}$ for $z \in \mathbb{C} \setminus \mathbb{R}$. It is an isometry which maps $\operatorname{dom} V_z = \operatorname{ran}(S - \bar{z})$ bijectively onto $\operatorname{ran} V_z = \operatorname{ran}(S - z)$. Hence the signature of the subspace $\operatorname{ran}(S - \bar{z})$ coincides with the signature of the subspace $\operatorname{ran}(S - z)$, and then the same is true for the signatures of the orthogonal complements of these subspaces. By Lemma 2.1.(d) the orthogonal complements are spanned by $\varphi(z)$ and $\varphi(\bar{z})$, respectively, and therefore $\operatorname{sign} \delta(z) = \operatorname{sign} \delta(\bar{z})$. This proves (c). It follows that

$$\operatorname{sign} \delta(z_j) = \operatorname{sign} \delta(\bar{z}_j) = \operatorname{sign} \pi_{jj},$$

which implies (d). □

Recall that if A is a selfadjoint linear relation in a Pontryagin space $\mathcal{P}$, the nonreal spectrum $\sigma_0(A) = \sigma(A) \cap (\mathbb{C} \setminus \mathbb{R})$ of A is the set of all nonreal eigenvalues of A and is symmetric with respect to the real axis; see, for example, [DS, Corollary to Theorem 4.6]. If $\rho(A) \neq \emptyset$ then $\sigma_0(A)$ contains at most 2κ points, where $\kappa = \operatorname{ind}_- \mathcal{P}$. The next theorem describes the location of the nonreal spectrum of a regular minimal selfadjoint extension $\tilde{A}$ of S relative to the partition (1) of $\mathbb{C}$. The proof is taken partly from the proofs of [KL1, Theorem 3.1 and Corollary 3.1] and [KL3, Behauptung 1.2(f)].

Theorem 2.5. (a) *The spectral inclusion $\sigma_0(\tilde{A}) \subseteq \Delta_0$ holds for every canonical selfadjoint extension $\tilde{A}$ of S. For every $z \in \Delta_0 \cap (\mathbb{C} \setminus \mathbb{R})$ there exists a unique regular minimal selfadjoint extension $\tilde{A}$ of S with $z \in \sigma_0(\tilde{A})$, and this extension is canonical.*
(b) *The spectral inclusion $\sigma_0(\tilde{A}) \subseteq \Delta_-$ holds for every noncanonical regular minimal selfadjoint extension $\tilde{A}$ of S. For every $z \in \Delta_- \cap (\mathbb{C} \setminus \mathbb{R})$ there exist infinitely many noncanonical regular minimal selfadjoint extensions $\tilde{A}$ of S with $z \in \sigma_0(\tilde{A})$.*
(c) *The intersection $\sigma_0(\tilde{A}) \cap \Delta_+$ is empty and hence the inclusion $\Delta_+ \cap (\mathbb{C} \setminus \mathbb{R}) \subseteq \rho(\tilde{A})$ holds for every regular minimal selfadjoint extension $\tilde{A}$ of S.*

Proof. Let $\tilde{A}$ in $\tilde{\mathcal{P}}$ be a regular minimal selfadjoint extension of S in $\mathcal{P}$. Assume that z belongs to $\sigma_0(\tilde{A})$, denote by $\tilde{\varphi} \in \tilde{\mathcal{P}}$ the corresponding eigenvector, and set $\varphi = P_{\mathcal{P}}\tilde{\varphi}$. From $S \subseteq \tilde{A}$ follows that φ belongs to $\operatorname{ran}(S-\bar{z})^{\langle\perp\rangle_{\mathcal{P}}}$ and hence $\varphi = \lambda\varphi(z)$ for some $\lambda \in \mathbb{C}$. The minimality of $\tilde{A}$ implies that $\varphi \neq 0$ and hence $\lambda \neq 0$. By the regularity of A,

$$|\lambda|^2\langle\varphi(z),\varphi(z)\rangle_{\mathcal{P}} \leq \langle\varphi,\varphi\rangle_{\mathcal{P}} + \langle\tilde{\varphi}-\varphi,\tilde{\varphi}-\varphi\rangle_{\tilde{\mathcal{P}}\ominus\mathcal{P}} = \langle\tilde{\varphi},\tilde{\varphi}\rangle_{\tilde{\mathcal{P}}} = 0.$$

If $\tilde{A}$ is canonical equality prevails (since then $\tilde{\varphi} = \lambda\varphi(z)$) and hence $z \in \Delta_0$; if $\tilde{A}$ is noncanonical the inequality is strict and hence $z \in \Delta_-$. This proves the first statements in (a) and (b), and (c) is a simple consequence of these statements.

To prove the second statement in (a), assume that z belongs to $\Delta_0 \cap (\mathbb{C}\setminus\mathbb{R})$. Then

$$A = S \dot{+} \operatorname{span}\{\{\varphi(z), z\varphi(z)\}\}$$

is symmetric. By Lemma 2.1.(c), the righthand side is a direct sum in $\mathcal{P}\oplus\mathcal{P}$, and so, by Lemma 2.1.(b), A is a canonical selfadjoint extension of S. Evidently, z belongs to $\sigma_0(A)$. To prove uniqueness, assume that $\tilde{A}$ in $\tilde{\mathcal{P}}$ is a regular minimal selfadjoint extension of S in $\mathcal{P}$ for which z is a nonreal eigenvalue. Let $\tilde{\varphi}$ be the corresponding eigenvector. Again we find that $\varphi = P_{\mathcal{P}}\tilde{\varphi} = \lambda\varphi(z)$ for some $\lambda \in \mathbb{C}$. Since z belongs to Δ_0, $\langle\varphi,\varphi\rangle_{\mathcal{P}} = |\lambda|^2\langle\varphi(z),\varphi(z)\rangle_{\mathcal{P}} = 0$, and hence

$$\langle\tilde{\varphi}-\varphi,\tilde{\varphi}-\varphi\rangle_{\tilde{\mathcal{P}}\ominus\mathcal{P}} = \langle\tilde{\varphi},\tilde{\varphi}\rangle_{\tilde{\mathcal{P}}} - \langle\varphi,\varphi\rangle_{\mathcal{P}} = 0.$$

The space $\tilde{\mathcal{P}}\ominus\mathcal{P}$ is a Hilbert space, so $\tilde{\varphi} = \varphi = \lambda\varphi(z)$. It follows that $A \subseteq \tilde{A}$, that is, for all $z \in \rho(A)\cap\rho(\tilde{A})$,

$$(\tilde{A}-z)^{-1}|_{\mathcal{P}} = (A-z)^{-1}.$$

By the minimality of A and $\tilde{A}$, we have that $\tilde{\mathcal{P}} = \mathcal{P}$ and then also that $\tilde{A} = A$. This proves the uniqueness statement in (a).

To prove the second statement in (b), assume that z belongs to $\Delta_-\cap(\mathbb{C}\setminus\mathbb{R})$. Let $\mathcal{P}_1$ stand for the Pontryagin space $\mathcal{P}\oplus\{\psi\}$, the orthogonal sum of $\mathcal{P}$ and the one-dimensional Hilbert space spanned by an abstract element denoted by ψ with self inner product $\langle\psi,\psi\rangle = -\langle\varphi(z),\varphi(z)\rangle_{\mathcal{P}} > 0$. Then

$$S_1 = S \dot{+} \operatorname{span}\{\{\varphi(z)+\psi, z(\varphi(z)+\psi)\}\}$$

is a symmetric extension in $\mathcal{P}_1$ of S in $\mathcal{P}$. The operator S_1 has infinitely many regular minimal selfadjoint extensions. Clearly, their nonreal spectrum contains the point z and they are regular selfadjoint extensions of S. It suffices to show that they are minimal as well: Denote by $\tilde{A}$ in $\tilde{\mathcal{P}}$ a regular minimal selfadjoint extension of S_1 in $\mathcal{P}_1$. Then

$$\overline{\operatorname{span}}\{\{(I+(w-\mu)(\tilde{A}-\mu)^{-1})\mathcal{P}_1 : w \in \rho(\tilde{A})\}\} = \tilde{\mathcal{P}}, \tag{2}$$

where μ is any point in $\rho(\tilde{A})$. We must show that

$$\overline{\operatorname{span}}\{\{(I+(w-\mu)(\tilde{A}-\mu)^{-1})\mathcal{P} : w \in \rho(\tilde{A})\}\} = \tilde{\mathcal{P}}. \tag{3}$$

Denote the space on the lefthand side of (3) by $\mathcal{L}$. Then $\mathcal{L}$ is $(\tilde{A}-w)^{-1}$-invariant for some and hence every $w \in \rho(\tilde{A})$; if $\tilde{A}$ is an operator, this is equivalent to $\mathcal{L}$ being $\tilde{A}$-invariant. The orthogonal complement of $\mathcal{L}$ in $\tilde{\mathcal{P}}$ is a Hilbert space and is also $(\tilde{A}-w)^{-1}$-invariant. This readily implies that $\varphi(z)+\psi$ belongs to $\mathcal{L}$, and hence

$$(I+(w-\mu)(\tilde{A}-\mu)^{-1})\psi = \frac{z-\mu}{z-w}(\varphi(z)+\psi) - (I+(w-\mu)(\tilde{A}-\mu)^{-1})\varphi(z)$$

also belongs to $\mathcal{L}$. Therefore

$$\operatorname{span}\{\{(I+(w-\mu)(\tilde{A}-\mu)^{-1})\mathcal{P}_1 : w \in \rho(\tilde{A})\}\} \subseteq \mathcal{L} \subseteq \tilde{\mathcal{P}}.$$

These inclusions and (2) imply that $\mathcal{L} = \tilde{\mathcal{P}}$, that is, (3) holds true. □

Theorem 2.5.(a) implies that if A is any canonical selfadjoint extension of S, then for every regular minimal selfadjoint extension $\tilde{A}$ of S with $\tilde{A} \neq A$,

$$\sigma_0(\tilde{A}) \cap \sigma_0(A) = \emptyset.$$

Indeed, if the intersection is not empty and contains the point z, then by the first statement in (a), z belongs to Δ_0, and by the uniqueness statement in (a), $\tilde{A} = A$, which is excluded.

3 Selfadjoint extensions of the model and the solutions of the interpolation problem

The elements $e_j - e_k$, $j, k = 1, 2 \ldots, n$, belong to $\operatorname{dom} S$ and hence for all $z \in \mathbb{C}$,

$$(S-z)(e_j - e_k) = (z_j - z)e_j - (z_k - z)e_k.$$

It follows that for every selfadjoint extension $\tilde{A}$ in $\tilde{\mathcal{P}}$ of S with nonempty resolvent set

$$e_j - e_k = (z_j - z)(\tilde{A}-z)^{-1}e_j - (z_k - z)(\tilde{A}-z)^{-1}e_k, \quad z \in \rho(\tilde{A}).$$

Hence the n-vector function

$$e(z) = e_k + (z - z_k)(\tilde{A}-z)^{-1}e_k, \quad z \in \rho(\tilde{A}),$$

is independent of $k \in \{1, 2, \ldots, n\}$. The same is true for the scalar function N defined by

$$\begin{aligned} N(z) &= \bar{w}_k + (z - \bar{z}_k)\langle e(z), e_k\rangle_{\tilde{\mathcal{P}}} \\ &= \bar{w}_k + (z - \bar{z}_k)\frac{\operatorname{Im} w_k}{\operatorname{Im} z_k} + (z - \bar{z}_k)(z - z_k)\langle (\tilde{A}-z)^{-1}e_k, e_k\rangle_{\tilde{\mathcal{P}}}, \; z \in \rho(\tilde{A}). \end{aligned}$$

This follows easily from the resolvent formula

$$(\tilde{A}-z)^{-1}-(\tilde{A}-w)^{-1}=(z-w)(\tilde{A}-z)^{-1}(\tilde{A}-w)^{-1}$$

and the formula $(\tilde{A}-z)^{-*}=(\tilde{A}-\bar{z})^{-1}$, $z,w\in\rho(\tilde{A})$. We mention these well known formulas, because they are frequently used in the sequel.

Theorem 3.1. *The function $N\in\mathbf{N}_\kappa$ is a solution of the Nevanlinna-Pick interpolation problem if and only if for some (and hence for each) $k\in\{1,2,\cdots,n\}$, N has the representation:*

$$N(z)=\bar{w}_k+(z-\bar{z}_k)\frac{\operatorname{Im}w_k}{\operatorname{Im}z_k}+(z-\bar{z}_k)(z-z_k)\langle(\tilde{A}-z)^{-1}e_k,e_k\rangle_{\tilde{\mathcal{P}}},\tag{1}$$

where $\tilde{A}$ in $\tilde{\mathcal{P}}$ is a regular minimal selfadjoint extension of S in $\mathcal{P}$ (or of an isomorphic copy of S and $\mathcal{P}$) whose resolvent set contains all interpolation points. The minimality of the extension $\tilde{A}$ of S is equivalent to the e_k-minimality of the representation (1) of N.

Proof. Sufficiency: Since the interpolation points belong to $\rho(\tilde{A})$, the formula (1) implies that N is holomorphic at these points. From the relations

$$\frac{N(z_j)-\bar{w}_k}{z_j-\bar{z}_k}=\langle e(z_j),e_k\rangle_{\tilde{\mathcal{P}}}=\langle e_j,e_k\rangle_{\mathcal{P}}=p_{kj}=\frac{w_j-\bar{w}_k}{z_j-\bar{z}_k}$$

it follows that $N(z_j)=w_j$ for each $j\in\{1,2,\cdots,n\}$ and that the Nevanlinna kernel K_N has at least κ negative squares. That K_N has at most and, consequently, precisely κ negative squares follows from the regularity of $\tilde{A}$ and the equality

$$\frac{N(z)-\overline{N(w)}}{z-\bar{w}}=\langle e(z),e(w)\rangle_{\tilde{\mathcal{P}}}.$$

Necessity: Let $N\in\mathbf{N}_\kappa$ be a solution of the interpolation problem, and let

$$N(z)=\overline{N(\mu)}+(z-\bar{\mu})\langle(I+(z-\mu)(\tilde{A}-z)^{-1})u,u\rangle_{\tilde{\mathcal{P}}}$$

be the u-minimal representation of N. By, for example, [DLS2, Theorem 1.1], we have that $\mathcal{D}_N=\rho(\tilde{A})$, hence the interpolation points $z_1,z_2,\ldots,z_n$ belong to $\rho(\tilde{A})$. We set $\tilde{e}(z)=(I+(z-\mu)(\tilde{A}-z)^{-1})u$ and

$$\mathcal{P}_1=\operatorname{span}\{\tilde{e}(z_j):j=1,2,\ldots,n\}.$$

Then

$$\langle\tilde{e}(z_j),\tilde{e}(z_k)\rangle_{\tilde{\mathcal{P}}}=\frac{N(z_j)-\overline{N(z_k)}}{z_j-\bar{z}_k}=p_{kj}$$

and it follows that $\mathcal{P}_1$ is a Pontryagin subspace of $\tilde{\mathcal{P}}$ of index κ and hence $\tilde{\mathcal{P}}\ominus\mathcal{P}_1$ is a Hilbert space. Since $\{(\tilde{A}-z)^{-1}u,z(\tilde{A}-z)^{-1}u+u\}$ belongs to $\tilde{A}$, we have that

$\{\tilde{e}(z) - u, z\tilde{e}(z) - \mu u\}$ also belongs to $\tilde{A}$. Therefore if $x = (x_1 \ x_2 \ \cdots \ x_n)^{\rm t}$ is an n-vector with $e^*x = 0$, then

$$\{\sum_{j=1}^{n} \tilde{e}(z_j)x_j, \sum_{j=1}^{n} z_j\tilde{e}(z_j)x_j\} = \sum_{j=1}^{n}\{\tilde{e}(z_j) - u, z_j\tilde{e}(z_j) - \mu u\}x_j \in \tilde{A}.$$

We define the operator S_1 in $\mathcal{P}_1$ by

$$S_1 \sum_{j=1}^{n} \tilde{e}(z_j)x_j = \sum_{j=1}^{n} z_j\tilde{e}(z_j)x_j,$$

where $x = (x_1 \ x_2 \ \cdots \ x_n)^{\rm t}$ varies over all n-vectors with $e^*x = 0$. Since $S_1 \subseteq \tilde{A}$, we have that S_1 is symmetric in the space $\mathcal{P}_1$ and $\tilde{A}$ is a regular selfadjoint extension of S_1. Clearly, the mapping $e_j \mapsto \tilde{e}(z_j)$ defines an isomorphism between the Pontryagin spaces $\mathcal{P}$ and $\mathcal{P}_1$ under which S and S_1 are isomorphic. Finally, it follows from the identities

$$\frac{N(z) - \overline{N(w)}}{z - \bar{w}} = \langle (I + (z-\mu)(\tilde{A} - z)^{-1})u, (I + (w-\mu)(\tilde{A} - w)^{-1})u \rangle_{\tilde{\mathcal{P}}}$$

and

$$(I + (z-\mu)(\tilde{A} - z)^{-1})u = (I + (z - z_k)(\tilde{A} - z)^{-1})\tilde{e}(z_k),$$

that N admits the representation (1). As $\tilde{A}$ is u-minimal, the last relation implies that $\tilde{A}$ is also $\tilde{e}(z_k)$-minimal.

To prove the last statement, assume that $\tilde{A}$ in $\tilde{\mathcal{P}}$ is a minimal extension of S, and let

$$\mathcal{L} = \operatorname{span}\{e_k + (z - z_k)(\tilde{A} - z)^{-1}e_k : z \in \rho(\tilde{A})\}.$$

Since $e_k + (z - z_k)(\tilde{A} - z)^{-1}e_k$ is independent of $k \in \{1, 2, \cdots, n\}$ and the interpolation points belong to $\rho(\tilde{A})$, we have that $\mathcal{P} \subseteq \mathcal{L}$ and that $(\tilde{A} - z)^{-1}\mathcal{P} \subseteq \mathcal{L}$ for every $z \in \rho(\tilde{A})$, $z \neq z_1, z_2, \ldots, z_n$. This readily implies that $\mathcal{L}$ is dense in $\tilde{\mathcal{P}}$, that is, $\tilde{A}$ is e_k-minimal for every $k \in \{1, 2, \cdots, n\}$. The converse implication is obvious.

4 A parametrization

In this section we give a parametrization of the expression on the righthand side of (1) when $\tilde{A}$ varies over the class of all regular minimal selfadjoint extensions of S; we do not require that the interpolation points lie in the resolvent sets of these extensions. The parameters in the parametrization are the Nevanlinna functions and the constant ∞.

We first recall some of the theory developed by M.G. Kreĭn and the second author. Their results also apply to the case where the underlying symmetric operator is nondensely defined. Again we present the theory not in full generality but in relation to the model.

a. *Q*-functions (see [KL2]). The Q-function associated with S and $\tilde{A}_\infty$ is given by

$$Q_\infty(z) = c - i\mathrm{Im}\,\mu\langle\varphi(\mu),\varphi(\mu)\rangle_{\mathcal{P}} + (z-\bar{\mu})\langle(I+(z-\mu)(\tilde{A}_\infty - z)^{-1})\varphi(\mu),\varphi(\mu)\rangle_{\mathcal{P}}$$

where c is some real constant and μ is a point in $\rho(\tilde{A}_\infty)$. If, for example,

$$e^*Z^{-1}\mathbb{P}^{-1}e = \sum_{j,k=1}^{n}\frac{1}{z_j}\pi_{jk} \neq 0,$$

then by Lemma 2.2.(a) we may take $\mu = 0$ and, choosing $Q_\infty(0) = 0$, we have that

$$Q_\infty(z) = z\frac{(e^*(Z-z)^{-1}\mathbb{P}^{-1}Z^{-1}e)(e^*Z^{-1}\mathbb{P}^{-1}e)}{e^*(Z-z)^{-1}\mathbb{P}^{-1}e}.$$

According to Lemma 2.2.(b) we may also choose $\mu = \bar{z}_j$ for some $j \in \{1,2,,\dots,n\}$, and then with $c = (\mathrm{Re}\,z_j)\pi_{jj}$,

$$Q_\infty(z) = z\pi_{jj} + (z-\bar{z}_j)\left(1-(e_j^*\mathbb{P}^{-1}e)\frac{e^*(Z-z)^{-1}\mathbb{P}^{-1}e_j}{e^*(Z-z)^{-1}\mathbb{P}^{-1}e}\right).$$

We set

$$\gamma(z) = (I+(z-\mu)(\tilde{A}_\infty - z)^{-1})\varphi(\mu), \quad z \in \mathbb{C}\setminus\mathbb{R}.$$

Up to the real constant c, Q_∞ is uniquely determined by the property that the corresponding Nevanlinna kernel is given by

$$K_{Q_\infty}(z,w) = \langle\gamma(z),\gamma(w)\rangle_{\mathcal{P}}.$$

It is easily verified that

$$\gamma(z) = \lambda(z)\varphi(z),$$

for some nonzero complex number $\lambda(z)$. This implies that

$$K_{Q_\infty}(z,w) = \lambda(z)\overline{\lambda(w)}\langle\varphi(z),\varphi(w)\rangle_{\mathcal{P}} \tag{1}$$

and hence, on account of Lemma 2.3., the kernel $K_{Q_\infty}(z,w)$ has $n-\kappa$ positive and κ negative squares. The Q-function plays an important role in the parametrization of all regular minimal selfadjoint extensions $\tilde{A}$ of S:

Theorem 4.1. *The formula*

$$P_{\mathcal{P}}(\tilde{A}-z)^{-1}\,|_{\mathcal{P}} = (\tilde{A}_\infty - z)^{-1} - \frac{\langle\,\cdot\,,\gamma(\bar{z})\rangle_{\mathcal{P}}}{Q_\infty(z)+T'(z)}\gamma(z), \quad z \in \rho(\tilde{A})\cap\rho(\tilde{A}_\infty),$$

establishes a one-to-one correspondence between all regular minimal selfadjoint extensions $\tilde{A}$ of S and all functions $T' \in \widetilde{\mathbf{N}}_0$, with $T' \neq -Q_\infty$ if $\mathbb{P}$ is negative definite. Under this correspondence the canonical extensions correspond to the real constants T' or $T' \equiv \infty$, and the extensions with a finite dimensional exit space correspond to the rational functions T' in $\widetilde{\mathbf{N}}_0$. If the selfadjoint extension corresponding to T' is denoted by $\tilde{A}_{T'}$, then

$$\sigma_0(\tilde{A}_{T'}) = \{z \in \mathbb{C}\setminus\mathbb{R} : Q_\infty(z)+T'(z) = 0\}, \quad T' \neq \infty. \tag{2}$$

The function $-Q_\infty$ is a Nevanlinna function if and only if $\kappa = n$. So the parameter $T' = -Q_\infty$ has to be excluded from the parametrization only if the Pick matrix $\mathbb{P}$ is negative definite.

The formula in Theorem 4.1. is Kreĭn 's formula (see [K]) for the case where the underlying symmetric operator S acts in a Pontryagin space and the selfadjoint extensions have an exit space which is a Hilbert space. The new feature here is that S is not densely defined. For other cases, Kreĭn 's formula appears in [KL1] and [LT]; see also [Br1] and [ABDS]. The proof of Kreĭn 's formula is similar to the proof of [DLS1, Theorem 6.1]; we omit the details. The equality (2) follows from Kreĭn 's formula and the fact that

$$\sigma_0(\tilde{A}) \cap \sigma_0(\tilde{A}_\infty) = \emptyset.$$

The last relation follows from the remark at the end of Section 2.

b. u-resolvent matrices (see [KL3]). If $W = (w_{jk})_{j,k=1}^2$ is a 2×2 matrix function and T is a Nevanlinna function we denote by $W_T(z)$ the fractional linear transform

$$W_T(z) = \frac{w_{11}(z)T(z) + w_{12}(z)}{w_{21}(z)T(z) + w_{22}(z)}.$$

For $T(z) \equiv \infty$ this expression reduces to $w_{11}(z)/w_{21}(z)$.

Since the righthand side of (1) is independent of $k \in \{1, 2, \ldots, n\}$, we take $k = 1$ and set $u = e_1$. Then u is a **module element** for S in the model. This means that the set of $z \in \mathbb{C}$ for which

$$\mathcal{P} = \operatorname{ran}(S - z) \dotplus \operatorname{span}\{u\}, \quad \text{direct sum in } \mathcal{P}, \tag{3}$$

is not empty; here this set is $\mathbb{C} \setminus \{z_2, z_3, \ldots, z_n\}$. It follows from Theorem 4.1. that

$$\langle (\tilde{A} - z)^{-1} u, u \rangle_{\tilde{\mathcal{P}}} = W_{T'}(z), \tag{4}$$

where $W = (w_{jk})_{j,k=1}^2$ is the 2×2 matrix function given by

$$w_{11}(z) = \frac{\langle (\tilde{A}_\infty - z)^{-1} u, u \rangle_{\mathcal{P}}}{\langle u, \gamma(\bar{z}) \rangle_{\mathcal{P}}},$$

$$w_{12}(z) = \frac{\langle (\tilde{A}_\infty - z)^{-1} u, u \rangle_{\mathcal{P}} Q_\infty(z) - \langle u, \gamma(\bar{z}) \rangle_{\mathcal{P}} \langle \gamma(z), u \rangle_{\mathcal{P}}}{\langle u, \gamma(\bar{z}) \rangle_{\mathcal{P}}},$$

and

$$w_{21}(z) = \frac{1}{\langle u, \gamma(\bar{z}) \rangle_{\mathcal{P}}}, \quad w_{22}(z) = \frac{Q_\infty(z)}{\langle u, \gamma(\bar{z}) \rangle_{\mathcal{P}}}.$$

The matrix function W is a so called **u-resolvent matrix**. This means that (4) determines a one-to-one correspondence between the extensions $\tilde{A}$ on the lefthand side and the parameters T' on the righthand side of the kind described in Theorem 4.1. All u-resolvent matrices are of the form

$$\beta(z)W(z)U,$$

where β is holomorphic on $\mathbb{C}\setminus\{z_1,\dots,z_n,\bar{z}_1,\dots,\bar{z}_n\}$ and U is a $(i\hat{J})$-unitary matrix: $U\hat{J}U^*=\hat{J}$. In [KL3, pp. 410, 411] it is shown that for $a\in\mathbb{R}$, $W(a)$ is $(i\hat{J})$-unitary and the u-resolvent matrix $W^a(z)=W(z)W(a)^{-1}$ takes the form

$$W^a(z)=I-(z-a)\begin{pmatrix}\mathsf{Q}(z)\\ -\mathsf{P}(z)\end{pmatrix}\begin{pmatrix}\mathsf{Q}(a)^* & -\mathsf{P}(a)^*\end{pmatrix}\hat{J}.$$

Here for $z\neq z_1,z_2,\dots,z_n$, $\mathsf{P}(z)$ and $\mathsf{Q}(z)$ are the linear mappings from the Pontryagin space $\mathcal{P}$ to the (one-dimensional Hilbert) space $\mathbb{C}$ defined as follows: According to the direct sum decomposition (3) every $x\in\mathcal{P}$ can be uniquely written as $x=y_z+\lambda_z u$, where y_z belongs to $\operatorname{ran}(S-z)$ and λ_z is a complex number. We set $\mathsf{P}(z)x=\lambda_z$ and

$$\mathsf{Q}(z)x=\langle(\tilde{A}_\infty-z)^{-1}y_z,u\rangle_{\mathcal{P}}.$$

It is easy to check that these linear functionals do not change if we replace $\tilde{A}_\infty$ by any other canonical selfadjoint extension of S. Explicit formulas for P and Q can be found in [KL3, pp. 404, 405, formulas (3.3)–(3.5)]. In our model the formulas for P and Q can be calculated directly:

$$\begin{array}{rclrcl}\mathsf{P}(z) & = & (z_1-z)e^*(Z-z)^{-1}, & \mathsf{P}(z)^* & = & \overline{(z_1-z)}\mathbb{P}^{-1}(Z-z)^{-*}e,\\ \mathsf{Q}(z) & = & (\mathbb{P}u-p_{11}e)^*(Z-z)^{-1}, & \mathsf{Q}(z)^* & = & \mathbb{P}^{-1}(Z-z)^{-*}(\mathbb{P}u-p_{11}e).\end{array}$$

The adjoints of P and Q are taken with respect to the indefinite inner product of the Pontryagin space $\mathcal{P}$ and the usual definite inner product of $\mathbb{C}$. To formulate the main theorem of this section we set for $a\in\mathbb{R}$,

$$\Theta^a(z)=I-(z-a)\begin{pmatrix}w_1 & \cdots & w_n\\ 1 & \cdots & 1\end{pmatrix}(Z-z)^{-1}\mathbb{P}^{-1}(Z-a)^{-*}\begin{pmatrix}w_1 & \cdots & w_n\\ 1 & \cdots & 1\end{pmatrix}^*\hat{J}$$

and for $a=\infty$,

$$\Theta^\infty(z)=I-\begin{pmatrix}w_1 & \cdots & w_n\\ 1 & \cdots & 1\end{pmatrix}(Z-z)^{-1}\mathbb{P}^{-1}\begin{pmatrix}w_1 & \cdots & w_n\\ 1 & \cdots & 1\end{pmatrix}^*\hat{J}.$$

These matrices are 2×2 matrix functions whose components we denote by $\theta^a_{jk}(z)$, $j,k=1,2$. Note that $\Theta^0(z)$ coincides with $\Theta(z)$ in (2).

Theorem 4.2. *For each $a\in\mathbb{R}\cup\{\infty\}$, the formula*

$$\bar{w}_k+(z-\bar{z}_k)\frac{\operatorname{Im}w_k}{\operatorname{Im}z_k}+(z-\bar{z}_k)(z-z_k)\langle(\tilde{A}-z)^{-1}e_k,e_k\rangle_{\tilde{\mathcal{P}}}=\Theta^a_T(z) \qquad (5)$$

gives a one-to-one correspondence between the set of all regular minimal selfadjoint extensions $\tilde{A}$ in Pontryagin spaces $\tilde{\mathcal{P}}$ of S in $\mathcal{P}$ in the model and all functions $T \in \widetilde{\mathbf{N}}_0$; if $\mathcal{P}$ is an anti-Hilbert space one parameter T has to be excluded. More specifically, there exists an $(i\hat{J})$-unitary matrix U^a, such that if via Kreĭn's formula $\tilde{A}$ on the lefthand side of (5) is given by $\tilde{A} = \tilde{A}_{T'}$ for some $T' \in \widetilde{\mathbf{N}}_0$, then T on the righthand side of (5) is given by $T = U^a_{T'}$; and the excluded parameter then is $T = U^a_{-Q_\infty}$. Hence the canonical selfadjoint extensions $\tilde{A}$ in (5) correspond to the constants $T \in \widetilde{\mathbf{N}}_0$, and the selfadjoint extensions $\tilde{A}$ with finite dimensional exit spaces to the rational functions $T \in \widetilde{\mathbf{N}}_0$.

Proof. We first consider the case $a \in \mathbb{R}$ and use the results and notation of the foregoing paragraphs. The lefthand side of (5) (where we set $k = 1$) can be written as $\tilde{\Theta}^a_{T'}(z)$, $T' \neq -Q_\infty$, where

$$\tilde{\Theta}^a(z) = X(z)W(z) = X(z)W^a(z)W(a).$$

Here the 2×2 matrix function X is defined by

$$X(z) = \begin{pmatrix} (z - z_1)(a - \bar{z}_1) & \bar{w}_1 + (z - \bar{z}_1)p_{11} \\ 0 & 1 \end{pmatrix}.$$

After some calculations as in [Br1] and [ABDS] using the above formulas for P and Q, we obtain that

$$\tilde{\Theta}^a(z) = (z - z_1)\Theta^a(z)YW(a),$$

where Y is the $(i\hat{J})$-unitary matrix

$$Y(z) = \frac{1}{z_1 - a}\begin{pmatrix} |z_1 - a|^2 & w_1 - (z_1 - a)p_{11} \\ 0 & 1 \end{pmatrix}.$$

Thus the lefthand side of (5) can be written as

$$\tilde{\Theta}^a_{T'}(z) = \Theta^a_T(z),$$

where $T(z) = U^a_{T'}(z)$ with $U^a = YW(a)$. Since U^a is $(i\hat{J})$-unitary, the last equality defines a bijection $T' \mapsto T$ from $\widetilde{\mathbf{N}}_0$ onto itself. If the Pick matrix $\mathbb{P}$ is negative the parameter $T(z) = U^a_{-Q_\infty}(z)$ has to be excluded.

It remains to prove the parametrization for $a = \infty$. We use the fact that for all $b \in \mathbb{R} \cup \{\infty\}$ and all $z, w \neq z_1, z_2, \dots, z_n$,

$$\Theta^b(z)\hat{J}\Theta^b(w)^* =$$
$$\hat{J} + (z - \bar{w})\begin{pmatrix} w_1 & \cdots & w_n \\ 1 & \cdots & 1 \end{pmatrix}(Z - z)^{-1}\mathbb{P}^{-1}(Z - w)^{-*}\begin{pmatrix} w_1 & \cdots & w_n \\ 1 & \cdots & 1 \end{pmatrix}^*$$

(see [Br1, p. 17, formula (1.35)]). Note that the righthand side is independent of $b \in \mathbb{R} \cup \{\infty\}$. It follows that for any $b \in \mathbb{R}$,

$$\Theta^\infty(z) = \Theta^b(z)V^b,$$

where $V^b = -\hat{J}\Theta^b(b)^*\hat{J}\Theta^\infty(b)$ is an $(i\hat{J})$-unitary matrix. The theorem now follows with $U^\infty = (V^b)^{-1}U^b$. □

The Theorems 3.1. and 4.2. imply the following result, which provides a parametrization of all solutions of our problem.

Theorem 4.3. *The function* $N(z)$ *is a solution of the interpolation problem if and only if* $N(z) = \Theta_T^a(z)$ *for some* $T \in \widetilde{\mathbf{N}}_0$ *for which the selfadjoint extension* $\tilde{A}_{T'}$ *of* S, $T = U_{T'}^a$, *has the property that all interpolation points* $z_1, z_2, \ldots, z_n$ *belong to* $\rho(\tilde{A}_{T'})$.

5 Excluded parameters

We call $T \in \widetilde{\mathbf{N}}_0$ an **excluded** parameter if the function $\Theta_T^a(z)$ is not a solution of the interpolation problem. Note that Theorem 4.3. characterizes the nonexcluded parameters in terms of the selfadjoint extensions of the symmetry S of the model. The following theorem characterizes the excluded parameters by a simple equation without reference to selfadjoint extensions.

Theorem 5.1. *The function* $T \in \widetilde{\mathbf{N}}_0$ *is an excluded parameter if and only if for some interpolation point* z_j

$$T(z_j) = -\theta_{22}^a(z_j)/\theta_{21}^a(z_j). \tag{1}$$

If T *is an excluded parameter then* Θ_T *belongs to a class* $\mathbf{N}_{\kappa'}$ *for some* κ' *with* $0 \le \kappa' < \kappa$.

Observe that the relation (1) just means that in the fractional linear description

$$N(z) = \Theta_T^a(z) = \frac{\theta_{11}^a(z)T(z) + \theta_{12}^a(z)}{\theta_{21}^a(z)T(z) + \theta_{22}^a(z)} \tag{2}$$

of the solutions the denominator vanishes.

Proof of Theorem 5.1.. The first statement follows from the equivalence

$$z_j \notin \rho(\tilde{A}_{T'}) \Longleftrightarrow z_j \in \sigma_0(\tilde{A}_{T'}) \Longleftrightarrow T'(z_j) = -Q_\infty(z_j)(= -w_{22}(z_j)/w_{21}(z_j))$$

$$\Longleftrightarrow T(z_j) = U_{T'}^a(z_j) = -\theta_{22}^a(z_j)/\theta_{21}^a(z_j).$$

For the proof of the second statement we first suppose that $n \ge 2$. It suffices to prove that if z_j belongs to the point spectrum of a regular minimal extension $\tilde{A}$ of S in the space $\tilde{\mathcal{P}}$, then the Nevanlinna kernel $K_N(z, w)$ of the function

$$N(z) = \bar{w}_k + (z - \bar{z}_k)\frac{\operatorname{Im} w_k}{\operatorname{Im} z_k} + (z - \bar{z}_k)(z - z_k)\langle(\tilde{A} - z)^{-1}e_k, e_k\rangle_{\tilde{\mathcal{P}}}$$

has less then κ negative squares. For that we use the relation

$$K_N(z, w) = \langle e(z), e(w)\rangle_{\tilde{\mathcal{P}}},$$

where $e(z) = e_k + (z - z_k)(\tilde{A} - z)^{-1}e_k$. Assume that z_j belongs to the point spectrum of $\tilde{A}$. We claim that then $e(z)$ belongs to $\operatorname{ran}(\tilde{A} - z_j)$. To see this, let $k \neq j$; then from $S \subseteq \tilde{A}$ and

$$(z_k - z_j)e_k = (S - z_j)(e_k - e_j)$$

it follows that $e_k \in \operatorname{ran}(\tilde{A} - z_j)$. Since $\operatorname{ran}(\tilde{A} - z_j)^{\langle\perp\rangle_{\tilde{\mathcal{P}}}} = \ker(\tilde{A} - \bar{z}_j)$, we have

$$\langle(\tilde{A} - z)^{-1}e_k, \ker(\tilde{A} - \bar{z}_j)\rangle_{\tilde{\mathcal{P}}} = \langle e_k, (\bar{z}_j - \bar{z})^{-1}\ker(\tilde{A} - \bar{z}_j)\rangle_{\tilde{\mathcal{P}}} = \{0\},$$

and hence $(\tilde{A} - z)^{-1}e_k$ belongs to $\operatorname{ran}(\tilde{A} - z_j)$. The claim now follows from the definition of $e(z)$. As $\ker(\tilde{A} - \bar{z}_j) \subseteq \operatorname{ran}(\tilde{A} - z_j)$, we have that

$$\ker(\tilde{A} - \bar{z}_j) = \operatorname{ran}(\tilde{A} - z_j) \cap \operatorname{ran}(\tilde{A} - z_j)^{\langle\perp\rangle_{\tilde{\mathcal{P}}}},$$

and hence $\ker(\tilde{A} - \bar{z}_j)$ is the isotropic part of $\operatorname{ran}(\tilde{A} - z_j)$. Since with z_j also $\bar{z}_j$ belongs to the point spectrum of $\tilde{A}$, this subspace is nontrivial. By, for example, [DS, Proposition 4.3(ii)], $\operatorname{ran}(\tilde{A} - z_j)$ is closed in $\tilde{\mathcal{P}}$. It follows that the quotient space

$$\hat{\mathcal{P}} = \ker(\tilde{A} - \bar{z}_j)^{\langle\perp\rangle_{\tilde{\mathcal{P}}}}/\ker(\tilde{A} - \bar{z}_j) = \operatorname{ran}(\tilde{A} - z_j)/\ker(\tilde{A} - \bar{z}_j)$$

is a Pontryagin space with $\hat{\kappa} = \operatorname{ind}_-\hat{\mathcal{P}} = \kappa - \dim\ker(\tilde{A} - \bar{z}_j) < \kappa$. If for $x \in \operatorname{ran}(\tilde{A} - z_j)$ we denote by $\hat{x}$ the equivalence class in $\hat{\mathcal{P}}$ which contains x, then, on account of the claim just proved, we have that

$$K_N(z, w) = \langle\widehat{e(z)}, \widehat{e(w)}\rangle_{\hat{\mathcal{P}}}.$$

This implies that the kernel $K_N(z, w)$ has at most $\hat{\kappa}$ negative squares.

If $n = 1$, we can suppose that $w_1 \in \mathbb{C}^-$. Then $\mathbb{P} = \dfrac{w_1 - \bar{w}_1}{z_1 - \bar{z}_1} < 0$ and hence $\kappa = 1$. So we look for solutions N in the class $\mathbf{N}_1$. The matrix function $\Theta^a(z)$ is given by

$$\Theta^a(z) = \begin{pmatrix} 1 & 0 \\ 0 & 1 \end{pmatrix} - \frac{(z-a)(z_1 - \bar{z}_1)}{(z_1 - z)(w_1\bar{w}_1)(\bar{z}_1 - a)} \begin{pmatrix} w_1 & -w_1\bar{w}_1 \\ 1 & -\bar{w}_1 \end{pmatrix}.$$

$\Theta^\infty(z)$ is obtained by letting $a \to \infty$. We consider the case $a = 0$; the other cases can treated in the same way. We find that the solutions of the interpolation are of the form:

$$N(z) = \frac{(\bar{z}_1(w_1 - \bar{w}_1)(z_1 - z) - zw_1(z_1 - \bar{z}_1))\,T(z) + zw_1\bar{w}_1(z_1 - \bar{z}_1)}{-z(z_1 - \bar{z}_1)T(z) + (\bar{z}_1(w_1 - \bar{w}_1)(z_1 - z) + z\bar{w}_1(z_1 - \bar{z}_1))}, \tag{3}$$

where T varies over $\widetilde{\mathbf{N}}_0$. (For $z_1 = i$ and $w_1 = -i$ this formula simply becomes $N(z) = (T(z) - z)/(zT(z) + 1)$.) The model associated with the interpolation problem is as follows: The Pontryagin space $\mathcal{P}$ is the space $\mathbb{C}$ provided with the negative inner product

$$\langle x, y\rangle = \bar{y}\frac{\operatorname{Im} w_1}{\operatorname{Im} z_1}x$$

and the symmetric operator S is the zero operator with domain equal to the trivial subspace $\{0\}$. The selfadjoint linear relation $\tilde{A}_\infty$ maps 0 to every element in $\mathbb{C}$; its resolvent operator is the zero operator defined on the whole space $\mathbb{C}$. The Q-function corresponding to S and $\tilde{A}_\infty$ is given by

$$Q_\infty(z) = c + (z - \operatorname{Re}\mu)\frac{w_1 - \bar{w}_1}{z_1 - \bar{z}_1}\frac{1}{|z_1 - \mu|^2} = -\alpha z + \beta,$$

where α and β are real numbers with $\alpha > 0$. Hence Q_∞ is an $\mathbf{N}_1$-function. Thus there is an excluded parameter due to the fact that $\mathbb{P}$ is negative: the denominator of the expression on the righthand side of (3) may not be identically equal to zero. The other parameters T which have to be excluded are those for which this denominator is equal to zero at z_1: $T(z_1) = \bar{w}_1$. These functions are given by

$$T(z) = \frac{(\bar{z}_1(w_1 - \bar{w}_1)(z_1 - z) + z\bar{w}_1(z_1 - \bar{z}_1))\,T'(z) - zw_1\bar{w}_1(z_1 - \bar{z}_1)}{z(z_1 - \bar{z}_1)T'(z) + (\bar{z}_1(w_1 - \bar{w}_1)(z_1 - z) - zw_1(z_1 - \bar{z}_1))},$$

where T' varies over $\widetilde{\mathbf{N}}_0$. If we substitute the expression for T in the formula (3) for N we obtain $N = T'$; evidently, these functions are Nevanlinna functions, hence they do not solve the interpolation problem. □

An immediate consequence of Theorem 5.1. is the following result.

Theorem 5.2. (a) *If z_j belongs to Δ_+ or, equivalently, $\pi_{jj} > 0$, then the relation (1) is satisfied for no $T \in \widetilde{\mathbf{N}}_0$, and hence for all parameters $T \in \widetilde{\mathbf{N}}_0$, the function $N(z)$ defined by (2) is holomorphic at z_j and satisfies $N(z_j) = w_j$.*
(b) *If z_j belongs to Δ_0 or, equivalently, $\pi_{jj} = 0$, then $\theta_{22}^a(z_j)/\theta_{21}^a(z_j) \in \mathbb{R} \cup \{\infty\}$ and the relation (1) is satisfied for exactly one $T \in \widetilde{\mathbf{N}}_0$, and hence for all other parameters T, the function $N(z)$ in (2) is holomorphic at z_j and satisfies $N(z_j) = w_j$.*
(c) *If z_j belongs to Δ_- or, equivalently, $\pi_{jj} < 0$, then $\operatorname{Im}\theta_{22}^a(z_j)/\theta_{21}^a(z_j) < 0$, and the relation (1) is satisfied for infinitely many $T \in \widetilde{\mathbf{N}}_0$ (all nonconstant), and hence for all other parameters T, the function $N(z)$ in (2) is holomorphic in z_j and satisfies $N(z_j) = w_j$.*

Lemma (2.4.)(d) implies that the properties "$z_j \in \Delta_+$," "$z_j \in \Delta_0$," or, "$z_j \in \Delta_-$," can be read off from the sign of the diagonal element π_{jj} of $\mathbb{P}^{-1}$. Theorem 5.2. implies that in the general problem if $z_1, z_2, \ldots, z_n$ are given, then for each point $z_j \in \Delta_0$ one constant parameter T and for each point $z_j \in \Delta_-$ infinitely many nonconstant parameters T with the property (1) have to be excluded; the other elements $T \in \widetilde{\mathbf{N}}_0$ describe by means of formula (2) all solutions of the interpolation problem.

Recall that the order of a rational function is the number of its poles including infinity and counted according to their multiplicities. We call a solution of the interpolation problem **canonical** if it is rational and of order n, where n is the number of interpolation points. A canonical solution corresponds to a canonical selfadjoint extension of the symmetric operator S in the model. As a consequence of the above theorems we formulate the following theorem.

Theorem 5.3. *If no interpolation point belongs to Δ_0 or, equivalently, all the diagonal entries of $\mathbb{P}^{-1}$ are nonzero, then the relation (2) gives a bijective correspondence between all canonical solutions of the interpolation problem and all $T \in \mathbb{R} \cup \{\infty\}$.*

We conclude this section with three simple examples. We always take $n = 2$ and calculate the function $\Theta^a(z)$ for the case $a = 0$, which corresponds to Potapov's formulas if $\kappa = 0$; we write $\Theta(z)$ instead of $\Theta^0(z)$.

Example 5.4. Take $z_1 = i$, $z_2 = 2i$, $w_1 = 1$ and $w_2 = 2$. Then

$$\mathbb{P} = \frac{i}{3}\begin{pmatrix} 0 & -1 \\ 1 & 0 \end{pmatrix}, \qquad \det \mathbb{P} = -\frac{1}{9},$$

$\mathbb{P}$ is invertible and has one negative and one positive eigenvalue, and

$$\Theta(z) = \frac{-1}{2(z-i)(z-2i)} \begin{pmatrix} 7z^2+4 & -6z^2 \\ 3z^2 & 4-2z^2 \end{pmatrix}.$$

The solutions of the interpolation problem in the class $\mathbf{N}_1$ are given by the formula

$$N(z) = \frac{(7z^2+4)T(z) - 6z^2}{3z^2 T(z) + (4-2z^2)},$$

where T runs through $\widetilde{\mathbf{N}}_0$ except for the values $T(z) \equiv 1$ and $T(z) \equiv 2$. For these exceptional values of $T(z)$ we have that $N(z) \equiv 1$ and $N(z) \equiv 2$, respectively. These functions are holomorphic at the interpolation points, but are only partial solutions of the problem and belong to the class $\mathbf{N}_0$, that is, have a nonnegative Nevanlinna kernel. Note that the diagonal entries of the matrix

$$\mathbb{P}^{-1} = 3i\begin{pmatrix} 0 & -1 \\ 1 & 0 \end{pmatrix}$$

are both zero. □

Example 5.5. Take $z_1 = i$, $z_2 = 2i$, $w_1 = i$ and $w_2 = 1$. Then

$$\mathbb{P} = \frac{1}{3}\begin{pmatrix} 3 & 1-i \\ 1+i & 0 \end{pmatrix}, \qquad \det \mathbb{P} = -\frac{2}{9},$$

$\mathbb{P}$ has one negative and one positive eigenvalue, and

$$\Theta(z) = \frac{-1}{4(z-i)(z-2i)} \begin{pmatrix} 5z^2-3z+8 & 3z^2+3z \\ 3z^2-3z & 5z^2+3z+8 \end{pmatrix}.$$

The righthand side of (2) is given by

$$\Theta_T(z) = \frac{(5z^2-3z+8)T(z)+(3z^2+3z)}{(3z^2-3z)T(z)+(5z^2+3z+8)}.$$

The function $N(z) = \Theta_T(z)$ is a solution of the interpolation problem in $\widetilde{\mathbf{N}}_1$ if and only if T belongs to the class $\widetilde{\mathbf{N}}_0$ and does *not* satisfy

$$T(i) = 1 \quad \text{and} \quad T(2i) = \frac{-3+4i}{5}.$$

The first equation has a unique solution: $T \equiv 1$, and for this T we have $\Theta_T(z) \equiv 1$. The second equation (represents a Nevanlinna-Pick problem in $\mathbf{N}_0$ with one interpolation point and hence) has infinitely many solutions T, and they are not constant. For each such T we find that Θ_T is holomorphic near $z = 2i$ and $\Theta_T(2i) \neq 1$. On account of Theorem 5.1., $\Theta_T \in \mathbf{N}_0$. Note that the inverse of the Pick matrix is given by

$$\mathbb{P}^{-1} = \frac{3}{2} \begin{pmatrix} 0 & 1-i \\ 1+i & -3 \end{pmatrix}$$

and the above conclusions can also be drawn from the sign of the diagonal entries of this matrix. □

Example 5.6. Take $z_1 = i$, $z_2 = 2i$, $w_1 = -i$ and $w_2 = 2i$. Then

$$\mathbb{P} = \frac{1}{3} \begin{pmatrix} -3 & 1 \\ 1 & 3 \end{pmatrix}, \qquad \det \mathbb{P} = -\frac{10}{9},$$

$\mathbb{P}$ is invertible and has one negative and one positive eigenvalue, and

$$\Theta(z) = \frac{-1}{20(z-i)(z-2i)} \begin{pmatrix} 25z^2+40 & -24z \\ 15z & 16z^2+40 \end{pmatrix}.$$

The righthand side of (2) is given by

$$\Theta_T(z) = \frac{(25z^2+40)T(z)-24z}{15zT(z)+(16z^2+40)}.$$

The function $N(z) = \Theta_T(z)$ is a solution of the interpolation problem in $\widetilde{\mathbf{N}}_1$ if and only if T belongs to $\widetilde{\mathbf{N}}_0$ and does not satisfy $T(i) = \frac{8i}{5}$. This equation has infinitely many solutions T and for each such T we find that Θ_T belongs to $\mathbf{N}_0$ and $\Theta_T(i) \neq -i$. For all parameters $T \in \widetilde{\mathbf{N}}_0$, $\Theta_T(2i) = 2i$. We have that

$$\mathbb{P}^{-1} = \frac{3}{10}\begin{pmatrix} -3 & 1 \\ 1 & 3 \end{pmatrix},$$

and the results just mentioned are in accordance with the sign of the diagonal entries of this matrix. □

References

[ABDS] D. Alpay, P. Bruinsma, A. Dijksma, and H.S.V. de Snoo, *Interpolation problems, extensions of symmetric operators and reproducing kernels I*, Operator Theory: Adv. Appl., vol 50, Birkhäuser Verlag, Basel, 1991, 35–82.

[AI] T.Ya. Azizov and I.S. Iohkvidov, *Foundations of the theory of linear operators in spaces with an indefinite metric*, "Nauka", Moscow, 1986; English transl.: *Linear operators in spaces with indefinite metric*, Wiley, New York, 1989.

[Ba] J.A. Ball, *Interpolation problems of Pick-Nevanlinna and Loewner types for meromorphic matrix functions*, Integral Equations and Operator Theory, **6** (1983), 804–840.

[BGR] J.A. Ball, I. Gohberg, and L. Rodman, *Interpolation of rational matrix functions*, Operator Theory: Adv. Appl., vol 45, Birkhäuser Verlag, Basel, 1990.

[Bo] J. Bognár, *Indefinite inner product spaces*, Springer Verlag, Berlin, 1974.

[Br1] P. Bruinsma, *Interpolation problems for Schur and Nevanlinna pairs*, Dissertation University of Groningen, 1991.

[Br2] P. Bruinsma, *Degenerate interpolation problems for Nevanlinna pairs*, Indag. Mathem., N.S. **2** (1991), 179–200.

[D] H. Dym, *J contractive matrix functions, reproducing kernel Hilbert spaces and interpolation*, Regional conference series in mathematics, **71**, Amer. Math. Soc., Providence, R.I., 1989.

[DGK] P. Delsarte, Y. Genin, and Y. Kamp, *Pseudo-Carathéodory functions and hermitian Toeplitz matrices*, Philips J. Res., **41** (1986), 1–54.

[DLS1] A. Dijksma, H. Langer, and H.S.V. de Snoo, *Selfadjoint Π_κ-extensions of symmetric subspaces: an abstract approach to boundary value problems with spectral parameter in the boundary conditions*, Integral Equations and Operator Theory, **7** (1984), 460–515.

[DLS2] A. Dijksma, H. Langer, and H.S.V. de Snoo, *Hamiltonian systems with eigenvalue depending boundary conditions*, Operator Theory: Adv. Appl., vol 35, Birkhäuser Verlag, Basel, 1988, 37–83.

[DLS3] A. Dijksma, H. Langer, and H.S.V. de Snoo, *Eigenvalues and pole functions of Hamiltonian systems with eigenvalue depending boundary conditions*, Math. Nachr., **161** (1993), 107–154.

[DS] A. Dijksma and H.S.V. de Snoo, *Symmetric and selfadjoint relations in Kreĭn spaces I*, Operator Theory: Adv. Appl., vol 24, Birkhäuser Verlag, Basel, 1987, 145–166.

[G] L.B. Golinskii, *A generalization of the Nevanlinna-Pick matrix problem*, Izv. Akad. Nauk Arm. SSR Math., **18** (1983), 187–205 (Russian). English transl. in: *Soviet Journal of Contemporary Mathematical Analysis* (Allerton Press), **18**:3 (1983), 22–39.

[IKL] I.S. Iohkvidov, M.G. Kreĭn , and H. Langer. *Introduction to the spectral theory of operators in spaces with an indefinite metric*, Akademie Verlag, Berlin, 1982.

[K] M.G. Kreĭn , *On the resolvents of an Hermitian operator with defect index* (m, m), Dokl. Akad. Nauk. SSSR, **52** (1946), 657–660 (Russian).

[KL1] M.G. Kreĭn and H. Langer, *The defect subspaces and generalized resolvents of a Hermitian operator in the space* Π_κ, Funkcional. Anal. i Prilozen., **5** (1971), no 2, 59–71; Conclusion, **5** (1971), no 3, 54–69; English transl. in Functional Anal. Appl., **5** (1971), 136–146 and 217–228.

[KL2] M.G. Kreĭn and H. Langer, *Über die Q-Funktion eines* π*-hermiteschen Operators im Raume* Π_κ, Acta Sci. Math. (Szeged), **34** (1973), 191–230.

[KL3] M.G. Kreĭn and H. Langer, *Über einige Fortsetzungsprobleme, die eng mit der Theorie hermitescher Operatoren im Raume* Π_κ *zusammenhängen II. Verallgemeinerte Resolventen, u-Resolventen und ganze Operatoren*, J. Funct. Anal., **30** (1978), 390-447.

[KP] I.V. Kovalishina and V.P. Potapov, *An indefinite metric in the Nevanlinna-Pick problem*, Akad. Nauk. Arm. SSR Dokl., **59** (1974), 17-22; English transl. in *Seven papers translated from the Russian* by I.V. Kovalishina and V.P. Potapov, Amer. Math. Soc. Transl., vol 138, Amer. Math. Soc., Providence, R.I., 1988, 15–19.

[LT] H. Langer and B. Textorius, *On generalized resolvents and Q-functions of symmetric linear relations (subspaces) in Hilbert space*, Pacific J. Math., **72** (1977), 135–165.

[N1] A.A. Nudelman, *On a generalization of classical interpolation problems*, Soviet Math. Dokl., **23** (1981), 125–128.

[N2] A.A. Nudelman, Lecture at the Leipzig conference "Recent developments in Schur Analysis", September 22–26, 1994.

[S] Yu.L. Shmul'yan, *On a class of holomorphic operator functions*, Matem. Zametki, **5** (1969), 351–359; English transl. in Mathematical Notes of the Academy of Sciences of the USSR, vol 5, Plenum, New York, 1969, 212–216.

[W1] H. Woracek, *Das verallgemeinerte Nevanlinna-Pick Problem im entarteten Fall*, Dissertation Technical University of Vienna, 1993.

[W2] H. Woracek, *An operator theoretic approach to degenerated Nevanlinna-Pick interpolation*, preprint.

Aad Dijksma
Department of Mathematics
Unversity of Groningen
P.O. Box 800
9700 AV Groningen
The Netherlands
E-mail: a.dijksma@math.rug.nl

Heinz Langer
Department of Mathematics
Technical University of Vienna
Wiedner Hauptstrasse 8–10
A-1040 Vienna
Austria
E-mail: hlanger@email.tuwien.ac.at

AMS Subject Classification Numbers 47A57, 47B50

Operator Theory
Advances and Applications, Vol. 95
© 1997 Birkhäuser Verlag Basel/Switzerland

The indefinite metric in the Schur interpolation problem for analytic functions. IV[1]

V.K. Dubovoj

Abstract. A complete description is given of the set of all contractive matrix valued holomorphic functions in the disc whose first $n+1$ matrix coefficients in their power series expansion about zero are specified when the associated information block (or Pick matrix) is positive semidefinite and singular. In an earlier part of this series,[2] a complete description was given when the associated information block is positive and nonsingular. [Abstract added by editors.]

10 Solving the degenerate Schur problem

In this section we draw attention to a method of solving the Fundamental Matrix Inequalities (FMI) (S) and $(\tilde{S})$ (see §1, [Dub1]) with a *degenerate* information block.[3] In order to split off the kernel (the zero-subspace) of the matrix A_n we introduce the notion of a *K-type subspace* (see Section 1). After splitting off this kernel the solution is representable in the form of a fractional linear transformation whose argument is a matrix-function $\omega(\zeta) \in \mathcal{S}_{p,q}$ (see Section 1; Lemmas 10.1, 10.2). The coefficient matrix of this fractional linear transformation is an elementary factor of nonfull rank [Dub4]. The presence of the kernel imposes some constraints on the parameter $\omega(\zeta)$. The nature of these constraints is clarified in Section 2. The final results are formulated in Theorems 10.1 and 10.2. Finally, an example of a K-type subspace is given in Section 3.

Interpolation problems with a degenerate information block (or Pick matrix) were considered by A.A. Nudelman [Nud], I.P. Fedćina [Fed] and L.A. Galstyan [Gal]. However, our method differs from the methods used in these works.

---

1) Translated from: Teor. Funkciĭ, Funkcional. Anal. i Prilozen **42** (1984), 46–57, (The Kharkov University publishing house, Marchenko, V.A., – ed) [**MR 86c**:47008].

2) The first three parts of this work were published in [Dub1], [Dub2],[Dub3]. In order to read this paper, however, it is enough to be familiar with [Dub1].

3) In the western mathematical literature it is more common to use the terminology *Pick matrix* instead of *information block* for $A_n = I - C_nC_n^*$.

10.1. Now we will identify the functions $\omega(\zeta) \in \mathcal{S}_{p,q}$ with operator-valued functions acting from a space E_- ($\dim E_- = q$) into a space E_+ ($\dim E_+ = p$).

In view of the block-structure of the operator C_k we consider the operator A_k as an operator acting in the space

$$E_+^{(k)} = E_+ \oplus E_+ \oplus \ldots \oplus E_+, \tag{10.1}$$

where the number of E_+ is $k+1$. This allows us to embed $E_+^{(k-1)}$ into $E_+^{(k)}$, i.e., $E_+^{(k)} = E_+^{(k-1)} \oplus E_+$. Thus,

$$E_+ = E_+^{(0)} \subset E_+^{(1)} \subset \ldots \subset E_+^{(n)}. \tag{10.2}$$

Definition 10.1. *A subspace $L \subset E_+^{(n)}$ is a K-type subspace if*
(1) L is a complement to $\mathrm{Ker} A_n$, *i.e., if $L \dot{+} \mathrm{Ker} A_n = E_+^{(n)}$.*
(2) L is invariant with respect to the operator $V_{p,n}^*$, where

$$V_{p,n} = \begin{bmatrix} 0 & & & & \\ I_p & 0 & & & \\ & I_p & & & \\ & & \ddots & & \\ & & & I_p & 0 \end{bmatrix}.$$

In Subsection 10.3 we will give an example of a K-type subspace. From this example it will turn out that infinitely many K-type subspaces exist.

Let L be an arbitrary K-type subspace and let P be the orthoprojector onto L.
Clearly,

$$V_{p,n}^* \, P = P \, V_{p,n}^* \, P. \tag{10.3}$$

Let

$$\tilde{C}_n = P\, C_n \quad \text{and} \quad A_n^{(1)} = (P - \tilde{C}_n \tilde{C}_n^*)_{|L} = P\, A_n\, P_{|L}.$$

In view of the orthogonal decomposition $E_+^{(n)} = L \oplus \tilde{L}$, we can consider the block representation of A_n

$$A_n = \begin{bmatrix} A_n^{(1)} & B \\ B^* & C \end{bmatrix} = \begin{bmatrix} I & 0 \\ X^* & I \end{bmatrix} \cdot \begin{bmatrix} A_n^{(1)} & 0 \\ 0 & 0 \end{bmatrix} \cdot \begin{bmatrix} I & X \\ 0 & I \end{bmatrix}, \tag{10.4}$$

where X is a solution of the equation $A_n^{(1)} \, X = B$. From (10.4) it follows that $\mathrm{Ker} A_n$ consists of vectors $f = f_1 + f_2$ with $f_1 \in L$, $f_2 \in \tilde{L}$ and $f_1 = -X\, f_2$, i.e., $\mathrm{Ker} A_n = \mathrm{Im} \begin{bmatrix} -X \\ I \end{bmatrix}$. Hence,

$$\mathrm{Ker}\,[-X^*, I] = \mathrm{Im} A_n. \tag{10.5}$$

The following properties of the operator $\tilde{C}_n$ will be significant in our future considerations:

$$\tilde{C}_n\, V_{q,n} = P\, V_{p,n}\, \tilde{C}_n\ , \tag{10.6}$$

$$P - \tilde{C}_n\, \tilde{C}_n^* \geq 0 \quad \text{and} \qquad \operatorname{rank}(P - \tilde{C}_n\, \tilde{C}_n^*) = \operatorname{rank} P\ . \tag{10.7}$$

To obtain (10.6) we multiply the equality $C_n\, V_{q,n} = V_{p,n}\, C_n$ by P from the left side and use (10.3). To obtain the relation (10.7) we use the fact that L is a K-type subspace and that $P - \tilde{C}_n\, \tilde{C}_n^* = P\, A_n\, P$. The block representation of the operator $P - \tilde{C}_n\, \tilde{C}_n^*$ corresponding to the decomposition $E_+^{(n)} = L \oplus \tilde{L}$ has the form

$$\begin{bmatrix} A_n^{(1)} & 0 \\ 0 & 0 \end{bmatrix}.$$

In this connection we denote the operator

$$\begin{bmatrix} (A_n^{(1)})^{-1} & 0 \\ 0 & 0 \end{bmatrix}.$$

by the symbol $(P - \tilde{C}_n\, \tilde{C}_n^*)^{-1}$.

Let us turn to the solution of the inequalities (S) and $(\tilde{S})$.

Writing down (S) in the form (see §1, (1.4))

$$\begin{bmatrix} I - C_n C_n^* & B^{(1)}(\zeta, n) \\ B^{(1)*}(\zeta, n) & \dfrac{I - \Theta(\zeta)\Theta^*(\zeta)}{1 - |\zeta|^2} \end{bmatrix} \geq 0,$$

where $B^{(1)}(\zeta, n) = \Lambda_{p,n}^*(\zeta) - C_n \Lambda_{q,n}^*(\zeta)\Theta^*(\zeta)$, we multiply this inequality by

$$T = \begin{bmatrix} \begin{bmatrix} I & -X \\ 0 & I \end{bmatrix} & 0 \\ 0 & I_p \end{bmatrix}$$

from the right and by T^* from the left. Taking into account (10.4), we obtain the equivalent inequality

$$\begin{bmatrix} \begin{bmatrix} A_n^{(1)} & 0 \\ 0 & 0 \end{bmatrix} & \begin{bmatrix} I & 0 \\ -X^* & I \end{bmatrix} B^{(1)}(\zeta, n) \\ * & \dfrac{I - \Theta(\zeta)\Theta^*(\zeta)}{1 - |\zeta|^2} \end{bmatrix} \geq 0.$$

The last inequality is equivalent to the following pair of relations

$$[-X^*, I] B^{(1)}(\zeta, n) = 0, \tag{10.8}$$

$$\frac{I - \Theta(\zeta)\Theta^*(\zeta)}{1 - |\zeta|^2} - B^{(1)*}(\zeta, n)(P - \tilde{C}_n \tilde{C}_n^*)^{-1} B^{(1)}(\zeta, n) \geq 0 \tag{10.9}$$

(see for example [EfPot], p.88 of the Russian original or pp. 93–94 of the English translation where the appropriate statement is referred to as *Lemma on a non-negative block matrix*). Now we will first solve the inequality (10.9). Then we shall select the solutions of the inequality (10.9) which satisfy the condition (10.8). Thus, the set of all solutions of the inequality (S) will be described. We solve the inequality (10.9) in the same way as in the nondegenerate case. Namely, by taking into account the formulas

$$\frac{I-\Theta(\zeta)\Theta^*(\zeta)}{1-|\zeta|^2} = [\Theta(\zeta), I]\frac{j}{1-|\zeta|^2}\begin{bmatrix}\Theta^*(\zeta)\\ I\end{bmatrix}$$

and

$$B^{(1)}(\zeta,n) = [-C_n, I]\begin{bmatrix}\Lambda^*_{q,n}(\zeta) & 0\\ 0 & \Lambda^*_{p,n}(\zeta)\end{bmatrix}\begin{bmatrix}\Theta^*(\zeta)\\ I\end{bmatrix}, \tag{10.10}$$

we rewrite (10.9) in the form

$$[\Theta(\zeta),I]\left\{\frac{j}{1-|\zeta|^2} - j\begin{bmatrix}\Lambda_{q,n}(\zeta) & 0\\ 0 & \Lambda_{p,n}(\zeta)\end{bmatrix}H_n\begin{bmatrix}\Lambda^*_{q,n}(\zeta) & 0\\ 0 & \Lambda^*_{p,n}(\zeta)\end{bmatrix}j\right\}\begin{bmatrix}\Theta^*(\zeta)\\ I\end{bmatrix}\geq 0, \tag{10.11}$$

where

$$H_n = \begin{bmatrix}\tilde{C}^*_n\\ P\end{bmatrix}(P-\tilde{C}_n\tilde{C}^*_n)^{-1}[\tilde{C}_n, P].$$

From (10.3), (10.6) and (10.7) it follows (just as was shown in [Dub4]) that the matrix-function

$$B_n(\zeta) = I + \frac{1-\zeta}{\zeta}j\begin{bmatrix}\Lambda_{q,n}(1) & 0\\ 0 & \Lambda_{p,n}(1)\end{bmatrix}H_n\begin{bmatrix}\Lambda^*_{q,n}\left(1/\overline{\zeta}\right) & 0\\ 0 & \Lambda^*_{p,n}\left(1/\overline{\zeta}\right)\end{bmatrix} \tag{10.12}$$

is a j-expansive elementary multiple factor. Moreover,

$$B^*_n(\zeta)jB_n(\zeta)-j = \frac{1-|\zeta|^2}{|\zeta|^2}\begin{bmatrix}\Lambda_{q,n}\left(1/\overline{\zeta}\right) & 0\\ 0 & \Lambda_{p,n}\left(1/\overline{\zeta}\right)\end{bmatrix}H_n\begin{bmatrix}\Lambda^*_{q,n}\left(1/\overline{\zeta}\right) & 0\\ 0 & \Lambda^*_{p,n}\left(1/\overline{\zeta}\right)\end{bmatrix}.$$

Since $(B^*_n)^{-1}(\zeta) = jB_n\left(1/\overline{\zeta}\right)j$, it follows that

$$\frac{j-B_n^{-1}(\zeta)j(B^*_n)^{-1}(\zeta)}{1-|\zeta|^2} = j\begin{bmatrix}\Lambda_{q,n}(\zeta) & 0\\ 0 & \Lambda_{p,n}(\zeta)\end{bmatrix}H_n\begin{bmatrix}\Lambda^*_{q,n}(\zeta) & 0\\ 0 & \Lambda^*_{p,n}(\zeta)\end{bmatrix}j.$$

Therefore (10.11) can be rewritten in the following way:

$$[\Theta(\zeta), I]\frac{B_n^{-1}(\zeta)j(B^*_n)^{-1}(\zeta)}{1-|\zeta|^2}\begin{bmatrix}\Theta^*(\zeta)\\ I\end{bmatrix}\geq 0. \tag{10.13}$$

Now we define a pair of matrix-functions by the formula

$$[u(\zeta), v(\zeta)] = [\Theta(\zeta), I]B_n^{-1}(\zeta). \tag{10.14}$$

Just as in the nondegenerate case we see that $[u(\zeta), v(\zeta)]$ is a holomorphic, j-expansive and nondegenerate pair in the unit circle, i.e., the quotient

$$\omega(\zeta) := v^{-1}(\zeta)u(\zeta) \in \mathcal{S}_{p,q} \tag{10.15}$$

is well defined. Let us consider the block decomposition

$$B_n(\zeta) = \begin{bmatrix} a(\zeta) & b(\zeta) \\ c(\zeta) & d(\zeta) \end{bmatrix}$$

of the matrix $B_n(\zeta)$ which corresponds to the block decomposition of the matrix j. Taking into account (10.15), we obtain the following formula from (10.14):

$$\Theta(\zeta) = [\omega(\zeta)b(\zeta) + d(\zeta)]^{-1}[\omega(\zeta)a(\zeta) + c(\zeta)]. \tag{10.16}$$

Conversely, if $\omega(\zeta) \in \mathcal{S}_{p,q}$ then the matrix-function $\omega(\zeta)b(\zeta) + d(\zeta)$ is invertible everywhere in the unit disc, and the matrix-function $\Theta(\zeta)$ satisfies (10.13) and hence (10.9). Thus, we have proved:

Lemma 10.1. *The general solution $\Theta(\zeta)$ of the inequality* (10.9) *can be represented in the form of the fractional linear transformation* (10.16), *where the parameter $\omega(\zeta)$ is an arbitrary function from the class $\mathcal{S}_{p,q}$. The coefficient matrix of this fractional linear transformation is an elementary j-expansive multiple factor* (10.12) *constructed from the matrix C_n and from a K-type subspace.*

Let us clarify what constraints on the parameter $\omega(\zeta)$ are imposed by the condition (10.8). From (10.16) it follows that

$$\begin{bmatrix} \Theta^*(\zeta) \\ I \end{bmatrix} = B_n^*(\zeta) \begin{bmatrix} \omega^*(\zeta) \\ I \end{bmatrix} q^*(\zeta),$$

where $q(\zeta) = (w(\zeta)b(\zeta) + d(\zeta))^{-1}$. Taking into account (10.10) we obtain

$$B^{(1)}(\zeta, n) = [-C_n, I] \begin{bmatrix} \Lambda_{q,n}^*(\zeta) & 0 \\ 0 & \Lambda_{p,n}^*(\zeta) \end{bmatrix} B_n^*(\zeta) \begin{bmatrix} \omega^*(\zeta) \\ I \end{bmatrix} q^*(\zeta). \tag{10.17}$$

A simple calculation gives

$$\Lambda_{p,n}^*(\zeta)\Lambda_{p,n}^*\left(1/\overline{\zeta}\right) = D_{p,n}(\overline{\zeta}) + D_{p,n}^*\left(1/\zeta\right) + I,$$

where

$$D_{p,n}(\zeta) = \zeta V_{p,n} + \zeta^2 V_{p,n}^2 + \ldots + \zeta^n V_{p,n}^n.$$

Let

$$\Phi_n(\zeta) = \begin{bmatrix} D_{q,n}(\overline{\zeta}) + D_{q,n}^*\left(1/\zeta\right) + I & 0 \\ 0 & D_{p,n}(\overline{\zeta}) + D_{p,n}^*\left(1/\zeta\right) + I \end{bmatrix}.$$

In view of (10.12) we can rewrite (10.17) in the following way:

$$B^{(1)}(\zeta,n)=[-C_n,I]\left\{\begin{bmatrix}\Lambda^*_{q,n}(\zeta) & 0\\ 0 & \Lambda^*_{p,n}(\zeta)\end{bmatrix}\right.$$
$$\left.+\frac{1-\overline{\zeta}}{\overline{\zeta}}\Phi_n(\zeta)H_n\begin{bmatrix}\Lambda^*_{q,n}(1) & 0\\ 0 & \Lambda^*_{p,n}(1)\end{bmatrix}j\right\}\begin{bmatrix}\omega^*(\zeta)\\ I\end{bmatrix}q^*(\zeta). \quad (10.18)$$

Since

$$\Lambda^*_{p,n}(\zeta)=\left(I-\frac{1-\overline{\zeta}}{\overline{\zeta}}D_{p,n}(\overline{\zeta})\right)\Lambda^*_{p,n}(1),$$

we have

$$[-C_n,I]\begin{bmatrix}\Lambda^*_{q,n}(\zeta) & 0\\ 0 & \Lambda^*_{p,n}(\zeta)\end{bmatrix}$$

$$=[-C_n,I]\left(I-\frac{1-\overline{\zeta}}{\overline{\zeta}}\begin{bmatrix}D_{q,n}(\overline{\zeta}) & 0\\ 0 & D_{p,n}(\overline{\zeta})\end{bmatrix}\begin{bmatrix}\Lambda^*_{q,n}(1) & 0\\ 0 & \Lambda^*_{p,n}(1)\end{bmatrix}\right)$$

$$=\left(I-\frac{1-\overline{\zeta}}{\overline{\zeta}}D_{p,n}(\overline{\zeta})\right)[-C_n,I]\begin{bmatrix}\Lambda^*_{q,n}(1) & 0\\ 0 & \Lambda^*_{p,n}(1)\end{bmatrix}. \quad (10.19)$$

Further, upon taking the formula $\tilde{C}_n=P_nC_n$ into account, we obtain

$$[-C_n,I]\Phi_n(\zeta)H_n$$
$$=[-C_n,I]\Phi_n(\zeta)\begin{bmatrix}C^*_n\\ I\end{bmatrix}(P-\tilde{C}_n\tilde{C}^*_n)^{-1}[C_n,I]$$
$$=\left\{A_n\left(I+D^*_{p,n}\left(\frac{1}{\zeta}\right)\right)+D_{p,n}\left(\overline{\zeta}\right)A_n\right\}(P-\tilde{C}_n\tilde{C}^*_n)^{-1}[C_n,I]$$
$$=\left\{A_n\left(I+D^*_{p,n}\left(\frac{1}{\zeta}\right)\right)(P-\tilde{C}_n\tilde{C}^*_n)^{-1}+D_{p,n}(\overline{\zeta})\begin{bmatrix}I & 0\\ X^* & 0\end{bmatrix}\right\}[C_n,I].$$

Thus, from (10) and (10.19) it follows that

$$B^{(1)}(\zeta,n)$$
$$=\left\{I-\frac{1-\overline{\zeta}}{\overline{\zeta}}D_{p,n}(\zeta)\begin{bmatrix}0 & 0\\ -X^* & I\end{bmatrix}+\frac{1-\overline{\zeta}}{\overline{\zeta}}A_n\left(I+D_{p,n}\left(\frac{1}{\zeta}\right)\right)(P-\tilde{C}_n\tilde{C}^*_n)^{-1}\right\}$$
$$\cdot[-C_n,I]\begin{bmatrix}\Lambda^*_{q,n}(1) & 0\\ 0 & \Lambda^*_{p,n}(1)\end{bmatrix}\begin{bmatrix}\omega^*(\zeta)\\ I\end{bmatrix}q^*(\zeta).$$

From this and from (10.5) we conclude that condition (10.8) is equivalent to the equality

$$[-X^*,I]\left(I-\frac{1-\overline{\zeta}}{\overline{\zeta}}D_{p,n}(\overline{\zeta})\begin{bmatrix}0 & 0\\ -X^* & I\end{bmatrix}\right)[-C_n,I]\begin{bmatrix}\Lambda^*_{q,n}(1) & 0\\ 0 & \Lambda^*_{p,n}(1)\end{bmatrix}\begin{bmatrix}w^*(\zeta)\\ I\end{bmatrix}=0,$$

or

$$\begin{bmatrix} 0 & 0 \\ -X^* & I \end{bmatrix}\left(I-\frac{1-\overline{\zeta}}{\overline{\zeta}}D_{p,n}(\overline{\zeta})\begin{bmatrix} 0 & 0 \\ -X^* & I \end{bmatrix}\right)[-C_n,I]\begin{bmatrix} \Lambda_{q,n}^*(1) & 0 \\ 0 & \Lambda_{p,n}^*(1) \end{bmatrix}\begin{bmatrix} \omega^*(\zeta) \\ I \end{bmatrix}=0,$$

i.e.,

$$\left(I-\frac{1-\overline{\zeta}}{\overline{\zeta}}\begin{bmatrix} 0 & 0 \\ -X^* & I \end{bmatrix}D_{p,n}(\overline{\zeta})\right)\begin{bmatrix} 0 & 0 \\ -X^* & I \end{bmatrix}[-C_n,I]\begin{bmatrix} \Lambda_{q,n}^*(1) & 0 \\ 0 & \Lambda_{p,n}^*(1) \end{bmatrix}\begin{bmatrix} \omega^*(\zeta) \\ I \end{bmatrix}=0.$$

Now it follows that (10.8) is equivalent to the condition

$$[-X^*,I][-C_n,I]\begin{bmatrix} \Lambda_{q,n}^*(1) & 0 \\ 0 & \Lambda_{p,n}^*(1) \end{bmatrix}\begin{bmatrix} \omega^*(\zeta) \\ I \end{bmatrix}=0.$$

Denoting the orthoprojector onto $\mathrm{Ker}A_0$ by P_0, we can rewrite the last condition in the form

$$P_0[-C_n,I]\begin{bmatrix} \Lambda_{q,n}^*(1) & 0 \\ 0 & \Lambda_{p,n}^*(1) \end{bmatrix}\begin{bmatrix} \omega^*(\zeta) \\ I \end{bmatrix}=0. \tag{10.20}$$

Now let us turn to the inequality $(\tilde{S})$ (see subsection 10.1):

$$\begin{bmatrix} I-C_nC_n^* & \tilde{B}^{(1)}(\zeta,n) \\ \tilde{B}^{(1)*}(\zeta,n) & \dfrac{1-\Theta^*(\zeta)\Theta(\zeta)}{1-|\zeta|^2} \end{bmatrix}\geq 0,$$

where

$$\tilde{B}^{(1)}(\zeta,n)=\frac{1}{\zeta}\Lambda_{p,n}^*\left(\frac{1}{\overline{\zeta}}\right)\Theta(\zeta)-\frac{1}{\zeta}C_n\Lambda_{q,n}^*\left(\frac{1}{\zeta}\right).$$

Multiplying the inequality $(\tilde{S})$ from the right by

$$\tilde{T}=\begin{bmatrix} \begin{bmatrix} I & -X \\ 0 & I \end{bmatrix} & 0 \\ 0 & I_q \end{bmatrix}$$

and from the left by $(\tilde{T})^*$ (just as was done with the inequality (S)) we obtain the equivalent inequality

$$\begin{bmatrix} \begin{bmatrix} A_n^{(1)} & 0 \\ 0 & 0 \end{bmatrix} & \begin{bmatrix} I & 0 \\ -X^* & I \end{bmatrix}\tilde{B}^{(1)}(\zeta,n) \\ * & \dfrac{I-\Theta(\zeta)\Theta^*(\zeta)}{1-|\zeta|^2} \end{bmatrix}\geq 0.$$

Just as in the case of the inequality (S) we see that the last inequality is equivalent to the following pair of conditions:

$$[-X^*,I]\tilde{B}^{(1)}(\zeta,n)=0 \tag{10.21}$$

and

$$\frac{I-\Theta(\zeta)\Theta^*(\zeta)}{1-|\zeta|^2} - \tilde{B}^{(1)*}(\zeta,n)(P-\tilde{C}_n\tilde{C}_n^*)^{-1}\tilde{B}^{(1)}(\zeta,n) \geq 0. \tag{10.22}$$

Let us first solve the inequality (10.22). For this goal we consider the $\tilde{\jmath}$-expansive elementary multiple factors ([Dub4])

$$\tilde{B}_n(\zeta) = I + (1-\zeta)\begin{bmatrix} \Lambda_{p,n}(\zeta) & 0 \\ 0 & \Lambda_{q,n}(\zeta) \end{bmatrix} \tilde{H}_n \begin{bmatrix} \Lambda_{p,n}^*(1) & 0 \\ 0 & \Lambda_{q,n}^*(1) \end{bmatrix} \tilde{\jmath}, \tag{10.23}$$

where

$$\tilde{H}_n = \begin{bmatrix} P \\ \tilde{C}_n^* \end{bmatrix} (P-\tilde{C}_n\tilde{C}_n^*)^{-1}[P,\tilde{C}_n] \text{ and } \tilde{\jmath} = \begin{bmatrix} -I_p & 0 \\ 0 & I_q \end{bmatrix}.$$

Now it is possible, just as in the case of the inequality (S), to show that (10.22) is equivalent to the inequality

$$[\Theta^*(\zeta), I]\frac{(\tilde{B}_n^*)^{-1}(\zeta)\tilde{\jmath}\tilde{B}_n^{-1}(\zeta)}{1-|\zeta|^2}\begin{bmatrix} \Theta(\zeta) \\ I \end{bmatrix} \geq 0.$$

The last inequality leads to the following statement:

Lemma 10.2. *The general solution $\Theta(\zeta)$ of the inequality* (10.22) *can be represented in the form of a fractional linear transformation*

$$\Theta(\zeta) = [\tilde{a}(\zeta)\tilde{\omega}(\zeta) + \tilde{b}(\zeta)][\tilde{c}(\zeta)\tilde{\omega}(\zeta) + \tilde{d}(\zeta)]^{-1},$$

where the parameter $\tilde{\omega}(\zeta)$ is an arbitrary function from the class $\mathcal{S}_{p,q}$. The coefficient matrix

$$\tilde{B}_n(\zeta) = \begin{bmatrix} \tilde{a}(\zeta) & \tilde{b}(\zeta) \\ \tilde{c}(\zeta) & \tilde{d}(\zeta) \end{bmatrix}$$

of this fractional linear transformation is a $\tilde{\jmath}$-expansive elementary multiple factor (10.23) *constructed from the matrix C_n and from a K-type subspace.*

Just as before we obtain further that the condition (10.21) is equivalent to the equality

$$P_0[I, -C_n]\begin{bmatrix} \Lambda_{p,n}^*(1) & 0 \\ 0 & \Lambda_{q,n}^*(1) \end{bmatrix}\begin{bmatrix} \tilde{\omega}(\zeta) \\ I \end{bmatrix} = 0. \tag{10.24}$$

10.2. Let us clarify the significance of the conditions (10.20) and (10.24). For this purpose we rewrite them in the form

$$-P_0C_n\Lambda_{q,n}^*(1)\omega^*(\zeta) + P_0\Lambda_{p,n}^*(1) \quad = \quad 0$$

and

$$P_0\Lambda_{p,n}^*(1)\tilde{\omega}(\zeta) - P_0C_n\Lambda_{q,n}^*(1) \quad = \quad 0,$$

respectively. Passing to the conjugate equalities we obtain

$$\omega(\zeta)\Lambda_{q,n}(1)C_n^*P_0 \quad = \quad \Lambda_{p,n}(1)P_0 \tag{10.25}$$

and

$$\tilde{\omega}^*(\zeta)\Lambda_{p,n}(1)P_0 \quad = \quad \Lambda_{q,n}(1)C_n^*P_0. \tag{10.26}$$

Now we note that

$$\begin{aligned}
P_0C_n\Lambda_{q,n}^*(1)\Lambda_{q,n}(1)C_n^*P_0 &= P_0C_n(D_{q,n}(1) + D_{q,n}^*(1) + I)C_n^*P_0 \\
&= P_0D_{p,n}(1)C_nC_n^*P_0 + P_0C_nC_n^*D_{p,n}(1)P_0 + P_0C_nC_n^*P_0 \\
&= P_0(D_{p,n}(1) + D_{p,n}^*(1) + I)P_0 \\
&= P_0\Lambda_{p,n}^*(1)\Lambda_{p,n}(1)P_0.
\end{aligned}$$

Let M_0 be the image of the operator $\Lambda_{q,n}(1)C_n^*P_0$ and N_0 be the image of the operator $\Lambda_{p,n}(1)P_0$. From the last equalities it follows that the mapping U defined by

$$U\Lambda_{q,n}(1)C_n^*P_0 = \Lambda_{p,n}(1)P_0$$

is a unitary operator from M_0 onto N_0.

Thus, the conditions (10.25) and (10.26) are equivalent to the equalities

$$\omega(\zeta)_{|M_0} = \tilde{\omega}(\zeta)_{|M_0} = U,$$

i.e., the parameter $\omega(\zeta) : E_- \to E_+$ has the block decomposition

$$\omega(\zeta) = \begin{bmatrix} U & 0 \\ 0 & \omega_1(\zeta) \end{bmatrix}$$

with respect to the space decompositions $E_- = M_0 \oplus M_1$ and $E_+ = N_0 \oplus N_1$. Here $\omega_1(\zeta)$ is an arbitrary contractive map from M_1 into N_1 which is holomorphic in the unit disc. The parameter $\tilde{w}$ can be described in analogous way.

Thus, the following statements are proved.

Theorem 10.1. *The general solution $\Theta(\zeta)$ of the inequality (S) is representable in the form of a fractional linear transformation*

$$\Theta(\zeta) = [\omega(\zeta)b(\zeta) + d(\zeta)]^{-1}[\omega(\zeta)a(\zeta) + c(\zeta)], \tag{10.27}$$

where $\omega(\zeta) \in \mathcal{S}_{p,q}$ is a matrix-function on which certain constraints are imposed. Let $\Theta(\zeta)$ act from the space E_- into the space E_+ and let these spaces are be decomposed in the following way:

$$E_- = M_0 \oplus M_1, \qquad E_+ = N_0 \oplus N_1,$$

where M_0 is the image of $\Lambda_{q,n}(1)C_n^*P_0$, N_0 is the image of $\Lambda_{p,n}(1)P_0$, P_0 is the orthoprojector onto $\mathrm{Ker}(I - C_nC_n^*)$ and $\Lambda_{p,n}(\zeta) = [I_p, \zeta I_p, \dots, \zeta^n I_p]$. Then the parameter w has the following block decomposition

$$\omega(\zeta) = \begin{bmatrix} U & 0 \\ 0 & \omega_1(\zeta) \end{bmatrix},$$

where U is a unitary map from M_0 onto N_0 which is uniquely determined from the interpolation data by the equality

$$U\Lambda_{q,n}(1)C_n^*P_0 = \Lambda_{p,n}(1)P_0.$$

The function $\omega_1(\zeta)$ is an arbitrary function which is holomorphic in the unit disc whose values are contractive operators acting from M_1 into N_1. The coefficient matrix

$$B_n(\zeta) = \begin{bmatrix} a(\zeta) & b(\zeta) \\ c(\zeta) & d(\zeta) \end{bmatrix}$$

of the fractional linear transformation (10.27) is a j-expansive elementary multiple factor of the form (10.12) which is constructed from the matrix C_n and from a K-type subspace.

Theorem 10.2. *The general solution $\Theta(\zeta)$ of the inequality $(\tilde{S})$ is representable in the form of a fractional linear transformation*

$$\Theta(\zeta) = [\tilde{a}(\zeta)\tilde{w}(\zeta) + \tilde{b}(\zeta)][\tilde{c}(\zeta)\tilde{w}(\zeta) + \tilde{d}(\zeta)]^{-1},$$

where the parameter $\tilde{w}(\zeta)$ is described as in the previous theorem and the coefficient matrix

$$\tilde{B}_n(\zeta) = \begin{bmatrix} \tilde{a}(\zeta) & \tilde{b}(\zeta) \\ \tilde{c}(\zeta) & \tilde{d}(\zeta) \end{bmatrix}$$

is a $\tilde{j}$-expansive elementary multiple factor of the form (10.23) which is constructed from the matrix C_n and from a K-type subspace.

Remark. *It is possible to show that*

$$\dim M_0 = \dim N_0 = \dim (\mathrm{Ker}A_n \setminus \mathrm{Ker}A_{n-1}).$$

10.3. Now we give an example of a K-type subspace. Let $\tilde{F}_k$ denote the orthogonal projection of $\mathrm{KerA_n}$ onto $E_+^{(k)} \ominus E_+^{(k-1)}$ $(k = 0, 1, \dots, n)$; $E_+^{(-1)} = 0$ (see (10.1)) and let $\tilde{F} = \tilde{F}_0 \oplus \tilde{F}_1 \oplus \dots \oplus \tilde{F}_n$.

Lemma 10.3. *The subspace $\tilde{F}$ is invariant with respect to the operator $V_{p,n}$.*

Proof. Since $V_{p,n}\tilde{F}_n = 0$ and the embeddings (10.2) are in force, it remains to show that

$$V_{p,k+1}\tilde{F}_k \subset \tilde{F}_{k+1}, \qquad k = 0, 1, \dots, n-1. \tag{10.28}$$

Let $h \in \tilde{F}_k$. According to (10.1) there exists an $f \in \mathrm{Ker}A_k \subset E_+^{(k)}$, $f = \{f_0, f_1, \dots, f_k\}$, $f_k = h$. Let $\tilde{f} = \{f, 0\} \in E_+^{(k+1)}$ and $g = V_{p,k+1}\tilde{f} = \{0, f_0, f_1, \dots, f_k\} = \{0, \tilde{f}\}$. Then, since

$$C_{k+1} = \begin{bmatrix} c_0 & 0 \dots 0 \\ c_1 & \\ \vdots & C_k \\ c_{k+1} & \end{bmatrix},$$

we obtain

$$\langle A_{k+1}g, g\rangle = -\|c_1^* f_0 + c_2^* f_1 + \ldots + c_{k+1}^* f_k\| \leq 0.$$

Hence, $g \in \mathrm{Ker} A_{k+1}, h = f_k \in \tilde{F}_{k+1}$, and the embeddings (10.28) are established. □

Lemma.10.4 *The equality*

$$\dim F = \dim \mathrm{Ker} A_n$$

holds true.

Proof. For $n = 0$ this statement is evident. Let us assume that the statement of the lemma holds true for the index $n-1$. We remark that

$$\begin{aligned} A_n &= \begin{bmatrix} A_{n-1} & B_n \\ B_n^* & I - \sum\limits_{k=0}^{n} c_k c_k^* \end{bmatrix} \\ &= \begin{bmatrix} I & 0 \\ Y^* & I \end{bmatrix} \begin{bmatrix} A_{n-1} & 0 \\ 0 & I - \sum\limits_{k=0}^{n} c_k c_k^* - Y^* A_{n-1} Y \end{bmatrix} \begin{bmatrix} I & Y \\ 0 & I \end{bmatrix}, \end{aligned}$$

where Y is a solution of the equation $A_{n-1}Y = B_n$. Hence, $\mathrm{Ker} A_n$ consists of those vectors $\{f, g\}$ such that $f + Yg \in \mathrm{Ker} A_{n-1}, g \in \mathrm{Ker}(I - \sum\limits_{k=0}^{n} c_k c_k^* - Y^* A_{n-1} Y)$ (g belongs to $\tilde{F}_n$).
From here it follows that

$$\dim \mathrm{Ker} A_n = \dim (\tilde{F}_n \setminus \mathrm{Ker} A_{n-1}).$$

The lemma is proved. □

Let $F = E_+^{(n)} \ominus \tilde{F}$. Then

$$F = F_0 \oplus F_1 \oplus \ldots \oplus F_n, \quad F_k = E_+ \ominus \tilde{F}_k, \quad k = 0, 1, \ldots, n.$$

Now let us prove that F is a K-type subspace. From Lemma 10.3 and Lemma 10.4 it follows that it remains to show that $F \cap \mathrm{Ker} A_n = 0$. If $A_n f = 0$, where $f = \{f_0, f_1, \ldots, f_n\} \in F$, then $f \in F_n$ and $f_n \in \tilde{F}_n$. Hence, $f_n = 0$. In the same way we can prove that

$$f_0 = f_1 = \ldots = f_{n-1} = 0.$$

Thus, $f = 0$.

References

[Dub1] Dubovoj, V.K. : *Indefinite metric in the interpolation problem of Schur for analytic matrix functions. I* (in Russian), Teor. Funkciĭ, Funkcional. Anal. i Prilozen **37** (1982), 14–26, (The Kharkov University publishing house, Marchenko, V.A. -ed) [**MR 85f:**30059a]. Engl. Transl. in: Amer. Math. Soc. Transl. (ser. **2**) 144 (1989) (*Thirteen Papers Translated from Russian*), 47–60.

[Dub2] Dubovoj, V.K. : *Indefinite metric in the interpolation problem of Schur for analytic matrix functions. II* (in Russian), Teor. Funkciĭ, Funkcional. Anal. i Prilozen **38** (1982), 32–40, (The Kharkov University publishing house, Marchenko, V.A. -ed) [**MR 85f:**30059b]. Engl. Transl. in: Amer. Math. Soc. Transl. (ser. **2**) 144 (1989) (*Thirteen Papers Translated from Russian*), 61–70.

[Dub3] Dubovoj, V.K. : *Indefinite metric in the interpolation problem of Schur for analytic matrix functions. III* (in Russian), Teor. Funkciĭ, Funkcional. Anal. i Prilozen **41** (1984), 55–64, (The Kharkov University publishing house, Marchenko, V.A. -ed) [**MR 86c:**47008].

[Dub4] Dubovoj, V.K. : *Parametrization of multiple elementary factors of non-full rank* (in Russian), in: Analysis in Indefinite Dimensional Spaces and Operator Theory (Ed. V.A. Marchenko), Nauk. Dumka, Kiev 1983, pp. 54–68.

[EfPot] Efimov, A.V. and V.P. Potapov, *J-expansive matrix-functions and their role in the theory of electrical circuits* (in Russian), Uspekhi Matem. Nauk **28**:1 (1973), 65–130 [**MR 81f:**15023]. English transl.: Russian Math. Surveys **28**:1 (1973), 69–140.

[Fed] Fedćina, I.P. : *Description of solutions of the tangent Nevanlinna-Pick tangent problem* (in Russian), Doklady Akad. Nauk Armjan. SSR **60**:1 (1975), pp. 37–42 [**MR 52**#5974].

[Gal] Galstyan, L.A. : *Analytic j-expansive matrix functions and the Fejér's problem* (in Russian), Doklady Akad. Nauk Armjan. SSR **63**:1 (1976), 22–26 [**MR 55**#8365].

[Nud] Nudelman, A.A. : *On one new moment-type problem* (in Russian), Doklady. Akad. Nauk SSSR **233**:5 , 792–795 [**MR 57**#10379]. Engl. Transl. in: Soviet Math. Dokl. **18**:2 (1977), 507–510.

Faculty of Mathematics
The Kharkov State University
Independence Square 4
31077 Kharkov
Ukraine

AMS Subject Classification: 47A45, 30D50, 30E05, 47B50.

Operator Theory
Advances and Applications, Vol. 95
© 1997 Birkhäuser Verlag Basel/Switzerland

Bitangential interpolation for upper triangular operators

Harry Dym* and Boris Freydin

Abstract. This paper deals with bitangential interpolation problems of the Nevanlinna-Pick type in the general setting of upper triangular operators. (In this setting upper triangular operators play the role of analytic functions and the classical cases emerge by restricting these operators to be Toeplitz.) The approach is based largely on adapting ideas which were introduced by Katsnelson, Kheifets and Yuditskii, and then further refined by Kheifets, (to establish the existence of and representation formulas for the solutions to a number of interpolation problems in settings based on the usual notion of analyticity) to the setting of upper triangular operators.

One advantage of this approach is that it yields a description of all the solutions to the interpolation problem under consideration in terms of a linear fractional representation of the Redheffer type even when the Pick operator associated with the problem is only positive semidefinite.

1 Introduction

In this paper we shall solve a two-sided interpolation problem of the Nevanlinna-Pick type in a general setting of upper triangular operators. Although such problems have been considered earlier by Ball, Gohberg and Kaashoek [BGK2], the methods considered here are quite different. In particular, we shall follow the general strategy which was introduced by Katsnelson, Kheifets and Yuditskii [KKY] for solving the "Abstract Interpolation Problem"; for a good expository account and additional references see [KY]. One advantage of this approach is that it allows one to give a description of all the solutions to the problem under consideration under the assumption that the associated Pick operator is positive semidefinite (and not just when it is uniformly positive). The description is formulated in terms of the characteristic functions of two coupled unitary colligations, the first of which is defined in terms of the data of the problem whereas the second is arbitrary. This leads to a description of all solutions in terms of a linear fractional transformation of the Redheffer type.

*) The author wishes to express his thanks to Renee and Jay Weiss for endowing the chair which supports his research.

In order to explain the problem under study it is necessary to introduce some notation. The basic setting, which is described in more detail in [ADD] and [DD], is the space $\mathcal{X}(\ell^2_{\mathcal{N}};\ell^2_{\mathcal{M}})$ of bounded linear operators from the Hilbert space

$$\ell^2_{\mathcal{N}} = \bigoplus_{i=-\infty}^{\infty} \mathcal{N}_i \text{ into the Hilbert space } \ell^2_{\mathcal{M}} = \bigoplus_{i=-\infty}^{\infty} \mathcal{M}_i ,$$

where the coordinate spaces $\mathcal{N}_i$ and $\mathcal{M}_i$, $i = 0, \pm 1, \ldots$, are themselves each separable Hilbert spaces. For the moment we shall assume that $\mathcal{N}_i = \mathcal{N}_0$ and $\mathcal{M}_i = \mathcal{M}_0$ for $i = \pm 1, \pm 2, \ldots$, in order to simplify the exposition. The more general setting will be considered a little later. Every $A \in \mathcal{X}(\ell^2_{\mathcal{N}};\ell^2_{\mathcal{M}})$ has a block matrix representation

$$\begin{bmatrix} & & \vdots & & \\ & & A_{-1,0} & & \\ \cdots & A_{0,-1} & \boxed{A_{00}} & A_{01} & \cdots \\ & & A_{10} & & \\ & & \vdots & & \end{bmatrix} ,$$

where $A_{ij} : \mathcal{N}_j \to \mathcal{M}_i$. Correspondingly, if we write $g \in \ell^2_{\mathcal{N}}$ and $f \in \ell^2_{\mathcal{M}}$ as infinite column vectors:

$$g = \begin{bmatrix} \vdots \\ g_{-1} \\ \boxed{g_0} \\ g_1 \\ \vdots \end{bmatrix} \quad \text{and } f = \begin{bmatrix} \vdots \\ f_{-1} \\ \boxed{f_0} \\ f_1 \\ \vdots \end{bmatrix} ,$$

respectively, then the components of $f = Ag$ follow the usual rules of matrix multiplication:

$$f_j = \sum_{k=-\infty}^{\infty} A_{jk} g_k .$$

Let $\mathcal{U}, \mathcal{L}$ and $\mathcal{D}$ denote the subspaces of $\mathcal{X}$ consisting of upper triangular, lower triangular and diagonal operators, respectively, in the indicated matrix representation, and let $Z \in \mathcal{U}(\ell^2_{\mathcal{B}};\ell^2_{\mathcal{B}})$ denote the shift operator which is specified by the rule

$$(Zf)_j = f_{j+1} \text{ for } f \in \ell^2_{\mathcal{B}} = \bigoplus_{i=-\infty}^{\infty} \mathcal{B}_i , \tag{1.1}$$

with $\mathcal{B}_i = \mathcal{B}_0$ a fixed separable Hilbert space for $i = \pm 1, \pm 2, \ldots$. In this setting, the embedding operators

$$\pi_i : u \in \mathcal{B}_i \longrightarrow f \in \ell^2_{\mathcal{B}} , \text{ where } \begin{cases} f_i = u \\ f_j = 0 \text{ for } j \neq i , \end{cases} \tag{1.2}$$

and their adjoints

$$\pi_i^* : f \in \ell_{\mathcal{B}}^2 \longrightarrow f_i \in \mathcal{B}_i \ , \tag{1.3}$$

can all be expressed in terms of π_0 and the shift operator Z:

$$\pi_i = Z^{*i} \pi_0 \ . \tag{1.4}$$

We shall use the letter Z for the forward shift in other spaces too; the dependence of Z on the space in question will not be indicated explicitly in order to avoid overburdening the notation (time for that later). Thus for example, we shall let

$$F^{(j)} = (Z^*)^j F Z^j \ , \ j = 0, \pm 1, \dots, \tag{1.5}$$

for $F \in \mathcal{X}(\ell_{\mathcal{N}}^2; \ell_{\mathcal{M}}^2)$ even though Z^* acts from $\ell_{\mathcal{M}}^2 \to \ell_{\mathcal{M}}^2$ and Z acts from $\ell_{\mathcal{N}}^2 \to \ell_{\mathcal{N}}^2$ in (1.5). It is useful to note that $F^{(j)}$ slides the entries in each diagonal of F j units in the South East direction: $(F^{(j)})_{st} = F_{s-j,t-j}$. Thus each of the spaces $\mathcal{U}, \mathcal{L}$ and $\mathcal{D}$ is invariant under the mapping $F \to F^{(j)}$. The same holds true for the spaces

$$\mathcal{U}' = Z\mathcal{U} \text{ and } \mathcal{L}' = Z^* \mathcal{L}$$

of strictly upper and strictly lower triangular operators, respectively.

Now let $\mathcal{X}_2(\ell_{\mathcal{N}}^2; \ell_{\mathcal{M}}^2)$ denote the set of operators in $\mathcal{X}(\ell_{\mathcal{N}}^2; \ell_{\mathcal{M}}^2)$ which are Hilbert-Schmidt and let

$$\mathcal{U}_2 = \mathcal{U} \cap \mathcal{X}_2 \ , \ \mathcal{U}_2' = \mathcal{U}' \cap \mathcal{X}_2 \ , \ \mathcal{D}_2 = \mathcal{D} \cap \mathcal{X}_2, \ , \ \mathcal{L}_2 = \mathcal{L} \cap \mathcal{X}_2 \text{ and } \mathcal{L}_2' = \mathcal{L}' \cap \mathcal{X}_2.$$

All of these spaces are Hilbert spaces with respect to the inner product

$$\langle F, G \rangle_{HS} = \text{trace } G^* F \ .$$

It is readily checked that

$$\langle Z^j D, Z^k E \rangle_{HS} = \begin{cases} \text{trace } E^* D & \text{if } j = k \\ 0 & \text{if } j \neq k \ , \end{cases}$$

for every choice of D and E in $\mathcal{D}_2(\ell_{\mathcal{N}}^2; \ell_{\mathcal{M}}^2)$, and hence that the spaces $\mathcal{L}_2', \mathcal{D}_2$ and $\mathcal{U}_2'$ are orthogonal with respect to this inner product:

$$\mathcal{X}_2 = \mathcal{U}_2' \oplus \mathcal{D}_2 \oplus \mathcal{L}_2' = \mathcal{U}_2 \oplus \mathcal{L}_2' = \mathcal{U}_2' \oplus \mathcal{L}_2 \ .$$

Let

$\underline{\underline{p}}$ = the orthogonal projection of $\mathcal{X}_2$ onto $\mathcal{U}_2$,

$\underline{\underline{q}}$ = the orthogonal projection of $\mathcal{X}_2$ onto $\mathcal{L}_2$,

$\underline{\underline{q}}'$ = the orthogonal projection of $\mathcal{X}_2$ onto $\mathcal{L}_2'$,

and let $\mathcal{S}(\ell_{\mathcal{N}}^2; \ell_{\mathcal{M}}^2)$ denote the set of $S \in \mathcal{U}(\ell_{\mathcal{N}}^2; \ell_{\mathcal{M}}^2)$ with operator norm $\|S\| \leq 1$.

We shall say that $S \in \mathcal{S}(\ell^2_{\mathcal{N}};\ell^2_{\mathcal{M}})$ is a solution of the BIP (basic interpolation problem) based on a given set of bounded linear operators

$$f_i = \begin{bmatrix} g_i \\ h_i \end{bmatrix} , \ i = 1, \dots, n ,$$

from $\ell^2_{\mathcal{B}}$ to $\ell^2_{\mathcal{M}} \oplus \ell^2_{\mathcal{N}}$ with components

$$g_i \in \mathcal{U}(\ell^2_{\mathcal{B}};\ell^2_{\mathcal{M}}) \text{ and } h_i \in \mathcal{U}(\ell^2_{\mathcal{B}};\ell^2_{\mathcal{N}}) \text{ for } i = 1, \dots, \mu ,$$

and

$$g_i \in \mathcal{L}'(\ell^2_{\mathcal{B}};\ell^2_{\mathcal{M}}) \text{ and } h_i \in \mathcal{L}'(\ell^2_{\mathcal{B}};\ell^2_{\mathcal{N}}) \text{ for } i = \mu+1, \dots, n ,$$

if, for every choice of $E \in \mathcal{D}_2(\ell^2_{\mathcal{B}};\ell^2_{\mathcal{B}})$,

$$\underline{\underline{p}}S^* g_i E = h_i E \text{ for } i = 1, \dots, \mu \tag{1.6}$$

and

$$\underline{\underline{q}}' S h_i E = g_i E \text{ for } i = \mu+1, \dots, n . \tag{1.7}$$

This formulation is modelled on the treatment of the classical case in [D2], [D3]. In the present setting, upper triangular operators play the role of functions which are analytic inside the unit disc in the classical case and diagonal operators play the role of scalars. The classical case emerges upon choosing all the operators to be constant along diagonals, i.e., to be block Toeplitz.

The special choice

$$g_i = \xi_i(I - ZV_i^*)^{-1} \text{ and } h_i = \eta_i(I - ZV_i^*)^{-1} \text{ for } i = 1, \dots, \mu , \tag{1.8}$$

$$g_i = \xi_i(Z - V_i)^{-1} \text{ and } h_i = \eta_i(Z - V_i)^{-1} , \text{ for } i = \mu+1, \dots, n , \tag{1.9}$$

with $\xi_i \in \mathcal{D}(\ell^2_{\mathcal{B}};\ell^2_{\mathcal{M}})$, $\eta_i \in \mathcal{D}(\ell^2_{\mathcal{B}};\ell^2_{\mathcal{N}})$, $V_i \in \mathcal{D}(\ell^2_{\mathcal{B}};\ell^2_{\mathcal{B}})$ and $r_{sp}(ZV_i^*) = r_{sp}(Z^*V_i) < 1$ for $i = 1, \dots, n$, corresponds to the two-sided Nevanlinna-Pick problem in the present setting. Here r_{sp} designates the spectral radius of the indicated operator.

For this choice of g_i and h_i, the interpolation conditions embodied in (1.6) and (1.7) can be reformulated in terms of a pair of diagonal transforms $(\xi_i^* S)^\wedge(V_i)$ and $(S\eta_i)^\Delta(V_i)$ as follows:

$$(\xi_i^* S)^\wedge(V_i) = \eta_i^* \text{ for } i = 1, \dots, \mu \tag{1.10}$$

and

$$(S\eta_i)^\Delta(V) = \xi_i \text{ for } i = \mu+1, \dots, n . \tag{1.11}$$

Here, for $V \in \mathcal{D}(\ell^2_{\mathcal{B}};\ell^2_{\mathcal{B}})$ with $r_{sp}(ZV^*) < 1$, $F \in \mathcal{U}(\ell^2_{\mathcal{N}};\ell^2_{\mathcal{B}})$ and $G \in \mathcal{U}(\ell^2_{\mathcal{B}};\ell^2_{\mathcal{M}})$, the diagonal transform $F^\wedge(V)$ may be characterized as the unique element in $\mathcal{D}(\ell^2_{\mathcal{N}};\ell^2_{\mathcal{B}})$ such that $(Z - V)^{-1}\{F - F^\wedge(V)\} \in \mathcal{U}(\ell^2_{\mathcal{N}};\ell^2_{\mathcal{B}})$, whereas the diagonal transform $G^\Delta(V)$ may be characterized as the unique element in $\mathcal{D}(\ell^2_{\mathcal{B}};\ell^2_{\mathcal{M}})$ such that $\{G - G^\Delta(V)\}(Z - V)^{-1} \in \mathcal{U}(\ell^2_{\mathcal{B}};\ell^2_{\mathcal{M}})$; see Theorems 3.3 and 3.4 of [ADD].

The formulas (1.10) and (1.11) serve to display this NP problem in a form which resembles the bitangential NP problem in the classical setting. Indeed, the latter emerges from the former by suitably restricting the data.

Other choices of g_i and h_i lead to assorted analogues of classical interpolation problems; see e.g., [DD], [D1], [Ko1], [Ko2], [DvV], [BGK1] and [BGK2]. We shall not make any further explicit use of these transforms for the remainder of this paper.

We remark that interpolations of the type considered here can be formulated in the general setting of nest algebras, and solved with the help of the commutant lifting theorem for nest algebras which is due to Paulsen and Power [PP]. J. Ball [Ba] used this theorem to establish the existence of solutions to the NP problem under the assumption that the associated Pick operator is positive and that the appropriate observability operators are uniformly positive.

In formula (1.7), the symbol S should be understood as the operator M_S of multiplication by S on the left acting from $\mathcal{X}_2(\ell^2_{\mathcal{B}};\ell^2_{\mathcal{N}})$ to $\mathcal{X}_2(\ell^2_{\mathcal{B}};\ell^2_{\mathcal{M}})$. Similarly, in formula (1.6) the symbol S^* should be understood as the operator M_{S^*} of multiplication by S^*, from $\mathcal{X}_2(\ell^2_{\mathcal{B}};\ell^2_{\mathcal{M}})$ to $\mathcal{X}_2(\ell^2_{\mathcal{B}};\ell^2_{\mathcal{N}})$.

We turn now to the more general setting in which the coordinate spaces $\mathcal{B}_i, \mathcal{M}_i, \mathcal{N}_i, \dots$ of $\ell^2_{\mathcal{B}}, \ell^2_{\mathcal{M}}, \ell^2_{\mathcal{N}}, \dots$, respectively, are allowed to vary with the "time" index i. The main effect of this relaxation is that the shift operator Z applied to the space

$$\ell^2_{\mathcal{B}(t)} = \cdots \oplus \mathcal{B}_{t-1} \oplus \boxed{\mathcal{B}_t} \oplus \mathcal{B}_{t+1} \oplus \cdots$$

maps it onto the "displaced" space $\ell^2_{\mathcal{B}(t+1)}$; here as usual, the box designates the zero'th coordinate space, and $\ell^2_{\mathcal{B}(0)} = \ell^2_{\mathcal{B}}$. The BIP can also be formulated in this more general setting, but this requires extra caution. Thus, for example, in order for the operators $Z - V$ and $\rho_V = I - ZV^*$ with $V \in \mathcal{D}$ to be meaningful, the diagonal operator V must belong to the same $\mathcal{U}$ space as Z. Thus, if $Z \in \mathcal{U}(\ell^2_{\mathcal{B}(-1)}, \ell^2_{\mathcal{B}})$, then V must belong to $\mathcal{D}(\ell^2_{\mathcal{B}(-1)}; \ell^2_{\mathcal{B}})$; then the powers $(ZV^*)^k$, $k = 1, 2, \dots$, all belong to $\mathcal{U}(\ell^2_{\mathcal{B}}; \ell^2_{\mathcal{B}})$, as does ρ_V^{-1} when $r_{sp}(ZV^*) < 1$.

Particular care must now be taken with all definitions involving powers: if, for example, $W \in \mathcal{X}(\ell^2_{\mathcal{C}}; \ell^2_{\mathcal{B}})$ and $Z_j^{\mathcal{C}}$ denotes the shift operator which maps $\ell^2_{\mathcal{C}(j)}$ onto $\ell^2_{\mathcal{C}(j+1)}$, then strictly speaking $W^{(1)} = Z^*WZ$ [resp. $W^{(-1)} = ZWZ^*$] should be written as

$$W^{(1)} = (Z^{\mathcal{B}}_{-1})^* W (Z^{\mathcal{C}}_{-1}) \text{ [resp. } W^{(-1)} = (Z^{\mathcal{B}}_0) W (Z^{\mathcal{C}}_0)^*] \,,$$

from which in turn it follows that

$$W^{(t)} = (Z^{\mathcal{B}}_{-t})^* \cdots (Z^{\mathcal{B}}_{-2})^* (Z^{\mathcal{B}}_{-1})^* W (Z^{\mathcal{C}}_{-1})(Z^{\mathcal{C}}_{-2}) \cdots (Z^{\mathcal{C}}_{-t})$$

and

$$W^{(-t)} = (Z^{\mathcal{B}}_{t-1})(Z^{\mathcal{B}}_{t-2}) \cdots (Z^{\mathcal{B}}_0) W (Z^{\mathcal{C}}_0)^* \cdots (Z^{\mathcal{C}}_{t-2})^* (Z^{\mathcal{C}}_{t-1})^*$$

for $t \geq 2$. From these formulas, supplemented by $W^{(0)} = W$, it is readily checked that for every choice of the integers t and s,

$$\{W^{(t)}\}^{(s)} = W^{(t+s)} \ , \ \{W^{(t)}\}^* = \{W^*\}^{(t)} \ ,$$

and that $W^{(-t)} \in \mathcal{X}(\ell^2_{\mathcal{C}(t)}; \ell^2_{\mathcal{B}(t)})$ for every integer t. The last fact implies in particular that if $\mathcal{C} = \mathcal{B}^{(-1)}$, then the product $W^{(t)}W^{(t+1)}$ is meaningful even though the ordinary multiplication $W \cdot W$ is not (unless $\mathcal{B}^{(-1)} = \mathcal{B}$). For additional discussion, see [D1].

In what follows we will always assume, without special mention, that these changes are made whenever we deal with spaces in which the coordinate spaces are allowed to vary with the index i.

In principle, the difficulties inherent in time varying spaces can be avoided by imbedding each coordinate space in a (possibly new) sufficiently large coordinate space which is the same for each index i. But this makes the coordinate spaces artificially large. Thus, for example, if each coordinate space is finite dimensional but these dimensions are not uniformly bounded, then the new coordinate spaces will each be infinitely dimensional. An advantage of time varying coordinate spaces is that one can incorporate finite dimensional problems by setting most of the coordinate spaces equal to zero. A number of applications of time varying coordinate spaces to assorted operator and matrix completion of norm approximation may be found in the dissertations of Kos [Ko2] and van der Veen [vV].

2 The basic interpolation problem

In this section we shall establish necessary conditions for the existence of a solution to the BIP (which is easy) and then begin the more arduous task of establishing the sufficiency of this condition for an appropriately restricted class of $f_i, \ i = 1, \dots, n$. The first step is to define the form

$$\Lambda^S(D, E) = \sum_{i,j=1}^{n} \Lambda^S_{ij}(D_j, E_i) \tag{2.1}$$

on $\{\mathcal{D}_2(\ell^2_{\mathcal{B}}; \ell^2_{\mathcal{B}})\}^n \times \{\mathcal{D}_2(\ell^2_{\mathcal{B}}; \ell^2_{\mathcal{B}})\}^n$ with components

$$\Lambda^S_{ij}(D_j, E_i) = \left\langle \begin{bmatrix} I & -S \\ -S^* & I \end{bmatrix} f_j D_j, f_i E_i \right\rangle_{HS} \tag{2.2}$$

for $S \in \mathcal{S}(\ell^2_{\mathcal{N}}; \ell^2_{\mathcal{M}})$ and $D_j, E_i \in \mathcal{D}_2(\ell^2_{\mathcal{B}}; \ell^2_{\mathcal{B}})$.

Since

$$\begin{bmatrix} I & -S \\ -S^* & I \end{bmatrix} \geq 0$$

for $S \in \mathcal{S}(\ell^2_{\mathcal{N}}; \ell^2_{\mathcal{M}})$, it is selfevident that

$$\Lambda^S(E,E) = \sum_{i,j=1}^{n} \Lambda^S_{ij}(E_j, E_i) \geq 0$$

for every choice of $E_1, \ldots, E_n$ in $\mathcal{D}_2(\ell^2_{\mathcal{B}}; \ell^2_{\mathcal{B}})$.

Theorem 2.1. *If $S \in \mathcal{S}(\ell^2_{\mathcal{N}}; \ell^2_{\mathcal{M}})$ is a solution of the BIP based on $f_1, \ldots, f_n$, then*

$$\Lambda^S_{ij}(E_j, E_i) = \begin{cases} \langle Jf_jE_j, f_iE_i\rangle_{HS} & for \quad i,j = 1,\ldots,\mu\ , \\ -\langle Sh_jE_j, g_iE_i\rangle_{HS} & for \quad i = 1,\ldots,\mu\ ,\ j = \mu+1,\ldots,n\ , \\ -\langle Jf_jE_j, f_iE_i\rangle_{HS} & for \quad i,j = \mu+1,\ldots,n\ , \end{cases} \tag{2.3}$$

where

$$J = \begin{bmatrix} I_{\ell^2_{\mathcal{M}}} & 0 \\ 0 & -I_{\ell^2_{\mathcal{N}}} \end{bmatrix} \tag{2.4}$$

and $E_i \in \mathcal{D}_2(\ell^2_{\mathcal{B}}; \ell^2_{\mathcal{B}})$, for $i = 1, \ldots, n$.

Proof. By definition,

$$\begin{aligned} \Lambda^S_{ij}(E_j, E_i) &= \left\langle \begin{bmatrix} I & -S \\ -S^* & I \end{bmatrix} \begin{bmatrix} g_j \\ h_j \end{bmatrix} E_j, \begin{bmatrix} g_i \\ h_i \end{bmatrix} E_i \right\rangle_{HS} \\ &= \langle (g_j - Sh_j)E_j, g_iE_i\rangle_{HS} - \langle (S^*g_j - h_j)E_j, h_iE_i\rangle_{HS} \\ &= ① - ② . \end{aligned}$$

Moreover, since S is presumed to be a solution of the BIP,

$$(g_j - Sh_j)E_j \in \mathcal{U}_2 \text{ and } (-S^*g_j + h_j)E_j \in \mathcal{L}_2'$$

for every choice of j, $j = 1, \ldots, n$. The rest of the calculation is split into cases:

If $i, j = 1, \ldots, \mu$, then ② $= 0$ since $h_i \in \mathcal{U}$, and

$$\begin{aligned} ① &= \langle g_jE_j, g_iE_i\rangle_{HS} - \langle h_jE_j, \underline{\underline{p}}S^*g_iE_i\rangle_{HS} \\ &= \langle g_jE_j, g_iE_i\rangle_{HS} - \langle h_jE_j, h_iE_i\rangle_{HS} \end{aligned}$$

by the interpolation conditions. This is equivalent to the first set of stated formulas.

Next, if $i = 1, \ldots, \mu$ and $j = \mu+1, \ldots, n$, then ② $= 0$ and

$$① = -\langle Sh_jE_j, g_iE_i\rangle_{HS}\ ,$$

which yields the second set of formulas.

Finally, if $i, j = \mu+1, \ldots, n$, then ① $= 0$ since $g_i \in \mathcal{L}'$, and

$$② = \langle g_j E_j, \underline{\underline{q}}' S h_i E_i\rangle_{HS} - \langle h_j E_j, h_i E_i\rangle_{HS}$$

$$= \langle g_j E_j, g_i E_i\rangle_{HS} - \langle h_j E_j, h_i E_i\rangle_{HS} \,,$$

because of the interpolation conditions. This yields the final set of formulas. □

This is as far as one can go unless additional structure is imposed upon the data. Upon adapting the development of the BIP in the classical setting in [D2] and [D3] to the present environment, it seems reasonable to restrict the operators $f_1, \ldots, f_n$ to the following form (when the coordinate spaces do not vary with the index i):

$$[f_1 \cdots f_n] = \begin{bmatrix} C_1 \\ C_2 \end{bmatrix} (M - \underset{\sim}{Z} N)^{-1} \,, \tag{2.5}$$

where[1]

$$C_1 = [C_{11}\ C_{12}] = [\xi_1 \cdots \xi_\mu\ \xi_{\mu+1} \cdots \xi_n], \ (\xi_i \in \mathcal{D}(\ell^2_{\mathcal{B}}; \ell^2_{\mathcal{M}}))$$

$$C_2 = [C_{21}\ C_{22}] = [\eta_1 \cdots \eta_\mu\ \eta_{\mu+1} \cdots \eta_n], \ (\eta_i \in \mathcal{D}(\ell^2_{\mathcal{B}}; \ell^2_{\mathcal{N}}))$$

$$M = \begin{bmatrix} I & 0 \\ 0 & -A_2 \end{bmatrix}, \ N = \begin{bmatrix} A_1 & 0 \\ 0 & -I \end{bmatrix},$$

$$A_1 = \begin{bmatrix} (A_1)_{11} & \cdots & (A_1)_{1\mu} \\ \vdots & & \vdots \\ (A_1)_{\mu 1} & \cdots & (A_1)_{\mu\mu} \end{bmatrix},$$

$$A_2 = \begin{bmatrix} (A_2)_{11} & \cdots & (A_2)_{1\nu} \\ \vdots & & \vdots \\ (A_2)_{\nu 1} & \cdots & (A_2)_{\nu\nu} \end{bmatrix}, \ ((A_i)_{st} \in \mathcal{D}(\ell^2_{\mathcal{B}}; \ell^2_{\mathcal{B}}))$$

$$\underset{\sim}{Z} = Z_1 \oplus Z_2 \,,$$

$$Z_1 = \operatorname{diag}(Z, \ldots, Z)\ \mu \text{ times} \,,$$

$$Z_2 = \operatorname{diag}(Z, \ldots, Z)\ \nu \text{ times} \,,$$

$$\mu + \nu = n$$

and it is assumed that

$$r_{sp}(Z_1 A_1) < 1 \text{ and } r_{sp}(A_2 Z_2^*) < 1 \,. \tag{2.6}$$

This ensures that

$$F_{11} = C_{11}(I - Z_1 A_1)^{-1} \text{ and } F_{21} = C_{21}(I - Z_1 A_1)^{-1}$$

1) When the coordinate spaces do vary with the index i, then the domains of the operators ξ_i and η_i should be $\ell^2_{\mathcal{B}(-1)}$ for $i = \mu+1, \ldots, n$, $(A_1)_{st} \in \{\mathcal{D}(\ell^2_{\mathcal{B}}; \ell^2_{\mathcal{B}(-1)})\}$ for $s, t = 1, \ldots, \mu$, and $(A_2)_{st} \in \{\mathcal{D}(\ell^2_{\mathcal{B}(-1)}; \ell^2_{\mathcal{B}})\}$ for $s, t = 1, \ldots, \nu$.

are bounded upper triangular operators, whereas

$$F_{12} = C_{12}Z_2^*(I - A_2Z_2^*)^{-1} \text{ and } F_{22} = C_{22}Z_2^*(I - A_2Z_2^*)^{-1}$$

are bounded strictly lower triangular operators.

It is convenient to express the form Λ^S in terms of the operator P which is expressed in the block form

$$P = \begin{bmatrix} P_{11} & P_{12} \\ P_{21} & P_{22} \end{bmatrix},$$

corresponding to the indicated decomposition of M and N, i.e., P maps $(\ell_{\mathcal{B}}^2)^\mu \oplus (\ell_{\mathcal{B}}^2)^\nu$ into itself with components $P_{11} \in \{\mathcal{D}(\ell_{\mathcal{B}}^2; \ell_{\mathcal{B}}^2)\}^{\mu\times\mu}$, $P_{12} \in \{\mathcal{D}(\ell_{\mathcal{B}}^2; \ell_{\mathcal{B}}^2)\}^{\mu\times\nu}$, $P_{21} = P_{12}^*$ and $P_{22} \in \{\mathcal{D}(\ell_{\mathcal{B}}^2; \ell_{\mathcal{B}}^2)\}^{\nu\times\nu}$. To this end, let $Q_{ij} \in \mathcal{D}(\ell_{\mathcal{B}}^2; \ell_{\mathcal{B}}^2)$ be defined by the rule

$$\Lambda_{ij}^S(D, E) = \langle Q_{ij}D, E\rangle_{HS} \tag{2.7}$$

for every choice of D and E in $\mathcal{D}_2(\ell_{\mathcal{B}}^2; \ell_{\mathcal{B}}^2)$. Then

$$P_{11} = \begin{bmatrix} Q_{11} & \cdots & Q_{1\mu} \\ \vdots & & \vdots \\ Q_{\mu 1} & \cdots & Q_{\mu\mu} \end{bmatrix}, \qquad P_{12} = \begin{bmatrix} Q_{1,\mu+1} & \cdots & Q_{1n} \\ \vdots & & \vdots \\ Q_{\mu,\mu+1} & \cdots & Q_{\mu n} \end{bmatrix},$$

$$P_{21} = \begin{bmatrix} Q_{\mu+1,1} & \cdots & Q_{\mu+1,\mu} \\ \vdots & & \vdots \\ Q_{n1} & \cdots & Q_{n\mu} \end{bmatrix} \text{ and } P_{22} = \begin{bmatrix} Q_{\mu+1,\mu+1} & \cdots & Q_{\mu+1,n} \\ \vdots & & \vdots \\ Q_{n,\mu+1} & \cdots & Q_{nn} \end{bmatrix}. \tag{2.8}$$

Theorem 2.2. *If $S \in \mathcal{S}(\ell_{\mathcal{N}}^2; \ell_{\mathcal{M}}^2)$ is a solution of the BIP with data of the form (2.5) which is subject to the restrictions (2.6), then the operator P which is defined by (2.7) and (2.8) is a positive semidefinite solution of the equation*

$$M^*PM - N^*\underset{\sim}{Z}^*P\underset{\sim}{Z}N = C_1^*C_1 - C_2^*C_2\,. \tag{2.9}$$

Moreover, the block diagonal components of P are uniquely specified by the formulas

$$P_{11} = \sum_{j=0}^{\infty}(A_1^*Z_1^*)^j(C_{11}^*C_{11} - C_{21}^*C_{21})(Z_1A_1)^j \tag{2.10}$$

and

$$P_{22} = -\sum_{j=0}^{\infty}(Z_2A_2^*)^jZ_2(C_{12}^*C_{12} - C_{22}^*C_{22})Z_2^*(A_2Z_2^*)^j\,, \tag{2.11}$$

respectively.

Proof. In order to establish (2.9), it suffices to check that the block components P_{11}, P_{22} and P_{12} of P are solutions of the equations

$$P_{11} - A_1^* Z_1^* P_{11} Z_1 A_1 = C_{11}^* C_{11} - C_{21}^* C_{21}\,, \tag{2.12}$$

$$A_2^* P_{22} A_2 - Z_2^* P_{22} Z_2 = C_{12}^* C_{12} - C_{22}^* C_{22}\,, \tag{2.13}$$

and

$$P_{12} A_2 - A_1^* Z_1^* P_{12} Z_2 = C_{21}^* C_{22} - C_{11}^* C_{12}\,, \tag{2.14}$$

respectively. But this may be carried out with the help of Theorem 2.1 much as in the classical case; see e.g., [D2] and [D3].

For the sake of completeness we shall sketch the verification of (2.14).

By formulas (2.3), (2.7) and (2.8),

$$\langle P_{12} D_2, D_1\rangle_{HS} = -\langle S C_{22} Z_2^* (I - A_2 Z_2^*)^{-1} D_2, C_{11}(I - Z_1 A_1)^{-1} D_1\rangle_{HS}$$

for every choice of $D_1 \in \{\mathcal{D}_2(\ell_{\mathcal{B}}^2; \ell_{\mathcal{B}}^2)\}^{\mu\times 1}$ and $D_2 \in \{\mathcal{D}_2(\ell_{\mathcal{B}}^2; \ell_{\mathcal{B}}^2)\}^{\nu\times 1}$. Therefore,

$$\begin{aligned}
&\langle P_{12} A_2 E_2, E_1\rangle_{HS} - \langle P_{12} Z_2 E_2,\ Z_1 A_1 E_1\rangle_{HS}\\
&= -\langle S C_{22} Z_2^* (I - A_2 Z_2^*)^{-1} A_2 E_2,\ C_{11}(I - Z_1 A_1)^{-1} E_1\rangle_{HS}\\
&\quad + \langle S C_{22} Z_2^* (I - A_2 Z_2^*)^{-1} Z_2 E_2,\ C_{11}(I - Z_1 A_1)^{-1} Z_1 A_1 E_1\rangle_{HS}
\end{aligned}$$

for every choice of $E_1 \in \{\mathcal{D}(\ell_{\mathcal{B}}^2; \ell_{\mathcal{B}}^2)\}^{\mu\times 1}$ and $E_2 \in \{\mathcal{D}(\ell_{\mathcal{B}}^2; \ell_{\mathcal{B}(-1)}^2)\}^{\nu\times 1}$. But, by adding and subtracting like terms, this last expression can be rewritten as ① − ②, where

$$\begin{aligned}
① &= \langle S C_{22} Z_2^* (I - A_2 Z_2^*)^{-1} (Z_2 - A_2) E_2,\ C_{11}(I - Z_1 A_1)^{-1} E_1\rangle_{HS}\\
&= \langle S C_{22} E_2,\ C_{11}(I - Z_1 A_1)^{-1} E_1\rangle_{HS}\\
&= \langle C_{22} E_2,\ C_{21}(I - Z_1 A_1)^{-1} E_1\rangle_{HS}\\
&= \langle C_{22} E_2,\ C_{21} E_1\rangle_{HS}
\end{aligned}$$

and

$$\begin{aligned}
② &= \langle S C_{22} Z_2^* (I - A_2 Z_2^*)^{-1} Z_2 E_2,\ C_{11}(I - Z_1 A_1)^{-1}(I - Z_1 A_1) E_1\rangle_{HS}\\
&= \langle S C_{22} (Z_2 - A_2)^{-1} Z_2 E_2,\ C_{11} E_1\rangle_{HS}\\
&= \langle S C_{22} (Z_2 - A_2)^{-1} E_2^{(-1)},\ C_{11} E_1 Z^*\rangle_{HS}\\
&= \langle C_{12} (Z_2 - A_2)^{-1} E_2^{(-1)},\ C_{11} E_1 Z^*\rangle_{HS}\\
&= \langle C_{12} Z_2^* (I - A_2 Z_2^*)^{-1} Z_2 E_2,\ C_{11} E_1\rangle_{HS}\\
&= \langle C_{12} E_2,\ C_{11} E_1\rangle_{HS}
\end{aligned}$$

The asserted result now drops out easily by combining formulas. □

Warning. Now that the statement of the problem is clear, we shall use the symbol Z in place of $\underset{\sim}{Z}$ in the sequel in order to keep the notation simple.

3 The augmented BIP

Theorem 2.2 implies that every solution $S \in \mathcal{S}(\ell^2_{\mathcal{N}};\ell^2_{\mathcal{M}})$ of the BIP generates a solution

$$P = \begin{bmatrix} Q_{11} & \cdots & Q_{1n} \\ \vdots & & \vdots \\ Q_{n1} & \cdots & Q_{nn} \end{bmatrix} = \begin{bmatrix} P_{11} & P_{12} \\ & \\ P_{21} & P_{22} \end{bmatrix} \geq 0$$

of equation (2.9) with unique diagonal blocks $P_{11} \in \{\mathcal{D}(\ell^2_{\mathcal{B}};\ell^2_{\mathcal{B}})\}^{\mu\times\mu}$ and $P_{22} \in \{\mathcal{D}(\ell^2_{\mathcal{B}};\ell^2_{\mathcal{B}})\}^{\nu\times\nu}$.

In general, the off diagonal block P_{12} is not uniquely determined by the equation. Following the usage in [D5], we shall say that S is a solution of the augmented BIP based on each solution $P \geq 0$ of (2.9) as well as on the data, if, in addition to conditions (1.6) and (1.7), the condition

$$-\langle Sh_jE_j, g_iE_i\rangle_{HS} = \langle Q_{ij}E_j, E_i\rangle_{HS} \tag{3.1}$$

is satisfied for every choice of E_i and E_j in $\mathcal{D}_2(\ell^2_{\mathcal{B}};\ell^2_{\mathcal{B}})$ with $i = 1,\ldots,\mu$, $j = \mu+1,\ldots,n$. The preceding analysis shows that every solution S of the BIP based on $f_1,\ldots,f_n$ is a solution of the augmented BIP based on $f_1,\ldots,f_n$ and the particular solution $P = P^S$ of (2.9) corresponding to the form Λ^S.

It is perhaps well to point out, that in terms of the notation introduced between formula (2.5) and the constraints (2.6), the solutions of the augmented BIP can be characterized as the set of operators $S \in \mathcal{S}(\ell^2_{\mathcal{N}};\ell^2_{\mathcal{M}})$ such that

(1) $(C_1 - SC_2)(M - ZN)^{-1}E \in \mathcal{U}_2(\ell^2_{\mathcal{B}};\ell^2_{\mathcal{M}})$

(2) $(-S^*C_1 + C_2)(M - ZN)^{-1}E \in \mathcal{L}_2'(\ell^2_{\mathcal{B}};\ell^2_{\mathcal{N}})$

(3) $$-\left\langle SC_2(M-ZN)^{-1}\begin{bmatrix} 0 \\ E_2 \end{bmatrix}, C_1(M-ZN)^{-1}\begin{bmatrix} E_1 \\ 0 \end{bmatrix}\right\rangle_{HS} = \operatorname{trace}[E_1^* \ 0]P\begin{bmatrix} 0 \\ E_2 \end{bmatrix}$$

for every choice of $E \in \{\mathcal{D}_2(\ell^2_{\mathcal{B}};\ell^2_{\mathcal{B}})\}^{n\times 1}$, $E_1 \in \{\mathcal{D}_2(\ell^2_{\mathcal{B}};\ell^2_{\mathcal{B}})\}^{\mu\times 1}$ and $E_2 \in \{\mathcal{D}_2(\ell^2_{\mathcal{B}};\ell^2_{\mathcal{B}})\}^{\nu\times 1}$.

Our next objective is to show that for every positive semidefinite solution $P \in \{\mathcal{D}(\ell^2_{\mathcal{B}};\ell^2_{\mathcal{B}})\}^{n\times n}$ of equation (2.9), there exists a solution S of the augmented BIP. We shall follow the strategy which was initiated by Katsnelson, Kheifets and Yuditskii [KKY], and further developed by Kheifets in [Kh1] and [Kh2].

The starting point is to rewrite the basic identity (2.9) as

$$M^*PM + C_2^*C_2 = N^*P^{(1)}N + C_1^*C_1 \ . \tag{3.2}$$

It is convenient to let[2]

$$\mathcal{H} = (\ell^2_{\mathcal{B}})^n \ , \ \mathcal{E}_1 = \ell^2_{\mathcal{N}} \text{ and } \mathcal{E}_2 = \ell^2_{\mathcal{M}} \ .$$

Then, because of formula (3.2), the mapping

$$V : \begin{bmatrix} P^{\frac{1}{2}}M \\ C_2 \end{bmatrix} x \longrightarrow \begin{bmatrix} (P^{(1)})^{\frac{1}{2}}N \\ C_1 \end{bmatrix} x \tag{3.3}$$

is well defined. In fact, it is an isometry from

$$d_V = \text{closure}\left\{ \begin{bmatrix} P^{\frac{1}{2}}M \\ C_2 \end{bmatrix} x : \ x \in \mathcal{H} \right\} \ \text{ in } \mathcal{H} \oplus \mathcal{E}_1$$

onto

$$\Delta_V = \text{closure}\left\{ \begin{bmatrix} (P^{(1)})^{\frac{1}{2}}N \\ C_1 \end{bmatrix} x : \ x \in \mathcal{H} \right\} \ \text{ in } \mathcal{H}^{(-1)} \oplus \mathcal{E}_2 \ .$$

Both of these spaces are "ℓ^2 spaces", and $V \in \mathcal{D}(d_V; \Delta_V)$. This follows easily from the form of d_V and Δ_V and the fact that

$$P^{\frac{1}{2}}, N, M \in \{\mathcal{D}(\ell^2_{\mathcal{B}}; \ell^2_{\mathcal{B}})\}^{n\times n}, \ C_1 \in \{\mathcal{D}(\ell^2_{\mathcal{B}}; \ell^2_{\mathcal{M}})\}^{1\times n} \text{ and } C_2 \in \{\mathcal{D}(\ell^2_{\mathcal{B}}; \ell^2_{\mathcal{N}})\}^{1\times n} \ .$$

To clarify this further, let us note that if $\mathcal{H} = \bigoplus_{j=-\infty}^{\infty} \mathcal{H}_j$ and $x_j \in \mathcal{H}_j$, then

$$\text{the } j\text{'th coordinate of } \begin{bmatrix} P^{\frac{1}{2}}M \\ C_2 \end{bmatrix} x \text{ is equal to } \begin{bmatrix} P_j^{\frac{1}{2}}M_j \\ (C_2)_j \end{bmatrix} x_j \in \mathcal{H}_j \oplus \mathcal{N}_j \ ,$$

whereas

$$\text{the } j\text{'th coordinate of } \begin{bmatrix} \{P^{(1)}\}^{\frac{1}{2}}N \\ C_1 \end{bmatrix} x \text{ is equal to } \begin{bmatrix} P_{j-1}^{\frac{1}{2}}N_j \\ (C_1)_j \end{bmatrix} x_j \in \mathcal{H}_{j-1} \oplus \mathcal{M}_j \ .$$

We shall refer to $\mathcal{H}$ as the state space of the isometry V and to the spaces $\mathcal{E}_1$ and $\mathcal{E}_2$ as the input space and the output space, respectively.

Let $\mathcal{K} = \{\ell^2_{\mathcal{C}}\}^n$ be a complex separable Hilbert space which contains $\mathcal{H}$, i.e., assume that $\mathcal{C}_i \supset \mathcal{B}_i$, where the spaces $\mathcal{C}_i$ are allowed to vary with i, and let $U \in \mathcal{D}(\mathcal{K}\oplus\mathcal{E}_1; \mathcal{K}^{(-1)}\oplus\mathcal{E}_2)$ be a unitary extension of V from $\mathcal{K}\oplus\mathcal{E}_1$ onto $\mathcal{K}^{(-1)}\oplus\mathcal{E}_2$:

$$U|_{d_V} = V \ .$$

The entries in the block decomposition

$$U = \begin{bmatrix} U_{11} & U_{12} \\ & \\ U_{21} & U_{22} \end{bmatrix} : \begin{matrix} \mathcal{K} \\ \oplus \\ \mathcal{E}_1 \end{matrix} \longrightarrow \begin{matrix} \mathcal{K}^{(-1)} \\ \oplus \\ \mathcal{E}_2 \end{matrix}$$

2) If the coordinate spaces vary with the index i, then one should take $\mathcal{H} = \{\ell^2_{\mathcal{B}}\}^{\mu} \oplus \{\ell^2_{\mathcal{B}(-1)}\}^{\nu}$.

of U with respect to the given spaces taken together with the spaces themselves:

$$\left\{U_{11}, U_{12}, U_{21}, U_{22}; \mathcal{K}, \mathcal{E}_1, \mathcal{K}^{(-1)}, \mathcal{E}_2\right\}$$

is usually termed a colligation. It is said to be a unitary colligation if, as in the present case, U is unitary. It is important to bear in mind that because of the prevailing assumptions on U, each of the components U_{ij}, $i, j = 1, 2$, is a diagonal operator.

We shall refer to the operator valued function

$$W(r) = U_{22} + U_{21}(I - rZU_{11})^{-1} rZU_{12} \ , \quad 0 \le r < 1 \ , \tag{3.4}$$

which maps $\mathcal{E}_1$ into $\mathcal{E}_2$ as the characteristic (scattering or transfer) function of the unitary colligation which was described above. In this formulation, the unitary operator Z (which maps $\mathcal{K}^{(-1)}$ into $\mathcal{K}$) plays the role of the phase factor $e^{i\theta}$ in the complex variable $re^{i\theta}$. Since U is unitary, the right hand side of formula (3.4) defines a bounded operator even if $r \in [0, 1)$ is replaced by any complex variable $\zeta \in \mathbb{D}$. $\Bigg($Then in fact $W(\zeta)$ coincides with the characteristic function of the unitary colligation

$$\begin{bmatrix} Z & 0 \\ 0 & I \end{bmatrix} \begin{bmatrix} U_{11} & U_{12} \\ U_{21} & U_{22} \end{bmatrix} . \Bigg)$$

However, we shall not make use of this extra freedom except in a few isolated remarks.

It is convenient to note here two elementary properties of characteristic functions, which we state as lemmas for ease of future reference.

Lemma 3.1. *If $D \in \mathcal{D}(\mathcal{K}; \mathcal{K})$ is unitary and if*

$$\widetilde{U} = \begin{bmatrix} D^{(1)} & 0 \\ 0 & I_{\mathcal{E}_2} \end{bmatrix} \begin{bmatrix} U_{11} & U_{12} \\ U_{21} & U_{22} \end{bmatrix} \begin{bmatrix} D^* & 0 \\ 0 & I_{\mathcal{E}_1} \end{bmatrix} ,$$

then

$$W_{\widetilde{U}}(r) = W_U(r) \ .$$

Lemma 3.2. *If $\widetilde{U} \in \mathcal{D}(\widetilde{\mathcal{K}} \oplus \mathcal{E}_1; \widetilde{\mathcal{K}}^{(-1)} \oplus \mathcal{E}_2)$ is a unitary extension of U with the same input and outspaces as U, i.e., if*

$$\widetilde{U} = \begin{bmatrix} U_{00} & 0 & 0 \\ 0 & U_{11} & U_{12} \\ 0 & U_{21} & U_{22} \end{bmatrix} ,$$

with $U_{00} \in \mathcal{D}(\widetilde{\mathcal{K}} \ominus \mathcal{K}; (\widetilde{\mathcal{K}} \ominus \mathcal{K})^{(-1)})$ and unitary, then

$$W_{\widetilde{U}}(r) = W_U(r) \ .$$

Since U is unitary, it is readily checked that for every choice of $r \in (0,1)$ and $u \in \mathcal{E}_1$, there exists a unique choice of $x_r \in \mathcal{K}$ and $y_r \in \mathcal{E}_2$ such that

$$\begin{bmatrix} r^{-1}Z^*x_r \\ y_r \end{bmatrix} = \begin{bmatrix} U_{11} & U_{12} \\ U_{21} & U_{22} \end{bmatrix} \begin{bmatrix} x_r \\ u \end{bmatrix} \tag{3.5}$$

and that

$$W(r):\ u \in \mathcal{E}_1 \longrightarrow y_r = W(r)u \in \mathcal{E}_2 \ .$$

Lemma 3.3. *The operator $W(r)$ belongs to $\mathcal{U}(\mathcal{E}_1;\mathcal{E}_2)$ and is contractive: $\|W(r)\| \le 1$, for every choice of $r \in [0,1)$.*

Proof. Since U and Z^* are unitary,

$$\|r^{-1}x_r\|^2 + \|y_r\|^2 = \|x_r\|^2 + \|u\|^2 \ ,$$

which in turn clearly implies that

$$\begin{aligned} \|u\|^2 - \|W(r)u\|^2 &= \|u\|^2 - \|y_r\|^2 \\ &= \left(\frac{1}{r^2} - 1\right)\|x_r\|^2 \\ &\ge 0 \ . \end{aligned}$$

Therefore $W(r)$ is contractive. It also clearly belongs to the space $\mathcal{U}(\mathcal{E}_1, \mathcal{E}_2)$. □

We remark that the contractivity of $W(r)$ also follows easily from the identity

$$I - W(\beta)^*W(\alpha) = (1-\alpha\beta^*)U_{12}^*(I - \beta^*Z^*U_{11}^*)^{-1}(I - \alpha U_{11}Z)^{-1}U_{12} \tag{3.6}$$

which is valid for every choice of α and β in the open unit disc $\mathbb{D}$.

The next step is to investigate the properties of the operator

$$L(r) = U_{21}(I - rZU_{11})^{-1} \ . \tag{3.7}$$

Lemma 3.4. *The inequalities*

$$\|L(r)E\|_{HS}^2 \le \sum_{j=0}^{\infty} \mathrm{trace}\{E^*(U_{11}^*Z^*)^j(I - U_{11}^*U_{11})(ZU_{11})^jE\} \tag{3.8}$$

and

$$\sum_{j=0}^{\infty} \mathrm{trace}\{E^*(U_{11}^*Z^*)^j(I - U_{11}^*U_{11})(ZU_{11})^jE\} \le \mathrm{trace}\ E^*E \tag{3.9}$$

hold for every choice $r \in (0,1)$ and $E \in \mathcal{D}_2(\ell_{\mathbb{C}}^2;\ell_{\mathbb{C}}^2)\}^{n\times 1}$.

Proof. By definition,

$$\begin{aligned}\langle L(r)E, L(r)E\rangle_{HS} &= \text{trace}\{E^*(I - rU_{11}^*Z^*)^{-1}U_{21}^*U_{21}(I - rZU_{11})^{-1}E\} \\ &= \text{trace}\{E^*(I - rU_{11}^*Z^*)^{-1}(I - U_{11}^*U_{11})(I - rZU_{11})^{-1}E\}\ ,\end{aligned}$$

since U is unitary. Therefore

$$\begin{aligned}\langle L(r)E, L(r)E\rangle_{HS} &= \sum_{j=0}^{\infty} \text{trace}\{E^*(rU_{11}^*Z^*)^j(I - U_{11}^*U_{11})(rZU_{11})^jE\} \\ &= \lim_{k\uparrow\infty} \sum_{j=0}^{k} \text{trace}\{E^*(rU_{11}^*Z^*)^j(I - U_{11}^*U_{11})(rZU_{11})^jE\} \\ &\le \lim_{k\uparrow\infty} \sum_{j=0}^{k} \text{trace}\{E^*(U_{11}^*Z^*)^j(I - U_{11}^*Z^*ZU_{11})(ZU_{11})^jE\} \\ &= \lim_{k\uparrow\infty} \text{trace}\{E^*[I - (U_{11}^*Z^*)^{k+1}(ZU_{11})^{k+1}]E\} \\ &\le \text{trace } E^*E\ .\end{aligned}$$

This completes the proof of both of the asserted inequalities. □

Let

$$L_k(r) = U_{21}\sum_{j=0}^{k}(rU_{11}Z)^j$$

and

$$W_k = U_{22} + L_{k-1}(1)ZU_{12}\ .$$

Lemma 3.5. *Let $E \in \{\mathcal{D}_2(\ell_{\mathcal{C}}^2; \ell_{\mathcal{C}}^2)\}^n$ and let $\varepsilon > 0$ be given. Then:*

(1) *There exists an $r_0 \in [0, 1)$ such that*

$$\|\{L(r_2) - L(r_1)\}E\|_{HS} \le \varepsilon$$

for every choice of r_1 and r_2 in the open interval $(r_0, 1)$.

(2) *There exists a positive integer k_0 such that for every choice of $k \ge k_0$ and $r \in [1 - 1/k_0, 1)$,*

$$\|\{rL(r) - L_k(1)\}E\|_{HS} \le \varepsilon\ .$$

Proof. Clearly

$$\|\{L(r)-L_k(r)\}E\|_{HS}^2 = \sum_{j=k+1}^{\infty} \operatorname{trace}\{E^*(rU_{11}^*Z^*)^j(I-U_{11}^*U_{11})(rZU_{11})^jE\}$$

$$\leq \sum_{j=k+1}^{\infty} \operatorname{trace}\{E^*(U_{11}^*Z^*)^j(I-U_{11}^*U_{11})(ZU_{11})^jE\}\,.$$

In view of (3.9), this last sum can be made smaller than $\varepsilon^2/9$ by choosing k large enough. Thus, for such k,

$$\begin{aligned}\|\{L(r_2)-L(r_1)\}E\|_{HS} &\leq \|\{L(r_2)-L_k(r_2)\}E\|_{HS}\\ &\quad+\|\{L_k(r_2)-L_k(r_1)\}E\|_{HS}\\ &\quad+\|\{L_k(r_1)-L(r_1)\}E\|_{HS}\\ &\leq \frac{2}{3}\varepsilon+\|\{L_k(r_2)-L_k(r_1)\}E\|_{HS}\,.\end{aligned}$$

Moreover,

$$\begin{aligned}\|\{L_k(r_2)-L_k(r_1)\}E\|_{HS} &= \|\sum_{j=0}^{k} U_{21}(ZU_{11})^j(r_2^j-r_1^j)E\|_{HS}\\ &\leq \sum_{j=0}^{k}|r_2^j-r_1^j|\;\|U_{21}(ZU_{11})^jE\|_{HS}\,,\end{aligned}$$

which can also be made arbitrarily small by trapping r_1 and r_2 in an interval $(r_0,1)$ of small enough width, since the sum is finite. This completes the proof of (1).

The proof of (2) now follows easily from the inequality

$$\begin{aligned}\|\{rL(r)-L_k(1)\}E\|_{HS} &\leq \|r\{L(r)-L_k(r)\}E\|_{HS}\\ &\quad+(r-1)\|L_k(r)E\|_{HS}+\|\{L_k(r)-L_k(1)\}E\|_{HS}\end{aligned}$$

and the preceding bounds. □

Lemma 3.6. *Let $E\in\mathcal{D}_2(\ell_{\mathcal{B}}^2;\ell_{\mathcal{B}}^2)$ and let $\varepsilon>0$ be given. Then (for the operator $W(r)$ defined in (3.4)):*

(1) There exists an $r_0\in[0,1)$ such that

$$\|\{W(r_2)-W(r_1)\}E\|_{HS}\leq\varepsilon$$

for every choice of r_1 and r_2 in the open interval $(r_0,1)$.

(2) *There exists a* k_0 *such that*

$$\|\{W(r) - W_k\}E\|_{HS} \leq \varepsilon$$

for $k \geq k_0$ *and* $r \in [1 - 1/k_0, 1)$.

(3) *There exists a* k_0 *such that*

$$\|\{W_j - W_k\}E\|_{HS} \leq \varepsilon$$

for $j \geq k \geq k_0$.

Proof. This is an easy consequence of the last two lemmas since

$$\{W(r_2) - W(r_1)\}E = \{L(r_2)(r_2 - r_1) + [L(r_2) - L(r_1)]r_1\}ZU_{12}E\ . \qquad \square$$

Let $\mathring{W}$ denote the operator which is defined on $\mathcal{E}_1 = \ell^2_{\mathcal{N}}$ by the rule

$$\mathring{W}x = \lim_{r \uparrow 1} W(r)x = \lim_{k \uparrow \infty} W_k x\ . \tag{3.10}$$

This rule is meaningful because for any $x \in \mathcal{E}_1$ of norm one with zero entries in all of the coordinate spaces except one, the orthogonal projection of $\mathcal{E}_1$ onto span$\{x\} \subset \mathcal{E}_1$ belongs to $\mathcal{D}_2(\ell^2_{\mathcal{N}}; \ell^2_{\mathcal{N}})$. Therefore, by the preceding analysis, the indicated limit exists: If E denotes the projection, then

$$\|\{W(r_2) - W(r_1)\}x\|_{\mathcal{E}_2} = \|\{W(r_2) - W(r_1)\}E\|_{HS}\ .$$

Thus (3.10) is well defined for every $x \in \mathcal{E}_1$ with nonzero entries in only one coordinate space, and hence also for every $x \in \mathcal{E}_1$ with nonzero entries in finitely many coordinate spaces. The existence of the limit for arbitrary $x \in \mathcal{E}_1$ now follows easily by standard estimates, since $\|W(r_2) - W(r_1)\| \leq 2$ for every choice of $r_1, r_2 \in [0, 1)$. It follows easily from definition (3.4) that $\mathring{W} \in \mathcal{S}(\ell^2_{\mathcal{N}}; \ell^2_{\mathcal{M}})$.

We remark that formula (3.10) implies that

$$(\mathring{W} - W_j)x = \sum_{i=j}^{\infty} (W_{i+1} - W_i)x$$

and hence, in view of the definition of W_i, that the "k'th diagonal" of $\mathring{W}$ agrees with the "k'th diagonal" of W_j for $j \geq k$, i.e.,

$$\mathring{W} - W_j \in Z^{k+1}\mathcal{U}$$

for $j \geq k$. The operator increments $W_0, W_1 - W_0, W_2 - W_1, \ldots$ are orthogonal in the sense that

$$\langle (W_i - W_{i-1})E_i, (W_j - W_{j-1})E_j \rangle_{HS} = 0$$

for every choice of $E_i, E_j \in \mathcal{D}_2$ with $i \neq j$. These increments determine the operator $\mathring{W}$ in much the same way that the Fourier coefficients of a function serve to determine that function in classical settings. Nevertheless, it is often more convenient to work with the operator $W(r)$, which is an analogue in the present setting of representing a function $f \in H_\infty$ which is given on the boundary by its values in the interior.

The next objective is to show that $\mathring{W}$ is a solution of the augmented interpolation problem which is formulated in terms of the conditions (1.6), (1.7) and (3.1) with f_j as in (2.5). The proof rests on a number of preliminary lemmas.

Lemma 3.7. *The identity*

$$(C_1 - W(r)C_2)(M - rZN)^{-1} = L(r)P^{\frac{1}{2}} \tag{3.11}$$

holds for every choice of $r \in [0,1)$ for which the indicated inverse exists (and so in particular for all $r \in (1-\varepsilon, 1)$ for sufficiently small $\varepsilon > 0$).

Proof. Since

$$\begin{bmatrix} U_{11} & U_{12} \\ U_{21} & U_{22} \end{bmatrix} \begin{bmatrix} P^{\frac{1}{2}}M \\ C_2 \end{bmatrix} x = \begin{bmatrix} \{P^{(1)}\}^{\frac{1}{2}}N \\ C_1 \end{bmatrix} x$$

for $x \in (\ell^2_{\mathcal{B}})^n$, $P^{(1)} = Z^*PZ$, and $\{P^{(1)}\}^{\frac{1}{2}} = \{P^{\frac{1}{2}}\}^{(1)}$ (since P is a diagonal operator), it is readily seen that

$$\begin{aligned} C_1 - W(r)C_2 &= C_1 - U_{22}C_2 - U_{21}(I - rZU_{11})^{-1}rZU_{12}C_2 \\ &= U_{21}P^{\frac{1}{2}}M - U_{21}(I - rZU_{11})^{-1}rZ(\{P^{(1)}\}^{\frac{1}{2}}N - U_{11}P^{\frac{1}{2}}M) \\ &= U_{21}(I - rZU_{11})^{-1}P^{\frac{1}{2}}(M - rZN)\ . \end{aligned}$$

This is equivalent to (3.11). □

At this point it is convenient to introduce the operator

$$K(r) = U_{12}^*Z^*(I - rU_{11}^*Z^*)^{-1} \tag{3.12}$$

which will play a role in the development of formulas involving U^* which is similar to that played earlier by $L(r)$.

Lemma 3.8. *The inequalities*

$$\|K(r)E\|_{HS}^2 \leq \sum_{j=0}^{\infty} \text{trace}\{E^*(ZU_{11})^j(I - ZU_{11}U_{11}^*Z^*)(U_{11}^*Z^*)^jE\} \tag{3.13}$$

and

$$\sum_{j=0}^{\infty} \text{trace}\{E^*(ZU_{11})^j(I - ZU_{11}U_{11}^*Z^*)(U_{11}^*Z^*)^jE\} \leq \text{trace}\, E^*E \tag{3.14}$$

hold for every choice of $r \in [0,1)$ and $E \in \{\mathcal{D}_2(\ell^2_{\mathcal{C}}; \ell^2_{\mathcal{C}})\}^n$. Moreover, for any such fixed choice of E and any $\varepsilon > 0$, there exists an $r_0 \in [0,1)$ such that

$$\|\{K(r_2) - K(r_1)\}E\|_{HS} \leq \varepsilon$$

for every choice of r_1 and r_2 in the open interval $(r_0, 1)$.

Proof. The proof is easily modelled on that of Lemma 3.4. □

Lemma 3.9. *The identity*

$$(W(r)^*C_1 - C_2)(rM - ZN)^{-1} = K(r)P^{\frac{1}{2}} \tag{3.15}$$

holds for every choice of $r \in [0,1)$ for which the indicated inverse exists (and so in particular for all $r \in (1-\varepsilon, 1)$ for sufficiently small $\varepsilon > 0$)

Proof. The formula

$$\begin{bmatrix} U_{11}^* & U_{21}^* \\ U_{12}^* & U_{22}^* \end{bmatrix} \begin{bmatrix} \{P^{(1)}\}^{\frac{1}{2}}Nx \\ C_1x \end{bmatrix} = \begin{bmatrix} P^{\frac{1}{2}}Mx \\ C_2x \end{bmatrix} \tag{3.16}$$

for $x \in \{\ell_{\mathcal{B}}^2\}^n$ implies that

$$\begin{aligned} W(r)^*C_1 - C_2 &= U_{22}^*C_1 + U_{12}^*rZ^*(I - rU_{11}^*Z^*)^{-1}U_{21}^*C_1 - C_2 \\ &= -U_{12}^*Z^*P^{\frac{1}{2}}ZN + U_{12}^*rZ^*(I - rU_{11}^*Z^*)^{-1}U_{21}^*C_1 \\ &= -U_{12}^*Z^*(I - rU_{11}^*Z^*)^{-1}\{(I - rU_{11}^*Z^*)Z\{P^{(1)}\}^{\frac{1}{2}}N - rU_{21}^*C_1\} \\ &= K(r)\{-(I - rU_{11}^*Z^*)P^{\frac{1}{2}}ZN + rU_{21}^*C_1\} \\ &= K(r)P^{\frac{1}{2}}(rM - ZN) . \end{aligned}$$

But this is clearly equivalent to (3.15). □

Theorem 3.1. *Let $P \in \{\mathcal{D}(\ell_{\mathcal{B}}^2;\ell_{\mathcal{B}}^2)\}^{n\times n}$ be a given positive semidefinite solution of (2.9) and let*

$$U = \begin{bmatrix} U_{11} & U_{12} \\ U_{21} & U_{22} \end{bmatrix}$$

be any unitary colligation with components $U_{11} \in \{\mathcal{D}(\ell_{\mathcal{C}}^2;\ell_{\mathcal{C}(-1)}^2)\}^{n\times n}$, $U_{12} \in \{\mathcal{D}(\ell_{\mathcal{N}}^2;\ell_{\mathcal{C}(-1)}^2)\}^{n\times 1}$, $U_{21} \in \{\mathcal{D}(\ell_{\mathcal{C}}^2;\ell_{\mathcal{M}}^2)\}^{1\times n}$ and $U_{22} \in \mathcal{D}(\ell_{\mathcal{N}}^2;\ell_{\mathcal{M}}^2)$ which extends the isometry V which is defined by (3.3). Then the operator $\mathring{W}$, which is defined in terms of the limit (3.10) of the characteristic function $W(r)$ of U, is a solution of the augmented interpolation problem, i.e.,

(1) $\mathring{W} \in \mathcal{S}(\ell_{\mathcal{N}}^2;\ell_{\mathcal{M}}^2)$

(2) $(C_1 - \mathring{W}C_2)(M - ZN)^{-1}E \in \mathcal{U}_2(\ell_{\mathcal{B}}^2;\ell_{\mathcal{M}}^2)$

(3) $(-\mathring{W}^*C_1 + C_2)(M - ZN)^{-1}E \in \mathcal{L}_2'(\ell_{\mathcal{B}}^2;\ell_{\mathcal{N}}^2)$

(4) $$-\left\langle \mathring{W}C_2(M - ZN)^{-1}\begin{bmatrix} 0 \\ E_2 \end{bmatrix}, C_1(M - ZN)^{-1}\begin{bmatrix} E_1 \\ 0 \end{bmatrix}\right\rangle_{HS} = \mathrm{trace}[E_1^* \ 0]P\begin{bmatrix} 0 \\ E_2 \end{bmatrix}$$

for every choice of $E \in \{\mathcal{D}_2(\ell_{\mathcal{B}}^2;\ell_{\mathcal{B}}^2)\}^{n\times 1}$, $E_1 \in \{\mathcal{D}_2(\ell_{\mathcal{B}}^2;\ell_{\mathcal{B}}^2)\}^{\mu\times 1}$ and $E_2 \in \{\mathcal{D}_2(\ell_{\mathcal{B}}^2;\ell_{\mathcal{B}}^2)\}^{\nu\times 1}$.

Proof. We have already verified (1).

Next, with the help of (3.11), it is readily seen that

$$(C_1 - \mathring{W}C_2)(M - ZN)^{-1}x - (C_1 - W(r)C_2)(M - rZN)^{-1}x$$

$$= \{C_1 - \mathring{W}C_2 - (C_1 - W(r)C_2)\}(M - ZN)^{-1}x$$

$$+ (C_1 - W(r)C_2)\{(M - ZN)^{-1} - (M - rZN)^{-1}\}x$$

$$= (W(r) - \mathring{W})C_2(M - ZN)^{-1}x + (1 - r)L(r)P^{\frac{1}{2}}ZN(M - ZN)^{-1}x$$

tends to zero as $r \uparrow 1$ for every choice of $x \in \{\ell^2_{\mathcal{B}}\}^n$. This serves to establish (2) since $(C_1 - W(r)C_2)(M - rZN)^{-1}E \in \mathcal{U}_2$ and is uniformly bounded in norm for r close to one for every $E \in \mathcal{D}_2$.

The proof of (3) follows in much the same way with the help of (3.15) from the identity

$$(\mathring{W}^*C_1 - C_2)(M - ZN)^{-1} - (W(r)^*C_1 - C_2)(rM - ZN)^{-1}$$

$$= \{\mathring{W}^*C_1 - C_2 - (W(r)^*C_1 - C_2)\}(M - ZN)^{-1}$$

$$+ (W(r)^*C_1 - C_2)\{(M - ZN)^{-1} - (rM - ZN)^{-1}\}$$

$$= \{\mathring{W}^* - W(r)^*\}C_1(M - ZN)^{-1} + (r - 1)K(r)P^{\frac{1}{2}}M(M - ZN)^{-1}$$

and definition (3.10).

Finally, to obtain (4), notice first that the left hand side of the asserted identity is equal to

$$\left\langle (C_1 - \mathring{W}C_2)(M - ZN)^{-1}\begin{bmatrix} 0 \\ E_2 \end{bmatrix}, C_1(M - ZN)^{-1}\begin{bmatrix} E_1 \\ 0 \end{bmatrix}\right\rangle_{HS}$$

$$= \lim_{r\uparrow 1}\left\langle \{C_1 - W(r)C_2\}(M - rZN)^{-1}\begin{bmatrix} 0 \\ E_2 \end{bmatrix}, C_1(M - ZN)^{-1}\begin{bmatrix} E_1 \\ 0 \end{bmatrix}\right\rangle_{HS}.$$

But, in view of (3.11), the last inner product is equal to

$$\alpha(r) = \left\langle L(r)P^{\frac{1}{2}}\begin{bmatrix} 0 \\ E_2 \end{bmatrix}, C_1(M - ZN)^{-1}\begin{bmatrix} E_1 \\ 0 \end{bmatrix}\right\rangle_{HS}$$

$$= \left\langle (I - rZU_{11})^{-1}P^{\frac{1}{2}}\begin{bmatrix} 0 \\ E_2 \end{bmatrix}, U_{21}^*C_1(M - ZN)^{-1}\begin{bmatrix} E_1 \\ 0 \end{bmatrix}\right\rangle_{HS}.$$

By (3.16),

$$U_{21}^*C_1 = -U_{11}^*Z^*P^{\frac{1}{2}}ZN + P^{\frac{1}{2}}M$$

$$= P^{\frac{1}{2}}(M - ZN) + (I - U_{11}^*Z^*)P^{\frac{1}{2}}ZN\,.$$

Therefore $\alpha(r) = \textcircled{1} + \textcircled{2}$, where

$$\begin{aligned}\textcircled{1} &= \left\langle (I - rZU_{11})^{-1} P^{\frac{1}{2}} \begin{bmatrix} 0 \\ E_2 \end{bmatrix}, P^{\frac{1}{2}} \begin{bmatrix} E_1 \\ 0 \end{bmatrix} \right\rangle_{HS} \\ &= \left\langle P^{\frac{1}{2}} \begin{bmatrix} 0 \\ E_2 \end{bmatrix}, P^{\frac{1}{2}} \begin{bmatrix} E_1 \\ 0 \end{bmatrix} \right\rangle_{HS}\end{aligned}$$

and

$$\begin{aligned}\textcircled{2} &= \left\langle (I - ZU_{11})(I - rZU_{11})^{-1} P^{\frac{1}{2}} \begin{bmatrix} 0 \\ E_2 \end{bmatrix}, P^{\frac{1}{2}} ZN(M - ZN)^{-1} \begin{bmatrix} E_1 \\ 0 \end{bmatrix} \right\rangle_{HS} \\ &= \left\langle P^{\frac{1}{2}} \begin{bmatrix} 0 \\ E_2 \end{bmatrix}, P^{\frac{1}{2}} ZN(M - ZN)^{-1} \begin{bmatrix} E_1 \\ 0 \end{bmatrix} \right\rangle_{HS} \\ &\quad + (r-1) \left\langle ZU_{11}(I - rZU_{11})^{-1} P^{\frac{1}{2}} \begin{bmatrix} 0 \\ E_2 \end{bmatrix}, P^{\frac{1}{2}} ZN(M - ZN)^{-1} \begin{bmatrix} E_1 \\ 0 \end{bmatrix} \right\rangle_{HS} .\end{aligned}$$

The first inner product in $\textcircled{2}$ is equal to zero; the second can be evaluated as follows:

$$\langle\ ,\ \rangle_{HS} = \sum_{j=0}^{\infty} r^j \left\langle (ZU_{11})^{j+1} P^{\frac{1}{2}} \begin{bmatrix} 0 \\ E_2 \end{bmatrix}, P^{\frac{1}{2}} (ZN)^{j+1} \begin{bmatrix} E_1 \\ 0 \end{bmatrix} \right\rangle_{HS} ,$$

since

$$(M - ZN)^{-1} \begin{bmatrix} E_1 \\ 0 \end{bmatrix} = \begin{bmatrix} (I - Z_1 A_1)^{-1} E_1 \\ 0 \end{bmatrix} = \sum_{j=0}^{\infty} (ZN)^j \begin{bmatrix} E_1 \\ 0 \end{bmatrix} .$$

Therefore,

$$\begin{aligned}|\textcircled{2}| &\leq (1-r) \sum_{j=0}^{\infty} r^j \left\| (ZU_{11})^{j+1} P^{\frac{1}{2}} \begin{bmatrix} 0 \\ E_2 \end{bmatrix} \right\|_{HS} \left\| P^{\frac{1}{2}} (ZN)^{j+1} \begin{bmatrix} E_1 \\ 0 \end{bmatrix} \right\|_{HS} \\ &\leq (1-r) \sum_{j=0}^{\infty} \left\| P^{\frac{1}{2}} \begin{bmatrix} 0 \\ E_2 \end{bmatrix} \right\|_{HS} \left\| P^{\frac{1}{2}} \right\| \left\| (Z_1 A_1)^{j+1} E_1 \right\|_{HS} \cdot\end{aligned}$$

which tends to zero as $r \uparrow 1$ in view of (2.6). □

4 Coupling

In this section we shall show that every unitary colligation U

$$U = \begin{bmatrix} U_{11} & U_{12} \\ U_{21} & U_{22} \end{bmatrix} : \mathcal{K} \oplus \mathcal{E}_1 \longrightarrow \mathcal{K}^{(-1)} \oplus \mathcal{E}_2$$

with input space $\mathcal{E}_1$, output space $\mathcal{E}_2$ (and components $U_{ij} \in \mathcal{D}$) which extends V (i.e., $\mathcal{K} \supset \mathcal{H}$ and $U|_{d_V} = V$) can be expressed as the "coupling" of a pair of unitary colligations A and B, the first of which is based on the data of the interpolation problem and the second of which is arbitrary. The strategy is adapted from Kheifets [Kh1] and [Kh2]; for additional information on coupling, see Arov and Grossman [AG1] and [AG2].

Recall that $V \in \mathcal{D}(d_V; \Delta_V)$ is an isometry from

$$d_V \subset \mathcal{H} \oplus \mathcal{E}_1$$

onto

$$\Delta_V \subset \mathcal{H}^{(-1)} \oplus \mathcal{E}_2$$

and let

$$d_V^{\perp} = (\mathcal{H} \oplus \mathcal{E}_1) \ominus d_V$$

and

$$\Delta_V^{\perp} = (\mathcal{H}^{(-1)} \oplus \mathcal{E}_2) \ominus \Delta_V$$

denote the defect spaces of the domain and range of V, respectively. Then, it follows readily from the definition of V that

$$d_V^{\perp} = \ker[M^* P^{\frac{1}{2}} \; C_2^*]$$

and

$$\Delta_V^{\perp} = \ker[N^* \{P^{(1)}\}^{\frac{1}{2}} \; C_1^*] \; .$$

Moreover, since $\mathcal{H} = \{\ell_{\mathcal{B}}^2\}^n$, $\mathcal{E}_1 = \ell_{\mathcal{N}}^2$ and $\mathcal{E}_2 = \ell_{\mathcal{M}}^2$, both of these defect spaces are "ℓ^2 spaces":

$$d_V^{\perp} = \ell_{\mathcal{G}}^2 \text{ and } \Delta_V^{\perp} = \ell_{\mathcal{F}}^2 \; ,$$

where

$$\mathcal{G} = \bigoplus_{j=-\infty}^{\infty} \mathcal{G}_j \; , \; \mathcal{F} = \bigoplus_{j=-\infty}^{\infty} \mathcal{F}_j \; ,$$

$$\mathcal{G}_j = \ker \; \pi_j^*[M^* P^{\frac{1}{2}} \; C_2^*]\pi_j = \ker[M_j^* P_j^{\frac{1}{2}} \; (C_2^*)_j]$$

and

$$\mathcal{F}_j = \ker \; \pi_j^*[N^* \{P^{(1)}\}^{\frac{1}{2}} \; C_1^*]\pi_j = \ker[N_j^* P_{j-1}^{\frac{1}{2}} \; (C_1^*)_j] \; .$$

Let[3] $\mathcal{N}_1 = \alpha d_V^{\perp}$ be unitarily equivalent to $d_V^{\perp}$ by a unitary diagonal operator $\alpha \in \mathcal{D}(d_V^{\perp}; \mathcal{N}_1)$ and let $\mathcal{N}_2 = \beta \Delta_V^{\perp}$ be unitarily equivalent to $\Delta_V^{\perp}$ by a unitary diagonal operator $\beta \in \mathcal{D}(\Delta_V^{\perp}; \mathcal{N}_2)$ and define the operator A from the space $\mathcal{H} \oplus \mathcal{E}_1 \oplus \mathcal{N}_2$ onto the space $\mathcal{H}^{(-1)} \oplus \mathcal{E}_2 \oplus \mathcal{N}_1$ by the rule

$$A|_{d_V} = V \; , \; A|_{d_V^{\perp}} = \alpha \; , \text{ and } A|_{\mathcal{N}_2} = \beta^* \; . \tag{4.1}$$

3) The choice of the symbols $\mathcal{N}_1$ and $\mathcal{N}_2$ for the orthogonal complements conflicts with the usage of $\mathcal{N}_j$ as the coordinate spaces of $\mathcal{E}_1 = \ell_{\mathcal{N}}^2$. However, since the intention is always clear from the context, this should not cause any trouble.

Lemma 4.1. *Let A denote the operator which is defined by rule (4.1). Then:*
(1) $A \in \mathcal{D}(\mathcal{H} \oplus \mathcal{E}_1 \oplus \mathcal{N}_2; \mathcal{H}^{(-1)} \oplus \mathcal{E}_2 \oplus \mathcal{N}_1)$.
(2) *A is unitary.*
(3) $P_{\mathcal{N}_1} A|_{\mathcal{N}_2} = 0$. (4.2)
(4) $P_{\mathcal{H}^{(-1)} \oplus \mathcal{E}_2} A|_{d_V^\perp} = 0$. (4.3)

Proof. The proof is selfevident from the definition of A and the preceding discussion. □

We shall refer to A as a unitary colligation with state space $\mathcal{H}$, input space $\mathcal{E}_1 \oplus \mathcal{N}_2$ and output space $\mathcal{E}_2 \oplus \mathcal{N}_1$, even though "$A_{11}$" maps $\mathcal{H}$ into $\mathcal{H}^{(-1)}$.

Now let $B \in \mathcal{D}(\mathcal{G} \oplus \mathcal{N}_1; \mathcal{G}^{(-1)} \oplus \mathcal{N}_2)$ be a second unitary colligation with state space $\mathcal{G} = \ell^2_{\mathcal{R}}$, input space $\mathcal{N}_1$ and output space $\mathcal{N}_2$. Then

$$A(h + e_1 + n_2) = h' + e_2 + n_1 \tag{4.4}$$

and

$$B(g + n_1) = g' + n_2 \ , \tag{4.5}$$

where h and h' stand for elements in $\mathcal{H}$, g and g' for elements in $\mathcal{G}$, $e_j \in \mathcal{E}_j$ and $n_j \in \mathcal{N}_j$ for $j = 1, 2$, imbedded as needed for the formulas to make sense.

Because of (4.2) and (4.3), this pair of relations serves to define a unitary colligation $U \in \mathcal{D}(\mathcal{G} \oplus \mathcal{H} \oplus \mathcal{E}_1; \mathcal{G}^{(-1)} \oplus \mathcal{H}^{(-1)} \oplus \mathcal{E}_2)$ from $\mathcal{G} \oplus \mathcal{H} \oplus \mathcal{E}_1$ onto $\mathcal{G}^{(-1)} \oplus \mathcal{H}^{(-1)} \oplus \mathcal{E}_2$ such that

$$U(g + h + e_1) = g' + h' + e_2 \tag{4.6}$$

and

$$U|_{d_V} = V \ .$$

In particular, given $g \in \mathcal{G}$, $h \in \mathcal{H}$ and $e_1 \in \mathcal{E}_1$, define

$$n_1 = P_{\mathcal{N}_1} A(h + e_1) \tag{4.7a}$$

and

$$g' + n_2 = B(g + n_1) \ . \tag{4.7b}$$

These two formulas are consistent because, in view of (4.2),

$$P_{\mathcal{N}_1} A(h + e_1 + n_2) = P_{\mathcal{N}_1} A(h + e_1) \ .$$

Now set

$$g' = P_{\mathcal{G}^{(-1)}} B(g + n_1) \ , \tag{4.8a}$$

$$h' = P_{\mathcal{H}^{(-1)}} A(h + e_1 + n_2) \ , \tag{4.8b}$$

$$e_2 = P_{\mathcal{E}_2} A(h + e_1 + n_2) \tag{4.8c}$$

and define U as the mapping from

$$g + h + e_1 \longrightarrow g' + h' + e_2 \ ;$$

this is the meaning of (4.6).

Lemma 4.2. *Let U denote the mapping from $\mathcal{G} \oplus \mathcal{H} \oplus \mathcal{E}_1$ into $\mathcal{G}^{(-1)} \oplus \mathcal{H}^{(-1)} \oplus \mathcal{E}_2$ which is defined by* (4.7) *and* (4.8). *Then:*

(1) *U is linear and isometric.*

(2) *U maps $\mathcal{G} \oplus \mathcal{H} \oplus \mathcal{E}_1$ onto $\mathcal{G}^{(-1)} \oplus \mathcal{H}^{(-1)} \oplus \mathcal{E}_2$ (and hence is in fact unitary).*

(3) *$U \in \mathcal{D}(\mathcal{G} \oplus \mathcal{H} \oplus \mathcal{E}_1; \mathcal{G}^{(-1)} \oplus \mathcal{H}^{(-1)} \oplus \mathcal{E}_2)$.*

(4) *$U|_{d_V} = V$.*

Proof. U is clearly linear. Since A and B are unitary, it follows from (4.4) and (4.5) that

$$\|h\|^2 + \|e_1\|^2 + \|n_2\|^2 = \|h'\|^2 + \|e_2\|^2 + \|n_1\|^2$$

and

$$\|g\|^2 + \|n_1\|^2 = \|g'\|^2 + \|n_2\|^2 ,$$

respectively. The asserted isometry in (1) now drops out easily from the sum of these two formulas.

Next, to obtain (2), let $g' \in \mathcal{G}^{(-1)}$, $h' \in \mathcal{H}^{(-1)}$ and $e_2 \in \mathcal{E}_2$ be given. Then

$$h' + e_2 = c + d$$

with $c \in \Delta_V$ and $d \in \Delta_V^{\perp} = \beta^* \mathcal{N}_2$. Therefore, since B is unitary and hence onto, there is a unique choice of $g \in \mathcal{G}$ and $n_1 \in \mathcal{N}_1$ such that

$$B(g + n_1) = g' + \beta d .$$

Now, having h', e_2 and n_1, there is a unique choice of $h \in \mathcal{H}$, $e_1 \in \mathcal{E}_1$ and $n_2 \in \mathcal{N}_2$ such that

$$A(h + e_1 + n_2) = h' + e_2 + n_1$$

because A is also unitary and hence onto. Moreover, by the definition of A,

$$n_2 = \beta d .$$

Assertion (3) is immediate from the fact that $A \in \mathcal{D}$ and $B \in \mathcal{D}$, whereas assertion (4) follows from (4.7) and the fact that $A|_{d_V} = V$. □

The preceding analysis exhibits a method of constructing a unitary colligation $U \in \mathcal{D}(\mathcal{G} \oplus \mathcal{H} \oplus \mathcal{E}_1; \mathcal{G}^{(-1)} \oplus \mathcal{H}^{(-1)} \oplus \mathcal{E}_2)$ with state space $\mathcal{G} \oplus \mathcal{H}$, input space $\mathcal{E}_1$ and output space $\mathcal{E}_2$ such that

$$U|_{d_V} = V .$$

We shall refer to such a U as the coupling of (the "standard" colligation) A with (the arbitrary colligation) B. Our next objective is to show that every unitary colligation $U \in \mathcal{D}(\mathcal{G} \oplus \mathcal{H} \oplus \mathcal{E}_1; \mathcal{G}^{(-1)} \oplus \mathcal{H}^{(-1)} \oplus \mathcal{E}_2)$ such that $U|_{d_V} = V$ arises this way.

Theorem 4.1. *Let* $U \in \mathcal{D}(\mathcal{G} \oplus \mathcal{H} \oplus \mathcal{E}_1; \mathcal{G}^{(-1)} \oplus \mathcal{H}^{(-1)} \oplus \mathcal{E}_2)$ *be a unitary colligation (with state space* $\mathcal{G} \oplus \mathcal{H}$*, input space* $\mathcal{E}_1$ *and output space* $\mathcal{E}_2$*) such that* $U|_{d_V} = V$*. Let* $\mathcal{N}_1 = \alpha d_V^{\perp}$ *and let* $\mathcal{N}_2 = \beta \Delta_V^{\perp}$*, where* α *and* β *are unitary diagonal operators (i.e.,* $\alpha \in \mathcal{D}(d_V^{\perp}; \mathcal{N}_1)$ *and* $\beta \in \mathcal{D}(\Delta_V^{\perp}; N_2)$*). Then* U *can be expressed as the coupling of a standard unitary colligation* $A \in \mathcal{D}(\mathcal{H} \oplus \mathcal{E}_1 \oplus \mathcal{N}_2; \mathcal{H}^{(-1)} \oplus \mathcal{E}_2 \oplus \mathcal{N}_1)$ *(with state space* $\mathcal{H}$*, input space* $\mathcal{E}_1 \oplus \mathcal{N}_2$ *and output space* $\mathcal{E}_2 \oplus \mathcal{N}_1$*) which is defined as in (4.1) and a unitary colligation* $B \in \mathcal{D}(\mathcal{G} \oplus \mathcal{N}_1; \mathcal{G}^{(-1)} \oplus \mathcal{N}_2)$ *(with state space* $\mathcal{G}$*, input space* $\mathcal{N}_1$ *and output space* $\mathcal{N}_2$*).*

Proof. Recall first that we have defined $d_V^{\perp} = (\mathcal{H} \oplus \mathcal{E}_1) \ominus d_V$ and $\Delta_V^{\perp} = (\mathcal{H}^{(-1)} \oplus \mathcal{E}_2) \ominus \Delta_V$. Therefore, since U is a unitary extension of V, it is readily checked that U maps $\mathcal{G} \oplus d_V^{\perp}$ onto $\mathcal{G}^{(-1)} \oplus \Delta_V^{\perp}$. Thus if

$$U(g + h + e_1) = g' + h' + e_2 ,$$

where $g \in \mathcal{G},\ g' \in \mathcal{G}^{(-1)},\ h \in \mathcal{H},\ h' \in \mathcal{H}^{(-1)}$ and $e_j \in \mathcal{E}_j$ for $j = 1, 2$, and if

$$h + e_1 = a + b \ , \text{ with } a \in d_V \text{ and } b \in d_V^{\perp} ,$$

and

$$h' + e_2 = c + d \ , \text{ with } c \in \Delta_V \text{ and } d \in \Delta_V^{\perp} ,$$

then

$$U(g + b) = g' + d .$$

Let

$$n_1 = \alpha b \text{ and } n_2 = \beta d$$

and set

$$B(g + n_1) = g' + n_2 .$$

Then $B \in \mathcal{D}(\mathcal{G} \oplus \mathcal{N}_1; \mathcal{G}^{(-1)} \oplus \mathcal{N}_2)$ is a unitary colligation with state space $\mathcal{G}$, input space $\mathcal{N}_1$ and output space $\mathcal{N}_2$.

Consider next the element

$$\begin{aligned} h + e_1 + n_2 &= a + b + n_2 \\ &= a + \alpha^* n_1 + \beta d \end{aligned}$$

and recall that

$$\begin{aligned} A(h + e_1 + n_2) &= Ua + d + n_1 \\ &= c + d + n_1 \\ &= h' + e_2 + n_1 . \end{aligned}$$

These calculations exhibit U as the coupling of A and B. □

Our next step is to express the characteristic function of the coupling of the unitary colligations A and B through their characteristic functions.

Theorem 4.2. *Let*

$$\Sigma(r) = W_A(r) : \mathcal{E}_1 \oplus \mathcal{N}_2 \longrightarrow \mathcal{E}_2 \oplus \mathcal{N}_1 ,$$

$$\Omega(r) = W_B(r) : \mathcal{N}_1 \longrightarrow \mathcal{N}_2 ,$$

and

$$S(r) = W_U(r) : \mathcal{E}_1 \longrightarrow \mathcal{E}_2$$

denote the characteristic functions of the unitary colligations $A \in \mathcal{D}(\mathcal{H} \oplus \mathcal{E}_1 \oplus \mathcal{N}_2; \mathcal{H}^{(-1)} \oplus \mathcal{E}_2 \oplus \mathcal{N}_1)$ which is defined by (4.1), $B \in \mathcal{D}(\mathcal{G} \oplus \mathcal{N}_1; \mathcal{G}^{(-1)} \oplus \mathcal{N}_2)$ and their coupling $U \in \mathcal{D}(\mathcal{G} \oplus \mathcal{H} \oplus \mathcal{E}_1; \mathcal{G}^{(-1)} \oplus \mathcal{H}^{(-1)} \oplus \mathcal{E}_2)$, respectively. Then

$$\|\Sigma_{22}(r)\| \leq r$$

and

$$S(r) = \Sigma_{11}(r) + \Sigma_{12}(r)\Omega(r)\{I - \Sigma_{22}(r)\Omega(r)\}^{-1}\Sigma_{21}(r) \tag{4.9}$$

for every choice of $r \in [0, 1)$.

Proof. The proof is broken into steps.

Step 1. $\|\Sigma_{22}(r)\| \leq r$.

Proof of Step 1. Since $\Sigma_{22}(r) = P_{\mathcal{N}_1}\Sigma(r)|_{\mathcal{N}_2}$, it follows from Lemma 3.3 that $\|\Sigma_{22}(r)\| \leq 1$. Moreover, by (4.2), $\Sigma_{22} \in U'(\ell^2_{\mathcal{N}_2}; \ell^2_{\mathcal{N}_1})$. Thus $(\Sigma_{22}Z^*)(\omega)$ (which is defined by replacing r by ω) is an analytic operator valued function of the complex variable ω. Therefore, by the maximum modulus principle for analytic operator valued functions (see e.g., Theorem 2.3 in Chapter 3 of [SzNF]),

$$\max\{\|(\Sigma_{22}Z^*)(\omega)\| : |\omega| = r\}$$

is a monotone increasing function of r. However, since

$$(\Sigma_{22}Z^*)(re^{i\theta}) = \Lambda_1(e^{i\theta})^*(\Sigma_{22}Z^*)(r)\Lambda_2(e^{i\theta}) ,$$

where the $\Lambda_k(e^{i\theta})$, $k = 1, 2$, are block diagonal operators with $e^{ij\theta} \times$ an appropriate identity in the j'th block, it follows that

$$\|(\Sigma_{22}Z^*)(\omega)\| = \|(\Sigma_{22}Z^*)(|\omega|)\| ,$$

and hence, since

$$\frac{\|\Sigma_{22}(r)\|}{r} = \|(\Sigma_{22}Z^*)(r)\| ,$$

that $\|\Sigma_{22}(r)\|/r$ is also a monotone increasing function of r. But this in turn leads easily to the stated conclusion.

Step 2. *Formula (4.9) is valid.*

Proof of Step 2. Fix $r \in (0,1)$ and $e_1 \in \mathcal{E}_1$. Then, since U is unitary, there exists a unique choice of $g_r \in \mathcal{G}$, $h_r \in \mathcal{H}$ and $e_2(r) \in \mathcal{E}_2$ such that

$$U(g_r + h_r + e_1) = r^{-1}Z^*g_r + r^{-1}Z^*h_r + e_2(r) \ .$$

Next, since

$$h_r + e_1 = a_r + b_r \ , \ \text{ with } a_r \in d_V \text{ and } b_r \in d_V^\perp \ ,$$

and

$$r^{-1}Z^*h_r + e_2(r) = c_r + d_r \ , \ \text{ with } c_r \in \Delta_V \text{ and } d_r \in \Delta_V^\perp \ ,$$

it follows that

$$\begin{aligned} A(h_r + e_1 + \beta d_r) &= A(a_r + b_r + \beta d_r) \\ &= c_r + d_r + \alpha b_r \\ &= r^{-1}Z^*h_r + e_2(r) + \alpha b_r \ . \end{aligned}$$

Thus, since $\Sigma(r)$ is the characteristic function of A,

$$\Sigma(r) \begin{bmatrix} e_1 \\ \beta d_r \end{bmatrix} = \begin{bmatrix} e_2(r) \\ \alpha b_r \end{bmatrix} \ . \tag{4.10}$$

At the same time, the formula

$$B(g_r + \alpha b_r) = r^{-1}Z^*g_r + \beta d_r$$

implies that

$$\Omega(r)\alpha b_r = \beta d_r \ . \tag{4.11}$$

Therefore, upon writing Σ in block form and substituting (4.11) into (4.10), we obtain

$$\begin{aligned} \Sigma_{11}(r)e_1 + \Sigma_{12}(r)\Omega(r)\alpha b_r &= e_2(r) \\ \Sigma_{21}(r)e_1 + \Sigma_{22}(r)\Omega(r)\alpha b_r &= \alpha b_r \ . \end{aligned}$$

Since $\|\Sigma_{22}(r)\| \leq r$ by Step 1,

$$\alpha b_r = \{I - \Sigma_{22}(r)\Omega(r)\}^{-1}\Sigma_{21}(r)e_1$$

and the asserted formula for $S(r)$ now drops out easily upon expressing $e_2(r)$ in terms of e_1. □

We remark that since $\|\Sigma_{22}(r)\| \leq 1$ and $\Sigma_{22}(0) = 0$, the inequality $\|\Sigma_{22}(r)\| \leq r$ is a particular case of a generalization of the Schwarz lemma to upper triangular operators which will be established below in Theorem 5.5. We also wish to emphasize that here too (just as in the preceding sections) the calculations are valid when the coordinate spaces depend upon the index, providing that Z is interpreted properly.

5 Linear fractional representation of all solutions

In this section we shall show that $S \in \mathcal{S}(\mathcal{E}_1;\mathcal{E}_2)$ is a solution of the augmented BIP if and only if it can be expressed as a linear fractional transformation in the Redheffer form:

$$S(r) = \Sigma_{11}(r) + \Sigma_{12}(r)\{I - \Omega(r)\Sigma_{22}(r)\}^{-1}\Omega(r)\Sigma_{21}(r) \ , \tag{5.1}$$

where Σ is the characteristic function of the standard unitary colligation

$$A \in \mathcal{D}(\mathcal{H} \oplus \mathcal{E}_1 \oplus \mathcal{N}_2; \mathcal{H}^{(-1)} \oplus \mathcal{E}_2 \oplus \mathcal{N}_1)$$

which is specified by (4.1) and $\Omega \in \mathcal{S}(\mathcal{N}_1, \mathcal{N}_2)$.

We need the following preliminary result.

Theorem 5.1. *There exists a choice of operators $\xi_j \in \mathcal{D}(\ell^2_{\mathcal{B}};\ell^2_{\mathcal{M}})$ and $W_j \in \mathcal{D}(\ell^2_{\mathcal{B}};\ell^2_{\mathcal{B}})$ for $j = n+1, n+2, \ldots$ with*

$$r_{sp}(ZW_j^*) \leq q < 1$$

such that $S = 0$ is the only operator in $\mathcal{U}(\mathcal{E}_1;\mathcal{E}_2)$ for which

$$\underline{\underline{p}}S^*\xi_j(I - ZW_j^*)^{-1}E = 0$$

for $j = n+1, n+2, \ldots$ and every choice of $E \in \mathcal{D}_2(\ell^2_{\mathcal{B}};\ell^2_{\mathcal{B}})$.

Proof. This is an easy consequence of Theorem 3.2 of [ADD]. □

Theorem 5.2. *Let $S \in \mathcal{U}(\mathcal{E}_1;\mathcal{E}_2)$ be a solution of the augmented BIP based on a given choice of data M, N, C_1 and C_2 and a given positive semidefinite solution $P \in \{\mathcal{D}(\ell^2_{\mathcal{B}};\ell^2_{\mathcal{B}})\}^{n\times n}$ of the generalized Lyapunov equation (2.9). Then:*

(1) S is the characteristic function of a unitary colligation U which extends the isometry V which is defined in terms of the data and P by (3.3).

(2) $S(r)$ can be expressed in the form (5.1), where $\Sigma(r)$ is the characteristic function of the standard unitary colligation $A \in \mathcal{D}(\mathcal{H} \oplus \mathcal{E}_1 \oplus \mathcal{N}_2; \mathcal{H}^{(-1)} \oplus \mathcal{E}_2 \oplus \mathcal{N}_1)$ which is specified by (4.1) and $\Omega \in \mathcal{S}(\mathcal{N}_1;\mathcal{N}_2)$.

Proof. Let $\xi_j \in \mathcal{D}(\ell^2_{\mathcal{B}};\ell^2_{\mathcal{M}})$ and $W_j \in \mathcal{D}(\ell^2_{\mathcal{B}};\ell^2_{\mathcal{B}})$ be chosen for $j = n+1, n+2, \ldots$ so that $r_{sp}(ZW_j^*) \leq q < 1$ and the conclusions of Theorem 5.1 are in force. Next define $\eta_j \in \mathcal{D}(\ell^2_{\mathcal{B}};\ell^2_{\mathcal{N}})$ via the formula

$$\underline{\underline{p}}S^*\xi_j(I - ZW_j^*)^{-1}E = \eta_j(I - ZW_j^*)^{-1}E \tag{5.2}$$

for $E \in \mathcal{D}_2(\ell^2_{\mathcal{B}};\ell^2_{\mathcal{B}})$ (the calculation is justified in Theorem 7.2 of [ADD]) and set

$$C_3 = [\xi_{n+1}\ \xi_{n+2}\ \cdots] \ ,$$

$$C_4 = [\eta_{n+1}\ \eta_{n+2}\ \cdots] \ ,$$

$$A_3 = \mathrm{diag}\{W^*_{n+1}, W^*_{n+2}\cdots\} \ ,$$

and

$$\mathcal{K} = \bigoplus_{j=-\infty}^{\infty} \mathcal{K}_j \ ,$$

where

$$\mathcal{K}_j \ = \ \left\{ \begin{bmatrix} u_{n+1,j} \\ u_{n+2,j} \\ \vdots \end{bmatrix} \ : \ u_{n+i,j} \in \mathcal{B}_j \text{ for } i = 1, 2, \ldots \right\} .$$

Then $C_3 \in \mathcal{D}(\mathcal{K}; \mathcal{E}_2)$, $C_4 \in \mathcal{D}(\mathcal{K}; \mathcal{E}_1)$, $A_3 \in \mathcal{D}(\mathcal{K}; \mathcal{K})$, $r_{sp}(ZA_3) \le q < 1$ and, by construction, S is a solution of the BIP based on the data

$$\widetilde{C}_1 \ = \ [C_1 \ C_3] \ , \qquad\qquad \widetilde{C}_2 \ = \ [C_2 \ C_4]$$

$$\widetilde{M} \ = \ \begin{bmatrix} M & 0 \\ 0 & I_{\mathcal{K}} \end{bmatrix} \quad \text{and} \quad \widetilde{N} \ = \ \begin{bmatrix} N & 0 \\ 0 & A_3 \end{bmatrix} .$$

Therefore, by Theorem 2.2, the operator $\widetilde{P} \in \mathcal{D}(\mathcal{H} \oplus \mathcal{K}; \mathcal{H} \oplus \mathcal{K})$ which is defined by the formula

$$\langle \widetilde{P} D, E \rangle_{HS} = \left\langle \begin{bmatrix} I_{\mathcal{E}_2} & -S \\ -S^* & I_{\mathcal{E}_1} \end{bmatrix} \begin{bmatrix} \widetilde{C}_1 \\ \widetilde{C}_2 \end{bmatrix} (\widetilde{M} - Z\widetilde{N})^{-1} D, \begin{bmatrix} \widetilde{C}_1 \\ \widetilde{C}_2 \end{bmatrix} (\widetilde{M} - Z\widetilde{N})^{-1} E \right\rangle_{HS}$$

(where, as usual, Z is chosen to fit the space) is a solution of the generalized Lyapunov equation

$$\widetilde{M}^* \widetilde{P} \widetilde{M} - \widetilde{N}^* \widetilde{P}^{(1)} \widetilde{N} = \widetilde{C}_1^* \widetilde{C}_1 - \widetilde{C}_2^* \widetilde{C}_2 \ . \tag{5.3}$$

Further, since S is a solution of the augmented BIP based on P, it follows readily from the form of $\widetilde{C}_1, \widetilde{C}_2, \widetilde{M}$ and $\widetilde{N}$ that $\widetilde{P}$ is of the form

$$\widetilde{P} \ = \ \begin{bmatrix} P & * \\ * & * \end{bmatrix}$$

with respect to the block decomposition $\mathcal{H} \oplus \mathcal{K} \longrightarrow \mathcal{H} \oplus \mathcal{K}$.

Let $\widetilde{V}$ denote the isometry which is induced by the Lyapunov identity (5.3), i.e.,

$$\widetilde{V} : \begin{bmatrix} \widetilde{P}^{\frac{1}{2}} \widetilde{M} \\ \widetilde{C}_2 \end{bmatrix} y \longrightarrow \begin{bmatrix} (\widetilde{P}^{\frac{1}{2}})^{(1)} \widetilde{N} \\ \widetilde{C}_1 \end{bmatrix} y$$

and let

$$\widetilde{U} \in \mathcal{D}(\widetilde{\mathcal{G}} \oplus \mathcal{H} \oplus \mathcal{K} \oplus \mathcal{E}_1; \widetilde{\mathcal{G}}^{(-1)} \oplus \mathcal{H}^{(-1)} \oplus \mathcal{K}^{(-1)} \oplus \mathcal{E}_2)$$

be a unitary colligation which extends $\widetilde{V}$. Then, by Theorem 3.1, the characteristic function $W_{\widetilde{U}}$ of $\widetilde{U}$ is a solution of the BIP based on the data $\widetilde{C}_1, \widetilde{C}_2, \widetilde{M}$ and $\widetilde{N}$. Therefore, since S is the only solution of this problem by construction, we see that

$$S \ = \ \mathring{W}_{\widetilde{U}} \ .$$

The next step is to construct a unitary operator U which extends the isometry V with the same characteristic function as $\widetilde{U}$, so that

$$\mathring{W}_U = S\ .$$

We shall take advantage of Lemmas 3.1 and 3.2.

To this end, observe first that if $y \in \mathcal{H} \oplus \mathcal{K}$ is chosen to be of the form

$$y = \begin{bmatrix} x \\ 0 \end{bmatrix}$$

with $x \in \mathcal{H}$ and $0 \in \mathcal{K}$, then

$$\begin{bmatrix} \widetilde{P}^{\frac{1}{2}}\widetilde{M} \\ \widetilde{C}_2 \end{bmatrix} y = \begin{bmatrix} \widetilde{P}^{\frac{1}{2}} \begin{bmatrix} \widetilde{M}x \\ 0 \end{bmatrix} \\ C_2 x \end{bmatrix}$$

and

$$\left\| \widetilde{P}^{\frac{1}{2}} \begin{bmatrix} x \\ 0 \end{bmatrix} \right\| = \|P^{\frac{1}{2}}x\|\ .$$

Assume that

$$\widetilde{\mathcal{G}}_i \oplus \mathcal{H}_i \oplus \mathcal{K}_i \ominus \text{ closure} \left\{ \widetilde{P}_i^{\frac{1}{2}} \begin{bmatrix} x_i \\ 0 \end{bmatrix} \ : \ x_i \in \mathcal{H} \right\}$$

and

$$\widetilde{\mathcal{G}}_i \oplus \mathcal{H}_i \oplus \mathcal{K}_i \ominus \text{ closure} \left\{ P_i x_i \ : \ x_i \in \mathcal{H} \right\}$$

have the same dimension for every integer i. If not, then this can always be achieved by enlarging $\widetilde{\mathcal{G}}_i$ as in Lemma 3.2. Then without affecting the characteristic function of the resulting colligation, the operator which is defined on elements of the indicated form by the rule

$$\gamma : \ \widetilde{P}^{\frac{1}{2}} \begin{bmatrix} x \\ 0 \end{bmatrix} \longrightarrow P^{\frac{1}{2}}x$$

is isometric and extends to a unitary operator in $\mathcal{D}(\widetilde{\mathcal{G}} \oplus \mathcal{H} \oplus \mathcal{K}\ ,\ \widetilde{\mathcal{G}} \oplus \mathcal{H} \oplus \mathcal{K})$ which we shall continue to call γ. Let

$$U = \begin{bmatrix} \gamma^{(1)} & 0 \\ 0 & I_{\mathcal{E}_2} \end{bmatrix} \widetilde{U} \begin{bmatrix} \gamma^* & 0 \\ 0 & I_{\mathcal{E}_1} \end{bmatrix} .$$

Then U is clearly a unitary diagonal operator with $U|_{d_V} = V$. Therefore, by Lemma 3.1,

$$\mathring{W}_U = \mathring{W}_{\widetilde{U}} = S\ .$$

This completes the proof of the first assertion. The second assertion now follows from Theorems 4.1 and 4.2. □

The preceding analysis serves to prove that every contractive upper triangular operator is the characteristic function of a unitary colligation, at practically no extra cost:

Theorem 5.3. *Let $S \in \mathcal{S}(\mathcal{E}_1;\mathcal{E}_2)$. Then S is the characteristic function of a unitary colligation $U \in \mathcal{D}(\mathcal{G}\oplus\mathcal{E}_1;\mathcal{G}\oplus\mathcal{E}_2)$.*

Proof. Choose $\xi_j \in \mathcal{D}(\ell^2_{\mathcal{B}};\ell^2_{\mathcal{M}})$, and $W_j \in \mathcal{D}(\ell^2_{\mathcal{B}};\ell^2_{\mathcal{B}})$ with $r_{sp}(ZW_j^*) < 1$ for $j = n+1, n+2, \dots$, as in Theorem 5.1. and then define the $\eta_j \in \mathcal{D}(\ell^2_{\mathcal{B}};\ell^2_{\mathcal{N}})$ via (5.2) for $j = n+1, n+2, \dots$ as in Theorem 5.2. In the present setting $n = 0$ because S is not presumed to be the solution of an interpolation problem to begin with. Thus, in terms of the notation in Theorem 5.2, $\widetilde{C}_1 = C_3$, $\widetilde{C}_2 = C_4$, $\widetilde{M} = I_{\mathcal{K}}$ and $\widetilde{N} = A_3$ and $S = W_{\widetilde{U}}$, the characteristic function of the unitary colligation $\widetilde{U}$ which was introduced in the proof of that theorem. □

We are now ready to state and prove the main theorem of this paper:

Theorem 5.4. *An operator $S \in \mathcal{S}(\mathcal{E}_1;\mathcal{E}_2)$ is a solution of the augmented BIP if and only if it can be expressed in the form*

$$S(r) = \Sigma_{11}(r) + \Sigma_{12}(r)\left\{I - \Omega(r)\Sigma_{22}(r)\right\}^{-1}\Omega(r)\Sigma_{21}(r) \ ,$$

for $0 \le r < 1$, where

$$\Sigma = \begin{bmatrix} \Sigma_{11} & \Sigma_{12} \\ \Sigma_{21} & \Sigma_{22} \end{bmatrix}$$

is the characteristic function of the standard unitary colligation A which was defined by (4.1) and Ω is any contractive upper triangular operator in the class $\mathcal{S}(\mathcal{N}_1;\mathcal{N}_2)$.

Proof. Theorem 5.2 guarantees that every solution S of the augmented BIP can be expressed in the indicated form for some choice of $\Omega \in \mathcal{S}(\mathcal{N}_1;\mathcal{N}_2)$.

Conversely if S can be expressed in this form, then, by Theorem 5.3, Ω is the characteristic function of a unitary colligation $B \in \mathcal{D}(\mathcal{G}\oplus\mathcal{N}_1;\ \mathcal{G}^{(-1)}\oplus\mathcal{N}_2)$ for some "ℓ_2 space" $\mathcal{G}$. Therefore, since Σ is the characteristic function of the standard unitary colligation $A \in \mathcal{D}(\mathcal{H}\oplus\mathcal{E}_1\oplus\mathcal{N}_2;\ \mathcal{H}^{(-1)}\oplus\mathcal{E}_2\oplus\mathcal{N}_1)$, the right hand side of (5.4) is equal to the characteristic function W_U of the unitary colligation U which corresponds to the coupling of A and B. This does the trick since by Theorem 3.1, $\mathring{W}_U$ is a solution of the augmented BIP. □

Now, as an application of Theorem 5.3, we shall establish a generalization of the maximum principle and the Schwarz lemma in the setting of upper triangular operators. These results will be expressed in terms of the symbol $F(r)$ which is defined for each upper triangular operator $F \in \mathcal{U}(\ell^2_{\mathcal{M}};\ell^2_{\mathcal{N}})$ as the operator in $\mathcal{U}(\ell^2_{\mathcal{M}};\ell^2_{\mathcal{N}})$ with blocks

$$F(r)_{ij} = r^{j-i}F_{ij}$$

for $j \ge i$. This notation is consistent with the notation introduced in formula (3.4) because $\mathring{W}(r)$ in this new sense (for the operator $\mathring{W}$ which is defined by (3.10)) is equal to $W(r)$ as defined by (3.4), thanks to Lemma 3.6 and formula (3.10).

Theorem 5.5. *Let $F \in \mathcal{U}(\ell^2_{\mathcal{M}}; \ell^2_{\mathcal{N}})$. Then*

$$\|F(r)\| \;\leq\; \|F\| \tag{5.4}$$

for every $r \in [0,1)$. If $F \in \mathcal{U}'(\ell^2_{\mathcal{M}}; \ell^2_{\mathcal{N}})$, i.e., if F is strictly upper triangular, then

$$\|F(r)\| \;\leq\; r\|F\| \tag{5.5}$$

for every choice of $r \in [0,1)$.

Proof. Let $F \in \mathcal{U}(\ell^2_{\mathcal{M}}; \ell^2_{\mathcal{N}})$. Then without loss of generality we can assume that $\|F\| \leq 1$ and hence, by Theorem 5.3, that

$$F = \lim_{r\uparrow 1} W_U(r) \text{ (strongly)}$$

for some unitary colligation U with input space $\mathcal{E}_1 = \ell^2_{\mathcal{M}}$ and output space $\mathcal{E}_2 = \ell^2_{\mathcal{N}}$. Therefore

$$\|F(r)\| = \|W_U(r)\| \leq 1 \,,$$

by Lemma 3.3.

Suppose next that $F \in U'(\ell^2_{\mathcal{M}}; \ell^2_{\mathcal{N}})$. Then for any $x \in \ell^2_{\mathcal{M}}$,

$$\begin{aligned}
\|F(r)x\|^2 &= \sum_{i=-\infty}^{\infty} \left\| \sum_{j=i+1}^{\infty} F_{ij} r^{j-i} x_j \right\|^2 \\
&= r^2 \sum_{i=-\infty}^{\infty} \left\| \sum_{j=i+1}^{\infty} F_{ij} r^{j-i-1} x_j \right\|^2 \\
&= r^2 \sum_{i=-\infty}^{\infty} \left\| \sum_{j=i+1}^{\infty} (Z^*F)_{i+1,j} \cdot r^{j-1-i} x_j \right\|^2 \\
&= r^2 \|(Z^*F)(r)x\|^2 \\
&\leq r^2 \|Z^*F\|^2 \|x\|^2 \text{ (by 5.4)} \\
&= r^2 \|F\|^2 \|x\|^2 \,,
\end{aligned}$$

which serves to complete the proof. □

We remark that $F(r)$ is of the form $\Lambda(r)F\Lambda(r)^{-1}$, where $\Lambda(r)$ is a block diagonal operator with $\Lambda_{ii}(r) = r^i I_{e_i}$ and $e_i = \mathcal{M}_i$ for the Λ which is applied on the right and $e_i = \mathcal{N}_i$ for the Λ which is applied on the left. An interesting application of estimates similar to the last one may be found in Katsnelson [Ka].

6 Formulas for the standard unitary colligation A

In this section we shall derive formulas for the block entries in the unitary colligation A which is defined by (4.1) when $P \geq 0$, under the added assumption that

$$d_V = \left\{ \begin{bmatrix} P^{\frac{1}{2}}Mx \\ C_2x \end{bmatrix} : x \in \mathcal{H} \right\} \text{ and } \Delta_V = \left\{ \begin{bmatrix} Z^*P^{\frac{1}{2}}ZNx \\ C_1x \end{bmatrix} : x \in \mathcal{H} \right\} . \tag{6.1}$$

Since $A : \mathcal{H} \oplus \mathcal{E}_1 \oplus \mathcal{N}_2 \longrightarrow \mathcal{H}^{(-1)} \oplus \mathcal{E}_2 \oplus \mathcal{N}_1$ it is convenient to express A as a 3×3 matrix of operators:

$$A = \begin{bmatrix} A_{11} & A_{12} & A_{13} \\ A_{21} & A_{22} & A_{23} \\ A_{31} & A_{32} & A_{33} \end{bmatrix} : \begin{matrix} \mathcal{H} \\ \oplus \\ \mathcal{E}_1 \\ \oplus \\ \mathcal{N}_2 \end{matrix} \longrightarrow \begin{matrix} \mathcal{H}^{(-1)} \\ \oplus \\ \mathcal{E}_2 \\ \oplus \\ \mathcal{N}_1 \end{matrix} . \tag{6.2}$$

It is also convenient to let

$$T_0 = M^*PM + C_2^*C_2 = N^*Z^*PZN + C_1^*C_1 .$$

Lemma 6.1. *If $P \geq 0$ and assumption (6.1) is in force, then:*
(1) *T_0 maps $\mathcal{H}$ onto* range $[M^*P^{\frac{1}{2}} \; C_2^*]$
(2) *The orthogonal projection Π_{d_V} of $\mathcal{H} \oplus \mathcal{E}_1$ onto d_V is given by the formula*

$$\Pi_{d_V} = \begin{bmatrix} P^{\frac{1}{2}}M \\ C_2 \end{bmatrix} T_0^{[-1]}[M^*P^{\frac{1}{2}} \; C_2^*] , \tag{6.4}$$

where $T_0^{[-1]}$ denotes the Moore-Penrose inverse of T_0.

Proof. Let $h \in \mathcal{H}$ and $e_1 \in \mathcal{E}_1$. Then by assumption (6.1), there exists at least one $x \in \mathcal{H}$ such that

$$\Pi_{d_V} \begin{bmatrix} h \\ e_1 \end{bmatrix} = \begin{bmatrix} P^{\frac{1}{2}}M \\ C_2 \end{bmatrix} x .$$

This in turn implies that for such a choice of x

$$\begin{bmatrix} h \\ e_1 \end{bmatrix} - \begin{bmatrix} P^{\frac{1}{2}}M \\ C_2 \end{bmatrix} x \in d_V^{\perp} ,$$

and hence that

$$M^*P^{\frac{1}{2}}h + C_2^*e_1 = T_0x .$$

This serves to establish (1); (2) now follows easily. □

Similar considerations lead easily to the following lemma, which is stated without proof.

Lemma 6.2. *If $P \geq 0$ and assumption (6.1) is in force, then*

$$T_0 \text{ maps } \mathcal{H} \text{ onto range } [N^* Z^* P^{\frac{1}{2}} Z \; C_1^*]$$

and the orthogonal projection Π_{Δ_V} of $\mathcal{H}^{(-1)} \oplus \mathcal{E}_2$ onto Δ_V is given by the formula

$$\Pi_{\Delta_V} = \begin{bmatrix} Z^* P^{\frac{1}{2}} Z N \\ C_1 \end{bmatrix} T_0^{[-1]} [N^* Z^* P^{\frac{1}{2}} Z \; C_1^*] \,. \tag{6.5}$$

Theorem 6.1. *If $P \geq 0$ and assumption (6.1) is in force, then the entries A_{ij} in the block decomposition of the unitary colligation A which is defined by (4.1), are given by the following formulas:*

$$A_{11} = \Pi_{\mathcal{H}^{(-1)}} A|_{\mathcal{H}} = Z^* P^{\frac{1}{2}} Z N T_0^{[-1]} M^* P^{\frac{1}{2}} \,,$$

$$A_{21} = \Pi_{\mathcal{E}_1} A|_{\mathcal{H}} = C_1 T_0^{[-1]} M^* P^{\frac{1}{2}} \,,$$

$$A_{31} = \Pi_{\mathcal{N}_1} A|_{\mathcal{H}} = \alpha \begin{bmatrix} I_{\mathcal{H}} - P^{\frac{1}{2}} M T_0^{[-1]} M^* P^{\frac{1}{2}} \\ -C_2 T_0^{[-1]} M^* P^{\frac{1}{2}} \end{bmatrix} ,$$

$$A_{12} = \Pi_{\mathcal{H}^{(-1)}} A|_{\mathcal{E}_1} = Z^* P^{\frac{1}{2}} Z N T_0^{[-1]} C_2^* \,,$$

$$A_{22} = \Pi_{\mathcal{E}_2} A|_{\mathcal{E}_1} = C_1 T_0^{[-1]} C_2^* \,,$$

$$A_{32} = \Pi_{\mathcal{N}_1} A|_{\mathcal{E}_1} = \alpha \begin{bmatrix} -P^{\frac{1}{2}} M T_0^{[-1]} C_2^* \\ I_{\mathcal{E}_1} - C_2 T_0^{[-1]} C_2^* \end{bmatrix} ,$$

$$\begin{aligned} A_{13} &= \Pi_{\mathcal{H}^{(-1)}} A|_{\mathcal{N}_2} = \Pi_{\mathcal{H}^{(-1)}} \beta^* \\ &= [I_{\mathcal{H}^{(-1)}} - Z^* P^{\frac{1}{2}} Z N T_0^{[-1]} N^* Z^* P^{\frac{1}{2}} Z \mid - Z^* P^{\frac{1}{2}} Z N T_0^{[-1]} C_1^*] \beta^* \\ &= [I_{\mathcal{H}^{(-1)}} \;\; 0] \beta^* \,, \end{aligned}$$

$$\begin{aligned} A_{23} &= \Pi_{\mathcal{E}_2} A|_{\mathcal{N}_2} = \Pi_{\mathcal{E}_2} \beta^* \\ &= [-C_1 T_0^{[-1]} N^* Z^* P^{\frac{1}{2}} Z \mid I_{\mathcal{E}_2} - C_1 T_0^{[-1]} C_1^*] \beta^* \\ &= [0 \;\; I_{\mathcal{E}_2}] \beta^* \end{aligned}$$

and

$$A_{33} = \Pi_{\mathcal{N}_1} A|_{\mathcal{N}_2} = 0 \,.$$

Proof. Clearly

$$\begin{aligned} A_{11} &= \Pi_{\mathcal{H}^{(-1)}} A|_{\mathcal{H}} \\ &= \Pi_{\mathcal{H}^{(-1)}} A(\Pi_{d_V} + \Pi_{d_V^\perp})|_{\mathcal{H}} \\ &= \Pi_{\mathcal{H}^{(-1)}} V \Pi_{d_V}|_{\mathcal{H}} \,. \end{aligned}$$

Therefore, by Lemma 6.1,

$$\begin{aligned} A_{11}h &= \Pi_{\mathcal{H}^{(-1)}} V \Pi_{d_V} \begin{bmatrix} h \\ 0 \end{bmatrix} \\ &= \Pi_{\mathcal{H}^{(-1)}} \begin{bmatrix} Z^* P^{\frac{1}{2}} Z N \\ C_1 \end{bmatrix} T_0^{[-1]} M^* P^{\frac{1}{2}} h \,, \end{aligned}$$

which yields the stated formula for A_{11}. The formula for A_{21} is obtained in much the same way. Next, by another application of 6.1,

$$\begin{aligned} A_{31} &= \Pi_{\mathcal{N}_1} A|_{\mathcal{H}} \\ &= \alpha \Pi_{d_V^\perp}|_{\mathcal{H}} \\ &= \alpha \left\{ \begin{bmatrix} I_{\mathcal{H}} \\ 0 \end{bmatrix} - \begin{bmatrix} P^{\frac{1}{2}} M \\ C_2 \end{bmatrix} T_0^{[-1]} M^* P^{\frac{1}{2}} \right\} \,, \end{aligned}$$

as claimed. The verification of the formulas for the entries in the next column is much the same. Thus, e.g.,

$$\begin{aligned} A_{12} &= \Pi_{\mathcal{H}^{(-1)}} A|_{\mathcal{E}_1} \\ &= \Pi_{\mathcal{H}^{(-1)}} A(\Pi_{d_V} + \Pi_{d_V^\perp})|_{\mathcal{E}_1} \\ &= \Pi_{\mathcal{H}^{(-1)}} V \Pi_{d_V}|_{\mathcal{E}_1} \\ &= \Pi_{\mathcal{H}^{(-1)}} V \begin{bmatrix} P^{\frac{1}{2}} M \\ C_2 \end{bmatrix} T_0^{[-1]} C_2^* \,, \end{aligned}$$

whereas

$$\begin{aligned} A_{32} &= \Pi_{\mathcal{N}_1} A|_{\mathcal{E}_1} \\ &= \Pi_{\mathcal{N}_1} A(\Pi_{d_V} + \Pi_{d_V^\perp})|_{\mathcal{E}_1} \\ &= \alpha \Pi_{d_V^\perp}|_{\mathcal{E}_1} \\ &= \alpha \left\{ \begin{bmatrix} I_{\mathcal{H}} & 0 \\ 0 & I_{\mathcal{E}_1} \end{bmatrix} - \Pi_{d_V} \right\} \Bigg|_{\mathcal{E}_1} \,. \end{aligned}$$

Finally, the stated formulas for A_{13} and A_{23} follow easily from Lemma 6.2 and the observation that

$$A_{13}^* = \Pi_{\mathcal{N}_2} A^*|_{\mathcal{H}^{(-1)}}$$

$$= \Pi_{\mathcal{N}_2} A^* (\Pi_{\Delta_V} + \Pi_{\Delta_V^\perp})|_{\mathcal{H}^{(-1)}}$$
$$= \Pi_{\mathcal{N}_2} A^* \Pi_{\Delta_V^\perp}|_{\mathcal{H}^{(-1)}}$$
$$= \beta \left\{ \begin{bmatrix} I_{\mathcal{H}^{(-1)}} & 0 \\ 0 & I_{\mathcal{E}_2} \end{bmatrix} - \Pi_{\Delta_V} \right\} \Bigg|_{\mathcal{H}^{(-1)}}$$

and

$$A_{23}^* = \Pi_{\mathcal{N}_2} A^*|_{\mathcal{E}_2} = \beta \left\{ \begin{bmatrix} I_{\mathcal{H}^{(-1)}} & 0 \\ 0 & I_{\mathcal{E}_2} \end{bmatrix} - \Pi_{\Delta_V} \right\} \Bigg|_{\mathcal{E}_2},$$

whereas $A_{33} = 0$, by the definition of A. □

The characteristic function of the unitary colligation A can be written as follows, with respect to the decomposition (6.2):

$$\Sigma(r) = \begin{bmatrix} A_{22} & A_{23} \\ A_{32} & A_{33} \end{bmatrix} + \begin{bmatrix} A_{21} \\ A_{31} \end{bmatrix} (I - rZA_{11})^{-1} [rZA_{12} \; rZA_{13}] \tag{6.6}$$
$$: \begin{bmatrix} \mathcal{E}_1 \\ \mathcal{N}_2 \end{bmatrix} \longrightarrow \begin{bmatrix} \mathcal{E}_2 \\ \mathcal{N}_1 \end{bmatrix}.$$

It is important to bear in mind that it is possible for either $d_V^\perp$ or $\Delta_V^\perp$ or even both of these spaces to be trivial. If $\mathcal{N}_1 = \{0\}$, then $A_{31} = A_{32} = A_{33} = 0$; if $\mathcal{N}_2 = \{0\}$, then $A_{13} = A_{23} = A_{33} = 0$. In both of these cases the augmented BIP has a unique solution (since in formula (4.9) we must have $\Omega = 0$). In the sequel we shall consider the augmented BIP under the added assumption that the Pick operator P is strictly positive. In this case the spaces d_V and Δ_V satisfy assumption (6.1). Moreover. $d_V^\perp$ and $\Delta_V^\perp$ are both nontrivial and there are infinitely many solutions to the augmented BIP.

It is also important to note that although we cannot always let $r = 1$ in formula (6.6), we can let $r = 1$ in each term

$$\begin{bmatrix} A_{21} \\ A_{22} \end{bmatrix} (rZA_{11})^k [rZA_{12} \; rZA_{13}]$$

of the power series for $\Sigma(r)$. This serves to determine the "k-th diagonal" of $\mathring{\Sigma}$.

Finally we remark that the circle of ideas discussed in this paper is applicable to interpolation problems with infinitely many interpolation conditions [Fr].

Added in proof. We remark that a byproduct of the preceding analysis is that if $S \in \mathcal{U}(\ell^2_{\mathcal{N}}; \ell^2_{\mathcal{M}})$, then

$$\sup\{\|\underline{\underline{q}}SA\|_{HS} : A \in \mathcal{L}_2 \,,\; \|A\|_{HS} \le 1\} = \sup\{\|Sa\|_{\ell^2_{\mathcal{M}}} : a \in \ell^2_{\mathcal{N}} \,,\; \|a\|_{\ell^2_{\mathcal{N}}} \le 1\} \,,$$

or, in terms of the operator M_S of multiplication by S on $\mathcal{X}_2$, $\|(M_{S^*}|_{\mathcal{L}_2})^*\| = \|(M_S|_{\mathcal{U}_2})^*\| = \|S\|$. This is easily verified directly.

References

[ADD] D. Alpay, P. Dewilde and H. Dym, Lossless inverse scattering and reproducing kernels for upper triangular operators, in: *Extension and Interpolation of Linear Operators and Matrix Functions* (I. Gohberg, ed.), *Operator Theory: Advances and Applications*, **OT47**, Birkhäuser Verlag, Basel, 1990, pp. 61–135.

[ArG1] D.Z. Arov and L.Z Grossman, Scattering matrices in the theory of unitary extensions of isometric operators, *Soviet Math. Dokl.* **270** (1983), 17–20.

[ArG2] D.Z. Arov and L.Z Grossman, Scattering matrices in the theory of unitary extensions of isometric operators, *Math. Nachr.* **157** (1992), 105–123.

[Ba] J.A. Ball, Commutant lifting and interpolation: the time varying case, *Integral Equations Operator Theory* **25** (1996), 377–405.

[BGK1] J.A. Ball, I. Gohberg and M.A. Kaashoek, Nevanlinna-Pick interpolation for time-varying input-output maps: The discrete case, in: *Time-variant Systems and Interpolation* (I. Gohberg, ed.), *Operator Theory: Advances and Applications*, **OT56**, Birkhäuser Verlag, Basel, 1992, pp. 1–51.

[BGK2] J.A. Ball, I. Gohberg and M.A. Kaashoek, Two-sided Nudelman interpolation for input-output operators of discrete time-varying systems, *Integral Equations Operator Theory* **21** (1994), 174-211.

[DD] P. Dewilde and H. Dym, Interpolation for upper triangular operators, in: *Time-variant Systems and Interpolation* (I. Gohberg, ed.), *Operator Theory: Advances and Applications*, **OT56**, Birkhäuser Verlag, Basel, 1992, pp. 153–260.

[DvV] P. Dewilde and A. van der Veen, On the Hankel-norm approximation of upper triangular operators and matrices, *Integral Equations Operator Theory*, **17** (1993), 1–45.

[D1] H. Dym, Remarks on interpolation for upper triangular operators, in: *Challenges of a Generalized System Theory*, (P. Dewilde, M.A. Kaashoek and M. Verhaegen, eds.), North-Holland, Amsterdam, 1993, 9–24.

[D2] H. Dym, Shifts, realizations and interpolation, Redux, in: *Nonselfadjoint Operators and Related Topics* (A. Feintuch and I. Gohberg, eds.), *Operator Theory: Advances and Applications* **OT73**, Birkhäuser Verlag, Basel, 1994, pp. 182–243.

[D3] H. Dym, A basic interpolation problem, in: *Holomorphic Spaces* (S. Axler, J. McCarthy and D. Sarason, eds.), Cambridge University Press, in preparation.

[Fr] B. Freydin, BIP with infinitely many interpolation nodes, Preprint.

[Ka] V.E. Katsnelson, Remark on canonic factorization in certain analytic function spaces, *J. Soviet Math.* **4** (1975), 444–445.

[KKY] V.E. Katsnelson, A.Ya. Kheifets and P.M. Yuditskii, An abstract interpolation problem and the theory of extensions of isometric operators, in: *Operators in Function Spaces and Problems in Function Theory* (V.A. Marchenko, ed.), **146**, *Naukova Dumka*, Kiev, 1987, pp. 83–96. (For an English translation, see this volume.)

[Kh1] A.Ya. Kheifets, Parseval equality in abstract interpolation problems and coupling of open systems, *J. of Soviet Math.*, **49**, No. 4 (1990), 1114–1120; **49**, No. 6 (1990), 1307–1310.

[Kh2] A.Ya. Kheifets, The generalized bitangential Schur-Nevanlinna-Pick problem and the related Parseval equality, *J. of Soviet Math.*, **58**, No. 4 (1992), 358–364.

[KY] A.Ya. Kheifets and P.M. Yuditskii, An analysis and extension of V.P. Potapov's approach to interpolation problems with applications to the generalized bitangential Schur-Nevanlinna-Pick problem and J-inner-outer factorization, in: *Matrix and Operator Valued Functions* (I. Gohberg and L.A. Sakhnovich, eds.), *Operator Theory: Advances and Application* **OT72**, Birkhäuser Verlag, Basel, 1994, pp. 133–161

[Ko1] J. Kos, Higher order time-varying Nevanlinna-Pick interpolation, in: *Challenges of a Generalized System Theory,* (P. Dewilde, M.A. Kaashoek, and M. Verhaegen, eds.), North Holand, Amsterdam, 1993, pp. 59–72.

[Ko2] J. Kos, *Time-dependent Problems in Linear Operator Theory*, Ph.D. Thesis, Amsterdam (1995).

[PP] V.I. Paulsen and S. Power, Lifting theorem for nest algebras, *J. Funct. Anal.* **80** (1980), 76–87.

[SzNF] B. Sz.-Nagy and C. Foias, *Harmonic Analysis of Operator on Hilbert Space*, North Holland, Amsterdam, 1970.

[vV] A. van der Veen, *Time-varying Theory and Computational Modeling*, Ph.D. Thesis, Delft (1993).

Harry Dym
Dept. of Theoretical Mathematics
The Weizmann Institute of Science
Rehovot 76100, Israel

E-mail: dym@wisdom.weizmann.ac.il

Boris Freydin
Dept. of Theoretical Mathematics
The Weizmann Institute of Science
Rehovot 76100, Israel

E-mail: fred@wisdom.weizmann.ac.il

AMS Subject Classification: 47A57, 47A48

Operator Theory
Advances and Applications, Vol. 95
© 1997 Birkhäuser Verlag Basel/Switzerland

Bitangential interpolation for triangular operators when the Pick operator is strictly positive

Harry Dym* and Boris Freydin

Abstract. This paper is a sequel to [DF]. It includes a number of applications and explicit formulas which are developed under the extra assumption that the Pick operator P which is associated with the basic interpolation problem under study is strictly positive (i.e., $P \geq \varepsilon I$ for some $\varepsilon > 0$) rather than positive semidefinite. In order to both minimize the length and simplify the referencing, we shall refer to the six sections of [DF] as if they were an integral part of this paper and shall begin this paper with Section 7. The applications include a discussion of chain scattering operators, a maximum entropy problem and a bitangential interpolation problem in the class of upper triangular operators with strictly positive real part.

7 Strictly positive Pick operators

In the preceding six sections (i.e., in [DF]) we obtained a representation for the set of all solutions to a two sided interpolation problem in the setting of upper triangular operators when the underlying Pick operator was positive semidefinite. This representation was obtained by extending some of the methods which were introduced by Katsnelson, Kheifets and Yuditskii [KKY] and further refined in [Kh1], [Kh2] and [KY] to this setting.

In this section we consider the augmented BIP under the added assumption that it is based on a solution $P \in \{\mathcal{D}(\ell^2_{\mathcal{B}}; \ell^2_{\mathcal{B}})\}^{n\times n}$ of the Lyapunov–Stein equation (2.9) which is strictly positive, i.e., we assume that $P \geq \varepsilon I$ for some $\varepsilon > 0$. Under this assumption, the characteristic function $\Sigma(r)$ of the standard unitary colligation A which was introduced in Section 4 is defined for all r, $0 \leq r \leq 1$, since (as will be shown below) $r_{sp}(ZA_{11}) < 1$. Moreover, $\mathring{\Sigma} = \Sigma(1)$ is unitary, $\Sigma_{12}(r)$ and $\Sigma_{21}(r)$ are invertible for r close to one and hence $\Sigma_{11}(1)$ and $\Sigma_{22}(1)$ are strictly contractive.

---

*) The author wishes to express his thanks to Renee and Jay Weiss for endowing the chair which supports his research.

7.1 Preliminaries

The first step is to show that the extra assumption implies that the real part of the operator

$$T_\omega = M^*P(M - \omega ZN) + C_2^*C_2 = (N^*Z^* - \omega M^*)PZN + C_1^*C_1 \qquad (7.1)$$

is strictly positive for every point $\omega \in \overline{\mathbb{D}}$. We begin with T_0.

Lemma 7.1. *If $P \geq \varepsilon I$, then $T_0 \geq (\varepsilon/2)I$.*

Proof. Let $x \in \{\ell_{\mathcal{B}}^2\}^\mu$ and $y \in \{\ell_{\mathcal{B}}^2\}^\nu$. Then (because of the special form of M and N; see Section 2)

$$\left\| M \begin{bmatrix} x \\ y \end{bmatrix} \right\|^2 = \|x\|^2 + \|A_2 y\|^2 \geq \|x\|^2 ,$$

whereas

$$\left\| N \begin{bmatrix} x \\ y \end{bmatrix} \right\|^2 = \|A_1 x\|^2 + \|y\|^2 \geq \|y\| .$$

Thus,

$$\begin{aligned} 2\langle T_0 \begin{bmatrix} x \\ y \end{bmatrix}, \begin{bmatrix} x \\ y \end{bmatrix} \rangle &\geq \langle PM \begin{bmatrix} x \\ y \end{bmatrix}, M \begin{bmatrix} x \\ y \end{bmatrix} \rangle + \langle P^{(1)} N \begin{bmatrix} x \\ y \end{bmatrix}, N \begin{bmatrix} x \\ y \end{bmatrix} \rangle \\ &\geq \varepsilon\{\|x\|^2 + \|y\|^2\} , \end{aligned}$$

which does the trick. □

The next lemma extends Theorem 4.1 of [D3]. (The operator T_1 in the present setting corresponds to $\Delta_0(\lambda)$ in [D3].)

Lemma 7.2. *If $P \geq \varepsilon I$ for some $\varepsilon > 0$, then there exists a finite positive constant γ_1 (which is independent of ω) such that*

$$T_\omega + T_\omega^* \geq \gamma_1 I \qquad (7.2)$$

and

$$\|T_\omega^{-1}\| \leq 2/\gamma_1 \qquad (7.3)$$

for every point $\omega \in \overline{\mathbb{D}}$.

Proof. It follows readily from the Lyapunov-Stein identity that

$$\begin{aligned} T_\omega + T_\omega^* &= (M - \omega ZN)^*P(M - \omega ZN) + M^*PM + 2C_2^*C_2 - |\omega|^2 N^*Z^*PZN \\ &= (M - \omega ZN)^*P(M - \omega ZN) + (1 - |\omega|^2)T_0 + |\omega|^2 C_1^*C_1 + C_2^*C_2 . \end{aligned}$$

Thus, in view of Lemma 7.1, this clearly implies that

$$T_\omega + T_\omega^* \geq (1-|\omega|^2)T_0$$
$$\geq (1-r_0^2)(\varepsilon/2)I$$

for $|\omega| \leq r_0 < 1$. Because of the special form of M and N and the assumption that $r_{sp}(Z_1A_1) < 1$ and $r_{sp}(Z_2^*A_2) < 1$, there exists a choice of r_0, $0 < r_0 < 1$, and a finite positive constant γ such that

$$\|(M-\omega ZN)^{-1}\| \leq \gamma \text{ for } 1-r_0 \leq |\omega| \leq 1 \ .$$

Therefore,

$$\langle (T_\omega + T_\omega^*)x, x\rangle \geq \langle P(M-\omega ZN)x, (M-\omega ZN)x\rangle$$
$$\geq \varepsilon\|(M-\omega ZN)x\|^2$$
$$\geq (\varepsilon/\gamma)\|x\|^2$$

for $1-r_0 \leq |\omega| \leq 1$ and $x \in \{\ell_{\mathcal{B}}^2\}^n$. The lower bound (7.2) now drops out easily upon combining estimates and setting

$$\gamma_1 = \min\{(1-r_0^2)(\varepsilon/2)\ ,\ \varepsilon/\gamma\} \ .$$

Finally, the inequality (7.3) follows from (7.2) by a standard argument. □

The next lemma is a straightforward application of the well known fact that if E, F and G are bounded operators between appropriately defined spaces such that E, and $E-FG$ are invertible, then $I - GE^{-1}F$ is invertible and

$$(I - GE^{-1}F)^{-1} = I + G(E-FG)^{-1}F \ .$$

Lemma 7.3. *If $P \geq \varepsilon I$ for some $\varepsilon > 0$, then, for every choice of $\omega \in \overline{\mathbb{D}}$, the operator $I - \omega P^{\frac{1}{2}}ZNT_0^{-1}M^*P^{\frac{1}{2}}$ has a (uniformly) bounded inverse which is given by the formula*

$$\{I - \omega P^{\frac{1}{2}}ZNT_0^{-1}M^*P^{\frac{1}{2}}\}^{-1} = I + \omega P^{\frac{1}{2}}ZNT_\omega^{-1}M^*P^{\frac{1}{2}} \ . \tag{7.4}$$

Proof. With the help of the identity

$$T_\omega = T_0 - \omega M^*PZN \ , \tag{7.5}$$

it is readily checked by direct calculation that

$$(I - \omega P^{\frac{1}{2}}ZNT_0^{-1}M^*P^{\frac{1}{2}})(I + \omega P^{\frac{1}{2}}ZNT_\omega^{-1}M^*P^{\frac{1}{2}})$$
$$= (I + \omega P^{\frac{1}{2}}ZNT_\omega^{-1}M^*P^{\frac{1}{2}})(I - \omega P^{\frac{1}{2}}ZNT_0^{-1}M^*P^{\frac{1}{2}})$$
$$= I$$

for every choice of $\omega \in \overline{\mathbb{D}}$. This proves that the operator of interest is invertible and serves to establish formula (7.4). Lemma 7.2 guarantees that the inverse is uniformly bounded for $\omega \in \overline{\mathbb{D}}$. □

7.2 Calculation of the characteristic function of A

In this subsection we shall calculate explicit formulas for the blocks $\Sigma_{ij}(\omega)$ of the characteristic function $\Sigma(\omega)$ of the unitary colligation $A \in \mathcal{D}(\mathcal{H}\oplus\mathcal{E}_1\oplus\mathcal{N}_2; \mathcal{H}^{(-1)}\oplus \mathcal{E}_2\oplus\mathcal{N}_1)$ (which was introduced in Section 4) for general points $\omega \in \overline{\mathbb{D}}$. The main cases of interest are when $\omega = r$ and $0 \le r \le 1$. Indeed, the operators $\Sigma(\omega)$ and $\Sigma(r)$ are unitarily equivalent when $|\omega| = r$. We shall make extensive use of Theorem 6.1.

In view of Lemma 7.1, the assumption that $P \ge \varepsilon I$ for some $\varepsilon > 0$, guarantees that $T_0 \ge (\varepsilon/2)I$ and hence that assumption (6.1) is in force. Thus Theorem 6.1 is applicable. Moreover, in all the formulas for the blocks of A, the Moore-Penrose inverse $T_0^{[-1]}$ is just the ordinary inverse T_0^{-1}.

Lemma 7.4. *If $P \ge \varepsilon I$, then $r_{sp}(ZA_{11}) < 1$ and for every choice of $e_1 \in \mathcal{E}_1$, $n_2 \in \mathcal{N}_2$ and $\omega \in \overline{\mathbb{D}}\setminus\{0\}$, there exists a unique choice of $h_\omega \in \mathcal{H}$, $e_2(\omega) \in \mathcal{E}_2$ and $n_1(\omega) \in \mathcal{N}_1$ such that*

$$A\begin{bmatrix} h_\omega \\ e_1 \\ n_2 \end{bmatrix} = \begin{bmatrix} Z^*h_\omega/\omega \\ e_2(\omega) \\ n_1(\omega) \end{bmatrix}. \tag{7.6}$$

Moreover, the "transfer function"

$$\Sigma(\omega) = \begin{bmatrix} \Sigma_{11}(\omega) & \Sigma_{12}(\omega) \\ \Sigma_{21}(\omega) & \Sigma_{22}(\omega) \end{bmatrix} : \begin{bmatrix} e_1 \\ n_2 \end{bmatrix} = \begin{bmatrix} e_2(\omega) \\ n_1(\omega) \end{bmatrix}$$

is given by the formula

$$\begin{bmatrix} \Sigma_{11}(\omega) & \Sigma_{12}(\omega) \\ \Sigma_{21}(\omega) & \Sigma_{22}(\omega) \end{bmatrix} = \begin{bmatrix} A_{22} & A_{23} \\ A_{32} & 0 \end{bmatrix} + \begin{bmatrix} A_{21} \\ A_{31} \end{bmatrix} (I-\omega ZA_{11})^{-1}[\omega ZA_{12} \;\; \omega ZA_{13}]\,. \tag{7.7}$$

It is upper triangular for every choice of $\omega \in \overline{\mathbb{D}}$, contractive for $\omega \in \mathbb{D}$ and unitary for $\omega \in \mathbb{T}$.

Proof. By Lemma 7.3 and the formula for A_{11} which is given in Theorem 6.1, the operator $I - \omega ZA_{11}$ has a bounded inverse for every choice of $\omega \in \overline{\mathbb{D}}$. Therefore, since $\|ZA_{11}\| \le 1$, $r_{sp}(ZA_{11}) < 1$, and hence the system equation (7.6) is uniquely solvable, and $\Sigma(\omega)$ can be expressed in the form (7.7) by standard calculations. Since A is unitary,

$$\begin{aligned} \|e_2(\omega)\|^2 + \|n_1(\omega)\|^2 &= \|e_1\|^2 + \|n_2\|^2 - \left(\frac{1}{|\omega|^2} - 1\right)\|h_\omega\|^2 \\ &\le \|e_1\|^2 + \|n_2\|^2 \end{aligned}$$

for $0 < |\omega| \le 1$, with equality if $|\omega| = 1$. This proves that

$$\|\Sigma(\omega)\| \le 1$$

for $\omega \in \mathbb{D}$ and that $\Sigma(\omega)$ is isometric for $|\omega| = 1$.

A similar calculation based on the fact that the transfer function of the adjoint equation

$$A^* \begin{bmatrix} h_\omega \\ e_1 \\ n_2 \end{bmatrix} = \begin{bmatrix} Zh_\omega/\omega^* \\ e_2(\omega) \\ n_1(\omega) \end{bmatrix}$$

is equal to $\Sigma(\omega)^*$ leads to the supplementary conclusion that $\Sigma(\omega)^*$ is also isometric for $\omega \in \mathbb{T}$. Therefore $\Sigma(\omega)$ is unitary for $\omega \in \mathbb{T}$. □

Formula (7.7) serves to identify $\Sigma(\omega)$ with the characteristic function of the unitary colligation A. It is now relatively easily to compute the block entries in $\Sigma(\omega)$ in terms of the data of the augmented BIP.

Lemma 7.5. *If $P \geq \varepsilon I$, then*

$$\Sigma_{11}(\omega) = C_1 T_\omega^{-1} C_2^*$$

$$\Sigma_{12}(\omega) = [C_1 T_\omega^{-1}(\omega M^* - N^* Z^*) P^{\frac{1}{2}} Z \mid I_{\mathcal{E}_2} - C_1 T_\omega^{-1} C_1^*]\beta^*$$

$$= [C_1 T_\omega^{-1} \omega M^* P^{\frac{1}{2}} Z \; I_{\mathcal{E}_2}]\beta^*$$

$$\Sigma_{21}(\omega) = \alpha \begin{bmatrix} -P^{\frac{1}{2}}(M - \omega Z N) T_\omega^{-1} C_2^* \\ I_{\mathcal{E}_1} - C_2 T_\omega^{-1} C_2^* \end{bmatrix}$$

$$\Sigma_{22}(\omega) = \alpha \begin{bmatrix} I_{\mathcal{H}} - P^{\frac{1}{2}}(M - \omega Z N) T_\omega^{-1} M^* P^{\frac{1}{2}} \\ -C_2 T_\omega^{-1} M^* P^{\frac{1}{2}} \end{bmatrix} \omega Z [I_{\mathcal{H}^{(-1)}} \; 0]\beta^*$$

for $\omega \in \overline{\mathbb{D}}$.

Proof. The proofs are fairly straightforward consequences of substituting the formulas for the A_{ij} which were obtained in Theorem 6.1 into formula (7.7) for the blocks of $\Sigma(\omega)$.

The evaluations

$$T_0^{-1} M^* P^{\frac{1}{2}} (I - \omega P^{\frac{1}{2}} Z N T_0^{-1} M^* P^{\frac{1}{2}})^{-1} = T_\omega^{-1} M^* P^{\frac{1}{2}} \tag{7.8}$$

and

$$(I - \omega P^{\frac{1}{2}} Z N T_0^{-1} M^* P^{\frac{1}{2}})^{-1} P^{\frac{1}{2}} Z N T_0^{-1} = P^{\frac{1}{2}} Z N T_\omega^{-1} \tag{7.9}$$

are useful for calculating $\Sigma_{12}(\omega)$ and $\Sigma_{21}(\omega)$, respectively, for $\omega \in \overline{\mathbb{D}}$; the second formula for $\Sigma_{12}(\omega)$ rests on the fact that

$$[N^* Z^* P^{\frac{1}{2}} Z \; C_1^*]\beta^* = 0 ,$$

since β^* maps $\mathcal{N}_2$ onto $\Delta_V^{\perp}$. Formula (7.4) comes into play in the calculation of $\Sigma_{22}(\omega)$. □

As a partial confirmation of these formulas, it is readily checked that

$$[M^* P^{\frac{1}{2}} \; C_2^*]\alpha^* \Sigma_{21}(\omega) = 0$$

and hence that $\alpha^* \Sigma_{21}(\omega)$ maps $\mathcal{E}_1$ into $d_V^{\perp}$; see Section 4 for the definition of $d_V^{\perp}$.

Lemma 7.6. *If $P \geq \varepsilon I$, for some $\varepsilon > 0$, then the operator $\Sigma_{21}(\omega)$ is a contractive invertible map from $\mathcal{E}_1$ onto $\mathcal{N}_1$ for every $\omega \in \overline{\mathbb{D}}$ at which the operator $M - \omega ZN$ has a bounded inverse. Moreover, the operator $\Sigma_{21}(\omega)^{-1}$ is uniformly bounded in every subset of $\overline{\mathbb{D}}$ in which the operator $(M - \omega ZN)^{-1}$ is uniformly bounded and is given by the formula*

$$\Sigma_{21}(\omega)^{-1}\alpha = \left[-C_2(M - \omega ZN)^{-1}P^{-\frac{1}{2}} \; I_{\mathcal{E}_1}\right]\alpha \,. \tag{7.10}$$

Proof. It suffices to show that $\alpha^*\Sigma_{21}(\omega)$ is a bounded invertible map from $\mathcal{E}_1$ onto $\mathcal{H} \oplus \mathcal{E}_1 \ominus d_V$.

Suppose first that $\alpha^*\Sigma_{21}(\omega)e_1 = 0$ for some $e_1 \in \mathcal{E}_1$. Then

$$P^{\frac{1}{2}}(M - \omega ZN)T_\omega^{-1}C_2^*e_1 = 0$$

and

$$C_2T_\omega^{-1}C_2^*e_1 = e_1 \,.$$

But this clearly shows that $e_1 = 0$ and hence that ker $\Sigma_{21}(\omega) = 0$ when $(M - \omega ZN)$ is invertible.

Finally, we need to show that $\alpha^*\Sigma_{21}(\omega)$ maps $\mathcal{E}_1$ onto $d_V^\perp$, i.e., that for every choice of $\begin{bmatrix} b_1 \\ b_2 \end{bmatrix}$ in $d_V^\perp$ there exists a choice of $e_1 \in \mathcal{E}_1$ such that

$$\begin{bmatrix} b_1 \\ b_2 \end{bmatrix} = \begin{bmatrix} -P^{\frac{1}{2}}(M - \omega ZN)T_\omega^{-1}C_2^*e_1 \\ e_1 - C_2T_\omega^{-1}C_2^*e_1 \end{bmatrix} .$$

Now if $M - \omega ZN$ is invertible, then the top entry in the last formula suggests that

$$T_\omega^{-1}C_2^*e_1 = -(M - \omega ZN)^{-1}P^{-\frac{1}{2}}b_1$$

and hence, upon substituting this into the second entry, that

$$e_1 = \left[-C_2(M - \omega ZN)^{-1}P^{-\frac{1}{2}} \,\vert\, I_{\mathcal{E}_1}\right]\begin{bmatrix} b_1 \\ b_2 \end{bmatrix} .$$

It remains only to check that this rule works, i.e., if $\begin{bmatrix} b_1 \\ b_2 \end{bmatrix} \in d_V^\perp$ and e_1 is defined by the last formula, then

$$\alpha^*\Sigma_{21}(\omega)e_1 = \begin{bmatrix} b_1 \\ b_2 \end{bmatrix} .$$

But this is a straightforward calculation which uses the fact that

$$M^*P^{\frac{1}{2}}b_1 + C_2^*b_2 = 0 \,.$$

This completes the proof that $\alpha^*\Sigma_{21}(\omega)$ is a one to one map of $\mathcal{E}_1$ onto $d_V^\perp$. Therefore, since both of these spaces are closed, a well known theorem of Banach guarantees that the operator $\alpha^*\Sigma_{21}(\omega)$ has a bounded inverse.

Finally, if $(M - \omega ZN)^{-1}$ is uniformly bounded on some subset Δ of $\overline{\mathbb{D}}$, then there exists a positive constant γ_1 which is independent of ω such that

$$\|\Sigma_{21}(\omega)e_1\|^2 \geq \gamma_1 \|C_2^* e_1\|^2 + \|(I - C_2 T_\omega^{-1} C_2^*)e_1\|^2$$

for every choice of $e_1 \in \mathcal{E}_1$ and $\omega \in \Delta$. But this in turn implies that $\|\Sigma_{21}(\omega)^{-1}\|$ is uniformly bounded on Δ, by elementary estimates. □

Lemma 7.7. *If $P \geq \varepsilon I$, then the operator $\Sigma_{12}(\omega)$ is a contractive invertible map from $\mathcal{N}_2$ onto $\mathcal{E}_2$ for every point $\omega \in \overline{\mathbb{D}}$ at which the operator $\omega^* M - ZN$ has a bounded inverse. Moreover, the operator $\Sigma_{12}(\omega)^{-1}$ is uniformly bounded in every subset of $\overline{\mathbb{D}}$ in which the operator $(\omega^* M - ZN)^{-1}$ is uniformly bounded.*

Proof. The first step is to show that

$$\beta^* \Sigma_{12}(\omega)^* \;=\; \begin{bmatrix} Z^* P^{\frac{1}{2}} (\omega^* M - ZN) T_\omega^{-1} C_1^* \\ I - C_1 T_\omega^{-1} C_1^* \end{bmatrix}$$

is a contractive invertible map from $\mathcal{E}_2$ onto $d_V^{\perp}$ for every point $\omega \in \overline{\mathbb{D}}$ at which the operator $\omega^* M - ZN$ is invertible. But this goes through in much the same way as the proof of Lemma 7.6, as do the remaining assertions. □

Lemma 7.8. *If $P \geq \varepsilon I$ for some $\varepsilon > 0$, then*

$$\|\Sigma_{22}(\omega)\| \leq \gamma < 1 \tag{7.11}$$

for every point $\omega \in \overline{\mathbb{D}}$.

Proof. The bound on $\Sigma_{22}(\omega)$ for $\omega \in \mathbb{T}$ follows from the equality

$$\Sigma_{22}(\omega)^* \Sigma_{22}(\omega) = I - \Sigma_{12}(\omega)^* \Sigma_{12}(\omega)$$

and Lemma 7.7. The rest follows from the maximum principle for upper triangular operators which was formulated in Theorem 5.5. □

Lemma 7.9. *If $P \geq \varepsilon I$ for some $\varepsilon > 0$ and if $S(\omega)$ is given by formula (4.9), then*

$$\begin{aligned} I - S(\omega)^* S(\omega) \geq & \Sigma_{21}(\omega)^* \{I - \Omega(\omega)^* \Sigma_{22}(\omega)^*\}^{-1} \{I - \Omega(\omega)^* \Omega(\omega)\} \\ & \times \{I - \Sigma_{22}(\omega)\Omega(\omega)\}^{-1} \Sigma_{21}(\omega) \end{aligned} \tag{7.12}$$

for every point $\omega \in \overline{\mathbb{D}}$ at which $M - \omega ZN$ is boundedly invertible. In particular, $M - \omega ZN$ is boundedly invertible and equality prevails in (7.12) for every point $\omega \in \mathbb{T}$.

Proof. The inequality (7.12) is a standard computation which depends upon the fact that $\Sigma(\omega)$ is contractive for $\omega \in \mathbb{D}$ and unitary for $\omega \in \mathbb{T}$. The explicit calculations are perhaps carried out most easily by setting

$$X(\omega) = \Sigma_{21}(\omega)^{-1} - \Sigma_{21}(\omega)^{-1}\Sigma_{22}(\omega)\Omega(\omega)$$

and checking that

$$\begin{aligned} & X(\omega)^*\{I - S(\omega)^*S(\omega)\}X(\omega) \\ &= \begin{bmatrix} X(\omega) \\ \Omega(\omega) \end{bmatrix}^* \{I - \Sigma(\omega)^*\Sigma(\omega)\} \begin{bmatrix} X(\omega) \\ \Omega(\omega) \end{bmatrix} + I - \Omega(\omega)^*\Omega(\omega) \\ &\geq I - \Omega(\omega)^*\Omega(\omega) \,, \end{aligned}$$

for every point $\omega \in \overline{\mathbb{D}}$ at which the indicated inverse exists, and taking note that the inequality is an equality for $\omega \in \mathbb{T}$. □

The main conclusions of this section can now be summarized by the following supplement to Theorem 5.4:

Theorem 7.1. *If $P \geq \varepsilon I$ for some $\varepsilon > 0$, then for every solution $S \in \mathcal{S}(\mathcal{E}_1;\mathcal{E}_2)$ of the augmented BIP, there exists a unique choice of $\Omega \in \mathcal{S}(\mathcal{N}_1;\mathcal{N}_2)$ such that*

$$S(\omega) = \Sigma_{11}(\omega) + \Sigma_{12}(\omega)\{I - \Omega(\omega)\Sigma_{22}(\omega)\}^{-1}\Omega(\omega)\Sigma_{21}(\omega) \tag{7.13}$$

for every point $\omega \in \overline{\mathbb{D}}$. Moreover, S is strictly contractive if and only if Ω is strictly contractive.

Proof. The asserted uniqueness is immediate from Theorem 5.4 and Lemmas 7.6 and 7.7. The fact that $\|S\| < 1$ if and only if $\|\Omega\| < 1$ then follows from Lemmas 7.9 and 7.8. □

We remark that Lemmas 7.6 and 7.7 guarantee that the spaces $\mathcal{N}_1$ and $\mathcal{N}_2$ are nontrivial when $P \geq \varepsilon I$ for some $\varepsilon > 0$ and hence, via Theorem 7.1, that there are infinitely many solutions of the augmented BIP in this case (because there are infinitely many $\Omega \in \mathcal{S}(\mathcal{N}_1;\mathcal{N}_2)$).

8 Chain scattering operators

In this section we shall show that when $P \geq \varepsilon I$ for some $\varepsilon > 0$, then the set of solutions S to the augmented BIP can be expressed as a linear fractional transformation

$$S = (\Theta_{11}\Omega + \Theta_{12})(\Theta_{21}\Omega + \Theta_{22})^{-1}$$

in terms of the block entries of an operator

$$\Theta = \begin{bmatrix} \Theta_{11} & \Theta_{12} \\ & \\ \Theta_{21} & \Theta_{22} \end{bmatrix} : \begin{matrix} \mathcal{N}_2 \\ \oplus \\ \mathcal{N}_1 \end{matrix} \longrightarrow \begin{matrix} \mathcal{E}_2 \\ \oplus \\ \mathcal{E}_1 \end{matrix}$$

which is $(J_{\mathcal{E}}, J_{\mathcal{N}})$ inner, in a sense that will be explained below.

Representations of this form were considered in a number of earlier papers on tangential interpolation problems in the setting of triangular operators; see e.g., [BGK2] for a bitangential problem with finite dimensional coordinate spaces $\mathcal{B}_i$, $\mathcal{M}_i$ and $\mathcal{N}_i$, and [BGK1], [D1], [vV], [Ko1], [DD] and [Ko2] for assorted variants of one-sided tangential interpolation problems. In the last two papers, the coordinate spaces $\mathcal{B}_i$, $\mathcal{M}_i$ and $\mathcal{N}_i$ are permitted to be infinite dimensional. A closely related problem was considered in [ADD].

In all the cited references the operator Θ referred to earlier was obtained by first embedding the Lyapunov-Stein identity into an appropriately defined J isometric operator and then calculating its characteristic function. We shall obtain this representation in another way, i.e., by applying the Potapov-Ginzburg transform to the operator $\Sigma(\omega)$.

8.1 General considerations

The basic observation is that elementary manipulations of formula (4.9) (with ω in place of r) lead readily to the conclusion that

$$\begin{aligned} S(\omega) &= \left\{\Sigma_{11}(\omega)\Sigma_{21}(\omega)^{-1} + \Sigma_{12}(\omega)\Omega(\omega)(I - \Sigma_{22}(\omega)\Omega(\omega))^{-1}\right\}\Sigma_{21}(\omega) \\ &= \left\{(\Sigma_{12}(\omega) - \Sigma_{11}(\omega)\Sigma_{21}(\omega)^{-1}\Sigma_{22}(\omega))\Omega(\omega)\right. \\ &\quad \left.+\Sigma_{11}(\omega)\Sigma_{21}(\omega)^{-1}\right\}\{I - \Sigma_{22}(\omega)\Omega(\omega)\}^{-1}\Sigma_{21}(\omega) \end{aligned}$$

for every point $\omega \in \overline{\mathbb{D}}$ at which $\Sigma_{21}(\omega)$ is invertible. Let

$$\Theta(\omega) = \begin{bmatrix} \Theta_{11}(\omega) & \Theta_{12}(\omega) \\ \Theta_{21}(\omega) & \Theta_{22}(\omega) \end{bmatrix}$$

be given by the formula

$$\Theta(\omega) = \begin{bmatrix} \Sigma_{12}(\omega) - \Sigma_{11}(\omega)\Sigma_{21}(\omega)^{-1}\Sigma_{22}(\omega) & \Sigma_{11}(\omega)\Sigma_{21}(\omega)^{-1} \\ -\Sigma_{21}(\omega)^{-1}\Sigma_{22}(\omega) & \Sigma_{21}(\omega)^{-1} \end{bmatrix}. \tag{8.1}$$

Then

$$S(\omega) = \{\Theta_{11}(\omega)\Omega(\omega) + \Theta_{12}(\omega)\}\{\Theta_{21}(\omega)\Omega(\omega) + \Theta_{22}(\omega)\}^{-1} \tag{8.2}$$

for every point $\omega \in \overline{\mathbb{D}}$ at which $\Sigma_{21}(\omega)$ is invertible. Let

$$J_{\mathcal{N}} = \begin{bmatrix} I_{\mathcal{N}_2} & 0 \\ 0 & -I_{\mathcal{N}_1} \end{bmatrix} \text{ and } J_{\mathcal{E}} = \begin{bmatrix} I_{\mathcal{E}_2} & 0 \\ 0 & -I_{\mathcal{E}_1} \end{bmatrix}. \tag{8.3}$$

Lemma 8.1. *The inequalities*

$$\Theta(\omega)^* J_{\mathcal{E}} \Theta(\omega) \leq J_{\mathcal{N}} \tag{8.4}$$

and

$$\Theta(\omega) J_{\mathcal{N}} \Theta(\omega)^* \leq J_{\mathcal{E}} \tag{8.5}$$

hold for every point $\omega \in \overline{\mathbb{D}}$ at which $\Sigma_{21}(\omega)$ is boundedly invertible. Equality prevails in both (8.4) and (8.5) for $\omega \in \mathbb{T}$.

Proof. Let

$$\Psi_1(\omega) = \begin{bmatrix} \Sigma_{12}(\omega) & \Sigma_{11}(\omega) \\ 0 & I_{\mathcal{E}_1} \end{bmatrix}, \quad \Psi_2(\omega) = \begin{bmatrix} I_{\mathcal{N}_2} & 0 \\ \Sigma_{22}(\omega) & \Sigma_{21}(\omega) \end{bmatrix},$$

$$\Psi_3(\omega) = \begin{bmatrix} I_{\mathcal{E}_2} & -\Sigma_{11}(\omega) \\ 0 & -\Sigma_{21}(\omega) \end{bmatrix} \quad \text{and } \Psi_4(\omega) = \begin{bmatrix} \Sigma_{12}(\omega) & 0 \\ \Sigma_{22}(\omega) & -I_{\mathcal{N}_2} \end{bmatrix}.$$

It is then readily checked that

$$\Theta(\omega) = \Psi_1(\omega)\Psi_2(\omega)^{-1} = \Psi_3(\omega)^{-1}\Psi_4(\omega)$$

at every point $\omega \in \overline{\mathbb{D}}$ at which $\Sigma_{21}(\omega)$ is invertible. Therefore,

$$J_{\mathcal{N}} - \Theta(\omega)^* J_{\mathcal{E}} \Theta(\omega) = \Psi_2(\omega)^{-*}\{\Psi_2(\omega)^* J_{\mathcal{N}} \Psi_2(\omega) - \Psi_1(\omega)^* J_{\mathcal{E}} \Psi_1(\omega)\}\Psi_2(\omega)^{-1} \geq 0$$

for such points ω because the term in curly brackets is equal to

$$\begin{bmatrix} I_{\mathcal{N}_2} & 0 \\ 0 & I_{\mathcal{E}_1} \end{bmatrix} - \begin{bmatrix} \Sigma_{22}(\omega)^* & \Sigma_{12}(\omega)^* \\ \Sigma_{21}(\omega)^* & \Sigma_{11}(\omega)^* \end{bmatrix} \begin{bmatrix} \Sigma_{22}(\omega) & \Sigma_{21}(\omega) \\ \Sigma_{12}(\omega) & \Sigma_{11}(\omega) \end{bmatrix}$$

$$= \begin{bmatrix} 0 & I_{\mathcal{N}_2} \\ I_{\mathcal{E}_1} & 0 \end{bmatrix} \left\{ \begin{bmatrix} I_{\mathcal{E}_1} & 0 \\ 0 & I_{\mathcal{N}_2} \end{bmatrix} - \Sigma(\omega)^*\Sigma(\omega) \right\} \begin{bmatrix} 0 & I_{\mathcal{E}_1} \\ I_{\mathcal{N}_2} & 0 \end{bmatrix}$$

and $\Sigma(\omega)$ is contractive for $\omega \in \overline{\mathbb{D}}$.

This serves to establish (8.4). Similar considerations based on the formula

$$J_{\mathcal{E}} - \Theta(\omega) J_{\mathcal{N}} \Theta(\omega)^* = \Psi_3(\omega)^{-1}\{\Psi_3(\omega) J_{\mathcal{E}} \Psi_3(\omega)^* - \Psi_4(\omega) J_{\mathcal{N}} \Psi_4(\omega)^*\}\Psi_3(\omega)^{-*}$$

$$= \Psi_3(\omega)^{-1} \left\{ \begin{bmatrix} I_{\mathcal{E}_2} & 0 \\ 0 & I_{\mathcal{N}_1} \end{bmatrix} - \Sigma(\omega)\Sigma(\omega)^* \right\} \Psi_3(\omega)^{-*}$$

lead readily to (8.5). The asserted equality in both (8.4) and (8.5) for $\omega \in \mathbb{T}$ follows from the fact that $\Sigma(\omega)$ is unitary for $\omega \in \mathbb{T}$. □

To this point we have established the following:

Theorem 8.1. *The Potapov–Ginzburg transform Θ, of the characteristic function Σ of the unitary colligation A associated with the augmented BIP based on a strictly positive Pick operator P, (which is defined by (8.1)) is $(J_{\mathcal{N}}, J_{\mathcal{E}})$ unitary and satisfies conditions (8.4) and (8.5). Moreover, every solution of the augmented BIP can be represented by formula (8.2).*

Lemma 8.1 suggests that the operator Θ which is defined by (8.1) is $(J_{\mathcal{N}}, J_{\mathcal{E}})$ inner in some sense. We shall make this precise in a moment. To this end, let Q_t denote the diagonal truncation operator which is defined by the rule

$$(Q_t x)_j = \begin{cases} 0 & \text{for} \quad j < t \\ x_j & \text{for} \quad j \geq t, \end{cases}$$

regardless of which ℓ_2 space x belongs to. Then, since $\Sigma(\omega) \in \mathcal{U}(\mathcal{E}_1 \oplus \mathcal{N}_2;\ \mathcal{E}_2 \oplus \mathcal{N}_1)$, it is readily checked that

$$Q_t \Sigma(\omega) = Q_t \Sigma(\omega) Q_t \tag{8.6}$$

and hence if

$$\begin{bmatrix} e_2(\omega) \\ n_1(\omega) \end{bmatrix} = \Sigma(\omega) \begin{bmatrix} e_1 \\ n_2 \end{bmatrix},$$

that

$$\begin{aligned} \left\| Q_t \begin{bmatrix} e_2(\omega) \\ n_1(\omega) \end{bmatrix} \right\|^2 &= \left\| Q_t \Sigma(\omega) \begin{bmatrix} e_1 \\ n_2 \end{bmatrix} \right\|^2 \\ &= \left\| Q_t \Sigma(\omega) Q_t \begin{bmatrix} e_1 \\ n_2 \end{bmatrix} \right\|^2 \\ &\leq \left\| Q_t \begin{bmatrix} e_1 \\ n_2 \end{bmatrix} \right\|^2 . \end{aligned}$$

But this in turn implies that

$$\|Q_t e_2\|^2 - \|Q_t e_1\|^2 \leq \|Q_t n_2\|^2 - \|Q_t n_1\|^2$$

or equivalently that

$$\langle J_{\mathcal{E}} Q_t \Theta \begin{bmatrix} n_2 \\ n_1 \end{bmatrix} , Q_t \Theta \begin{bmatrix} n_2 \\ n_1 \end{bmatrix} \rangle \leq \langle J_{\mathcal{N}} Q_t \begin{bmatrix} n_2 \\ n_1 \end{bmatrix} , Q_t \begin{bmatrix} n_2 \\ n_1 \end{bmatrix} \rangle ,$$

i.e.,

$$\Theta^* Q_t J_{\mathcal{E}} Q_t \Theta \leq Q_t J_{\mathcal{N}} Q_t . \tag{8.7}$$

We shall say that an operator $\Theta \in \mathcal{X}(\mathcal{N}_2 \oplus \mathcal{N}_1; \mathcal{E}_2 \oplus \mathcal{E}_1)$ is $(J_{\mathcal{N}}, J_{\mathcal{E}})$ inner if it meets the following three conditions:

(1) $\Theta^* J_{\mathcal{E}} \Theta = J_{\mathcal{N}}$.

(2) $\Theta J_{\mathcal{N}} \Theta^* = J_{\mathcal{E}}$.

(3) $\Theta^* Q_t J_{\mathcal{E}} Q_t \Theta \leq Q_t J_{\mathcal{N}} Q_t$ for every integer t.

This definition was introduced for some more restricted classes of Θ in [BGK2] and [Ko2].

Theorem 8.2. *Let $\Theta \in \mathcal{X}(\mathcal{N}_2 \oplus \mathcal{N}_1; \mathcal{E}_2 \oplus \mathcal{E}_1)$. Then Θ is the Potapov-Ginzburg transform of a unitary operator $\Sigma \in \mathcal{U}(\mathcal{E}_1 \oplus \mathcal{N}_2; \mathcal{E}_2 \oplus \mathcal{N}_1)$ if and only if Θ is $(J_{\mathcal{N}}, J_{\mathcal{E}})$ inner.*

Proof. The preceding calculations show that the Potapov-Ginzburg transform of a unitary operator Σ in the indicated class is $(J_{\mathcal{N}}, J_{\mathcal{E}})$ inner.

Conversely, if Θ is $(J_{\mathcal{N}}, J_{\mathcal{E}})$ inner, then the same set of calculations run backwards from (8.7) show that the Potapov-Ginzburg transform

$$\Sigma = \begin{bmatrix} \Theta_{12}\Theta_{22}^{-1} & \Theta_{11} - \Theta_{12}\Theta_{22}^{-1}\Theta_{21} \\ \Theta_{22}^{-1} & -\Theta_{22}^{-1}\Theta_{21} \end{bmatrix}$$

of Θ is subject to the inequality

$$\Sigma^* Q_t \Sigma \leq Q_t$$

for every integer t. But this implies that

$$(I - Q_t)\Sigma^* Q_t \Sigma (I - Q_t) = 0$$

and hence that

$$Q_t \Sigma (I - Q_t) = 0$$

for every integer t. Therefore Σ is upper. This completes the proof since Σ is clearly unitary. □

A similar development is presented as Proposition 7.1 in [BGK2]; see also pp. 180–181 of [DD].

Lemma 8.2. *If $P \geq \varepsilon I$ and if the BIP is a left interpolation problem only (i.e., if $\mu = n$ and $M = I$ in the formulation of the problem which is given in the third paragraph of Section* 3), *then:*

(1) *The operator $\Theta(\omega)$ which is defined by formula (8.1) belongs to the class $\mathcal{U}(\mathcal{N}_2 \oplus \mathcal{N}_1; \mathcal{E}_1 \oplus \mathcal{E}_2)$ for every point $\omega \in \overline{\mathbb{D}}$.*
(2) *There exists a unitary diagonal operator α_1 which maps $\mathcal{E}_1$ onto $\mathcal{N}_1$.*
(3) *The operator $\Sigma_{21}(0)$ is invertible.*

Proof. By formula (8.1), $\Theta(\omega) \in \mathcal{U}$ for every point $\omega \in \overline{\mathbb{D}}$ for which $\Sigma_{21}(\omega)$ is invertible and $\Sigma_{21}(\omega)^{-1} \in \mathcal{U}(\mathcal{N}_1, \mathcal{E}_1)$. The formula

$$\Sigma_{21}(\omega)^{-1}\alpha : \begin{bmatrix} b_1 \\ b_2 \end{bmatrix} \in d_V^{\perp} \longrightarrow [-C_2(I - \omega Z N)^{-1} P^{-\frac{1}{2}} \; I_{\mathcal{E}_1}] \begin{bmatrix} b_1 \\ b_2 \end{bmatrix}$$

clearly displays the fact that $\Sigma_{21}(\omega)^{-1} \in \mathcal{U}(\mathcal{N}_1; \mathcal{E}_1)$ for every point $\omega \in \overline{\mathbb{D}}$ since $r_{sp}(ZN) = r_{sp}(ZA_1) < 1$ by (2.6).

Next, it is readily checked that if $M = I_{\mathcal{H}}$, then

$$d_V^{\perp} = \left\{ \begin{bmatrix} -P^{-\frac{1}{2}} C_2^* \\ I_{\mathcal{E}_1} \end{bmatrix} e_1 : e_1 \in \mathcal{E}_1 \right\}$$

and hence that (2) is valid.

Finally, (3) is clear from Lemma 7.6. □

We remark that the same arguments serve to establish (2) and (3) of the last lemma for the (two-sided) augmented BIP when the operator M is boundedly invertible. In fact, by Lemma 7.6, $\Sigma_{21}(0)$ is boundedly invertible if and only if M is boundedly invertible.

In [DD] a one-sided BIP was studied under the added assumption that there exists a diagonal unitary operator α_2 which maps $\mathcal{E}_2$ onto $\mathcal{N}_2$. This assumption was referred to as the kernel condition. In view of the preceding lemma, it follows that the kernel condition guarantees that the operators $J_{\mathcal{E}}$ and $J_{\mathcal{N}}$ are unitarily equivalent via diagonal unitary operators.

Next we observe that

$$\begin{bmatrix} A_{11} & A_{13} & A_{12} \\ A_{21} & A_{23} & A_{22} \\ A_{31} & 0 & A_{32} \end{bmatrix} = \begin{bmatrix} A_{11} & A_{12} & A_{13} \\ A_{21} & A_{22} & A_{23} \\ A_{31} & A_{32} & 0 \end{bmatrix} \begin{bmatrix} I & 0 & 0 \\ 0 & 0 & I \\ 0 & I & 0 \end{bmatrix}$$

is a unitary colligation from $\mathcal{H} \oplus \mathcal{N}_2 \oplus \mathcal{E}_1$ onto $\mathcal{H}^{(-1)} \oplus \mathcal{E}_2 \oplus \mathcal{N}_1$ and set

$$\begin{bmatrix} D_{11} & D_{12} \\ D_{21} & D_{22} \end{bmatrix} = \left[\begin{array}{cc|c} A_{11} & A_{13} & A_{12} \\ A_{21} & A_{23} & A_{22} \\ \hline A_{31} & 0 & A_{32} \end{array}\right].$$

Now if $P \geq \varepsilon I$ for some $\varepsilon > 0$ and $M = I_{\mathcal{H}}$, then $\Sigma_{21}(0) = D_{22}$ is invertible by Lemma 6.6. Thus the Potapov-Ginzburg transform of D:

$$\begin{bmatrix} D_{11} - D_{12}D_{22}^{-1}D_{21} & D_{12}D_{22}^{-1} \\ -D_{22}^{-1}D_{21} & D_{22}^{-1} \end{bmatrix},$$

is well defined and is equal to

$$\tau = \left[\begin{array}{c|c|c} A_{11} - A_{12}A_{32}^{-1}A_{31} & A_{13} & A_{12}A_{32}^{-1} \\ A_{21} - A_{22}A_{32}^{-1}A_{31} & A_{23} & A_{22}A_{32}^{-1} \\ -A_{32}^{-1}A_{31} & 0 & A_{32}^{-1} \end{array}\right]. \tag{8.8}$$

Moreover, it is readily checked that the characteristic operator function of this colligation:

$$\begin{bmatrix} A_{23} & A_{22}A_{32}^{-1} \\ 0 & A_{32}^{-1} \end{bmatrix} + \begin{bmatrix} A_{21} - A_{22}A_{32}^{-1}A_{31} \\ -A_{32}^{-1}A_{31} \end{bmatrix} \{I - \omega Z(A_{11} - A_{12}A_{32}^{-1}A_{31})\}^{-1}$$
$$\times [\omega Z A_{13} \; wZA_{12}A_{32}^{-1}]$$

agrees with $\Theta(\omega)$.

Thus we have established the following fact:

Theorem 8.3. *If $P \geq \varepsilon I$ for some $\varepsilon > 0$ and if the BIP is a left interpolation problem only, then the $(J_{\mathcal{N}}, J_{\mathcal{E}})$ inner operator Θ which is defined by formula (8.1) is the characteristic operator of the $(\widehat{J_{\mathcal{N}}}, \widehat{J_{\mathcal{E}}})$ unitary colligation which is defined by (8.8), where*

$$\widehat{J_{\mathcal{N}}} = \operatorname{diag}(I_{\mathcal{H}}, J_{\mathcal{N}}) \text{ and } \widehat{J_{\mathcal{E}}} = \operatorname{diag}(I_{\mathcal{H}^{(-1)}}, J_{\mathcal{E}}) .$$

We remark that since the proof of Theorem 2.3 is based on the fact that $D_{22} = \Sigma_{21}(0)$ is boundedly invertible, it follows from Lemma 7.6 that Theorem 2.3 is valid also for the special case of the augmented BIP for which the operator M is boundedly invertible.

8.2 Formulas in terms of the data

Our next objective is to express the block entries of the chain scattering operator in terms of the data.

Theorem 8.4. *If $P \geq \varepsilon I$ for some $\varepsilon > 0$, then the block entries in the chain scattering operator $\Theta(\omega) : \mathcal{N}_2 \oplus \mathcal{N}_1 \longrightarrow \mathcal{E}_1 \oplus \mathcal{E}_1$, which is given by (8.1) can be expressed explicitly in terms of the data of the augmented BIP by means of the formulas*

$$\Theta_{11}(\omega) = [C_1(M - \omega Z N)^{-1} P^{-\frac{1}{2}} \omega Z \ I_{\mathcal{E}_2}]\beta^* , \tag{8.9}$$

$$\Theta_{12}(\omega) = C_1 T_\omega^{-1} C_2^* [-C_2(M - \omega Z N)^{-1} P^{-\frac{1}{2}} \ I_{\mathcal{E}_1}]\alpha^* , \tag{8.10}$$

$$\Theta_{21}(\omega) = C_2(M - \omega Z N)^{-1} P^{-\frac{1}{2}} \omega Z [I_{\mathcal{H}^{(-1)}} \ 0]\beta^* , \tag{8.11}$$

$$\Theta_{22}(\omega) = [-C_2(M - \omega Z N)^{-1} P^{-\frac{1}{2}} \ I_{\mathcal{E}_1}]\alpha^* , \tag{8.12}$$

for every point $\omega \in \overline{\mathbb{D}}$ at which the indicated inverse exists.

Proof. The proof is a straightforward calculation which is based on substituting the formula (7.10) and the formulas given in Lemma 7.5 into (8.1). □

Corollary. *If $P \geq \varepsilon I$ for some $\varepsilon > 0$ and if the operator M is boundedly invertible, then the entries in the second block column of $\Theta(\omega)$ can be expressed even more explicitly as*

$$\Theta_{12}(\omega) = C_1(M - \omega Z N)^{-1} P^{-1} M^{-*} C_2^* \{I_{\mathcal{E}_1} + C_2 M^{-1} P^{-1} M^{-*} C_2^*\}^{-\frac{1}{2}} \tag{8.13}$$

$$\begin{aligned}\Theta_{22}(\omega) = {} & \{I_{\mathcal{E}_1} + C_2(M - \omega Z N)^{-1} P^{-1} M^{-*} C_2^*\} \\ & \times \{I_{\mathcal{E}_1} + C_2 M^{-1} P^{-1} M^{-*} C_2^*\}^{-\frac{1}{2}}\end{aligned} \tag{8.14}$$

for every point $\omega \in \overline{\mathbb{D}}$ at which the indicated inverse exists.

Proof. If M is invertible, then $\mathcal{E}_1$ is unitarily equivalent to $d_v^{\perp}$ by a diagonal operator and hence we may choose

$$\alpha^* = \begin{bmatrix} -P^{-\frac{1}{2}}M^{-*}C_2^* \\ I_{\mathcal{E}_1} \end{bmatrix} \{I_{\mathcal{E}_1} + C_2M^{-1}P^{-1}M^{-*}C_2^*\}^{-\frac{1}{2}} .$$

The given formulas follow easily from (8.10) and (8.12) and this choice of α. □

In particular, if $M = I$ and $\omega = 1$, then formulas (8.9), (8.11), (8.13) and (8.15) yield the formulas which appear in Theorem 3.2.2 of [Ko2]. The same formulas appear as (4.17)–(4.20) of [DD], but under the added assumption that $\mathcal{E}_2$ is unitarily equivalent to $\Delta_v^{\perp}$ by a diagonal operator. This condition, which was referred to as the kernel condition in [DD], was not imposed in [Ko2] nor is it imposed in the present analysis.

Similar reductions are possible if N is boundedly invertible because then $\mathcal{E}_2$ is unitarily equivalent to $\Delta_v^{\perp}$ (by a diagonal operator) and we may choose

$$\beta^* = \begin{bmatrix} -Z^*P^{-\frac{1}{2}}ZN^{-*}C_1^* \\ I_{\mathcal{E}_2} \end{bmatrix} (I_{\mathcal{E}_2} + C_1N^{-1}Z^*P^{-1}ZN^{-*}C_1^*)^{-\frac{1}{2}} .$$

Thus, if M and N are both boundedly invertible and $\omega = 1$, we can write

$$\Theta(1) = \left\{ \begin{bmatrix} I_{\mathcal{E}_2} & 0 \\ 0 & I_{\mathcal{E}_1} \end{bmatrix} + \begin{bmatrix} C_1 \\ C_2 \end{bmatrix} (M - ZN)^{-1}P^{-1}[-N^{-*}C_1^* \; M^{-*}C_2^*] \right\} L^{-\frac{1}{2}}$$

where

$$L = \mathrm{diag}\{I_{\mathcal{E}_2} + C_1N^{-1}Z^*P^{-1}ZN^{-*}C_1^* \, , \, I_{\mathcal{E}_1} + C_2M^{-1}P^{-1}M^{-1}C_2^*\} .$$

Some similar formulas for finite dimensional coordinate spaces are given in [BGK2].

9 The maximum entropy problem

In this section we shall apply the results of the preceding two sections to formulate and solve a maximum entropy problem in the general setting of upper triangular operators. The analysis is much the same as in the classical case (see e.g., Chapter 11 of [D1]; examples in the setting of upper triangular operators may be found in the work of Iglesias [Ig] (which is also described in the lecture notes of Feintuch [Fe]) and Constantinescu, Sayed and Kailath [CSK]. For another approach to entropy problems, see the paper of Gohberg, Kaashoek and Woerdeman [GKW].

We continue to assume that $P \geq \varepsilon I$ for some $\varepsilon > 0$ and shall let $\mathcal{K}$ denote the set of all strictly contractive solutions $S \in \mathcal{S}(\ell_{\mathcal{N}}^2; \ell_{\mathcal{M}}^2)$ to the one-sided BIP which is discussed in Lemma 8.2. Then by a theorem of Arveson [Ar], the operator $I - S^*S$ admits a spectral factorization, i.e., it can be expressed in the form

$$I - S^*S = F_S^*F_S \; ,$$

where $F_S \in \mathcal{U}(\ell^2_{\mathcal{N}};\ell^2_{\mathcal{N}'})$ is a bounded invertible operator with $F_S^{-1} \in \mathcal{U}(\ell^2_{\mathcal{N}'};\ell^2_{\mathcal{N}})$ for some suitable choice of $\ell^2_{\mathcal{N}'}$. Let $\Delta(F_S)$ denote the diagonal of F_S, i.e., in terms of the notation which was introduced just before Theorem 5.5,

$$\Delta(F_S) = F_S(0)$$

(or alternatively, in terms of the diagonal transforms which were discussed briefly in the introduction just after formula (1.11),

$$\Delta(F_S) = F_S^{\wedge}(0) = F_S^{\Delta}(0) \ .)$$

The maximum entropy problem in this setting is to (1) evaluate

$$\sup\{\Delta(F_S)^*\Delta(F_S) : \ S \in \mathcal{K}\} \ ,$$

and (2) find those S which achieve this supremum, if any exist.

Theorem 9.1. *If $P \geq \varepsilon I$, then*

$$\Delta(F_S)^*\Delta(F_S) \leq I_{\mathcal{E}_1} - C_2(P + C_2^*C_2)^{-1}C_2^* \tag{9.1}$$

with equality if and only if

$$S = \Sigma_{11} \ .$$

Proof. Under the given assumptions, every $S \in \mathcal{K}$ can be expressed in the form (4.9) with ω in place of r and hence by formula (7.12) with $\omega = 1$ and then dropped from the notation,

$$I_{\mathcal{N}_1} - S^*S = \Psi^*(I_{\mathcal{N}_1} - \Omega^*\Omega)\Psi \ ,$$

where

$$\Psi = (I_{\mathcal{N}_1} - \Sigma_{22}\Omega)^{-1}\Sigma_{21}$$

belongs to $\mathcal{U}(\mathcal{E}_1;\mathcal{N}_2)$ and is invertible with $\Psi^{-1} \in \mathcal{U}(\mathcal{N}_1;\mathcal{E}_1)$. (It is perhaps reassuring to recall here that for the one-sided problem with $M = I$, $\mathcal{E}_1$ is unitarily equivalent to $\mathcal{N}_2$.) Therefore,

$$F_S = F_\Omega\Psi$$

and

$$\Delta(F_S) = \Delta(F_\Omega)\Delta(\Psi) \ ,$$

up to left unitary diagonal factors which do not affect the final calculation. Moreover, since $\Delta(\Sigma_{22}) = \Sigma_{22}(0) = 0$,

$$\begin{aligned}\Delta(F_S) &= \Delta(F_\Omega)\Delta(\Sigma_{21}) \\ &= \Delta(F_\Omega)\Sigma_{21}(0)\end{aligned}$$

and hence

$$\Delta(F_S)^*\Delta(F_S) = \Sigma_{21}(0)^*\Delta(F_\Omega)^*\Delta(F_\Omega)\Sigma_{21}(0) \ .$$

Therefore, since

$$\|\Delta(F_\Omega)\| \leq \|F_\Omega\| ,$$

it follows that

$$\Delta(F_\Omega)^*\Delta(F_\Omega) \leq F_\Omega^*F_\Omega \leq I$$

and hence that

$$\Delta(F_S)^*\Delta(F_S) \leq \Sigma_{21}(0)^*\Sigma_{21}(0)$$

with equality if and only if $\Omega = 0$. By formula (7.13) the condition $\Omega = 0$ is equivalent to the statement that $S = \Sigma_{11}$. The proof is now completed by checking that

$$\Sigma_{21}(0)^*\Sigma_{21}(0) = I_{\mathcal{E}_1} - C_2(P + C_2^*C_2)^{-1}C_2^* , \tag{9.2}$$

which is a straightforward calculation. □

10 Positive real interpolants

In this section we shall consider the BIP in the class of operators $\Phi \in \mathcal{U}(\ell^2_{\mathcal{M}};\ell^2_{\mathcal{M}})$ with strictly positive real part, i.e., with $\Phi + \Phi^* \geq \varepsilon I$ for some $\varepsilon > 0$. A one-sided version of this problem was considered in Kos [Ko2]. It will be convenient to let $\mathcal{C}^\circ(\ell^2_{\mathcal{M}};\ell^2_{\mathcal{M}})$ denote this class of operators and correspondingly to let $\mathcal{S}^\circ(\ell^2_{\mathcal{M}};\ell^2_{\mathcal{M}})$ denote the set of operators in $\mathcal{S}(\ell^2_{\mathcal{M}};\ell^2_{\mathcal{M}})$ which are strictly contractive. We shall take advantage of the fact that the Cayley transform $T \longrightarrow (I - T)(I + T)^{-1}$ defines a one to one correspondence between the class of operators with strictly positive real part and the class of strictly contractive operators. More is true:

Lemma 10.1. *If $\Phi \in \mathcal{C}^\circ(\ell^2_{\mathcal{M}};\ell^2_{\mathcal{M}})$, then $S = (I-\Phi)(I+\Phi)^{-1}$ belongs to $\mathcal{S}^\circ(\ell^2_{\mathcal{M}};\ell^2_{\mathcal{M}})$. Moreover (since the Cayley transform is its own inverse), every such S may be obtained as the Cayley transform of such a Φ.*

Proof. The main task is to show that $(I + \Phi)^{-1} \in \mathcal{U}$; the rest is easy. To this end, let

$$\mathcal{C}_\gamma = \{\Phi \in \mathcal{U}(\ell^2_{\mathcal{M}};\ell^2_{\mathcal{M}}) : \ \Phi + \Phi^* \geq \varepsilon I \text{ for some } \varepsilon > 0 \text{ and } \|\Phi\| < \gamma\}$$

and suppose for the moment that $(I + \Phi)^{-1} \in \mathcal{U}$ for every $\Phi \in \mathcal{C}_\gamma$ for some fixed $\gamma > 0$ and choose a $\Phi \in \mathcal{C}_{2\gamma}$. Then, since $\Phi/2 \in \mathcal{C}_\gamma$,

$$\begin{aligned} I + \Phi &= I + \Phi/2 + \Phi/2 \\ &= (I + \Phi/2)(I + \Psi) \end{aligned}$$

where

$$\Psi = (I + \Phi/2)^{-1}\Phi/2$$

is readily seen to belong to $\mathcal{C}_\gamma$. Therefore

$$(I + \Phi)^{-1} = (I + \Psi)^{-1}(I + \Phi/2)^{-1}$$

belongs to $\mathcal{U}$ as claimed. Since $(I+\Phi)^{-1} \in \mathcal{U}$ for every $\Phi \in \mathcal{U}$ with $\|\Phi\| < 1$ (and so in particular for $\Phi \in \mathcal{C}_\gamma$ with $0 \leq \gamma < 1$), it follows readily by induction that $(I+\Phi)^{-1} \in \mathcal{U}$ for every $\Phi \in \mathcal{U}$ with strictly positive real part. □

Our next objective is to show that if S is a solution of the augmented BIP which is expressed in terms of P, A_1, A_2, C_{11}, C_{12}, C_{21} and C_{22} as in the third paragraph of Section 3 (between (3.1) and (3.2)), then Φ is the solution of a related problem which will be expressed in terms of the same P, A_1 and A_2 but with

$$\widetilde{C}_{11} = \frac{C_{11}+C_{21}}{\sqrt{2}} \quad , \quad \widetilde{C}_{12} = \frac{C_{22}-C_{12}}{\sqrt{2}} \ ,$$
$$\widetilde{C}_{21} = \frac{C_{11}-C_{21}}{\sqrt{2}} \quad \text{and} \quad \widetilde{C}_{22} = \frac{C_{22}+C_{12}}{\sqrt{2}} \ .$$

(Note that $\widetilde{\widetilde{C}}_{ij} = C_{ij}$.)

Theorem 10.1. *Let $S \in \mathcal{S}^\circ(\ell^2_{\mathcal{M}};\ell^2_{\mathcal{M}})$ be a solution of the augmented BIP based on P, A_1, A_2, C_{11}, C_{12}, C_{21} and C_{22} as set forth in the third paragraph of Section 3, then*

$$\Phi = (I-S)(I+S)^{-1}$$

belongs to $\mathcal{C}^\circ(\ell^2_{\mathcal{M}};\ell^2_{\mathcal{M}})$ and is a solution of the following augmented BIP:

$$\underline{\underline{p}}\Phi^*\widetilde{C}_{11}(I-Z_1A_1)^{-1}E_2 = \widetilde{C}_{21}(I-Z_1A_1)^{-1}E_1 \ , \tag{10.1}$$

$$\underline{\underline{q}}'\Phi\widetilde{C}_{22}(Z_2-A_2)^{-1}E_2 = \widetilde{C}_{12}(Z_2-A_2)^{-1}E_2 \ , \tag{10.2}$$

and

$$\langle \Phi\widetilde{C}_{22}(Z_2-A_2)^{-1}E_2, \widetilde{C}_{11}(I-Z_1A_1)^{-1}E_1\rangle_{HS} = \operatorname{trace}[E_1^* \ 0]P\begin{bmatrix} 0 \\ E_2 \end{bmatrix} \tag{10.3}$$

for every choice of $E_1 \in \{\mathcal{D}_2(\ell^2_{\mathcal{B}};\ell^2_{\mathcal{B}})\}^{\mu\times 1}$ and $E_2 \in \{\mathcal{D}_2(\ell^2_{\mathcal{B}};\ell^2_{\mathcal{B}})\}^{\nu\times 1}$.

Conversely, if $\Phi \in \mathcal{C}^\circ(\ell^2_{\mathcal{M}};\ell^2_{\mathcal{M}})$ is a solution of the second stated problem, then $S = (I-\Phi)(I+\Phi)^{-1}$ belongs to $\mathcal{S}^\circ(\ell^2_{\mathcal{M}};\ell^2_{\mathcal{M}})$ and is a solution of the original augmented BIP.

Proof. Since

$$\begin{bmatrix} I & -S \\ -S^* & I \end{bmatrix}\begin{bmatrix} C_{11} & C_{12} \\ C_{21} & C_{22} \end{bmatrix}(M-ZN)^{-1}$$

$$= \frac{1}{2}\begin{bmatrix} I & -S \\ -S^* & I \end{bmatrix}\begin{bmatrix} -I & I \\ I & I \end{bmatrix}\begin{bmatrix} -I & I \\ I & I \end{bmatrix}\begin{bmatrix} C_{11} & C_{12} \\ C_{21} & C_{22} \end{bmatrix}(M-ZN)^{-1}$$

$$= \frac{1}{2}\begin{bmatrix} -I-S^* & I-S \\ I+S^* & I-S^* \end{bmatrix}\begin{bmatrix} C_{21}-C_{11} & C_{22}-C_{12} \\ C_{21}+C_{11} & C_{22}+C_{12} \end{bmatrix}(M-ZN)^{-1}$$

$$= \frac{1}{2}\begin{bmatrix} I+S & 0 \\ 0 & I+S^* \end{bmatrix}\begin{bmatrix} -I & \Phi \\ I & \Phi^* \end{bmatrix}\begin{bmatrix} C_{21}-C_{11} & C_{22}-C_{12} \\ C_{21}+C_{11} & C_{22}+C_{12} \end{bmatrix}(M-ZN)^{-1} \ ,$$

it is readily checked that S is a solution of the BIP (not the augmented BIP!) if and only if

$$[-I\ \Phi]\begin{bmatrix} C_{22}-C_{12} \\ C_{22}+C_{12} \end{bmatrix}(Z_2-A_2)^{-1}$$

is upper, and

$$[I\ \ -\Phi^*]\begin{bmatrix} C_{11}-C_{21} \\ C_{11}+C_{21} \end{bmatrix}(I-Z_1A_1)^{-1}$$

is strictly lower, i.e., if and only if (10.1) and (10.2) hold for every choice of E_1 and E_2 in the indicated classes.

Next, in order to establish (10.3), it suffices to verify the identity

$$\begin{aligned} &\langle \Phi\widetilde{C}_{22}(Z_2-A_2)^{-1}E_2, \widetilde{C}_{11}(I-Z_1A_1)^{-1}E_1\rangle_{HS} \\ &= -\langle SC_{22}(Z_2-A_2)^{-1}E_2, C_{11}(I-Z_1A_1)^{-1}E_1\rangle_{HS}\ . \end{aligned} \tag{10.4}$$

It is convenient to let $\Psi_- = (Z_2-A_2)^{-1}E_2$ and $\Psi_+ = (I-Z_1A_1)^{-1}E_1$. Then $\sqrt{2}$ times the left hand side of the asserted identity is equal to

$$\begin{aligned} \sqrt{2}\langle \Phi\widetilde{C}_{22}\Psi_-, \widetilde{C}_{11}\Psi_+\rangle_{HS} &= \langle \Phi\widetilde{C}_{22}\Psi_-, (C_{11}+C_{21})\Psi_+\rangle_{HS} \\ &= \langle \Phi\widetilde{C}_{22}\Psi_-, \underline{\underline{p}}(I+S^*)C_{11}\Psi_+\rangle_{HS} \\ &= \langle (I+S)\Phi\widetilde{C}_{22}\Psi_- - (I+S)\underline{\underline{q}}'\Phi\widetilde{C}_{22}\Psi_-, C_{11}\Psi_+\rangle_{HS} \\ &= \langle (I-S)\widetilde{C}_{22}\Psi_- - (I+S)\widetilde{C}_{12}\Psi_-, C_{11}\Psi_+\rangle_{HS} \\ &= -\langle S(\widetilde{C}_{22}+\widetilde{C}_{12})\Psi_-, C_{11}\Psi_+\rangle_{HS} \\ &= -\sqrt{2}\langle SC_{22}\Psi_-, C_{11}\Psi_+\rangle_{HS}\ , \end{aligned}$$

as needed. □

The preceding analysis shows that the new augmented BIP based on (10.1)–(10.3) admits a solution $\Phi \in \mathcal{C}^\circ(\ell^2_{\mathcal{M}};\ell^2_{\mathcal{M}})$ for each strictly positive solution P of the Lyapunov–Stein equation (2.9). Conversely, the inequality

$$\begin{aligned} \langle PE, E\rangle_{HS} &= \left\langle \begin{bmatrix} I & -S \\ -S^* & I \end{bmatrix} FE, FE \right\rangle_{HS} \\ &\geq \gamma\langle FE, FE\rangle_{HS}\ , \end{aligned}$$

shows that the condition that P be strictly positive is necessary for the existence of such solutions providing that

$$\langle FE, FE\rangle_{HS} \geq \varepsilon_1\langle E, E\rangle_{HS}$$

for some choice of $\varepsilon_1 > 0$.

A number of facts are now easily transferred from the setting of the original augmented BIP in the class $\mathcal{S}^\circ$ to the class of the new augmented BIP based on (10.1), (10.2) and (10.3) in the class $\mathcal{C}^\circ$.

Theorem 10.2. *If $P \geq \varepsilon I$ for some $\varepsilon > 0$, then Φ is a solution of the new augmented BIP based on (10.1), (10.2) and (10.3) in the class $\mathcal{C}^\circ(\ell^2_{\mathcal{M}};\ell^2_{\mathcal{M}})$ if and only if it can be expressed in the form*

$$\Phi = (\widetilde{\Theta}_{11}\Omega + \widetilde{\Theta}_{12})(\widetilde{\Theta}_{21}\Omega + \widetilde{\Theta}_{22})^{-1} \tag{10.5}$$

for some choice of $\Omega \in \mathcal{S}^\circ(\mathcal{N}_1;\mathcal{N}_2)$, where

$$\widetilde{\Theta} = \frac{1}{\sqrt{2}}\begin{bmatrix} -I_{\mathcal{E}} & I_{\mathcal{E}} \\ I_{\mathcal{E}} & I_{\mathcal{E}} \end{bmatrix}\Theta\,,$$

$\mathcal{E} = \mathcal{E}_1 = \mathcal{E}_2$ and Θ is given by formula (8.1).

Formula (10.5) is meaningful for $\Omega \in \mathcal{S}^\circ(\mathcal{N}_1;\mathcal{N}_2)$ because $\widetilde{\Theta}_{22}$ is invertible (as is $\widetilde{\Theta}_{12}$) and $\widetilde{\Theta}_{22}^{-1}\widetilde{\Theta}_{21}$ is unitary (as is $\widetilde{\Theta}_{12}^{-1}\widetilde{\Theta}_{11}$). Moreover, the Potapov-Ginzburg transform

$$\widetilde{\Sigma} = \begin{bmatrix} \widetilde{\Theta}_{11} - \widetilde{\Theta}_{12}\widetilde{\Theta}_{22}^{-1}\widetilde{\Theta}_{21} & \widetilde{\Theta}_{12}\widetilde{\Theta}_{22}^{-1} \\ -\widetilde{\Theta}_{22}^{-1}\widetilde{\Theta}_{21} & \widetilde{\Theta}_{22}^{-1} \end{bmatrix}$$

belongs to $\mathcal{U}(\mathcal{N}_2 \oplus \mathcal{E};\mathcal{E} \oplus \mathcal{N}_1)$ and, by the usual manipulations, formula (10.5) can be reexpressed in terms of $\widetilde{\Sigma}$:

$$\Phi = \widetilde{\Sigma}_{12} + \widetilde{\Sigma}_{11}\Omega(I - \widetilde{\Sigma}_{21}\Omega)^{-1}\widetilde{\Sigma}_{22}\,.$$

Finally, we remark that when the operator M is boundedly invertible, then (as follows from the discussion leading to and from Theorem 8.3) $\widetilde{\Theta}$ is the characteristic operator function of the $(\widehat{J}_{\mathcal{N}}, \widehat{J}'_{\mathcal{E}})$ unitary colligation

$$\widehat{\tau} = 2^{-\frac{1}{2}}\begin{bmatrix} 2^{\frac{1}{2}} I_{\mathcal{H}(-1)} & 0 & 0 \\ 0 & -I_{\mathcal{E}} & I_{\mathcal{E}} \\ 0 & I_{\mathcal{E}} & I_{\mathcal{E}} \end{bmatrix}\tau\,,$$

where τ is defined by (8.8), $\widehat{J}'_{\mathcal{E}} = \operatorname{diag}(I_{\mathcal{H}(-1)}, J')$ and

$$J' = \begin{bmatrix} 0 & -I_{\mathcal{E}} \\ -I_{\mathcal{E}} & 0 \end{bmatrix}.$$

For additional information on such formulas, see [ArG1] and [ArG2].

References

[ADD] D. Alpay, P. Dewilde and H. Dym, Lossless inverse scattering and reproducing kernels for upper triangular operators, in: *Extension and Interpolation of Linear Operators and Matrix Functions* (I. Gohberg, ed.), *Operator Theory: Advances and Applications*, **OT47**, Birkhäuser Verlag, Basel, 1990, pp. 61–135.

[ArG1] D.Z. Arov and L.Z Grossman, Scattering matrices in the theory of unitary extensions of isometric operators, *Soviet Math. Dokl.* **270** (1983), 17–20.

[ArG2] D.Z. Arov and L.Z Grossman, Scattering matrices in the theory of unitary extensions of isometric operators, *Math. Nachr.* **157** (1992), 105–123.

[Arv] W.B. Arveson, Interpolation problems in nest algebras, *J. Funct. Anal.* **20** (1975), 208–233.

[BGK1] J.A. Ball, I. Gohberg and M.A. Kaashoek, Nevanlinna-Pick interpolation for time-varying input-output maps: The discrete case, in: *Time-variant Systems and Interpolation* (I. Gohberg, ed.), *Operator Theory: Advances and Applications*, **OT56**, Birkhäuser Verlag, Basel, 1992, pp. 1–51.

[BGK2] J.A. Ball, I. Gohberg and M.A. Kaashoek, Two-sided Nudelman interpolation for input-output operators of discrete time-varying systems, *Integral Equations Operator Theory* **21** (1994), 174-211.

[CSK] T. Constantinescu, A.H. Sayed and T. Kailath, Displacement structure and maximum entropy, preprint, 1996.

[DD] P. Dewilde and H. Dym, Interpolation for upper triangular operators, in: *Time-variant Systems and Interpolation* (I. Gohberg, ed.), *Operator Theory: Advances and Applications*, **OT56**, Birkhäuser Verlag, Basel, 1992, pp. 153–260.

[D1] H. Dym, *J Contractive Matrix Functions, Reproducing Kernel Hilbert Spaces and Interpolation*, CBMS Reg. Conf. Ser. in Math., **71**, Amer. Math. Soc., Providence, RI, 1989.

[D2] H. Dym, Remarks on interpolation for upper triangular operators, in: *Challenges of a Generalized System Theory,* (P. Dewilde, M.A. Kaashoek and M. Verhaegen, eds.), North Holland, Amsterdam, 1993, pp. 9–24.

[D3] H. Dym, More on maximum entropy interpolants and maximum determinant completions of associated Pick matrice, *Integral Equations Operator Theory,* **24** (1996) 188–229.

[DF] H. Dym and B. Freydin, Bitangential interpolation for upper triangular operators, in: this volume.

[Fe] A. Feintuch, *Robust Control Theory in Hilbert Space*, Lecture notes, (1995).

[GKW] I. Gohberg, M.A. Kaashoek and H.J. Woerdeman, A maximum entropy principle in the general framework of the band method, *J. Functional Analysis* **95** (1991), 231–254.

[Ig] P. Iglesias, An entropy formula for time-varying discrete-time control system, *SIAM J. Control Optim.*, in press.

[KKY] V.E. Katsnelson, A.Ya. Kheifets and P.M. Yuditskii, An abstract interpolation problem and the theory of extensions of isometric operators, in: *Operators in Function Spaces and Problems in Function Theory* (V.A. Marchenko, ed.), **146**, *Naukova Dumka*, Kiev, 1987, pp. 83–96; English transl., in: this volume.

[Kh1] A.Ya. Kheifets, Parseval equality in abstract interpolation problems and coupling of open systems, *J. Soviet Math.* **49**, No. 4 (1990), 1114–1120; **49**, No. 6 (1990), 1307–1310.

[Kh2] A.Ya. Kheifets, The generalized bitangential Schur-Nevanlinna-Pick problem and the related Parseval equality, *Journal of Soviet Mathematic*, **58**, No. 4 (1992), 358–364.

[KY] A.Ya. Kheifets and P.M. Yuditskii, An analysis and extension of V.P. Potapov's approach to interpolation problems with applications to the generalized bitangential Schur-Nevanlinna-Pick problem and J-inner-outer factorization. in: Matrix and Operator Valued Functions (I. Gohberg and L.A. Sakhnovich, eds.), *Operator Theory: Advances and Application* **OT72**, Birkhäuser Verlag, Basel, 1994, pp. 133–161

[Ko1] J. Kos, Higher order time-varying Nevanlinna-Pick interpolation, in: *Challenges of a Generalized System Theory*, (P. Dewilde, M.A. Kaashoek and M. Verhaegen, eds.), North Holland, Amsterdam, 1993, pp. 59–72.

[Ko2] J. Kos, *Time-dependent problems in linear operator theory*, Ph.D. Thesis, Amsterdam (1995).

[Vv] A. van der Veen, *Time-varying theory and computational modeling*, Ph.D. Thesis, Delft (1993).

Harry Dym
Dept. of Theoretical Mathematics
The Weizmann Institute of Science
Rehovot 76100, Israel

E-mail: dym@wisdom.weizmann.ac.il

Boris Freydin
Dept. of Theoretical Mathematics
The Weizmann Institute of Science
Rehovot 76100, Israel

E-mail: fred@wisdom.weizmann.ac.il

AMS Subject Classification: 47A57, 47A48

Operator Theory
Advances and Applications, Vol. 95
© 1997 Birkhäuser Verlag Basel/Switzerland

Integral representations of a pair of nonnegative operators and interpolation problems in the Stieltjes class

Yu.M. Dyukarev

Abstract. An abstract moment problem (which includes the Stieltjes moment problem and the Nevanlinna-Pick problem as typical representatives) is formulated as an integral representation problem. Potapov's method is used to show that this problem is solvable if and only if two associated Pick operators are nonnegative, or equivalently if and only if a pair of fundamental matrix inequalities are satisfied. The general solution is described in terms of a linear fractional transformation in the nonsingular case. [Abstract added by editors.]

1 Introduction

Definition 1. A function $w(z)$, which is defined and holomorphic on the half-plane $\mathrm{Im}\, z > 0$ and has its values in the set of the bounded operators, acting in some Hilbert space H, is called a Nevanlinna function, if $(w(z) - w^*(z))/2i \geq 0$.

Let us denote the class of all such functions by **N**.

Definition 2. A Nevanlinna function $s(z)$ is called Stieltjes, if it is defined and continuous for $x < 0$ and $s(x) \geq 0$ for $x < 0$.

Let us denote the class of all such functions by **S**. It is obvious, that $\mathbf{S} \subset \mathbf{N}$.

Let us consider the Nevanlinna-Pick problem in the classes **N** and **S**: Given the complex numbers $z_1, \ldots, z_n$ with $\mathrm{Im}\, z_j > 0$ and $z_j \neq z_k$ for $j \neq k$ (the interpolation points) and the bounded operators $s_1, \ldots, s_n$ in H (the interpolation values) the objective is to find the solutions of the following problems:

Problem **A**. Find $w(z) \in \mathbf{N}$ such that $w(z_j) = s_j,\ j = 1, \ldots, n$.

Problem **B**. Find $s(z) \in \mathbf{S}$ such that $s(z_j) = s_j,\ j = 1, \ldots, n$.

Problem (**A**) and analogues to it were used to develop the Potapov method for solving classical interpolation problems [KP], [Ko]. Generalizations of this method were proposed in the papers [Nu], [IS], [KKY].

The Potapov method was extended to problem (**B**) in the paper [DK]. Now that a number of concrete interpolation problems have been solved, it seems to be an appropriate time to search for a general abstract formulation for solving such problems in the class **S**. This paper exhibits one approach. Our constructions will be closed to similar constructions for interpolation problems in the class **N** that were proposed in the paper [IS]. But the special features of the class **S** will constantly appear. For example, instead of a problem about the integral representation of one nonnegative operator we have a problem about consistent integral representations of a pair of nonnegative operators.

2 The formulation of the problem about consistent representation of the pair of nonnegative operators

Let X be a Hilbert space and let $K_1 \geq 0$ and $K_2 \geq 0$ be two nonnegative and bounded operators in X; let T be a bounded operator in X, the spectrum of which consists of not more than a countable set of points, such that the points of spectrum in the half-planes Im $z > 0$ and Im $z < 0$ are isolated. Let $R(t) = (T - tI)^{-1}$ denote the resolvent of the operator T; let H be a Hilbert space and let $v : H \to X$, $u : H \to X$ be bounded operators; $\sigma(t)$ is a nondecreasing function defined on the interval $[0, +\infty)$ and having values in the set of bounded Hermitian operators acting in H; $\gamma \geq 0$ is a bounded operator in H.

Let the operators be connected by the *Fundamental Identity*

$$K_2 - TK_1 = vu^*. \tag{1}$$

We need to find conditions under which the operators K_1 and K_2 possess consistent integral representations of the form:

$$K_1 = \int_0^\infty R(t)vd\sigma(t)v^*R^*(t), \quad K_2 = \int_0^\infty R(t)vtd\sigma(t)v^*R^*(t) + v\gamma v^*, \tag{2}$$

and the operator u has an integral representation of the form

$$u = -\int_0^\infty R(t)vd\sigma(t) + v\gamma. \tag{3}$$

The integrals used in the representations (2) and (3) are understood in the weak sense. We also suppose that $\int_0^\infty (1+t)^{-1}d\sigma(t) < \infty$.

Definition 3. The function $\sigma(t)$ and the operator γ which were introduced in the representations in (2), are called the solution of the problem on consistent integral representation of the operators K_1 and K_2.

Two more identities can be obtained from (1) and the fact that K_1 and K_2 are Hermitian:

$$\begin{aligned} TK_1 - K_1T^* &= uv^* - vu^*, &(4)\\ TK_2 - K_2T^* &= Tuv^* - vu^*T^*. &(5) \end{aligned}$$

Identities having the form (4) (or (5)) play a fundamental role in the solution of Problem **A** and problems which are similar to it (see, for example [IS]). If only these two identities take place, then K_1 and K_2 possess the representation indicated in (2). But if in the representation of K_1 the measure $d\sigma(t)$ is involved, then in the representation K_2 a measure $d\tau(t)$ is involved, which is not connected to the measure $d\sigma(t)$.

The identity (1) is characteristic for interpolation in the class **S** and does not have an analogue when solving problems of class (**A**). This identity leads to the consistency of the representations K_1 and K_2 in formulas (2).

Theorem 1. *If the integral representations (2) and (3) take place, then the Fundamental Identity (1) holds.*

Proof. We have

$$\begin{aligned} K_2 - K_1 T^* &= \int_0^\infty R(t) v t d\sigma(t) v^* R^*(t) + v\gamma v^* - \int_0^\infty R(t) v d\sigma(t) v^* R^*(t) T^* \\ &= \int_0^\infty R(t) v d\sigma(t) v^* (tI - T^*) R^*(t) + v\gamma v^* \\ &= -\int_0^\infty R(t) v d\sigma(t) v^* + v\gamma v^* = uv^*. \end{aligned}$$

The theorem is proved. □

It turns out that when solving problem **B** and analogues to it we always come to the problem formulated above about consistent representation of the pair of nonnegative operators.

Remark. Let us give the form of the operators introduced for problem **B** when the space X is of the form

$$X = \underbrace{H \oplus \ldots \oplus H}_{n}.$$

In the obvious matrix representations, we have

$$K_1 = \begin{bmatrix} \frac{s_1 - s_1^*}{z_1 - \bar{z}_1} & \cdots & \frac{s_1 - s_n^*}{z_1 - \bar{z}_n} \\ \vdots & \ddots & \vdots \\ \frac{s_n - s_1^*}{z_n - \bar{z}_1} & \cdots & \frac{s_n - s_n^*}{z_n - \bar{z}_n} \end{bmatrix}, \quad K_2 = \begin{bmatrix} \frac{z_1 s_1 - \bar{z}_1 s_1^*}{z_1 - \bar{z}_1} & \cdots & \frac{z_1 s_1 - \bar{z}_n s_n^*}{z_1 - \bar{z}_n} \\ \vdots & \ddots & \vdots \\ \frac{z_n s_n - \bar{z}_1 s_1^*}{z_n - \bar{z}_1} & \cdots & \frac{z_n s_n - \bar{z}_n s_n^*}{z_n - \bar{z}_n} \end{bmatrix},$$

$$T = \begin{bmatrix} z_1 I & 0 & \ldots & 0 \\ 0 & z_2 I & \ldots & 0 \\ \vdots & \vdots & \ddots & \vdots \\ 0 & 0 & \ldots & z_n I \end{bmatrix}, \quad u = \begin{bmatrix} s_1 \\ s_2 \\ \vdots \\ s_n \end{bmatrix} \quad \text{and} \quad v = \begin{bmatrix} I \\ I \\ \vdots \\ I \end{bmatrix}.$$

The identity (1) can be checked immediately.

3 The system of fundamental matrix inequalities of V.P. Potapov

In this section the problem about the integral representation will be transformed to a function theoretic problem. After that the integral representation of the operator u will be obtained.

Definition 4. If the function $\sigma(t)$ and operator γ are the solutions of the problem on consistent integral representation of the operators K_1 and K_2, then the Stieltjes function

$$s(z) = \gamma + \int_0^\infty \frac{d\sigma(t)}{t-z}$$

is said to be associated with the problem on consistent integral representation.

If the function $s(z)$ is known, then, using the inversion formula, we can find $\sigma(t)$ and γ.

Theorem 2. *If $s(z)$ is associated with problem on integral representation, then it satisfies the system of Fundamental Matrix Inequalities (FMI) of V.P. Potapov:*

$$\left[\begin{array}{c|c} K_1 & R(z)\,[u - vs(z)] \\ \hline * & \frac{s(z)-s^*(z)}{z-\bar{z}} \end{array}\right] \geq 0 \text{ and } \left[\begin{array}{c|c} K_2 & R(z)\,[Tu - vzs(z)] \\ \hline * & \frac{zs(z)-\bar{z}s^*(z)}{z-\bar{z}} \end{array}\right] \geq 0. \quad (6)$$

The left hand sides of the inequalities are understood as matrix representations of operators in the Hilbert space $X \oplus H$.

Proof. We have

$$\begin{aligned} &R(z)(u - vs(z)) \\ &\quad = R(z)\left[-\int_0^\infty R(t)vd\sigma(t) + v\gamma - v\gamma - v\int_0^\infty d\sigma(t)/(t-z)\right] \\ &\quad = -\int_0^\infty [R(z)R(t) + R(z)/(t-z)]vd\sigma(t) \\ &\quad = -\int_0^\infty \left[\frac{R(t)-R(z)}{t-z} + \frac{R(z)}{t-z}\right] vd\sigma(t) = -\int_0^\infty \frac{R(t)v}{t-z}d\sigma(t), \\ &\frac{s(z)-s^*(z)}{z-\bar{z}} = \int_0^\infty \frac{d\sigma(t)}{(t-z)(t-\bar{z})} \quad \text{and } K_1 = \int_0^\infty R(t)vd\sigma(t)v^*R^*(t). \end{aligned}$$

Thus

$$\left[\begin{array}{c|c} K_1 & R(z)\,[u - vs(z)] \\ \hline * & \frac{s(z)-s^*(z)}{z-\bar{z}} \end{array}\right] = \int_0^\infty \left[\begin{array}{c} R(t)v \\ \frac{-I}{t-z} \end{array}\right] d\sigma(t) \left[v^*R^*(t); \frac{-I}{t-\bar{z}}\right] \geq 0.$$

Therefore, the first FMI is proved. In the same way

$$\left[\begin{array}{c|c} K_2 & R(z)\left[Tu - vzs(z)\right] \\ \hline * & \dfrac{zs(z)-\bar{z}s^*(z)}{z-\bar{z}} \end{array}\right]$$

$$= \int_0^\infty \begin{bmatrix} R(t)v \\ \frac{-I}{t-\bar{z}} \end{bmatrix} td\sigma(t)\left[v^*R^*(t); \frac{-I}{t-z}\right] + \begin{bmatrix} v \\ I \end{bmatrix}\gamma[v^*I] \geq 0.$$

The theorem is proved. □

Definition 5. Let $\mathcal{N}$ denote the class of all bounded operators T in X such that there exists a region Ω in the half-plane $\operatorname{Im} z > 0$ such that for each solution $s(z)$ of the FMI and for each vector $g \in X$ with $\|g\| = 1$ we have

1. $\|(s(z) - s^*(z))/(z - \bar{z})\|$ is bounded for $z \in \Omega$.
2. $|(R(z)g, e)|$ is bounded, $z \in \Omega$, $\forall e \in X$, $\|e\| = 1 \Leftrightarrow g = 0$.

Remark. For example for the operator T of Problem **B** we can take any bounded region in the half-plane $\operatorname{Im} z > \varepsilon$, $\varepsilon > 0$, that includes the spectrum of the operator T. Condition 1 of Definition 5 is obvious. Let us check Condition 2. Let $e = g$ and in accordance with the representation

$$X = \underbrace{H \oplus \ldots \oplus H}_{n},$$

let us write

$$g = g_1 + \ldots + g_n.$$

Then Condition 2 takes the form

$$\frac{\|g_1\|^2}{|z_1 - z|} + \ldots + \frac{\|g_n\|^2}{|z_n - z|} = O(1) \quad \text{for} \quad z \in \Omega.$$

Because $z_1, \ldots, z_n \in \Omega$, it follows that we must have $g_1 = \ldots = g_n = 0$. Therefore $g = 0$.

Theorem 3. *Let $s(z)$ be holomorphic in the half plane $\operatorname{Im} z > 0$ and satisfy the system FMI (6). Then*

1. $s(z) \in \mathbf{S}$, $s(z) = \gamma + \int_0^\infty \frac{d\sigma(t)}{t-z}$ *and* $\int_0^\infty \frac{d\sigma(t)}{1+t} < \infty$.
2. *If $T \in \mathcal{N}$ and integrals*

$$\int_0^\infty R(t)vd\sigma(t)v^*R^*(t) \quad \text{and} \quad \int_0^\infty R(t)vd\sigma(t)$$

are understood in the weak sense, then $u = -\int_0^\infty R(t)vd\sigma(t) + v\gamma$.

Proof.

1. We have $(s(z)-s^*(z))/(z-\bar z)\geq 0$, and $(zs(z)-\bar z s^*(z))/(z-\bar z)\geq 0$. Now the result of Theorem 1 comes out from the results of [KN].

2. Under the conditions of the theorem, we can consider the operators

$$K_1^\sigma=\int_0^\infty R(t)v\,d\sigma(t)v^*R^*(t)\quad\text{and}\quad u^\sigma=-\int_0^\infty R(t)v\,d\sigma(t)+v\gamma.$$

From the condition and Theorem 2 we have

$$\left[\begin{array}{c|c} K_1 & R(z)\,[u-vs(z)] \\ \hline * & \dfrac{s(z)-s^*(z)}{z-\bar z}\end{array}\right]\geq 0\ \text{ and }\ \left[\begin{array}{c|c} K_1^\sigma & R(z)\,[u^\sigma-vs(z)] \\ \hline * & \dfrac{s(z)-s^*(z)}{z-\bar z}\end{array}\right]\geq 0.$$

From this it follows that

$$\left[\begin{array}{c|c} (K_1e,e) & (R(z)\,[u-vs(z)]f,e) \\ \hline * & \left(\dfrac{s(z)-s^*(z)}{z-\bar z}f,f\right)\end{array}\right]\geq 0,\qquad \begin{array}{c}\forall e\in X,\ \forall f\in H,\\ \|e\|=\|f\|=1.\end{array}$$

Therefore

$$\left[\begin{array}{c|c} \|K_1\| & (R(z)\,[u-vs(z)]f,e) \\ \hline * & \left\|\dfrac{s(z)-s^*(z)}{z-\bar z}\right\|\end{array}\right]\geq 0.$$

Thus

$$|(R(z)\,[u-vs(z)]f,e)|^2\leq\|K_1\|\left\|\frac{s(z)-s^*(z)}{z-\bar z}\right\|.$$

Therefore

$$|(R(z)\,[u-vs^*(z)]f,e)|=O(1),\quad z\in\Omega,\quad \|e\|=\|f\|=1.$$

In the same way,

$$|(R(z)\,[u^\sigma-vs(z)]f,e)|=O(1),\quad z\in\Omega,\quad \|e\|=\|f\|=1.$$

From these asymptotic expressions we obtain

$$|(R(z)\,(u-u^\sigma)f,e)|=O(1),\quad z\in\Omega,\quad \forall e\in X,\ \forall f\in H,\ \|e\|=\|f\|=1.$$

Therefore $u=u^\sigma$, because $T\in\mathcal{N}$.

The theorem is proved. □

Remark. For Problem **B** we have

$$u=\begin{bmatrix} s_1\\ \vdots\\ s_n\end{bmatrix}\quad\text{and}\quad u^\sigma=-\int_0^\infty\begin{bmatrix}\frac{I}{z_1-t}\\ \vdots\\ \frac{I}{z_n-t}\end{bmatrix}d\sigma(t)+\begin{bmatrix}\gamma\\ \vdots\\ \gamma\end{bmatrix}=\begin{bmatrix}s(z_1)\\ \vdots\\ s(z_n)\end{bmatrix}.$$

It follows from $u=u^\sigma$ that $s(z_1)=s_1,\ldots,s(z_n)=s_n$.

So for problem **B**, the integral representation of the operator u is equivalent to the solution of the Nevanlinna-Pick problem.

4 The transformation of the FMI of V.P. Potapov

In order to obtain the integral representation of the operators K_1 and K_2, we shall make use of a special transformation [Ka].

Theorem 4. *The system FMI of V.P. Potapov is equivalent to the system of the Transformed Fundamental Matrix Inequalities (TFMI) of V.P. Potapov*

$$\left[\begin{array}{c|c} K_1 & S_1^*(z) \\ \hline S_1(z) & \frac{S_1(z)-S_1^*(z)}{z-\bar{z}} \end{array}\right] \geq 0 \text{ and } \left[\begin{array}{c|c} K_2 & S_2^*(z) \\ \hline S_2(z) & \frac{S_2(z)-S_2^*(z)}{z-\bar{z}} \end{array}\right] \geq 0\,, \tag{7}$$

where

$$S_1(z) = R(z)K_1 - R(z)v(u^* - s(z)v^*)R^*(\bar{z}) \tag{8}$$

and

$$S_2(z) = R(z)K_2 - R(z)v(u^*T^* - zs(z)v^*)R^*(\bar{z}). \tag{9}$$

The functions $S_1(z)$ and $S_2(z)$ are connected by the equation

$$S_2(z) = K_1 + zS_1(z). \tag{10}$$

Proof. In the first inequality of the FMI system we will denote the independent variable by $\bar{z}$. Using the symmetry principle $s^*(\bar{z}) = s(z)$, we will have

$$\left[\begin{array}{c|c} K_1 & R(\bar{z})\,[u - vs^*(z)] \\ \hline * & \frac{s(z)-s^*(z)}{z-\bar{z}} \end{array}\right] \geq 0.$$

Multiplying this inequality from the left and the right sides by the matrices

$$\left[\begin{array}{c|c} I & 0 \\ \hline R(z) & -R(z)v \end{array}\right] \text{ and } \left[\begin{array}{c|c} I & R^*(z) \\ \hline 0 & -v^*R^*(z) \end{array}\right],$$

respectively, we see that

$$\left[\begin{array}{c|c} I & 0 \\ \hline R(z) & -R(z)v \end{array}\right] \left[\begin{array}{c|c} K_1 & R(\bar{z})\,[u - vs^*(z)] \\ \hline * & \frac{s(z)-s^*(z)}{z-\bar{z}} \end{array}\right] \left[\begin{array}{c|c} I & R^*(z) \\ \hline 0 & -v^*R^*(z) \end{array}\right]$$

$$= \left[\begin{array}{c|c} K_1 & K_1R^*(z) - R(\bar{z})[u - vs^*(z)]v^*R^*(z) \\ \hline * & \begin{array}{c} R(z)K_1R^*(z) - R(z)v[u^* - s(z)v^*]R^*(\bar{z})R^*(z) - R(z) \\ \times R(\bar{z})[u - vs^*(z)]v^*R^*(z) + R(z)v\frac{s(z)-s^*(z)}{z-\bar{z}}v^*R^*(z) \end{array} \end{array}\right]$$

$$\geq 0.$$

Let

$$S_1(z) = R(z)K_1 - R(z)v(u^* - s(z)v^*)R^*(\bar{z}).$$

Then the last inequality takes the form

$$\left[\begin{array}{c|c} K_1 & S_1^*(z) \\ \hline S_1(z) & \dfrac{S_1(z)-S_1^*(z)}{z-\bar z} \end{array}\right] \geq 0.$$

Indeed,

$$\begin{aligned}
&R(z)K_1R^*(z) - R(z)v[u^* - s(z)v^*]R^*(\bar z)R^*(z) \\
&\quad -R(z)R(\bar z)[u - vs^*(z)]v^*R^*(z) + R(z)v\frac{s(z)-s^*(z)}{z-\bar z}v^*R^*(z) \\
&\quad = R(z)K_1R^*(z) - R(z)v[u^* - s(z)v^*]\frac{R^*(\bar z)-R^*(z)}{z-\bar z} \\
&\quad -\frac{R(z)-R(\bar z)}{z-\bar z}[u - vs^*(z)]v^*R^*(z) + R(z)v\frac{s(z)-s^*(z)}{z-\bar z}v^*R^*(z) \\
&\quad = R(z)K_1R^*(z) - R(z)v[u^* - s(z)v^*]\frac{R^*(\bar z)}{z-\bar z} \\
&\quad +\frac{R(\bar z)}{z-\bar z}[u - vs^*(z)]v^*R^*(z) + \frac{1}{z-\bar z}\{R(z)v[u^* - s(z)v^*]R^*(z) \\
&\quad -R(z)[u - vs^*(z)]v^*R^*(z) \\
&\quad +R(z)vs(z)v^*R^*(z) - R(z)vs^*(z)v^*R^*(z)\} \\
&\quad = R(z)K_1R^*(z) - \frac{R(z)v[u^* - s(z)v^*]R^*(\bar z)}{z-\bar z} \\
&\quad +\frac{R(\bar z)[u - vs^*(z)]v^*R^*(z)}{z-\bar z} + \frac{R(z)[vu^* - uv^*]R^*(z)}{z-\bar z} \\
&\quad = -\frac{R(z)v[u^* - s(z)v^*]R^*(\bar z)}{z-\bar z} + \frac{R(\bar z)[u - vs^*(z)]v^*R^*(z)}{z-\bar z} \\
&\quad +\frac{R(z)[zK_1 - \bar zK_1 + vu^* - uv^*]R^*(z)}{z-\bar z} \\
&\quad = -\frac{R(z)v[u^* - s(z)v^*]R^*(\bar z)}{z-\bar z} + \frac{R(\bar z)[u - vs^*(z)]v^*R^*(z)}{z-\bar z} \\
&\quad +\frac{R(z)[zK_1 - \bar zK_1 - TK_1 + K_1T^*]R^*(z)}{z-\bar z} \\
&\quad = \frac{R(z)K_1 - R(z)v(u^* - s(z)v^*)R^*(\bar z)}{z-\bar z} \\
&\quad -\frac{K_1R^*(z) - R(\bar z)(u - vs^*(z))v^*R^*(z)}{z-\bar z} = (S_1(z) - S_1^*(z))/(z-\bar z).
\end{aligned}$$

The first of the inequalities (7) is proved. The second inequality is proved in an analogous way. Obviously the FMI follows from the TFMI.

Let us now show that (10) is valid. We have

$$S_2(z) = R(z)K_2 - R(z)v(u^*T^* - zs(z)v^*)R^*(\bar z)$$

$$
\begin{aligned}
&= R(z)TK_1 + R(z)vu^* - R(z)vu^*T^*R^*(\bar{z}) + R(z)vzs(z)v^*R^*(\bar{z}) \\
&= R(z)TK_1 + R(z)vu^*(T^* - zI - T^*)R^*(\bar{z}) + R(z)vzs(z)v^*R^*(\bar{z}) \\
&= R(z)TK_1 - zR(z)v(u^* - s(z)v^*)R^*(\bar{z}) \\
&= R(z)(T - zI + zI)K_1 - zR(z)v(u^* - s(z)v^*)R^*(\bar{z}) \\
&= K_1 + z\{R(z)K_1 - R(z)v(u^* - s(z)v^*)R^*(\bar{z})\} \\
&= K_1 + zS_1(z).
\end{aligned}
$$

The theorem is proved. □

Theorem 5. *Let the functions $S_1(z)$ and $S_2(z) = K_1 + zS_1(z)$ be solutions of the TFMI system. Then there exists a monotonically increasing function $\Sigma(t)$ such that*

$$S_1(z) = \int_0^\infty \frac{d\Sigma(t)}{t - z}, \quad S_2(z) = \int_0^\infty \frac{td\Sigma(t)}{t - z}\ , \tag{11}$$

$$K_1 = \int_0^\infty d\Sigma(t) \quad \text{and} \quad \int_0^\infty td\Sigma(t) \le K_2\ , \tag{12}$$

where all the integrals are understood in the weak sense.

Proof. The function $S_1(z)$ has only isolated singularities in Im $z > 0$. At all other points the inequalities $(S_1(z) - S_1^*(z))/2i \ge 0$ and $(zS_1(z) - \bar{z}S_1^*(z))/2i \ge 0$ hold. Therefore all the singularities of $S_1(z)$ in the half-plane Im $z > 0$ are removable and we can assume that $S_1(z) \in \mathbf{S}$. Now put $z = iy$ with $y > 0$ in the first of the inequalities (7). We obtain

$$\left[\begin{array}{c|c} (K_1e, e) & (S_1^*(iy)g, e) \\ \hline * & \left(\frac{S_1(iy) - S_1^*(iy)}{2yi}g, g\right) \end{array}\right] \ge 0, \quad \forall e, g \in X, \quad \|e\| \le 1, \quad \|g\| \le 1.$$

From the inequalities

$$(K_1e, e) \le \|K_1\| \quad \text{and} \quad \left(\frac{S_1(iy) - S_1^*(iy)}{2yi}g, g\right) \le \frac{\|S_1(iy)\|}{y}$$

we obtain

$$\left[\begin{array}{c|c} \|K_1\| & (S_1^*(iy)g, e) \\ \hline * & \|S_1(iy)\|/y \end{array}\right] \ge 0, \quad \forall e, g \in X, \quad \|e\| \le 1, \quad \|g\| \le 1,$$

and

$$|(S_1(iy)g, e)|^2 \le \|K_1\|\frac{\|S_1(iy)\|}{y} \quad \text{and} \quad y\|S_1(iy)\| \le \|K_1\| \quad \text{for} \quad y > 0.$$

Therefore [KN]

$$S_1(z)=\int_0^\infty \frac{d\Sigma(t)}{t-z} \quad\text{and}\quad K_1^\Sigma=\int_0^\infty d\Sigma(t) \text{ is a bounded operator.}$$

Moreover,

$$-\lim_{y\to+\infty} iyS_1(iy)=\lim_{y\to+\infty} iyS_1^*(iy)=\lim_{y\to+\infty} y(S_1(iy)-S_1^*(iy))/2i=K_1^\Sigma.$$

Let us multiply the first of the inequalities (7) from the left and the right side by the matrices

$$\left[\begin{array}{c|c} I & 0 \\ \hline 0 & -iyI \end{array}\right] \quad\text{and}\quad \left[\begin{array}{c|c} I & 0 \\ \hline 0 & iyI \end{array}\right],$$

respectively. Upon passing to the limit $y\to+\infty$, we obtain

$$\left[\begin{array}{c|c} K_1 & K_1^\Sigma \\ \hline K_1^\Sigma & K_1^\Sigma \end{array}\right] \geq 0\,.$$

From this it follows that $K_1^\Sigma \leq K_1$.

By similar considerations, the second of the inequalities (7) leads to the conclusion that

$$S_2(z)=\int_0^\infty \frac{d\Sigma_2(t)}{t-z} \quad\text{and}\quad \int_0^\infty d\Sigma_2(t) \text{ is a bounded operator.}$$

Moreover,

$$S_2(z)=K_1+zS_1(z)=(K_1-K_1^\Sigma)+\int_0^\infty \frac{td\Sigma(t)}{t-z}$$

and

$$\int_0^\infty \frac{d\Sigma_2(t)}{t-z}=(K_1-K_1^\Sigma)+\int_0^\infty \frac{td\Sigma(t)}{t-z}.$$

From this it follows that

$$K_1=K_1^\Sigma \quad\text{and}\quad S_2(z)=\int_0^\infty \frac{td\Sigma(t)}{t-z}.$$

Finally, by paralleling the proof of the inequality $K_1^\Sigma \leq K_1$, we obtain

$$\int_0^\infty td\Sigma(t)\leq K_2.$$

The theorem is proved. □

Definition 6. Let $\mathcal{N}_0$ denote the class of all bounded operators $T\in\mathcal{N}$ such that there exists an $\varepsilon>0$ and $0<\delta<\pi/2$ such that the region Ω which is used in Definition 5 can be chosen in the form

$$\Omega_\varepsilon^\delta=\{z:\delta<\arg(z)<\pi-\delta, |z|>\varepsilon\}.$$

Theorem 6. *Let $s(z)$ satisfy the system FMI (6) and admit the representation*

$$s(z) = \gamma + \int_0^\infty \frac{d\sigma(t)}{t-z}.$$

Then,

1. $K_1 = \int_0^\infty R(t)vd\sigma(t)v^*R^*(t)$ *and* $\int_0^\infty R(t)vtd\sigma(t)v^*R^*(t) \leq K_2$.

2. *If* $T \in \mathcal{N}_0$, *then* $K_2 = \int_0^\infty R(t)vtd\sigma(t) + v\gamma v^*$.

Proof. 1. Let $0 < \alpha < \beta < +\infty$ and let $R(z)$ be holomorphic in a region that includes the segment $[\alpha, \beta]$. Then using formulas (8), (9), (11) and the formula of inversion from [KN] and [IS], we obtain

$$\int_\alpha^\beta d\Sigma(t) = \int_\alpha^\beta R(t)vd\sigma(t)v^*R^*(t) \quad \text{and} \quad \int_\alpha^\beta td\Sigma(t) = \int_\alpha^\beta R(t)vtd\sigma(t)v^*R^*(t).$$

Therefore, by (12),

$$K_1 = \int_0^\infty R(t)vd\sigma(t)v^*R^*(t) \quad \text{and} \quad \int_0^\infty R(t)vtd\sigma(t)v^*R^*(t) \leq K_2.$$

2. Let $\tilde{K}_2 = \int_0^\infty R(t)vtd\sigma(t)v^*R^*(t) + v\gamma v^*, \quad \tilde{u} = -\int_0^\infty R(t)vtd\sigma(t) + v\gamma,$

$$S_2(z) = R(z)K_2 - R(z)v(u^*T^* - zs(z)v^*)R^*(\bar{z})$$

and

$$\tilde{S}_2(z) = R(z)\tilde{K}_2 - R(z)v(\tilde{u}^*T^* - zs(z)v^*)R^*(\bar{z}).$$

Using the condition and Theorem 2, we have

$$\left[\begin{array}{c|c} K_2 & S_2^*(z) \\ \hline S_2(z) & \dfrac{S_2(z) - S_2^*(z)}{z - \bar{z}} \end{array}\right] \geq 0 \text{ and } \left[\begin{array}{c|c} \tilde{K}_2 & \tilde{S}_2^*(z) \\ \hline \tilde{S}_2(z) & \dfrac{\tilde{S}_2(z) - \tilde{S}_2^*(z)}{z - \bar{z}} \end{array}\right] \geq 0\ .$$

Let $\Omega_\varepsilon^\delta$ be the region from Definition 6. Using the last inequalities we have

$$|(S_2(z)f, e)| = O(1) \text{ and } |(\tilde{S}_2(z)f, e)| = O(1), \quad z \in \Omega_\varepsilon^\delta,\ \forall f, e \in X,\ \|f\| = \|e\| = 1.$$

From this it follows that

$$|((S_2(z) - \tilde{S}_2(z))f, e)| = O(1), \quad z \in \Omega_\varepsilon^\delta,\ \forall f, e \in X,\ \|f\| = \|e\| = 1.$$

From Theorem 3, $u = \tilde{u}$ and therefore,

$$S_2(z) - \tilde{S}_2(z) = R(z)(K_2 - \tilde{K}_2).$$

Therefore

$$|(R(z)(K_2 - \tilde{K}_2)f, e)| = O(1), \quad z \in \Omega_\varepsilon^\delta,\ \forall f, e \in X,\ \|f\| = \|e\| = 1.$$

From this it follows that $K_2 = \tilde{K}_2$ because $T \in \mathcal{N}_0$.
The theorem is proved. □

Remark. For problem **B**, when the space X is as in the remark following Theorem 1, we have

$$\begin{aligned} K_1 &= \left[\frac{s_j - s_k^*}{z_j - \bar{z}_k}\right] = \left[\frac{s(z_j) - s^*(z_k)}{z_j - \bar{z}_k}\right] \\ &= \left[\frac{1}{z_j - \bar{z}_k}\left(\int_0^\infty \frac{d\sigma(t)}{t - z_j} - \int_0^\infty \frac{d\sigma(t)}{t - \bar{z}_k}\right)\right] \\ &= \left[\int_0^\infty \frac{d\sigma(t)}{(t - z_j)(t - \bar{z}_k)}\right] \\ &= \int_0^\infty \begin{bmatrix} \frac{I}{z_1 - t} \\ \vdots \\ \frac{I}{z_n - t} \end{bmatrix} d\sigma(t) \left[\frac{I}{\bar{z}_1 - t} \cdots \frac{I}{\bar{z}_n - t}\right] \\ &= \int_0^\infty R(t) v d\sigma(t) v^* R^*(t). \end{aligned}$$

In an analogous way,

$$\begin{aligned} K_2 &= \left[\frac{z_j s_j - \bar{z}_k s_k^*}{z_j - \bar{z}_k}\right] = \left[\frac{z_j s(z_j) - \bar{z}_k s^*(z_k)}{z_j - \bar{z}_k}\right] \\ &= \int_0^\infty \begin{bmatrix} \frac{I}{z_1 - t} \\ \vdots \\ \frac{I}{z_n - t} \end{bmatrix} t d\sigma(t) \left[\frac{I}{\bar{z}_1 - t} \cdots \frac{I}{\bar{z}_n - t}\right] + \begin{bmatrix} I \\ \vdots \\ I \end{bmatrix} \gamma \left[I \dots I\right] \\ &= \int_0^\infty R(t) v t d\sigma(t) v^* R^*(t) + v \gamma v^*. \end{aligned}$$

Let us show one remarkable solution of the system of FMI of V.P. Potapov.

Theorem 7. *If the function*

$$s^\mu(z) = u^*(K_2 - z\,K_1)^{-1} u,$$

is defined and holomorphic on the half-plane $\operatorname{Im} z > 0$, *then it is a solution of the system of FMI's (6).*

Proof. We only need to prove that the functions $S_1^\mu(z)$, which is built using formula (8), and $S_2^\mu(z) = K_1 + z S_1^\mu(z)$ satisfy the transformed system (7). We have

$$\begin{aligned} S_1^\mu(z) &= R(z) K_1 - R(z) v [u^* - s^\mu(z) v^*] R^*(\bar{z}) \\ &= R(z) K_1 - R(z) v [u^* - u^*(K_2 - z\,K_1)^{-1} u v^*] R^*(\bar{z}) \\ &= R(z) \{K_1 - v u^* (K_2 - z\,K_1)^{-1} [K_2 - z\,K_1 - u v^*] R^*(\bar{z})\} \end{aligned}$$

$$
\begin{aligned}
&= R(z)\{K_1 - vu^*(K_2 - z K_1)^{-1}[K_1T^* - z K_1]R^*(\bar{z})\}\\
&= R(z)\{K_1 - vu^*(K_2 - z K_1)^{-1}K_1\}\\
&= R(z)\{K_2 - z K_1 - vu^*\}(K_2 - z K_1)^{-1}K_1\\
&= R(z)\{TK_1 - z K_1\}(K_2 - z K_1)^{-1}K_1\\
&= K_1(K_2 - z K_1)^{-1}K_1.
\end{aligned}
$$

Therefore

$$S_1^{\mu}(z) = K_1(K_2 - z K_1)^{-1}K_1.$$

In an analogous way

$$S_2^{\mu}(z) = K_1(K_2 - z K_1)^{-1}K_2,$$

$$\frac{S_1^{\mu}(z) - S_1^{\mu *}(z)}{z - \bar{z}} = K_1(K_2 - z K_1)^{-1}K_1(K_2 - z K_1)^{-1*}K_1$$

and

$$\frac{S_2^{\mu}(z) - S_2^{\mu}(z)^*}{z - \bar{z}} = K_1(K_2 - z K_1)^{-1}K_2(K_2 - z K_1)^{-1*}K_1.$$

Bearing in mind these equalities, the system (7) can be written in the following way:

$$\left[\begin{array}{c|c} I & 0 \\ \hline 0 & K_1(K_2 - zK_1)^{-1} \end{array}\right]\left[\begin{array}{c|c} K_1 & K_1 \\ \hline K_1 & K_1 \end{array}\right]\left[\begin{array}{c|c} I & 0 \\ \hline 0 & (K_2 - zK_1)^{-1*}K_1 \end{array}\right] \geq 0,$$

$$\left[\begin{array}{c|c} I & 0 \\ \hline 0 & K_1(K_2 - zK_1)^{-1} \end{array}\right]\left[\begin{array}{c|c} K_2 & K_2 \\ \hline K_2 & K_2 \end{array}\right]\left[\begin{array}{c|c} I & 0 \\ \hline 0 & (K_2 - zK_1)^{-1*}K_1 \end{array}\right] \geq 0.$$

The theorem is proved. □

5 J-Expansive functions associated with a nondegenerate system of FMI's

Definition 7. The FMI system of V.P. Potapov is termed nondegenerate if the operators K_1^{-1} and K_2^{-1} exist and are bounded.

In this case the FMI system is associated with two functions that are J-contractive in the half-plane $\operatorname{Im} z > 0$ and J-expansive in the half-plane $\operatorname{Im} z < 0$, with respect to the J-operator in $H \oplus H$ that is given by block matrix

$$J = \begin{bmatrix} 0 & -iI \\ iI & 0 \end{bmatrix},$$

where I is the unit operator in H.

Theorem 8. *Let*

$$U_1(z) = \left\{ I_2 + i \begin{bmatrix} u^* \\ v^* \end{bmatrix} R_{T^*}(z) K_1^{-1}[u,v]J \right\} \begin{bmatrix} I & 0 \\ M_1 & I \end{bmatrix} \tag{13}$$

and

$$U_2(z) = \left\{ I_2 + i \begin{bmatrix} u^*T^* \\ v^* \end{bmatrix} R_{T^*}(z) K_2^{-1}[Tu,v]J \right\} \begin{bmatrix} I & -M_2 \\ 0 & I \end{bmatrix}, \tag{14}$$

where I_2 is the unit operator in $H \oplus H$,

$$R_{T^*}(z) = (T^* - zI)^{-1}, \quad M_1 = v^* K_2^{-1} v \text{ and } M_2 = u^* K_1^{-1} u.$$

Then

$$\frac{U_1(z)JU_1^*(z) - J}{i(\bar{z} - z)} = - \begin{bmatrix} u^* \\ v^* \end{bmatrix} R_{T^*}(z) K_1^{-1} R_{T^*}^*(z)[u,v], \tag{15}$$

$$\frac{U_2(z)JU_1^*(z) - J}{i(\bar{z} - z)} = - \begin{bmatrix} u^*T^* \\ v^* \end{bmatrix} R_{T^*}(z) K_2^{-1} R_{T^*}^*(z)[Tu,v] \tag{16}$$

and the functions $U_1(z)$ and $U_2(z)$ are connected by relation

$$U_2(z) = \begin{bmatrix} zI & 0 \\ 0 & I \end{bmatrix} U_1(z) \begin{bmatrix} z^{-1}I & 0 \\ 0 & I \end{bmatrix}. \tag{17}$$

Proof. We have

$$\begin{bmatrix} I & 0 \\ M_1 & I \end{bmatrix} J \begin{bmatrix} I & M_1 \\ 0 & I \end{bmatrix} = J.$$

Therefore,

$$\begin{aligned}
&U_1(z)JU_1^*(z) - J \\
&= \left\{ I_2 + i \begin{bmatrix} u^* \\ v^* \end{bmatrix} R_{T^*}(z) K_1^{-1}[u,v]J \right\} \cdot J \left\{ I_2 - iJ \begin{bmatrix} u^* \\ v^* \end{bmatrix} K_1^{-1} R_{T^*}^*(z)[u,v] \right\} - J \\
&= i \begin{bmatrix} u^* \\ v^* \end{bmatrix} R_{T^*}(z) K_1^{-1}[u,v] - i \begin{bmatrix} u^* \\ v^* \end{bmatrix} K_1^{-1} R_{T^*}^*(z)[u,v] \\
&\quad + \begin{bmatrix} u^* \\ v^* \end{bmatrix} R_{T^*}(z) K_1^{-1}[u,v]J \begin{bmatrix} u^* \\ v^* \end{bmatrix} K_1^{-1} R_{T^*}^*(z)[u,v] \\
&= i \begin{bmatrix} u^* \\ v^* \end{bmatrix} \{ -K_1^{-1} R_{T^*}^*(z) + R_{T^*}(z) K_1^{-1} + R_{T^*}(z) K_1^{-1}(-uv^* + vu^*) \\
&\quad \times K_1^{-1} R_{T^*}^*(z) \} [u,v]
\end{aligned}$$

$$= i \begin{bmatrix} u^* \\ v^* \end{bmatrix} \{-K_1^{-1} R_{T^*}^*(z) + R_{T^*}(z) K_1^{-1} + R_{T^*}(z) K_1^{-1}(-TK_1 + K_1 T^*)$$
$$\times K_1^{-1} R_{T^*}^*(z)\}[u, v]$$
$$= i \begin{bmatrix} u^* \\ v^* \end{bmatrix} \{-K_1^{-1} R_{T^*}^*(z) + R_{T^*}(z) K_1^{-1} - R_{T^*}(z) K_1^{-1} T R_{T^*}^*(z)$$
$$+ R_{T^*}(z) T^* K_1^{-1} R_{T^*}^*(z)\}[u, v]$$
$$= i \begin{bmatrix} u^* \\ v^* \end{bmatrix} \{R_{T^*}(z)(-T^* + zI + T^*) K_1^{-1} R_{T^*}^*(z)$$
$$+ R_{T^*}(z) K_1^{-1}(T - \bar{z} - T) R_{T^*}^*(z)\}[u, v]$$
$$= i(z - \bar{z}) \begin{bmatrix} u^* \\ v^* \end{bmatrix} R_{T^*}(z) K_1^{-1} R_{T^*}^*(z)[u, v].$$

The equality (15) is proved. The equality (16) may be proved in an analogous way.

Now let us show that (17) holds. We have

$$U_1(z) = \begin{bmatrix} I - u^* R_{T^*}(z) K_1^{-1} v & u^* R_{T^*}(z) K_1^{-1} u \\ -v^* R_{T^*}(z) K_1^{-1} v & I + v^* R_{T^*}(z) K_1^{-1} u \end{bmatrix} \begin{bmatrix} I & 0 \\ M_1 & I \end{bmatrix}$$

$$= \begin{bmatrix} I - u^* R_{T^*}(z) K_1^{-1} v + u^* R_{T^*}(z) K_1^{-1} u M_1 & u^* R_{T^*}(z) K_1^{-1} u \\ -v^* R_{T^*}(z) K_1^{-1} v + (I + v^* R_{T^*}(z) K_1^{-1} u) M_1 & I + v^* R_{T^*}(z) K_1^{-1} u \end{bmatrix}.$$

Moreover,

$$I - u^* R_{T^*}(z) K_1^{-1} v + u^* R_{T^*}(z) K_1^{-1} u v^* K_2^{-1} v$$
$$= I - u^* R_{T^*}(z) K_1^{-1} v + u^* R_{T^*}(z) K_1^{-1}(K_2 - K_1 T^*) K_2^{-1} v$$
$$= I - u^* R_{T^*}(z) K_1^{-1} v + u^* R_{T^*}(z) K_1^{-1} v - u^* R_{T^*}(z) T^* K_2^{-1} v$$
$$= I - u^* T^* R_{T^*}(z) K_2^{-1} v.$$

In an analogous way,

$$-v^* R_{T^*}(z) K_1^{-1} v + (I + v^* R_{T^*}(z) K_1^{-1} u) v^* K_2^{-1} v$$
$$= -v^* R_{T^*}(z) K_1^{-1} v + v^* K_2^{-1} v + v^* R_{T^*}(z) K_1^{-1}(K_2 - K_1 T^*) K_2^{-1} v$$
$$= -v^* R_{T^*}(z) K_1^{-1} v + v^* K_2^{-1} v + v^* R_{T^*}(z) K_1^{-1} v - v^* R_{T^*}(z) T^* K_2^{-1} v$$
$$= v^* R_{T^*}(z)(T^* - zI - T^*) K_2^{-1} v = -z v^* R_{T^*}(z) K_2^{-1} v.$$

Therefore

$$U_1(z) = \begin{bmatrix} I - u^*T^*R_{T^*}(z)K_2^{-1}v & u^*R_{T^*}(z)K_1^{-1}u \\ -zv^*R_{T^*}(z)K_2^{-1}v & I + v^*R_{T^*}(z)K_1^{-1}u \end{bmatrix}.$$

Now

$$\begin{bmatrix} zI & 0 \\ 0 & I \end{bmatrix} U_1(z) \begin{bmatrix} z^{-1}I & 0 \\ 0 & I \end{bmatrix}$$

$$= \begin{bmatrix} I - u^*T^*R_{T^*}(z)K_2^{-1}v & zu^*R_{T^*}(z)K_1^{-1}u \\ -v^*R_{T^*}(z)K_2^{-1}v & I + v^*R_{T^*}(z)K_1^{-1}u \end{bmatrix}$$

$$= \begin{bmatrix} I - u^*T^*R_{T^*}(z)K_2^{-1}v & zu^*R_{T^*}(z)K_1^{-1}u \\ -v^*R_{T^*}(z)K_2^{-1}v & I + v^*R_{T^*}(z)K_1^{-1}u \end{bmatrix} \begin{bmatrix} I & M_2 \\ 0 & I \end{bmatrix} \begin{bmatrix} I & -M_2 \\ 0 & I \end{bmatrix}$$

$$= \begin{bmatrix} I - u^*T^*R_{T^*}(z)K_2^{-1}v & (I - u^*T^*R_{T^*}(z)K_2^{-1}v)M_2 + zu^*R_{T^*}(z)K_1^{-1}u \\ -v^*R_{T^*}(z)K_2^{-1}v & -v^*R_{T^*}(z)K_2^{-1}vM_2 + I + v^*R_{T^*}(z)K_1^{-1}u \end{bmatrix}$$

$$\times \begin{bmatrix} I & -M_2 \\ 0 & I \end{bmatrix}.$$

Moreover,

$$\begin{aligned}
&(I - u^*T^*R_{T^*}(z)K_2^{-1}v)u^*K_1^{-1}u + zu^*R_{T^*}(z)K_1^{-1}u \\
&\quad = u^*K_1^{-1}u - u^*T^*R_{T^*}(z)K_2^{-1}(K_2 - TK_1)K_1^{-1}u + zu^*R_{T^*}(z)K_1^{-1}u \\
&\quad = u^*K_1^{-1}u - u^*T^*R_{T^*}(z)K_1^{-1}u + u^*T^*R_{T^*}(z)K_2^{-1}Tu + zu^*R_{T^*}(z)K_1^{-1}u \\
&\quad = u^*K_1^{-1}u - u^*R_{T^*}(z)(T^* - zI)K_1^{-1}u + u^*T^*R_{T^*}(z)K_2^{-1}Tu \\
&\quad = u^*T^*R_{T^*}(z)K_2^{-1}Tu.
\end{aligned}$$

In an analogous way

$$\begin{aligned}
&-v^*R_{T^*}(z)K_2^{-1}vu^*K_1^{-1}u + I + v^*R_{T^*}(z)K_1^{-1}u \\
&\quad = -v^*R_{T^*}(z)K_2^{-1}(K_2 - TK_1)K_1^{-1}u + I + v^*R_{T^*}(z)K_1^{-1}u \\
&\quad = -v^*R_{T^*}(z)K_1^{-1}u + v^*R_{T^*}(z)K_2^{-1}Tu + I + v^*R_{T^*}(z)K_1^{-1}u \\
&\quad = I + v^*R_{T^*}(z)K_2^{-1}Tu.
\end{aligned}$$

Finally,

$$\begin{bmatrix} zI & 0 \\ 0 & I \end{bmatrix} U_1(z) \begin{bmatrix} z^{-1}I & 0 \\ 0 & I \end{bmatrix}$$

$$= \begin{bmatrix} I - u^*T^*R_{T^*}(z)K_2^{-1}v & u^*T^*R_{T^*}(z)K_2^{-1}Tu \\ -v^*R_{T^*}(z)K_2^{-1}v & I + v^*R_{T^*}(z)K_2^{-1}Tu \end{bmatrix} \begin{bmatrix} I & -M_2 \\ 0 & I \end{bmatrix}$$

$$= \left\{ I_2 + \begin{bmatrix} -u^*T^* \\ -v^* \end{bmatrix} R_{T^*}(z)K_2^{-1}[v, Tu] \right\} \begin{bmatrix} I & -M_2 \\ 0 & I \end{bmatrix}$$

$$= \left\{ I_2 + i \begin{bmatrix} u^*T^* \\ v^* \end{bmatrix} R_{T^*}(z)K_2^{-1}[Tu, v]J \right\} \begin{bmatrix} I & -M_2 \\ 0 & I \end{bmatrix} = U_2(z).$$

The theorem is proved. □

6 The solution of the FMI system in the nondegenerate case

For the sake of simplicity we shall assume that $\dim H < +\infty$ in this section.

Theorem 9. *In the nondegenerate case, the FMI system of V.P. Potapov (6) is equivalent to the system*

$$[s^*(z), I] \frac{U_1^{-1^*}(z) J U_1^{-1}(z)}{i(\bar{z} - z)} \begin{bmatrix} s(z) \\ I \end{bmatrix} \leq 0 \tag{18}$$

and

$$[\bar{z}s^*(z), I] \frac{U_2^{-1^*}(z) J U_2^{-1}(z)}{i(\bar{z} - z)} \begin{bmatrix} zs(z) \\ I \end{bmatrix} \leq 0 \, , \tag{19}$$

where the functions $U_1(z)$ and $U_2(z)$ are defined by formulas (13) and (14).

Proof. From the lemma about the non-negative block-matrix, it follows that the FMI system (6) is equivalent to the system

$$\frac{s(z) - s^*(z)}{z - \bar{z}} - \{R(z)[u - vs(z)]\}^* K_1^{-1} \{R(z)[u - vs(z)]\} \geq 0$$

and

$$\frac{zs(z) - \bar{z}s^*(z)}{z - \bar{z}} - \{R(z)[Tu - vs(z)]\}^* K_2^{-1} \{R(z)[Tu - vs(z)]\} \geq 0.$$

This system can be written in the form

$$[s^*(z), I] \left\{ \frac{-J}{i(\bar{z} - z)} - J \begin{bmatrix} u^* \\ v^* \end{bmatrix} R^*(z) K_1^{-1} R(z)[u, v] J \right\} \begin{bmatrix} s(z) \\ I \end{bmatrix} \geq 0 \tag{20}$$

and

$$[\bar{z}s^*(z), I]\left\{\frac{-J}{i(\bar{z}-z)} - J\begin{bmatrix} u^* \\ v^* \end{bmatrix} R^*(z)K_2^{-1}R(z)[u,v]J\right\}\begin{bmatrix} zs(z) \\ I \end{bmatrix} \geq 0. \tag{21}$$

Let us denote the independent variable in (15) as $\bar{z}$ and multiply this equality from the left and the right by J. We obtain

$$\frac{JU_1(\bar{z}))JJJU_1^*(\bar{z})J - J}{i(z-\bar{z})} = -J\begin{bmatrix} u^* \\ v^* \end{bmatrix} R_{T^*}(\bar{z}))K_1^{-1}R_{T^*}^*(\bar{z}))[u,v]J\,.$$

Using the symmetry principle $JU_1^*(\bar{z})J = U_1^{-1}(z)$ and the obvious equality $R_{T^*}^*(\bar{z}) = R_T(z) = R(z)$, we obtain

$$\frac{U_1^{-1^*}(z)JU_1^{-1}(z) - J}{i(\bar{z}-z)} = J\begin{bmatrix} u^* \\ v^* \end{bmatrix} R^*(z)K_1^{-1}R(z)[u,v]J.$$

Having transformed the core of the inequality (20) with this relation, we obtain (18).

In an analogous way we obtain inequality (19) from inequality (21).

The theorem is proved. □

Definition 8. A pair $\begin{bmatrix} p(z) \\ q(z) \end{bmatrix}$ of operator-valued functions which is meromorphic in $\mathbf{C} \setminus [\mathbf{0}, +\infty)$ is called a Stieltjes pair if

1. $\Delta(z) = \det\{p^*(z)p(z) + q^*(z)q(z)\} \neq 0.$

2. $[p^*(z), q^*(z)]\frac{J}{i(\bar{z}-z)}\begin{bmatrix} p(z) \\ q(z) \end{bmatrix} \leq 0, \ \operatorname{Im} z \neq 0.$

3. $[\bar{z}p^*(z), q^*(z)]\frac{J}{i(\bar{z}-z)}\begin{bmatrix} zp(z) \\ q(z) \end{bmatrix} \leq 0, \ \operatorname{Im} z \neq 0.$

We introduce an equivalence relation on the set of Stieltjes pairs: a pair $\begin{bmatrix} p(z) \\ q(z) \end{bmatrix}$ is said to be equivalent to the pair $\begin{bmatrix} p_1(z) \\ q_1(z) \end{bmatrix}$ if there exists an operator-valued function $Q(z)(\det Q(z) \neq 0)$ which is meromorphic in $\mathbf{C} \setminus [\mathbf{0}, +\infty)$ such that $p_1(z) = p(z)Q(z)$ and $q_1(z) = q(z)Q(z)$. The relation we have introduced is obviously an equivalence relation and, consequently, the set of Stieltjes pairs breaks up into equivalence classes. We identify the pairs in a single class. The meaning of this identification is clear, since equivalent pairs $\begin{bmatrix} p(z) \\ q(z) \end{bmatrix}$ and $\begin{bmatrix} p_1(z) \\ q_1(z) \end{bmatrix}$ lead to the same operator-valued function $s(z)$ as a result of the linear fractional transformation.

Theorem 10.

1. *Suppose that the function $s(z) \in \mathbf{S}$ is a solution of the FMI of V.P. Potapov (6). Then $s(z)$ can be represented as the linear fractional transformation*

$$s(z) = [\alpha(z)p(z) + \beta(z)q(z)][\gamma(z)p(z) + \delta(z)q(z)]^{-1} \tag{22}$$

of a Stieltjes pair $\begin{bmatrix} p(z) \\ q(z) \end{bmatrix}$. *The coefficient matrix of this transformation is constructed according to (13), namely,*

$$\begin{bmatrix} \alpha(z) & \beta(z) \\ \gamma(z) & \delta(z) \end{bmatrix} = \left\{ I_2 + i \begin{bmatrix} u^* \\ v^* \end{bmatrix} R_{T^*}(z) K_1^{-1}[u, v] J \right\} \begin{bmatrix} I & 0 \\ M_1 & I \end{bmatrix}.$$

2. *If* $\begin{bmatrix} p(z) \\ q(z) \end{bmatrix}$ *is an arbitrary Stieltjes pair, then the linear fractional transformation (22) defines a function $s(z) \in \mathbf{S}$ that is a solution of the FMI's (6).*

3. *The transformation (22) applied to the pairs* $\begin{bmatrix} p(z) \\ q(z) \end{bmatrix}$ *and* $\begin{bmatrix} p_1(z) \\ q_1(z) \end{bmatrix}$ *leads to the same function $s(z)$ if and only if these pairs are equivalent.*

The proof of this theorem follows the scheme of the proof of Theorem 3.5 in [DK]. In carrying this out, the inequalities (18) and (19) are used.

References

[DK] Dyukarev Yu.M. and Katsnelson V.E., Multiplicative and additive classes of Stieltjes analytic matrix-valued functions, and interpolation problems associated with them. I (in Russian), *Teor. Funkt., Funktsional. Anal. i Prilozhen* **36** (1981) 13–27.

[IS] Ivanchenko T.S. and Sahnovich L.A., *Operator-theoretic approach to the V.P. Potapov scheme of investigation of interpolation problem*, (in Russian), *Ukr. Math. Journal* **39:5** (1987) 573–578.

[Ka] Katsnelson V.E., Continuous analogies of Hamburger-Nevanlinna theorem and fundamental matrix inequality of classical problems, (in Russian), *Teor. Funkt., Funktsional. Anal. i Prilozhen* **40** (1984) 79–90.

[KKY] Katsnelson V.E., Kheifets A. Ya. and Yuditskii P.M., The abstract interpolation problem and extension theory of isometric operators, (in Russian), in: *Operator in Spaces of Functions and Problems in Function Theory: Collected scientific papers*, Kiev, Naukova Dumka, 1987, 83–96.

[KN] Krein M.G., and Nudelman A.A., *The Markov moment problem and extremal problems*, (in Russian), "Nauka", Moscow, 1973; Engl. transl.in: *Translation Math. Monographs* **50** Amer. Math. Soc., Providence, Rhode Island, 1977.

[Ko] Kovalishina I.V., Analytic theory of a class of interpolation problems, (in Russian), *Izv. Akad. Nauk SSSR Ser. Mat.* **47** (1983) 455–497. Engl. transl. in: *Math. USSR Izv.* **22** (1983) 419–463.

[KP] Kovalishina I.V. and Potapov V.P., An indefinite matrix in the Nevanlinna-Pick problem, (in Russian), *Akad. Nauk Armyan. SSR Dokl.* **59** (1974) 17–22. Engl. transl. in: *Collected Papers of V.P. Potapov*, Sapporo. Japan: Private translation and edition by T. Ando 1982, pp. 67–99.

[Nu] Nudelman A.A., On a generalization of classical interpolation problems, (in Russian), *Dokl. Akad. Nauk SSSR* **256:4** (1981) 290–293. Engl. transl. in: *Sov. Math. Dokl.* **23** (1981) 125–128.

Department of Physics
Kharkov State University
Svoboda Square 4
Kharkov 310077, Ukraine

AMS Subject Classification: 47A57, 42A82.

Operator Theory
Advances and Applications, Vol. 95
© 1997 Birkhäuser Verlag Basel/Switzerland

On recovering a multiplicative integral from its modulus*

Yu.P. Ginzburg

Abstract. An improvement and generalization of V.P. Potapov's theory of the J-modulus is established. The results, in the special case $J = I$, are applied to obtain the representation by a multiplicative integral of an operator-valued outer function f from the values at some interior point of the disc of the moduli of certain divisors of f, or from the moduli of the boundary values of f. [Abstract added by editors.]

0 Preface

In the proof of the classical theorem on the multiplicative representation of a j-contractive analytic matrix function the theory of the j-modulus created by V.P. Potapov [1] plays an important role. A considerably improved presentation of this deep theory has been developed in [2, 3].

The formulas which express a summable j-hermitian nonnegative function $H(s)$ in terms of the j-modulus $R(s)$ of the multiplicative integral

$$W(s) = \overset{s}{\underset{0}{\int^{\curvearrowleft}}} \exp\{-H(t)dt\}$$

(formulas (6.4)–(6.6) from [3]) fill a highly important place in Potapov's theory of the j-modulus.

The present paper is dedicated to some generalizations and refinements of these results. We also apply these results (in the case $J = I$) to the reconstruction of the multiplicative representation of an outer operator function from the moduli of its boundary values (or from the values of the moduli of its special divisors at some point). The results which we obtain can be reformulated in terms of canonical differential systems by standard methods [1].

---

*) Translated from: Teor. Funktsiĭ Funktsional. Anal. i Prilozhen. No.41 (1984), 135–143. MR 85k:47063

1 On one factorization of an absolutely continuous operator function

Let $W(s)$ $(0 \leq s \leq \ell)$ be an operator valued function which is uniformly continuous with respect to the uniform operator metric. Then the weak derivative $F(s) = W'(s)$ exists at almost every point in the interval. This derivative $F(s)$ has the following properties:

(1) $F(s)$ is weakly measurable on $[0, \ell]$.

(2) $|F| \in L^{(1)}[0, \ell]$.

Operator functions which possess the properties (1) and (2) are said to be *summable on* $[0, \ell]$.

If the function $W(s)$ is invertible for each point $s \in (0, \ell]$ and its inverse $W^{-1}(s)$ is a bounded operator, then the function

$$\overset{\curvearrowleft}{\dot{W}}(s) \equiv W'(s)W^{-1}(s)$$

is said to be a *multiplicative derivative* of the function $W(s)$. The function $W(s)$ can be recovered from the function $C(s) \equiv \overset{\curvearrowleft}{\dot{W}}(s)$:

$$W(s) = W_0 \overset{\curvearrowleft}{\int_0^s} \exp\{C(t)dt\} \ , \tag{1}$$

where $\overset{\curvearrowleft}{\int_0^s}$ is the multiplicative Lebesgue integral [1; 4] and $W_0 = W(0)$.

Conversely, if $W(s)$ is defined by (1), where $C(t)$ is a summable operator function, then $\overset{\curvearrowleft}{\dot{W}}(s)$ exists almost everywhere on $[0, \ell]$ and $\overset{\curvearrowleft}{\dot{W}}(s) = C(s)$. In what follows the letter j will denote an operator such that $j = j^* = j^{-1}$.

Lemma. *Let $W(s)$ $(0 \leq s \leq \ell)$ be an absolutely continuous operator function, $W(0) = I$. In order that the operator $W(s)$ be j-unitary for almost every $s \in [0, \ell]$ it is necessary and sufficient that the operator $B(s) = i\overset{\curvearrowleft}{\dot{W}}(s)$ be j-hermitian almost everywhere on $[0, \ell]$.*

Proof. Since $W'(s) = -B(s)W(s)$, it follows that

$$(W^*(s))' = iW^*(s)(B^*(s)j - jB(s))W(s) \ .$$

If $B(s)$ is j-hermitian, then $(W^*(s)jW(s))' = 0$ almost everywhere on $[0, \ell]$. From $W(0) = I$ it follows that $W^*(s)jW(s) = j$ $(s \in [0, \ell])$. Therefore, since the operator $W(s)$ is boundedly invertible, it is j-unitary. Conversely, if $W(s)$ is a j-unitary operator, then $(W^*(s)jW(s))' \equiv 0$. Hence, the operator $B(s)$ is j-unitary for almost all $s \in [0, \ell]$. $\square$

Theorem 1. *Let $W(s)$ be an absolutely continuous operator function on $[0,\ell]$ with $W(0)=I$. Then $W(s)$ admits a unique representation of the form*

$$W(s)=U(s)\overset{\curvearrowleft}{\int_0^s}\exp\{H(t)dt\}\ , \tag{2}$$

where $U(s)$ takes j-unitary values and $H(s)$ takes j-hermitian values and is summable. Moreover,

$$U(s)=\overset{\curvearrowleft}{\int_0^s}\exp\{iB(t)dt\}\ ,\qquad B(s)=\frac{1}{2i}\left(\overset{\curvearrowleft}{W}(s)-j\overset{\curvearrowleft}{W^*}(s)j\right)\ , \tag{3}$$

$$H(s)=U^{-1}(s)A(s)U(s)\quad\text{and}\quad A(s)=\frac{1}{2}\overset{\curvearrowleft}{W}(s)+j\overset{\curvearrowleft}{W^*}j\ . \tag{4}$$

Proof. Let $U(s)$ be of the form (3). Because of the lemma, $U(s)$ is a j-unitary operator for almost all $s\in[0,\ell]$. Clearly, the function $Z(s)\equiv U^{-1}(s)W(s)$ is absolutely continuous and $Z(0)=I$. Let

$$H(s)=\overset{\curvearrowleft}{Z}(s)\quad\text{and}\quad Z(s)=\overset{\curvearrowleft}{\int_0^{s}}\exp\{H(t)dt\}\ .$$

Then

$$W'=iBUZ+UHZ\quad\text{and}\quad\overset{\curvearrowleft}{W}=iB+UMU^{-1}\ .$$

Hence (4) holds.

The uniqueness of the representation (2) follows immediately from the equality $\overset{\curvearrowleft}{W}=\overset{\curvearrowleft}{U}+UHU^{-1}$ which can be obtained by differentiating (2) and invoking the fact that the operators $I\overset{\curvearrowleft}{U}$ and UHU^{-1} are j-hermitian. □

2 The modulus of the multiplicative integral of a Hermitian-valued function

Let M_j be a set of all boundedly invertible operators A such that the j-hermitian operator $B\equiv jA^*jA$ has a positive spectrum. Clearly, every boundedly invertible operator belongs to the class M_I. It is known that for an arbitrary j, every boundedly invertible doubly j-expansive operator (as well as every j-contractive operator) belongs to the class M_j. For each $A\in M_j$, let us denote by R_A the unique operator with the positive spectrum such that $R_A^2=B$. It is easy to see that the operator R_A is j-hermitian. This operator R_A is said to be *the j-modulus of the operator A*.

Clearly, $A=V_AR_A$, where V_A is a j-unitary operator.

Let $H(s)$ $(0 \leq s \leq \ell)$ be an arbitrary j-hermitian-valued summable function. Let us consider the function

$$Z(s) = \overset{s}{\underset{0}{\curvearrowleft\!\!\!\!\int}} \exp\{H(t)dt\} .$$

If[1] $Z(s) \in M_j$ for all $s \in [0, \ell]$, then $R_{Z(s)}$ is an absolutely continuous function [1; 3; 4].

The following converse theorem is an immediate consequence of Theorem 1.

Theorem 2. *Let $R(s)$ $(0 \leq s \leq \ell)$ be an absolutely continuous function whose values are j-hermitian operators with positive spectrum such that $R(0) = I$. Then there exists a unique j-hermitian-valued function $H(s)$ which is summable on $[0, \ell]$ such that $R(s)$ is the j-modulus of the operator $Z(s) := \overset{s}{\curvearrowleft\!\!\!\!\int_0} \exp\{H(t)dt\}$. (Note that $Z(s)$ belongs to M_j.) Moreover, V.P. Potapov's formulas hold:*

$$H(s) = \frac{1}{2} U^{-1}(s)(R'(s)R^{-1}(s) + R^{-1}(s)R'(s))U(s) , \tag{5}$$

$$U(s) = \overset{s}{\underset{0}{\curvearrowleft\!\!\!\!\int}} \exp\left\{\frac{1}{2}(R'(t)R^{-1}(t) - R^{-1}(t)R'(t)dt\right\} \tag{6}$$

and

$$Z(s) = U^{-1}(s)R(s) .$$

3 On some classes of analytic operator functions and their multiplicative representations

We consider functions $X(z)$ which are holomorphic in $|z| < 1$ whose values are boundedly invertible operators. Such a function is said to belong to:

the class A, if $\sup_{0<r<1} \int_0^{2\pi} ln^+ \|X(re^{it})\| dt < \infty$.

the class D if the family $\{ln^+ \|X(re^{it})\|\}_{0<r<1}$ is uniformly integrable on $[0, \ell]$.

the class C if $\|X(z)\| \leq 1$ for $|z| < 1$.

Clearly, $C \subset D \subset A$. The scalar analogs of the classes C, D and A are denoted by C_L, D_L and A_L, respectively. The simple statements which are formulated below, were proved in [5].

1°. $X \in A \Leftrightarrow X = y^{-1}Y$ ($y \in C_1$, $y \neq 0$ for $|z| < 1$).

2°. $X \in D \Leftrightarrow X = y^{-1}Y$ (y is an outer function from C_1, $Y \in C$).

3°. If $X \in A$ then $X(z)$ and $X^*(z)$ have strong radial boundary values $X(e^{it})$ and $X^*(e^{it})$ at almost every point e^{it} of the unit circle.

---

1) This will be satisfied if there exists a scalar summable function $h(s)$ such that the operator $j(H(s) + W(s)I)$ is nonnegative on $[0, \ell]$ (or nonpositive on $[0, \ell]$).

Let K' and U'' be any two of the classes of operator functions which were introduced above. We shall say that an operator $X \in (K' : K'')$ if $X \in K'$ and $X^{-1} \in K''$. From 1° and 2° it follows that the study of any class $(K' : K'')$ can be reduced to the study of the class $(C : A) \equiv P$. In the matrix (i.e., the finite-dimensional) case each class $(K' : A)$ consists of all these matrix functions $X \in K'$ which satisfy the condition $\det X(z) \neq 0$ for $|z| < 1$.

4°. For each function $X \in P$ there exists a function $m \in P_1 \equiv (C_1 : A_1)$ such that
   (a) $mX^{-1} \in P$;
   (b) The condition $m_1 m^{-1} \in P_1$ is satisfied for every function $m_1 \in P_1$ which meets the condition $m_1 X^{-1} \in P$. Such a function $m(z)$ is said to be *the best minorant of the function* $X \in P$. The function $m(z)$ is unique up to a scalar unimodular factor.

5°. The following assertions are equivalent:
   (a) $X \in (C : D)$.
   (b) X is an outer function which belongs to the class P.
   (c) X is extremal in P, i.e., if $X \in P$, $X_1 \in P$ and $X_1^*(t)X_1(t) \leq X^*(t)X(t)$ almost everywhere on $[0, 2\pi]$, then $X_1^*(z)X_1(z) \leq X^*(z)X(z)$ for $|z| < 1$.
   (d) $X \in P$, and the best minorant of the function X is an outer function.

6°. If $X_1, X_2 \in (D : D)$ and $X_1^*(e^{it})X_1(e^{it}) = X_2^*(e^{it})X_2(e^{it})$ almost everywhere on $[0, 2\pi]$, then $X_2(z) = UX_2(z)$ for $|z| < 1$, where U is a constant unitary operator.

The following statement is the extension of Potapov's fundamental theorem to the class P.

7°. Let the best minorant m of the function $X \in P$ be representable in the form

$$m(z) = e^{i\alpha}\exp\left\{\int_0^{2\pi} \frac{z + e^{it}}{z - e^{it}} d\sigma(t)\right\} ,$$

where $\sigma(t)$ is a nondecreasing function on $[0, 2\pi]$, $\sigma(0) = 0$, $\sigma(t - 0) = \sigma(t)$ on the interval $0 < t \leq 2\pi$ and $\operatorname{Im} \alpha = 0$. Let us define

$$\varphi(s) = \operatorname{mes}\{t : \sigma(t) < s \ , \quad 0 \leq s \leq \ell := \sigma(2\pi)\} .$$

Then there exists a hermitian nondecreasing operator function $E(s)$ with $\|E(s'') - E(s')\| \leq |s'' - s'|$ for every $s', s'' \in [0, \ell]$ such that

$$X(z) = U \overset{\ell}{\underset{0}{\int^{\curvearrowleft}}} \exp\left\{\frac{z + e^{i\varphi(s)}}{z - e^{i\varphi(s)}} dE(s)\right\} , \tag{7}$$

where U is a constant unitary operator.

Now let $X \in (C : D)$. In view of 5°, the function $\sigma(t)$ is absolutely continuous. In particular, it follows from this that the function φ does not have intervals of constancy. This allows us to change variables by the formulas $\varphi(s) = t$ and $s = \sigma(t)$ in the integral (7).

Theorem 3. *A function $X(z)$ ($|z| < 1$) belongs to the class $(C : D)$ (i.e., is an outer function from the class $P = (C : A)$) if and only if it is representable in the form*

$$X(z) = U \overset{2\pi}{\underset{0}{\curvearrowleft\!\!\!\!\!\int}} \exp\left\{\frac{z+e^{it}}{z-e^{it}}G(t)dt\right\} , \tag{8}$$

where U is a constant unitary operator and $G(t)$ is a summable function on $[0, 2\pi]$ whose values are nonnegative hermitian operators.

Proof. Using (7) and denoting the weak derivative of the absolutely continuous function $E(\sigma(t))$ by $G(t)$, we obtain the necessity of the stated condition. Next, to establish sufficiency, we observe first that the inclusion $X \in C$ follows immediately from the condition $\operatorname{Re}\left\{\frac{z+e^{it}}{z-e^{it}}G(t)\right\} \leq 0$ ($|z| < 1$, $t \in [0, 2\pi]$). Finally, the membership of X^{-1} in the class D is immediate from the next theorem. □

Theorem 4. *A function $X(z)$ ($|z| < 1$) belongs to $(D : D)$ if and only if it admits a representation of the form (8) with a summable hermitian valued function $G(t)$.*

Proof. Let $X \in (D : D)$. Since $X \in D$ and $X^{-1} \in D$, it follows from 2° that $X = y^{-1}Y$ and $X^{-1} = y_1^{-1}Y_1$, where $Y, Y_1 \in C$, and y, y_1 are outer functions from the class C_1. Thus, $Y^{-1} = (yy_1)^{-1}Y_1$. Therefore, since yy_1 is an outer function from the class C_1, it follows that $Y^{-1} \in D$. Thus, $Y \in (C : D)$. By Theorem 3,

$$Y(z) = V \overset{2\pi}{\underset{0}{\curvearrowleft\!\!\!\!\!\int}} \exp\left\{\frac{z+e^{it}}{z-e^{it}}F(t)dt\right\} , \tag{9}$$

where V is a constant unitary operator and F is a summable function on $[0, 2\pi]$ with $F(t) = 0$.

Since

$$X^{-1}(z) = \overset{2\pi}{\underset{0}{\curvearrowleft\!\!\!\!\!\int}} \exp\left\{\frac{z+e^{it}}{z-e^{it}}(-G(t)dt\right\} U^{-1} , \tag{10}$$

it follows that $X^{-1} \in D$, and hence that $X \in (D : D)$. The theorem is proved. □

Theorems 3 and 4 give grounds for referring to functions from the class $(D : D)$ as outer functions of the class $(D : A)$. It is possible to show that $X \in (D : D)$ if and only if $X(z)$ is extremal in $(D : A)$ (see 5°c).

The uniqueness of the representation of the form (8) follows from Theorem 6 which will be proved below.

First we prove

Theorem 5. *For each function $X \in (D:D)$ and every $s \in [0, 2\pi]$ there exists a unique function $X_s \in (D:D)$ (up to a constant unitary left factor) such that the functions $X_s(z)$ and $X(z)X_s^{-1}(z)$ are holomorphic and take unitary values on the arcs*

$$\alpha_s = \{e^{it} : s < t < 2\pi\} \quad \text{and} \quad \widetilde{\alpha}_s = \{e^{it} : 0 < t < s\} ,$$

respectively. If (8) is the representation of the function X (which exists by Theorem 4), then

$$X_s(z) = \overset{s}{\underset{0}{\curvearrowleft\!\!\int}} \exp\left\{\frac{z + e^{it}}{z - e^{it}} G(t)dt\right\} . \tag{11}$$

Proof. It is clear that the function $X_s(z)$ defined by (11) and the function

$$X(z)X_s^{-1}(z) = \overset{2\pi}{\underset{s}{\curvearrowleft\!\!\int}} \exp\left\{\frac{z + e^{it}}{z - e^{it}} G(t)dt\right\}$$

are holomorphic on the arcs α_s and $\widetilde{\alpha}_s$, respectively. The unitarity follows directly from the lemma.

Now let the functions $\widetilde{X}_s(z) (\in (D:D))$ and $X(z)\widetilde{X}_s^{-1}(z)$ be holomorphic and unitary on the arcs α_s and $\widetilde{\alpha}_s$, respectively. It is clear that

$$\widetilde{X}_s^*(e^{it})\widetilde{X}_s(e^{it}) = X^*(t)X(t) \ (0 < t < s) , \ \widetilde{X}_s^*(e^{it})X_s(e^{it}) = I \ (s < t < 2\pi) .$$

Since the function X_s satisfies the same conditions, it follows from 6° that $\widetilde{X}_s(z) = UX_s(z)$, where U is a unitary operator. □

Theorem 6. *Let $Q(t)$ $(0 \le t \le 2\pi)$ be an absolutely continuous function whose values are nonnegative hermitian operators with $Q(0) = 1$ and let z_0 be an arbitrary point with $|z_0| < 1$. Then there exists a unique function $X \in (D:D)$ (up to a constant unitary left factor) such that $X_s^*(z_0)X_s(z_0) = Q(s)$ for $0 \le s \le 2\pi$. For $z_0 = 0$, the function $X(z)$ can be constructed from the absolutely continuous function $R(s) = \sqrt{Q(s)}$ by the formulas (8), (5) (where one should put $G = -H$) and (6).*

Proof. Without loss of generality, we may assume that $z_0 = 0$. Then the proof follows immediately from Theorems 2, 4 and 5. □

Remark. From the obvious equality

$$\frac{d}{ds}(X_s^*(0)X_s(0)) = -2X_s^*(0)G(s)X_s(0)$$

and Theorem 3, it follows that the function X which appears in Theorem 6 belongs to the class $(C:D)$ if and only if the function $Q(s)$ is hermitian-nonincreasing.

Corollary. *The function G from the multiplicative decomposition (8) of the function* $X \in (D : D)$ *is defined uniquely.*[2]

The following theorem shows that $G(t)$ depends on the moduli of the boundary values of the function $X(z)$.

Theorem 7. *Let* $N(s)$ $(0 \leq s \leq 2\pi)$ *be a weakly measurable function whose values are nonnegative hermitian operators which satisfy the conditions* $ln\|N(s)\| \in L^{(1)}[0, 2\pi]$ *and* $ln\|N^{-1}(s)\| \in L^{(1)}[0, 2\pi]$. *Then there exists a function* $G(t)$ *which is summable on* $[0, 2\pi]$ *whose values are hermitian operators such that for the function*

$$X(z) := \int_0^{2\pi} \exp\left\{\frac{z + e^{it}}{z - e^{it}} G(t)d(t)\right\}$$

the equality

$$X^*(e^{is})X(e^{is}) = N(s)^2 \tag{12}$$

holds almost everywhere on $[0, 2\pi]$.

Proof. Using 6° and the corollary, we reduce the original problem to the problem of establishing the existence of a function $X \in (D : D)$ which satisfies condition (12). To attain this goal we consider the function

$$w(z) = \exp\left\{\frac{1}{2\pi}\int_0^{2\pi} \frac{e^{it} + z}{e^{it} - z} ln\|N(t)\|dt\right\} \in (D_1 : D_1)$$

It is well known that $|w(e^{is})| = \|N(s)\|$ almost everywhere on $[0, 2\pi]$. Put $M(s) = \|N(s)\|^{-1}N(s)$. The function $M(s)$ is weakly measurable, $\|M(s)\| \leq 1$, and $ln\|M^{-1}(s)\| \in L^{(1)}[0, 2\pi]$ (because $0 \leq ln\|M^{-1}(s)\| \leq ln\|N(s)\| + ln\|N^{-1}(s)\|$).

According to 5° and a well known theorem of Devinatz on the factorization of hermitian nonnegative operator-valued functions, there exists a function $Y \in (C : D)$ which satisfies the condition $Y^*(e^{is})Y(e^{is}) = M(s)^2$ almost everywhere on $[0, 2\pi]$.

Let us consider the function $X = wY$. Since $X \in (D : D)$ and

$$X^*(e^{is})X(e^{is}) = |w(e^{is})|^2 Y^*(e^{is})Y(e^{is}) = \|N(s)\|M(s)^2 = N(s)^2 ,$$

the theorem is proved. □

For $0 \leq s \leq 2\pi$, let us put $N_s(t) = \begin{cases} N(t) & (0 \leq t < s) \\ I & (s \leq t \leq 2\pi) \end{cases}$.

---

2) Some results on the uniqueness of the multiplicative decomposition (for other classes of operator functions) are established in [4; 6]. In the same places references are given to other works which are related to problems discussed in the present paper.

It is clear that $ln\|N_s\| \in L^1[0,2\pi]$, $ln\|N_s^{-1}\| \in L^1[0,2\pi]$, and (by Theorem 5) that the function $X_s(z)$ satisfies the condition

$$X_s^*(e^{it})X_s(e^{it}) = N_s(t)^2 \quad (\text{for a.e. } t \in [0,2\pi]) \ . \tag{13}$$

Thus, the problem of recovering the function G from the function N is reduced to the problem of resolving the family of boundary problems (13) and to using Potapov's formulas (5) and (6) applied to the function $R(s) = \sqrt{X_s^*(0)X_s(0)}$.

Finally, we remark that since the function $W(s) := \overset{\curvearrowleft}{\int_0^s} \exp\{A(t)dt\}$ is the solution of the equation $W'(s) = A(s)W(s)$ with initial condition $W(0) = I$, some results of the present paper can be reformulated as results on the construction of a canonical differential system from some information on the Wronskiĭ matrix (see the Introduction to [1]).

References

[1] Potapov, V.P., The multiplicative structure of J-contractive matrix-functions. *Trudy Mosk. Mat. Obshch.* **4** (1955) 125–236 (Russian). English transl.: *Amer. Math. Soc. Transl.* (2) **15** (1960), 131–243.

[2] Potapov, V.P., A theorem on the modulus, I. Fundamental concepts. The modulus. *Teor. Functsiĭ Funksional Anal. i Prilozh.* **38** (1982), 91–101 (Russian). English transl.: *Amer. Math. Soc. Transl.* (2) **138** (1988), 55–66.

[3] Potapov, V.P., A theorem on the modulus, II. *Teor. Functsiĭ Funksional Anal. i Prilozh.* **39** (1983), 95–106 (Russian). English transl.: *Amer. Math. Soc. Transl.* (2) **138** (1988), 67–74.

[4] Ginzburg, Ju.P., On multiplicative representations on J-nonexpansive operator-functions. Part I. *Mat. Issled.* **2:2** (1967), 52–83 (Russian). English transl.: *Amer. Math. Soc. Transl.* (2) **96** (1970), 189–221.

[5] Ginzburg, Ju.P., On multiplicative representations on J-nonexpansive operator-functions. Part II. *Mat. Issled.* **2:3** (1967), 20–51 (Russian). English transl.: *Amer. Math. Soc. Transl.* (2) **96** (1971), 223–254.

[6] Ginzburg, Ju.P., Divisors and minorants of operator-valued functions of bounded form (Russian). *Mat. Issled.* **2:4** (1967), 47–72. English transl: *Amer. Math. Soc. Transl.* (2) **103** (1974), 153–169.

[7] Ginzburg, Ju.P., Multiplicative representations and minorants of bounded analytic operator-functions. *Teor. Functsiĭ Funksional Anal. i Prilozh.* **1:3** (1967), 9–23 (Russian). English transl.: *Func. Anal. Appl.* **1** (1967), 180–192.

C/o Professor L.A. Sakhnovich, pr. Dobrovolskogo 154, ap. 199,
Odessa 270111, Ukraine

AMS Subject Classification: 30D50, 30E20, 47A68.

Operator Theory
Advances and Applications, Vol. 95
© 1997 Birkhäuser Verlag Basel/Switzerland

On Schur functions and Szegő orthogonal polynomials

Leonid Golinskii*

Abstract. Properties of Schur functions in terms of their Schur parameters are investigated using Szegő orthogonal polynomials as a key tool. The relation between the summability of the Schur parameters and the Bernstein condition for the function to have an absolutely convergent Fourier series is discussed. It is shown that the summability of the Schur parameters implies the summability of the Taylor coefficients.

This note may be regarded as a supplement to the paper [7]. We study further relations between Schur functions, Schur parameters and orthogonal polynomials on the unit circle (Szegő polynomials).

Let us recall that by a *Schur function* we mean a function f which is analytic in the open unit disk $\mathbf{D} = \{|z| < 1\}$ with modulus not exceeding one:

$$||f|| = \sup_{|z|<1} |f(z)| \leq 1,$$

which is not a finite Blaschke product. Each Schur function $f = f_0$ generates (by means of the well-known Schur algorithm) an infinite sequence of Schur functions $f_n, n = 0, 1, \ldots$. The complex numbers $\gamma_n = f_n(0)$ satisfy $|\gamma_n| < 1$; they are generally called *Schur parameters*. A remarkable theorem of Schur asserts that there is one-to-one correspondence between Schur functions and infinite sequences of complex numbers with moduli strictly less than one. Thereby Schur parameters appear to be the natural characteristic of Schur functions. The problem we study here is that of finding the relationship between the Schur functions and their parameters.

It was Ya.L. Geronimus who first observed the intimate relation between Schur functions, Schur parameters and Szegő polynomials. His result plays a crucial role in our investigations.

Let f be a Schur function with parameters γ_n and put

$$F(z) = \frac{1 + zf(z)}{1 - zf(z)}.$$

$*$) The research described in this publication was made possible in part by Grant No. U9S000 from the NSF.

It can be easily checked that $\mathrm{Re}F > 0$ and $F(0) = 1$. According to the Riesz-Herglotz theorem there is a unique probability measure μ on the unit circle $\mathbf{T}$ such that

$$F(z) = \int_{\mathbf{T}} \frac{t+z}{t-z}\, d\mu(t). \tag{1}$$

Let $L^2(\mu, \mathbf{T})$ be the Hilbert space of measurable, square-integrable functions on the unit circle $\mathbf{T}$ with the scalar product and norm

$$(f,g)_\mu = \int_{\mathbf{T}} f(t)\overline{g(t)}\, d\mu, \qquad ||f||^2 = (f,f)_\mu.$$

Let $\Phi_n(z) = z^n + \dots$ be the uniquely determined system of monic polynomials which are orthogonal with respect to μ:

$$(\Phi_n, \Phi_m)_\mu = 0, \qquad n \neq m. \tag{2}$$

Geronimus revealed the relation between Schur parameters γ_n and the so-called reflection coefficients $\Phi_n(0)$:

$$\gamma_n = -\overline{\Phi_{n+1}(0)} \tag{3}$$

(cf. [3, Theorem 18.2], [7, pp. 459–460]). Therefore Szegő orthogonal polynomials can be applied to the problem in question.

Let $\{\varphi_n(z) = \kappa_n z^n + \dots,\ \kappa_n > 0\}_{n=0}^\infty$ be the corresponding orthonormal polynomials

$$(\varphi_n, \varphi_m)_\mu = \delta_{n,m}, \qquad \varphi_n(z) = \kappa_n \Phi_n(z).$$

These polynomials satisfy the Szegő recurrences (cf. [3, Chapter 1, formulas (1.2), (1.9)]:

$$\kappa_n \varphi_{n+1}(z) = z\kappa_{n+1}\varphi_n(z) + \varphi_{n+1}(0)\varphi_n^*(z), \quad n = 1, 2, \dots, \tag{4}$$

and

$$\kappa_n \varphi_n^*(z) = \sum_{i=0}^{n} \overline{\varphi_i(0)}\varphi_i(z), \quad n = 1, 2, \dots, \tag{5}$$

where the reversed *-polynomial of a polynomial P_n of degree n is given by $P_n^*(z) = z^n \bar{P}_n(1/z)$ [$= z^n \overline{P_n(1/\bar{z})}$, editorial note]. If we eliminate φ_n^* from (4) using (5), we obtain

$$z\varphi_n(z) = \frac{\kappa_n}{\kappa_{n+1}}\varphi_{n+1}(z) - \sum_{i=0}^{n} \frac{\kappa_i}{\kappa_n}\Phi_{n+1}(0)\overline{\Phi_i(0)}\varphi_i(z), \quad n = 1, 2, \dots, \tag{6}$$

which is the orthogonal Fourier expansion of $z\varphi_n(z)$ in terms of the orthonormal polynomial system $\{\varphi_n\}_{n=0}^\infty$. The latter formula is useful when considering the matrix representation of the multiplication operator in terms of $\{\varphi_n\}_{n=0}^\infty$.

We start with the following theorem due to Geronimus [4, Theorem 1′] concerning the polynomials with asymptotically constant recurrence coefficients. For the reader's convenience we bring here a proof (cf. [8, Theorem 3]); it is different from the original one.

Theorem 1. *Let μ be a probability measure on $\mathbf{T}$ with infinite support, and let $\{\Phi_n\}$ be the corresponding orthogonal polynomials* (2). *Suppose* $\lim_{n\to\infty}\Phi_n(0)=a$, *where a is a nonzero complex number, $0<|a|<1$. Let $\sin\frac{\alpha}{2}=|a|$ where $\alpha\in(0,\pi)$ and $\Delta_\alpha=\{e^{i\theta}:\alpha\le\theta\le 2\pi-\alpha\}$. Then $\Delta_\alpha\subset\operatorname{supp}(\mu)$ and $\operatorname{supp}(\mu)\backslash\Delta_\alpha$ is at most a countable set whose limit points (if any) belong to Δ_α.*

Proof. Orthogonal polynomials on the real line are intimately related to the Jacobi matrices containing the coefficients of the three-term recurrence relation for the orthonormal polynomials. Perturbation theory of self-adjoint operators allows the interpretation of spectral properties of the Jacobi matrix as properties of the orthogonality measure, including in particular information about its support.

For orthogonal polynomials on the unit circle there is a similar relationship with infinite matrices, but instead of a self-adjoint tridiagonal matrix (for determinate moment problems on the real line) one deals with a unitary Hessenberg matrix (for measures outside the Szegő class).

By a theorem of A.N. Kolmogorov and M.G. Krein, the system $\{\varphi_n\}_{n=0}^{\infty}$ forms an orthonormal basis in $L^2(\mu,\mathbf{T})$ if and only if

$$\log\mu'\notin L^1 \iff \sum_{n=0}^{\infty}|\Phi_n(0)|^2=\infty$$

(cf. [9, Theorem 3.3(a)], [5, Theorem 8.2]); that is obviously the case under the assumptions of the theorem.

In our investigations a key role is played by the unitary multiplication operator $U(\mu):L^2(\mu,\mathbf{T})\longrightarrow L^2(\mu,\mathbf{T})$ defined by

$$\big[U(\mu)f\big](t)=tf(t),\qquad t\in\mathbf{T},\ f\in L^2(\mu,\mathbf{T}),\tag{7}$$

and its representation $\widehat{U}(\mu)$ in the orthonormal basis $\{\varphi_n\}_{n=0}^{\infty}$.

By (6) and the well-known formulas (cf. [5, p. 7])

$$\frac{\kappa_{n-1}^2}{\kappa_n^2}=1-|\Phi_n(0)|^2 \text{ and } \kappa_n=\left[\prod_{k=1}^{n}\left(1-|\Phi_n(0)|^2\right)\right]^{-\frac{1}{2}},$$

we have

$$\widehat{U}(\mu)=\begin{pmatrix}u_{00} & u_{01} & \cdots\\ u_{10} & u_{11} & \cdots\\ \vdots & \vdots & \ddots\end{pmatrix},\qquad u_{kj}=\left(U(\mu)\varphi_j,\varphi_k\right)_\mu,\tag{8}$$

where for $k,j=0,1,\dots$

$$u_{kj}=\begin{cases}-\Phi_{j+1}(0)\overline{\Phi_k(0)}\prod_{p=k+1}^{j}\sqrt{1-|\Phi_p(0)|^2}, & \text{for } k=0,1,\dots,j,\\ \sqrt{1-|\Phi_{j+1}(0)|^2}, & \text{for } k=j+1,\\ 0, & \text{for } k\ge j+2.\end{cases}\tag{9}$$

Infinite matrices such as (8)–(9) in which all the entries below the first subdiagonal vanish are called (upper) Hessenberg matrices.

We can view the infinite matrix (8)–(9) as a unitary operator $\widehat{U}(\mu) : \ell^2 \longrightarrow \ell^2$ which is unitarily equivalent to the multiplication operator $U(\mu)$. In particular, $\operatorname{supp}(\mu)$ coincides with the spectrum of $\widehat{U}(\mu)$.

Along with the multiplication operator $U(\mu)$ of (7), consider the multiplication operator $U(\mu_a)$ on the space $L^2(\mu_a, \mathbf{T})$, where the measure μ_a corresponds to the constant reflection coefficients $\Phi_n(0) = a,\ n = 1, 2, \ldots$. Such measure exists by Favard's theorem on $\mathbf{T}$ (cf. [2]). To study both these operators simultaneously consider their matrix representation $\widehat{U}(\mu)$ of (8)–(9) and

$$\widehat{U}(\mu_a) = \begin{pmatrix} u_{00} & u_{01} & \ldots \\ u_{10} & u_{11} & \ldots \\ \vdots & \vdots & \ddots \end{pmatrix}, \qquad u_{kj} = (U(\mu_a)\varphi_j, \varphi_k)_{\mu_a},$$

where for $k, j = 0, 1, \ldots$

$$u_{kj} = \begin{cases} -a\sqrt{(1-|a|^2)^k}, & \text{for } k = 0, \\ -|a|^2\sqrt{(1-|a|^2)^{(j-k)}} & \text{for } k = 1, \ldots, j, \\ \sqrt{1-|a|^2}, & \text{for } k = j+1, \\ 0, & \text{for } k \geq j+2. \end{cases}$$

This follows from (8)–(9) applied with μ replaced by μ_a. Both $\widehat{U}(\mu)$ and $\widehat{U}(\mu_a)$ are unitary operators acting on the *same* Hilbert space ℓ^2.

Let $S : \ell^2 \longrightarrow \ell^2$ be the shift operator given by the matrix representation

$$S = \begin{pmatrix} 0 & 1 & 0 & 0 \ldots & \\ 0 & 0 & 1 & 0 \ldots & \\ 0 & 0 & 0 & 1 \ldots & \\ 0 & 0 & 0 & 0 \ldots & \\ \vdots & \vdots & \vdots & \vdots & \ddots \end{pmatrix}.$$

Then we can write $\widehat{U}(\mu)$ as

$$\widehat{U}(\mu) = D_{-1}(\mu)S^* + \sum_{k=0}^{\infty} D_k(\mu)S^k \tag{10}$$

where S^* is the adjoint of S and each $D_k(\mu) : \ell^2 \longrightarrow \ell^2$ is a diagonal matrix whose main diagonal is equal to the kth diagonal above the main diagonal of $\widehat{U}(\mu)$, that is,

$$\operatorname{diag} D_k = \begin{cases} \left[-\Phi_{k+i+1}(0)\overline{\Phi_i(0)} \prod_{j=i+1}^{i+k} \sqrt{1-|\Phi_j(0)|^2}\right]_{i=0}^{\infty}, & \text{for } k = 0, 1, \ldots \\ \left[\sqrt{1-|\Phi_i(0)|^2}\right]_{i=0}^{\infty}, & \text{for } k = -1. \end{cases}$$

This infinite series representation for $\widehat{U}(\mu)$ converges in the operator norm. To see this note that $||S|| = 1$ and

$$||D_k(\mu)|| = \sup_{i\geq 0} \Phi_{k+i+1}(0)\overline{\Phi_i(0)} \prod_{j=i+1}^{i+k} \sqrt{1-|\Phi_j(0)|^2} \leq \sup_{i\geq 0} \prod_{j=i+1}^{i+k} \sqrt{1-|\Phi_j(0)|^2}, \tag{11}$$

for $k = 0, 1, \ldots$, since $|\Phi_n(0)| < 1$ for positive integer n. If $\lim_{n\to\infty} \Phi_n(0) = a$ with $a \neq 0$, then $\{||D_k(\mu)||\}_{k=-1}^{\infty}$ decreases exponentially and therefore the series in (10) converges uniformly.

Similarly, we can write $\widehat{U}(\mu_a)$ as a uniformly convergent series

$$\widehat{U}(\mu_a) = D_{-1}(\mu_a)S^* + \sum_{k=0}^{\infty} D_k(\mu_a)S^k.$$

Now consider the difference $\widehat{U}(\mu) - \widehat{U}(\mu_a)$. Given $k = -1, 0, 1, \ldots$, the difference $D_k(\mu) - D_k(\mu_a)$ is a compact operator since it is a diagonal operator for which the entries converge to zero. Since the set of compact operators on a Hilbert space is a closed ideal, then $\widehat{U}(\mu) - \widehat{U}(\mu_a)$ is a compact operator. According to H. Weyl's theorem the essential spectra of $\widehat{U}(\mu)$ and $\widehat{U}(\mu_a)$ are the same. Geronimus [6, p. 94] (see also [8, Section 2]) showed that the operator $\widehat{U}(\mu_a)$ has purely absolutely continuous spectrum on the arc Δ_α and, in addition, $\widehat{U}(\mu_a)$ may have at most one eigenvalue which must be in $\mathbf{T}$ but is located outside this arc. Since the spectra of $\widehat{U}(\mu)$ and $\widehat{U}(\mu_a)$ are equal to the support of μ and μ_a, respectively, this is precisely what had to be proved. □

Our first remark concerns some results from the theory of Szegő polynomials, which can be reformulated in terms of Schur functions and parameters.

Theorem 2. *Let f be a Schur function with parameters γ_n.*

1. *If*

$$\lim_{n\to\infty} \gamma_n = \gamma, \quad 0 < |\gamma| < 1,$$

then f can be continued analytically to a meromorphic function in the domain $\overline{\mathbf{C}}\backslash\Delta_\alpha$, where $\Delta_\alpha = \{e^{ix} : \alpha \leq x \leq 2\pi - \alpha,\ \alpha = 2\arcsin|\gamma|\}$, and $|f(e^{ix})| = 1$ for $|x| < \alpha$.

2. *If $\limsup_{n\to\infty} |\gamma_n| = 1$, then f is an inner function.*
3. *The relation*

$$\lim_{n\to\infty} \gamma_{n-1}\overline{\gamma_n} = \zeta, \quad |\zeta| = 1\ ,$$

holds iff the Schur function f is an infinite Blaschke product with zeros tending to $-\zeta$.

Proof.

1. Since for the affiliated orthogonal polynomials Φ_n we have $\lim_{n\to\infty} \Phi_n(0) = -\overline{\gamma}$ by (3), it follows from Theorem 1 that $\Delta_\alpha \subset \mathrm{supp}(\mu)$ and $\mathrm{supp}(\mu)\backslash\Delta_\beta$ is finite for every $0 < \beta < \alpha$. Therefore the representation (1) takes the form

$$F(z) = \int_{\Delta_\alpha} \frac{t+z}{t-z}\, d\mu(t) + \sum_{k=1}^{\omega} \frac{t_k + z}{t_k - z}\mu_k,$$

where $\omega \le \infty$, $t_k \in \mathbf{T}$ and $\mu_k > 0$ (if $\omega = \infty$, then the sequence t_k may have no more then two limit points $e^{\pm i\alpha}$). The rest is immediate from the equalities

$$f(z) = \frac{1}{z}\frac{F(z)-1}{F(z)+1} \tag{12}$$

and

$$1 - |f(e^{ix})|^2 = \frac{4\mathrm{Re}F(e^{ix})}{|F(e^{ix})+1|^2}. \tag{13}$$

2. The relation $\limsup_{n\to\infty} |\Phi_n(0)| = 1$ forces the measure μ in (1) to be singular by Rakhmanov's lemma (cf. [10, Lemma 4]), so that $\mathrm{Re}F(e^{ix}) = 0$ a.e. and $|f(e^{ix})| = 1$ a.e. by (4).

3. We now have $\lim_{n\to\infty} \overline{\Phi_n(0)\Phi_{n+1}(0)} = \zeta$ by (3), and that obviously implies that

$$\lim_{n\to\infty} |\Phi_n(0)| = 1.$$

Consider the operator $U(\mu)$ of (7) and it matrix representation $\widehat{U}(\mu)$ which is given by (8)–(9). It can be readily seen that now the diagonal operators $D_k(\mu)$ in (10) for $k \ne 0$ as well as $D_0(\mu)+\zeta\mathbf{I}$ are compact operators. Hence, the operator $\widehat{U}(\mu)+\zeta\mathbf{I}$, acting in ℓ^2 is also compact.

According to a theorem of F. Riesz and J. Schauder (cf. [11, Theorem VI.15, p. 203]), the spectrum of the operator $\widehat{U}(\mu)+\zeta\mathbf{I}$ is an at most countable set with no limit point except possibly zero. Since $\mathrm{supp}(\mu)$ is an infinite set, the spectrum of the unitary operator $\widehat{U}(\mu) = (\widehat{U}(\mu)+\zeta\mathbf{I}) - \zeta\mathbf{I}$ is a countable set, it lies on the unit circle, and its only limit point is $-\zeta$.

Since the spectrum of $\widehat{U}(\mu)$ is equal to $\mathrm{supp}(\mu)$, the representation (1) takes the form

$$F(z) = \sum_{k=1}^{\omega} \frac{t_k + z}{t_k - z}\mu_k,$$

where $\mu_k > 0$ and $t_k \to -\zeta$. The rest is plain by (12). □

In his recent paper [13] W.H. Thron has shown that a Schur function f can be analytically continued to a meromorphic function in the entire plane except, possibly, for an essential singularity at $z=-1$, as long as $\lim_{n\to\infty}\gamma_n=\zeta,\ |\zeta|=1$ and

$$\sum_{n=0}^{\infty}\max_{k\geq n}|\gamma_n-\zeta|<\infty.$$

Note that this result is a special case of our Theorem 2.

Our next remark applies to Theorem 1 from [7].

Theorem 3 (see [7, Theorem 1]). *Let a Schur function f be continuous in the closed unit disk $\overline{\mathbf{D}}$ with $||f||<1$. If*

$$\sum_{n=1}^{\infty}\frac{\omega(\frac{1}{n},f)}{\sqrt{n}}<\infty, \tag{14}$$

where $\omega(x,f)$ is the modulus of continuity of the boundary value $f(e^{ix})$, then the Schur parameters γ_n satisfy

$$\sum_{n=0}^{\infty}|\gamma_n|<\infty. \tag{15}$$

The following example shows that in general (15) does not imply (14).

Example. Let W denote the set of absolutely convergent Fourier series. As in [7, Remark 2, p. 465] we start with the function

$$g(x)=\operatorname{Re}\sum_{n=1}^{\infty}n^{-1}\exp(in\log n+inx),$$

which is an element of $\mathrm{Lip}\frac{1}{2}$, but not of W. It is obvious that

$$\sum_{n=1}^{\infty}\frac{\omega(\frac{1}{n},g)}{\sqrt{n}}=\infty \tag{16}$$

(or otherwise g would belong to W by Bernstein theorem).

Consider the function

$$h(x)=1+\omega(x),\quad 0\leq x\leq\pi,\quad h(-x)=h(x),$$

where $\omega(x)=\omega(x,g)=O(x^{1/2}),\ x\to 0$. Since the modulus of continuity satisfies $\omega(x+y)\leq\omega(x)+\omega(y)$, then h is a continuous function with $\omega(x,h)=\omega(x,g)=\omega(x)$. Hence, by Privalov's theorem (cf. [14, Chapter 6, Theorem 13.29]) for the conjugate function $\tilde{h}$, we have $\tilde{h}\in\mathrm{Lip}\frac{1}{2}$. Therefore the function

$$F(z)=\frac{1}{2\pi M}\int_{-\pi}^{\pi}\frac{e^{ix}+z}{e^{ix}-z}h(x)\,dx,\quad M=\frac{1}{2\pi}\int_{-\pi}^{\pi}h(x)\,dx\ ,$$

is continuous in $\overline{\mathbf{D}}$, $F(0) = 1$ and $\mathrm{Re}F(z) \geq M^{-1}$. This means that the Schur function (12) is continuous in $\overline{\mathbf{D}}$ and $||f|| < 1$.

On the other hand, applying [14, Chapter 6, Theorem 3.6] we obtain $h \in W$ and by G. Baxter's theorem (cf. [1, Corollary 1.1]) we have $\sum_{n=0}^{\infty} |\Phi_n(0)| < \infty$, where Φ_n are the monic orthogonal polynomials with respect to the weight function $M^{-1}h$. The remark due to Geronimus mentioned at the begining of the paper leads to the relation (15). Nevertheless the inequalities $\omega(x,h) = \omega(x,g) \leq \omega(x,F) \leq C(f)\omega(x,f)$ (see [7, p. 464 after formula (4.2)]) along with (16) yields the divergence of the series (14).

Our final remark gives an affirmative answer on the following question, posed by H. Dym: whether the condition (15) ensures the summability of the Taylor coefficients of a Schur function f.

Theorem 4. *Let $f(z) = \sum_{k=0}^{\infty} f_k z^k$ be a Schur function with Schur parameters γ_n and let (15) be valid. Then*

$$\sum_{k=0}^{\infty} |f_k| < \infty.$$

Proof. Since for the associated system of monic orthogonal polynomials Φ_n we have

$$\sum_{n=0}^{\infty} |\Phi_n(0)| < \infty,$$

then, by the theorem of G. Baxter which was mentioned above, the measure μ in (1) is absolutely continuous with respect to Lebesgue measure, its derivative μ' is positive, continuous and belongs to W. This means that the function $F(z)$ (1) has an absolutely convergent Taylor series:

$$F(z) = \sum_{k=0}^{\infty} F_k z^k, \quad \sum_{k=0}^{\infty} |F_k| < \infty.$$

Since $\mathrm{Re}F(z) \geq 0$ and the function $\varphi(w) = (w-1)(w+1)^{-1}$ is analytic in the closed right halfplane, the Wiener-Levy theorem (cf. [14, Chapter 6, Theorem 5.2]) implies that the function $\varphi(F(z))$ has an absolutely convergent Taylor series, as does the function f of (12). □

The example $f(z) = (2-z)^{-1}$, $\gamma_n = (n+2)^{-1}$ (see [7, example 7, p. 463]) shows that the converse statement is false in general.

References

[1] Baxter G., A convergence equivalence related to polynomials on the unit circle, *Trans. Amer. Math. Soc.* **99** (1961) 471–487.

[2] Erdelyi T., Geronimo J.S., Nevai P. and Zhang J., Simple proof of "Favard's theorem" on the unit circle, *Ati. Sem. Mat. Fis Univ. Modena* **29** (1991) 41–46.

[3] Geronimus Ja.L., Polynomials orthogonal on a circle and their applications, *Amer. Math. Soc. Transl.* **3** (1962) 1–78.

[4] Geronimus Ja.L., On the character of the solution of the moment problem in the case of the periodic in the limit associated fraction, *Izv. Akad. Nauk SSSR* **5** (1941) 203–210.

[5] Geronimus Ja.L., *Orthogonal Polynomials*, Consultants Bureau, New York, 1961.

[6] Geronimus Ja.L, *Orthogonal Polynomials*, Appendix to the Russian translation of Szegő's book [12], *Amer. Math. Soc. Transl.* **108** (1977) 37–130.

[7] Golinskii L.B., Schur functions, Schur parameters and orthogonal polynomials on the unit circle, *Zeitschrift fur Analysis und ihre Anwendungen* **12** (1993) 457–469.

[8] Golinskii L.B., Nevai P.G. and van Assche W., Perturbation of orthogonal polynomials on an arc of the unit circle, to appear in *JAT*.

[9] Grenander U. and Szegő G., *Toeplitz Forms and Their Applications*, Chelsea Publishing Company, New York, 1984.

[10] Rakhmanov E.A., On the asymptotics of the ratio of orthogonal polynomials, II, *Math. USSR Sb* **46** (1983) 105–117.

[11] Reed M. and Simon B., *Methods of Modern Mathematical Physics, Vol. 1: Functional Analysis*, Academic Press, New York, 1972.

[12] Szegő G., *Orthogonal Polynomials*. Amer. Math. Soc. Colloq. Publ., Vol. 23, Amer. Math. Soc., Providence, RI, 1975 (4th edition).

[13] Thron W.J., Limit periodic Schur algorithms, the case $|\gamma| = 1, \sum d_n < \infty$, *Numerical Algorithms* **3** (1992) 441–450.

[14] Zygmund A., *Trigonometric Series, Vol. 1.*, Cambridge University Press, Cambridge, 1977.

Mathematical Division
Institute for Low Temperature Physics and Engineering
47 Lenin Avenue
Kharkov 310164, Ukraine

AMS Subject Classification: 30D50, 33C45

Operator Theory
Advances and Applications, Vol. 95
© 1997 Birkhäuser Verlag Basel/Switzerland

Hilbert spaces of entire functions as a J theory subject

Leonid Golinskii and Irina Mikhailova
(edited by V.P. Potapov)

Abstract. In this work a number of fundamental results in the de Branges theory of Hilbert spaces of entire functions are obtained from the point of view of J theory. Particular attention is focused on the set of measures which satisfy a Parseval equality in such a Hilbert space of entire functions.

A resolvent matrix for the solutions of this problem is studied. It is a J inner matrix valued meromorphic function. A number of theorems about its real representation, structure and parametrization are obtained.

0 Preface

The theory of Hilbert spaces of entire functions was created by L. de Branges about 20 years ago[1]. Its main results are summed up in the monograph [1]. Starting with the classical Paley and Wiener theory, that characterizes entire functions of exponential type which are square integrable on the real line, the author introduced and developed the theory of the class $H(E)$ of entire functions, generated by an entire function $E(z)$, which is a far reaching analog of the exponential function in the Paley-Wiener theory. This function is subject only to the restriction

$$|E(x+iy)| < |E(x-iy)|, \qquad y > 0 ;$$

it serves to characterize the behavior of the functions $H(E)$ in the upper and lower half planes. L. de Branges endowed $H(E)$ with a Hilbert space structure and established a number of fundamental identities, with highly nontrivial consequences. A generalized Fourier transformation and an analog of the Paley-Wiener theorem for the spaces $H(E)$ complete the author's construction. The general theory has been illustrated by a number of examples which prove useful for applications. In particular, L. de Branges used certain properties of the Hilbert spaces $H(E)$ to resolve an important uniqueness problem which is connected with the reconstruction of a canonical system of two differential equations from its monodromy matrix, a

1) The original version of this paper appeared (in Russian) as Preprint No. 28–80, Institute for Low Temperature Physics and Engineering, Kharkov, 1980.

problem which defied solution for a long time. That is why the de Branges theory must be regarded as one of the outstanding achievements of mathematical analysis of our century.

Nevertheless, in our opinion, the exposition in this monograph is far from being perfect. The main defect is a complete avoidance of the matrix nature of the problem, even though it underlies many of the formulas. As a matter of fact, all the essential theorems in de Branges' theory are statements concerning analytic matrix valued functions of second order. Similar problems can be formulated for matrix valued functions of arbitrary order. The exposition in [1] makes it difficult to envision such extensions. A cumbersome scalar technique often leads to nonconstructive considerations (cf. [1, Theorem 27]). The didactic principles also leave much to be desired: the reader is forced to solve more than 200 problems. A great number of these are not just exercises illustrating the basic theory, but components in the proof of the main results. The exposition abounds with purely technical details, and the principal ideas can hardly be singled out among them. All this taken together make the study and perception of this extremely important theory very difficult.

We are deeply convinced of the intimate relation between the theory of Hilbert spaces of entire functions and the theory of analytic J expansive matrix valued functions. Our main objective is to expose the basic propositions of de Branges' theory from this viewpoint. The present paper encompasses an essential part of the general theory, including its cornerstone: the Parseval equality in $H(E)$. It is perhaps well to remark at this point that a number of interpolation problems (Nevanlinna-Pick problem, extension of positive-definite functions etc.) can be formulated as a Parseval equality in an appropriately chosen space $H(E)$. The matrix techniques employed here serve to simplify de Branges' constructions and makes them much more transparent.

1 Preliminaries from J theory

Throughout this paper the symbol J will denote a constant $m \times m$ matrix which is both selfadjoint and unitary with respect to the standard inner product in the m dimensional complex linear space of column vectors $\mathbf{C}^m$:

$$J^* = J, \qquad J^*J = I_m.$$

We shall refer to such a J as a signature matrix.

We shall endow $\mathbf{C}^m$ with an indefinite metric by means of the indefinite inner product

$$[f, g] = g^*Jf.$$

A vector f is called a "+" vector, a "−" vector or a neutral vector with respect to this metric depending on whether $[f, f] > 0$, $[f, f] < 0$ or $[f, f] = 0$.

An $m \times m$ matrix $\mathcal{A}$ is said to be J expansive (resp. J contractive) if

$$[\mathcal{A}f, \mathcal{A}f] \geq [f, f] \qquad (\text{resp.}\ [\mathcal{A}f, \mathcal{A}f] \leq [f, f])$$

or equivalently if

$$\mathcal{A}^* J \mathcal{A} - J \geq 0 \qquad (\text{ resp. } \mathcal{A}^* J \mathcal{A} - J \leq 0)\,,$$

and J unitary if $[\mathcal{A}f, \mathcal{A}f] = [f, f]$ or $\mathcal{A}^* J \mathcal{A} - J = 0$. The expression $\mathcal{A}^* J \mathcal{A} - J$ which occurs frequently in what follows, will be called the J *form* of the matrix $\mathcal{A}$. The J form corresponding to the product of two matrices $\mathcal{A}_1$ and $\mathcal{A}_2$ can be expressed in terms of the J forms corresponding to each factor:

$$\mathcal{A}_2^* \mathcal{A}_1^* J \mathcal{A}_1 \mathcal{A}_2 - J = \mathcal{A}_2^* \left(\mathcal{A}_1^* J \mathcal{A}_1 - J\right) \mathcal{A}_2 + \mathcal{A}_2^* J \mathcal{A}_2 - J. \tag{1.1}$$

The relation (1.1) clearly displays the fact that the product $\mathcal{A}_1 \mathcal{A}_2$ is J expansive (contractive, unitary) when $\mathcal{A}_1$ and $\mathcal{A}_2$ are both J expansive (contractive, unitary).

A number of properties for J expansive matrices can be derived from well known properties of contractive matrices. To clarify this, let

$$P = \frac{I_m + J}{2} \quad \text{and} \quad Q = \frac{I_m - J}{2}.$$

It is easily checked that $P^* = P^2 = P$, $Q^* = Q^2 = Q$ and

$$PQ = QP = 0, \quad P + Q = I_m, \quad P - Q = J.$$

As a matter of fact, P and Q are the orthogonal projections of $\mathbf{C}^m$ onto the span of the eigenvectors of J corresponding to the eigenvalues $+1$ and -1, respectively.

Lemma 1. *For each J expansive matrix $\mathcal{A}$:*
1. *$Q - P\mathcal{A}$ and $Q + \mathcal{A}P$ are invertible and*

$$S = (P - Q\mathcal{A})\,(Q - P\mathcal{A})^{-1} = -\,(Q + \mathcal{A}P)^{-1}\,(P + \mathcal{A}Q) \tag{1.2}$$

is a contractive matrix.
2. *$Q - SP$ and $Q + PS$ are invertible and*

$$\mathcal{A} = (Q - SP)^{-1}\,(P - SQ) = -\,(P + QS)\,(Q + PS)^{-1}\ . \tag{1.3}$$

3. *The equalities*

$$I_m - S^* S = (Q - \mathcal{A}^* P)^{-1}\,(\mathcal{A}^* J \mathcal{A} - J)\,(Q - P\mathcal{A})^{-1} \tag{1.4}$$

$$I_m - S S^* = (Q + \mathcal{A}P)^{-1}\,(\mathcal{A} J \mathcal{A}^* - J)\,(Q + P\mathcal{A}^*)^{-1} \tag{1.5}$$

hold.

Proof. The nonnegativity of the J form is equivalent to the inequalities

$$0 \le P + \mathcal{A}^* Q \mathcal{A} \le Q + \mathcal{A}^* P \mathcal{A}. \tag{1.6}$$

For any vector h satisfying $(Q - P\mathcal{A})\, h = 0$ we have

$$(Q - \mathcal{A}^* P)\,(Q - P\mathcal{A})\, h = (Q + \mathcal{A}^* P \mathcal{A})\, h = 0$$

and by (1.6) $(P + \mathcal{A}^* Q \mathcal{A})\, h = 0$. Therefore $h = 0$. This means that $Q - P\mathcal{A}$ is invertible and hence that the first linear fractional transformation in (1.2) is well defined. Since $Q - \mathcal{A}^* P$ is also invertible, the identity

$$(P - \mathcal{A}^* Q)\,(Q + P\mathcal{A}^*) = -\,(Q - \mathcal{A}^* P)\,(P + Q\mathcal{A}^*) \tag{1.7}$$

implies that if $(Q + P\mathcal{A}^*)\, h = 0$, then also $(P + Q\mathcal{A}^*)\, h = 0$ holds, from whence it follows that $h = 0$, i.e., $Q + P\mathcal{A}^*$ and $Q + \mathcal{A}P$ are invertible. Taking the conjugate matrices in (1.7) we come to the equality

$$(Q + \mathcal{A}P)\,(P - Q\mathcal{A}) = -\,(P + \mathcal{A}Q)\,(Q - P\mathcal{A})$$

which proves (1.2). In particular, it follows readily from (1.2) that

$$Q - SP \;=\; Q + (Q + \mathcal{A}P)^{-1}\, P = (Q + \mathcal{A}P)^{-1}$$

and

$$Q + PS \;=\; Q + P\,(Q - P\mathcal{A})^{-1} = (Q - P\mathcal{A})^{-1}$$

are invertible and hence that (1.3) holds.

The equalities (1.4)–(1.5) are immediate from (1.2). □

The following basic statement is a straightforward consequence of Lemma 1.

Theorem 1. *A matrix $\mathcal{A}$ is J expansive (resp. J unitary) if and only if $\mathcal{A}^*$ is J expansive (resp. J unitary).*

Proof. The equivalence of the two inequalities $\mathcal{A}^* J \mathcal{A} - J \ge 0$ and $\mathcal{A} J \mathcal{A}^* - J \ge 0$ is immediate from (1.4)–(1.5) and the well known properties of contractive and unitary matrices. □

The main object of J theory is a J expansive invertible matrix valued function $\mathcal{A}(z)$, which is analytic in a region G except for isolated singularities. For reasons of symmetry, the unit disk, and the right and upper half planes are the most natural regions of the complex plane. We shall focus our attention on the latter one and denote it $\mathbf{C}_+$.

At every point z of analyticity of $\mathcal{A}(z)$ the linear fractional transformations (1.2) and (1.3) make sense

$$S(z) = (P - Q\mathcal{A}(z))\,(Q - P\mathcal{A}(z))^{-1}\ , \tag{1.8}$$

$$\mathcal{A}(z) = (Q - S(z)P)^{-1}(P - S(z)Q) \;, \tag{1.9}$$

where $S(z)$ is a contractive analytic matrix valued function, having only isolated singularities. But then $S(z)$ is analytic in the whole upper half plane. By (1.9), $\mathcal{A}(z)$ is meromorphic and of bounded type in $\mathbf{C}_+$ (i.e., each matrix entry can be represented as a ratio of bounded analytic functions). It follows from a well known theorem of Fatou that a J expansive matrix valued function $\mathcal{A}(z)$ admits nontangential one sided boundary limits at a.e. point $t \in \mathbf{R}$, so that $\mathcal{A}(t)$ is well determined a.e. on the real line and is clearly a J expansive matrix valued function.

We shall say that an $m \times m$ matrix valued function $\mathcal{U}(z)$ is J *inner* if it is J expansive and meromorphic in $\mathbf{C}_+$ and $\mathcal{U}(t)$ is J unitary a.e. on the real line. Let $\bar{\mathcal{A}}(z) = \mathcal{A}^*(\bar{z})$. Given a J expansive matrix valued function $\mathcal{A}(z)$, it seems natural enough to extend it into the lower half plane by means of the relation

$$\mathcal{A}(z) = J\left(\bar{\mathcal{A}}(z)\right)^{-1} J, \tag{1.10}$$

which is usually called a symmetry principle. It enables one to deal with a single function $\mathcal{A}(z)$ in the region $\{\Im z \neq 0\}\backslash K$, where K is an isolated set of points, instead of the pair of functions $\mathcal{A}(z)$ and $\mathcal{A}^*(z)$ in $\mathbf{C}_+$. Furthermore, if $\mathcal{A}(z)$ is J unitary and analytic on the real axis, then it admits an analytic continuation downwards and (1.10) holds. Such a matrix valued function $\mathcal{A}(z)$ is meromorphic and of bounded type in the lower half plane.

A J inner matrix valued function having a first order pole as its only singularity will be termed an *elementary* J inner factor. Unlike the situation for scalar contractive functions, this pole $z_0 = \sigma_0 + i\tau_0$ may be situated anywhere in the complex plane. We shall distinguish elementary J inner factors of the first, second and third kind depending on whether its pole is in the upper half plane, lower half plane or on the real line. Apart from a constant J unitary multiplier, they must be in one of the following forms:

$$b(z) = I_m + \frac{2i\tau_0}{z - z_0}P, \quad P^2 = P, \; PJ = JP^* \geq 0, \quad (\tau_0 > 0), \tag{I}$$

$$b(z) = I_m - \frac{2i\tau_0}{z - z_0}Q, \quad Q^2 = Q, \; QJ = JQ^* \geq 0, \quad (\tau_0 < 0), \tag{II}$$

$$b(z) = I_m + \frac{i}{z - \sigma_0}\epsilon, \quad \epsilon^2 = 0, \; \epsilon J = J\epsilon^* \geq 0, \quad (\tau_0 = 0). \tag{III}$$

Their J forms can be easily evaluated:

$$\begin{aligned}
b(z)Jb^*(z) - J &= \frac{z - \bar{z}}{i}\frac{2i}{|z - z_0|^2}PJ, \quad (\tau_0 > 0),\\
b(z)Jb^*(z) - J &= \frac{z - \bar{z}}{i}\frac{2i}{|z - z_0|^2}QJ, \quad (\tau_0 < 0),\\
b(z)Jb^*(z) - J &= \frac{z - \bar{z}}{i}\frac{1}{|z - \sigma_0|^2}\epsilon J, \quad (\tau_0 = 0).
\end{aligned}$$

Note, that for a pole on the real line, the elementary J inner factor of the third kind can be represented as

$$b(z) = \exp\left(\frac{i}{z-\sigma_0}\epsilon\right).$$

We do not exclude a pole at infinity:

$$b(z) = I_m - iz\epsilon.$$

For $m = 2n$ the matrices $P,\ Q,\ \epsilon$ in (I)–(III) can be parametrized as follows

$$P = \begin{bmatrix} x \\ y \end{bmatrix} [x^*,\ y^*]\, J, \qquad [x^*,\ y^*]\, J \begin{bmatrix} x \\ y \end{bmatrix} = 1;$$

$$Q = \begin{bmatrix} x \\ y \end{bmatrix} [x^*,\ y^*]\, J, \quad [x^*,\ y^*]\, J \begin{bmatrix} x \\ y \end{bmatrix} = -1;$$

$$\epsilon = \begin{bmatrix} \xi \\ \eta \end{bmatrix} [\xi^*,\ \eta^*]\, J, \qquad [\xi^*,\ \eta^*]\, J \begin{bmatrix} \xi \\ \eta \end{bmatrix} = 0;$$

where $x,\ y,\ \xi$ and η are $n \times n$ matrices.

Definition. Let $\mathcal{A}(z)$ and $\mathcal{A}_2(z)$ be J expansive matrix valued functions. $\mathcal{A}_2(z)$ is called a *right* (resp. *left*) *divisor* of $\mathcal{A}(z)$ if there exists a J expansive matrix valued function $\mathcal{A}_1(z)$ such that

$$\mathcal{A}(z) = \mathcal{A}_1(z)\mathcal{A}_2(z) \qquad (\text{resp. } \mathcal{A}(z) = \mathcal{A}_2(z)\mathcal{A}_1(z))\ .$$

Theorem 2. *A J expansive matrix valued function $\mathcal{A}_2(z)$ is a right divisor of $\mathcal{A}(z)$ if and only if*

$$\mathcal{A}^*(z)J\mathcal{A}(z) - J \geq \mathcal{A}_2^*(z)J\mathcal{A}_2(z) - J \geq 0\ . \tag{1.11}$$

Proof. If $\mathcal{A}_2(z)$ is a right divisor of $\mathcal{A}(z)$, then (1.11) is a straightforward consequence of (1.1). Conversely, it follows immediately from (1.11) that $\mathcal{A}_1(z) = \mathcal{A}(z)\mathcal{A}_2^{-1}(z)$ is a J expansive matrix valued function. □

For left divisors, the relation (1.11) must be replaced by

$$\mathcal{A}(z)J\mathcal{A}^*(z) - J \geq \mathcal{A}_2(z)J\mathcal{A}_2^*(z) - J \geq 0\ . \tag{1.11}$$

In general J theory, the explicit form and size of the signature matrix is not essential. Here, however, we shall restrict ourselves to 2×2 matrices and shall fix

$$J = \begin{pmatrix} 0 & i \\ -i & 0 \end{pmatrix}.$$

Such a choice is justified by the following proposition:

Theorem 3. *The linear fractional transformation*

$$w = \frac{\delta\omega + \gamma}{\beta\omega + \alpha}$$

based on the J expansive matrix

$$\mathcal{A} = \begin{pmatrix} \delta & \gamma \\ \beta & \alpha \end{pmatrix}$$

maps $\mathbf{C}_+ = \{\omega : \Im\omega > 0\}$ *into itself.*

Proof. Define the vector $\begin{bmatrix} u \\ v \end{bmatrix}$ by the equality

$$\begin{bmatrix} u \\ v \end{bmatrix} = \mathcal{A}\begin{bmatrix} \omega \\ 1 \end{bmatrix}.$$

Then, since

$$\begin{aligned} i\,(u^*v - v^*u) &= [u^*, v^*]\,J\begin{bmatrix} u \\ v \end{bmatrix} \\ &= [\omega^*, 1]\,\{\mathcal{A}^*J\mathcal{A} - J\}\begin{bmatrix} \omega \\ 1 \end{bmatrix} + [\omega^*, 1]\,J\begin{bmatrix} \omega \\ 1 \end{bmatrix} \\ &\geq [\omega^*, 1]\,J\begin{bmatrix} \omega \\ 1 \end{bmatrix} > 0\,, \end{aligned}$$

$v \neq 0$ and $w = v^{-1}u$. The rest is immediate from the relation

$$|v|^2\,\frac{w - w^*}{i} = [\omega^*, 1]\,\{\mathcal{A}^*J\mathcal{A} - J\}\begin{bmatrix} \omega \\ 1 \end{bmatrix} + \frac{\omega - \omega^*}{i}\,, \tag{1.12}$$

and the presumed J expansiveness of $\mathcal{A}$. □

A crucial role in what follows is played by the well known class of Nevanlinna functions, i.e., functions which are analytic in $\mathbf{C}_+$ with positive imaginary part there. Every such function $w(z)$ admits a representation of the form

$$w(z) = pz + q + \frac{1}{\pi}\int_{-\infty}^{\infty}\left(\frac{1}{t - z} - \frac{t}{1 + t^2}\right) d\,\mu(t), \quad \Im z > 0, \tag{1.13}$$

where $p \geq 0,\ q = q^*$ and $\mu(t)$ is a nondecreasing function such that

$$\mu(0) = 0, \quad \mu(t) = \frac{\mu(t + 0) + \mu(t - 0)}{2} \text{ and } \int_{-\infty}^{\infty}\frac{d\,\mu(t)}{1 + t^2} < \infty. \tag{1.14}$$

If, in particular, the function $\Im w(z)$ is continuous in the closed upper half plane, then the measure $d\mu(t)$ is absolutely continuous with respect to Lebesgue measure on the line and

$$d\,\mu(t) = \Im w(t)\,dt. \tag{1.15}$$

Conversely, every function satisfying (1.13)–(1.14) is a Nevanlinna function and

$$\frac{w(z) - w^*(z)}{z - \bar{z}} = p + \frac{1}{\pi}\int_{-\infty}^{\infty} \frac{d\,\mu(t)}{|t-z|^2}\,.$$

Keeping in mind formula (1.13) we shall extend the function $w(z)$ into the lower half plane by the equality $w(z) = w^*(\bar{z})$, $\Im z < 0$, which is called the symmetry continuation. If $w(z)$ is analytic and real valued on an interval (a, b), this continuation coincides with the analytic one. However, in some cases we shall consider the analytic continuation wherein the symmetry is violated. Note that if a Nevanlinna function $w(z)$ has a pole at a real point x_0, then necessarily

$$w(z) = \frac{a}{x_0 - z} + w_1(z), \qquad a > 0, \tag{1.16}$$

where $w_1(z)$ is analytic at x_0. As is well known, the parameters p and q in (1.13) are uniquely determined by the function $w(z)$. In particular,

$$p = \lim_{y\to+\infty} \frac{w(iy)}{iy} = \lim_{y\to+\infty} \frac{\Im w(iy)}{y}. \tag{1.17}$$

Let

$$\mathcal{A}(z) = \begin{pmatrix} \delta(z) & \gamma(z) \\ \beta(z) & \alpha(z) \end{pmatrix}$$

be a matrix valued function which is analytic and J expansive in the upper half plane and let $\omega(z)$ be an arbitrary Nevanlinna function. By Theorem 3,

$$w(z) = \frac{\delta(z)\omega(z) + \gamma(z)}{\beta(z)\omega(z) + \alpha(z)}$$

is also a Nevanlinna function. Such expressions, based on a J inner matrix $\mathcal{A}(z)$, occur extensively in representation formulas for the solutions of a wide variety of interpolation problems. In this instance, the matrix valued function $\mathcal{A}(z)$ is usually called a resolvent matrix of the problem.

2 de Branges matrices

We start with a J expansive matrix valued function

$$\mathcal{B}(z) = \begin{pmatrix} \delta(z) & \gamma(z) \\ \beta(z) & \alpha(z) \end{pmatrix} \tag{2.1}$$

and its J forms

$$\mathcal{B}(z)J\mathcal{B}^*(z)-J=\frac{1}{i}\begin{pmatrix}\gamma(z)\delta^*(z)-\delta(z)\gamma^*(z) & \gamma(z)\beta^*(z)-\delta(z)\alpha^*(z)+1\\ \alpha(z)\delta^*(z)-\beta(z)\gamma^*(z)-1 & \alpha(z)\beta^*(z)-\beta(z)\alpha^*(z)\end{pmatrix} \tag{2.2}$$

and

$$\mathcal{B}^*(z)J\mathcal{B}(z)-J=\frac{1}{i}\begin{pmatrix}\beta^*(z)\delta(z)-\delta^*(z)\beta(z) & \beta^*(z)\gamma(z)-\delta^*(z)\alpha(z)+1\\ \alpha^*(z)\delta(z)-\gamma^*(z)\beta(z)-1 & \alpha^*(z)\gamma(z)-\gamma^*(z)\alpha(z)\end{pmatrix}. \tag{2.3}$$

Definition. A matrix valued function which is meromorphic in the complex plane and J inner in the upper half plane is said to be a *de Branges matrix* if the Nevanlinna function

$$\Phi(z)=\Phi(z,\mathcal{B})=\frac{\delta(z)i+\gamma(z)}{\beta(z)i+\alpha(z)} \tag{2.4}$$

is analytic on the real line.

Assumption (2.4) is certainly fulfilled if the matrix valued function $\mathcal{B}(z)$ itself is analytic on the real line. Indeed, let $\Phi(z)$ have a pole at a point $x_0\in\mathbf{R}$. Then, since $\mathcal{B}(z)$ is analytic at x_0,

$$\beta(x_0)i+\alpha(x_0)=0.$$

The J unitary property of $\mathcal{B}(x_0)$ implies that

$$\alpha(x_0)\beta^*(x_0)=\beta(x_0)\alpha^*(x_0)=\alpha^*(x_0)\beta(x_0).$$

Comparing these two equalities we see that $\alpha(x_0)\alpha^*(x_0)=\beta(x_0)\beta^*(x_0)=0$ and hence that $\mathcal{B}(x_0)$ is a degenerate matrix, which is incompatible with its J unitary property.

Thus the elementary J inner factors (I) and (II) as well as their finite products are de Branges matrices. We now turn our attention to the factors of the third kind (III):

$$b(z)=I+\frac{i}{z-x_0}\begin{pmatrix}-i\xi\eta^* & i\xi\xi^*\\ -i\eta\eta^* & i\eta\xi^*\end{pmatrix}=\begin{pmatrix}1+\dfrac{\xi\eta^*}{z-x_0} & \dfrac{-\xi\xi^*}{z-x_0}\\ \dfrac{\eta\eta^*}{z-x_0} & 1-\dfrac{\eta\xi^*}{z-x_0}\end{pmatrix},$$

where

$$\Im x_0=0,\quad \xi\eta^*=\eta\xi^* \text{ and } \xi\xi^*+\eta\eta^*>0. \tag{2.5}$$

For the linear fractional transformation (2.4) we have in this instance

$$\Phi(z)=\frac{(z-x_0)i+\xi(\eta^*i-\xi^*)}{(z-x_0)+\eta(\eta^*i-\xi^*)}.$$

By (2.3) we obtain as above $\eta^* i - \xi^* \neq 0$ and hence that

$$\lim_{z \to x_0} \Phi(z) = \frac{\xi}{\eta}.$$

Therefore the function $\Phi(z)$ has a pole at the point x_0 if and only if $\eta = 0$. Thus the only factor of the third kind which is not a de Branges matrix takes the form

$$b_0(z) = I + \frac{ip}{z - x_0}\epsilon_0, \text{ where } \epsilon_0 = \begin{pmatrix} 0 & i \\ 0 & 0 \end{pmatrix} \text{ and } p > 0. \tag{2.6}$$

Lemma 2. *A matrix valued function $\mathcal{B}(z)$ which is meromorphic in the complex plane and J inner is not a de Branges matrix if and only if it possesses a left divisor $b_0(z)$ of the form* (2.6).

Proof. Let $\mathcal{B}(z) = b_0(z)\mathcal{A}(z)$, where $\mathcal{A}(z)$ is a matrix valued function which is meromorphic and J inner. Then

$$\Phi(z, \mathcal{B}) = \Phi(z, \mathcal{A}) + \frac{p}{x_0 - z}$$

and according to the remark (1.16) the function $\Phi(z, \mathcal{B})$ must have a pole at the point x_0 and so $\mathcal{B}(z)$ is not a de Branges matrix.

Conversely, if $\mathcal{B}(z)$ is not a de Branges matrix, then it necessarily has a pole at a real point x_0. By Potapov's Splitting-Off Theorem, there exists a uniquely determined elementary factor $b(z)$ of the third kind (III) such that $\mathcal{B}(z) = b(z)\mathcal{B}_1(z)$, where $\mathcal{B}_1(z)$ is a matrix valued function which is meromorphic and J inner. Therefore

$$\Phi(z, \mathcal{B}) = \frac{\left(1 + \frac{\xi\eta^*}{z - x_0}\right)\Phi(z, \mathcal{B}_1) - \frac{\xi\xi^*}{z - x_0}}{\frac{\eta\eta^*}{z - x_0}\Phi(z, \mathcal{B}_1) + \left(1 - \frac{\eta\xi^*}{z - x_0}\right)}.$$

It is actually not hard to see that for $\eta \neq 0$ the function $\Phi(z, \mathcal{B})$ is analytic at x_0 regardless of whether the Nevanlinna function $\Phi(z, \mathcal{B})$ is analytic at x_0 or has a pole. Thus, if $\Phi(z, \mathcal{B})$ is not analytic, then $\eta = 0$, i.e., $b(z)$ is of the form (2.6), which completes the proof. □

Corollary. *If $\mathcal{U}$ is a constant J inner matrix, then the meromorphic and J inner matrix valued functions $\mathcal{B}(z)$ and $\mathcal{B}(z)\mathcal{U}$ are de Branges matrices simultaneously.*

We now turn to the following statement:

Theorem 4. *Let $\mathcal{B}(z)$ be a de Branges matrix of the form* (2.1). *Then the entries $\alpha(z)$ and $\beta(z)$ are linearly dependent if and only if $\mathcal{B}(z)$ takes the form*

$$\mathcal{B}(z) = \begin{pmatrix} 1 & az \\ 0 & 1 \end{pmatrix}\mathcal{U}, \tag{2.7}$$

where $a \geq 0$ and $\mathcal{U}$ is a constant J unitary matrix.

Proof. For matrices of the form (2.7), the entries $\alpha(z)$ and $\beta(z)$ are obviously linearly dependent. Conversely, let $u\alpha(z) + v\beta(z) = 0$ for some complex numbers u and v. For appropriately chosen complex numbers x and y, the matrix

$$\mathcal{U}_1 = \begin{pmatrix} v & x \\ u & y \end{pmatrix}$$

is J unitary. Then, by the above corollary,

$$\mathcal{B}_1(z) = \mathcal{B}(z)\mathcal{U}_1 = \begin{pmatrix} \delta_1(z) & \gamma_1(z) \\ \beta_1(z) & \alpha_1(z) \end{pmatrix} = \begin{pmatrix} v\delta(z) + u\gamma(z) & x\delta(z) + y\gamma(z) \\ 0 & x\beta(z) + y\alpha(z) \end{pmatrix}$$

is a de Branges matrix. From (2.2) we infer that $\alpha_1(z)\delta_1^*(z) = 1$ and hence that α_1 and δ_1 are constants. Finally, put

$$\mathcal{B}_2(z) = \mathcal{B}_1(z)\mathcal{U}_2 = \begin{pmatrix} \delta_1(z) & \gamma_1(z) \\ \beta_1(z) & \alpha_1(z) \end{pmatrix} \begin{pmatrix} \alpha_1^* & 0 \\ 0 & \delta_1^* \end{pmatrix} = \begin{pmatrix} 1 & \gamma_2(z) \\ 0 & 1 \end{pmatrix},$$

where $\mathcal{U}_2$ is a J unitary matrix. By the corollary, $\mathcal{B}_2(z)$ is still a de Branges matrix. Therefore $\gamma_2(z)$ is a real Nevanlinna function (i.e., real valued at every real point of analyticity of γ_2) and analytic on the whole real line by (2.4). Consequently, $\gamma_2(z) = az + b$ with $a \geq 0$ and $b = b^*$, which yields (2.7). □

Throughout the rest of the paper we shall assume that the entries $\alpha(z)$ and $\beta(z)$ are linearly independent (in particular none of them vanishes identically).

The J innerness of the matrix $\mathcal{B}(z)$ in (2.1) enables one to deduce information about the structure of the functions $\delta(z)$, $\gamma(z)$, $\beta(z)$ and $\alpha(z)$. From (2.2)–(2.3) it follows that the functions $\alpha(z)\beta^{-1}(z)$, $\delta(z)\beta^{-1}(z)$ and $\gamma(z)\alpha^{-1}(z)$ are real Nevanlinna functions:

$$\frac{\alpha(z)\beta^*(z) - \beta(z)\alpha^*(z)}{i} = |\beta(z)|^2 \frac{\alpha(z)\beta^{-1}(z) - \beta^{-1*}(z)\alpha^*(z)}{i} > 0$$

for $\Im z > 0$ and is equal to zero for $\Im z = 0$. The same argument is applicable to the other two functions.

We might equally well have considered the matrix valued function

$$\bar{\mathcal{B}}(z) = \mathcal{B}^*(\bar{z}) = J\mathcal{B}^{-1}(z)J = \begin{pmatrix} \bar{\delta}(z) & \bar{\beta}(z) \\ \bar{\gamma}(z) & \bar{\alpha}(z) \end{pmatrix}, \tag{2.8}$$

which is J contractive in $\mathbf{C}_+$ and J unitary on the real line and made similar conclusions about its entries. The formula

$$\begin{pmatrix} \bar{\alpha}(z) & -\bar{\gamma}(z) \\ -\bar{\beta}(z) & \bar{\delta}(z) \end{pmatrix} \begin{pmatrix} \delta(z) & \gamma(z) \\ \beta(z) & \alpha(z) \end{pmatrix} = \begin{pmatrix} 1 & 0 \\ 0 & 1 \end{pmatrix}, \tag{2.9}$$

which is a straightforward consequence of (2.8), seems worth mentioning.

Theorem 5 (Real Representation of de Branges Matrices). *Every de Branges matrix admits a representation of the form*

$$\mathcal{B}(z) = \frac{1}{S(z)} \begin{pmatrix} D(z) & C(z) \\ B(z) & A(z) \end{pmatrix}, \tag{2.10}$$

where $D(z)$, $C(z)$, $B(z)$, and $A(z)$ are real entire functions and $S(z)$ is an entire function. The function $E(z) = A(z) + iB(z)$ has no zeros in $\mathbf{C}_+$ and satisfies the inequality

$$|E(\bar{z})| < |E(z)|, \qquad \Im z > 0. \tag{2.11}$$

Proof. For a complex number ζ and a function $f(z)$ which is meromorphic in the complex plane, let $n(\zeta, f)$ stand for the order of the pole of $f(z)$ at the point ζ, so that $n(\zeta, f) = 0, 1, \ldots$. Let

$$Z = \{z_k | \min \{n(z_k, \delta),\ n(z_k, \gamma),\ n(z_k, \beta),\ n(z_k, \alpha)\} > 0\}$$

and let

$$N(z_k) = \max \{n(z_k, \delta),\ n(z_k, \gamma),\ n(z_k, \beta),\ n(z_k, \alpha)\}\ .$$

If $S_1(z)$ is the Weierstrass product with the zeros z_k in Z of order $N(z_k)$, then

$$\mathcal{B}_1(z) = S_1(z)\mathcal{B}(z) = \begin{pmatrix} D_1(z) & C_1(z) \\ B_1(z) & A_1(z) \end{pmatrix}$$

is an entire matrix valued function and $\mathcal{B}_1(z_k) \neq 0$. Let $\Im\lambda \neq 0$ and $B_1(\lambda) = 0$. Since $D_1(z){B_1}^{-1}(z)$, $A_1(z)B_1^{-1}(z)$ and $C_1(z)A_1^{-1}(z)$ are real Nevanlinna functions, we conclude consecutively that

$$D_1(z) = 0, \quad A_1(z) = 0 \text{ and } C_1(z) = 0,$$

i.e., $\mathcal{B}_1(\lambda) = 0$. The latter equality means that $\lambda \notin Z$, so that $\mathcal{B}(\lambda) = 0$, which contradicts its J property. Hence $B_1(\lambda) \neq 0$ and $A_1(\lambda) \neq 0$ for nonreal λ.

Next, the function

$$E_1(z) = A_1(z) + iB_1(z) = \left(A_1(z){B_1}^{-1}(z) + i\right) B_1(z) \neq 0, \quad \Im z > 0\ ,$$

since $A_1(z){B_1}^{-1}(z)$ is a Nevanlinna function. Setting $r(z) = \bar{B}_1(z)$, consider the entire functions

$$\begin{aligned} B(z) &= r(z)B_1(z), \\ D(z) &= r(z)D_1(z) = B(z)\frac{\delta(z)}{\beta(z)}, \\ A(z) &= r(z)A_1(z) = B(z)\frac{\alpha(z)}{\beta(z)}, \\ C(z) &= r(z)C_1(z) = B(z)\frac{\gamma(z)}{\beta(z)}. \end{aligned}$$

The function $B(z)$ is obviously a real entire function. Since the functions $\alpha(z)\beta^{-1}(z)$, $\delta(z)\beta^{-1}(z)$ and $\gamma(z)\beta^{-1}(z)$ are real valued on the real line, we see that $D(z)$, $A(z)$ and $C(z)$ are real entire functions. Finally,

$$E(z) = A(z) + iB(z) = r(z)E_1(z) \neq 0, \qquad \Im z > 0. \tag{2.12}$$

We thus obtain the representation (2.10) with $S(z) = r(z)S_1(z)$.

To prove the inequality (2.11), note that

$$\left|\frac{\bar{E}(z)}{E(z)}\right| = \left|\frac{A(z) - iB(z)}{A(z) + iB(z)}\right| = \left|\frac{\alpha(z) - i\beta(z)}{\alpha(z) + i\beta(z)}\right|$$

and

$$\left|\frac{\bar{E}(z)}{E(z)}\right|^2 - 1 = 2\left\{\frac{\beta(z)\alpha^*(z) - \alpha(z)\beta^*(z)}{i|\alpha(z) + i\beta(z)|^2}\right\} \leq 0, \qquad \Im z > 0.$$

If the equality $|\bar{E}(z)/E(z)| = 1$ holds for some point $z_0 \in \mathbf{C}_+$, then, by Schwarz's lemma, $\bar{E}(z)/E(z) = e^{-2i\theta}$ i.e.,

$$A(z)\sin\theta - B(z)\cos\theta = 0 .$$

Thus the functions $A(z)$ and $B(z)$ (as well as $\alpha(z)$ and $\beta(z)$) are linearly dependent, which contradicts the aforementioned convention. Thus,

$$\left|\frac{\bar{E}(z)}{E(z)}\right| < 1, \qquad \Im z > 0 ,$$

which is equivalent to (2.11) in view of the relation $E(z) \neq 0$, $\Im z > 0$. The proof is complete. □

Remark. The representation of de Branges matrices given in Theorem 5 is called the *real representation*. It is not uniquely determined, as, starting from any such representation and multiplying each function in (2.10) by the same real entire function with real zeros we come to another one.

The theorem on real representation enables one to evaluate the measure $d\mu$ in the integral representation (1.13) for the Nevanlinna function $\Phi(z, \mathcal{B})$. Applying (1.12) and putting $\omega = i$ we obtain

$$|v(z)|^2 \frac{\Phi(z) - \Phi^*(z)}{i} = [-i, 1]\{\mathcal{B}^*(z)J\mathcal{B}(z) - J\}\begin{bmatrix} i \\ 1 \end{bmatrix} + 2, \tag{2.13}$$

where

$$v(z) = \alpha(z) + i\beta(z) = \frac{E(z)}{S(z)},$$

so that

$$\Im\Phi(z) \begin{cases} \geq |S(z)/E(z)|^2, & \text{if } \Im z > 0, \\ \leq |S(z)/E(z)|^2, & \text{if } \Im z < 0, \\ = |S(z)/E(z)|^2, & \text{if } \Im z = 0 \end{cases} \tag{2.14}$$

at every point of analyticity of $\mathcal{B}(z)$.

Much the same sort of argument may be applied to the matrix valued function $\bar{\mathcal{B}}(z)$ in (2.8). Using (2.9) it is not hard to check that

$$\frac{\bar{\delta}(z)i+\bar{\gamma}(z)}{\bar{\beta}(z)i+\bar{\alpha}(z)}=\frac{\delta(z)i+\gamma(z)}{\beta(z)i+\alpha(z)}=\Phi(z), \qquad \bar{\beta}(z)i+\bar{\alpha}(z)=\frac{E(z)}{\bar{S}(z)},$$

$$\frac{\Phi(z)-\bar{\Phi}(z)}{2i}=\frac{S(z)\bar{S}(z)}{E(z)\bar{E}(z)} \text{ and } \frac{\Phi(z)+\bar{\Phi}(z)}{2}=\Phi(z)-i\frac{S(z)\bar{S}(z)}{E(z)\bar{E}(z)}. \tag{2.15}$$

The matrix valued function

$$\tilde{\mathcal{B}}(z)=\begin{pmatrix} \bar{\delta}(z) & \bar{\gamma}(z) \\ \bar{\beta}(z) & \bar{\alpha}(z) \end{pmatrix}$$

is readily seen to be J expansive in $\mathbf{C}_+$. Therefore the system

$$\Im\Phi(z)\begin{cases} \geq \left|\bar{S}(z)/E(z)\right|^2, & \text{if } \Im z>0, \\ \leq \left|\bar{S}(z)/E(z)\right|^2, & \text{if } \Im z<0, \\ = \left|\bar{S}(z)/E(z)\right|^2, & \text{if } \Im z=0\,, \end{cases} \tag{2.14$'$}$$

which is analogous to (2.14), is also valid.

We are now in a position to obtain the representation (1.13). The condition (2.4) guarantees the continuity of the function $\Phi(z)$ up to the real line, so that (cf. (1.15))

$$d\mu(t)=\left|\frac{S(t)}{E(t)}\right|^2 dt$$

and hence, in the upper half plane,

$$\Phi(z)=\frac{D(z)i+C(z)}{B(z)i+A(z)}=pz+q+\frac{1}{\pi}\int_{-\infty}^{\infty}\left\{\frac{1}{t-z}-\frac{t}{1+t^2}\right\}\left|\frac{S(t)}{E(t)}\right|^2 dt. \tag{2.16}$$

The first equality in (2.15) makes it possible to find a representation for the function $\Phi(z)$ in the lower half plane:

$$\Phi(z)=pz+q+\frac{1}{\pi}\int_{-\infty}^{\infty}\left\{\frac{1}{t-z}-\frac{t}{1+t^2}\right\}\left|\frac{S(t)}{E(t)}\right|^2 dt+2i\frac{S(z)\bar{S}(z)}{E(z)\bar{E}(z)}. \tag{2.17}$$

We now recall the main definitions and basic facts concerning functions of bounded type which are analytic in $\mathbf{C}_+$. An excellent source of information on the subject is Chapter 1 of de Branges' monograph [1].

By Nevanlinna's factorization theorem (cf. [1, Chapter 1, Theorem 9]) every such function $N(z)$, which does not have the origin as a limit point of zeros, may be represented as follows:

$$N(z)=B_0(z)\exp(-ihz)\exp G(z), \tag{2.18}$$

where $B_0(z)$ is a Blaschke product, h is a real number, and $G(z)$ is a function which is analytic in the upper half plane such that

$$\Re G(x+iy) = \frac{y}{\pi}\int_{-\infty}^{\infty} \frac{d\mu(t)}{(t-x)^2+y^2}$$

for some real valued function $\mu(x)$ such that

$$\int_{-\infty}^{\infty} \frac{|d\mu(t)|}{1+t^2} < \infty.$$

If $\mu(t)$ is normalized by the conditions

$$\mu(0) = 0 \text{ and } \mu(t) = \frac{\mu(t+0)+\mu(t-0)}{2},$$

then the Stieltjes inversion formula

$$\mu(b)-\mu(a) = \lim_{y\to 0+0}\int_a^b \Re G(x+iy)\,dx$$

holds. The number h, which is called the *mean type*, is closely related to the concept of exponential type in the theory of entire functions. It can be evaluated by means of the formula

$$h = \limsup_{y\to+\infty} \frac{\log|N(iy)|}{y}. \tag{2.19}$$

A function $F(z)$ of bounded type which is analytic in $\mathbf{C}_+$ is called an *outer* function if $F(z) \neq 0$ for $\Im z > 0$ and

$$\log|F(i)| = \frac{1}{\pi}\int_{-\infty}^{\infty} \frac{\log|F(t)|}{1+t^2}\,dt.$$

Recall that the de Branges matrices $\mathcal{B}(z)$ and $\bar{\mathcal{B}}(z)$ are matrix valued functions of bounded type in $\mathbf{C}_+$. Hence the functions

$$\frac{S(z)}{E(z)} = \frac{1}{\beta(z)i+\alpha(z)} \quad\text{and}\quad \frac{\bar{S}(z)}{E(z)} = \frac{1}{\bar{\beta}(z)i+\bar{\alpha}(z)}$$

are analytic in $\mathbf{C}_+$ and of bounded type. The top inequalities in (2.14)–(2.14′) show that these functions (as well as $\Phi(z)$) are analytic on the real line and grow slowly along the positive imaginary axis. Since for each Nevanlinna function $w(z)$ we have $w(iy) = O(y)$ as $y \to +\infty$, it follows that

$$\left|\frac{S(iy)}{E(iy)}\right| + \left|\frac{\bar{S}(iy)}{E(iy)}\right| = O(y^{1/2}), \quad y\to+\infty\,,$$

so that these functions have a nonpositive mean type. Therefore the following statement is verified.

Theorem 6. *The functions $S(z)$ and $E(z) = A(z)+iB(z)$ in the real representation (2.10) of the de Branges matrix $\mathcal{B}(z)$ are subject to the following conditions:*

(1) *The functions $S(z)/E(z)$ and $\bar{S}(z)/E(z)$ are analytic in the closed upper half plane and of nonpositive mean type.*

(2) *The relation*

$$\int_{-\infty}^{\infty} \left|\frac{S(t)}{E(t)}\right|^2 \frac{dt}{1+t^2} < \infty \tag{2.20}$$

holds.

3 Parametrization of de Branges matrices

One of the main objectives of the present paper is to prove the converse statement: given entire functions $S(z)$ and $E(z)$ which obey (2.11) and the assumptions of Theorem 6, there exists a de Branges matrix $\mathcal{B}(z)$ of the form (2.10). We also describe the set of all such matrices. To this end, note that we can determine the functions $D(z)$ and $C(z)$ via $S(z)$, $E(z)$, $\Phi(z)$ and the parameters $(p, q) : p \geq 0$ and $q = q^*$ by means of the formulas (2.16)–(2.17). Indeed, putting

$$A(z) = \frac{E(z) + \bar{E}(z)}{2} \text{ and } B(z) = \frac{E(z) - \bar{E}(z)}{2i},$$

we have

$$\begin{aligned} D(z)i + C(z) &= \Phi(z)\,(B(z)i + A(z))\,, \\ -D(z)i + C(z) &= \bar{\Phi}(z)\,(-B(z)i + A(z))\,, \end{aligned}$$

and hence

$$\begin{aligned} C(z) &= \frac{\Phi(z) + \bar{\Phi}(z)}{2} A(z) - \frac{\Phi(z) - \bar{\Phi}(z)}{2i} B(z), \\ D(z) &= \frac{\Phi(z) - \bar{\Phi}(z)}{2i} A(z) + \frac{\Phi(z) + \bar{\Phi}(z)}{2} B(z). \end{aligned}$$

With the help of (2.15) we have finally

$$C(z) = A(z)\Phi(z) - i\frac{S(z)\bar{S}(z)}{E(z)} \text{ and } D(z) = B(z)\Phi(z) + \frac{S(z)\bar{S}(z)}{E(z)}. \tag{3.1}$$

The remaining steps are now quite clear: given $S(z)$ and $E(z)$ and arbitrary $p \geq 0$ and $q = q^*$, we define $\Phi(z)$ by (2.16)–(2.17), $C(z)$ and $D(z)$ by (3.1), and then show that the corresponding matrix valued function $\mathcal{B}(z)$ is a de Branges matrix.

The first step is plain. Due to (2.20), the function $\Phi(z)$ which is defined by (2.16)–(2.17) is meromorphic in the complex plane and has a positive imaginary part in $\mathbf{C}_+$. The set of its poles is contained in the set of zeros of the function

$E(z)$. The function $\Phi(z)$ obviously satisfies (2.15). Next we define the functions $C(z)$ and $D(z)$ by (3.1). They are both clearly analytic in the closed upper half plane and satisfy the equalities

$$C(z) - \bar{C}(z) = D(z) - \bar{D}(z) = 0 .$$

Therefore, $C(z)$ and $D(z)$ are real entire functions and the matrix valued function

$$\mathcal{B}(z) = \frac{1}{S(z)} \begin{pmatrix} D(z) & C(z) \\ B(z) & A(z) \end{pmatrix} \tag{3.2}$$

is meromorphic in the complex plane. The condition (2.4) is fulfilled automatically because of the identity

$$\Phi(z) = \frac{D(z)i + C(z)}{B(z)i + A(z)}.$$

It is a much more complicated problem to show that $\mathcal{B}(z)$ is J inner. Our reasoning relies on the method suggested by D.Z.Arov in his study of Darlington realization for matrix valued functions. Invoking the linear fractional transformation

$$\mathcal{S}(z) = (P - Q\mathcal{B}(z))\,(Q - P\mathcal{B}(z))^{-1}, \tag{3.3}$$

which we encountered earlier in Section 1, we will show that $\mathcal{S}(z)$ is an inner contractive matrix valued function. This is much easier to carry out because we can invoke the maximum modulus principle.

It is both convenient and instructive here to deal with another signature matrix:

$$j = \begin{pmatrix} -1 & 0 \\ 0 & 1 \end{pmatrix},$$

and the accompanying projectors

$$P_j = \frac{I+j}{2} = \begin{pmatrix} 0 & 0 \\ 0 & 1 \end{pmatrix} \text{ and } Q_j = \frac{I-j}{2} = \begin{pmatrix} 1 & 0 \\ 0 & 0 \end{pmatrix}.$$

The identity

$$J = VjV^*, \text{ based on the unitary matrix } V = \frac{1}{\sqrt{2}} \begin{pmatrix} 1 & i \\ i & 1 \end{pmatrix},$$

establishes a connection between the J properties of $\mathcal{B}(z)$ and the j properties of $\mathcal{A}(z) = V^*\mathcal{B}(z)V$. Therefore it suffices to show that $\mathcal{A}(z)$ is j inner or, in other words, that

$$\mathcal{S}(z) = (P_j - Q_j\mathcal{A}(z))\,(Q_j - P_j\mathcal{A}(z))^{-1} \tag{3.4}$$

is an inner contractive matrix valued function.

For the matrix valued function $\mathcal{B}(z)$ in (3.2) we have, upon setting $E_0(z) = D(z)i + C(z)$,

$$\mathcal{A}(z) = \begin{pmatrix} a_{11}(z) & a_{12}(z) \\ a_{21}(z) & a_{22}(z) \end{pmatrix} = \frac{1}{2\mathcal{S}(z)} \begin{pmatrix} \bar{E}(z) + i\bar{E}_0(z) & E_0(z) - iE(z) \\ \bar{E}_0(z) + i\bar{E}(z) & E(z) - iE_0(z) \end{pmatrix}$$

and

$$\mathcal{S}(z) = \begin{pmatrix} s_{11}(z) & s_{12}(z) \\ s_{21}(z) & s_{22}(z) \end{pmatrix} = \begin{pmatrix} a_{12}a_{22}^{-1}a_{21} - a_{11} & a_{12}a_{22}^{-1} \\ -a_{22}^{-1}a_{21} & -a_{22}^{-1} \end{pmatrix}. \tag{3.5}$$

Note that $2\mathcal{S}(z)a_{22}(z) = B(z)i + A(z) - i\,(D(z)i + C(z)) \neq 0$ in the closed upper half plane apart from (at most) an isolated set of real points.

Consider the j form of the matrix $\mathcal{A}(x)$ for $x \in \mathbf{R}$:

$$\mathcal{A}(x)j\mathcal{A}^*(x) - j = \begin{pmatrix} -|a_{11}|^2 + |a_{12}|^2 + 1 & -a_{11}a_{21}^* + a_{12}a_{22}^* \\ -a_{21}a_{11}^* + a_{22}a_{12}^* & -|a_{21}|^2 + |a_{22}|^2 - 1 \end{pmatrix}.$$

For real x such that $S(x) \neq 0$, the equalities

$$\begin{aligned} |a_{11}(x)|^2 &= \frac{|\bar{E}(x) + i\bar{E}_0(x)|^2}{4|S(x)|^2} = \frac{|E(x) - iE_0(x)|^2}{4|S(x)|^2} = |a_{22}(x)|^2, \\ |a_{12}(x)|^2 &= \frac{|E_0(x) - iE(x)|^2}{4|S(x)|^2} = \frac{|\bar{E}_0(x) + i\bar{E}(x)|^2}{4|S(x)|^2} = |a_{21}(x)|^2 \end{aligned}$$

and

$$a_{11}(x)a_{21}^*(x) = \frac{\left(\bar{E}(x) + i\bar{E}_0(x)\right)\left(E_0(x) - iE(x)\right)}{4|S(x)|^2} = a_{12}(x)a_{22}^*(x)$$

are checked by direct calculation. Upon taking into account the relations

$$E_0(z) = \Phi(z)E(z) \text{ and } \frac{\Phi(x) - \Phi^*(x)}{2i} = \left|\frac{S(x)}{E(x)}\right|^2, \tag{3.6}$$

we get

$$\begin{aligned} |a_{11}(x)|^2 - |a_{12}(x)|^2 &= |a_{22}(x)|^2 - |a_{21}(x)|^2 \\ &= \frac{|E(x)|^2}{4|S(x)|^2}\left(|\Phi(x) + i|^2 - |\Phi(x) - i|^2\right) = 1, \end{aligned}$$

so that $\mathcal{A}(x)j\mathcal{A}^*(x) - j = 0$ apart from an isolated set of real points. Therefore $\mathcal{S}(x)$ is unitary on the whole real line.

A function $H(z)$ which is analytic and of bounded type in $\mathbf{C}_+$ is said to belong to the (V.I.) Smirnov class $\mathbf{D}$ if

$$H(z) = \frac{H_1(z)}{H_2(z)}, \text{ for } \Im z > 0,$$

where $H_1(z)$ is a bounded analytic function and $H_2(z)$ is a bounded outer function. A matrix valued function $\mathcal{S}(z)$ belongs to $\mathbf{D}$ if all of its entries belong to $\mathbf{D}$.

The class $\mathbf{D}$ is known as the widest subclass of the class of bounded type functions wherein the maximum modulus principle is still valid: for $H \in \mathbf{D}$ the assumption $|H(x)| \leq 1$ a.e. on the real line implies that $|H(z)| \leq 1$ for $\Im z > 0$. The following elementary lemma provides an effective means of invoking this principle to matrix valued functions (cf. [2], Lemma 2.2).

Lemma 3. *If a matrix valued function $\mathcal{S}(z) \in \mathbf{D}$ and $I - \mathcal{S}^*(x)\mathcal{S}(x) \geq 0$ a.e. on the real line then $I - \mathcal{S}^*(z)\mathcal{S}(z) \geq 0$ in the upper half plane.*

Proof. Consider the bilinear form

$$(\mathcal{S}(z)\varphi, \psi) = \sum_{k,j=1}^{2} s_{kj}(z)\varphi_j\psi_k^*,$$

where φ and ψ are unit vectors in 2-dimensional unitary space. The function $(\mathcal{S}(z)\varphi, \psi)$ clearly lies in $\mathbf{D}$. By the assumptions of the lemma,

$$|(\mathcal{S}(z)\varphi, \psi)| \leq (\mathcal{S}(z)\varphi,\ \mathcal{S}(z)\varphi)\,(\psi,\ \psi) \leq 1.$$

The rest is immediate from the scalar maximum modulus principle and the arbitrariness of the vectors φ, ψ. □

The next step is to show that the matrix valued function $\mathcal{S}(z)$ of (3.4) belongs to $\mathbf{D}$. The following lemma furnishes a key tool.

Lemma 4. *A function $N(z)$ which is analytic in the closed upper half plane and is of nonpositive mean type belongs to $\mathbf{D}$.*

Proof. To begin with, let $N(z)$ be of the form $N(z) = \exp G(z)$, where

$$\Re G(x+iy) = \log|N(x+iy)| = \frac{y}{\pi}\int_{-\infty}^{\infty} \frac{d\mu(t)}{(t-x)^2+y^2}.$$

By the Stieltjes inversion formula,

$$\mu(b) - \mu(a) = \lim_{y\to 0+} \int_a^b \log|N(x+iy)|\,dx.$$

Then, since $N(z)$ is analytic on the real line,

$$\lim_{y\to 0+} \int_a^b \log|N(x+iy)|\,dx = \int_a^b \log|N(x)|\,dx$$

and hence

$$\log|N(x+iy)| = \frac{y}{\pi}\int_{-\infty}^{\infty} \frac{\log|N(t)|dt}{(t-x)^2+y^2},$$

i.e., $N(z)$ is an outer function.

In the general situation we have

$$N(z) = B_0(z)\exp(-ihz)\exp G(z),$$

where $B_0(z)$ is a Blaschke product over the zeros of $N(z)$ and $h \le 0$. Since $N(z)$ is analytic on the real line, infinity is the only permissible limit point of its zeros. Therefore $B_0(z)$ is analytic on the real line and so is the function $N_1(z) = \exp G(z)$. The above argument shows that

$$N(z) = B_0(z)\exp(-ihz)N_1(z)$$

with an outer factor $N_1(z)$. Therefore the desired conclusion is obtained. □

It is reasonable to expect that the entries $s_{kj}(z)$ of the matrix valued function $\mathcal{S}(z)$ in (3.5) are subject to the conditions of the previous lemma. Observe first that the function

$$s(z) = \frac{\Phi(z) - i}{\Phi(z) + i} = \frac{i\Phi(z) + 1}{i\Phi(z) - 1}$$

is a contractive function: $|s(z)| \le 1$, $\Im z > 0$. Bearing in mind the first equalities in (3.6) and (2.15) we have the following conclusions for the entries $s_{kj}(z)$ of (3.5):

$$s_{12}(z) = \frac{E_0(z) - iE(z)}{E(z) - iE_0(z)} = \frac{\Phi(z) - i}{1 - i\Phi(z)} = is(z),$$

so that $|s_{12}(z)| \le 1$ and a fortiori $s_{12}(z) \in \mathbf{D}$;

$$s_{22}(z) = \frac{2S}{iE_0(z) - E(z)} = \frac{S(z)}{E(z)}(s(z) - 1)$$

belongs to $\mathbf{D}$, since $S(z)/E(z)$ is in $\mathbf{D}$ by the lemma and the function $s(z) - 1$ is bounded;

$$\begin{aligned}
s_{11}(z) &= \frac{1}{2S(z)}\left\{\frac{(E_0(z) - iE(z))\left(\bar{E}_0(z) + i\bar{E}(z)\right)}{E(z) - iE_0(z)} - \left(\bar{E}(z) + i\bar{E}_0(z)\right)\right\} \\
&= \frac{i}{S(z)}\,\frac{E_0(z)\bar{E}(z) - E(z)\bar{E}_0(z)}{E(z) - iE_0(z)} = \frac{\bar{E}(z)}{S(z)}\,\frac{\Phi(z) - \bar{\Phi}(z)}{i}\,\frac{1}{i\Phi(z) - 1} \\
&= \frac{\bar{E}(z)}{S(z)}\,\frac{S(z)\bar{S}(z)}{E(z)\bar{E}(z)}\,\frac{2}{i\Phi(z) - 1} = \frac{\bar{S}(z)}{E(z)}(s(z) - 1)
\end{aligned}$$

belongs to $\mathbf{D}$ by the same argument;

$$s_{21}(z) = \frac{\bar{E}_0(z) + i\bar{E}(z)}{iE_0(z) - E(z)} = \frac{\bar{E}(z)}{E(z)}\,\frac{\bar{\Phi}(z) + i}{i\Phi(z) - 1}$$

$$= \frac{\bar{E}(z)}{E(z)}\left\{-i - \frac{\Phi(z)-\bar{\Phi}(z)}{2i}\,\frac{2i}{i\Phi(z)-1}\right\}$$

$$= -i\frac{\bar{E}(z)}{E(z)}\left\{1 + \frac{S(z)\bar{S}(z)}{E(z)\bar{E}(z)}\,(s(z)-1)\right\}$$

$$= -i\frac{\bar{E}(z)}{E(z)} - i\frac{S(z)}{E(z)}\,\frac{\bar{S}(z)}{E(z)}\,(s(z)-1)$$

belongs to $\mathbf{D}$, since $|\bar{E}(z)| < |E(z)|$ for $\Im z > 0$ and the class $\mathbf{D}$ is closed with respect to addition and multiplication.

Thus the matrix valued function $\mathcal{S}(z)$ is inner and hence $\mathcal{B}(z)$ is J inner, as needed.

We remark that given $S(z)$ and $E(z)$, the de Branges matrix $\mathcal{B}(z)$ is completely determined by the parameters $(p,\, q)$. Furthermore, if the de Branges matrices

$$\mathcal{B}_k(z) = \frac{1}{S(z)}\begin{pmatrix} D_k(z) & C_k(z) \\ B(z) & A(z) \end{pmatrix}, \quad k = 1, 2\ ,$$

correspond to the parameters $(p_k,\, q_k)$, then, upon taking into account the equalities (cf. (3.1))

$$C_2(z) - C_1(z) = (\Phi_2(z) - \Phi_1(z))\,A(z) = ((p_2 - p_1)z + (q_2 - q_1))\,A(z)$$

and

$$D_2(z) - D_1(z) = (\Phi_2(z) - \Phi_1(z))\,B(z) = ((p_2 - p_1)z + (q_2 - q_1))\,B(z)\ ,$$

we see that

$$\mathcal{B}_2(z) = \begin{pmatrix} 1 & (p_2 - p_1)z + (q_2 - q_1) \\ 0 & 1 \end{pmatrix}\mathcal{B}_1(z). \tag{3.7}$$

The results obtained in this section can be summed up in the following statement:

Theorem 7 (Parametrization of de Branges Matrices). *Let $\mathcal{B}(z)$ be a de Branges matrix of the form* (3.2). *Then the function $E(z) = A(z) + iB(z)$ meets the constraint* (2.11)*, the function $S(z)$ satisfies the conditions of Theorem 6 and the Nevanlinna function*

$$\Phi(z) = \frac{D(z)i + C(z)}{B(z)i + A(z)}$$

admits the representation (2.16) – (2.17) *with parameters $(p,\, q) : p \geq 0$ and $q = q^*$. Conversely, starting with an arbitrary such pair $(p,\, q)$ and entire functions $E(z)$ and $S(z)$ which satisfy* (2.11) *and the assumptions of Theorem 6, there exists a uniquely determined de Branges matrix $\mathcal{B}(z)$ of the form* (3.2) *such that* (2.16) *and* (2.17) *hold. For two different pairs $(p_k,\, q_k)$ the corresponding de Branges matrices are related by* (3.7).

Definition. A de Branges matrix $\mathcal{B}(z)$ is said to be *perfect* if $p = 0$ in formula (2.16).

The following statement is a direct consequence of the last assertion of Theorem 7.

Theorem 8. *Every de Branges matrix $\mathcal{B}(z)$ admits a representation of the form*

$$\mathcal{B}(z) = \exp(-izp\epsilon_0)\,\mathcal{B}_0(z), \text{ where } \epsilon_0 = \begin{pmatrix} 0 & i \\ 0 & 0 \end{pmatrix}, \tag{3.8}$$

$p \geq 0$ and $\mathcal{B}_0(z)$ is a perfect de Branges matrix.

The perfect de Branges matrices are exactly those which correspond to the parameters $(0,\ q)$ in the Parametrization Theorem. In order for a de Branges matrix to be perfect it is necessary and sufficient that

$$\lim_{y\to+\infty} \frac{1}{iy}\,\frac{D(iy)i + C(iy)}{B(iy)i + A(iy)} = \lim_{y\to+\infty} \frac{1}{iy}\,\frac{\delta(iy)i + \gamma(iy)}{\beta(iy)i + \alpha(iy)} = 0.$$

The following matrix serves as an example of a perfect de Branges matrix:

$$\mathcal{B}_E(z) = \frac{1}{E(z)} \begin{pmatrix} A(z) & -B(z) \\ B(z) & A(z) \end{pmatrix}.$$

Here $S(z) = E(z)$ and, since

$$\Phi_E(z) = i = \frac{1}{\pi}\int_{-\infty}^{\infty} \left\{\frac{1}{t-z} - \frac{t}{1+t^2}\right\} dt, \quad \Im z > 0, \tag{3.9}$$

we have $p = q = 0$. The matrix $\mathcal{B}_0(z)$ is called *special.*

4 de Branges spaces

Our main objective is to investigate the multiplicative structure of de Branges matrices. The key tool for this purpose is the theory of de Branges spaces.

Definition. Let $E(z) = A(z) + iB(z)$ be an entire function which satisfies the condition

$$|\bar{E}(z)| < |E(z)|, \qquad \Im z > 0, \tag{4.1}$$

and for which $A(z)$ and $B(z)$ are real entire functions. The set $H(E)$ of entire functions $F(z)$ such that

1) the functions $F(z)/E(z)$ and $\bar{F}(z)/E(z)$ are analytic in the upper half plane and of nonpositive mean type, and

2) the constraint $\displaystyle\int_{-\infty}^{\infty} \left|\frac{F(t)}{E(t)}\right|^2 \frac{dt}{1+t^2} < \infty$ holds

will be called a *de Branges space*.

For example, if $E(z) = \exp(-i\sigma z)$, $\sigma > 0$, then $H(E)$ is the well known Paley-Wiener space of entire functions of exponential type at most σ which are square integrable on the real line.

$H(E)$ is a Hilbert space with inner product

$$\langle F, G\rangle = \int_{-\infty}^{\infty} \frac{F(t)G^*(t)}{|E(t)|^2}\, dt$$

and involution $F(z) \to \bar{F}(z)$. In what follows a crucial role is played by the Phragmén-Lindelöf principle and Cauchy's formula for functions of bounded type.

Theorem 9 (cf. [1, Chapter 1, Theorem 1]). *Assume that $f(z)$ is analytic in the open upper half plane, that $|f(z)|$ has a continuous extension to the closed half plane, that*

$$\liminf_{r\to\infty} \frac{1}{r}\int_0^{\pi} \log^+ |f(re^{i\theta})| \sin\theta \, d\theta = 0,$$

where $\log^+ x = \max(0, \log x)$ for $x \geq 0$ and that $|f(t)| \leq 1$ for $t \in \mathbf{R}$. Then $|f(z)| \leq 1$ everywhere in the upper half plane.

Theorem 10 (cf. [1, Chapter 1, Theorem 12]). *Assume that $f(z)$ is analytic and of nonpositive mean type in the open upper half plane and has a continuous extension to the closed half plane. If*

$$\int_{-\infty}^{\infty} |f(t)|^2 \, dt < \infty ,$$

then Cauchy's formula holds:

$$\frac{1}{2\pi i}\int_{-\infty}^{\infty} \frac{f(t)\, dt}{t-z} = \begin{cases} f(z), & \Im z > 0, \\ 0, & \Im z < 0. \end{cases}$$

Observe that for $F(z) \in H(E)$, the function $F(z)/E(z)$ satisfies the hypotheses of the last theorem and so

$$\frac{1}{2\pi i}\int_{-\infty}^{\infty} \frac{F(t)\, dt}{E(t)(t-z)} = \begin{cases} F(z)/E(z), & \Im z > 0, \\ 0, & \Im z < 0. \end{cases} \tag{4.2}$$

Cauchy's formula guarantees the existence of a reproducing kernel in $H(E)$: the function

$$K(z,\omega) = \frac{\bar{E}(z)\bar{E}^*(\omega) - E(z)E^*(\omega)}{2\pi i(z-\bar{\omega})} = \frac{A(z)B^*(\omega) - B(z)A^*(\omega)}{\pi(z-\bar{\omega})} \tag{4.3}$$

belongs to $H(E)$ for every fixed ω and posseses the reproducing property

$$\langle F(\cdot),\ K(\cdot,\omega)\rangle = F(\omega) \tag{4.4}$$

for every function $F \in H(E)$ and every complex ω.

The equality (4.4) implies the completeness of the system $\{K(z,\omega_k)\}_k$ for every sequence of complex numbers $\{\omega_k\}$ having a finite limit point. It is clear that $K^*(z,\omega) = K(\omega,z)$ and (cf. (4.1))

$$K(\omega,\omega) = K(\bar{\omega},\bar{\omega}) > 0, \qquad \Im z \neq 0.$$

An important consequence of (4.4) is the inequality

$$|F(\omega)|^2 \leq \|F\|^2 \, K(\omega,\omega)$$

which means that point evaluation is a bounded linear functional.

Let $S(z)$ be an entire function. Given a complex number ω consider the linear operator

$$R_\omega \, F(z) = \frac{F(z)S(\omega) - F(\omega)S(z)}{z-\omega}.$$

The function $S(z)$ is said to be *associated* with $H(E)$ if the operator R_ω acts in $H(E)$ for every complex ω. It is known (cf. [1, Chapter 2, p. 68]) that R_ω is a bounded linear operator in $H(E)$ and that the identity

$$(z-\omega)R_z R_\omega = S(\omega)R_z - S(z)R_\omega,$$

which may be recognized as a generalized Hilbert identity for the resolvent, holds. Clearly, $R_z R_\omega = R_\omega R_z$.

The associated functions form a linear space which is invariant under involution. The class of functions which are associated with the space $H(E)$ can be identified with the help of the next theorem.

Theorem 11 (cf. [1, Chapter 2, Theorem 25]). *A function $S(z)$ is associated with the space $H(E)$ if and only if:*

1. *The functions $S(z)/E(z)$ and $\bar{S}(z)/E(z)$ are analytic in the closed upper half plane and of nonpositive mean type.*

2. *The relation*

$$\int_{-\infty}^{\infty} \left| \frac{S(t)}{E(t)} \right|^2 \frac{dt}{1+t^2} < \infty$$

holds.

The proof amounts to matching the conditions of the theorem with the definition of a de Branges space.

The functions $F \in H(E)$ are clearly associated with $H(E)$. The functions $S(z) = E(z)$ and $S(z) = zF(z)$ for $F \in H(E)$ serve as examples of associated functions which do not in general belong to $H(E)$.

Lemma 5. *If $F(z)$ and $G(z)$ are arbitrary functions from $H(E)$ and $S(z)$ is an associated function, then the inner product*

$$\langle R_\omega \, F(\cdot), \; G(\cdot)\rangle = \int_{-\infty}^{\infty} \frac{F(t)S(\omega) - F(\omega)S(t)}{(t-\omega)|E(t)|^2} G^*(t) \, dt$$

is an entire function.

Proof. Putting $a < b$ we have

$$\langle R_\omega F(\cdot),\ G(\cdot)\rangle = \left\{\int_{-\infty}^{a} + \int_{a}^{b} + \int_{b}^{\infty}\right\} \frac{F(t)S(\omega) - F(\omega)S(t)}{(t-\omega)|E(t)|^2} G^*(t)\,dt,$$

where the first integral on the right hand side is analytic in the halfplane $\Re\omega > a$, the last one is analytic in the halfplane $\Re\omega < b$ and the middle one is an entire function. The rest is plain. □

5 Structure of perfect de Branges matrices and basic formulas

We are going to clarify the close connection between de Branges matrices and de Branges spaces. Namely, given a de Branges matrix

$$\mathcal{B}(z) = \frac{1}{S(z)} \begin{pmatrix} D(z) & C(z) \\ B(z) & A(z) \end{pmatrix}, \tag{5.1}$$

consider the corresponding de Branges space $H(E)$ which is generated by the function $E(z) = A(z) + iB(z)$. In this instance $S(z)$ is associated with $H(E)$, as is readily seen by comparing Theorem 6 and Theorem 11 above. Conversely, given de Branges space $H(E)$ and an associated function $S(z)$, the de Branges matrix (5.1) arises naturally from the Parametrization Theorem. We shall focus here primarily on perfect de Branges matrices.

Let $\mathcal{B}(z)$ be a not necessarily perfect de Branges matrix of the form (5.1). Then its bilinear J form

$$\mathcal{H}(\xi,\eta) = \mathcal{B}(\xi)J\mathcal{B}^*(\bar\eta) - J = \frac{\xi-\eta}{i}\,\frac{\pi}{S(\xi)\bar S(\eta)} \begin{pmatrix} Q(\xi,\bar\eta) & -L^*(\bar\eta,\xi) \\ -L(\xi,\bar\eta) & K(\xi,\bar\eta) \end{pmatrix},$$

where $K(z,\omega)$ is the reproducing kernel (4.3) of the corresponding de Branges space $H(E)$,

$$L(z,\omega) = \frac{S(z)S^*(\omega) - A(z)D^*(\omega) + B(z)C^*(\omega)}{\pi(z-\bar\omega)}$$

and

$$Q(z,\omega) = \frac{C(z)D^*(\omega) - D(z)C^*(\omega)}{\pi(z-\bar\omega)}.$$

Putting

$$T(\xi,\eta) = I + \mathcal{H}(\xi,\eta)J\ , \tag{5.2}$$

we have

$$T(\xi,\eta) = \mathcal{B}(\xi)J\bar{\mathcal{B}}(\eta)J = \mathcal{B}(\xi)\mathcal{B}^{-1}(\eta)$$

by the symmetry principle, from whence it follows that

$$T(\xi,\eta)T(\eta,\zeta)=T(\xi,\zeta), \tag{5.3}$$

or

$$\mathcal{H}(\xi,\eta)J\mathcal{H}(\eta,\zeta)=\mathcal{H}(\xi,\zeta)-\mathcal{H}(\xi,\eta)-\mathcal{H}(\eta,\zeta). \tag{5.4}$$

The identities (5.3) and (5.4) are called *chain identities*.

The aim of this section is to express the matrix $\mathcal{H}(\xi,\eta)$ in terms of the inner product in the related space $H(E)$. The desired representation for $K(\xi,\bar\eta)$ is obvious:

$$K(\xi,\bar\eta)=\langle K(\cdot,\bar\eta),\, K(\cdot,\xi)\rangle\,.$$

Turning next to the function $L(z,\omega)$, we show first that it belongs to $H(E)$ for every fixed complex number ω. By (3.1),

$$A(z)D(z)-B(z)C(z)=S(z)\bar S(z)$$

and hence $L(\cdot,\omega)$ is an entire function.

Moreover, both of the functions $L(z,\omega)/E(z)$ and $\bar L(z,\omega)/E(z)$ are clearly of nonpositive mean type. Next, for nonreal ω,

$$\int_{-\infty}^{\infty}\left|\frac{L(t,\omega)}{E(t)}\right|^2 dt=\int_{-\infty}^{\infty}\left|\frac{S(t)S^*(\omega)-A(t)D^*(\omega)+B(t)C^*(\omega)}{\pi(t-\bar\omega)E(t)}\right|^2 dt<\infty.$$

The same is true for real ω as well because of the continuity at the point ω and the appropriate decrease at infinity of the expression under the integral sign. Thus, $L(\cdot,\omega)\in H(E)$, as claimed, and

$$L(\xi,\bar\eta)=\langle L(\cdot,\bar\eta),\, K(\cdot,\xi)\rangle\,. \tag{5.5}$$

It is a much more delicate problem to cope with the function $Q(z,\omega)$ which is not in general an element of $H(E)$. To obtain a representation similar to (5.5) for $Q(z,\omega)$ we proceed in several steps. Now the presumed 'perfectness' of the de Branges matrix will play an essential role.

Step 1. Let us first reexpress $L(z,\omega)$ with the help of (3.1) and (4.3):

$$L(z,\omega)=\frac{S(z)S^*(\omega)}{\pi(z-\bar\omega)}-K(z,\omega)\Phi^*(\omega)-\frac{\bar E(z)}{E^*(\omega)}\,\frac{S^*(\omega)S(\bar\omega)}{\pi(z-\bar\omega)}.$$

Hence for nonreal ω,

$$\begin{aligned}\langle F(\cdot),\, L(\cdot,\omega)\rangle &= \int_{-\infty}^{\infty}\frac{F(t)L^*(t,\omega)dt}{|E(t)|^2}\\ &= \frac{S(\omega)}{\pi}\int_{-\infty}^{\infty}\frac{F(t)\bar S(t)dt}{(t-\omega)|E(t)|^2}-\Phi(\omega)F(\omega)-\frac{S(\omega)\bar S(\omega)}{\pi E(\omega)}\int_{-\infty}^{\infty}\frac{F(t)dt}{(t-\omega)\bar E(t)}.\end{aligned}$$

Cauchy's formula (4.2) gives

$$\frac{1}{2\pi i}\int_{-\infty}^{\infty}\frac{F(t)dt}{(t-\omega)\bar{E}(t)}=\begin{cases}0, & \Im\omega>0,\\ -F(\omega)/\bar{E}(\omega), & \Im\omega<0,\end{cases}$$

so that

$$\langle F(\cdot),\, L(\cdot,\omega)\rangle=\frac{S(\omega)}{\pi}\int_{-\infty}^{\infty}\frac{F(t)\bar{S}(t)dt}{(t-\omega)|E(t)|^2}-F(\omega)\Phi_0(\omega),$$

where

$$\Phi_0(\omega)=\begin{cases}\Phi(\omega), & \Im\omega>0,\\ \Phi(\omega)-2i\;S(\omega)\bar{S}(\omega)/E(\omega)\bar{E}(\omega), & \Im\omega<0.\end{cases}$$

Taking into account formulas (2.16) and (2.17) for the function $\Phi(\omega)$ with $p=0$, we obtain

$$\langle F(\cdot),\, L(\cdot,\omega)\rangle=\frac{S(\omega)}{\pi}\int_{-\infty}^{\infty}\frac{F(t)\bar{S}(t)dt}{(t-\omega)|E(t)|^2}$$
$$-F(\omega)\left[q+\frac{1}{\pi}\int_{-\infty}^{\infty}\left\{\frac{1}{t-\omega}-\frac{t}{1+t^2}\right\}\left|\frac{S(t)}{E(t)}\right|^2 dt\right]. \tag{5.6}$$

Step 2. For an entire function $F(z)$ such that

$$R_z\,F(t) \text{ and } R_\omega\,F(t)\in H(E)$$

the important permutability relation

$$\langle R_zF(\cdot),\, L(\cdot,\omega)\rangle=\langle R_\omega F(\cdot),\, L(\cdot,z)\rangle \tag{5.7}$$

can be derived from (5.6). Consider first the case of nonreal z, ω and replace $F(t)$ by $R_z\,F(t)$ in (5.6):

$$\langle R_zF(\cdot),\, L(\cdot,\omega)\rangle=\frac{S(\omega)}{\pi}\int_{-\infty}^{\infty}\frac{R_zF(t)\bar{S}(t)dt}{(t-\omega)|E(t)|^2}$$
$$-\,R_zF(\omega)\left[q+\frac{1}{\pi}\int_{-\infty}^{\infty}\left\{\frac{1}{t-\omega}-\frac{t}{1+t^2}\right\}\left|\frac{S(t)}{E(t)}\right|^2 dt\right].$$

Interchanging z and ω we get

$$\langle R_\omega F(\cdot),\, L(\cdot,z)\rangle=\frac{S(z)}{\pi}\int_{-\infty}^{\infty}\frac{R_\omega F(t)\bar{S}(t)dt}{(t-z)|E(t)|^2}$$
$$-R_\omega F(z)\left[q+\frac{1}{\pi}\int_{-\infty}^{\infty}\left\{\frac{1}{t-z}-\frac{t}{1+t^2}\right\}\left|\frac{S(t)}{E(t)}\right|^2 dt\right].$$

The relation (5.7) now drops out upon subtracting the latter equality from the former since $R_z\,F(\omega)=R_\omega\,F(z)$. Much the same sort of argument which was used to prove Lemma 5 shows that for $G \in H(E)$ the functions $\langle G(\cdot),\, L(\cdot,\omega)\rangle$ and $\langle R_\omega F(\cdot),\, G(\cdot)\rangle$ are both entire functions. Therefore the relation (5.7) is true for every complex z and ω.

A modified permutability relation:

$$\langle R_z F(\cdot),\, L(\cdot,\omega)\rangle = R_z \,\langle F(\cdot),\, L(\cdot,\omega)\rangle\,, \quad F \in H(E)\,, \tag{5.8}$$

will prove useful in the sequel. From the equality

$$R_z\,(tF(t)) = zR_z\,F(t) + S(z)F(t) \tag{5.9}$$

it follows that $R_z\,(tF(t)) \in H(E)$ for every fixed z. The relation (5.8) is now immediate from (5.7) with $F(t)$ replaced by $tF(t)$, and (5.9).

Step 3. The fundamental identity which we establish here is characteristic of V.P.Potapov's approach to interpolation. It is usually called the *principal identity.*

Lemma 6. *For every de Branges matrix (perfect or not) the identity*

$$\langle R_\omega F(\cdot),\, S(\bar\omega)G(\cdot)\rangle - \langle S(\omega)F(\cdot),\, R_{\bar\omega}G(\cdot)\rangle$$
$$= \pi\,\langle F(\cdot),\, L(\cdot,\omega)\rangle\,\langle K(\cdot,\bar\omega),\, G(\cdot)\rangle - \pi\,\langle F(\cdot),\, K(\cdot,\omega)\rangle\,\langle L(\cdot,\bar\omega),\, G(\cdot)\rangle \tag{5.10}$$

holds for every choice of F and G in $H(E)$.

Proof. The identity (5.10) can be extracted from the chain identity (5.4). For the (2,2) entries we have

$$\frac{\xi-\eta}{i}\,\frac{\pi}{S(\xi)\bar S(\eta)}[-L(\xi,\bar\eta),\, K(\xi,\bar\eta)]J\begin{bmatrix} -L^*(\bar\zeta,\eta) \\ K(\eta,\bar\zeta)\end{bmatrix}\frac{\pi}{S(\eta)\bar S(\zeta)}\,\frac{\eta-\zeta}{i}$$
$$= \frac{\xi-\zeta}{i}\,\frac{\pi K(\xi,\bar\zeta)}{S(\xi)\bar S(\zeta)} - \frac{\xi-\eta}{i}\,\frac{\pi K(\xi,\bar\eta)}{S(\xi)\bar S(\eta)} - \frac{\eta-\zeta}{i}\,\frac{\pi K(\eta,\bar\zeta)}{S(\eta)\bar S(\zeta)}\,, \tag{5.11}$$

or

$$\pi\left[K(\xi,\bar\eta)L^*(\bar\zeta,\eta) - L(\xi,\bar\eta)K(\eta,\bar\zeta)\right]$$
$$= S(\eta)\bar S(\eta)\frac{(\xi-\zeta)K(\xi,\bar\zeta)}{(\xi-\eta)(\eta-\zeta)} - S(\eta)\bar S(\zeta)\frac{K(\xi,\bar\eta)}{\eta-\zeta} - S(\xi)\bar S(\eta)\frac{K(\eta,\bar\zeta)}{\xi-\eta}.$$

On the other hand, a straightforward calculation leads to the equality

$$\left\langle R_\eta K(\cdot,\bar\zeta),\, S(\bar\eta)K(\cdot,\xi)\right\rangle - \left\langle S(\eta)K(\cdot,\bar\zeta),\, R_{\bar\eta}K(\cdot,\xi)\right\rangle$$
$$= S(\eta)\bar S(\eta)\frac{(\xi-\zeta)K(\xi,\bar\zeta)}{(\xi-\eta)(\eta-\zeta)} - S(\eta)\bar S(\zeta)\frac{K(\xi,\bar\eta)}{\eta-\zeta} - S(\xi)\bar S(\eta)\frac{K(\eta,\bar\zeta)}{\xi-\eta}.$$

Matching the left hand sides, we come to the conclusion

$$\begin{aligned} &\pi^{-1}\left\langle R_\eta K(\cdot,\bar\zeta),\, S(\bar\eta)K(\cdot,\xi)\right\rangle - \pi^{-1}\left\langle S(\eta)K(\cdot,\bar\zeta),\, R_{\bar\eta}K(\cdot,\xi)\right\rangle \\ = \;&\left[K(\xi,\bar\eta)L^*(\bar\zeta,\eta) - L(\xi,\bar\eta)K(\eta,\bar\zeta)\right] \\ = \;&\left\langle K(\cdot,\bar\zeta),\, L(\cdot,\eta)\right\rangle\left\langle K(\cdot,\bar\eta),\, K(\cdot,\xi)\right\rangle - \left\langle K(\cdot,\bar\zeta),\, K(\cdot,\eta)\right\rangle\left\langle L(\cdot,\bar\eta),\, K(\cdot,\xi)\right\rangle. \end{aligned}$$

The completeness of the system $\{K(\cdot,\tau)\}_{\tau\in\mathbb{C}}$ makes it possible to replace the reproducing kernels $K(t,\bar{\zeta})$ and $K(t,\xi)$ in the last identity by arbitrary functions F and $G \in H(E)$, respectively. Therefore, we reach the desired identity with $\omega = \eta$.

The principal identity (5.10) is a particular case of the identity

$$\langle R_\alpha F(\cdot),\, S(\beta)G(\cdot)\rangle - \langle S(\alpha)F(\cdot),\, R_\beta G(\cdot)\rangle - (\alpha-\bar{\beta})\,\langle R_\alpha F(\cdot),\, R_\beta G(\cdot)\rangle$$
$$= \pi\,\langle F(\cdot),\, L(\cdot,\alpha)\rangle\,\langle K(\cdot,\beta),\, G(\cdot)\rangle - \pi\,\langle F(\cdot),\, K(\cdot,\alpha)\rangle\,\langle L(\cdot,\beta),\, G(\cdot)\rangle\,, \qquad (5.10')$$

which has also been established by L. de Branges in [1, Chapter 2, §27].

We now deduce (5.10′) from (5.10) under the added assumption that the de Branges matrix is perfect. To this end replace F and G in (5.10) by $R_\alpha(t-\omega)F(t)$ and $R_\beta(t-\omega)G(t)$, respectively, assuming that

$$\omega = \bar{\omega}, \quad S(\omega) \neq 0, \quad \omega \neq \alpha, \text{ and } \omega \neq \beta.$$

Applying the identity

$$R_\beta(t-\omega)G(t) = S(\beta)G(t) + (\beta-\omega)R_\beta G(t)$$

(cf. (5.9)), transform the left hand side in (5.10):

$$\begin{aligned}
& \langle R_\omega R_\alpha(t-\omega)F(t),\, S(\omega)R_\beta(t-\omega)G(t)\rangle \\
- \;& \langle S(\omega)R_\alpha(t-\omega)F(t),\, R_\omega R_\beta(t-\omega)G(t)\rangle \\
= \;& \{\langle R_\alpha F(\cdot),\, S(\beta)G(\cdot)\rangle + (\bar{\beta}-\omega)\,\langle R_\alpha F(\cdot),\, R_\beta G(\cdot)\rangle\}\,|S(\omega)|^2 \\
- \;& \{\langle S(\alpha)F(\cdot),\, R_\beta G(\cdot)\rangle - (\alpha-\omega)\,\langle R_\alpha F(\cdot),\, R_\beta G(\cdot)\rangle\}\,|S(\omega)|^2,
\end{aligned}$$

which coincides with the left hand side in (5.10′) up to the positive factor $|S(\omega)|^2$.

Next, we apply (5.7) to the right hand side of (5.10), which is permissible because $R_\alpha tF(t) \in H(E)$ and the de Branges matrix is perfect, to obtain

$$\begin{aligned}
\langle R_\alpha(t-\omega)F(t),\, L(t,\omega)\rangle &= \langle R_\omega(t-\omega)F(t),\, L(t,\alpha)\rangle = S(\omega)\,\langle F(\cdot),\, L(\cdot,\alpha)\rangle\,, \\
\langle R_\alpha(t-\omega)F(t),\, K(t,\omega)\rangle &= S(\omega)F(\alpha) = S(\omega)\,\langle F(\cdot),\, K(\cdot,\alpha)\rangle\;,
\end{aligned}$$

and a pair of similar formulas for $R_\beta(t-\omega)G(t)$. The rest is plain.

It is worth pointing out once again that the identities (5.7) and (5.10′) are valid for perfect de Branges matrices. What is more, a de Branges matrix is perfect if and only if the permutability relation (5.7) holds.

Step 4. We are now in a position to express the function $Q(\eta,\bar{\zeta})$ in terms of the inner product in $H(E)$. For the (2,1) entries in (5.4) we have, much as in (5.11):

$$\begin{aligned}
&\frac{\xi-\eta}{i}\,\frac{\pi}{S(\xi)\bar{S}(\eta)}\,[-L(\xi,\bar{\eta}),\, K(\xi,\bar{\eta})]\, J \begin{bmatrix} Q(\eta,\bar{\zeta}) \\ -L(\eta,\bar{\zeta}) \end{bmatrix} \frac{\pi}{S(\eta)\bar{S}(\zeta)}\,\frac{\eta-\zeta}{i} \\
&\quad = -\frac{\xi-\zeta}{i}\,\frac{\pi L(\xi,\bar{\zeta})}{S(\xi)\bar{S}(\zeta)} + \frac{\xi-\eta}{i}\,\frac{\pi L(\xi,\bar{\eta})}{S(\xi)\bar{S}(\eta)} + \frac{\eta-\zeta}{i}\,\frac{\pi L(\eta,\bar{\zeta})}{S(\eta)\bar{S}(\zeta)}\,,
\end{aligned} \qquad (5.12)$$

or

$$\pi\left[K(\xi,\bar\eta)Q(\eta,\bar\zeta)-L(\xi,\bar\eta)L(\eta,\bar\zeta)\right]$$
$$=S(\eta)\bar S(\eta)\frac{(\xi-\zeta)L(\xi,\bar\zeta)}{(\xi-\eta)(\eta-\zeta)}-S(\eta)\bar S(\zeta)\frac{L(\xi,\bar\eta)}{\eta-\zeta}-S(\xi)\bar S(\eta)\frac{L(\eta,\bar\zeta)}{\xi-\eta}.$$

Now, upon setting $\omega=\eta$, $F(t)=L(t,\bar\zeta)$ and $G(t)=K(t,\xi)$ in the principal identity (5.10) and taking advantage of (5.8), we get first that the left hand side

$$\begin{aligned}
&\left\langle R_\eta L(\cdot,\bar\zeta),\, S(\bar\eta)K(\cdot,\xi)\right\rangle-\left\langle S(\eta)L(\cdot,\bar\zeta),\, R_{\bar\eta}K(\cdot,\xi)\right\rangle\\
=\;&\bar S(\eta)R_\eta L(\xi,\bar\zeta)-S(\eta)\left\{R_{\bar\eta}\left\langle K(\cdot,\xi),\, L(\cdot,\bar\zeta)\right\rangle\right\}^*\\
=\;&\bar S(\eta)\frac{L(\xi,\bar\zeta)S(\eta)-L(\eta,\bar\zeta)S(\xi)}{\xi-\eta}-S(\eta)\frac{L(\xi,\bar\zeta)\bar S(\eta)-L(\xi,\bar\eta)\bar S(\zeta)}{\zeta-\eta}\\
=\;&S(\eta)\bar S(\eta)\frac{(\xi-\zeta)L(\xi,\bar\zeta)}{(\xi-\eta)(\eta-\zeta)}-S(\eta)\bar S(\zeta)\frac{L(\xi,\bar\eta)}{\eta-\zeta}-S(\xi)\bar S(\eta)\frac{L(\eta,\bar\zeta)}{\xi-\eta}
\end{aligned}$$

and hence that

$$\begin{aligned}
&\left\langle R_\eta L(\cdot,\bar\zeta),\, S(\bar\eta)K(\cdot,\xi)\right\rangle-\left\langle S(\eta)L(\cdot,\bar\zeta),\, R_{\bar\eta}K(\cdot,\xi)\right\rangle\\
=\;&\pi\left[K(\xi,\bar\eta)Q(\eta,\bar\zeta)-L(\xi,\bar\eta)L(\eta,\bar\zeta)\right].
\end{aligned}$$

Next, the right hand side of (5.10) is easily seen to be equal to

$$\begin{aligned}
&\pi\left\langle L(\cdot,\bar\zeta),\, L(\cdot,\eta)\right\rangle\left\langle K(\cdot,\bar\eta),\, K(\cdot,\xi)\right\rangle-\pi\left\langle L(\cdot,\bar\zeta),\, K(\cdot,\eta)\right\rangle\left\langle L(\cdot,\bar\eta),\, K(\cdot,\xi)\right\rangle\\
=\;&\pi\left\langle L(\cdot,\bar\zeta),\, L(\cdot,\eta)\right\rangle K(\xi,\bar\eta)-\pi L(\xi,\bar\eta)L(\eta,\bar\zeta).
\end{aligned}$$

Thus, this lengthy and tedious calculation ends up with the simple formula

$$Q(\eta,\bar\zeta)=\left\langle L(\cdot,\bar\zeta),\, L(\cdot,\eta)\right\rangle \tag{5.13}$$

for every complex η and ζ.

According to generally accepted notation, it seems advisable to introduce the functions

$$b_z(t)=\sqrt{\pi}\,\frac{K(t,\bar z)}{\bar S(z)},\qquad c_z(t)=-\sqrt{\pi}\,\frac{L(t,\bar z)}{\bar S(z)} \tag{5.14}$$

and to summarize the results obtained in this section in the following statement:

Theorem 12 (Structure of perfect de Branges matrices). *Let $\mathcal{B}(z)$ be a perfect de Branges matrix of the form* (5.1)*. Then its bilinear J forms can be expressed as*

$$\mathcal{B}(z)J\mathcal{B}^*(\omega)-J=\frac{z-\bar\omega}{i}\begin{pmatrix}\langle c_{\bar\omega},\, c_{\bar z}\rangle & \langle b_{\bar\omega},\, c_{\bar z}\rangle\\ \langle c_{\bar\omega},\, b_{\bar z}\rangle & \langle b_{\bar\omega},\, b_{\bar z}\rangle\end{pmatrix} \tag{5.15}$$

and

$$J-\mathcal{B}^{-1*}(z)J\mathcal{B}^{-1}(\omega)=\frac{\omega-\bar z}{i}\begin{pmatrix}\langle b_\omega,\, b_z\rangle & -\langle c_\omega,\, b_z\rangle\\ -\langle b_\omega,\, c_z\rangle & \langle c_\omega,\, c_z\rangle\end{pmatrix}, \tag{5.16}$$

respectively. Moreover, the matrix $\mathcal{B}(z)$ admits a representation of the form

$$\mathcal{B}(z) = [I + H(z, x_0)J]\,\mathcal{B}(x_0)$$
$$= \left\{ I + \frac{z - x_0}{i} \begin{pmatrix} \langle c_{x_0}, c_{\bar z} \rangle & \langle b_{x_0}, c_{\bar z} \rangle \\ \langle c_{x_0}, b_{\bar z} \rangle & \langle b_{x_0}, b_{\bar z} \rangle \end{pmatrix} \right\} \mathcal{B}(x_0), \tag{5.17}$$

where x_0 is a real point of analyticity of $\mathcal{B}(z)$.

6 Parseval's equality

1. The relationship between perfect de Branges matrices and the spaces $H(E)$ is fully displayed in the following problem: given a de Branges space $H(E)$ and an associated function $S(z)$, describe the set of all measures $d\mu$ such that

$$\int_{-\infty}^{\infty} \left| \frac{F(t)}{S(t)} \right|^2 d\mu = \int_{-\infty}^{\infty} \left| \frac{F(t)}{E(t)} \right|^2 dt \tag{6.1}$$

holds for every $F \in H(E)$. The measure

$$d\mu(t) = \left| \frac{S(t)}{E(t)} \right|^2 dt$$

serves an example of such measure. Therefore the set is not empty. The equality

$$\int_{-\infty}^{\infty} \frac{F_1(t) F_2^*(t)}{|S(t)|^2} d\mu = \langle F_1, F_2 \rangle, \qquad F_1, F_2 \in H(E), \tag{6.2}$$

which stems from (6.1) in a standard way, should be noted.

The question formulated above underlies a wide class of interpolation problems. We refer to moment problems, extension problems for positive definite functions and others. These problems were thoroughly investigated from the viewpoint of J theory in the papers [4]–[9]. It was established that all the solutions of these problems are given by a linear fractional transformation of an arbitrary R-function, based on a certain matrix, which is called a resolvent matrix of the problem. By an R-function we understand here a function $\omega(z)$ which is analytic in the domain $\Im z \neq 0$ with positive imaginary part in $\mathbf{C}_+$ and which obeys the symmetry principle $\omega(\bar z) = \omega^*(z)$, as well as the improper element $\omega(z) = \infty$. In all these problems the resolvent matrix turns out to be a perfect de Branges matrix. Its real representation enables one to set up the Parseval equality problem (6.1) in the corresponding de Branges space.

On the other hand, Theorem 12 gives rise to the *principle matrix inequality*

$$\begin{bmatrix} \langle F(\cdot), F(\cdot) \rangle & \langle w(z) b_z(\cdot) - c_z(\cdot), F(\cdot) \rangle \\ * & \frac{w(z) - w^*(z)}{z - \bar z} \end{bmatrix} \geq 0, \quad \Im z \neq 0\,, \tag{6.3}$$

for every $F \in H(E)$, where b_z and c_z are given by (5.14). We shall show later on that all the solutions of (6.3) take the form

$$w(z) = \frac{\delta(z)\omega(z) + \gamma(z)}{\beta(z)\omega(z) + \alpha(z)}, \tag{6.4}$$

where $\omega(z)$ is an arbitrary R-function and

$$\begin{pmatrix} \delta(z) & \gamma(z) \\ \beta(z) & \alpha(z) \end{pmatrix} = \mathcal{B}(z).$$

The measures which appear in the integral representation of the functions $w(z)$ which are given by (6.4) are exactly those which occur in (6.1). Thus we obtain a unique interpretation of two distinct interpolation problems as the Parseval's equality problem in a de Branges space.

2. The remaining subsections are devoted to the realization of the program which was outlined in Item 1.

Let $E(z)$ be an entire function which satisfies (4.1) and let $S(z)$ be an entire function which is associated with the space $H(E)$. Then, by the Parametrization Theorem, there exists a perfect de Branges matrix

$$\mathcal{B}(z) = \frac{1}{S(z)} \begin{pmatrix} D(z) & C(z) \\ B(z) & A(z) \end{pmatrix}.$$

Now we consider the principal matrix inequality (6.3) with

$$b_z(t) = \frac{\sqrt{\pi}}{\bar{S}(z)} K(t, \bar{z}) = \frac{\sqrt{\pi}}{\bar{S}(z)} \frac{A(t)\bar{B}(z) - B(t)\bar{A}(z)}{\pi(t-z)},$$

and

$$c_z(t) = -\frac{\sqrt{\pi}}{\bar{S}(z)} L(t, \bar{z}) = \frac{\sqrt{\pi}}{\bar{S}(z)} \frac{-S(t)\bar{S}(z) + A(t)\bar{D}(z) - B(t)\bar{C}(z)}{\pi(t-z)}.$$

Theorem 13. *A function $w(z)$ which is analytic in $\Im z \neq 0$ solves (6.3) if and only if it can be expressed in the form*

$$w(z) = \frac{D(z)\omega(z) + C(z)}{B(z)\omega(z) + A(z)} \tag{6.5}$$

with an arbitrary R-function $\omega(z)$.

Proof. It is readily seen that (6.3) is equivalent to the inequality

$$\frac{w(z) - w^*(z)}{z - \bar{z}} \geq \langle w(z)b_z(\cdot) - c_z(\cdot), \, w(z)b_z(\cdot) - c_z(\cdot) \rangle \tag{6.6}$$

because, if $w(z)$ solves (6.3), then, upon setting

$$F(t) = w(z)b_z(t) - c_z(t) \ ,$$

we obtain (6.6). Conversely, (6.3) clearly follows from (6.6) and the Gram inequality

$$\begin{bmatrix} \langle F(\cdot), F(\cdot)\rangle & \langle w(z)b_z(\cdot) - c_z(\cdot), F(\cdot)\rangle \\ * & \langle w(z)b_z(\cdot) - c_z(\cdot), w(z)b_z(\cdot) - c_z(\cdot)\rangle \end{bmatrix} \geq 0.$$

Next, we can rewrite the two sides of (6.6) as

$$\frac{w(z) - w^*(z)}{z - \bar{z}} = [w^*(z), 1] \frac{J}{\frac{z-\bar{z}}{i}} \begin{bmatrix} w(z) \\ 1 \end{bmatrix}$$

and

$$\|w(z)b_z(t) - c_z(t)\|^2 = [w^*(z), 1] \begin{bmatrix} \langle b_z, b_z\rangle & -\langle c_z, b_z\rangle \\ -\langle b_z, c_z\rangle & \langle c_z, c_z\rangle \end{bmatrix} \begin{bmatrix} w(z) \\ 1 \end{bmatrix},$$

respectively. Applying (5.16), we come to the inequality

$$[w^*(z), 1] \frac{\mathcal{B}^{-1*}(z) J \mathcal{B}^{-1}(z)}{\frac{z-\bar{z}}{i}} \begin{bmatrix} w(z) \\ 1 \end{bmatrix} \geq 0.$$

It is readily checked that all the solutions of this inequality are exactly those functions which emerge from (6.5). □

3. Our next goal is to establish the relation between the measures $d\mu$ in (6.1) and the solutions of (6.3) (cf. [1, Chapter 2, §§29–32]).

Theorem 14. *Let the measure $d\mu$ be a solution of the Parseval equality* (6.1). *Then*

$$\int_{-\infty}^{\infty} \frac{d\mu(t)}{1+t^2} < \infty$$

and there exists a real number q such that

$$w(z) = q + \frac{1}{\pi} \int_{-\infty}^{\infty} \left\{ \frac{1}{t-z} - \frac{t}{1+t^2} \right\} d\mu(t), \quad \Im z \neq 0 , \tag{6.7}$$

satisfies (6.3).

Proof. To prove the theorem, we proceed in several steps.

Step 1. Putting

$$F(t) = R_z G(t) = \frac{G(t)S(z) - G(z)S(t)}{t - z}, \qquad G \in H(E),$$

in (6.1) with $\Im z \neq 0$ and $G(z) \neq 0$, we have

$$\frac{G(t)S(z) - G(z)S(t)}{(t-z)S(t)} \in L^2(d\mu).$$

The same is clearly true for the function $G(t)/S(t)$ and hence for the function $(t-z)^{-1}$ as well. Thus, the first statement is verified.

Step 2. Following de Branges, we define a function $w(z)$ which is analytic in $\Im z \neq 0$ by the relation

$$w(z)\,\langle b_z(\cdot),\, F(\cdot)\rangle - \langle c_z(\cdot),\, F(\cdot)\rangle = \frac{1}{\sqrt{\pi}}\int_{-\infty}^{\infty}\frac{\bar{F}(t)d\mu(t)}{\bar{S}(t)(t-z)}. \tag{6.8}$$

The objective is to show that $w(z) = \bar{w}(z)$ and that $w(z)$ does not depend on the choice of $F \in H(E)$.

For $F_1 \in H(E)$, let

$$w_1(z)\,\langle b_z(\cdot),\, F_1(\cdot)\rangle - \langle c_z(\cdot),\, F_1(\cdot)\rangle = \frac{1}{\sqrt{\pi}}\int_{-\infty}^{\infty}\frac{\bar{F}_1(t)d\mu(t)}{\bar{S}(t)(t-z)}.$$

Next, replace z by $\bar{z}$ and take complex conjugates of both sides to get

$$\bar{w}_1(z)\,\langle F_1(\cdot),\, b_{\bar{z}}(\cdot)\rangle - \langle F_1(\cdot),\, c_{\bar{z}}(\cdot)\rangle = \frac{1}{\sqrt{\pi}}\int_{-\infty}^{\infty}\frac{F_1(t)d\mu(t)}{S(t)(t-z)}\,. \tag{6.9}$$

Multiplying (6.8) [resp. (6.9)] through by

$$\langle F_1(\cdot),\, b_{\bar{z}}(\cdot)\rangle = \sqrt{\pi}\,\frac{F_1(z)}{S(z)} \qquad [\text{resp. } \langle b_z(\cdot),\, F(\cdot)\rangle = \sqrt{\pi}\,\frac{\bar{F}(z)}{\bar{S}(z)}]$$

and subtracting the latter from the former we conclude that

$$\begin{aligned}
&[w(z) - \bar{w}_1(z)]\,\langle F_1(\cdot),\, b_{\bar{z}}(\cdot)\rangle\,\langle b_z(\cdot),\, F(\cdot)\rangle \\
&+ \langle F_1(\cdot),\, c_{\bar{z}}(\cdot)\rangle\,\langle b_z(\cdot),\, F(\cdot)\rangle - \langle F_1(\cdot),\, b_{\bar{z}}(\cdot)\rangle\,\langle c_z(\cdot),\, F(\cdot)\rangle \\
&= \frac{F_1(z)}{S(z)}\int_{-\infty}^{\infty}\frac{\bar{F}(t)d\mu(t)}{\bar{S}(t)(t-z)} - \frac{\bar{F}(z)}{\bar{S}(z)}\int_{-\infty}^{\infty}\frac{F_1(t)d\mu(t)}{S(t)(t-z)}.
\end{aligned} \tag{6.10}$$

Next, we shall appeal to the principal identity (5.10) in terms of the functions $b_z(t)$ and $c_z(t)$:

$$\begin{aligned}
&\left\langle \frac{R_z F_1(\cdot)}{S(z)},\, F(\cdot)\right\rangle - \left\langle F_1(\cdot),\, \frac{R_{\bar{z}}F(\cdot)}{S(\bar{z})}\right\rangle \\
&= -\langle F_1(\cdot),\, c_{\bar{z}}(\cdot)\rangle\,\langle b_z(\cdot),\, F(\cdot)\rangle + \langle F_1(\cdot),\, b_{\bar{z}}(\cdot)\rangle\,\langle c_z(\cdot),\, F(\cdot)\rangle\ .
\end{aligned}$$

The left hand side may be evaluated with the help of (6.2):

$$\begin{aligned}
&\left\langle \frac{R_z F_1(\cdot)}{S(z)},\, F(\cdot)\right\rangle - \left\langle F_1(\cdot),\, \frac{R_{\bar{z}}F(\cdot)}{S(\bar{z})}\right\rangle \\
&= \int_{-\infty}^{\infty}\frac{S(z)F_1(t) - S(t)F_1(z)}{|S(t)|^2 S(z)(t-z)}F^*(t)\,d\mu(t) - \int_{-\infty}^{\infty}F_1(t)\frac{\bar{S}(z)F^*(t) - S^*(t)\bar{F}(z)}{|S(t)|^2\bar{S}(z)(t-z)}\,d\mu(t) \\
&= -\frac{F_1(z)}{S(z)}\int_{-\infty}^{\infty}\frac{\bar{F}(t)d\mu(t)}{\bar{S}(t)(t-z)} + \frac{\bar{F}(z)}{\bar{S}(z)}\int_{-\infty}^{\infty}\frac{F_1(t)d\mu(t)}{S(t)(t-z)}.
\end{aligned}$$

It now follows from (6.10) that

$$[w(z) - \bar{w}_1(z)] \langle F_1(\cdot), b_{\bar{z}}(\cdot)\rangle \langle b_z(\cdot), F(\cdot)\rangle = 0$$

and hence that $w(z) = \bar{w}_1(z)$. Putting $F_1 = F$ implies that $w(z) = \bar{w}(z)$, so that $w(z) = w_1(z)$, as needed.

Step 3. As we have just shown, the function $w(z)$ is meromorphic in $\Im z \neq 0$. Our next step is to ascertain that $w(z)$ is an R-function of the form (6.7) which satisfies (6.3). To this end consider two copies of the formula (6.8) and set $z = \eta$ and $F(t) = R_{\bar{\zeta}}G(t)/S(\bar{\zeta})$ in the first, and $z = \zeta$ and $F(t) = R_{\bar{\eta}}G(t)/S(\bar{\eta})$ in the second, where $G \in H(E)$. Subtracting the latter from the former, we get

$$w(\eta)\left\langle b_\eta(\cdot), \frac{R_{\bar{\zeta}}G(\cdot)}{S(\bar{\zeta})}\right\rangle - w(\zeta)\left\langle b_\zeta(\cdot), \frac{R_{\bar{\eta}}G(\cdot)}{S(\bar{\eta})}\right\rangle - \left\langle c_\eta(\cdot), \frac{R_{\bar{\zeta}}G(\cdot)}{S(\bar{\zeta})}\right\rangle + \left\langle c_\zeta(\cdot), \frac{R_{\bar{\eta}}G(\cdot)}{S(\bar{\eta})}\right\rangle$$

$$= \frac{1}{\sqrt{\pi}}\int_{-\infty}^{\infty} \frac{\{R_{\bar{\zeta}}G(t)\}^* d\mu(t)}{\bar{S}(\zeta)\bar{S}(t)(t-\eta)} - \frac{1}{\sqrt{\pi}}\int_{-\infty}^{\infty} \frac{\{R_{\bar{\eta}}G(t)\}^* d\mu(t)}{\bar{S}(\eta)\bar{S}(t)(t-\zeta)}.$$

Let us turn to the permutability relations

$$\left\langle b_\eta(\cdot), \frac{R_{\bar{\zeta}}G(\cdot)}{S(\bar{\zeta})}\right\rangle = \left\langle b_\zeta(\cdot), \frac{R_{\bar{\eta}}G(\cdot)}{S(\bar{\eta})}\right\rangle = \sqrt{\pi}\left\{\frac{R_{\bar{\zeta}}G(\bar{\eta})}{S(\bar{\zeta})S(\bar{\eta})}\right\}^*,$$

and

$$\left\langle c_\eta(\cdot), \frac{R_{\bar{\zeta}}G(\cdot)}{S(\bar{\zeta})}\right\rangle = \left\langle c_\zeta(\cdot), \frac{R_{\bar{\eta}}G(\cdot)}{S(\bar{\eta})}\right\rangle.$$

The first one is just the reproducing property and the second one is proved in Section 5 (see (5.7)). Therefore, upon making use of these identities in the preceding formula, we see that

$$(w(\eta) - w(\zeta))\left\{\frac{R_{\bar{\zeta}}G(\bar{\eta})}{S(\bar{\zeta})S(\bar{\eta})}\right\}$$

$$= \frac{1}{\pi}\int_{-\infty}^{\infty}\left\{\frac{\bar{S}(\zeta)\bar{G}(t) - \bar{S}(t)\bar{G}(\zeta)}{\bar{S}(\zeta)} - \frac{\bar{S}(\eta)\bar{G}(t) - \bar{S}(t)\bar{G}(\eta)}{\bar{S}(\eta)}\right\}\frac{d\mu(t)}{\bar{S}(t)(t-\zeta)(t-\eta)}$$

$$= \left\{\frac{\bar{G}(\eta)}{\bar{S}(\eta)} - \frac{\bar{G}(\zeta)}{\bar{S}(\zeta)}\right\}\int_{-\infty}^{\infty}\frac{d\mu(t)}{(t-\eta)(t-\zeta)},$$

or finally that

$$\frac{w(\eta) - w(\zeta)}{\eta - \zeta} = \frac{1}{\pi}\int_{-\infty}^{\infty}\frac{d\mu(t)}{(t-\eta)(t-\zeta)} \tag{6.11}$$

for every nonreal η and ζ. In the special case that $\eta = \zeta = z$ we have

$$\frac{w(z) - w^*(z)}{z - \bar{z}} = \frac{1}{\pi}\int_{-\infty}^{\infty}\frac{d\mu(t)}{|t-z|^2} \tag{6.12}$$

by the symmetry property of $w(z)$. The representation (6.7) drops out immediately from (6.12).

Step 4. To show that $w(z)$ satisfies (6.3), consider the Gram inequality in the space $L^2(d\mu)$:

$$\begin{bmatrix} \int_{-\infty}^{\infty} \left|\frac{F(t)}{S(t)}\right|^2 d\mu & \frac{1}{\sqrt{\pi}} \int_{-\infty}^{\infty} \frac{\bar{F}(t)}{\bar{S}(t)} \frac{d\mu(t)}{t-z} \\ * & \frac{1}{\pi} \int_{-\infty}^{\infty} \frac{d\mu(t)}{|t-z|^2} \end{bmatrix} \geq 0, \quad \Im z \neq 0\,. \tag{6.13}$$

By Parseval's equality and the relations (6.8) and (6.12) the inequality (6.13) is readily recognized as the principal matrix inequality (6.3), and the theorem follows. □

4. Theorems 13 and 14 assert that given an associated function $S(z)$, every measure $d\mu = d\mu_S$ which solves the Parseval equality (6.1), is the spectral measure of the linear fractional transformation

$$w_S(z) = \frac{D(z)\omega(z) + C(z)}{B(z)\omega(z) + A(z)} \tag{6.14}$$

based on the perfect de Branges matrix

$$\mathcal{B}(z) = \mathcal{B}_S(z) = \frac{1}{S(z)} \begin{pmatrix} D(z) & C(z) \\ B(z) & A(z) \end{pmatrix} \tag{6.15}$$

and some R-function $\omega(z)$. The following question arises naturally: what can one say about the spectral measure, corresponding to the linear fractional transformation (6.14)–(6.15) with an *arbitrary* parameter $\omega(z)$? It will turn out that in general the Bessel inequality

$$\int_{-\infty}^{\infty} \left|\frac{F(t)}{S(t)}\right|^2 d\mu_S(t) \leq \int_{-\infty}^{\infty} \left|\frac{F(t)}{E(t)}\right|^2 dt \tag{6.16}$$

holds. In particular cases Bessel's inequality may turn into Parseval's equality for every $w_S(z)$ and every $F \in H(E)$. We shall examine this phenomenon in detail in the sequel and shall show that it is an intimate property of the space $H(E)$ which does not depend on the choice of the associated function $S(z)$.

Along with the linear fractional transformation (6.14)–(6.15)

$$w_S(z) = \frac{D(z)\omega(z) + C(z)}{B(z)\omega(z) + A(z)}$$

$$= p_S z + q_S + \frac{1}{\pi} \int_{-\infty}^{\infty} \left\{ \frac{1}{t-z} - \frac{t}{1+t^2} \right\} d\mu_S(t)\,, \tag{6.17}$$

consider the linear fractional transformation

$$w_E(z) = \frac{A(z)\omega(z) - B(z)}{B(z)\omega(z) + A(z)}$$

$$= p_E z + q_E + \frac{1}{\pi}\int_{-\infty}^{\infty}\left\{\frac{1}{t-z} - \frac{t}{1+t^2}\right\} d\mu_E(t) \tag{6.18}$$

which is based on the special de Branges matrix (cf. (3.9))

$$\mathcal{B}_E(z) = \frac{1}{E(z)}\begin{pmatrix} A(z) & -B(z) \\ B(z) & A(z) \end{pmatrix},$$

which corresponds to the associated function $S(z) = E(z)$.

Theorem 15. *The functions $w_S(z)$ in (6.17) and $w_E(z)$ in (6.18) are connected by the relation*

$$w_S(z) = \Phi(z) + \frac{S(z)\bar{S}(z)}{E(z)\bar{E}(z)}(w_E(z) - i), \qquad \Im z \neq 0, \tag{6.19}$$

where $\Phi(z)$ is given by (2.16), and their spectral measures are connected by the relation

$$d\mu_S(t) = \left|\frac{S(t)}{E(t)}\right|^2 d\mu_E(t)\ . \tag{6.20}$$

Proof. We make use of the representation (3.1) for the elements $D(z)$ and $C(z)$ in order to express $w_S(z)$ in terms of $w_E(z)$:

$$\begin{aligned} w_S(z) &= \frac{D(z)\omega(z) + C(z)}{B(z)\omega(z) + A(z)} \\ &= \frac{\left(B(z)\Phi(z) + S(z)\bar{S}(z)/E(z)\right)\omega(z) + \left(A(z)\Phi(z) - i\ S(z)\bar{S}(z)/E(z)\right)}{B(z)\omega(z) + A(z)} \\ &= \Phi(z) + \frac{S(z)\bar{S}(z)}{E(z)\bar{E}(z)}\ \frac{\bar{E}(z)\omega(z) - i\bar{E}(z)(z)}{B(z)\omega(z) + A(z)} \\ &= \Phi(z) + \frac{S(z)\bar{S}(z)}{E(z)\bar{E}(z)}\left\{\frac{A(z)\omega(z) - B(z)}{B(z)\omega(z) + A(z)} - i\right\}, \end{aligned}$$

as needed.

In (6.19) $w_S(z) = \psi(z) + \varphi(z)w_E(z)$, where

$$\psi(z) = \Phi(z) - i\frac{S(z)\bar{S}(z)}{E(z)\bar{E}(z)} \text{ and } \varphi(z) = \frac{S(z)\bar{S}(z)}{E(z)\bar{E}(z)}$$

are analytic on the real line, $\varphi(x) \geq 0$ and $\psi(x) = \psi^*(x)$ (cf.(2.14′). According to the generalized Stieltjes inversion formula [10, Lemma DI.2.1, p. 634], for every pair of points a and b of continuity of the measure $d\mu_E$,

$$\lim_{y\to 0+0} \int_a^b \Im w_S(x+iy)\,dx = \int_a^b \varphi(x)\,d\mu_E(x),$$

which implies (6.20). The theorem follows. □

A crucial consequence of Theorem 15 is that

$$\int_{-\infty}^{\infty} \left|\frac{F(t)}{S(t)}\right|^2 d\mu_S(t) = \int_{-\infty}^{\infty} \left|\frac{F(t)}{E(t)}\right|^2 d\mu_E(t),$$

which means that given $F \in H(E)$ and an R-function $\omega(z)$ in (6.17)–(6.18), Parseval's equality or Bessel's inequality are valid regardless of the choice of the associated function $S(z)$. That is why it suffices to prove Bessel's inequality for the special associated function $S(z) = E(z)$.

Theorem 16. *For the special linear fractional transformation* (6.18) *of an arbitrary R-function $\omega(z)$, the Bessel inequality*

$$\int_{-\infty}^{\infty} \left|\frac{F(t)}{E(t)}\right|^2 d\mu_E(t) \leq \int_{-\infty}^{\infty} \left|\frac{F(t)}{E(t)}\right|^2 dt \tag{6.21}$$

holds for every $F \in H(E)$.

Proof. It is clear from (6.20) that for every associated function $S(z)$

$$\int_{-\infty}^{\infty} \left|\frac{S(t)}{E(t)}\right|^2 \frac{d\mu_E(t)}{1+t^2} = \int_{-\infty}^{\infty} \frac{d\mu_S(t)}{1+t^2} < \infty.$$

Therefore, since the functions $S(z) = zF(z)$ and $S(z) = F(z)$ are associated with $H(E)$ whenever $F \in H(E)$ by Theorem 11, it follows that

$$\int_{-\infty}^{\infty} \left|\frac{F(t)}{E(t)}\right|^2 d\mu_E(t) = \int_{-\infty}^{\infty} \left|\frac{tF(t)}{E(t)}\right|^2 \frac{d\mu_E(t)}{1+t^2} + \int_{-\infty}^{\infty} \left|\frac{F(t)}{E(t)}\right|^2 \frac{d\mu_E(t)}{1+t^2} < \infty$$

for every $f \in H(E)$. To prove (6.21), express $w_S(z)$ and $\Phi(z)$ in (6.19) in terms of formulas (6.17), (6.20) and (2.16), bearing in mind that in the latter formula $p = 0$ (since we are dealing with perfect de Branges matrices):

$$p_S z + q_S + \frac{1}{\pi}\int_{-\infty}^{\infty} \left\{\frac{1}{t-z} - \frac{t}{1+t^2}\right\} \left|\frac{S(t)}{E(t)}\right|^2 d\mu_E(t)$$

$$= q + \frac{1}{\pi}\int_{-\infty}^{\infty} \left\{\frac{1}{t-z} - \frac{t}{1+t^2}\right\} \left|\frac{S(t)}{E(t)}\right|^2 dt + \frac{S(z)\bar{S}(z)}{E(z)\bar{E}(z)}\,(w_E(z) - i)\,, \quad \Im z > 0. \tag{6.22}$$

This equality exhibits a simple relation between the Nevanlinna functions with spectral measures $|S(t)/E(t)|^2 d\mu_E(t)$ and $|S(t)/E(t)|^2 dt$. We shall exploit this formula for two particular associated functions. First, for $S(z) = zF(z)$, we have

$$p_{zF}z + q_{zF} + \frac{1}{\pi}\int_{-\infty}^{\infty}\left\{\frac{1}{t-z} - \frac{t}{1+t^2}\right\}\left|\frac{tF(t)}{E(t)}\right|^2 d\mu_E(t)$$

$$= q_1 + \frac{1}{\pi}\int_{-\infty}^{\infty}\left\{\frac{1}{t-z} - \frac{t}{1+t^2}\right\}\left|\frac{tF(t)}{E(t)}\right|^2 dt + \frac{z^2F(z)\bar{F}(z)}{E(z)\bar{E}(z)}\left(w_E(z) - i\right), \quad (6.23)$$

whence it follows by (1.17) that

$$p_{zF} = i \lim_{y\to+\infty} y\left\{\frac{F(iy)\bar{F}(iy)}{E(iy)\bar{E}(iy)}\left(w_E(iy) - i\right)\right\}. \quad (6.24)$$

Next, for $S(z) = F(z)$, we have

$$p_F z + q_F + \frac{1}{\pi}\int_{-\infty}^{\infty}\left\{\frac{1}{t-z} - \frac{t}{1+t^2}\right\}\left|\frac{F(t)}{E(t)}\right|^2 d\mu_E(t)$$

$$= q_2 + \frac{1}{\pi}\int_{-\infty}^{\infty}\left\{\frac{1}{t-z} - \frac{t}{1+t^2}\right\}\left|\frac{F(t)}{E(t)}\right|^2 dt + \frac{F(z)\bar{F}(z)}{E(z)\bar{E}(z)}\left(w_E(z) - i\right). \quad (6.25)$$

Since

$$\int_{-\infty}^{\infty}\left|\frac{F(t)}{E(t)}\right|^2 d\mu_E(t) < \infty \text{ and } \int_{-\infty}^{\infty}\left|\frac{F(t)}{E(t)}\right|^2 dt < \infty,$$

we can rewrite (6.25) as

$$\begin{aligned} p_F z + q_3 + \frac{1}{\pi}\int_{-\infty}^{\infty}\frac{1}{t-z}\left|\frac{F(t)}{E(t)}\right|^2 d\mu_E(t) \\ = \frac{1}{\pi}\int_{-\infty}^{\infty}\frac{1}{t-z}\left|\frac{F(t)}{E(t)}\right|^2 dt + \frac{F(z)\bar{F}(z)}{E(z)\bar{E}(z)}\left(w_E(z) - i\right). \end{aligned} \quad (6.26)$$

Now letting $z = iy$ and $y \to +\infty$, we obtain

$$\lim_{y\to+\infty} p_F \cdot iy + q_3 = 0$$

by (6.24) and hence that $p_F = q_3 = 0$. Thus the equality (6.26) takes the form

$$\frac{1}{\pi}\int_{-\infty}^{\infty}\frac{1}{t-z}\left|\frac{F(t)}{E(t)}\right|^2 d\mu_E(t) = \frac{1}{\pi}\int_{-\infty}^{\infty}\frac{1}{t-z}\left|\frac{F(t)}{E(t)}\right|^2 dt + \frac{F(z)\bar{F}(z)}{E(z)\bar{E}(z)}\left(w_E(z) - i\right). \quad (6.27)$$

Multiplying (6.27) with $z = iy$ through by $-iy$ and letting $y \to +\infty$ we get

$$\frac{1}{\pi}\int_{-\infty}^{\infty}\left|\frac{F(t)}{E(t)}\right|^2 d\mu_E(t) = \frac{1}{\pi}\int_{-\infty}^{\infty}\left|\frac{F(t)}{E(t)}\right|^2 dt - p_{zF}\,, \quad (6.28)$$

by (6.24). Thus the Bessel inequality (6.21) is verified. □

Equality prevails in (6.21) for $F \in H(E)$ if and only if $p_{zF} = 0$.

Corollary. *If for some $F \in H(E)$ the relation $zF(z) \in H(E)$ holds, then the measure $d\mu_E$ in (6.18) satisfies the Parseval equality*

$$\frac{1}{\pi}\int_{-\infty}^{\infty}\left|\frac{F(t)}{E(t)}\right|^2 d\mu_E(t) = \frac{1}{\pi}\int_{-\infty}^{\infty}\left|\frac{F(t)}{E(t)}\right|^2 dt \tag{6.29}$$

for every R-function $\omega(z)$.

Indeed, as we have just proved, $p_G = 0$ for every $G \in H(E)$. The equality (6.29) is now immediate from (6.28) and the assumption that $G = zF \in H(E)$

Theorem 17. *The measure $d\mu_E$ in* (6.18) *furnishes the Parseval equality on the whole space $H(E)$ if and only if $p_E = 0$.*

Proof. To prove the sufficiency, we begin with the equality (6.27) and show that under the assumption $p_E = 0$ we have

$$\lim_{y\to+\infty} y\left\{\frac{F(iy)\bar{F}(iy)}{E(iy)\bar{E}(iy)}(w_E(iy) - i)\right\} = 0$$

for the set of functions of the form

$$F(z) = \sum_{k=1}^{n} \alpha_k K(z, \zeta_k). \tag{6.30}$$

This set of functions is dense in $H(E)$. By Theorem 13, the function $w_E(z)$ solves the principal matrix inequality (6.3), corresponding to the special de Branges matrix $\mathcal{B}_E(z)$ in (3.9). Now

$$\begin{aligned} L(t,z) &= \frac{E(t)E^*(z) - A(t)A^*(z) - B(t)B^*(z)}{\pi(t-z)} \\ &= -iK(t,z), \end{aligned}$$

so that $c_z(t) = ib_z(t)$,

$$\langle w_E(z)b_z(\cdot) - c_z(\cdot),\, F(\cdot)\rangle = (w_E(z) - i)\,\frac{\sqrt{\pi}\bar{F}(z)}{\bar{E}(z)}$$

and (6.3) takes the form

$$\begin{bmatrix} \|F\|^2 & (w_E(z) - i)\,\dfrac{\sqrt{\pi}\bar{F}(z)}{\bar{E}(z)} \\ * & \dfrac{w_E(z) - w_E^*(z)}{z - \bar{z}} \end{bmatrix} \geq 0, \quad \Im z \neq 0. \tag{6.31}$$

Multiplying (6.31) from the left by the matrix

$$T = \begin{pmatrix} \frac{F(z)}{E(z)} & 0 \\ 0 & 1 \end{pmatrix},$$

and from the right by the matrix T^*, we obtain the inequality

$$\begin{bmatrix} \|F\|^2 \left|\frac{F(z)}{E(z)}\right|^2 & (w_E(z) - i)\frac{\sqrt{\pi}\bar{F}(z)F(z)}{\bar{E}(z)E(z)} \\ * & \frac{w_E(z) - w_E^*(z)}{z - \bar{z}} \end{bmatrix} \geq 0.$$

Therefore.,

$$y^2\|F\|^2 \left|\frac{F(z)}{E(z)}\right|^2 \frac{w_E(z) - w_E^*(z)}{z - \bar{z}} \geq y^2 \left|(w_E(z) - i)\frac{\sqrt{\pi}\bar{F}(z)F(z)}{\bar{E}(z)E(z)}\right|^2. \tag{6.32}$$

We wish to show that the left hand side in (6.32) tends to zero when $z = iy$ and $y \to +\infty$. To this end, put $F(z) = K(z, \zeta)$ for some fixed complex number ζ. Then, since $|\bar{E}(z)| < |E(z)|$ for $\Im z > 0$,

$$y\left|\frac{K(iy,\zeta)}{E(iy)}\right| = y\left|\frac{E(iy)E^*(\zeta) - \bar{E}(iy)\bar{E}^*(\zeta)}{2\pi E(iy)(iy - \zeta)}\right|$$

$$= y\left|\frac{E^*(\zeta) - \frac{\bar{E}(z)}{E(z)}\bar{E}^*(\zeta)}{2\pi(z - \zeta)}\right| = O(1), \quad \text{as } y \to +\infty. \tag{6.33}$$

The same is also true for every function $F(z)$ of the form (6.30). Since $p_E = 0$ implies that

$$\lim_{y\to+\infty} \frac{w_E(iy) - w_E^*(iy)}{2iy} = 0,$$

we conclude that

$$\lim_{y\to+\infty} y^2\|F\|^2 \left|\frac{F(iy)}{E(iy)}\right|^2 \frac{w_E(iy) - w_E^*(iy)}{2iy} = 0,$$

and so a fortiori that

$$\lim_{y\to+\infty} y\left|(w_E(iy) - i)\frac{\sqrt{\pi}\bar{F}(iy)F(iy)}{\bar{E}(iy)E(iy)}\right| = 0.$$

To prove (6.29) for functions of the form (6.30), it remains only to multiply (6.27) with $z = iy$ through by $-iy$ and let $y \to +\infty$. As the set (6.30) is dense in $H(E)$, the Parseval equality holds on the whole space $H(E)$.

Conversely, we shall now show that if $p_E > 0$, then Bessel's inequality is strict for the function

$$N(z) = \frac{K(z,\zeta)E(\zeta) - K(\zeta,\zeta)E(z)}{z-\zeta} \in H(E)$$

when ζ is not real. In fact, $N(z)$ is readily seen to satisfy the limit relation

$$\lim_{y\to+\infty} y^2 \frac{N(iy)\bar{N}(iy)}{E(iy)\bar{E}(iy)} = -K^2(\zeta,\zeta)$$

(cf. (6.33)). Now, from (6.27) for $F = N$, after multiplication through by $-iy$ and passage to the limit, we get as above

$$\frac{1}{\pi}\int_{-\infty}^{\infty} \left|\frac{N(t)}{E(t)}\right|^2 d\mu_E(t) = \frac{1}{\pi}\int_{-\infty}^{\infty} \left|\frac{N(t)}{E(t)}\right|^2 dt - p_E K^2(\zeta,\zeta),$$

so that

$$\frac{1}{\pi}\int_{-\infty}^{\infty} \left|\frac{N(t)}{E(t)}\right|^2 d\mu_E(t) < \frac{1}{\pi}\int_{-\infty}^{\infty} \left|\frac{N(t)}{E(t)}\right|^2 dt,$$

as was to be proved. □

5. We now raise the following question: how to distinguish the situation when Parseval's equality holds on the whole of $H(E)$ *for every* measure $d\mu_E$ (6.18)? Theorem 17 enables one to reformulate this problem: when is $p_E = 0$ in (6.18) *for every* parameter $\omega(z)$ in the special linear fractional transformation? Now the operator of multiplication by z moves from the wings to center stage.

Let $\mathcal{D}$ denote the domain of this operator. Following de Branges, we show that its closure $\overline{\mathcal{D}}$ either coincides with $H(E)$ or has codimension 1 (cf. [1, Chapter 2, problem 85]).

Lemma 7. *The function $T(z) \in H(E)$ is of the form*

$$T(z) = A(z)\xi + B(z)\eta \tag{6.34}$$

for complex numbers ξ, η if and only if $\xi^*\eta = xi\eta^*$ and for every pair of complex numbers z and u the relation

$$\frac{T(z)K(u,u) - T(u)K(z,u)}{z-u} = \frac{T(z)K(\bar{u},\bar{u}) - T(\bar{u})K(z,\bar{u})}{z-\bar{u}} \tag{6.35}$$

holds.

Proof. If $\xi^*\eta \neq \xi\eta^*$, then

$$A(z) = \frac{\bar{T}(z)\eta - T(z)\eta^*}{\xi^*\eta - \xi\eta^*} \in H(E) \quad \text{and} \quad B(z) = \frac{T(z)\xi^* - \bar{T}(z)\xi}{\xi^*\eta - \xi\eta^*} \in H(E),$$

which is not the case. The relation (6.35) can be checked by a direct (though rather tedious) calculation.

To obtain the converse, solve (6.35) for $T(z)$ with the aid of formula (4.3) for the reproducing kernel to get

$$T(z) = A(z)\frac{T(u)B(\bar{u}) - T(\bar{u})B(u)}{\pi(u-\bar{u})K(u,u)} + B(z)\frac{T(\bar{u})A(u) - T(u)A(\bar{u})}{\pi(u-\bar{u})K(u,u)}, \tag{6.36}$$

as claimed. □

The next theorem is proved in [1, Chapter 2, Theorem 29].

Theorem 18. *The function $T(z)$ is orthogonal to the domain of multiplication by z if and only if $T(z)$ is of the form* (6.34).

Proof. Let $T(z)$ be a function of the form (6.34) which belongs to $H(E)$ and, for any $F \in \mathcal{D}$ and any nonreal number u, consider the inner product

$$\langle (t-u)F(t),\, T(t)\rangle\, K(u,u) = \langle (t-u)F(t),\, T(t)K(u,u) - T(u)K(t,u)\rangle\,.$$

By the previous Lemma,

$$T(t)K(u,u) - T(u)K(t,u) = \frac{t-u}{t-\bar{u}}\left\{T(t)K(\bar{u},\bar{u}) - T(\bar{u})K(t,\bar{u})\right\}.$$

Now, taking into account the specific form of the inner product in a de Branges spaces (cf. Section 4), we have

$$\begin{aligned} \langle (t-u)F(t),\, T(t)\rangle\, K(u,u) &= \langle (t-\bar{u})F(t),\, T(t)K(\bar{u},\bar{u}) - T(\bar{u})K(t,\bar{u})\rangle \\ &= \langle (t-\bar{u})F(t),\, T(t)\rangle\, K(\bar{u},\bar{u}). \end{aligned}$$

Since $K(u,u) = K(\bar{u},\bar{u}) > 0$, it follows that

$$\langle (t-u)F(t),\, T(t)\rangle = \langle (t-\bar{u})F(t),\, T(t)\rangle\,,$$

or equivalently that $(u-\bar{u})\langle F(t),\, T(t)\rangle = 0$ for every $F \in \mathcal{D}$, which proves the sufficiency.

To prove the necessity, we can modify the argument given above to attain the relation

$$\langle (t-u)F(t),\, G(t)\rangle = 0 \quad \text{and} \quad F \in \mathcal{D}, \tag{6.37}$$

where

$$G(z) = T(z)K(u,u) - T(u)K(z,u) - \frac{z-u}{z-\bar{u}}\{T(z)K(\bar{u},\bar{u}) - T(\bar{u})K(z,\bar{u})\,.$$

Putting

$$F(z) = \frac{G(z)}{z-u} \in \mathcal{D}$$

in (6.37), we obtain $\|G\|^2 = 0$, i.e., the equality (6.35) holds for $T(z)$. By Lemma 7 (cf. (6.36)) $T(z)$ is of the form (6.34), as was to be proved. □

Corollary. *If $\mathcal{D}$ is not dense in $H(E)$, then $\overline{\mathcal{D}}$ is a hyperplane.*

6. Going back to Parseval's equality we can now state the following result:

Theorem 19. *Parseval's equality* (6.29) *is valid for every measure $d\mu_E$ in* (6.18) *if and only if $F \in \overline{\mathcal{D}}$.*

Proof. Let us first presume that $p_E = 0$ for every function $w_E(z)$ in (6.18). Then by Theorem 17, Parseval's equality holds on the full space $H(E)$ for every measure $d\mu_E$. We have to show that now $\overline{\mathcal{D}} = H(E)$.

As we know from Theorem 18, the assumption $\overline{\mathcal{D}} \neq H(E)$ entails the existence of complex numbers ξ, η such that $T(z) = A(z)\xi + B(z)\eta \in H(E)$ and $T(z)$ is orthogonal to $\overline{\mathcal{D}}$. By Lemma 7, the numbers ξ, η satisfy $\xi^*\eta = \xi\eta^*$. Therefore we may regard them to be real numbers. Substituting the parameter $\omega(z) = \eta/\xi$ (the case $\xi = 0$ is not excluded) into (6.18), we have

$$w_E(z) = \frac{A(z)\eta - B(z)\xi}{B(z)\eta + A(z)\xi} = (\eta - i\xi)\frac{\bar{E}(z)}{T(z)} + i.$$

Since $T(z) = \bar{T}(z)$ and $w_E(z) = \bar{w}_E(z)$, it is easily seen that

$$w_E(z) = (\eta + i\xi)\frac{E(z)}{T(z)} - i. \tag{6.38}$$

It now follows from (6.27) with $F = T$ that

$$\frac{1}{\pi}\int_{-\infty}^{\infty} \frac{1}{t-z}\left|\frac{T(t)}{E(t)}\right|^2 d\mu_E(t) = \frac{1}{\pi}\int_{-\infty}^{\infty} \frac{1}{t-z}\left|\frac{T(t)}{E(t)}\right|^2 dt + (\eta - i\xi)\frac{\bar{T}(z)}{E(z)} \tag{6.39}$$

and hence that

$$y\left|\frac{\bar{T}(iy)}{E(iy)}\right| = y\left|\frac{T(iy)}{E(iy)}\right| \leq C < \infty, \quad \text{for} \quad y > 1.$$

This in turn yields (see (6.38))

$$\frac{|w_E(iy)|}{y} \geq c > 0 \quad \text{for} \quad y > 1,$$

which contradicts the hypothesis that $p_E = 0$ for every $w_E(z)$.

Note that the converse statement is also true: if $\overline{\mathcal{D}} = H(E)$, then $p_E = 0$ for every $w_E(z)$ (6.18). This is a straightforward consequence of Theorem 17 and the corollary to Theorem 16.

Next, assume that $p_E > 0$ for some $w_E(z)$ of the form (6.18), so that $\overline{\mathcal{D}} \neq H(E)$. We shall show that for the corresponding measure $d\mu_E$ the strict Bessel inequality prevails for every function $F \in H(E)\backslash\overline{\mathcal{D}}$ (for $F \in \overline{\mathcal{D}}$ Parseval's equality holds for every $d\mu_E$ regardless of whether $p_E = 0$ or not).

By the corollary to Theorem 18, $\overline{\mathcal{D}}$ is now a hyperplane, i.e.,

$$F(z) = bT(z) + G(z), \qquad b \neq 0, \quad G \in \overline{\mathcal{D}}\,,$$

for every $F \in H(E)\backslash\overline{\mathcal{D}}$. Let us point out that if F_1 and F_2 belong to $H(E)$ and meet Parseval's equality, then so does their sum $F_1 + F_2$. In fact, given a complex number b consider the identity

$$\|bF_1 + F_2\|^2_{\mu_E} - \|bF_1 + F_2\|^2 = 2\Re\left\{b\left(\langle F_1(\cdot),\, F_2(\cdot)\rangle_{\mu_E} - \langle F_1(\cdot),\, F_2(\cdot)\rangle\right)\right\}, \quad (6.40)$$

where

$$\langle F_1(\cdot),\, F_2(\cdot)\rangle_{\mu_E} = \int_{-\infty}^{\infty} \frac{F_1(t)F_2^*(t)}{|E(t)|^2}\, d\mu_E(t)$$

and

$$\|F\|^2_{\mu_E} = \langle F(\cdot),\, F(\cdot)\rangle_{\mu_E}\,.$$

According to Bessel's inequality (which is always valid), the right hand side in (6.40) is nonpositive for every choice of the complex number b. This means that both sides in (6.40) must be equal to zero, and and hence that Parseval's equality holds for $bF_1 + F_2$, as needed.

Therefore, if at least one function $F = bT + G$ with $b \neq 0$ meets Parseval's equality, then so does every $F \in H(E)$. But then by Theorem 17, $p_E = 0$, which contradicts the assumption above. The theorem follows. □

7. It is time to sum up the results obtained in this section in the following statement.

Theorem 20 (Parseval's Equality). *Let $H(E)$ be the de Branges space which is generated by an entire function $E(z) = A(z) + iB(z)$, let $S(z)$ be an associated function and consider the perfect de Branges matrix*

$$\mathcal{B}(z) = \frac{1}{S(z)}\begin{pmatrix} D(z) & C(z) \\ B(z) & A(z) \end{pmatrix}.$$

If a measure $d\mu_S$ furnishes Parseval's equality on the whole of $H(E)$:

$$\int_{-\infty}^{\infty} \left|\frac{F(t)}{S(t)}\right|^2 d\mu_S(t) = \int_{-\infty}^{\infty} \left|\frac{F(t)}{E(t)}\right|^2 dt, \quad \textit{for every} \quad F \in H(E), \qquad (6.41)$$

then $d\mu_S$ is the spectral measure of the linear fractional transformation based on $\mathcal{B}(z)$ of some R-function $\omega(z)$:

$$\begin{aligned} w_S(z) &= \frac{D(z)\omega(z)+C(z)}{B(z)\omega(z)+A(z)} \\ &= p_S z + q_S + \frac{1}{\pi}\int_{-\infty}^{\infty}\left\{\frac{1}{t-z}-\frac{t}{1+t^2}\right\} d\mu_S(t)\ . \end{aligned} \tag{6.42}$$

Next, consider the operator of multiplication by z in $H(E)$ with domain $\mathcal{D}$.

1. *If $\overline{\mathcal{D}} = H(E)$, then (conversely) for any R-function $\omega(z)$, the measure $d\mu_S$ in (6.42) furnishes Parseval's equality (6.41) on the whole of $H(E)$.*
2. *The case $\overline{\mathcal{D}} \neq H(E)$ is more delicate. An arbitrary measure $d\mu_S$ in (6.42) provides only Bessel's inequality*

$$\int_{-\infty}^{\infty}\left|\frac{F(t)}{S(t)}\right|^2 d\mu_S(t) \le \int_{-\infty}^{\infty}\left|\frac{F(t)}{E(t)}\right|^2 dt, \quad \text{for every} \quad F\in H(E). \tag{6.43}$$

The measure $d\mu_S$ corresponding to the parameter $\omega(z)$ in (6.42) satisfies (6.41) for every $f \in H(E)$ if and only if

$$p_E = \lim_{y\to+\infty}\frac{1}{iy}\frac{A(iy)\omega(iy)-B(iy)}{B(iy)\omega(iy)+A(iy)} = 0.$$

For parameters $\omega(z)$ with $p_E > 0$ the relations

$$\int_{-\infty}^{\infty}\left|\frac{F(t)}{S(t)}\right|^2 d\mu_S(t) \quad = \quad \int_{-\infty}^{\infty}\left|\frac{F(t)}{E(t)}\right|^2 dt, \quad \text{for} \quad F\in\overline{\mathcal{D}}$$

and

$$\int_{-\infty}^{\infty}\left|\frac{F(t)}{S(t)}\right|^2 d\mu_S(t) \quad < \quad \int_{-\infty}^{\infty}\left|\frac{F(t)}{E(t)}\right|^2 dt, \quad \text{for} \quad F\in H(E)\backslash\overline{\mathcal{D}}$$

hold.

References

[1] L. de Branges, *Hilbert Spaces of Entire Functions*, Prentice Hall, NY, 1968.

[2] D.Z. Arov, Darlington realization of matrix valued functions, *Math. USSR Izvestija* **7** (1973) 1295–1326.

[3] D.Z. Arov, Realization of the canonical system with the dissipative boundary condition at one end of the segment according to dynamic compliance coefficients, *Sib. Math. Zhurn.*, **16** (1975) 440–463.

[4] V.P. Potapov, The multiplicative structure of J-contractive matrix functions, *Trudy Moskov. Mat. Obshch.* **4** (1955) 125–136; English transl. in *Amer. Math. Soc. Transl.* **15** (1960) 131–243.

[5] V.P. Potapov, Fractional linear transformations of matrices, in: *Studies in the Theory of Operators and their Applications*, Kiev (1979), pp. 75–91.

[6] A.V. Efimov and V.P. Potapov, J-expanding matrix valued functions and their role in the analytical theory of electrical circuits, *Russian Math. Surveys* **28** (1973) 69–140.

[7] I.V. Kovalishina and V.P. Potapov, An indefinite metric in the Nevanlinna Pick problem, *Akad. Nauk Arm. SSR Dokl.* **59** (1974) 17–22.

[8] I.V. Kovalishina, J expansive matrix valued functions and Caratheodory problem, *Akad. Nauk Arm. SSR Dokl.* **59** (1974) 129–135.

[9] I.V. Kovalishina, J expansive matrix valued functions and moment problem, *Akad. Nauk Arm. SSR Dokl.* **60** (1975) 3–10.

[10] I.S. Kac and M.G. Krein, R-functions – analytic functions, mapping the upper half plane into itself, Supplement I to Russian edition of the book: F. Atkinson, *Discrete and Continuous boundary problems*, Moscow, Mir, 1968, pp. 629–647. English transl. in *Amer. Math. Soc. Transl.* **103** (1974), 1–102.

Golinskii Leonid
Mathematical Division
Institute for Low Temperature
Physics and Engineering
47 Lenin Avenue,
Kharkov 310164, Ukraine

Mikhailova Irina
Meierstrasse 1A
Steinbergkirche 29472
Deutschland

AMS Subject Classification: 30D20, 30E10, 30E15, 30E20, 42A82, 46E22.

Operator Theory
Advances and Applications, Vol. 95
© 1997 Birkhäuser Verlag Basel/Switzerland

On transformations of Potapov's fundamental matrix inequality

V.E. Katsnelson

Abstract. According to V.P. Potapov, a classical interpolation problem can be reformulated in terms of a so-called Fundamental Matrix Inequality (FMI). To show that every solution of the FMI satisfies the interpolation problem, we usually have to transform the FMI in some special way. In this paper a number of the transformations of the FMI which come into play are motivated and demonstrated by simple, but typical examples.

0 Preface

V.P. Potapov's approach to classical interpolation problems is based on the following strategy. Instead of the original interpolation problem (or problem on integral representation), an inequality for analytic functions is considered in an appropriate domain. This inequality is said to be *the Fundamental Matrix Inequality* (FMI) for the considered interpolation problem. Here two problems appear. The first problem is how to "solve" this inequality. The second problem is to prove that this inequality is equivalent to the original interpolation problem.

The study of the second problem consist of two parts. First, we have to prove that any function which is a solution of the original problem is also a solution of the FMI. Usually this part is not difficult. Secondly, we have to extract the full interpolation information from the FMI. This means that we have to prove that any analytic function which satisfies the FMI is also a solution of of the original interpolation problem. In simple situations it is not difficult to obtain the interpolation information from the FMI. However, in the general case this is not easy, and we have to apply a special transformation to the FMI. Such a transformation can be applied to every FMI. However, in the simplest situations it is possible to do without such a transformation. The development of Potapov's method began with consideration of the simplest interpolation problem, i.e., the Nevanlinna-Pick ($\mathcal{NP}$) problem. The equivalence of the $\mathcal{NP}$ problem to its FMI is clear. Because of this, this transform was camouflaged in the beginning of the theory. However, in the study of the power moment problem we already can not do without it. In the paper [KKY] such a transform was used in the very general setting of the so called *Abstract Interpolation Problem.* Namely, such a transformation was used in considerations related to Theorem 1 of that paper. Of course, the authors of [KKY] took into account the experience which was accumulated by previous work with concrete problems. However, this transformation was introduced in [KKY] in a formal way, without any motivation. As result, the proof of Theorem 1 of [KKY] looks like a trick. This is not satisfactory, because the transformation of the FMI lies at the heart of the FMI business. The main goal of the present paper is to motivate and to demonstrate the transformation of the FMI by the simplest but typical example of the power moment problem. For contrast, the $\mathcal{NP}$ problem and the FMI for it are considered as well. We would like to demonstrate the algebraic side of the matter. Therefore, we will avoid the entourage of general vector spaces and Hilbert spaces in the generality of the paper [KKY]. All our spaces are finite-dimensional. Instead of abstract kernels and operators, we will consider matrices.

1 The FMI and its structure

Classical interpolation problems can be considered for various function classes in various domains. Here we consider two function classes related to the unit disc $\mathbb{D}$ and to the upper half plane $\mathbb{H}$.

Definition 1.1. *I.* The class $\mathfrak{C}\,(\mathbb{D})$ *is the class of functions w which are holomorphic outside the unit circle $\mathbb{T}$ and satisfy the symmetry condition*

$$w(z) = -w^*(1/\overline{z}) \qquad (z \in \mathbb{C} \setminus \mathbb{T}) \tag{1.1}$$

and the positivity condition

$$\frac{w(z) + w^*(z)}{1 - |z|^2} \geq 0 \qquad (z \in \mathbb{C} \setminus \mathbb{T}). \tag{1.2}$$

II. The class $\mathfrak{R}\,(\mathbb{H})$ *is the class of functions w which are holomorphic outside the real axes $\mathbb{R}$ and satisfy the symmetry condition*

$$w(z) = w^*(\overline{z}) \quad (z \in \mathbb{C} \setminus \mathbb{R}) \tag{1.3}$$

and the positivity condition

$$\frac{w(z) - w^*(z)}{z - \overline{z}} \geq 0 \qquad (z \in \mathbb{C} \setminus \mathbb{R}). \tag{1.4}$$

III. The class $\mathfrak{R}_0\,(\mathbb{H})$ *is the subclass of the class $\mathfrak{R}\,(\mathbb{H})$ which is singled out by the condition*

$$\overline{\lim_{y \uparrow \infty}}\, y\,|w(iy)| < \infty. \tag{1.5}$$

The FMI of a classical interpolation problem has the form

$$\begin{bmatrix} A & B_w(z) \\ B_w^*(z) & C_w(z) \end{bmatrix} \geq 0, \tag{1.6}$$

where A is some hermitian matrix, constructed from the interpolation data (interpolation points and interpolating values) only. It is nonnegative if and only if the considered interpolation problem is solvable. The entry $C_w(z)$ contains the function w only, but not the interpolation data. Its form depend on the function class to which the function w belongs. For an interpolation problem in the class $\mathfrak{C}(\mathbb{D})$ the entry $C_w(z)$ has the form

$$C_w(z) = \frac{w(z) + w^*(z)}{1 - |z|^2}. \tag{1.7}$$

For an interpolation problem in the class $\mathfrak{R}(\mathbb{H})$ the entry $C_w(z)$ has the form

$$C_w(z) = \frac{w(z) - w^*(z)}{z - \bar{z}}. \tag{1.8}$$

In the entry $B_w(z)$ both the interpolation data and the function w are combined. This entry looks like

$$B_w(z) = (zI - T)^{-1}(u \cdot w(z) - v), \tag{1.9}$$

or like

$$B_w(z) = T(I - zT)^{-1}(u \cdot w(z) - v) . \tag{1.10}$$

To each classical interpolation problem the following objects are related:

1. The hermitian matrix A, which is nonnegative iff the problem is solvable.
2. The matrix T which "determines" the interpolation nodes.
3. The vectors u and v which determine the interpolation values.

The terms A, T, u and v satisfy the so called *Fundamental Identity* (FI). The form of the FI depends on the function class in which the interpolation problem is considered. For the function class $\mathfrak{C}(\mathbb{D})$, the FI has the form

$$A - TAT^* = uv^* + vu^*. \tag{1.11}$$

For the class $\mathfrak{R}(\mathbb{H})$, the FI has the form

$$TA - AT^* = uv^* - vu^*. \tag{1.12}$$

If the FMI (1.6) is satisfied (for some z), and if M is a matrix of appropriate size, then the inequality

$$M \begin{bmatrix} A & B_w(z) \\ B_w^*(z) & C_w(z) \end{bmatrix} M^* \geq 0 \tag{1.13}$$

holds as well. If the matrix M is invertible, then the inequalities (1.6) and (1.13) are equivalent.

2 The FMI for the Nevanlinna-Pick problem

Now we obtain the FMI for the $\mathcal{NP}$ problem in the function class $\mathfrak{C}(\mathbb{D})$.

Definition 2.1. *Given n points $z_1, z_2, \ldots, z_n$ in the unit disc $\mathbb{D}$* (interpolation nodes) *and n complex numbers $w_1, w_2, \ldots, w_n$* (interpolation values). *A holomorphic function $w(z)$ from the class $\mathfrak{C}(\mathbb{D})$ is said to be* a solution of the Nevanlinna-Pick problem with interpolation data $\{z_1, w_1\}, \{z_2, w_2\}, \ldots, \{z_n, w_n\}$, *if the* interpolation conditions

$$w(z_k) = w_k \qquad (k = 1, 2, \ldots, n) \tag{2.1}$$

are satisfied.

Let us associate with the $\mathcal{NP}$ problem two $n \times 1$ vectors, which characterize the interpolation values:

$$u = \begin{bmatrix} 1 \\ \vdots \\ 1 \end{bmatrix} \qquad \text{and} \qquad v = \begin{bmatrix} w_1 \\ \vdots \\ w_n \end{bmatrix}. \tag{2.2}$$

The matrix T, which characterize the interpolation nodes, has the form

$$T = \operatorname{diag}[\, z_1, \, z_2\,, \cdots, z_n\,]\,. \tag{2.3}$$

The matrix A, the so called *Pick matrix* for the problem, has the form

$$A = \|a_{kl}\|_{1 \le k, l \le n}\,, \qquad a_{kl} = \frac{w_k + \bar{w}_l}{1 - z_k \bar{z}_l}. \tag{2.4}$$

The Fundamental Identity (1.11) for this choice of u, v, T, and$A$ can be checked directly.

The Fundamental Matrix Inequality for the Nevanlinna-Pick problem (FMI($\mathcal{NP}$)) has the form (1.6) with A from (2.4), $C_w(z)$ from (1.7) and $B_w(z)$ from (1.9), (2.2) and (2.3).

Theorem 2.1 (from FMI($\mathcal{NP}$) to interpolation conditions). *Let $w(z)$ be a function which is holomorphic in the unit disc $\mathbb{D}$ and which satisfies the FMI($\mathcal{NP}$) for every $z \in \mathbb{D}$. Then the function w satisfies the condition $w(z) + w^*(z) \ge 0$ $(z \in \mathbb{D})$ and the interpolation conditions* (2.1).

Proof. Since the entry $C_w(z)$ must be nonnegative for $z \in \mathbb{D}$, the real part of the function w is nonnegative in[1] $\mathbb{D}$. Now we take into account the concrete form of the entry $B_w(z)$:

$$B_w(z) = \begin{bmatrix} b_{1,w}(z) \\ b_{2,w}(z) \\ \vdots \\ b_{n,w}(z) \end{bmatrix}, \tag{2.5}$$

where

$$b_{k,w}(z) = \frac{w(z) - w_k}{z - z_k} \quad (k = 1, 2, \cdots, n). \tag{2.6}$$

Because the "full" matrix (1.6) is nonnegative, its principal subblocks are also nonnegative:

$$\begin{bmatrix} a_{kk} & b_{w,k}(z) \\ b^*_{w,k}(z) & C_w(z) \end{bmatrix} \geq 0. \tag{2.7}$$

Since the function w is holomorphic in $\mathbb{D}$, the entry $C_w(z)$ is also locally bounded in $\mathbb{D}$. Thus, from (2.7) it follows, that the entry $b_{w,k}(z)$ is locally bounded in $\mathbb{D}$ as well. However, if the function b_k is bounded even near the point z_k, then the interpolation conditions (2.1) are satisfied. □

Thus, for the $\mathcal{NP}$ interpolation problem it is not difficult to extract the interpolation information from its FMI.

It is worth mentioning, that the inequality (2.7) can be consider as an inequality of the form (1.13), with

$$M = \begin{bmatrix} 0 & 0 & \cdots & \overset{k}{1} \cdots & 0 & \vdots & \overset{n+1}{0} \\ 0 & 0 & \cdots & 0 \cdots & 0 & \vdots & 1 \end{bmatrix}. \tag{2.8}$$

3 Derivation of the FMI ($\mathcal{NP}$)

A crucial role in the derivation of the FMI for the $\mathcal{NP}$ problem is played by the Riesz-Herglotz theorem. Given a nonnegative measure σ and a real number c, we associate with them the function $w_{\sigma,c}$:

$$w_{\sigma,c}(z) = ic + \frac{1}{2} \int\limits_{\mathbb{T}} \frac{t+z}{t-z} \, d\sigma(t), \qquad (z \in \mathbb{C} \setminus \mathbb{T}). \tag{3.1}$$

The function $w_{\sigma,c}$ belongs to the class $\mathfrak{C}(\mathbb{D})$.

---

1) If we continue the function w, which is defined originally in $\mathbb{D}$ only, into the exterior of the unit circle according to the symmetry (1.1), then the function which is continued in this way will satisfy the condition (1.2).

Theorem (Riesz-Herglotz). *Let w be a function which belongs to the class $\mathfrak{C}(\mathbb{D})$. Then this function w is of the form (3.1) for some σ and c, which are uniquely determined from the given w.*

Let us start to derive the FMI($\mathcal{NP}$). Given a measure $\sigma \geq 0$ on $\mathbb{T}$, a real number c and points $z_1, z_2, \cdots, z_n; z \in \mathbb{D}$,. Let u be defined by (2.2), T be defined by (2.3). Then the following inequality ($z_1, z_2, \cdots, z_n$ appear in T) holds:

$$\int\limits_{\mathbb{T}} \begin{bmatrix} (tI-T)^{-1}u \\ \cdots\cdots \\ \bar{t}(\bar{t}-\bar{z})^{-1} \end{bmatrix} \cdot d\sigma(t) \cdot \begin{bmatrix} u^*(\bar{t}I - T^*)^{-1} & \vdots & \dfrac{t}{t-z} \end{bmatrix} \geq 0. \tag{3.2}$$

This is a block-matrix inequality of the form

$$\begin{bmatrix} A_\sigma & B_\sigma(z) \\ B_\sigma^*(z) & C_\sigma(z) \end{bmatrix} \geq 0. \tag{3.3}$$

Now let us discuss the entries of the block-matrix on the right-hand side of the inequality (3.3). Originally these entries were defined by means of an integral representation. However, they can be expressed in terms of the function $w_{\sigma,c}$. Let us consider the block A_σ:

$$A_\sigma = \int\limits_{\mathbb{T}} (tI-T)^{-1}u \cdot d\sigma(t) \cdot u^*(\bar{t} - T^*)^{-1}, \tag{3.4}$$

or, for the entries $A_\sigma = \|a_{\sigma,kl}\|_{1\leq k,l\leq n}$:

$$a_{\sigma,kl} = \int\limits_{\mathbb{T}} (t-z_k)^{-1} \cdot d\sigma(t) \cdot (\bar{t} - \bar{z}_l)^{-1}, \qquad (1 \leq k, l \leq n).$$

According to the well-known identity for the *Schwarz kernel* $\frac{1}{2}(t+z)(t-z)^{-1}$,

$$A_\sigma = \left\| \frac{w_{\sigma,c}(z_k) + \overline{w_{\sigma,c}(z_l)}}{1 - z_k \bar{z}_l} \right\|_{1\leq k,l\leq n}. \tag{3.5}$$

(The constant c does not appear in (3.5).) The block B_σ has the following form:

$$B_\sigma = \int\limits_{\mathbb{T}} (tI-T)^{-1}u \cdot \frac{t}{t-z} \cdot d\sigma(t). \tag{3.6}$$

The block B_σ (which does not depend on c) can be transformed in the following way. Integrating the identity

$$(tI-T)^{-1}\frac{t}{t-z} = (zI-T)^{-1} \cdot \frac{1}{2}\frac{t+z}{t-z} - (zI-T)^{-1} \cdot \frac{1}{2}\frac{tI+T}{tI-T}$$

with respect the measure $d\sigma$, we obtain

$$B_\sigma = (zI - T)^{-1}\Big(uw_{\sigma,c}(z) - v_{\sigma,c}\Big), \tag{3.7}$$

where

$$v_{\sigma,c} = icu + \frac{1}{2}\int\limits_{\mathbb{T}} \frac{tI + T}{tI - T}\, d\sigma(t). \tag{3.8}$$

It can be checked that

$$A_\sigma - TA_\sigma T^* = u \cdot v_{\sigma,c}^* - v \cdot u_{\sigma,c}^*. \tag{3.9}$$

According to (3.1) and to (2.3),

$$v_{\sigma,c} = \begin{bmatrix} w_{\sigma,c}(z_1) \\ w_{\sigma,c}(z_2) \\ \cdots \\ w_{\sigma,c}(z_n) \end{bmatrix}. \tag{3.10}$$

Of course,

$$C_\sigma(z) = \int\limits_{\mathbb{T}} \frac{d\sigma(t)}{|t - z|^2} = \frac{w_{\sigma,c}(z) + \overline{w_{\sigma,c}(z)}}{1 - |z|^2}. \tag{3.11}$$

Now let the function $w_{\sigma,c}$ satisfy the interpolation conditions (2.1) , i.e., let

$$w_{\sigma,c}(z_k) = w_k \qquad (k = 1, 2, \ldots, n). \tag{3.12}$$

Comparing (3.5) and (2.4), we obtain that

$$A_\sigma = A. \tag{3.13}$$

From (3.10) and (2.2),

$$v_{\sigma,c} = v. \tag{3.14}$$

Comparing now (3.7) with (1.9), we obtain that

$$B_\sigma(z) = B_{w_{\sigma,c}}(z). \tag{3.15}$$

Of course, by (3.11), $C_\sigma(z) = C_{w_{\sigma,c}}(z)$. Thus, we obtain the following statement:

Lemma 3.1. *If the function $w_{\sigma,c}$, defined by (3.1), satisfies the interpolation conditions (3.12), then the* FMI *(1.6) (with w replaced by $w_{\sigma,c}$) is satisfied for every $z \in \mathbb{C} \setminus \mathbb{T}$, where A is defined by (2.4), B_w is defined by (1.9), (2.2), (2.3) and C_w is defined by (1.7).*

According to the Riesz-Herglotz theorem, each function w from the considered class has the representation $w = w_{\sigma,c}$. Thus, the following result holds:

Theorem 3.1 (from interpolation conditions to FMI($\mathcal{NP}$)). *Let interpolation data for $\mathcal{NP}$ problem be given. Let w be a function, which belongs to the class $\mathfrak{C}(\mathbb{D})$. If the function w satisfies the interpolation conditions* (2.1), *then the* FMI($\mathcal{NP}$) *for this function (with A and v constructed from the given interpolation data) is satisfied for every $z \in \mathbb{C} \setminus \mathbb{T}$.*

We have stated this (well-known) derivation of the FMI ($\mathcal{NP}$) because the formulas (3.4) and (3.6) are a very convenient starting point to guess formulas for transformations of the FMI.

4 The Hamburger moment problem as a classical interpolation problem

This problem can be considered as a classical interpolation problem in the class $\mathfrak{R}(\mathbb{H})$.

Formulation of the Hamburger moment problem. The data *of the Hamburger problem is a finite sequence $s_0, s_1, \dots, s_{2n-1}, s_{2n}$ of real numbers. A nonnegative measure σ on the real numbers is said to be* a solution of the Hamburger moment problem *(with these data), if its power moments*

$$s_k(\sigma) = \int_{\mathbb{R}} \lambda^k d\sigma(\lambda) \qquad (k = 0, 1, \dots, 2n-1, 2n) \tag{4.1}$$

exist and satisfy the moment conditions

$$\text{i) } s_k(\sigma) = s_k \qquad (k = 0, 1, \dots, 2n-1); \qquad \text{ii) } s_{2n}(\sigma) \le s_{2n}. \tag{4.2}$$

Measures σ satisfying these moment conditions are sought.

At first glance the formulated moment problem does not look like an interpolation problem. However, this problem can be reformulated as a classical interpolation problem.

Namely, let σ be a nonnegative measure on $\mathbb{R}$ which is finite: $s_0(\sigma) < \infty$. We associate with this measure σ the function w_σ :

$$w_\sigma(z) = \int_{\mathbb{R}} \frac{d\sigma(\lambda)}{\lambda - z} \qquad (z \in \mathbb{C} \setminus \mathbb{R}) \tag{4.3}$$

This function w_σ belongs to the class $\mathfrak{R}_0(\mathbb{H})$.

The following result is a version of the Riesz-Herglotz theorem for the upper half-plane.

Theorem (Nevanlinna). *Let w be a function from the class $\mathfrak{R}_0(\mathbb{H})$. Then this function w is representable in the form (4.3), with some finite nonnegative measure $\sigma : \sigma \geq 0$, $s_0(\sigma) < \infty$. This measure σ is determined from the function w uniquely.*

It turns out that if a measure σ solves the Hamburger moment problem (4.2), then the function w_σ, associated with this measure σ, satisfies some asymptotic relation. To obtain such a relation, we consider the functions $w_{\sigma,k}$:

$$w_{\sigma,k}(z) = \int_{\mathbb{R}} \lambda^k \frac{d\sigma(\lambda)}{\lambda - z} \qquad (k = 0, 1, 2, \ldots, 2n). \tag{4.4}$$

(In this notation, $w_\sigma = w_{\sigma,0}$.) Assume that a finite measure $\sigma \geq 0$ on $\mathbb{R}$ has a finite $2n$'th moment $s_{2n}(\sigma)$ (and hence, also finite moments $s_0(\sigma), \ldots, s_{2n-1}(\sigma)$). Integrating the identity

$$\frac{\lambda^k}{\lambda - z} = \frac{z^k}{\lambda - z} + \sum_{0 \leq j \leq k-1} z^{k-1-j} \lambda^j \tag{4.5}$$

with respect to the measure σ, we come to the equality

$$w_{\sigma,k}(z) = z^k \left(w_\sigma(z) + \sum_{0 \leq j \leq k-1} \frac{s_j(\sigma)}{z^{j+1}} \right) \qquad (k = 0, 1, 2, \ldots, 2n). \tag{4.6}$$

Since

$$w_{\sigma,2n}(z) = -\frac{s_{2n}(\sigma)}{z}(1 + o(1)) \qquad (|z| \to \infty,\ z = iy), \tag{4.7}$$

it follows from (4.6) (with $k = 2n$) that

$$z^{2n} \left(w_\sigma(z) + \sum_{0 \leq j \leq 2n-1} \frac{s_j(\sigma)}{z^{j+1}} \right) = -\frac{s_{2n}(\sigma)}{z}(1 + o(1)) \qquad (|z| \to \infty,\ z = iy). \tag{4.8}$$

The asymptotic relation (4.8), together with (4.2) and (4.6) suggests the following:

Given a function w of class $\mathfrak{R}(\mathbb{H})$ and a set of real numbers $s_0, s_1, \ldots, s_{2n-1}$, it ought to be profitable to consider the functions $b_{w,k}(z) = b_{w,k}(z; s_0, s_1, \ldots, s_{k-1})$:

$$b_{w,k}(z) = z^k w(z) + \sum_{0 \leq j \leq k-1} z^{k-1-j} s_j \qquad (k = 0, 1, 2, \ldots, 2n) \tag{4.9}$$

and the asymptotic relation of the form

$$|b_{w,k}(z)| = O(|z|^{-1}) \qquad (|z| \to \infty,\ z = iy). \tag{4.10}$$

In this notation the equality (4.6) means that

$$w_{\sigma,k}(z) = b_{w_\sigma,k}(z; s_0(\sigma), \ldots, s_{k-1}(\sigma)). \tag{4.11}$$

From (4.8) and (4.11) it follows that:

If a measure $\sigma \geq 0$ on $\mathbb{R}$ satisfies the moment conditions (4.2), then the asymptotic relation

$$|b_{w_\sigma,2n}(z; s_0, \ldots, s_{2n-1})| \leq \frac{s_{2n}}{|z|}(1 + o(1)) \qquad (|z| \to \infty, \ z = iy) \tag{4.12}$$

holds.

It is remarkable that the last statement can be inverted.

Theorem (Hamburger). *Let w be a function which belongs to the class $\mathfrak{R}(\mathbb{H})$ and let $s_0, s_1, \ldots, s_{2n-1}$ be real numbers. Assume that the function w satisfies the asymptotic condition*

$$|b_{w,2n}(z; s_0, \ldots, s_{2n-1})| = O(|z|^{-1}) \qquad (|z| \to \infty, \ z = iy) \tag{4.13}$$

(where $b_{w,2n}$ is defined in (4.9)). Then the function w has the representation of the form (4.3), with a finite nonnegative measure σ, which has a finite $2n$'th moment: $s_{2n}(\sigma) < \infty$. Moreover,

$$s_0(\sigma) = s_0, s_1(\sigma) = s_1, \ldots, s_{2n-1}(\sigma) = s_{2n-1}, \tag{4.14}$$

$$s_{2n}(\sigma) = \lim_{\substack{|z| \to \infty \\ z = iy}} (-z) b_{w,2n}(z; s_0, s_1, \ldots, s_{2n-1}) \tag{4.15}$$

This theorem was proved by Hamburger ([H], Theorem IX). It is reproduced in the monograph by N. Akhiezer ([A], Theorem 2.3.1). The proof which was presented by Hamburger is based on a "step by step" algorithm. Another proof of this theorem, and its far reaching generalizations, is presented in [K1].

Thus the Hamburger moment problem can be reformulated as the following interpolation problem:

Function class: *the class $\mathfrak{R}(\mathbb{H})$.*

Interpolation data: *a finite sequence $s_0, s_1, \ldots, s_{2n}$ of real numbers.*

The asymptotic relation

$$\left| z^{2n} \left(w(z) + \sum_{0 \leq j \leq 2n-1} \frac{s_j}{z^{j+1}} \right) \right| \leq \frac{s_{2n}}{|z|}(1 + o(1)) \qquad (|z| \to \infty, \ z = iy) \tag{4.16}$$

is considered as an interpolation condition. *(The point $z = \infty$ is a multiple interpolation node which lies on the boundary of the upper half-plane $\mathbb{H}$. Its multiplicity equals $2n$). We seek functions w from this class which satisfy the condition* (4.16).[2]

2) Strictly speaking, the considered problem has two interpolation nodes which are symmetric with respect to the real axis and are located at the points $+i \cdot \infty$ and $-i \cdot \infty$. The multiplicity of each of them equals n.

Remark 4.1. *i*). Assume that a function w from the class $\mathfrak{R}(\mathbb{H})$ satisfies the condition (4.13). Suppose that we also know (for example, from the Hamburger theorem), that $w = w_\sigma$, where $s_{2n}(\sigma) < \infty$. Then we can construct the function $w_{\sigma,2n}$ by (4.6). Comparing the asymptotics (4.13) and (4.7), we conclude, that $b_{w,2n} = w_{\sigma,2n}$. Hence, the moment condition (4.2.i) is satisfied, as well as the condition

$$\left| z^{2n} \left(w(z) + \sum_{0 \le j \le 2n-1} \frac{s_j}{z^{j+1}} \right) \right| \le \frac{s_{2n}(\sigma)}{|z|}(1 + o(1)) \qquad (\,|z| \to \infty,\; z = iy\,). \tag{4.17}$$

Moreover, the function $b_{w,2n}(z; s_0, \ldots, s_{2n-1})$ belongs to the class $\mathfrak{R}_0(\mathbb{H})$. (If $d\sigma(\lambda)$ is a measure which represents w, then the measure $\lambda^{2n} d\sigma(\lambda)$ represents the function $b_{w,2n}$.)

ii). Assume now that the function $b_{w,2n}(z; s_0, \ldots, s_{2n-1})$ belongs to the class $\mathfrak{R}_0(\mathbb{H})$. Then, by Nevanlinna's theorem, the function $b_{w,2n}$ has the form w_τ for some $d\tau \ge 0$ with $s_0(\tau) < \infty$. Thus,

$$\int_{\mathbb{R}} \frac{d\tau(\lambda)}{\lambda - z} = z^{2n} \int_{\mathbb{R}} \frac{d\sigma(\lambda)}{\lambda - z} + \sum_{0 \le j \le 2n-1} s_j z^{2n-1-j}.$$

Applying the generalized Stieltjes inversion formula ([KaKr],§2), we conclude that $d\tau(\lambda) = \lambda^{2n} d\sigma(\lambda)$. Hence, $\int_{\mathbb{R}} \lambda^{2n} d\sigma(\lambda) = \int_{\mathbb{R}} d\tau(\lambda) < \infty$. Thus, $b_{w,2n} = w_{\sigma,2n}$, and (4.17) is satisfied.

5 Derivation of the FMI ($\mathcal{H}$)

Given the Hamburger moment problem with data $s_0, s_1, \ldots, s_{2n}$, we associate with this problem *the Pick matrix*

$$A = \begin{bmatrix} s_0 & s_1 & \cdots & s_n \\ s_1 & s_2 & \cdots & s_{n+1} \\ \cdots & \cdots & \cdots & \cdots \\ s_{n-1} & s_n & \cdots & s_{2n-1} \\ s_n & s_{n+1} & \cdots & s_{2n} \end{bmatrix}, \tag{5.1}$$

and the vectors of the interpolation data

$$u = \begin{bmatrix} 1 \\ 0 \\ \vdots \\ 0 \\ 0 \end{bmatrix} \quad \text{and} \quad v = \begin{bmatrix} 0 \\ -s_0 \\ \vdots \\ -s_{n-2} \\ -s_{n-1} \end{bmatrix}. \tag{5.2}$$

The matrix, which is responsible for interpolation nodes (with multiplicity) is:

$$T = \left.\begin{bmatrix} 0 & 0 & \cdots & 0 & 0 & 0 \\ 1 & 0 & \cdots & 0 & 0 & 0 \\ 0 & 1 & \cdots & 0 & 0 & 0 \\ \cdots & \cdots & \cdots & \cdots & \cdots & \cdots \\ 0 & 0 & \cdots & 0 & 0 & 0 \\ 0 & 0 & \cdots & 1 & 0 & 0 \\ 0 & 0 & \cdots & 0 & 1 & 0 \end{bmatrix}\right\} {\scriptstyle (n+1)} \cdot \tag{5.3}$$

The Fundamental Identity (1.12) for this choice of u, v, T and A can be checked straightforwardly.

Now we derive the Fundamental Matrix Inequality for the Hamburger Moment Problem (FMI$(\mathcal{H})$). Let $d\sigma(\lambda)$ be a nonnegative measure on $\mathbb{R}$ for which the $2n$'th moment is finite: $s_{2n}(\sigma) < \infty$. The following inequality is clear:

$$\int_{\mathbb{R}} \begin{bmatrix} (I-\lambda T)^{-1}u \\ \cdot - \cdot - \cdot - \cdot \\ (\bar{\lambda}-\bar{z})^{-1} \end{bmatrix} \cdot d\sigma(\lambda) \cdot \begin{bmatrix} u^*(I-\bar{\lambda}T^*)^{-1} & \vdots & (\lambda - z)^{-1} \end{bmatrix} \geq 0. \tag{5.4}$$

This inequality has the form

$$\begin{bmatrix} A_\sigma & B_\sigma(z) \\ B_\sigma^*(z) & \dfrac{w_\sigma(z) - w_\sigma^*(z)}{z - \bar{z}} \end{bmatrix} \geq 0, \tag{5.5}$$

where the function w_σ is defined by (4.3). It is clear that

$$A_\sigma = \int_{\mathbb{R}} (I-\lambda T)^{-1}u \cdot d\sigma(\lambda) \cdot u^*(I-\lambda T^*)^{-1}, \tag{5.6}$$

where

$$A_\sigma = \|a_{\sigma,kl}\|_{0\leq k.l\leq n}, \qquad a_{\sigma,kl} = s_{k+l}(\sigma) \quad (0 \leq k,\, l \leq n). \tag{5.7}$$

It is also clear, that

$$B_\sigma(z) = \int_{\mathbb{R}} \frac{(I-\lambda T)^{-1}u}{\lambda - z}\, d\sigma(\lambda). \tag{5.8}$$

Since

$$\frac{(I-\lambda T)^{-1}}{\lambda - z} = (I - zT)^{-1}\left(\frac{1}{\lambda - z} + T(I-\lambda T)^{-1}\right), \tag{5.9}$$

it follows that

$$B_\sigma(z) = (I-zT)^{-1}\big(u \cdot w_\sigma(z) - v_\sigma\big), \tag{5.10}$$

where

$$v_\sigma = -\int_{\mathbb{R}} T(I-\lambda T)^{-1}u\, d\sigma(\lambda). \tag{5.11}$$

From the concrete expressions (5.2) and (5.3) for u and T it is not difficult to see that

$$v_\sigma = \begin{bmatrix} 0 \\ -s_0(\sigma) \\ \vdots \\ -s_{n-2}(\sigma) \\ -s_{n-1}(\sigma) \end{bmatrix}. \tag{5.12}$$

Assume now, that the measure σ satisfies the moment conditions (4.2). Then, according to (5.2) and (5.12), $v_\sigma = v$, and according to (5.1) and (5.7), $a_{\sigma,kl} = a_{kl}$ $(0 \le k+l < 2n$, $a_{\sigma,nn} \le a_{nn}$; hence, $A_\sigma \le A$. Thus, we obtain

Theorem 5.1 (from the moment conditions to the FMI $(\mathcal{H})$). *Let interpolation data for the Hamburger moment problem be given. Let w be a function of the form* (4.3), *where the measure σ satisfies the moment conditions* (4.2) *(or, what is the same according to Hamburger, the interpolation condition* (4.16) *is satisfied). Then the* FMI$(\mathcal{H})$ (1.6) *holds for this function w at every point $z \in \mathbb{C} \setminus \mathbb{R}$, where A is defined by* (5.1), *C_w is defined by* (1.8) *and B_w is defined by (1.10), (5.2), (5.3).*

6 Transformation of the FMI $(\mathcal{H})$

Let $s_0, \ldots, s_{2n}$ be interpolation data for the Hamburger moment problem. Then the Pick matrix A is defined by (5.1), the interpolation nodes matrix T is defined by (5.3) and the interpolation values vectors u and v are defined by (5.2). Given a function w, which is holomorphic in $\mathbb{C} \setminus \mathbb{R}$ and satisfies the symmetry conditions (1.3), assume that the FMI $(\mathcal{H})$

$$\begin{bmatrix} A & B_w(z) \\ B_w^*(z) & \dfrac{w(z) - w^*(z)}{z - \bar{z}} \end{bmatrix} \ge 0 \tag{6.1}$$

is satisfied for every $z \in \mathbb{C} \setminus \mathbb{R}$. Here B_w is defined by (1.10), (5.2) and (5.3), or in detail,

$$B_w(z) = \begin{bmatrix} 0 \\ b_{w,0}(z) \\ \cdots \\ b_{w,n-1}(z) \end{bmatrix}. \tag{6.2}$$

Our goal is to extract interpolation information from this FMI. Of course, from (6.1) it follows, that the function w satisfies the positivity condition (1.4). Proceeding in the same way, as in the Proof of Theorem 2.1, we have to consider the "subinequalities" (2.7) of the inequality (6.1). The most information which we can obtain in this way from (6.1) is contained in the subinequality

$$\begin{bmatrix} s_{2n} & b_{w,n-1} \\ b^*_{w,n-1} & \frac{w(z)-w^*(z)}{z-\bar{z}} \end{bmatrix} \geq 0. \tag{6.3}$$

First and foremost, from (6.3) we obtain the estimate (1.5) for w. By the Nevanlinna Theorem, the function w has the form w_σ for some nonnegative measure σ with $s_0(\sigma) < \infty$. Moreover, the estimate $|b_{w,n-1}(iy)| = O(|y|^{-1})$ as $y \uparrow \infty$ follows from (6.3). This is not enough since the function $b_{w,n-1}$ contains the interpolation data $s_0, s_1, \dots, s_{n-1}$ only, and does not contain the data $s_n, s_{n+1}, \dots, s_{2n-1}$ at all. We need to obtain the condition (4.16) from (6.1). Clearly, it is impossible to extract the condition (4.16) by considering "subinequalities" of the inequality (6.1). More generally, it is impossible to obtain (4.16) from any inequality of the form (1.13) when the framing matrix M does not depend on z because the data $s_n, s_{n+1}, \dots, s_{2n-1}, s_{2n}$ appear in the block A only, which does not depend on z.

Therefore, in order to extract (4.16) from (6.1) (if it is at all possible), we have to choose a matrix M in (1.13), which depends on z. To understand how to do this we return to the derivation of the FMI $(\mathcal{H})$. Let us consider the inequality (5.5). It contains the functions $w_{\sigma,k} = b_{w_{\sigma,k}}$ with $k = 0, 1, \dots, n-1$ only. However, we need the function $w_{\sigma,2n-1}$. The only information which is available for us is the block A_σ, which is defined by (5.6) and (5.7). The Hankel matrix A_σ is related to the Hankel matrix

$$W_\sigma(z) = \|w_{\sigma,kl}(z)\|_{0 \le k,l \le n}, \tag{6.4}$$

with entries

$$w_{\sigma,kl}(z) = \int_{\mathbb{R}} \lambda^k \cdot \frac{d\sigma(\lambda)}{\lambda - z} \cdot \lambda^l \qquad (0 \le k, l \le n). \tag{6.5}$$

The k,l-entries of the matrix W_σ with $k + l < n$ are the same functions which appear in the column B_σ. The entries with $n \le k + l \le 2n$ are exactly those which we need. Thus, the problem is to obtain the matrix W_σ from the matrix A_σ. According to (6.5), (5.2) and (5.3),

$$W_\sigma(z) = \int_{\mathbb{R}} (I - \lambda T)^{-1} u \cdot \frac{d\sigma(\lambda)}{\lambda - z} \cdot u^*(I - \lambda T^*). \tag{6.6}$$

Comparing (6.6) with (5.6) we see that we have to replace $(I - \lambda T)^{-1}$ with $\dfrac{(I - \lambda T)^{-1}}{\lambda - z}$ in (5.6). Let us turn to the identity (5.9):

$$T(I - zT)^{-1} \cdot (I - \lambda T)^{-1} u = \frac{(I - \lambda T)^{-1}}{\lambda - z} u - \frac{(I - zT)^{-1}}{\lambda - z} u. \tag{6.7}$$

From (6.6) and (6.7) it follows that

$$T(I - zT)^{-1} A_\sigma = W_\sigma(z) - (I - zT)^{-1} u \cdot \int_{\mathbb{R}} d\sigma(\lambda) \frac{u^*(I - \lambda T^*)^{-1}}{\lambda - z}. \tag{6.8}$$

Taking into account (5.8), we obtain the equality

$$W_\sigma(z) = T((I - zT)^{-1})\, A_\sigma + (I - zT)^{-1}u \cdot B^*_\sigma(\bar{z}). \tag{6.9}$$

The equality (6.9) provide us a heuristic reason for the following

Definition 6.1. *Given a Hermitian matrix A, a matrix T and vectors u and v, which satisfy the Fundamental Identity (1.12), we associate with each function w, which is holomorphic in $\mathbb{C} \setminus \mathbb{R}$ and satisfies the symmetry condition (1.3), the function W_w:*

$$W_w(z) = T(I - zT)^{-1}\, A + (I - zT)^{-1}u \cdot B^*_w(\bar{z}). \tag{6.10}$$

or, in detail,

$$W_w(z) = T(I - zT)^{-1}\, A - (I - zT)^{-1}u \cdot v^*(I - zT^*)^{-1}$$

$$+ (I - zT)^{-1}u \cdot w(z) \cdot u^*(I - zT^*)^{-1}. \tag{6.11}$$

Lemma 6.1. *The matrix function W_w satisfies the same symmetry condition as the function w:*

$$W_w(z) = W^*_w(\bar{z}) \qquad (z \in \mathbb{C} \setminus \mathbb{R}). \tag{6.12}$$

Straightforward calculation gives us an explicit expression for $W_w(z)$:

$$W_w(z) = \|b_{w,k+l}(z)\|_{0 \le k,l \le n}. \tag{6.13}$$

Thus, the matrix-function W_w is exactly what we need: it contains the function $b_{w,2n}$. In particular, from the formula it follows that the matrix $W_w(z)$ is a Hankel matrix. However, the Hankel structure of the matrix $W_w(z)$ can be obtained in a less special way, i.e., by using the FI (1.12) only:

Lemma 6.2. *The matrix $W_w(z)$ satisfies the identity*[3]

$$T\, W_w(z) - W_w(z)\, T^* = u \cdot \varphi^*_w(\bar{z}) - \varphi_w(z) \cdot u^*, \tag{6.14}$$

where

$$\varphi_w(z) = -T(I - zT)^{-1}\Big(u \cdot w(z) - v\Big). \tag{6.15}$$

Lemma 6.3. *For the Hamburger moment problem, the function $w(z)$ and the column $B_w(z)$ can be recovered from the matrix-function $W_w(z)$ in the following way:*

$$w(z) = e_0 \cdot W_w(z) \cdot e^*_0\,, \qquad B_w(z) = W_w(z) \cdot e^*_0\,, \tag{6.16}$$

where $e_0 = \begin{bmatrix} 1 & 0 & \cdots & 0 \end{bmatrix}$ is a $1 \times (n+1)$ vector.

3) The equality (6.14), considered as an equation with respect to the matrix $W_w(z)$, can be used to calculate this matrix.

Proof. The formulas in (6.16) follow from the equalities

$$e_0\,T = 0\,, \quad e_0\,u = 1 \quad \text{and} \quad e_0\,v = 0. \tag{6.17}$$

□

Remark 6.1. The proof of the lemma depends on the equalities (6.17), not on the FI (1.12). It is specific for the problem in question.

Let us turn to the FMI (6.1). It is clear that the matrix $W_w(\bar{z})$ appears in the product

$$\begin{bmatrix} T(I-\bar{z}T)^{-1} & \vdots & (I-\bar{z}T)^{-1}u \end{bmatrix} \cdot \begin{bmatrix} A & | & B_w(z) \\ -\cdot- & \cdot & -\cdot-\cdot-\cdot- \\ B_w^*(z) & | & \dfrac{w(z)-w^*(z)}{z-\bar{z}} \end{bmatrix}. \tag{6.18}$$

In order to to transform the FMI (6.1), we have to "frame" it according to (1.13), where now the matrix M depends on z. It is clear that the row

$$\begin{bmatrix} T(I-\bar{z}T^{-1}) & \vdots & (I-\bar{z}T)^{-1}u \end{bmatrix}$$

ought to be one of the rows of the matrix $M(z)$. There are two main possibilities. Either the mentioned row is the first row of the matrix M:

$$M_1(z) = \begin{bmatrix} T(I-\bar{z}T)^{-1} & (I-\bar{z}T)^{-1}u \\ 0 & 1 \end{bmatrix}, \tag{6.19}$$

or the mentioned row is the second row of the matrix M:

$$M_2(z) = \begin{bmatrix} I & 0 \\ T(I-\bar{z}T)^{-1} & (I-\bar{z}T)^{-1}u \end{bmatrix}. \tag{6.20}$$

Upon performing the matrix multiplications, we obtain (after some calculations with the matrix entries):

$$\begin{aligned} M_1(z) \cdot \begin{bmatrix} A & B_w(z) \\ & \\ B_w^*(z) & \dfrac{w(z)-w^*(z)}{z-\bar{z}} \end{bmatrix} \cdot M_1^*(z) \\ = \begin{bmatrix} \dfrac{W_w(z)-W_w^*(z)}{z-\bar{z}} & \dfrac{B_w(z)-B_w(\bar{z})}{z-\bar{z}} \\ & \\ \dfrac{B_w^*(\bar{z})-B_w^*(z)}{z-\bar{z}} & \dfrac{w(z)-w^*(z)}{z-\bar{z}} \end{bmatrix} \end{aligned} \tag{6.21}$$

and

$$M_2(z) \cdot \begin{bmatrix} A & B_w(z) \\ B_w^*(z) & \dfrac{w(z)-w^*(z)}{z-\bar{z}} \end{bmatrix} \cdot M_2^*(z) = \begin{bmatrix} A & W_w(z) \\ W_w^*(z) & \dfrac{W_w(z)-W_w^*(z)}{z-\bar{z}} \end{bmatrix}. \tag{6.22}$$

The calculations with the matrix entries are based essentially on the following consequence of the FI (1.12):

Lemma 6.4. *The identity*

$$\begin{aligned} &T(I-zT)^{-1} \cdot A \cdot (I-\bar{z}T^*)^{-1}T^* \\ &= \frac{T(I-zT)^{-1}A - A(I-\bar{z}T^*)^{-1}T^*}{z-\bar{z}} \\ &\quad - (I-zT)^{-1} \cdot \frac{uv^*-vu^*}{z-\bar{z}} \cdot (I-\bar{z}T^*)^{-1} \end{aligned} \tag{6.23}$$

holds.

7 On using the TFMI $(\mathcal{H})$ to obtain interpolation information from the FMI $(\mathcal{H})$

We consider two kinds of Transformed Fundamental Matrix Inequalities (for the Hamburger problem): $\text{TFMI}_{\text{I}}(\mathcal{H})$ and $\text{TFMI}_{\text{II}}(\mathcal{H})$.

The $\text{TFMI}_{\text{I}}(\mathcal{H})$ is of the form

$$\begin{bmatrix} \dfrac{W_w(z)-W_w^*(z)}{z-\bar{z}} & \dfrac{B_w(z)-B_w(\bar{z})}{z-\bar{z}} \\ \dfrac{B_w^*(\bar{z})-B_w^*(z)}{z-\bar{z}} & \dfrac{w(z)-w^*(z)}{z-\bar{z}} \end{bmatrix} \geq 0. \tag{7.1}$$

The $\text{TFMI}_{\text{II}}(\mathcal{H})$ is of the form

$$\begin{bmatrix} A & W_w(z) \\ W_w^*(z) & \dfrac{W_w(z)-W_w^*(z)}{z-\bar{z}} \end{bmatrix} \geq 0. \tag{7.2}$$

We see that both of the TFMI's contain the function $W_w(z)$. Now the problem of extracting interpolation information from the TFMI arises.

Now we will discuss the extent to which the FMI $(\mathcal{H})$ and the TFMI $(\mathcal{H})$ are equivalent. In view of (6.21) and (6.22), it is clear that

$$\text{FMI}\,(\mathcal{H}) \quad \Rightarrow \quad \text{TFMI}_{\text{I}}\,(\mathcal{H}) \tag{7.3}$$

and

$$\mathrm{FMI}\,(\mathcal{H}) \quad \Rightarrow \quad \mathrm{TFMI}_{\mathrm{II}}\,(\mathcal{H}). \tag{7.4}$$

More formally:

Lemma 7.1. *If the* FMI *($\mathcal{H}$) is satisfied for some $z \in \mathbb{C} \setminus \mathbb{R}$, then both* $\mathrm{TFMI}_{\mathrm{I}}\,(\mathcal{H})$ *and* $\mathrm{TFMI}_{\mathrm{II}}\,(\mathcal{H})$ *are satisfied for the same z as well.*

The opposite implications (with respect to (7.3), (7.4)) may be false, because the matrices $M_1(z)$ and $M_2(z)$ are not invertible: $e_0\,T = 0$, and the matrix $M_2(z)$ is not even square. Actually,

$$\mathrm{FMI}\,(\mathcal{H}) \quad \not\Rightarrow \quad \mathrm{TFMI}_{\mathrm{I}}\,(\mathcal{H}) \tag{7.5}$$

Indeed, the product in the left hand side does not contain the nn'th entry s_{2n} of the matrix A at all, and the positivity of the matrix A (and hence, the positivity of the matrix of the FMI ($\mathcal{H}$)) depends essentially on this entry. However, the FMI ($\mathcal{H}$) *and the* $\mathrm{TFMI}_{\mathrm{I}}$ *($\mathcal{H}$) are "almost equivalent"* : the matrix $M_1(z)$ (6.9) is "almost invertible". Since $T^*T = P$, where P is a projector matrix: $P = \mathrm{diag}\,[1, \ldots, 1, 0]$ ($p_{kk} = 1$, $k = 0, 1, \ldots, n-1$; $p_{nn} = 0$),

$$\begin{bmatrix} T^*(I - \bar{z}T)^{-1} & 0 \\ 0 & 1 \end{bmatrix} \cdot M_1(z) = \begin{bmatrix} P_{n-1} & 0 \\ 0 & 1 \end{bmatrix}. \tag{7.6}$$

Hence, *the inequality, which is obtained from the inequality (6.1) by replacing*[4] *the matrix A by the matrix PAP and the column $B_w(z)$ by the column $PB_w(z)$, holds.*

The inequalities FMI ($\mathcal{H}$) and $\mathrm{TFMI}_{\mathrm{II}}$ *($\mathcal{H}$) are equivalent,* because there exists a left inverse matrix to the matrix $M(z)$:

$$N(z) = \begin{bmatrix} I & 0 \\ 0 & e_0\,(I - \bar{z}T) \end{bmatrix}, \qquad N(z) \cdot M_2(z) = \begin{bmatrix} I & 0 \\ 0 & 1 \end{bmatrix}. \tag{7.7}$$

Thus, we have proved that

$$\mathrm{FMI}\,(\mathcal{H}) \quad \Leftrightarrow \quad \mathrm{TFMI}_{\mathrm{II}}\,(\mathcal{H}). \tag{7.8}$$

More formally:

Lemma 7.2. *The inequality* FMI ($\mathcal{H}$) *is satisfied at some point $z \in \mathbb{C} \setminus \mathbb{R}$ if and only if the inequality* $\mathrm{TFMI}_{\mathrm{II}}\,(\mathcal{H})$ *is satisfied for the same z.*

4) The last inequality is nothing more than the FMI of the form (6.1), which is constructed from the "truncated" date $s_0, s_1, \ldots, s_{n-2}$. (The FMI (6.1) is constructed from the data $s_0, s_1, \ldots, s_{2n}$.)

The matrix of the $\mathrm{TFMI_I}\,(\mathcal{H})$ is invariant with respect to the change $z \to \bar{z}$. Thus:

If the inequality $\mathrm{TFMI_I}\,(\mathcal{H})$ *is satisfied at some point* $z \in \mathbb{C} \setminus \mathbb{R}$, *than it is satisfied also at the conjugate point* $\bar{z}$.

The following statement is not so evident:

Lemma 7.3. *If the* FMI $(\mathcal{H})$ *is satisfied at some point* $z \in \mathbb{C} \setminus \mathbb{R}$, *than it is satisfied also at the conjugate point* $\bar{z}$.

Proof. The FMI $(\mathcal{H})$ can be written in the form

$$\begin{bmatrix} (I - zT)A(I - \bar{z}T^* & u \cdot w(z) - v \\ w^*(z) \cdot u^* - v^* & \dfrac{w(z) - w^*(z)}{z - \bar{z}} \end{bmatrix} \geq 0.$$

The claim of the lemma follows from the matrix identity

$$\begin{bmatrix} I & (\bar{z} - z)u \\ 0 & 1 \end{bmatrix} \begin{bmatrix} (I - zT)\,A\,(I - \bar{z}T^*) & u \cdot w - v \\ w^* \cdot u^* - v^* & \dfrac{w - w^*}{z - \bar{z}} \end{bmatrix} \begin{bmatrix} I & 0 \\ (z - \bar{z})u^* & 1 \end{bmatrix}$$
$$= \begin{bmatrix} (I - \bar{z}T)\,A\,(I - zT^*) & u \cdot w^* - v \\ w \cdot u^* - v^* & \dfrac{w - w^*}{z - \bar{z}} \end{bmatrix}, \tag{7.9}$$

where w is an arbitrary complex number; we have to put $w = w(z)$, then $w^* = w(\bar{z})$. To obtain the identity (7.9), we perform the matrix multiplication and use the identity

$$(I - \bar{z}T)\,A\,(I - zT^*) - (z - \bar{z})\,(u \cdot v^* - v \cdot u^*) = (I - zT)\,A\,(I - \bar{z}T^*), \tag{7.10}$$

which is equivalent to the Fundamental Identity (1.12). □

Now we turn to the extraction of interpolation information from the FMI $(\mathcal{H})$.

Theorem 7.1 (from the FMI $(\mathcal{H})$ to the moment conditions). *Let the interpolation data* $s_0, s_1, \ldots, s_{2n-1}, s_{2n}$ *for the Hamburger moment problem be given. Let* w *be a function of class* $\mathfrak{R}\,(\mathbb{H})$ *and let the* FMI $(\mathcal{H})$ (6.1) *for this* w *be satisfied at every point* z *in the upper half plane. Then the function* w *is representable in the form* $w = w_\sigma$ *for some (uniquely determined) measure* σ. *This measure satisfies the moment conditions (4.2); the interpolation conditions (4.16) are satisfied as well.*

Proof. According to Lemma 7.3, the FMI $(\mathcal{H})$ is satisfied for every $z \in \mathbb{C} \setminus \mathbb{R}$. By Lemma 7.2, the $\mathrm{TFMI_{II}}\,(\mathcal{H})$ is satisfied for every $z \in \mathbb{C} \setminus \mathbb{R}$. First, from the $\mathrm{TFMI_{II}}\,(\mathcal{H})$ we obtain the positivity condition

$$\frac{W_w(z) - W_w^*(z)}{z - \bar{z}} \geq 0 \qquad (\forall z \in \mathbb{C} \setminus \mathbb{R}). \tag{7.11}$$

Secondly, we derive the estimate

$$y\, W_w(iy) = O(1) \qquad (\text{as } \ y \uparrow \infty). \tag{7.12}$$

According to the matrix version of Nevanlinna's theorem, the matrix function $W_w(z)$ is representable in the form

$$W_w(z) = \int\limits_{\mathbb{R}} \frac{d\Sigma(\lambda)}{\lambda - z} \qquad (\forall z \in \mathbb{C} \setminus \mathbb{R}), \tag{7.13}$$

where $d\Sigma(\lambda)$ is a nonnegative matrix-valued measure and the integral

$$s_0\,(\Sigma) = \int\limits_{R} d\Sigma(\lambda) \tag{7.14}$$

exists in the proper sense. Moreover,

$$\lim_{y\,\uparrow\,\infty} -iy\, W_w(iy) = s_0\,(\Sigma). \tag{7.15}$$

From the $\mathrm{TFMI_{II}}\,(\mathcal{H})$ (7.2) (for $z = iy$, $y \to \infty$) and from (7.15) it now follows, that

$$A - s_0\,(\Sigma) \geq 0. \tag{7.16}$$

Of course, the condition (1.5) for w (see (6.16)) follows from the condition (7.12). Thus, $w = w_\sigma$ for some $\sigma :\ s_0\,(\sigma) < \infty$. Let us clarify the structure of the measure $d\Sigma$. We can expect that $W_w = W_\sigma$, and hence (see (6.6)) that

$$d\Sigma(\lambda) = (I - \lambda T)^{-1} u \cdot d\sigma(\lambda) \cdot u^* (I - \lambda T^*)^{-1}. \tag{7.17}$$

This is the case indeed. To prove (7.17), we turn to the formula (6.11). The functions $(I - zT)^{-1}$ and $(I - zT^*)^{-1}$ are holomorphic near the real axis (actually, these function are entire). Applying the generalized Sieltjes inversion formula ([KaKr], §2) to (6.11), we obtain (7.17). In particular (see (5.6) and (7.17)), the equality

$$s_0\,(\Sigma) = A_\sigma \tag{7.18}$$

holds. Now (7.16) takes the form

$$A - A_\sigma \geq 0. \tag{7.19}$$

The inequality (7.19) itself ensures the condition (4.2.ii), but it does not ensure the condition (4.2.i). However, we can also exploit the asymptotics (7.15). Taking into account the concrete structure (6.13) of the matrix-function W_w, we see that (7.15) together with (4.2.ii) leads to the condition (4.16). From (4.16) of course the moment condition (4.2.i) follows.

Another way to obtain these results is to multiply the equality (6.11) by $(I - zT)$ from the left and by $(I - zT^*)$ from the right and then upon comparing the asymptotics of both sides, we see that

$$T\,(A - A_\sigma)T^* = 0. \tag{7.20}$$

Thus, the nonnegative matrix $A - A_\sigma$ vanishes at all vectors in the image of the matrix T (applied to the right). The orthogonal complement to this image is generated by the $1 \times (n+1)$ vector

$$e_n = \begin{bmatrix} 0 & 0 & \cdots & 0 & 1 \end{bmatrix}. \tag{7.21}$$

Hence,

$$A = A_\sigma + \rho \cdot e_n^* e_n, \quad \text{where } \rho \text{ is a nonnegative number.} \tag{7.22}$$

In view of (5.1) and (5.7), the representation (7.22) is equivalent to the moment conditions (4.2). □

Remark 7.1. To obtain the estimate for the function $b_{w,2n}$, we could restrict ourself to the subinequality of the inequality (7.2):

$$\begin{bmatrix} s_{2n} & b_{w,2n}(z) \\ b^*_{w,2n}(z) & \dfrac{b_{w,2n}(z) - b^*_{w,2n}(z)}{z - \bar{z}} \end{bmatrix} \geq 0. \tag{7.23}$$

We can obtain this inequality from the inequality (7.2), by "framing" it with the matrix

$$\begin{bmatrix} 0 & \cdots & 0 & 1 & \vdots & 0 & \cdots & 0 & 0 \\ 0 & \cdots & 0 & 0 & \vdots & 0 & \cdots & 0 & 1 \end{bmatrix}.$$

Combining this with (6.22), we obtain the following "truncated" transformation:

$$m(z) \cdot \begin{bmatrix} A & B_w(z) \\ B^*_w(z) & \dfrac{w(z) - w^*(z)}{z - \bar{z}} \end{bmatrix} \cdot m^*(z) = \begin{bmatrix} s_{2n} & b_{w,2n}(z) \\ b^*_{w,2n}(z) & \dfrac{b_{w,2n}(z) - b^*_{w,2n}(z)}{z - \bar{z}} \end{bmatrix}, \tag{7.24}$$

where

$$m(z) = \begin{bmatrix} 0 & 0 & \cdots & 0 & 1 & \vdots & 0 \\ \bar{z}^{n-1} & \bar{z}^{n-2} & \cdots & 1 & 0 & \vdots & \bar{z}^n \end{bmatrix}. \tag{7.25}$$

A transformation of the FMI of approximately the form (7.25) appeared in the paper [Kov] by I. Kovalishina (see pages 460–461 of the Russian original or pages 424–425 of the English translation). (I. Kovalishina used a step by step algorithm, and did not introduce the matrix (7.25) explicitly, but it is possible to extract this matrix from her considerations.) Starting from[5] [Kov], the author considered transformations of the FMI for various problems on integral representations, both discrete and continuous in [K2]. The nontruncated transformation FMI $(\mathcal{H}) \rightarrow$ TFMI$_{\text{II}}$ $(\mathcal{H})$ was considered by the author in [K3]. Such a transformation was considered also by T. Ivanchenko and L. Sakhnovich [IS1], [IS2]. The nontruncated transformation FMI $(\mathcal{H}) \rightarrow$ TFMI$_{\text{I}}$ $(\mathcal{H})$ was considered (for other classes of functions and in different notation) in [KKY]. A systematic development of transformations of the FMI was also presented in the preprint [K4], but [K4] is not easily available.

8 Transformation of the FMI $(\mathcal{NP})$

It is very easy to extract interpolation information from the FMI $(\mathcal{NP})$. For this goal we need not transform the FMI. However, we have already learnd that such transformations and related structures are objects which are interesting in themselves. Therefore, we will discuss transformations of the FMI $(\mathcal{NP})$. (We know, that to a large extent such transformations depend only on the Fundamental Identity for the considered problem and not on the concret expression for the entries in this identity.) Thus, we consider a FMI of the form (1.6) with B_w and C_w of the forms (1.9), (2.2), (2.3) and (1.7), respectively, and we assume, that the Fundamental Identity (1.11) is satisfied.

Let the function w which appears in FMI $(\mathcal{NP})$ be of the form $w = w_{\sigma,c}$ as in (3.1). To guess formulas for transformations of the FMI, we first consider the matrix function

$$W_\sigma(z) = \int_{\mathbb{T}} (tI - T)^{-1} \cdot \frac{1}{2}\frac{t+z}{t-z}\, d\sigma(t) \cdot (\bar{t}I - T^*)^{-1}, \tag{8.1}$$

which is obtained by inserting the Schwarz kernel into the formula (3.4) for A_σ. We would like to obtain W_σ from A_σ. For this goal we use the identity

$$\frac{1}{2}\frac{T+zI}{T-zI}(tI-T)^{-1} = \frac{1}{2}\frac{t+z}{t-z}(tI-T)^{-1} + \frac{z}{z-t}(zI-T)^{-1}, \tag{8.2}$$

which was constructed with formulas (3.4) and (3.6) for A_σ and B_σ in mind. Now we multiply the identity (8.2) by $u \cdot d\sigma(t) \cdot u^*(\bar{t}I - T^*)^{-1}$ and integrate over $\mathbb{T}$. Taking into account (3.4) and (3.6), we obtain

$$W_\sigma(z) = \frac{1}{2}\frac{T+zI}{T-zI}A_\sigma - (zI-T)^{-1}u \cdot B^*_{\sigma,c}(1/\bar{z}). \tag{8.3}$$

5) The paper [Kov] was published in 1983, but the author was aware of its contents much earlier.

The last formula is a heuristic reason for the following

Definition 8.1. *Given a Hermitian matrix A, a matrix T and vectors u and v which satisfy the FI* (1.11), *we associate with each function w, which is holomorphic in $\mathbb{C} \setminus \mathbb{T}$ and satisfies the symmetry condition* (1.1), *the function W_w:*

$$W_w(z) = \frac{1}{2}\frac{T+zI}{T-zI}A - (zI-T)^{-1}u \cdot B^*_{w,c}(1/\bar{z}). \tag{8.4}$$

or, in detail,

$$\begin{aligned} W_w(z) = \frac{1}{2}\frac{T+zI}{T-zI}A + (zI-T)^{-1}u \cdot v^*\,(z^{-1}I-T^*)^{-1} \\ +(zI-T)^{-1}u \cdot w(z) \cdot u^*\,(z^{-1}I-T^*)^{-1}. \end{aligned} \tag{8.5}$$

Using the FI (1.11), we obtain also another representation for $W_w(z)$:

$$W_w(z) = \frac{1}{2}A\frac{I+zT}{I-zT} + B_w(z) \cdot u\frac{z}{I-zT^*}, \tag{8.6}$$

or, in detail,

$$\begin{aligned} W_w(z) = \frac{1}{2}A\frac{I+zT}{I-zT} - (zI-T)^{-1}v \cdot u^*\,(z^{-1}I-T^*)^{-1} \\ +(zI-T)^{-1}u \cdot w(z) \cdot u^*\,(z^{-1}I-T^*)^{-1}. \end{aligned} \tag{8.7}$$

In other words:

Lemma 8.1. *The matrix-function W_w satisfies the symmetry condition*

$$W_w(z) = -W_w^*(1/\bar{z}) \qquad (\forall z \in \mathbb{C} \setminus \mathbb{T}). \tag{8.8}$$

Using the FI (1.11), we obtain also the following result:

Lemma 8.2. *The matrix-function W_w satisfies the identity*

$$W_w(z) - TW_w(z)T^* = u \cdot \varphi_w^*(1/\bar{z}) - \varphi_w(z) \cdot u^*, \tag{8.9}$$

where

$$\varphi_w(z) = \frac{1}{2}\frac{T+zI}{T-zI}(u \cdot w(z) - v). \tag{8.10}$$

Remark 8.1. For $z = 0$, the expression on the left hand side of (8.9) is equal to $\frac{1}{2}(A - TAT^*)$, and the expression on the right hand side is equal to $\frac{1}{2}(u \cdot v^* + v \cdot u^*)$. Thus, the formula (8.9) is in some sense an analytic continuation of the FI (1.11).

Remark 8.2. The equality (8.9), considered as an equation with respect to the matrix $W_w(z)$, can be used to calculate this matrix.

Let us calculate the matrix $W_w(z)$ for the $\mathcal{NP}$ problem with data given by (2.2) and (2.3).

From the equation (8.9), we obtain the following formula:

$$W_w(z) = \frac{1}{2} \left\| \frac{\dfrac{z_k + z}{z_k - z}(w_k - w(z)) + \dfrac{1 + z\bar{z}_l}{1 - z\bar{z}_l}(w(z) + w_l^*)}{1 - z_k \bar{z}_l} \right\|_{1 \le k, l \le n} . \tag{8.11}$$

Let us introduce the matrices

$$M_1(z) = \begin{bmatrix} (I - \bar{z}T)^{-1} & \bar{z}(I - \bar{z}T)^{-1}u \\ 0 & 1 \end{bmatrix} \tag{8.12}$$

and

$$M_2(z) = \begin{bmatrix} I & 0 \\ (I - \bar{z}T)^{-1} & \bar{z}(I - \bar{z}T)^{-1}u \end{bmatrix} . \tag{8.13}$$

Performing the matrix multiplication, we obtain (after some calculations with the entries):

$$\begin{aligned} M_1(z) \cdot \begin{bmatrix} A & B_w(z) \\ B_w^*(z) & \dfrac{w(z) - w^*(z)}{z - \bar{z}} \end{bmatrix} \cdot M_1^*(z) \\ = \begin{bmatrix} \dfrac{W_w(z) + W_w^*(z)}{1 - z\bar{z}} & \dfrac{B_w(z) - B_w(1/\bar{z})}{1 - z\bar{z}} \\ \dfrac{B_w^*(z) - B_w^*(1/\bar{z})}{1 - z\bar{z}} & \dfrac{w(z) + w^*(z)}{1 - z\bar{z}} \end{bmatrix} \end{aligned} \tag{8.14}$$

and

$$\begin{aligned} M_2(z) \cdot \begin{bmatrix} A & B_w(z) \\ B_w^*(z) & \dfrac{w(z) - w^*(z)}{z - \bar{z}} \end{bmatrix} \cdot M_2^*(z) \\ = \begin{bmatrix} A & W_w(z) + \dfrac{A}{2} \\ W_w^*(z) + \dfrac{A}{2} & \dfrac{W_w(z) + W_w^*(z)}{1 - z\bar{z}} \end{bmatrix} . \end{aligned} \tag{8.15}$$

The calculations mentioned above are based essentially on the following consequence of the FI (1.11):

$$\begin{aligned}
&(z-T)^{-1}A(\bar{z}-T^*)^{-1}\\
&\quad=\frac{1}{1-z\bar{z}}\left(\frac{1}{2}\frac{T+zI}{T-zI}A+\frac{1}{2}A\frac{T^*+\bar{z}I}{T^*-\bar{z}I}\right)\\
&\quad+(zI-T)^{-1}\cdot\frac{u\,v^*+v\,u^*}{1-z\bar{z}}\cdot(\bar{z}I-T^*)^{-1}.
\end{aligned}\tag{8.16}$$

We consider two variants of the Transformed Fundamental Matrix Inequality (for the Nevanlinna-Pick problem): the The $\mathrm{TFMI_I}(\mathcal{NP})$ and the TFMI $_{\mathrm{II}}(\mathcal{NP})$.

The $\mathrm{TFMI_I}(\mathcal{NP})$ has the form

$$\begin{bmatrix}
\dfrac{W_w(z)+W_w^*(z)}{1-z\bar{z}} & \dfrac{B_w(z)-B_w(1/\bar{z})}{1-z\bar{z}}\\[2ex]
\dfrac{B_w^*(z)-B_w^*(1/\bar{z})}{1-z\bar{z}} & \dfrac{w(z)+w^*(z)}{1-z\bar{z}}
\end{bmatrix}\geq 0.\tag{8.17}$$

The $\mathrm{TFMI_{II}}(\mathcal{NP})$ has the form

$$\begin{bmatrix}
A & W_w(z)+\dfrac{A}{2}\\[2ex]
W_w^*(z)+\dfrac{A}{2} & \dfrac{W_w(z)+W_w^*(z)}{1-z\bar{z}}
\end{bmatrix}\geq 0.\tag{8.18}$$

We see that both of these TFMI's contain the function $W_w(z)$.

Definition 8.2. *Given a $\mathcal{NP}$ problem with interpolation nodes $z_1, z_2, \ldots, z_n$ in the unit disc $\mathbb{D}$, the point $z\in\mathbb{C}\setminus\mathbb{T}$ is said to be* nonsingular, *if $z\neq 0, \infty$; z_1, z_2, …, z_n; $\bar{z}_1^{-1}$, $\bar{z}_2^{-1}$, …, $\bar{z}_n^{-1}$.*

If z is a nonsingular point, then the matrices $(zI-T)^{-1}$, $(I-\bar{z}T)^{-1}$ are defined (and, of course, invertible). (Strictly speaking, we can define the matrices $W_w(z)$, $M_1(z)$ and $M_2(z)$ for nonsingular z only). *For nonsingular z, the matrix $M_1(z)$ is invertible and the matrix $M_2(z)$ has a left inverse.*

Lemma 8.3. *Let $z\in\mathbb{C}\setminus\mathbb{T}$ be a nonsingular point. Then the FMI $(\mathcal{NP})$ is satisfied at this point z if and only if each of two inequalities $\mathrm{TFMI_I}(\mathcal{H})$ and $\mathrm{TFMI_{II}}(\mathcal{H})$ is satisfied at this point.*

Lemma 8.4. *Let $z\in\mathbb{C}\setminus\mathbb{T}$ be a nonsingular point. Then the FMI $(\mathcal{NP})$ is satisfied at this point z if and only if it is satisfied at the "symmetric" point $\bar{z}^{-1}$ as well.*

Proof. The FMI $(\mathcal{NP})$ is equivalent to the inequality

$$\begin{bmatrix} (zI-T)\,A(\bar{z}I-T^*) & u\cdot w(z)-v \\ w^*(z)\cdot u^*-v^* & \dfrac{w_w(z)+w_w^*(z)}{1-z\bar{z}} \end{bmatrix} \geq 0. \tag{8.19}$$

The claim of the lemma follows from the matrix identity

$$\begin{bmatrix} I & -(1-z\bar{z})u \\ 0 & 1 \end{bmatrix} \begin{bmatrix} (I-zT)A(I-\bar{z}T^*) & u\cdot w-v \\ w^*\cdot u^*-v^* & \dfrac{w(z)+w^*(z)}{1-z\bar{z}} \end{bmatrix} \begin{bmatrix} I & 0 \\ -(1-z\bar{z})u^* & 1 \end{bmatrix}$$
$$= \begin{bmatrix} (I-\bar{z}T)A(I-zT^*) & u\cdot w^*-v \\ w\cdot u^*-v^* & \dfrac{w(z)+w^*(z)}{1-z\bar{z}} \end{bmatrix}, \tag{8.20}$$

where $w = w(z)$ and $w^* = -w(1/\bar{z})$. To obtain the identity (8.20), we perform the matrix multiplication and use the identity

$$(zI-T)\,A\,(\bar{z}I-T^*) - (1-z\bar{z})\,(u\cdot v^* + v\cdot u^*) = (I-\bar{z}T)\,A\,(I-zT^*), \tag{8.21}$$

which is equivalent to the FI (1.11). □

Lemma 8.5. *The* $\mathrm{TFMI}_{\mathrm{II}}(\mathcal{NP})$ *(8.18) holds for every point* $z \in \mathbb{D}$ *if and only if the function* $W_w(z)$ *satisfies the positivity condition:*

$$W_w(z) + W_w^*(z) \geq 0 \qquad (\forall z \in \mathbb{D}). \tag{8.22}$$

Proof. The implication $\mathrm{TFMI}_{\mathrm{II}} \Rightarrow$ (8.22) is evident. The opposite implication is nothing more that the Schwarz-Pick inequality for the function $W_w(z)$ for the points: 0 and z (because $W_w(0) = 2^{-1}A$). □

From Lemmas 8.3 and 8.5 we obtain the following conclusion:

Theorem 8.1. *A function* w*, holomorphic in* $\mathbb{C}\setminus\mathbb{T}$ *and satisfying the symmetry condition (1.1), satisfies the FMI* $(\mathcal{NP})$ *for all* $z \in \mathbb{D}$ *(or, what is the same, for all* $z \in \mathbb{C}\setminus\mathbb{T}$*) if and only if the function* $W_w(z)$ *which is defined by (8.4) satisfies the positivity condition (8.22).*

Taking into account the concrete form (8.11) of the matrix W for the $\mathcal{NP}$ problem, we obtain:

Theorem 8.2. *Let the interpolation data for the* $\mathcal{NP}$ *problem (2.1) in the class* $\mathfrak{C}\,(\mathbb{D})$ *be given by (2.2) and (2.3). A function* w*, which is holomorphic in* $\mathbb{D}$*, is a solution of the* $\mathcal{NP}$ *problem (with these data) if and only if the real part of the matrix on the right hand side of (8.11) is nonnegative for every* $z \in \mathbb{D}$*.*

Remark 8.3. The matrix in (8.11) is an orthogonal projection of the operator $\frac{1}{2}(I + zU)(I - zU)^{-1}$, where U is a generalised unitary extension of some isometric operator, related to the considered problem.

This is a consequence of the TFMI$_{\mathrm{II}}(\mathcal{NP})$. A consecuence of the TFMI$_{\mathrm{I}}(\mathcal{NP})$ also may be interesting. The inequality (8.17) is equivalent to the inequality

$$\begin{bmatrix} W_w(z) + W_w^*(z) & B_w(z) - B_w(1/\bar{z}) \\ B_w^*(z) - B_w^*(1/\bar{z}) & w(z) + w^*(z) \end{bmatrix} \geq 0 \qquad (\forall z \in \mathbb{D}) . \tag{8.23}$$

The matrix function on the left hand side of (8.23) is harmonic and nonnegative in $\mathbb{D}$ and hence it admits a Riesz-Herglotz representation. Let

$$\begin{bmatrix} d\Sigma(t) & d\mu(t) \\ d\mu^*(t) & d\sigma(t) \end{bmatrix} \tag{8.24}$$

be the block decomposition of the representing measure.

Now we can apply Šmul'yan's results from[6] [S], to obtain the inequality

$$\int_{\mathbb{T}} d\mu(t)\,(d\sigma(t))^{-1}\,d\mu^*(t) \leq \int_{\mathbb{T}} d\Sigma(t)\,, \tag{8.25}$$

where the integral on the left hand side is the so called *Operator Hellinger Integral.* Because

$$W_w(0) + W_w^*(0)) = A, \quad \text{it follows that} \int_{\mathbb{T}} d\Sigma = A.$$

Thus

$$\int_{\mathbb{T}} d\mu(t)\,(d\sigma(t))^{-1}\,d\mu^*(t) \leq A\,. \tag{8.26}$$

It is not difficult to show that in the case under consideration (the $\mathcal{NP}$ problem with finitely many interpolation nodes located inside $\mathbb{D}$) equality holds in (8.26). In the general situation, A is a nonnegative Hermitian form in some vector space. Then, the TFMI$_{\mathrm{I}}(\mathcal{NP})$ leads to the representation of a nonnegative Hermitian form by the Hellinger Integral. It is worth mentioning that it was the Hellinger integral, which was used for the integral representation of Hermitian kernels early in the development of the theory. In more recent times, the Stieltjes integral ousted the Hellinger integral from this circle of problems. However, the use of the Stieltjes integral leads to difficulties. It may not exist, and we have to use rigged Hilbert spaces and all that. The Hellinger integral always exists (and under some conditions it may be reduced to the Stieltjes integral). But in our opinion, the use of the Hellinger integral lies at the heart of the matter. The moral is clear:

$\Longrightarrow$ GO BACK TO THE CLASSICS. $\Longleftarrow$

6) The paper [S] by Yu.L. Šmul'yan looks as if it was written especially to be used in this paper.

References

[A] Akhiezer, N.I., *The Classical Moment Problem.* (Russian) Moscow: Fizmatgiz 1961. English translation: Edinburg and London: Oliver & Boyd 1965.

[H] Hamburger, H., *Über eine Erweiterung des Stieltjesschen Momentenproblems. I.* (German) Math. Annalen **81**:3 (1920), 235–319.

[IS1] Ivanchenko, T.S.; Sakhnovich, L.A., *Operator identities in the theory of interpolation problems.* (Russian) Izv. Akad. Nauk Armyan. SSR Ser. Mat., Ser. Mat. **22**:3 (1987), 298–308, Engl. translation: Soviet. J. Contemporary Math. Anal. **22**:3 (1987), 84–94.

[IS2] Ivanchenko, T.S.; Sakhnovich, L.A., *An operator approach to V.P. Potapov's scheme for the investigation of interpolation problems.* (Russian), Ukrain. Mat. Zh. **39**:5 (1987), 573–578. Engl. translation: Ukrainian Math. J. **39**:5 (1987), 464–469.

[KaKr] Kac, I.S. and M.G. Krein, *R-functions – analytic functions mapping the upper half-plane into itself*, Supplement I to the Russian transl. of F.V. Atkinson, *Discrete and Continuous Boundary Problems*, Moscow: Mir 1968, 629–647. English translation: Amer. Math. Soc. Transl. (**2**), **103**, 1973, pp. 1–18, 99–102.

[K1] Katsnelson, V., *Continuous analogues of the Hamburger-Nevanlinna theorem and fundamental matrix inequalities for classical problems. III.* (Russian) Teoriya Funktsiĭ, Funktsional'nyĭ Analiz i Ikh Prilozheniya, **39**, (1983), 61–73. English translation: Amer. Math. Soc. Transl. (**2**) **136** (1987), 85–96.

[K2] Katsnelson, V., *Continuous analogues of the Hamburger-Nevanlinna theorem and fundamental matrix inequalities for classical problems. IV.* (Russian), Teoriya Funktsiĭ, Funktsional'nyĭ Analiz i Ikh Prilozheniya, **40**, (1983), 79–90. English translation: Amer. Math. Soc. Transl. (**2**) **136** (1987), 97–108.

[K3] Katsnelson, V., *An integral representation of hermitian positive kernels of mixed type and the generalized Nehari problem.I.* (Russian), Teoriya Funktsiĭ, Funktsional'nyĭ Analiz i Ikh Prilozheniya, **43**, (1985), 54–70. English translation: Journ. of Soviet. Math. **48**:2 (1990), 162–176.

[K4] Katsnelson, V., The fundamental matrix inequality of the problem of the decomposition of a positive definite kernel into elementary kernels. (Russian), Deposited in UkrNIINTI. 10.7.1984. No. 1184 Uk Dep.

[KKY] Katsnelson, V.; A. Kheifets and P. Yuditskii, *An abstract interpolation problem and the extension theory of isometric operators.* (Russian), in: *Operators in Function Spaces and Problems in Function Theory*, Kiev: Naukova Dumka 1987 (V.A. Marchenko – editor), 83–96. English translation – this Volume.

[Kov] Kovalishina, I.V., *Analytic theory of a class of interpolation problems.* (Russian), Izvestiya Akad. Nauk SSR Ser. Mat. **47**:3 (1983), 455–497. Engl. translation: Math. USSR Izvestiya **22**:3 (1984), 419–463.

[S] Šmul'yan, Yu.L., *A Hellinger operator integral.* (Russian), Matem. Sbornik **47**((**91**):4 (1959), 381–430. Engl. translation: Amer. Math. Soc. Transl., (**2**) **22** (1962), 289–337.

[W] Weyl, H., *Singuläre Integralgleichungen.* (German), Math. Annalen (1908), 273–324. *Reprinted in:* Weyl, H. *Gesammelte Abhandlungen. Band 1.* Berlin·Heidelberg·New-York: Springer-Verlag 1968, 102–153.

Victor Katsnelson
Department of Theoretical Mathematics
The Weizmann Institute of Science
Rehovot, 76100, Israel
E-mail: katze@wisdom.weizmann.ac.il

AMS Subject Classification: 30D50, 46E10.

Operator Theory
Advances and Applications, Vol. 95
© 1997 Birkhäuser Verlag Basel/Switzerland

An abstract interpolation problem and the extension theory of isometric operators*

V.E. Katsnelson, A.Ya. Kheifets and P.M. Yuditskii

Abstract. The algebraic structure of V.P. Potapov's Fundamental Matrix Inequality (FMI) is discussed and its interpolation meaning is analyzed. Functional model spaces are involved. A general Abstract Interpolation Problem is formulated which seems to cover all the classical and recent problems in the field and the solution set of this problem is described using the Arov-Grossman formula. The extension theory of isometric operators is the proper language for treating interpolation problems of this type.

1 Interpolation data and examples

Let $\mathbb{L}$ and $\mathbb{L}'$ be Hilbert spaces. We shall denote by $B(\mathbb{L}, \mathbb{L}')$ the class of operator-valued functions which are holomorphic in the unit disk $|\zeta| < 1$ and whose values are contractive operators from $\mathbb{L}$ into $\mathbb{L}'$. Let $\mathbb{X}$ be a linear space. We do not suppose that $\mathbb{X}$ is endowed with a topological structure. Let D be a sesquilinear form in $\mathbb{X}$ and let T be a linear operator on $\mathbb{X}$. Let E and M be linear operators from $\mathbb{X}$ into $\mathbb{L}$ and $\mathbb{L}'$, respectively.

We assume that the operators and the sesquilinear form are linked through the so-called *Fundamental Identity* (F I):

$$D(x,y) - D(Tx,Ty) = \langle Ex, Ey\rangle_{\mathbb{N}} - \langle Mx, My\rangle_{\mathbb{N}'} \ . \qquad (FI)$$

The Fundamental Identity must be fulfilled for arbitrary x and y in $\mathbb{X}$. This framework arose from the study of a number of interpolation problems. We shall give some examples.

Example 1. (The Nevanlinna-Pick Problem.)
Interpolation Data: A sequence $\{\zeta_k\}_{1\le k\le\infty}$ of complex numbers ($|\zeta_k| < 1$) and a sequence of contractive operators $\{s_k\}_{1\le k\le\infty}$, acting from $\mathbb{L}$ into $\mathbb{L}'$.
A solution $s(\zeta)$ of this problem is an arbitrary holomorphic function from the class $B(\mathbb{L}, \mathbb{L}')$ which satisfies the interpolation conditions $s(\zeta_k) = s_k$. *It is required* to

*) Translated from "Operators in Function Spaces and Problems in Function Theory", pp. 83–96. (Naukova Dumka, Kiev, 1987, Edited by V.A. Marchenko.)

give a criteria for the existence of solutions and to describe the set of all solutions of the interpolation problem.

In this example, $\mathbb{X}$ is the space of all infinite sequences whose entries are vectors from $\mathbb{L}$ such that only a finite number of them do not vanish:

$$x = \{\ell_1, \ell_2, \cdots, \ell_n, 0, \cdots, 0, \cdots\} .$$

The operators T, E, M and the sesquilinear form D are defined by the formulas

$$Tx = \{\zeta_1\ell_1, \zeta_2\ell_2, \cdots, \zeta_n\ell_n, 0, \cdots, 0, \cdots\} ,$$

$$Ex = \sum_{1\le k<\infty} \ell_k , \quad Mx = \sum_{1\le k<\infty} s_k\ell_k ,$$

$$D(x,x) = \sum_{1\le j,k<\infty} \left\langle \frac{I - s_j^* s_k}{1-\bar{\zeta}_j\zeta_k} \ell_k, \ell_j \right\rangle_{\mathbb{N}} .$$

The Fundamental Identity can be verified directly.

The condition $D(x,x) \ge 0$ ($\forall\ x \in \mathbb{X}$) is necessary and sufficient for the solvability of this problem.

Example 2. (The Sarason Problem.) Let θ be an inner function, let $\mathbb{K}_\theta := H^2 \ominus \theta H^2$, and let P_θ be the orthoprojection from H^2 onto $\mathbb{K}_\theta$.
The interpolation data for this problem are the operator $T = P_\theta t_{|\mathbb{K}_\theta}$ i.e., the compressed shift on $\mathbb{K}_\theta$, and a contractive operator W which commutes with T.
A solution of this problem is an arbitrary holomorphic function $w(\zeta)$ from the class $B(\mathbb{L}, \mathbb{L}')$ which satisfies the *interpolation condition* $W = P_\theta w|_{\mathbb{K}_\theta}$. *It is required* to describe the set of all solutions to this problem. The existence of solutions follows from the contractivity of W.

In this example $\mathbb{X}$ coincides with $\mathbb{K}_\theta$, the operator T is defined in the statement of the problem, $\mathbb{L} = \mathbb{L}' = \mathbb{C}$,

$$Ex \stackrel{\text{def}}{=} \langle x, e_* \rangle , \quad Mx \stackrel{\text{def}}{=} \langle Wx, e_* \rangle$$

and

$$D(x,x) \stackrel{\text{def}}{=} \langle (I - W^*W)x, x \rangle ,$$

where $\langle\ ,\ \rangle$ is the inner product in $\mathbb{K}_\theta$, the vector $e_* \in \mathbb{K}_\theta$ is defined by the formula $e_* = [\theta(t) - \theta(0)]/t$. The Fundamental Identity is a consequence of the fact that

$$(I - T^*T)x = e_* \langle x, e_* \rangle$$

for any $x \in \mathbb{K}_\theta$.

2 V.P. Potapov's fundamental matrix inequality and its transformation

We shall say that a function $s \in B(\mathbb{L}, \mathbb{L}')$ satisfies V.P. Potapov's Fundamental Matrix Inequality (the FMI) if for arbitrary $x \in \mathbb{X}$, $\ell \in \mathbb{L}$ and $|\zeta| < 1$,

$$\begin{bmatrix} D((I-T\bar{\zeta})x,(I-T\bar{\zeta})x) & \langle \ell,(E-s^*(\zeta)M)x\rangle_{\mathbb{N}} \\ \langle (E-s^*(\zeta)M)x,\ell\rangle_{\mathbb{N}} & \langle \frac{I_{\mathbb{N}}-s^*(\zeta)s(\zeta)}{1-\bar{\zeta}\zeta}\,\ell,\ell\rangle_{\mathbb{N}} \end{bmatrix} \geq 0\,. \qquad (FMI)$$

This inequality can be rewritten in another (equivalent) form:

$$\begin{bmatrix} D((\zeta I-T)x,(\zeta I-T)x) & \langle \ell',(s(\zeta)E-M)x,\ell'\rangle_{\mathbb{N}'} \\ \langle (s(\zeta)E-M)x,\ell'\rangle_{\mathbb{N}'} & \langle \frac{I_{\mathbb{N}'}-s(\zeta)s^*(\zeta)}{1-\zeta\bar{\zeta}}\,\ell',\ell'\rangle_{\mathbb{N}'} \end{bmatrix} \geq 0 \qquad (FMI')$$

for arbitrary $x \in \mathbb{X}$, $\ell' \in \mathbb{L}'$ and $|\zeta| < 1$.

It will be shown, that the equivalence of the FMI and the FMI$'$ is a consequence of the Fundamental Identity .

In the sequel we shall use several facts which we formulate here as propositions without proof.

Proposition 1 (The block-matrix lemma). *Let $\mathbb{H}$ be a Hilbert space and let A be a selfadjoint positive semidefinite ($A \geq 0$) operator acting on $\mathbb{H}$. Suppose that $h_0 \in \mathbb{H}$ and that there exists a constant $C \geq 0$ such that*

$$|\langle h_0, h\rangle|^2 \leq C\|\sqrt{A}h\|^2\,, \qquad \forall\, h \in D_{\sqrt{A}}\,,$$

where $D_{\sqrt{A}}$ is the domain of the selfadjoint operator $\sqrt{A}$. Then there exists a unique vector $g_0 \in (Ker A)^{\perp} \cap D_{\sqrt{A}}$ such that $\sqrt{A}g_0 = h_0$. Moreover,

$$\|g_0\|^2 \leq C.$$

We shall use the following notations:

$$g_0 = A^{[-\frac{1}{2}]}h_0 \quad \text{and} \quad \langle g_0, g_0\rangle \ = \langle A^{[-1]}h_0, h_0\rangle.$$

Remark. The converse statement is obvious: If $A \geq 0$, $g_0 \in D_{\sqrt{A}}$ and

$$\|g_0\|^2 \leq C,$$

then

$$|\langle \sqrt{A}g_0, h\rangle|^2 \leq C\|\sqrt{A}h\|^2$$

for arbitrary $h \in D_{\sqrt{A}}$.

Proposition 2. *Let s be a contractive operator from $\mathbb{L}$ into $\mathbb{L}'$, then*

$$s(I_{\mathbb{N}} - s^*s)^{[-\frac{1}{2}]} = (I_{\mathbb{N}'} - ss^*)^{[-\frac{1}{2}]}s$$

and

$$(I_{\mathbb{N}} - s^*s)^{[-\frac{1}{2}]}s^* = s^*(I_{\mathbb{N}'} - ss^*)^{[-\frac{1}{2}]} .$$

In particular, Proposition 2 contains the assertion that the domains of the operators on the right and left hand sides of these equalities coincide.

Proposition 3. *Let s be a contractive operator from $\mathbb{L}$ into $\mathbb{L}'$. Then the following statements are equivalent:*

(1) *$\ell' \oplus \ell$ belongs to the domain of the operator* $\begin{bmatrix} I_{\mathbb{N}'} & s \\ s^* & I_{\mathbb{N}} \end{bmatrix}^{[-\frac{1}{2}]}$.

(2) *$\ell - s^*\ell'$ belongs to the domain of the operator* $(I_{\mathbb{L}} - s^*s)^{[-\frac{1}{2}]}$.

(3) *$\ell' - s\ell$ belongs to the domain of the operator* $(I_{\mathbb{N}'} - ss^*)^{[-\frac{1}{2}]}$.

Moreover, if these properties are in force, then

$$\begin{aligned} &\langle \begin{bmatrix} I_{\mathbb{N}'} & s \\ s^* & I_{\mathbb{N}} \end{bmatrix}^{[-1]} (\ell' \oplus \ell), (\ell' \oplus \ell)\rangle_{\mathbb{N}'\oplus\mathbb{N}} \\ &= \langle \ell', \ell'\rangle_{\mathbb{N}'} + \langle (I_{\mathbb{N}} - s^*s)^{[-1]}(\ell - s^*\ell'), (\ell - s^*\ell')\rangle_{\mathbb{N}} \\ &= \langle \ell, \ell\rangle_{\mathbb{N}} + \langle (I_{\mathbb{N}'} - ss^*)^{[-1]}(\ell' - s\ell), (\ell' - s\ell)\rangle_{\mathbb{N}'} . \end{aligned}$$

The latter formulas inply the following:

Corollary. *The inequalities*

$$\langle \begin{bmatrix} I_{\mathbb{N}'} & s \\ s^* & I_{\mathbb{N}} \end{bmatrix}^{[-1]} (\ell' \oplus \ell), (\ell' \oplus \ell)\rangle_{\mathbb{N}'\oplus\mathbb{N}} \geq \langle \ell, \ell\rangle_{\mathbb{N}}$$

and

$$\langle \begin{bmatrix} I_{\mathbb{N}'} & s \\ s^* & I_{\mathbb{N}} \end{bmatrix}^{[-1]} (\ell' \oplus \ell), (\ell' \oplus \ell)\rangle_{\mathbb{N}'\oplus\mathbb{N}} \geq \langle \ell', \ell'\rangle_{\mathbb{N}'}$$

hold.

We are now ready to prove the equivalence of the FMI and the FMI$'$.

Proposition 4. *The FMI holds if and only if the FMI $'$ holds.*

Proof. In view of the block-matrix lemma, the FMI is equivalent to the following inequality

$$D((I - T\bar{\zeta})x, (I - T\bar{\zeta})x) \\ - \langle \left[\frac{I_{\mathbb{N}'} - s^*(\zeta)s(\zeta)}{1 - \bar{\zeta}\zeta} \right]^{[-1]} (E - s^*(\zeta)M)x, (E - s^*(\zeta)M)x \rangle \geq 0 \; .$$

Using Proposition 3 we get

$$D((I - T\bar{\zeta})x, (I - T\bar{\zeta})x) + (1 - \zeta\bar{\zeta})\{\langle Mx, Mx\rangle - \langle Ex, Ex\rangle\} \\ - \langle \left[\frac{I_{\mathbb{N}} - s(\zeta)s^*(\zeta)}{1 - \zeta\bar{\zeta}} \right]^{[-1]} (s(\zeta)E - M)x, (s(\zeta)E - M)x \rangle \geq 0 \; .$$

According to the Fundamental Identity

$$\langle Mx, Mx\rangle - \langle Ex, Ex\rangle \; = D(Tx, Tx) - D(x, x) \; .$$

Combining similar terms in the sesquilinear form D we obtain the inequality

$$D((\zeta I - T)x, (\zeta I - T)x) - \langle \left[\frac{I_{\mathbb{N}'} - s(\zeta)s^*(\zeta)}{1 - \zeta\bar{\zeta}} \right]^{[-1]} (s(\zeta)E - M)x, (s(\zeta)E - M)x \rangle \geq 0 \, ,$$

which is equivalent to the FMI because of the block-matrix lemma. Hence Proposition 4 is proved. □

Our aim now is to extract interpolation information from the FMI. We shall apply the method used earlier by one of the authors in [3].

We begin with an illustrative example: Let ζ_0 be an eigenvalue of T, let x_0 be the corresponding eigenvector, choose $\zeta = \zeta_0$ and $x = x_0$ in the FMI. Then

$$D((\zeta_0 I - T)x_0, (\zeta_0 I - T)x_0) = 0 \; ,$$

and hence $(s(\zeta_0)E - M)x_0 = 0$, that is $s(\zeta_0)Ex_0 = Mx_0$. Thus, the value $s(\zeta_0)Ex_0$ is the same for any solution $s(\zeta)$ of the FMI.

To formulate the theorem we need to define the Sz.-Nagy-Foias function space $\mathbb{K}_s$ (see [4], [5]).

For $s \in B(\mathbb{L}, \mathbb{L}')$ the space $\mathbb{K}_s$ is the set of all vector-valued functions $f = f_+ \oplus f_-$ which satisfy the following two conditions:

1) $f_+(\zeta) \in H^2_+(\mathbb{L}')$ and $f_-(\zeta) \in H^2_-(\mathbb{L})$, where $H^2_+(\mathbb{L}')$ and $H^2_-(\mathbb{L})$ are vector Hardy spaces of the disc $|\zeta| \leq 1$ with coefficients in the indicated space;

2) $\int_{\mathbb{T}} \langle \begin{bmatrix} I_{\mathbb{N}'} & s \\ s^* & I_{\mathbb{N}} \end{bmatrix}^{[-1]} f, f\rangle dm < \infty$, where $\mathbb{T}$ is the unit circle and dm is normalized Lebesgue measure on it.

This integral defines an inner product in $\mathbb{K}_s$. Sometimes it is convenient to use the following equivalent definition of the space $\mathbb{K}_s$ as the set of all vector valued functions $f = f_+ \oplus f_-$ such that

1′) $f_+(\zeta)$ is holomorphic in $|\zeta| < 1$, $f_-(\zeta)$ is anti-holomorphic in $|\zeta| < 1$, $f_-(0) = 0$,

2′) $\sup_{0<r<1} \int_{\mathbb{T}_r} \langle \begin{bmatrix} I_{\mathbb{N}'} & s \\ s^* & I_{\mathbb{N}} \end{bmatrix}^{[-1]} f, f\rangle dm < \infty$,

where $\mathbb{T}_r$ is the circle of radius r centered at the origin and dm is normalized Lebesgue measure on it.

Note that vector-functions from $\mathbb{K}_s$ have the following properties:

$$\langle \begin{bmatrix} I_{\mathbb{N}'} & s \\ s^* & I_{\mathbb{N}} \end{bmatrix}^{[-1]} f, f\rangle(rt) \to \langle \begin{bmatrix} I_{\mathbb{N}'} & s \\ s^* & I_{\mathbb{N}} \end{bmatrix}^{[-1]} f, f\rangle(t),$$

for almost all t, $|t| = 1$ (as $r \to 1$) and

$$\int_{\mathbb{T}_r} \langle \begin{bmatrix} I_{\mathbb{N}'} & s \\ s^* & I_{\mathbb{N}} \end{bmatrix}^{[-1]} f, f\rangle(dm) \to \int_{\mathbb{T}} \langle \begin{bmatrix} I_{\mathbb{N}'} & s \\ s^* & I_{\mathbb{N}} \end{bmatrix}^{[-1]} f, f\rangle dm ,$$

(as $r \to 1$). The interpolation sense of the FMI can be seen from the following:

Theorem 1. *Assume that the intersection of the spectrum of T with the set $|\zeta| \neq 1$ consists of isolated points only, that the vector-valued functions $E((\zeta I - T)^{-1}x)$ and $M((\zeta I - T)^{-1}x)$ are holomorphic for $|\zeta| \neq 1$ (where they are defined) and the function $D(x, \frac{I+T\overline{\zeta}}{I-T\zeta}\, x)$ is holomorphic in the disc $|\zeta| < 1$ (for all those ζ for which it is defined). Then every holomorphic operator-valued function $s(\zeta)$ which is a solution of the FMI has the following properties: $F_s x \in \mathbb{K}_s$ and*

$$\langle F_s x, F_s x\rangle_{\mathbb{K}_s} \le D(x,x), \ (\forall\, x \in \mathbb{X}) ,$$

where

$$F_s x \stackrel{\text{def}}{=} (F_s x)_+ \oplus (F_s x)_- \in \mathbb{K}_s,$$
$$(F_s x)_+(\zeta) \stackrel{\text{def}}{=} (s(\zeta)E - M)(\zeta I - T)^{-1}x, \ |\zeta| < 1 ,$$
$$(F_s x)_-(\zeta) \stackrel{\text{def}}{=} \overline{\zeta}(E - s^*(\zeta)M)(I - T\overline{\zeta})^{-1}x, \ |\zeta| < 1 .$$

Proof. Since $x \in \mathbb{X}$ and $\ell \in \mathbb{L}$ are chosen arbitrarily the FMI is equivalent to the inequality

$$D((I - T\overline{\zeta})x, (I - T\overline{\zeta})x) + 2Re\langle (E - s^*(\zeta)M)x, \ell\rangle + \Big\langle \frac{I_{\mathbb{N}} - s^*(\zeta)s(\zeta)}{1 - \overline{\zeta}\zeta}\, \ell, \ell\Big\rangle \ge 0 . \tag{1}$$

We now replace the vector ℓ in (1) by the vector $(\ell - E(I - T\overline{\zeta})x)$ and then, after multiplying the resulting expression through by $1 - \overline{\zeta}\zeta$, combine separately the terms which are quadratic with respect to x, the terms which are linear with respect to x and the terms which are independent of x.

The quadratic term has the form

$$\begin{aligned} C_2 =&(1-\zeta\overline{\zeta})D((I-T\overline{\zeta})x,(I-T\overline{\zeta})x) - (1-\zeta\overline{\zeta})\langle E(I-T\overline{\zeta})x,(E-s^*(\zeta)M)x\rangle \\ &- (1-\zeta\overline{\zeta})\langle (E-s^*(\zeta)M)x, E(I-T\overline{\zeta})x\rangle \\ &+ \langle (I_{\mathbb{N}} - s^*(\zeta)s(\zeta))E(I-T\overline{\zeta})x, E(I-T\overline{\zeta})x\rangle \ . \end{aligned}$$

The linear term is

$$\begin{aligned} C_1 = &(1-\zeta\overline{\zeta})\langle (E-s^*(\zeta)M)x,\ell\rangle \\ &- \langle (I - s^*(\zeta)s(\zeta))E(I-T\overline{\zeta})x,\ell\rangle \ . \end{aligned}$$

The constant term is

$$C_0 = \langle (I_{\mathbb{N}} - s^*(\zeta)s(\zeta))\ell,\ell\rangle \ .$$

In terms of this notation, inequality (1) can be expressed in the form

$$C_2 + 2ReC_1 + C_0 \geq 0 \ . \tag{2}$$

Using the arbitraryness of x and ℓ, we shall rewrite (2) in the form

$$\begin{bmatrix} C_2 & \overline{C}_1 \\ & \\ C_1 & C_0 \end{bmatrix} \geq 0 \ . \tag{3}$$

From the obvious identity

$$(1-\zeta\overline{\zeta})I = (I - T\overline{\zeta}) - \overline{\zeta}(\zeta I - T), \tag{4}$$

we have

$$\begin{aligned} &(1-\zeta\overline{\zeta})(E - s^*(\zeta)M)x \\ &= (E - s^*(\zeta)M)(I-T\overline{\zeta})x - \overline{\zeta}(E - s^*(\zeta)M)(\zeta I - T)x \ . \end{aligned} \tag{5}$$

It follows from (5) that

$$\begin{aligned} &(1-\zeta\overline{\zeta})(E - s^*(\zeta)M)x - (I_{\mathbb{N}} - s^*(\zeta)s(\zeta))E(I-T\overline{\zeta})x \\ &= s^*(\zeta)(s(\zeta)E - M)(I-T\overline{\zeta})x - \overline{\zeta}(E - s^*(\zeta)M)(\zeta I - T)x \ , \end{aligned} \tag{6}$$

and hence that

$$C_1 = \langle s^*(\zeta)(s(\zeta)E - M)(I-T\overline{\zeta})x - \overline{\zeta}(E - s^*(\zeta)M)(\zeta I - T)x,\ell\rangle \ . \tag{7}$$

Next, we transform the second term in C_2 with the help of (5) and the sum of the third and fourth terms with the help of (6) to obtain:

$$\begin{aligned} C_2 &= (1-\zeta\overline{\zeta})D((I-T\overline{\zeta})x,(I-T\overline{\zeta})x) \\ &\quad - \langle E(I-T\overline{\zeta})x,(E-s^*(\zeta)M)(I-T\overline{\zeta})x-\overline{\zeta}(E-s^*(\zeta)M)(\zeta I-T)x\rangle \\ &\quad - \langle s^*(\zeta)(s(\zeta)E-M)(I-T\overline{\zeta})x-\overline{\zeta}(E-s^*(\zeta)M)(\zeta I-T)x,E(I-T\overline{\zeta})x\rangle \\ &= (1-\zeta\overline{\zeta})D((I-T\overline{\zeta})x,(I-T\overline{\zeta})x) \\ &\quad + 2Re\langle E(I-T\overline{\zeta})x,\overline{\zeta}(E-s^*(\zeta)M)(\zeta I-T)x\rangle \\ &\quad - \langle E(I-T\overline{\zeta})x,(E-s^*(\zeta)M)(I-T\overline{\zeta})x\rangle \\ &\quad - \langle s^*(\zeta)(s(\zeta)E-M)(I-T\overline{\zeta})x,E(I-T\overline{\zeta})x\rangle\ . \end{aligned} \tag{8}$$

Since

$$\begin{aligned} &\langle s^*(\zeta(s(\zeta)E-M)(I-T\overline{\zeta})x,E(I-T\overline{\zeta})x\rangle \\ &\qquad = \langle (s(\zeta)E-M)(I-T\overline{\zeta})x,s(\zeta)E(I-T\overline{\zeta})x\rangle \end{aligned}$$

and

$$\begin{aligned} &\langle E(I-T\overline{\zeta})x,(E-s^*(\zeta)M)(I-T\overline{\zeta})x\rangle \\ &\qquad = \langle E(I-T\overline{\zeta})x,E(I-T\overline{\zeta})x\rangle - \langle E(I-T\overline{\zeta})x,s^*(\zeta)M(I-T\overline{\zeta})x\rangle \\ &\qquad = \langle E(I-T\overline{\zeta})x,E(I-T\overline{\zeta})x\rangle - \langle s(\zeta)E(I-T\overline{\zeta})x,M(I-T\overline{\zeta})x\rangle \\ &\qquad = \langle E(I-T\overline{\zeta})x,E(I-T\overline{\zeta})x\rangle - \langle M(I-T\overline{\zeta})x,M(I-T\overline{\zeta})x\rangle \\ &\qquad\quad - \langle (s(\zeta)E-M)(I-T\overline{\zeta})x,M(I-T\overline{\zeta})x\rangle\ , \end{aligned}$$

the expression (8) for C_2 takes the form

$$\begin{aligned} C_2 &= (1-\zeta\overline{\zeta})D((I-T\overline{\zeta})x,(I-T\overline{\zeta})x) \\ &\quad + 2Re\langle E(I-T\overline{\zeta})x,\overline{\zeta}(E-s^*(\zeta)M)(\zeta I-T)x\rangle \\ &\quad - \langle E(I-T\overline{\zeta})x,E(I-T\overline{\zeta})x\rangle + \langle M(I-T\overline{\zeta})x,M(I-T\overline{\zeta})x\rangle \\ &\quad - \langle s(\zeta)E-M)(I-T\overline{\zeta})x,(s(\zeta)E-M)(I-T\overline{\zeta})x\rangle\ . \end{aligned} \tag{9}$$

Using the Fundamental Identity and grouping similar terms in the quadratic form D we can reexpress (9) as

$$\begin{aligned} C_2 &= ReD((T+\zeta I)(I-T\overline{\zeta})x,(T-\zeta I)(I-T\overline{\zeta})x) \\ &\quad + 2Re\langle E(I-T\overline{\zeta})x,\overline{\zeta}(E-s^*(\zeta)M)(\zeta I-T)x\rangle \\ &\quad - \langle (s(\zeta)E-M)(I-T\overline{\zeta})x,(s(\zeta)E-M)(I-T\overline{\zeta})x\rangle\ . \end{aligned} \tag{10}$$

Next, upon taking into account expression (7) for C_1 and the block-matrix lemma, one can transform inequality (3) to the following equivalent inequality:

$$\begin{aligned}C_2 \geq \|&(I_{\mathbb{N}} - s^*(\zeta)s(\zeta))^{[-1/2]}\\ &\cdot [s^*(\zeta)(s(\zeta)E - M)(I - T\overline{\zeta})x - \overline{\zeta}(E - s^*(\zeta)M)(\zeta I - T)x]\|^2.\end{aligned} \tag{11}$$

Then, upon substituting formula (10) for C_2 into (11) and using Proposition 3 we obtain

$$\begin{aligned}&ReD((T + \zeta I)(I - T\overline{\zeta})x,\ (T - \zeta I)(I - T\overline{\zeta})x)\\ &\quad + 2Re\langle E(I - T\overline{\zeta})x, \overline{\zeta}(E - s^*(\zeta)M)(\zeta I - T)x\rangle\\ &- \left\langle \begin{bmatrix} I_{\mathbb{N}'} & s(\zeta) \\ s^*(\zeta) & I_{\mathbb{N}} \end{bmatrix}^{[-1]} \begin{bmatrix} (s(\zeta)E - M)(I - T\overline{\zeta})x \\ \overline{\zeta}(E - s^*(\zeta)M)(\zeta I - T)x \end{bmatrix}, \right.\\ &\qquad \left. \begin{bmatrix} (s(\zeta)E - M)(I - T\overline{\zeta})x \\ \overline{\zeta}(E - s^*(\zeta)M)(\zeta I - T)x \end{bmatrix} \right\rangle \geq 0 .\end{aligned} \tag{12}$$

The inequality (12) can be considered as the final form of the transformed FMI. We emphasize that all the transformations are based on identities and do not use any spectral properties of the operator T.

The left hand side of the inequality (12) admits a dual representation. It can be obtained from expression (12) by regrouping the entries in the first two terms and invoking the identity

$$\begin{aligned}&\frac{1}{2}D((T + \zeta I)y_1, (I - T\overline{\zeta})y_2) - \zeta\langle Ey_1, (E - s^*(\zeta)M)y_2\rangle\\ &\quad = \frac{1}{2}\ D((T - \zeta I)y_1, (I + T\overline{\zeta})y_2) + \zeta\langle(s(\zeta)E - M)y_1, My_2\rangle\ ,\end{aligned} \tag{13}$$

which follows directly from the FI for arbitrary y_1 and $y_2 \in \mathbb{X}$. Inserting $y_1 = (I - T\overline{\zeta})x$ and $y_2 = (T - \zeta I)y$ into (13), we obtain

$$\begin{aligned}&\frac{1}{2}\ D((T + \zeta I)(I - T\overline{\zeta})x, (T - \zeta I)(I - T\overline{\zeta})y)\\ &\quad + \zeta\langle E(I - T\overline{\zeta})x,\ (E - s^*(\zeta)M)(\zeta I - T)y\rangle\\ &= \frac{1}{2}\ D((I - T\overline{\zeta})(\zeta I - T)x, (I + T\overline{\zeta})(\zeta I - T)y)\\ &\quad - \zeta\langle(s(\zeta)E - M)(I - T\overline{\zeta})x, M(\zeta I - T)y\rangle\ .\end{aligned} \tag{14}$$

for arbitrary $x, y \in \mathbb{X}$. Substituting (14) into (12), we obtain

$$\begin{aligned} &ReD((I - T\overline{\zeta})(\zeta I - T)x, (I + T\overline{\zeta})(\zeta I - T)x) \\ &\quad - 2Re\zeta\langle (s(\zeta)E - M)(I - T\overline{\zeta})x, M(\zeta I - T)x\rangle \\ &\quad - \left\langle \begin{bmatrix} I_{\mathbb{N}'} & s(\zeta) \\ s^*(\zeta) & I_{\mathbb{N}} \end{bmatrix}^{[-1]} \begin{bmatrix} (s(\zeta)E - M)(I - T\overline{\zeta})x \\ \overline{\zeta}(E - s^*(\zeta)M)(\zeta I - T)x \end{bmatrix}, \right. \\ &\quad \left. \begin{bmatrix} (s(\zeta)E - M)(I - T\overline{\zeta})x \\ \overline{\zeta}(E - s^*(\zeta)M)(\zeta I - T)x \end{bmatrix} \right\rangle \geq 0 . \end{aligned} \tag{12'}$$

Finally, we shall use the spectral properties of operator T. Substituting the vector $(I - T\overline{\zeta})^{-1}(\zeta I - T)^{-1}x$ in place of x in (12′) we obtain

$$\begin{aligned} &ReD(x, \frac{I + T\overline{\zeta}}{I - T\overline{\zeta}}\, x) - 2Re\zeta\langle (s(\zeta)E - M)(\zeta I - T)^{-1}x, M(I - T\overline{\zeta})^{-1}x\rangle \\ &- \| \begin{bmatrix} I_{\mathbb{N}'} & s(\zeta) \\ s^*(\zeta) & I_{\mathbb{N}} \end{bmatrix}^{[-1/2]} \begin{bmatrix} (s(\zeta)E - M)(\zeta I - T)^{-1}x \\ \overline{\zeta}(E - s^*(\zeta)M)(I - T\overline{\zeta})^{-1}x \end{bmatrix} \|^2 \geq 0. \end{aligned} \tag{15}$$

Let us now recall the notation:

$$\begin{aligned} (F_s x)_+(\zeta) &\stackrel{\text{def}}{=} (s(\zeta)E - M)(\zeta I - T)^{-1}x\ , \\ (F_s x)_-(\zeta) &\stackrel{\text{def}}{=} \overline{\zeta}(E - s^*(\zeta)M)(I - T\overline{\zeta})^{-1}x\ , \\ F_s x &\stackrel{\text{def}}{=} (F_s x)_+ \oplus (F_s x)_-\ , \end{aligned}$$

and define

$$P_\zeta(x, y) = \frac{1}{2}\, D(x, \frac{I + T\overline{\zeta}}{I - T\overline{\zeta}}\, y) - \zeta\langle (F_s x)_+(\zeta),\ M(I - T\overline{\zeta})^{-1}y\rangle\ . \tag{16}$$

Then, in view of (14), we also have

$$P_\zeta(x, y) = \frac{1}{2}\, D(\frac{T + \zeta I}{T - \zeta I}\, x, y) + \langle E(\zeta I - T)^{-1}x, (F_s y)_-(\zeta)\rangle\ . \tag{16'}$$

In terms of these notations inequality (15) can be expressed in the following form:

$$P_\zeta(x, x) + \overline{P_\zeta(x, x)} - \left\langle \begin{bmatrix} I_{\mathbb{N}'} & s(\zeta) \\ s^*(\zeta) & I_{\mathbb{N}} \end{bmatrix}^{[-1]} (F_s x)(\zeta), (F_s x)(\zeta) \right\rangle \geq 0\ . \tag{17}$$

By assumption, the function $P_\zeta(x,x)$ is holomorphic everywhere in $|\zeta|<1$ with the possible exception of a set of isolated points. It follows from (17) that the real part of the function $P_\zeta(x,x)$ is nonnegative. Hence, all the singularities of this function in the disk $|\zeta|<1$ are removable. Furthermore, in view of (17) and the corollary to Proposition 3, the functions $(F_s x)_\pm(\zeta)$ possess a harmonic majorant:

$$P_\zeta(x,x)+\overline{P_\zeta(x,x)} \geq \|(F_s x)_\pm(\zeta)\|^2.$$

This implies that these functions are in $H^2_+(\mathbb{L}')$ and $H^2_-(\mathbb{L})$ respectively. In particular, all their singularities are removable. Moreover, it can be seen directly from the definition that $(F_s x)_-(0)=0$.

It follows from formula (16) and from the regularity of the functions $(F_s x)(\zeta)$ on $|\zeta|<1$ that

$$P_0(x,y)=\frac{1}{2}\,D(x,y)\ .$$

Integrating the inequality (17) over the circle $\mathbb{T}_r$ of radius r centered at the origin with respect to normalized Lebesgue measure $dm(\zeta)=\frac{1}{2\pi i}\frac{d\zeta}{\zeta}$ we get

$$\begin{aligned}\int_{\mathbb{T}_r}\left\langle\begin{bmatrix} I_{\mathbb{N}'} & s(\zeta)\\ s(\zeta)^* & I_{\mathbb{N}}\end{bmatrix}^{[-1]}(F_s x)(\zeta),(F_s x)(\zeta)\right\rangle dm(\zeta) &\leq \int_{\mathbb{T}_r}[P_\zeta(x,x)+\overline{P_\zeta(x,x)}]dm(\zeta)\\ &= P_0(x,x)+\overline{P_0(x,x)}=D(x,x)\ .\end{aligned}$$

Thus $F_s x\in\mathbb{K}_s$ and $\langle F_s x,F_s x\rangle_{\mathbb{K}_s}\leq D(x,x)$. The theorem is proved. □

The following proposition shows how the action of the operator T (which was introduced in Theorem 1) is changed by the transformation F_s which acts from the space $\mathbb{X}$ into the space $\mathbb{K}_s$.

Proposition 5. *Let T and F_s be the same as in Theorem 1, then*

$$(F_s T x)(t)\overset{a.e.}{=}t(F_s x)(t)-\begin{bmatrix} I_{\mathbb{N}'} & s(t)\\ s^*(t) & I_{\mathbb{N}}\end{bmatrix}\begin{bmatrix}-Mx\\ Ex\end{bmatrix}\ ,\quad (|t|=1)\ . \tag{18}$$

Proof. This follows from the definition of F_s by a straightforward calculation. □

Remark. If the spectral condition for T which was formulated in Theorem 1 is satisfied, then the transformation F_s from $\mathbb{X}$ in $\mathbb{K}_s$ is defined uniquely by the relation (18).

The following proposition is a converse of Theorem 1.

Proposition 6. *Let $\mathbb{L}$, $\mathbb{L}'$, $\mathbb{X}$, T, D, E and M be the objects occurring*[1] *in Theorem 1, let $s \in B(\mathbb{L}, \mathbb{L}')$ and let $(F_s x)(t)$ be a family of functions in the variable $t \in \mathbb{T}$ which depends linearly on $x \in \mathbb{X}$ and is defined by the following (generically implicit) formula:*

$$\text{i)} \quad (F_s Tx)(t) \overset{a.e.}{=} t(F_s x)(t) - \begin{bmatrix} I_{\mathbb{N}'} & s(t) \\ s^*(t) & I_{\mathbb{N}} \end{bmatrix} \begin{bmatrix} -Mx \\ Ex \end{bmatrix}, \quad (|t| = 1).$$

Assume further that

ii) $F_s x \in \mathbb{K}_s$, $\forall\, x \in \mathbb{X}$

and

iii) $\langle F_s x, F_s x \rangle_{\mathbb{K}_s} \leq D(x, x)$, $\forall\, x \in \mathbb{X}$.

Then $s(\zeta)$ is a solution of the FMI.

Proof. Fix a point ζ with $|\zeta| < 1$ and consider the pair of vectors

$$(F_s x)(t) \quad \text{and} \quad \begin{bmatrix} I_{\mathbb{N}'} & s(t) \\ s^*(t) & I_{\mathbb{N}} \end{bmatrix} \begin{bmatrix} I_{\mathbb{N}'} \\ -s^*(\zeta) \end{bmatrix} \frac{\ell'}{1 - t\bar{\zeta}}\,, \quad \ell' \in \mathbb{N}'\,,$$

both of which belong to $\mathbb{K}_s$. Then, upon calculating all pairwise scalar products (in $\mathbb{K}_s$) formed from them and writing out the nonnegativity condition for the Gram matrix we obtain

$$\begin{bmatrix} \langle F_s x, F_s x \rangle_{\mathbb{K}_s} & \overline{\langle (F_s x)_+(\zeta), \ell' \rangle} \\ \langle (F_s x)_+(\zeta), \ell' \rangle & \langle \frac{I_{\mathbb{N}'} - s(\zeta)s^*(\zeta)}{1 - \zeta\bar{\zeta}}\, \ell', \ell' \rangle \end{bmatrix} \geq 0\,. \tag{19}$$

Substituting the vector $(\zeta I - T)x$ in place of the vector x in (19) and invoking the analytic continuation of $F_s x$ (which is defined on the boundary in i)) and the linearity of $F_s x$ in x, we have

$$(F_s(\zeta I - T)x)_+(\zeta) = (s(\zeta)E - M)x\,, \tag{20}$$

and, in view of iii),

$$\langle F_s(\zeta I - T)x, F_s(\zeta I - T)x \rangle_{\mathbb{K}_s} \leq D((\zeta I - T)x, (\zeta I - T)x)\,. \tag{21}$$

Inserting (20) and (21) into (19) we obtain the FMI′.

The FMI can be obtained analogously by considering the Gram matrix of the pair of vectors

$$(F_s x)(t) \quad \text{and} \quad \zeta \begin{bmatrix} I_{\mathbb{N}'} & s(t) \\ s(t)^* & I_{\mathbb{N}} \end{bmatrix} \begin{bmatrix} -s(\zeta) \\ I_{\mathbb{N}} \end{bmatrix} \frac{\ell}{t - \zeta}\,, \quad \ell \in \mathbb{N}\,.$$ □

1) We do not impose any spectral conditions on T here.

3 The abstract interpolation problem

Let $\mathbb{X}$ be a linear space, T a linear operator on $\mathbb{X}$, D a nonnegative sesquilinear form in $\mathbb{X}$ and let E and M be linear operators from $\mathbb{X}$ into the Hilbert spaces $\mathbb{L}$ and $\mathbb{L}'$, respectively. Suppose, moreover, that the identity

$$D(x,x) - D(Tx,Tx) = \langle Ex, Ex\rangle_{\mathbb{N}} - \langle Mx, Mx\rangle_{\mathbb{N}'}$$

is satisfied.

Let $(F_s x)(t)$ be a family of functions in the variable $t \in \mathbb{T}$ which depends linearly on $x \in \mathbb{X}$ and is defined by the following (generically implicit) formula:

$$\text{i) } (F_s Tx)(t) \overset{a.e.}{=} t(F_s x)(t) - \begin{bmatrix} I_{\mathbb{N}'} & s(t) \\ s^*(t) & I_{\mathbb{N}} \end{bmatrix} \begin{bmatrix} -Mx \\ Ex \end{bmatrix}, \quad (|t|=1).$$

The operator-valued function $s(\zeta) \in B(\mathbb{L}, \mathbb{L}')$ which is holomorphic in the disc $|\zeta| < 1$ is said to be a solution of the Abstract Interpolation Problem if
ii) $F_s x \in \mathbb{K}_s$, $\forall\, x \in \mathbb{X}$
and
iii) $\langle F_s x, F_s x\rangle_{\mathbb{K}_s} \leq D(x,x)$, $\forall\, x \in \mathbb{X}$.
Our objective is to describe all the solutions of the Abstract Interpolation Problem.[2]

The remaining discussion depends on the paper [6]. To study the problem in question, we need some objects connected with the given spaces and operators. Let Dx denotes the conjugate linear functional defined by the formula

$$Dx(y) \overset{\text{def}}{=} D(x,y) .$$

The scalar product is naturally defined on the set $\{Dx\}_{x\in\mathbb{N}}$ by the rule

$$\langle Dx_1, Dx_2\rangle \overset{\text{def}}{=} D(x_1, x_2) .$$

Obviously this inner product is well-defined. Let us denote by $\mathbb{K}$ the completion of the space $\{Dx\}_{x\in\mathbb{N}}$ with respect to the inner product introduced above. Then, $\mathbb{K}$ is a Hilbert space. The Fundamental Identity enables us to define an isometric operator from the space $\mathbb{K}\oplus\mathbb{L}$ into the space $\mathbb{K}\oplus\mathbb{L}'$. Let us define an operator V by the formula

$$V(DTx \oplus Ex) \overset{\text{def}}{=} Dx \oplus Mx . \tag{22}$$

2) For some choices of data there exists a unique linear mapping F_s from the space $\mathbb{X}$ into the space $\mathbb{K}_s$ with properties i)–iii) for any solution $s(\zeta)$ of the Abstract Interpolation Problem; for some other data there might be many mappings for the same solution $s(\zeta)$. In any case all these mappings F_s can be described along with the description of the solutions (see [7], [12]).

The domain (D_V) of the operator V is the closure in $\mathbb{K} \oplus \mathbb{L}$ of the set of all the vectors $DTx \oplus Ex$, the range Δ_V is the closure in $\mathbb{K} \oplus \mathbb{L}'$ of the set of all the vectors $Dx \oplus Mx$.

Let $\mathbb{H}$ be a Hilbert space and let U be a unitary operator from $\mathbb{H} \oplus \mathbb{L}$ onto $\mathbb{H} \oplus \mathbb{L}'$. Following the paper [6] we define the scattering function $s(\zeta)$ of U with respect to the spaces $\mathbb{L}$ and $\mathbb{L}'$ in the following way:

$$s(\zeta) = P_{\mathbb{N}'} U (I_{\mathbb{H}\oplus\mathbb{N}} - \zeta P_{\mathbb{H}} U)^{-1} | \mathbb{N} .$$

Consider also the functional representation of the space $\mathbb{H}$ which is defined by formula

$$Gh = (Gh)_+ \oplus (Gh)_- ,$$

where

$$(Gh)_+(\zeta) = P_{\mathbb{N}'} U (I_{\mathbb{H}\oplus\mathbb{N}} - \zeta P_{\mathbb{H}} U)^{-1} h$$

and

$$(Gh)_-(\zeta) = \bar{\zeta} P_{\mathbb{N}} U^* (I_{\mathbb{H}\oplus\mathbb{N}'} - \bar{\zeta} P_{\mathbb{H}} U^*)^{-1} h . \qquad (|\zeta| < 1)$$

Proposition 7. *G maps $\mathbb{H}$ into $\mathbb{K}_s$, and*

$$\|Gh\|_{\mathbb{K}_s} \le \|h\|_{\mathbb{H}} .$$

The following statement yields a connection between all the solutions of the abstract interpolation problem and all the scattering functions of the unitary extensions of the isometry V (see (22)) with respect to the spaces $\mathbb{L}$ and $\mathbb{L}'$.

Proposition 8. *Let $\mathbb{H} \supset \mathbb{K}$, let U be a unitary operator from $\mathbb{H} \oplus \mathbb{L}$ onto $\mathbb{H} \oplus \mathbb{L}'$ which extends V and let $s(\zeta)$ be the scattering function of U with respect to the spaces $\mathbb{L}$ and $\mathbb{L}'$. Then the functional transformation F_s*

$$F_s x \stackrel{\text{def}}{=} GDx \qquad (x \in X)$$

has property (18), i.e.,

$$F_s T x = t F_s x - \begin{bmatrix} I_{\mathbb{N}'} & s \\ s^* & I_{\mathbb{N}} \end{bmatrix} \begin{bmatrix} -Mx \\ Ex \end{bmatrix} .$$

Corollary. *The scattering function of any unitary extension of an isometry V is a solution of the abstract interpolation problem.*

Proposition 9. *Let $s(\zeta) \in B(\mathbb{L}, \mathbb{L}')$, and let F_s be a mapping from $\mathbb{X}$ into $\mathbb{K}_s$ which satisfies the conditions i)–iii) of the abstract interpolation problem. Then $s(\zeta)$ is the scattering matrix of a unitary extension of the isometry V with respect to the spaces $\mathbb{L}$ and $\mathbb{L}'$.*

Corollary. *The set of all the solutions of the abstract interpolation problem admits the following description*

$$s(\zeta) = s_{12}(\zeta) + s_{11}(\zeta)\varepsilon(\zeta)[I - s_{21}(\zeta)\varepsilon(\zeta)]^{-1}s_{22}(\zeta), \ |\zeta| < 1 ,$$

where

$$S(\zeta) = \begin{bmatrix} s_{11}(\zeta) & s_{12}(\zeta) \\ s_{21}(\zeta) & s_{22}(\zeta) \end{bmatrix}$$

is the scattering matrix of the isometry V [6], *and* $\varepsilon(\zeta)$ *is an arbitrary holomorphic contractive operator-valued function which acts from* $\mathbb{M}_V = (\mathbb{K} \oplus \mathbb{L}) \ominus D_V$ *into* $\mathbb{N}_V = (\mathbb{K} \oplus \mathbb{L}') \ominus \Delta_V$.

References

[1] Kovalishina I.V. and Potapov V.P., *Indefinite metric in the Nevanlinna-Pick problem*, Dokl. Akad. Nauk Armjan. SSr. **59**, (1974), No.1, 17–22.

[2] Kheifets A.Ya. and Yuditskiĭ P.M., *Interpolation of operators commuting with truncated shift by functions of the Schur class*, Teoriya funktsiĭ, funktsional. analys i ikh prilošeniya. **40** (1983), 129–136.

[3] Katsnelson V.E., *Fundamental matrix inequality of the problem of decomposition of positive definite kernel on elementary kernel*, Kharkov, 1984; Deposited in UkrNIINTI 10.7.1984, No. 1184, Uk Dep.

[4] Sz.-Nagy B. and Foias C.,*Analyse harmonique des operateurs de l'espace de Hilbert*, Akádémiai Kiado, Budapest 1967.

[5] Pavlov B.S., *Selfadjoint dilatation of a dissipative operator and expansion by its eigenfunctions*, Matem. Sbornik **102**, (1977), No. 4, 511–536.

[6] Arov D.Z. and Grossman L.Z., *Scattering matrix in the extension theory of isometric operators*, Dokl. Akad. Nauk. SSSR. **270**, (1983), No. 1, 17–20.

Supplementary references added in translation

[7] A.Ya. Kheifets, *Parseval equality in abstract interpolation problem and coupling of open systems*, Teor. Funk., Funk. Anal. i ikh Prilozhen **49** (1988) 112–120, **50** (1988) 98–103, Russian. English transl., J. Sov. Math. **49**, 4 (1990) 1114–1120, **49**, 6 (1990) 1307–1310.

[8] A.Ya. Kheifets, *Generalized bitangential Schur-Nevanlinna-Pick problem, related Parseval equality and scattering operator*, deposited in VINITI, 11.05.1989, No. 3108–B89 Dep., 1–60, 1989, Russian.

[9] A.Ya. Kheifets, *Generalized bitangential Schur-Nevanlinna-Pick problem and the related Parseval equality*, Teor. Funk., Funk. Anal. i ikh Prilozhen. **54** (1990) 89–96, Russian. English transl., J. Sov. Math. **58**, 4 (1992) 358–364.

[10] A.Ya. Kheifets, *Nevanlinna-Adamjan-Arov-Krein theorem in semi-determinate case*, Teor. Funkt., Funkt. Anal. i ikh Prilozhen **56** (1991) 128–137, Russian. English transl., Journal of Mathematical Sciences **76**, 4 (1995) 2542–2549.

[11] A.Ya. Kheifets, *Scattering Matrices and Parseval Equality in Abstract Interpolation Problem*, Ph.D. Thesis, 1990, Kharkov, Russian

[12] A.Ya. Kheifets and P.M. Yuditskii, *An analysis and extension of V.P. Potapov's approach to interpolation problems with applications to the generalized bi-tangential Schur-Nevanlinna-Pick problem and j-inner-outer factorization*, in: Operator Theory: Advances and Applications, **72** (1994) 133–161, Birkhäuser Verlag, Basel.

[13] D.Z. Arov and L.Z. Grossman, *Scattering matrices in the theory of unitary extension of isometric operators*, Math. Nachr. **157** (199), 105–123.

V.E. Katsnelson
Department of Theoretical Mathematics
The Weizmann Institute of Science
Rehovot 76100, ISRAEL
E-mail: katze@wisdom.weizmann.ac.il

A.Ya. Kheifets
Mathematics Division
Institute for Low Temperature
Physics and Engineering
47 Lenin Avenue
Kharkov 310164, UKRAINE
E-mail: kheifets@ilt.kharkov.ua

P.M. Yuditskii
Mathematics Division
Institute for Low Temperature Physics and Engineering
47 Lenin Avenue
Kharkov 310164, UKRAINE
E-mail: yuditskii@ilt.kharkov.ua

AMS Subject Classification: Primary: 47A57, 47A20, 30D50; Secondary: 47A45, 47A48, 30C80.

Operator Theory
Advances and Applications, Vol. 95
© 1997 Birkhäuser Verlag Basel/Switzerland

On the theory of matrix-valued functions belonging to the Smirnov class

V.E. Katsnelson and B. Kirstein

Abstract. A theory of matrix-valued functions from the matricial Smirnov class $\mathfrak{N}_n^+(\mathbb{D})$ is systematically developed. In particular, the maximum principle of V.I.Smirnov, inner-outer factorization, the Smirnov-Beurling characterization of outer functions and an analogue of Frostman's theorem are presented for matrix-valued functions from the Smirnov class $\mathfrak{N}_n^+(\mathbb{D})$. We also consider a family $F_\lambda = F - \lambda I$ of functions belonging to the matricial Smirnov class which is indexed by a complex parameter λ. We show that with the exception of a "very small" set of such λ the corresponding inner factor in the inner-outer factorization of the function F_λ is a Blaschke-Potapov product.

The main goal of this paper is to provide users of analytic matrix-function theory with a standard source for references related to the matricial Smirnov class.

Notations

$\mathbb{C}$ – the complex plane.

$\mathbb{T} := \{t \in \mathbb{C} : |t| = 1\}$ – the unit circle.

$\mathbb{D} := \{z \in \mathbb{C} : |z| < 1\}$ – the unit disc.

$\mathfrak{B}_\mathbb{T}$ – the σ-algebra of Borel subsets of $\mathbb{T}$.

m – normalized Lebesgue measure on the measurable space $(\mathbb{T}, \mathfrak{B}_\mathbb{T})$.

$\mathbb{C}^n$ – the n-dimensional complex space equipped with the usual Euclidean norm, i.e., for $x = (\xi_1, \dots, \xi_n)^\top$ we define $\|x\|_{\mathbb{C}^n} := \left\{ \sum_{k=1}^{n} |\xi_k|^2 \right\}^{1/2}$.

$\mathfrak{M}_n$ – the set of all complex $n \times n$ matrices equipped with the standard matrix norm, namely if $M \in \mathfrak{M}_n$ then $\|M\| := \sup_{x \in \mathbb{C}^n \setminus \{0\}} \|Mx\|_{\mathbb{C}^n} / \|x\|_{\mathbb{C}^n}$.

$\mathfrak{C}_n := \{M \in \mathfrak{M}_n : \|M\| \leq 1\}$ – the subset of all contractive matrices in $\mathfrak{M}_n$.

I_n – the $n \times n$ unit matrix.

As usual for $r \in \mathbb{R}$ we set $r^+ := \max\{r, 0\}$ and $r^- := \max\{-r, 0\}$.
Hence, $r = r^+ - r^-$ and $|r| = r^+ + r^-$. In particular, if $a \in (0, \infty)$, then

$$\ln^+ a = \max\{\ln a, 0\} \quad , \quad \ln^- a = \max\left\{\ln \frac{1}{a}, 0\right\},$$
$$\ln a = \ln^+ a - \ln^- a \quad , \quad |\ln a| = \ln^+ a + \ln^- a.$$

If $A \in \mathbb{C}^{p \times q}$, then the symbol $A^\top$ stands for the transposed matrix.

0 Preface

In this paper, we discuss various aspects of a class of matrix-valued functions which is named after V.I. Smirnov who introduced it for the scalar case in his famous paper [Sm]. It should be mentioned that the scalar Smirnov class also appeared in early papers of Doob (see e.g. [Doo1], [Doo2] and the bibliographies in the monographs of Collingwood and Lohwater [CoLo] and Noshiro [No] which contain references to many other related works of Doob). For a collection of basic facts on the Smirnov class and the intimately related function spaces named after Nevanlinna and Hardy we refer the reader to the monographs of P.L. Duren [Dur], J.B. Garnett [G], K. Hoffman [Hoff], P. Koosis [Koo], I.I. Privalov [Pri] and M. Rosenblum and J. Rovnyak [RoRo2]. These books concentrate more or less on function-theoretic properties of functions belonging to some of the mentioned classes. In the last two decades much progress has been made in clearing up topological and functional-analytic questions connected with the structure of the Smirnov class (see e.g. Yanagihara [Y1]–[Y10], Yanagihara and Kawase [YK], Yanagihara and Nakamura [YN], Stoll [St1], [St2], Roberts [Rob], Roberts and Stoll [RoSt1], [RoSt2], Mochizuki [Mo1], [Mo2], Helson [Hel2]–[Hel4], McCarthy [McC], and Camera [Cam]).

A systematic study of the matricial Smirnov class was mainly promoted by the work of D.Z. Arov. In his paper [Ar1] on Darlington synthesis, a matricial generalization of V.I. Smirnov's important maximum principle was used in an essential way, namely with its aid a powerful criterion for proving the J-contractivity of a meromorphic matrix function was established. Moreover, D.Z. Arov's description of all Darlington representations of a given (pseudocontinuable) Schur function is based on the concept of denominators. *A pair* $[b_1, b_2]$ *of inner matrix-valued functions* of appropriate sizes is *called a denominator of a given meromorphic matrix-valued function* f of bounded characteristic if $b_1 f b_2$ belongs to the matricial Smirnov class.

Nehari interpolation and generalized bitangential Schur-Nevanlinna-Pick interpolation are other important problems which turned out to be closely related with the matricial Smirnov class. This is an immediate consequence of D.Z. Arov's work [Ar3]–[Ar9] (see also Nicolau [Nic1], [Nic2]). In his investigations on the corresponding inverse problem D.Z. Arov introduced particular subclasses of J-inner functions which are now called the classes of Arov-regular and Arov-singular J-inner functions. Here a J-inner function V is called Arov-singular if V and V^{-1} belong to the matricial Smirnov class. Furthermore, a J-inner function W is called left Arov-regular (resp. right Arov-regular) if it does not contain any nonconstant Arov-singular right (resp. left) divisors. D.Z. Arov (see [Ar3]–[Ar7]) proved that each J-inner function W admits (essentially unique) factorizations

$$W = W_{l,r} \cdot W_{l,s} = W_{r,s} \cdot W_{r,r} \ ,$$

where the J-inner functions $W_{l,s}$ and $W_{r,s}$ are Arov-singular whereas the J-inner functions $W_{l,r}$ and $W_{r,r}$ are left Arov-regular and right Arov-regular, respectively. Furthermore, D.Z. Arov proved that a J-inner function is a left (resp. right) resolvent matrix of a completely indeterminate bitangential Schur-Nevanlinna-Pick interpolation problem if and only if it is left Arov-regular (resp. right Arov-regular). For several connections between left and right Arov-regularity we refer the reader to the papers [Kats1], [Kats2] where essential connections between left and right Blaschke-Potapov products were established. In this way the first author (see [Kats3], [Kats4]) was led to a weighted approximation problems for pseudocontinuable functions belonging to the Smirnov class. The papers [Kats1] -[Kats3] laid the basis for the study of an inverse problem for Arov-singular J-inner functions which was considered in [AFK7]. The papers [Ar2], [AFK1]–[AFK6] deal with several completion problems for J-inner functions with particular emphasis on various subclasses of J-inner functions (Smirnov type, inverse Smirnov type, Arov-singular type). Using the concept of Arov-singularity and Arov-regularity of J-inner functions and the approximation method created in [Kats3], A. J. Kheifets [Kh] answered a question of D. Sarason [Sar1] (see also [Sar2]) on exposed points in the Hardy space $H^1(\mathbb{D})$. Prediction theory for multivariate stationary sequences formed an important source for the development of the theory of matrix-valued holomorphic functions (see Wiener and Masani [WM1], [WM2], Helson and Lowdenslager [HL1], [HL2], Rozanov [Roz1], [Roz2] and Masani [Ma1]–[Ma4]). In particular, the matricial Hardy class $H_n^2(\mathbb{D})$ (see Definition 5.1 below) became an essential tool. It turned out that the basic problems of prediction theory could be reformulated as analytic problems for appropriate functions belonging to the Hardy class $H_n^2(\mathbb{D})$. Using functional-analytic methods, Beurling's inner-outer factorization was generalized to $H_n^2(\mathbb{D})$ (see Masani [Ma2], Rozanov [Roz1]). Moreover classical results due to Szegö [Sz1]–[Sz3], Kolmogorov [Kol] and Krein [Kr] were extended to the multivariate case. Here, it turned out (see Devinatz [De]) that the matrix version of Szegö's factorization theorem and other results due to Wiener and Masani [WM1], [WM2] and Helson and Lowdenslager [HL1], [HL2] are not so much generalizations of Szegö's classical results as consequences of it. An algebraic treatment of this theory was given by Helson [Hel1].

Carrying on from the theory of matrix-valued functions belonging to the Hardy class $H_n^2(\mathbb{D})$, we will study various aspects of outer functions from the matricial Smirnov class in this paper. In particular, we will extend the theory of inner-outer factorization to the matricial Smirnov class. A central topic in our investigations is to describe the situation where the inner factor in the inner-outer factorization of a matrix-valued Smirnov class function is a Blaschke-Potapov product. Moreover, we will consider a family of functions belonging to the matricial Smirnov class which is indexed by a complex parameter λ. Then it will be shown that with exception of a "very small" set of such parameters λ the corresponding inner factor in the inner-outer factorization of the function F_λ is a Blaschke-Potapov product. Our methods to prove this use a matrix generalization of logarithmic potentials. In this way, we obtain a generalization of a classical theorem of Frostman [Fr] (see also Heins [Hei] and Rudin [Ru1], [Ru2]). It should be mentioned that it was Yu.P. Ginzburg who was a pioneer in matrix (and in operator) generalizations of Frostman's results (see [Gi6] and [GiTa1]–[GiTa3]).

1 On the matricial Nevanlinna and Smirnov classes

For $F : \mathbb{D} \to \mathfrak{M}_n$ and $r \in [0,1)$, we define the function $F_{[r]} : \mathbb{T} \to \mathfrak{M}_n$ via $t \to F(rt)$.

Definition 1.1. *A matrix-valued function $F : \mathbb{D} \to \mathfrak{M}_n$ is said to belong to the matricial Nevanlinna class $\mathfrak{N}_n(\mathbb{D})$ if F is holomorphic in $\mathbb{D}$ and if the family $\left(\ln^+ \|F_{[r]}\|\right)_{r\in[0,1)}$ is bounded in $\mathcal{L}^1(m)$, or more precisely, if*

$$\sup_{r\in[0,1)} \int_{\mathbb{T}} \ln^+ \|F_{[r]}(t)\| \; m(dt) < +\infty. \tag{1.1}$$

Remark 1.1. Let $F : \mathbb{D} \to \mathfrak{M}_n$ be a matrix-valued function which is holomorphic in $\mathbb{D}$. Then F belongs to $\mathfrak{N}_n(\mathbb{D})$ if and only if the (subharmonic) function $\ln^+ \|F\|$ has a harmonic majorant in $\mathbb{D}$.

The definition of the Smirnov class $\mathfrak{N}^+(\mathbb{D})$ and of its matricial analogue $\mathfrak{N}_n^+(\mathbb{D})$ are connected with the notion of uniform integrability. Since this notion is not used very often we give the definition.

Definition of uniform integrability. *Let $(\Omega, \mathfrak{A}, \mu)$ be a measure space.*
Then the family $(f_\alpha)_{\alpha\in A}$ belonging to $\mathcal{L}^1(\Omega, \mathfrak{A}, \mu; \mathbb{C})$ is called uniformly integrable with respect to μ *if the following conditions are satisfied:*

(i) $$\sup_{\alpha\in A} \int_{\Omega} |f_\alpha(t)| \; \mu(dt) < +\infty.$$

(ii) *For every $\epsilon \in (0,\infty)$ there exists a $\delta \in (0,\infty)$ (which depends only on ϵ) such that for all $\alpha \in A$ and for all $\Delta \in \mathfrak{A}$, with $\mu(\Delta) < \delta$, the inequality*

$$\int_{\Delta} |f_\alpha(t)| \; \mu(dt) < \epsilon$$

is fulfilled.

Remark 1.2. If $\mu(\Omega) < +\infty$ and if for each fixed $\delta \in (0, \infty)$ there exist an $N(\delta) \in \mathbb{N}$ and a sequence $(X_{k,\delta})_{k=1}^{N(\delta)}$ from Δ such that $\Omega = \bigcup_{k=1}^{N(\delta)} \Omega_{k,\delta}$ and $\mu(\Omega_{k,\delta}) \leq \delta$ for all $k \in \{1, 2, \ldots, N(\delta)\}$, then a family of functions for which condition (ii) in the preceding definition is fulfilled, automatically satisfies condition (i) . Consequently, in the case of a finite measure space $(\Omega, \mathfrak{A}, \mu)$ condition (i) can be omitted in the definition of uniform integrability. A special case of such a measure space is the Lebesgue space on $\mathbb{T}$, where $\mathfrak{A}$ is the σ-algebra of Borel subsets of $\mathbb{T}$ and m is the normalized Lebesgue measure on $\mathbb{T}$.

In the sequel we will repeatedly use the following theorem from measure theory which goes back to G. Vitali [Vit] (see also [Ru3, p.133, Exercise 10]).

Vitali's convergence theorem. *Let $(\Omega, \mathfrak{A}, \mu)$ be a finite measure space (i.e., $\mu(\Omega) < \infty$). Let $(f_n)_{n\in\mathbb{N}}$ be a sequence from $\mathcal{L}^1(\Omega, \mathfrak{A}, \mu; \mathbb{C})$ which is uniformly integrable with respect to μ and converges μ-a.e. to a Borel measurable function $f : \Omega \to \mathbb{C}$. Then $f \in \mathcal{L}^1(\Omega, \mathfrak{A}, \mu; \mathbb{C})$,*

$$\lim_{n\to\infty} \int_\Omega |f_n - f| \, d\mu = 0 \quad \textit{and} \quad \lim_{n\to\infty} \int_\Omega f_n \, d\mu = \int_\Omega f \, d\mu.$$

Proof. Let $\epsilon \in (0, \infty)$. In view of the uniform μ-integrability of $(f_n)_{n\in\mathbb{N}}$ there exists a number $\delta \in (0, \infty)$ such that for all $n \in \mathbb{N}$ and for all $\Delta \in \mathfrak{A}$, which satisfy $\mu(\Delta) < \delta$, the inequality

$$\int_\Delta |f_n| \, d\mu < \frac{\epsilon}{3} \tag{1.2}$$

is satisfied. Since $\mu(\Omega) < \infty$, Egorov's Theorem guarantees the existence of a set $B_\delta \in \mathfrak{A}$ such that

$$\mu(B_\delta) < \delta \tag{1.3}$$

and

$$\lim_{n\to\infty} \sup_{w\in\Omega\setminus B_\delta} |f_n(\omega) - f(\omega)| = 0. \tag{1.4}$$

Thus, there exists an $n_0 \in \mathbb{N}$ such that for all $n \geq n_0$ and all $\omega \in \Omega \setminus B_\delta$ the inequality

$$|f_n(\omega) - f(\omega)| < \frac{\epsilon}{3[1 + \mu(\Omega)]} \tag{1.5}$$

is satisfied. In view of (1.2) and (1.3) for $n \in \mathbb{N}$ we have

$$\int_{B_\delta} |f_n| \, d\mu < \frac{\epsilon}{3}. \tag{1.6}$$

From Fatou's Theorem and (1.6) we obtain

$$\int_{B_\delta} |f|\, d\mu \leq \varliminf_{n\to\infty} \int_{B_\delta} |f_n|\, d\mu \leq \frac{\epsilon}{3}. \tag{1.7}$$

Combining (1.5)–(1.7) we obtain the estimate

$$\begin{aligned}
\int_\Omega |f_n - f|\, d\mu &= \int_{\Omega\setminus B_\delta} |f_n - f|\, d\mu + \int_{B_\delta} |f_n - f|\, d\mu \\
&\leq \frac{\epsilon}{3[1+\mu(\Omega)]}\, \mu(\Omega \setminus B_\delta) + \int_{B_\delta} |f|\, d\mu + \int_{B_\delta} |f_n|\, d\mu \\
&< \frac{\epsilon}{3} + \frac{\epsilon}{3} + \frac{\epsilon}{3} = \epsilon
\end{aligned}$$

for $n \geq n_0$. Thus,

$$\lim_{n\to\infty} \int_\Omega |f_n - f|\, d\mu = 0.$$

From this, all the remaining assertions follow immediately. □

Definition 1.2. *A function $\varphi : \mathbb{R} \to \mathbb{R}$ is called* strongly convex *if it has the following properties:*

(i) *φ is convex.*

(ii) *φ is monotonically nondecreasing.*

(iii) *φ takes its values in $[0, \infty)$.*

(iv) $$\lim_{x\to\infty} \frac{\varphi(x)}{x} = \infty.$$

(v) *For some $c \in (0,\infty)$ there exist constants $M \in [0,\infty)$ and $a \in \mathbb{R}$ such that $\varphi(t+c) \leq M \cdot \varphi(t)$ for all $t \in [a,\infty)$.*

If (v) holds for just one value of $c \in (0,\infty)$, say $c = c_0$, then by (ii) it holds for all $c \in (0, c_0)$. By iteration it holds for $c = nc_0$, $n \in \mathbb{N}$ and hence it holds for all $c \in (0,\infty)$.

Theorem 1.1 (de la Vallée Poussin [LVP1], Nagumo [Na]). *Let $(\Omega, \mathfrak{A}, \mu)$ be a (finite or infinite) measure space, and let $(f_\alpha)_{\alpha\in A}$ be a family of functions belonging to $\mathcal{L}^1(\Omega, \mathfrak{A}, \mu; \mathbb{C})$. In case $\mu(\Omega) = +\infty$, we assume also that*

$$\sup_{\alpha\in A} \int_\Omega |f_\alpha|\, d\mu < \infty.$$

(i) *Suppose that there exists a function* $\varphi : [0,\infty) \to [0,\infty)$ *satisfying*

$$\lim_{x\to+\infty} \frac{\varphi(x)}{x} = +\infty \quad \text{and} \quad \sup_{\alpha\in A} \int_\Omega \varphi(|f_\alpha|)\, d\mu < +\infty.$$

Then the family $(f_\alpha)_{\alpha\in A}$ *is uniformly integrable with respect to* μ.

(ii) *Suppose that the family* $(f_\alpha)_{\alpha\in A}$ *is uniformly integrable with respect to* μ. *Then there exists a strongly convex function* $\varphi : \mathbb{R} \to \mathbb{R}$ *such that*

$$\sup_{\alpha\in A} \int_\Omega \varphi(|f_\alpha|)\, d\mu < +\infty.$$

For a modern proof of Theorem 1.1 we refer to [RoRo2, Theorem 3.10] (see also Theorem 3.1.2 in [Ru2]). This modern proof based on Vitali's Convergence Theorem.

Definition 1.3. *A matrix-valued function* $F : \mathbb{D} \to \mathfrak{M}_n$ *is said to belong to the matricial Smirnov class* $\mathfrak{N}_n^+(\mathbb{D})$ *if* F *is holomorphic in* $\mathbb{D}$ *and if the family* $\left(\ln^+ \|F_{[r]}\|\right)_{r\in[0,1)}$ *is uniformly integrable with respect to the normalized Lebesgue measure* m, *i.e., if for each* $\epsilon \in (0,\infty)$ *there exists a* $\delta \in (0,\infty)$ *(which depends only on* ϵ*) such that for all* $r \in [0,1)$ *and for all Borel subsets* Δ *of* $\mathbb{T}$ *satisfying* $m(\Delta) < \delta$ *the inequality*

$$\int_\Delta \ln^+ \|F_{[r]}(t)\| \;\; m(dt) < \epsilon \tag{1.8}$$

is fulfilled.

Remark 1.3. In view of Remark 1.2, each matrix-valued function $F \in \mathfrak{N}_n^+(\mathbb{D})$ satisfies condition (1.1). Hence, *the matricial Smirnov class* $\mathfrak{N}_n^+(\mathbb{D})$ *is a subclass of the matricial Nevanlinna class* $\mathfrak{N}_n(\mathbb{D})$:

$$\mathfrak{N}_n^+(\mathbb{D}) \subseteq \mathfrak{N}_n(\mathbb{D}). \tag{1.9}$$

For a matrix-valued function F belonging to $\mathfrak{N}_n(\mathbb{D})$ we denote by $\underline{F} : \mathbb{T} \to \mathfrak{M}_n$ a boundary limit function associated with F, i.e., $\underline{F}$ is a Borel measurable function and there exists a Borel subset Δ_0 of $\mathbb{T}$ satisfying $m(\Delta_0) = 0$ such that for all $t \in \mathbb{T} \setminus \Delta_0$ we have

$$\lim_{r\to1-0} F(rt) = \underline{F}(t).$$

Observe that in view of Vitali's theorem a function $F \in \mathfrak{N}_n^+(\mathbb{D})$ satisfies

$$\lim_{r\to1-0} \int_\mathbb{T} \ln^+ \|F(rt)\| \;\; m(dt) = \int_\mathbb{T} \ln^+ \|\underline{F}(t)\| \;\; m(dt). \tag{1.10}$$

According to Fatou's theorem,

$$\varliminf_{r\to 1-0} \int_{\mathbb{T}} \ln^- \|F(rt)\| \; m(dt) \geq \int_{\mathbb{T}} \ln^- \|\underline{F}(t)\| \; m(dt) \tag{1.11}$$

(where equality does not hold in general). Hence,

$$\varlimsup_{r\to 1-0} \int_{\mathbb{T}} \ln \|F(rt)\| \; m(dt) \leq \int_{\mathbb{T}} \ln \|\underline{F}(t)\| \; m(dt). \tag{1.12}$$

Lemma 1.1. *A matrix-valued function $F : \mathbb{D} \to \mathfrak{M}_n$ belongs to the matricial class $\mathfrak{N}_n(\mathbb{D})$ (resp. $\mathfrak{N}_n^+(\mathbb{D})$) if and only if each of its entries belongs to the scalar class $\mathfrak{N}(\mathbb{D})$ (resp. $\mathfrak{N}^+(\mathbb{D})$).*

As the determinant of a matrix is a polynomial of its elements and because each of the classes $\mathfrak{N}(\mathbb{D})$ and $\mathfrak{N}^+(\mathbb{D})$ is an algebra over $\mathbb{C}$ the following result holds true.

Lemma 1.2.
(i) *If $F \in \mathfrak{N}_n(\mathbb{D})$, then $\det F \in \mathfrak{N}(\mathbb{D})$.*
(ii) *If $F \in \mathfrak{N}_n^+(\mathbb{D})$, then $\det F \in \mathfrak{N}^+(\mathbb{D})$.*

As a special case of (1.10), (1.11) and (1.12) (corresponding to the scalar case) we obtain the following relations for a function $F \in \mathfrak{N}_n^+(\mathbb{D})$ from part (ii) of Lemma 1.2:

$$\lim_{r\to 1-0} \int_{\mathbb{T}} \ln^+ |\det[F(rt)]| \; m(dt) = \int_{\mathbb{T}} \ln^+ |\det[\underline{F}(t)]| \; m(dt), \tag{1.13}$$

$$\varliminf_{r\to 1-0} \int_{\mathbb{T}} \ln^- |\det[F(rt)]| \; m(dt) \geq \int_{\mathbb{T}} \ln^- |\det[\underline{F}(t)]| \; m(dt) \tag{1.14}$$

and, finally, that

$$\varlimsup_{r\to 1-0} \int_{\mathbb{T}} \ln |\det[F(rt)]| \; m(dt) \leq \int_{\mathbb{T}} \ln |\det[\underline{F}(t)]| \; m(dt). \tag{1.15}$$

In the following we will use the *Poisson kernel* $P : \mathbb{D} \times \mathbb{T} \to (0, \infty)$ which is defined by the formula

$$P(z,t) := \frac{1-|z|^2}{|t-z|^2}.$$

Theorem 1.2. *Let $F \in \mathfrak{N}_n(\mathbb{D})$ with $F \not\equiv 0$ and let u_F denote the least harmonic majorant of* $\log \|F\|$. *Then the following statements are equivalent:*

(i) $F \in \mathfrak{N}_n^+(\mathbb{D})$.

(ii) $u_F(z) \leq \int_{\mathbb{T}} \ln \|\underline{F}(t)\| \, \frac{1-|z|^2}{|t-z|^2} \, m(dt)$ *for every* $z \in \mathbb{D}$.

(iii) $\ln \|F(z)\| \leq \int_{\mathbb{T}} \ln \|\underline{F}(t)\| \, \frac{1-|z|^2}{|t-z|^2} \, m(dt)$ *for every* $z \in \mathbb{D}$.

(iv) *There exists a strongly convex function $\varphi : \mathbb{R} \to \mathbb{R}$ and a number $r_0 \in (0,1)$ such that* $\sup_{r \in [r_0,1)} \int_{\mathbb{T}} \varphi\left(\ln \|F_{[r]}(t)\|\right) \, m(dt) < +\infty$.

(v) *There exists a strongly convex function $\psi : \mathbb{R} \to \mathbb{R}$ such that*
$$\sup_{r \in [0,1)} \int_{\mathbb{T}} \psi\left(\ln^+ \|F_{[r]}(t)\|\right) \, m(dt) < +\infty.$$

Proof. Theorem 1.2 can be proved by a slight modification of the proof of Theorem 3.3.5 in [Ru2]. Here, Theorem 1.1 plays an essential role. □

For further results on matrix-valued functions belonging to one of the classes named after Nevanlinna, Smirnov and Hardy we refer the reader to chapter 4 in [RoRo1].

2 Matrix functions of the Smirnov class as multiples of contractive matrix functions

Recall that a scalar function $e : \mathbb{D} \to \mathbb{C}$ is said to be *outer* (in the sense of V.I. Smirnov) if there exist a unimodular constant $C \in \mathbb{T}$ and a function $w : \mathbb{T} \to [0,\infty)$ for which $\log w$ is m-integrable such that for $z \in \mathbb{D}$ the relation

$$e(z) = C \cdot \exp\left\{ \int_{\mathbb{T}} \frac{t+z}{t-z} \, \ln\left[w(t)\right] \, m(dt) \right\} \tag{2.1}$$

holds true. Let $\mathfrak{E}(\mathbb{D})$ denote the class of all outer functions. From its definition it is obvious, that the class $\mathfrak{E}(\mathbb{D})$ is multiplicative: If $e_1, e_2 \in \mathfrak{E}(\mathbb{D})$, then $e_1 \cdot e_2 \in \mathfrak{E}(\mathbb{D})$.

The following statement is well-known (see e.g. Theorem 4.29 in [RoRo2]).

Lemma 2.1. *Let $e : \mathbb{D} \to \mathbb{C}$ be some function. Then the following statements are equivalent:*

(i) $e \in \mathfrak{E}(\mathbb{D})$.

(ii) $e \in \mathfrak{N}^+(\mathbb{D})$, $e \not\equiv 0$ *and* $e^{-1} \in \mathfrak{N}^+(\mathbb{D})$.

In particular, a function e of type (2.1) belongs to the class $\mathfrak{N}(\mathbb{D})$. Consequently, it possesses a boundary function $\underline{e} : \mathbb{T} \to \mathbb{C}$. It is known that for almost all $t \in \mathbb{T}$ with respect to m,

$$|\underline{e}(t)| = w(t). \tag{2.2}$$

As the function $\ln |e|$ is harmonic in $\mathbb{D}$ we obtain

$$\int_{\mathbb{T}} \ln |e(rt)| \;\; m(dt) = \ln |e(0)| = \int_{\mathbb{T}} \ln [w(t)] \;\; m(dt)$$

for $r \in [0,1)$. Consequently, if $e \in \mathfrak{E}(\mathbb{D})$, then for $r \in [0,1)$ we obtain

$$\int_{\mathbb{T}} \ln |e(rt)| \;\; m(dt) = \int_{\mathbb{T}} \ln |\underline{e}(t)| \;\; m(dt). \tag{2.3}$$

Let us recall the following useful characterization of outer functions (see e.g. Corollaries 4.16 and 4.17 in [RoRo2]).

Lemma 2.2. *Let $e \in \mathfrak{N}(\mathbb{D})$ but $e \not\equiv 0$. Then the following statements are equivalent:*

(i) *e is outer*

(ii) *For all $z \in \mathbb{D}$, $\ln |e(z)| = \int_{\mathbb{T}} \operatorname{Re}\frac{t+z}{t-z} \;\ln |\underline{e}(t)| \; m(dt)$.*

(iii) *There is a $z_0 \in \mathbb{D}$ such that $\ln |e(z_0)| = \int_{\mathbb{T}} \operatorname{Re}\frac{t+z_0}{t-z_0} \;\ln |\underline{e}(t)| \; m(dt)$.*

(iv) *If $h \in \mathcal{N}^+(\mathbb{D})$ satisfies $|\underline{h}(t)| \le |\underline{e}(t)|$ for almost all $t \in \mathbb{T}$ with respect to m,m then for all $z \in \mathbb{D}$ the inequality $|h(z)| \le |e(z)|$ holds true.*

(v) *If $z_0 \in \mathbb{D}$ and if $h \in \mathcal{N}^+(\mathbb{D})$ satisfies $|\underline{h}(t)| \le |\underline{e}(t)|$ for almost all $t \in \mathbb{T}$ with respect to m, then the inequality $|h(z_0)| \le |e(z_0)|$ holds true.*

Observe that conditions (iii) and (v) of Lemma 2.2 are usually used with the choice $z_0 = 0$.

In the proof of Lemma 2.4 and also in further considerations we will use the following result which goes back to V.I. Smirnov [Sm].

The maximum principle of V.I. Smirnov. *Let $f \in \mathfrak{N}^+(\mathbb{D})$ be such that its boundary function $\underline{f}$ is m-essentially bounded. Then f is bounded in the unit disc and satisfies*

$$\sup_{z\in\mathbb{D}} |f(z)| = \operatorname*{ess\,sup}_{t\in\mathbb{T}} |\underline{f}(t)|$$

This result can be generalized to the matrix case.

The maximum principle of V.I. Smirnov for matrix functions.
Let $F \in \mathfrak{N}_n^+(\mathbb{D})$ be such that its boundary function $\underline{F}$ satisfies $\operatorname*{ess\,sup}_{t\in\mathbb{T}} \|\underline{F}(t)\| < \infty$. Then F is bounded in the unit disc and satisfies

$$\sup_{z\in\mathbb{D}} \|F(z)\| = \operatorname*{ess\,sup}_{t\in\mathbb{T}} \|\underline{F}(t)\|.$$

Proof. Let $F = (F_{j,k})_{j,k=1}^n$, and fix the indices $j,k \in \{1,\dots,n\}$. In view of the inequality $|F_{j,k}(z)| \leq \|F(z)\|$ $(z \in \mathbb{D})$, $F_{j,k} \in \mathfrak{N}^+(\mathbb{D})$ and

$$\operatorname*{ess\,sup}_{t\in\mathbb{T}} |\underline{F_{j,k}}(t)| \leq \operatorname*{ess\,sup}_{t\in\mathbb{T}} \|\underline{F}(t)\|.$$

According to the maximum principle for scalar functions we then have

$$\sup_{z\in\mathbb{D}} |F_{j,k}(z)| < +\infty.$$

Hence,

$$\sup_{z\in\mathbb{D}} \|F(z)\| < +\infty.$$

The bounded holomorphic matrix-valued function F admits the Poisson integral representation

$$F(z) = \int_{\mathbb{T}} \underline{F}(t) \cdot P(z,t)\, m(dt) \quad , \quad z \in \mathbb{D}.$$

Therefore, by the integral version of the triangle inequality, we obtain

$$\|F(z)\| \leq \int_{\mathbb{T}} \|\underline{F}(t)\| \cdot P(z,t)\, m(dt) \quad , \quad z \in \mathbb{D}.$$

But this in turn implies the inequality $\|F(z)\| \leq \operatorname*{ess\,sup}_{t\in\mathbb{T}} \|\underline{F}(t)\|$ $(z \in \mathbb{D})$. □

Since the function $\ln^+ \|F\|$ is subharmonic for an analytic matrix-valued function F the following result is true.

Lemma 2.3. *Let $F \in \mathfrak{N}_n^+(\mathbb{D})$. Then for all $z \in \mathbb{D}$ the inequality*

$$\|F(z)\| \leq \exp\left\{ \int_{\mathbb{T}} P(z,t) \ln \|\underline{F}(t)\|\ m(dt) \right\} \tag{2.4}$$

holds true.

For a proof of Lemma 2.3 we refer to Theorem 3.13 in [RoRo2].
Clearly, the maximum principle of V.I. Smirnov is a consequence of inequality (2.4).

Definition 2.1. *The set* $\mathfrak{S}_{n\times n}(\mathbb{D})$ *of all holomorphic matrix-valued functions* $S : \mathbb{D} \to \mathfrak{C}_n$ *is called* the $n \times n$ Schur class.

Lemma 2.4. *A matrix-valued function* $F : \mathbb{D} \to \mathfrak{M}_n$ *belongs to the Smirnov class* $\mathfrak{N}_n^+(\mathbb{D})$ *if and only if it admits a representation of the form*

$$F = \frac{1}{d} \cdot \Phi, \tag{2.5}$$

where $\Phi \in \mathfrak{S}_{n\times n}(\mathbb{D})$ *and* d *is an outer function which belongs to* $\mathfrak{S}(\mathbb{D})$.

Proof. I. Suppose that F admits a representation of the form (2.5). Then $\Phi \in \mathfrak{N}_n^+(\mathbb{D})$ and, as d is outer, we have $d^{-1} \in \mathfrak{N}^+(\mathbb{D})$. Thus, as $\mathfrak{N}^+(\mathbb{D})$ is an algebra over $\mathbb{C}$, we get $\Phi \cdot d^{-1} \in \mathfrak{N}_n^+(\mathbb{D})$.

II. Suppose that $F \in \mathfrak{N}_n^+(\mathbb{D})$. We can assume that F is not the null function in $\mathbb{D}$. Then $\ln \|\underline{F}\|$ is m-integrable. We define $d : \mathbb{D} \to \mathbb{C}$ via

$$d(z) := \exp\left\{ - \int\limits_{\mathbb{T}} \frac{t+z}{t-z} \ln \|\underline{F}\| \quad m(dt) \right\}.$$

Then, from our earlier considerations (see (2.1) - (2.4)), it is clear that d is a scalar outer function and that the corresponding boundary function $\underline{d}$ satisfies

$$|\underline{d}(t)| = \|\underline{F}(t)\|^{-1} \tag{2.6}$$

for almost all $t \in \mathbb{T}$ with respect to m. Now define $\Phi : \mathbb{D} \to \mathfrak{M}_n$ via

$$\Phi(z) := d(z) \cdot F(z). \tag{2.7}$$

Then, since $F \in \mathfrak{N}_n^+(\mathbb{D}), d \in \mathfrak{N}^+(\mathbb{D})$ and $\mathfrak{N}^+(\mathbb{D})$ is an algebra over $\mathbb{C}$, we see that

$$\Phi \in \mathfrak{N}_n^+(\mathbb{D}). \tag{2.8}$$

From (2.6) and (2.7) we get

$$\|\underline{\Phi}(t)\| = 1 \tag{2.9}$$

for almost all $t \in \mathbb{T}$ with respect to m. Finally, in view of (2.8) and (2.9), the maximum principle of V.I. Smirnov implies that for $z \in \mathbb{D}$ we obtain $\|\Phi(z)\| \leq 1$. Thus, $\Phi \in \mathfrak{S}_{n\times n}(\mathbb{D})$. □

3 Outer matrix-valued functions

The main goal of this section is to discuss outer matrix-valued functions which belong to the Smirnov class $\mathfrak{N}_n^+(\mathbb{D})$. The needs of prediction theory of multivariate stationary stochastic processes initiated an intensive study of matrix-valued outer functions belonging to the Hardy class $H_n^2(\mathbb{D})$ (see Definition 5.1 below) which is a subclass of $\mathfrak{N}_n^+(\mathbb{D})$. The formula for the best predictor of a multivariate stationary stochastic process of a given time in terms of its past depends in an essential manner on a particular outer matrix-valued function belonging to $H_n^2(\mathbb{D})$ (see Wiener and Masani [WM1], [WM2], Helson and Lowdenslager [HL1], [HL2], Rozanov [Roz1], [Roz2], Masani [Ma1]–[Ma4] and for operator-valued generalizations also Devinatz [De], Helson [Hel1], Sz.-Nagy and Foias [SZNF], Nikolskii [Nik2]).

Definition 3.1. *A matrix-valued function* $E : \mathbb{D} \to \mathfrak{M}_n$ *is called* outer *(in the sense of V.I. Smirnov) if* $E \in \mathfrak{N}_n^+(\mathbb{D})$ *and* $\det E$ *is outer. The class of all* $n \times n$ *matrix-valued outer functions will be denoted by* $\mathfrak{E}_n(\mathbb{D})$.

If $E \in \mathfrak{E}_n(\mathbb{D})$ then, in particular, for all $z \in \mathbb{D}$ we have

$$\det [E(z)] \neq 0.$$

Definition 3.1 is clearly an immediate generalization of the notion of a scalar outer function. This definition of an outer matrix-valued function enables us to avoid the study of the question of a matricial analogue of formula (2.1).

Remark 3.1. The class $\mathfrak{E}_n(\mathbb{D})$ is multiplicative: If $E_1, E_2 \in \mathfrak{E}_n(\mathbb{D})$ then $E_1 \cdot E_2 \in \mathfrak{E}_n(\mathbb{D})$.

Remark 3.2. Let $E \in \mathfrak{E}_n(\mathbb{D})$. Then $E^\top \in \mathfrak{E}_n(\mathbb{D})$.

Theorem 3.1 (Determinant characterization of outer matrix-valued functions).

(i) *Let* $E \in \mathfrak{E}_n(\mathbb{D})$. *Then* $\det [E(z)] \neq 0$ *for all* $z \in \mathbb{D}$ *and* $E^{-1} \in \mathfrak{N}_n^+(\mathbb{D})$.

(ii) *Let* E *be a function from* $\mathfrak{N}_n^+(\mathbb{D})$ *for which* $\det E$ *never vanishes in* $\mathbb{D}$ *and* E^{-1} *belongs to* $\mathfrak{N}_n^+(\mathbb{D})$. *Then* $E \in \mathfrak{E}_n(\mathbb{D})$.

Proof. (i) According to the rule for computing the inverse matrix we have the representation

$$E^{-1} = \frac{1}{\det E} \cdot A \tag{3.1}$$

where $A : \mathbb{D} \to \mathfrak{M}_n$ is a matrix-valued function the entries of which are polynomials of the elements of matrix E (namely, the cofactors of the corresponding elements). Since the class $\mathfrak{N}^+(\mathbb{D})$ is an algebra over $\mathbb{C}$, each entry of A belongs to $\mathfrak{N}^+(\mathbb{D})$. Hence, $A \in \mathfrak{N}_n^+(\mathbb{D})$. From the fact that $E \in \mathfrak{E}_n(\mathbb{D})$ and Lemma 2.1 it then follows that $(\det E)^{-1} \in \mathfrak{N}^+(\mathbb{D})$, and thus in view of (3.1), $E^{-1} \in \mathfrak{N}_n^+(\mathbb{D})$. Hence, (i) is proved.

(ii) By Lemma 1.2, $\det E \in \mathfrak{N}^+(\mathbb{D})$ and $\det (E^{-1}) \in \mathfrak{N}^+(\mathbb{D})$. Therefore, the function $\det E$ satisfies condition (ii) in Lemma 2.1. Thus, $\det E \in \mathfrak{E}(\mathbb{D})$, and so, in view of Definition 3.1, $E \in \mathfrak{E}_n(\mathbb{D})$. Hence (ii) is proved. □

The following result supplements the statement of Lemma 2.4.

Lemma 3.1. *Let $E \in \mathfrak{E}_n(\mathbb{D})$. Then E has a representation of the form*

$$E = \frac{1}{d} \cdot C, \tag{3.2}$$

where $C \in \mathfrak{S}_{n\times n}(\mathbb{D}) \cap \mathfrak{E}_n(\mathbb{D})$ and $d \in \mathfrak{E}(\mathbb{D})$.

Proof. In view of Lemma 2.4, the function E has a representation of the form

$$E = \frac{1}{d} \cdot C,$$

where $C \in \mathfrak{S}_{n\times n}(\mathbb{D})$ and $d \in \mathfrak{E}(\mathbb{D})$. Lemma 2.1 guarantees that $d^{-1} \in \mathfrak{N}^+(\mathbb{D})$. Since $E \in \mathfrak{E}_n(\mathbb{D})$, it follows from Theorem 3.1 that $E^{-1} \in \mathfrak{N}_n^+(\mathbb{D})$. Therefore, as $\mathfrak{N}^+(\mathbb{D})$ is an algebra over $\mathbb{C}$, from $C^{-1} = d^{-1} \cdot E^{-1}$ we see that $C^{-1} \in \mathfrak{N}_n^+(\mathbb{D})$. Thus, as $C \in \mathfrak{S}_{n\times n}(\mathbb{D}) \subseteq \mathfrak{N}_n^+(\mathbb{D})$ it follows from Theorem 3.1 that $C \in \mathfrak{E}_n(\mathbb{D})$. □

Let us recall the following notion.

Definition 3.2. *The class $H_n^\infty(\mathbb{D})$ consists of all matrix-valued functions $F : \mathbb{D} \to \mathfrak{M}_n$ which are holomorphic and bounded in $\mathbb{D}$, i.e.,*

$$\sup_{z\in\mathbb{D}} \|F(z)\| < \infty. \tag{3.3}$$

Theorem 3.2.

(i) *Let $E \in \mathfrak{E}_n(\mathbb{D})$. Then there exists a sequence $(F_k)_{k\in\mathbb{N}}$ from $H_n^\infty(\mathbb{D})$ with the following properties:*
  - (α) *For almost all $t \in \mathbb{T}$ with respect to m,* $\lim_{k\to\infty} \underline{E}(t) \cdot \underline{F_k}(t) = I_n$.
  - (β) *The family $\left(\ln^+ \|\underline{F_k}\|\right)_{k\in\mathbb{N}}$ is uniformly integrable with respect to* m.
  - (γ) *There exists a Borel subset B_0 of $\mathbb{T}$ with $m(B_0) = 0$ such that for all $k \in \mathbb{N}$ and all $t \in \mathbb{T} \setminus B_0$ the inequality $\|\underline{E}(t) \cdot \underline{F_k}(t)\| \leq 1$ holds true.*

(ii) *Let $E \in \mathfrak{N}_n^+(\mathbb{D})$ be such that there exists a sequence $(F_k)_{k\in\mathbb{N}}$ belonging to $H_n^\infty(\mathbb{D})$ satisfying the above conditions (α) and (β). Then $E \in \mathfrak{E}_n(\mathbb{D})$.*

Remark 3.3. Theorem 3.2 expresses in some sense a Smirnov class generalization of that characterization of the property that a function is outer which is formulated in terms of the shift-invariant subspace generated by this function. Sometimes the approximation property contained in Theorem 3.2 is called weak invertibility of the function E (see [Sh] or [Nik1,Ch.2]). For the spaces $H_n^\infty(\mathbb{D})$ or $H_n^2(\mathbb{D})$ this approximation property (weak invertibility) will be often used for defining the notion "outer function". Observe that in the scalar case ($n = 1$) it was already shown by V.I. Smirnov [Sm] that for an outer function e the linear subspace $e \cdot H^2(\mathbb{D})$ is dense in $H^2(\mathbb{D})$. Concerning several generalizations of this result of V.I. Smirnov we refer the reader to chapter 2 in [Nik1] (in particular, see Theorem 3 in Section 2.2.).

Proof of Theorem 3.2. (i) Since E is a matrix-valued outer function, Theorem 3.1 guarantees that $E^{-1} \in \mathfrak{N}_n^+(\mathbb{D})$. We fix a boundary function $\underline{E}$ of E such $\det[\underline{E}(t)] \neq 0$ for $t \in \mathbb{T}$. Then for $k \in \mathbb{N}$ we define $w_k : \mathbb{T} \to (0, \infty)$ via

$$w_k(t) := \begin{cases} 1 & , \text{ if } \ \|\underline{E}^{-1}(t)\| < k \\ \dfrac{1}{\|\underline{E}^{-1}(t)\|} & , \text{ if } \ \|\underline{E}^{-1}(t)\| \geq k. \end{cases} \tag{3.4}$$

Clearly

$$0 < w_1(t) \leq w_2(t) \leq w_3(t) \leq \ldots \leq 1 \tag{3.5}$$

for $t \in \mathbb{T}$ and

$$\lim_{k \to \infty} w_k(t) = 1. \tag{3.6}$$

From (3.5) we see that the inequality

$$w_1(t) \geq \|\underline{E}^{-1}(t)\|^{-1} \tag{3.7}$$

holds for $t \in \mathbb{T}$. Since $E^{-1} \in \mathfrak{N}_n^+(\mathbb{D})$, we infer that

$$\ln[\|\underline{E}^{-1}\|^{-1}] \in \mathcal{L}^1(\mathbb{T}, \mathfrak{B}_{\mathbb{T}}, m; \mathbb{C}). \tag{3.8}$$

From (3.5)–(3.7) we obtain

$$\int_{\mathbb{T}} \ln[w_k(t)] \; m(dt) > -\infty. \tag{3.9}$$

Hence, for $k \in \mathbb{N}$ the function $\varphi_k : \mathbb{D} \to \mathbb{C}$ which is given by

$$\varphi_k(z) := \exp\left\{ \int_{\mathbb{T}} \frac{t+z}{t-z} \ln[w_k(t)] \; m(dt) \right\}$$

is well defined. Moreover from its definition it is clear that $\varphi_k \in \mathfrak{N}^+(\mathbb{D})$ (or more precisely, that φ_k is even outer). In view of (3.5) and (3.6) the monotone convergence theorem guarantees that

$$\lim_{k \to \infty} \varphi_k(z) = 1 \,, \qquad z \in \mathbb{D}. \tag{3.10}$$

Since $|\underline{\varphi_k}(t)| = w_k(t)$ for almost all $t \in \mathbb{T}$ with respect to m, formula (3.6) yields

$$\lim_{k \to \infty} |\underline{\varphi_k}(t)| = 1$$

for almost all $t \in \mathbb{T}$ with respect to m. In view of (3.5) and (3.6), another application of the monotone convergence theorem gives us

$$\lim_{k \to \infty} \int_{\mathbb{T}} |\underline{\varphi_k}(t)|^2 \; m(dt) = \lim_{k \to \infty} \int_{\mathbb{T}} [w_k(t)]^2 \; m(dt) = \int_{\mathbb{T}} 1 \; dm = 1. \tag{3.11}$$

For $k \in \mathbb{N}$, we have

$$\int_{\mathbb{T}} |\underline{\varphi_k}(t) - 1|^2 \, m(dt) = \int_{\mathbb{T}} |\underline{\varphi_k}(t)|^2 \, m(dt) - 2\Re[\varphi_k(0)] + 1. \tag{3.12}$$

Combining (3.10)–(3.12) it follows that

$$\lim_{k\to\infty} \int_{\mathbb{T}} |\underline{\varphi_k}(t) - 1|^2 \, m(dt) = 0. \tag{3.13}$$

In view of (3.13), the F. Riesz-Fischer theorem yields a subsequence $(\underline{\varphi_{l_k}})_{k\in\mathbb{N}}$ of $(\underline{\varphi_k})_{k\in\mathbb{N}}$ such that

$$\lim_{k\to\infty} \underline{\varphi_{l_k}}(t) = 1 \tag{3.14}$$

for almost all $t \in \mathbb{T}$ with respect to m. Let $k \in \mathbb{N}$ and set

$$F_k := E^{-1} \cdot \varphi_{l_k}. \tag{3.15}$$

Then, since $E^{-1} \in \mathfrak{N}_n^+(\mathbb{D})$ and $\varphi_{l_k} \in \mathfrak{N}^+(\mathbb{D})$, we get $F_k \in \mathfrak{N}_n^+(\mathbb{D})$. Thus as $|\underline{\varphi_{l_k}}| = w_{l_k}$ almost everywhere with respect to m it follows from (3.15) and (3.4) that

$$\|\underline{F_k}(t)\| = w_{l_k}(t) \cdot \|\underline{E}^{-1}(t)\| \le l_k$$

for almost all $t \in \mathbb{T}$ with respect to m. Thus, the maximum principle of V.I. Smirnov implies that $\|F_k(z)\| \le l_k$ for all $z \in \mathbb{D}$. Consequently, $F_k \in H_n^\infty(\mathbb{D})$. From (3.15) it follows that

$$E \cdot F_k = \varphi_{l_k} \cdot I_n. \tag{3.16}$$

From (3.5) we obtain

$$|\underline{\varphi_{l_k}}(t)| = w_{l_k}(t) \le 1$$

and hence since $\varphi_{l_k} \in \mathfrak{N}^+(\mathbb{D})$, the maximum principle of V.I. Smirnov guarantees that that

$$|\varphi_{l_k}(z)| \le 1, \qquad z \in \mathbb{D}. \tag{3.17}$$

Thus, combining (3.16) and (3.17) we see that (γ) is fulfilled.
Moreover, from (3.16) and (3.14) we get that (α) is satisfied.
For almost all $t \in \mathbb{T}$ with respect to m we have $|\underline{\varphi_{l_k}}(t)| \le 1$ and, consequently, in view of (3.15), the inequality

$$\ln^+ \|\underline{F_k}(t)\| \le \ln^+ \|\underline{E}^{-1}(t)\|$$

holds for almost all $t \in \mathbb{T}$ with respect to m. Hence, the family $(\ln^+ \|\underline{F_k}(t)\|)_{k\in\mathbb{N}}$ has an m-integrable majorant. This implies that (β) is fulfilled.
Part (i) of Theorem 3.2 is now proved.

Before proving part (ii) of Theorem 3.2 we recall the following result (see [WM1, Lemma 3.12]).

The generalized Minkowski inequality. *Let $(\Omega, \mathfrak{A}, P)$ be a probability space and let $M : \Omega \to \mathfrak{M}_n$ be a P-integrable matrix function with nonnegative Hermitian values. Then*

$$\ln\left[\det\left(\int_\Omega M\, dP\right)\right] \geq \int_\Omega \ln\left[\det M\right] dP. \tag{3.18}$$

Proof of part (ii) of Theorem 3.2. For $k \in \mathbb{N}$ we define $v_k : \mathbb{T} \to [1, \infty)$ via the rule

$$v_k(t) := \begin{cases} \|\underline{E}(t) \cdot \underline{F_k}(t)\| & , \text{ if } \ \|\underline{E}(t) \cdot \underline{F_k}(t)\| \geq 1 \\ & , \text{ if } \ \|\underline{E}(t) \cdot \underline{F_k}(t)\| < 1. \end{cases} \tag{3.19}$$

For $k \in \mathbb{N}$ and $t \in \mathbb{T}$ we then have

$$\ln\left[v_k(t)\right] \in [0, \infty). \tag{3.20}$$

Combining (α) and (3.19) we infer that for almost all $t \in \mathbb{T}$ with respect to m,

$$\lim_{k\to\infty} \ln\left[v_k(t)\right] = 0. \tag{3.21}$$

For $k \in \mathbb{N}$ and $t \in \mathbb{T}$ we get the inequality

$$\ln\left[v_k(t)\right] \leq \ln^+ \|\underline{E}(t)\| + \ln^+ \|\underline{F_k}(t)\|$$

from (3.19), which together with (β) implies that the family $(\ln v_k)_{k\in\mathbb{N}}$ is uniformly m-integrable. Combining this fact with (3.20) and (3.21), an application of Vitali's Theorem yields

$$\lim_{k\to\infty} \int_\mathbb{T} \ln\left[v_k(t)\right]\, m(dt) = 0. \tag{3.22}$$

For $k \in \mathbb{N}$ we define $\Psi_k : \mathbb{D} \to \mathbb{C}$ via the formula

$$\Psi_k(z) := \exp\left\{ - \int_\mathbb{T} \ln\left[v_k(t)\right] \frac{t+z}{t-z}\, m(dt) \right\}. \tag{3.23}$$

Therefore, in view of (3.20), we obtain the inequality

$$\begin{aligned} |\Psi_k(z)| &= \exp\left\{ \Re\left[- \int_\mathbb{T} \ln\left[v_k(t)\right] \frac{t+z}{t-z}\, m(dt) \right] \right\} \\ &= \exp\left\{ - \int_\mathbb{T} \ln\left[v_k(t)\right] \frac{1-|z|^2}{|t-z|^2}\, m(dt) \right\} \leq \exp\{0\} = 1 \end{aligned} \tag{3.24}$$

for $z \in \mathbb{D}$. In view of (3.21), an application of Lebesgue's dominated convergence theorem yields

$$\lim_{k\to\infty} \Psi_k(z) = 1 \tag{3.25}$$

for all $z \in \mathbb{D}$. For almost all $t \in \mathbb{T}$ with respect to m we get

$$|\underline{\Psi_k}(t)| \leq 1 \tag{3.26}$$

from (3.24) and hence, since formula (3.23) implies that

$$|\underline{\Psi_k}(t)| = [v_k(t)]^{-1}, \tag{3.27}$$

we see from (3.19) and (α) that

$$\lim_{k\to\infty} |\underline{\Psi_k}(t)| = 1. \tag{3.28}$$

For $k \in \mathbb{N}$,

$$\int_{\mathbb{T}} |\underline{\Psi_k}(t) - 1|^2 \; m(dt) = \int_{\mathbb{T}} |\underline{\Psi_k}(t)|^2 \; m(dt) - 2\Re[\Psi_k(0)] + 1. \tag{3.29}$$

In view of (3.26) and (3.28), Lebesgue's dominated convergence theorem yields

$$\lim_{k\to\infty} \int_{\mathbb{T}} |\underline{\Psi_k}(t)|^2 \; m(dt) = m(\mathbb{T}) = 1. \tag{3.30}$$

Combining (3.25), (3.29) and (3.30) we obtain

$$\lim_{k\to\infty} \int_{\mathbb{T}} |\underline{\Psi_k}(t) - 1|^2 \; m(dt) = 0. \tag{3.31}$$

In view of (3.31), the F. Riesz-Fischer theorem provides a subsequence $(\underline{\Psi_{l_k}})_{k\in\mathbb{N}}$ of $(\underline{\Psi_k})_{k\in\mathbb{N}}$ such that

$$\lim_{k\to\infty} \underline{\Psi_{l_k}}(t) = 1 \tag{3.32}$$

for almost all $t \in \mathbb{T}$ with respect to m. Suppose that $k \in \mathbb{N}$ and define

$$\Phi_k := E \cdot F_k \cdot \Psi_k. \tag{3.33}$$

Then, since $E \in \mathfrak{N}_n^+(\mathbb{D})$, $F_k \in H_n^\infty(\mathbb{D})$ and (3.24) holds, we get

$$\Phi_k \in \mathfrak{N}_n^+(\mathbb{D}). \tag{3.34}$$

For almost all $t \in \mathbb{T}$ with respect to m it follows from (3.33), (3.19) and (3.27) that

$$\|\underline{\Phi_k}(t)\| = |\underline{\Psi_k}(t)| \cdot \|\underline{E}(t) \cdot \underline{F_k}(t)\| \leq |\underline{\Psi_k}(t)| \cdot v_k(t) = 1. \tag{3.35}$$

Therefore the maximum principle of V.I. Smirnov implies that

$$\|\Phi_k(z)\| \leq 1 \tag{3.36}$$

for all $z \in \mathbb{D}$. In particular,

$$\Phi_k \in H_n^\infty(\mathbb{D}). \tag{3.37}$$

From (3.34) and (3.35) it follows that

$$\underline{\Phi_k}^\star(t) \cdot \underline{\Phi_k}(t) \leq I_n \tag{3.38}$$

for almost all $t \in \mathbb{T}$ with respect to m. Combining (3.33), (α) and (3.28) we get

$$\lim_{k\to\infty} \underline{\Phi_k}^\star(t) \cdot \underline{\Phi_k}(t) = \lim_{k\to\infty} |\underline{\Psi_k}(t)|^2 [\underline{E}(t)\underline{F_k}(t)]^\star [\underline{E}(t)\underline{F_k}(t)] = I_n. \tag{3.39}$$

From (3.32), (3.33) and (α) we now obtain

$$\lim_{k\to\infty} \underline{\Phi_{l_k}}(t) = \lim_{k\to\infty} \underline{E}(t) \cdot \underline{F_{l_k}}(t) \cdot \underline{\Psi_{l_k}}(t) = I_n. \tag{3.40}$$

Using (3.37), (3.38), (3.40) and Lebesgue's dominated convergence theorem we get

$$\lim_{k\to\infty} \Phi_{l_k}(0) = \lim_{k\to\infty} \int_{\mathbb{T}} \underline{\Phi_{l_k}}(t)\, m(dt) = I_n. \tag{3.41}$$

Suppose that $k \in \mathbb{N}$. We define $M_k : \mathbb{T} \to \mathfrak{M}_n$ via the rule

$$M_k(t) := \underline{\Phi_k}^\star(t) \cdot \underline{\Phi_k}(t). \tag{3.42}$$

Then (3.42) and (3.38) imply that the inequality $0 \leq M_k(t) \leq I_n$ holds true for almost all $t \in \mathbb{T}$ with respect to m. Hence,

$$0 \leq \int_{\mathbb{T}} M_k(t)\, m(dt) \leq I_n. \tag{3.43}$$

Now we apply the Generalized Minkowski inequality to the M_k. (Note that normalized Lebesgue measure m is a probability measure.) From (3.43) we infer first that

$$\ln\left[\det\left(\int_{\mathbb{T}} M_k(t)\, m(dt)\right)\right] \leq \ln\left[\det I_n\right] = 0. \tag{3.44}$$

Hence, (3.44) and the Generalized Minkowski inequality guarantee that

$$\int_{\mathbb{T}} \ln\left(\det\left[M_k(t)\right]\right) m(dt) \leq 0. \tag{3.45}$$

Using (3.42) and (3.33) it follows that

$$\begin{aligned} \frac{1}{2}\ln(\det[M_k(t)]) &= \ln|\det[\underline{\Phi_k}(t)]| \\ &= \ln|\det[\underline{E}(t)]| + \ln|\det\{[\underline{F_k}(t)]\cdot[\underline{\Psi_k}(t)]\}| \end{aligned} \tag{3.46}$$

for almost all $t \in \mathbb{T}$ with respect to m. Thus, from (3.45) and (3.46) we see that

$$\int_{\mathbb{T}} \ln|\det[\underline{E}(t)]|\, m(dt) \leq -\int_{\mathbb{T}} \ln|\det\{[\underline{F_k}(t)]\cdot[\underline{\Psi_k}(t)]\}|\, m(dt). \tag{3.47}$$

By assumption, $F_k \in H_n^\infty(\mathbb{D})$. Using (3.23) and (3.24) we see that $\Psi_k \in H_n^\infty(\mathbb{D})$. Thus, $F_k \cdot \Psi_k \in H_n^\infty(\mathbb{D})$ and, consequently, $\det[F_k \cdot \Psi_k] \in H^\infty(\mathbb{D})$. Now Jensen's inequality gives

$$-\int_{\mathbb{T}} \ln|\det\{[\underline{F_k}(t)]\cdot[\underline{\Psi_k}(t)]\}|\, m(dt) \leq -\ln|\det\{[\underline{F_k}(0)]\cdot[\underline{\Psi_k}(0)]\}|. \tag{3.48}$$

From (3.47) and (3.48) it now follows that

$$\int_{\mathbb{T}} \ln|\det[\underline{E}(t)]|\, m(dt) \leq -\ln|\det\{[\underline{F_k}(0)]\cdot[\underline{\Psi_k}(0)]\}|. \tag{3.49}$$

From (3.33) and (3.41) we obtain

$$\lim_{k\to\infty} \ln|\det\{[F_{l_k}(0)]\cdot[\Psi_{l_k}(0)]\}| = -\ln|\det[E(0)]|. \tag{3.50}$$

Combining (3.49) and (3.50) we obtain

$$\int_{\mathbb{T}} \ln|\det[\underline{E}(t)]|\, m(dt) \leq \ln|\det[E(0)]|.$$

By assumption, $E \in \mathfrak{N}_n^+(\mathbb{D})$. Thus, $\det E \in \mathfrak{N}^+(\mathbb{D})$ and Jensen's inequality yields

$$\ln|\det[E(0)]| \leq \int_{\mathbb{T}} \ln|\det[\underline{E}(t)]|\, m(dt).$$

Hence,

$$\int_{\mathbb{T}} \ln|\det[\underline{E}(t)]|\, m(dt) = \ln|\det[E(0)]|. \tag{3.51}$$

From (3.51) and Lemma 2.1 we see that $\det E \in \mathfrak{E}(\mathbb{D})$. Therefore, by definition 3.1, $E \in \mathfrak{E}_n(\mathbb{D})$. Part (ii) of Theorem 3.2 is now proved. □

Theorem 3.3.

(i) Let $E \in \mathfrak{E}_n(\mathbb{D})$. Then there exists a sequence $(F_k)_{k \in \mathbb{N}}$ from $H_n^\infty(\mathbb{D})$ with the following properties:

(α) For almost all $t \in \mathbb{T}$ with respect to m, $\lim\limits_{k \to \infty} \underline{F_k}(t) \cdot \underline{E}(t) = I_n$.

(β) The family $\left(\ln^+ \|\underline{F_k}\| \right)_{k \in \mathbb{N}}$ is uniformly integrable with respect to m.

(γ) There exists a Borel subset B_0 of $\mathbb{T}$ with $m(B_0) = 0$ such that for all $k \in \mathbb{N}$ and all $t \in \mathbb{T} \setminus B_0$ the inequality $\|\underline{F_k}(t) \cdot \underline{E}(t)\| \leq 1$ holds.

(ii) Let $E \in \mathfrak{N}_n^+(\mathbb{D})$ be such that there exists a sequence $(F_k)_{k \in \mathbb{N}}$ which belongs to $H_n^\infty(\mathbb{D})$ and satisfies the above conditions (α) and (β). Then $E \in \mathfrak{E}_n(\mathbb{D})$.

Proof. Combine Theorem 3.2 and Remark 3.2. $\qquad\qquad\qquad\qquad\qquad$ □

It should be mentioned that Ginzburg [Gi1] obtained a multiplicative integral representation for outer functions which belong to $\mathfrak{E}_n(\mathbb{D})$.

4 Matrix-valued inner functions

In this section, we draw attention to a distinguished subclass of the Schur class $\mathfrak{S}_{n \times n}$ (compare Definition 2.1).

Definition 4.1. Let $\Theta \in \mathfrak{S}_{n \times n}(\mathbb{D})$. Then Θ is called *inner* if

$$I_n - \underline{\Theta}^\star(t) \cdot \underline{\Theta}(t) = \mathbb{O}_{n \times n} \qquad\qquad (4.1)$$

for almost all $t \in \mathbb{T}$ with respect to m. The class of all $n \times n$ matrix-valued inner functions will be denoted by $\mathfrak{I}_n(\mathbb{D})$.

Remark 4.1. Let $\Theta \in \mathfrak{I}_n(\mathbb{D})$. Then obviously $\det \Theta \not\equiv 0$.

Remark 4.2. Let $\Theta \in \mathfrak{I}_n(\mathbb{D})$. Then $\Theta^\top \in \mathfrak{I}_n(\mathbb{D})$.

The class $\mathfrak{I}_n(\mathbb{D})$ contains two important subclasses, namely the so-called singular inner functions and the Blaschke-Potapov products. Now we will formulate the corresponding definitions.

Definition 4.2. Let $S \in \mathfrak{I}_n(\mathbb{D})$. Then S is called *singular*, if $\det[S(z)] \neq 0$ for all $z \in \mathbb{D}$ (or in other words if S^{-1} is holomorphic in $\mathbb{D}$). The class of all $n \times n$ matrix-valued singular inner functions will be denoted by $\mathfrak{I}_{n,s}(\mathbb{D})$.

Remark 4.3. If $S \in \mathfrak{I}_{n,s}(\mathbb{D})$, then $S^{-1} \in \mathfrak{N}_n(\mathbb{D})$, be cause S^{-1} admits the representation $S^{-1} = L \cdot (\det S)^{-1}$ with bounded holomorphic functions L and $\det S$.

Lemma 4.1. Let $S \in \mathfrak{I}_{n,s}(\mathbb{D})$ be such that $S^{-1} \in \mathfrak{N}_n^+(\mathbb{D})$. Then S is constant.

Proof. . Since $\underline{S}(t)$ is unitary for a.e. $t \in \mathbb{T}$ it follows that

$$\|\underline{S^{-1}}(t)\| = 1.$$

Therefore, by the maximum principle of V.I. Smirnov, $\|S^{-1}(z)\| \leq 1$ for all $z \in \mathbb{D}$. Since $\|S(z)\| \leq 1$ then it follows that $S(z)$ is a unitary matrix for all $z \in \mathbb{D}$. However a holomorphic matrix function with unitary values is necessarily constant (see e.g. Corollary 2.3.2 in [DFK]). □

Now we are going to define Blaschke-Potapov products. For this reason, we recall first the notion of a scalar elementary Blaschke factor. Let $a \in \mathbb{D}$. Then we define $b_a : \mathbb{D} \to \mathbb{C}$ via the rule

$$b_a(z) := \begin{cases} \frac{|a|}{a} \cdot \frac{a - z}{1 - \overline{a}z} & , \text{ if } \quad a \in \mathbb{D} \setminus \{0\} \\ z & , \text{ if } \quad a = 0 \end{cases} . \tag{4.2}$$

Assume that $P \in \mathfrak{M}_n$ is a non-zero orthoprojection matrix, i.e., that the conditions

$$P^2 = P \quad \text{and} \quad P = P^* \tag{4.3}$$

are satisfied. Then the matrix-valued function $B_{a,P} : \mathbb{D} \to \mathfrak{M}_n$ which is defined by

$$B_{a,P}(z) := I_n + [b_a(z) - 1] \cdot P \tag{4.4}$$

is called *the Blaschke-Potapov elementary factor associated with a and P*.
From (4.3) and (4.4) it is clear that

$$\det [B_{a,P}] = (b_a)^{\text{rank } P}. \tag{4.5}$$

Suppose that $(z_k)_{k \in I}$ is a sequence from $\mathbb{D}$ and that $(P_k)_{k \in I}$ is a sequence of orthoprojection matrices for which the condition

$$\sum_{k \in I} (1 - |z_k|) \cdot \text{tr } P_k < +\infty \tag{4.6}$$

is fulfilled. (The index set I can be finite or infinite.) Then, according to a result due to V.P. Potapov [Pot], the product

$$\overset{\curvearrowright}{\prod_{k \in I}} B_{z_k,P_k}(z) \qquad \left(\text{resp. } \overset{\curvearrowleft}{\prod_{k \in I}} B_{z_k,P_k}(z) \right) \tag{4.7}$$

converges for all $z \in \mathbb{D}$. (V.P. Potapov has also shown that condition (4.6) is necessary for the convergence of the product in (4.7).)

Definition 4.3. *Let $B : \mathbb{D} \to \mathfrak{M}_n$. Then B is called* a left *(resp.* right*)* Blaschke-Potapov product *if B is a constant function with unitary value or if there exist a* *unitary matrix V, a set of orthoprojection matrices $(P_k)_{k\in I}$ and a sequence $(z_k)_{k\in I}$ which belongs to $\mathbb{D}$ such that (4.6) is satisfied and moreover the representation*

$$B = \left(\overset{\curvearrowright}{\prod_{k\in I}} B_{z_k,P_k}(z)\right)\cdot V \qquad \left(\text{resp.} \quad B = V\cdot\left(\overset{\curvearrowleft}{\prod_{k\in I}} B_{z_k,P_k}(z)\right)\right)$$

is valid. The set of left (resp. right) Blaschke-Potapov products will be denoted by $\mathfrak{I}_{n,B,l}(\mathbb{D})$ (resp. $\mathfrak{I}_{n,B,r}(\mathbb{D})$).

We will see below that each left Blaschke-Potapov product is also a right Blaschke-Potapov product and vice versa. Moreover, it will turn out that $\mathfrak{I}_{n,B,l}(\mathbb{D}) \subseteq \mathfrak{I}_n(\mathbb{D})$.

Lemma 4.2. *Let $A, B \in \mathfrak{C}_n$ be such that $A\cdot B$ is unitary. Then A and B are unitary too.*

Proof. Since $A, B \in \mathfrak{C}_n$ we have $I_n - AA^\star \geq \mathbb{O}_{n\times n}$ and $I_n - BB^\star \geq \mathbb{O}_{n\times n}$. Hence, $A(I_n - BB^\star)A^\star \geq \mathbb{O}_{n\times n}$. Therefore, the identity

$$\mathbb{O}_{n\times n} = I_n - (AB)(AB)^\star = (I_n - AA^\star) + A(I_n - BB^\star)A^\star$$

implies that $I_n - AA^\star = \mathbb{O}_{n\times n}$ and $A(I_n - BB^\star)A^\star = \mathbb{O}_{n\times n}$. Thus, A is unitary. In particular, we have $\det A \neq 0$. This implies that $I_n - BB^\star = \mathbb{O}_{n\times n}$ and hence that B is unitary too. □

Theorem 4.1. *Suppose that $\Theta \in \mathfrak{I}_n(\mathbb{D})$.*

(a) *There exist functions $B \in \mathfrak{I}_{n,B,l}(\mathbb{D})$ (resp. $C \in \mathfrak{I}_{n,B,r}(\mathbb{D})$) and $S \in \mathfrak{I}_{n,s}(\mathbb{D})$ (resp. $T \in \mathfrak{I}_{n,s}(\mathbb{D})$) such that the multiplicative representation*

$$\Theta = B\cdot S \qquad (\text{resp. } \Theta = T\cdot C) \tag{4.8}$$

holds true.

(b) *Suppose that the functions $B_1, B_2 \in \mathfrak{I}_{n,B,l}(\mathbb{D})$ (resp. $C_1, C_2 \in \mathfrak{I}_{n,B,r}(\mathbb{D})$) and $S_1, S_2 \in \mathfrak{I}_{n,s}(\mathbb{D})$ (resp. $T_1, T_2 \in \mathfrak{I}_{n,s}(\mathbb{D})$) satisfy $B_1S_1 = B_2S_2 = \Theta$ (resp. $T_1C_1 = T_2C_2 = \Theta$). Then there exist a unitary matrix $U \in \mathfrak{M}_n$ (resp. $V \in \mathfrak{M}_n$) such that $B_2 = B_1U$ and $S_2 = U^\star S_1$ (resp. $C_2 = VC_1$ and $T_2 = T_1V^\star$) are fulfilled.*

Proof. Theorem 4.1 is a special case of a much more general result due to V.P. Potapov [Pot]. The Potapov theory handles the case of meromorphic matrix-valued functions in $\mathbb{D}$ which have a nonidentically vanishing determinant and which are J-contractive where J is a signature matrix (i.e. $J = J^\star$ and $J^2 = I_n$). In the special case that $J = I$, V.P. Potapov's result (see [Pot] and also a series of papers by Ginzburg [Gi1]–[Gi5], [GiSh]) provides the existence of functions $B \in \mathfrak{I}_{n,B,l}(\mathbb{D})$ and $S \in \mathfrak{S}_{n\times n}(\mathbb{D})$ such that

$$\Theta = B \cdot S \tag{4.9}$$

and for all $z \in \mathbb{D}$,

$$\det [S(z)] \neq 0. \tag{4.10}$$

Since the boundary function $\underline{\Theta}$ has unitary values almost everywhere with respect to m we infer from Lemma 4.2 that the boundary functions $\underline{B}$ and $\underline{S}$ also have unitary values almost everywhere with respect to m. Taking into account (4.10) we obtain $S \in \mathfrak{I}_{n,s}(\mathbb{D})$. The uniqueness part goes back to V.P. Potapov [Pot] too. □

Lemma 4.3. *Let $M \in \mathfrak{C}_n$. Then*

(a) $|\det M| \leq 1$

(b) $|\det M| = 1$ *if and only if M is unitary.*

Proof. Let $(l_k(M^\star M))_{k=1}^n$ denote the system of eigenvalues of $M^\star M$. Then, since $M \in \mathfrak{C}_n$, $0 \leq l_k(M^\star M) \leq 1$ for all $k \in \{1,\dots,n\}$. Thus, as

$$|\det M|^2 = \det (M^\star M) = \prod_{k=1}^{n} l_k(M^\star M), \tag{4.11}$$

we see that $|\det M| \leq 1$ with equality if and only if $l_k(M^\star M) = 1$ for all $k \in \{1,\dots,n\}$. But $l_k(M^\star M) = 1$ for all $k \in \{1,\dots,n\}$ if and only if $M^\star M = I_n$. □

Now we recall a well-known characterization of Blaschke products (see e.g., Privalov [Pri, Ch.I, Sec.7.1]).

Lemma 4.4. *Let $\Theta \in \mathfrak{S}_{1\times 1}(\mathbb{D})$. Then Θ is a Blaschke product if and only if*

$$\lim_{r\to 1-0} \int_{\mathbb{T}} \ln |\det [\Theta(rt)]| \, m(dt) = 0.$$

Theorem 4.2. *Let $f \in \mathfrak{S}_{n\times n}(\mathbb{D})$. Then:*

(a) *The function* $\det f$ *belongs to* $\mathfrak{S}_{1\times 1}(\mathbb{D})$.

(b) $f \in \mathfrak{I}_n(\mathbb{D})$ *if and only if* $\det f \in \mathfrak{I}_1(\mathbb{D})$. *If* $f \in \mathfrak{I}_n(\mathbb{D})$ *then* $\det f \not\equiv 0$.

(c) $f \in \mathfrak{I}_{n,s}(\mathbb{D})$ *if and only if* $\det f \in \mathfrak{I}_{1,s}(\mathbb{D})$.

(d) *The following statements are equivalent:*

(i) $f \in \mathfrak{I}_{n,B,l}(\mathbb{D})$,

(ii) $f \in \mathfrak{I}_{n,B,r}(\mathbb{D})$,

(iii) $\det f$ *is a Blaschke product.*

(iv) *The limit relation* $\lim\limits_{r\to 1-0} \int\limits_{\mathbb{T}} \ln|\det[f(rt)]|\, m(dt) = 0$ *holds true.*

Proof. The assertions stated in part (a) and (b) are immediate consequences of Lemma 4.3. Part (c) follows from part (a) and the definition of a singular inner function.

It remains to prove part (d). From (a) and Lemma 4.4 we can immediately conclude the equivalence of statements (iii) and (iv). In view of (4.5), it is readily checked that each of the conditions (i) and (ii) implies (iii). Now suppose that (iii) holds. By virtue of part (b) we see that f is an inner function. From Theorem 4.1 we infer that there exist functions $B \in \mathfrak{I}_{n,B,l}(\mathbb{D})$ and $S \in \mathfrak{I}_{n,s}(\mathbb{D})$ satisfying the multiplicative decomposition $f = B \cdot S$. Hence, $\det f = \det B \cdot \det S$. The implication "(i) $\Rightarrow$ (iii)" which is already verified shows that $\det B$ is a Blaschke product. Part (c) yields that $\det S$ is a singular inner function. Therefore, the uniqueness part of Theorem 4.1 yields that $\det S$ is a constant inner function with unimodular value. Hence, we obtain from part (b) of Lemma 4.3 that the matrix $S(z)$ is unitary for each $z \in \mathbb{D}$. Since S belongs to $\mathfrak{S}_{n\times n}(\mathbb{D})$, the maximum modulus principle for matrix-valued Schur functions (see e.g. [DFK, Corollary 2.3.2]) implies that S is a constant function. From $f = B \cdot S$ we infer that (i) holds. The implication "(iii) $\Rightarrow$ (ii)" can be shown analogously. The theorem is proved. □

For further results on matrix-valued and operator-valued inner functions we refer the reader to the monographs Helson [Hel1], Sz.-Nagy and Foias [SZNF] and Nikolskii [Nik2].

5 Inner-outer factorization

This section is aimed at a Smirnov class generalization of the inner-outer factorization of matrix-valued functions belonging to the Hardy class $H_n^2(\mathbb{D})$.

Let us recall the following notions:

Definition 5.1. The Hardy class $H_n^2(\mathbb{D})$ *is the set of all matrix-valued functions* $F : \mathbb{D} \to \mathfrak{M}_n$ *which are holomorphic in* $\mathbb{D}$ *and satisfy*

$$\sup_{r\in[0,1)} \int\limits_{\mathbb{T}} \|F(rt)\|^2\, m(dt) < \infty.$$

Remark 5.1. Obviously, $H_n^\infty(\mathbb{D}) \subseteq H_n^2(\mathbb{D}) \subseteq \mathfrak{N}_n^+(\mathbb{D})$.

Remark 5.2. Define $\| \bullet \|_{H^2} : H_n^2(\mathbb{D}) \to [0, \infty)$ via

$$F \to \sqrt{\sup_{r \in [0,1)} \int_{\mathbb{T}} \|F(rt)\|^2 \, m(dt)}.$$

Then $(H_n^2(\mathbb{D}), \| \bullet \|_{H^2})$ is a complex Hilbert space.

Remark 5.3. Let $S \in \mathfrak{S}_{n \times n}(\mathbb{D})$ be such that $\det(I_n + S)$ does not identically vanish in $\mathbb{D}$. Then $(I_n + S) \in \mathfrak{E}_n(\mathbb{D}) \cap H_n^\infty(\mathbb{D})$ (see Arov [Ar1], Lemma 3.1).

The definition of a matrix-valued outer function given above (see Definition 3.1) is too rough for the purposes of prediction theory of stationary sequences. For this reason, P.R. Masani [Ma1, Ma2] introduced the following notion for the space $H_n^2(\mathbb{D})$ (compare Lemma 2.2).

Definition 5.2. *Let $E \in H_n^2(\mathbb{D})$. Then E is said to be* left optimal (*resp.* right optimal) *if E has the following property: If $F \in H_n^2(\mathbb{D})$ satisfies $[\underline{F}(t)] \cdot [\underline{F}(t)]^* = [\underline{E}(t)] \cdot [\underline{E}(t)]^*$ (resp. $[\underline{F}(t)]^* \cdot [\underline{F}(t)] = [\underline{E}(t)]^* \cdot [\underline{E}(t)]$), then $[F(0)] \cdot [F(0)]^* \leq [E(0)] \cdot [E(0)]^*$ (resp. $[F(0)]^* \cdot [F(0)] \leq [E(0)]^* \cdot [E(0)]$).*

Remark 5.4. Let $E \in H_n^2(\mathbb{D})$. Then E is left optimal if and only if $E^\top$ is right optimal.

This notion of optimality is closely related to the following definition which in the scalar case goes back to Beurling [Be].

Definition 5.3. *Let $E \in H_n^2(\mathbb{D})$. Then E is called* left Beurling-outer (*resp.* right Beurling outer) *if there exists a sequence $(f_k)_{k \in \mathbb{N}}$ from $H_n^\infty(\mathbb{D})$ which satisfies*

$$\lim_{k \to \infty} \int_{\mathbb{T}} \|\underline{F_k}(t) \cdot \underline{E}(t) - I_n\|^2 \, m(dt) = 0$$

$$\Big(\textit{resp.} \lim_{k \to \infty} \int_{\mathbb{T}} \|\underline{E}(t) \cdot \underline{F_k}(t) - I_n\|^2 \, m(dt) = 0 \Big).$$

The class of all $n \times n$ matrix-valued left Beurling-outer (resp. right Beurling outer) functions will be denoted by $\mathfrak{E}_{n,B,l}(\mathbb{D})$ (resp. $\mathfrak{E}_{n,B,r}(\mathbb{D})$).

Remark 5.5. Let $E \in H_n^2(\mathbb{D})$. Then $E \in \mathfrak{E}_{n,B,l}(\mathbb{D})$ if and only if $E^\top \in \mathfrak{E}_{n,B,r}(\mathbb{D})$.

Remark 5.6. Let $E \in H_n^2(\mathbb{D})$. Then it is readily checked that E is left Beurling outer (resp. right Beurling outer) if and only if the subspace $H_n^2(\mathbb{D}) \cdot E$ (resp. $E \cdot H_n^2(\mathbb{D})$) is dense in $(H_n^2(\mathbb{D}), \| \bullet \|_{H^2})$.

Remark 5.7. Let E be a function belonging to $\mathfrak{E}_{n,B,l}(\mathbb{D})$ or $\mathfrak{E}_{n,B,r}(\mathbb{D})$. Then for all $z \in \mathbb{D}$ the relation $\det[E(z)] \neq 0$ holds true.

Proof. Let us consider the case $E \in \mathfrak{E}_{n,B,r}(\mathbb{D})$. Then there exists a sequence $(F_k)_{k \in \mathbb{N}}$ from $H^{\infty}(\mathbb{D})$ such that

$$\lim_{k \to \infty} \int_{\mathbb{T}} \|\underline{E}(t) \cdot \underline{F_k}(t) - I_n\|^2 \, m(dt) = 0.$$

From this it follows by the Poisson integral representation for $H_n^2(\mathbb{D})$ functions that

$$\lim_{k \to \infty} E(z) \cdot F_k(z) = I_n$$

for $z \in \mathbb{D}$ and hence that

$$\lim_{k \to \infty} \det [E(z)] \cdot \det [F_k(z)] = 1.$$

Thus, $\det [E(z)] \neq 0$. If $E \in \mathfrak{E}_{n,B,l}(\mathbb{D})$, then the assertion follows from Remark 5.5 and the preceding analysis. □

The following result due to Masani [Ma2, Corollary 4.6] clarifies the relation between optimality and Beurling-outerness.

Theorem 5.1. *Let $E \in H_n^2(\mathbb{D})$. Then:*

(a) *If $\det E \not\equiv 0$ and E is left optimal (resp. right optimal), then $E \in \mathfrak{E}_{n,B,l}(\mathbb{D})$ (resp. $E \in \mathfrak{E}_{n,B,r}(\mathbb{D})$).*

(b) *If $E \in \mathfrak{E}_{n,B,l}(\mathbb{D})$ (resp. $E \in \mathfrak{E}_{n,B,r}(\mathbb{D})$), then E is left optimal (resp. right optimal).*

The notion of optimality is more general than the notion of Beurling – outer because it allows the functions in question to have identically vanishing determinants. In the theory of multivariate stationary stochastic processes this corresponds to the case of a singular prediction error matrix.

The following result plays a key role in the theory of holomorphic matrix-valued functions.

Theorem 5.2. *Let $F \in H_n^2(\mathbb{D})$ be such that $\det F \not\equiv 0$. Then:*

(i) *There exist functions $\Theta_r \in \mathfrak{I}_n(\mathbb{D})$ and $E_r \in \mathfrak{E}_{n,B,r}(\mathbb{D})$ such that the multiplicative decomposition*

$$F = \Theta_r \cdot E_r$$

is satisfied.

(ii) *Suppose that the functions $\Theta_{r1}, \Theta_{r2} \in \mathfrak{I}_n(\mathbb{D})$ and $E_{r1}, E_{r2} \in \mathfrak{E}_{n,B,r}(\mathbb{D})$ satisfy*

$$\Theta_{r1} \cdot E_{r1} = \Theta_{r2} \cdot E_{r2} = F.$$

Then there exists a unitary matrix $V \in \mathfrak{M}_n$ such that $\Theta_{r2} = \Theta_{r1} \cdot V$ and $E_{r2} = V^{\star} \cdot E_{r1}$.

(iii) *There exist functions* $\Theta_l \in \mathfrak{J}_n(\mathbb{D})$ *and* $E_l \in \mathfrak{E}_{n,B,l}(\mathbb{D})$ *such that the multiplicative decomposition*

$$F = E_l \cdot \Theta_l$$

is satisfied.

(iv) *Suppose that the functions* $\Theta_{l1}, \Theta_{l2} \in \mathfrak{J}_n(\mathbb{D})$ *and* $E_{l1}, E_{l2} \in \mathfrak{E}_{n,B,l}(\mathbb{D})$ *satisfy*

$$E_{l1} \cdot \Theta_{l1} = E_{l2} \cdot \Theta_{l2} = F.$$

Then there exists a unitary matrix $U \in \mathfrak{M}_n$ *such that* $\Theta_{l2} = U \cdot \Theta_{l1}$ *and* $E_{l2} = E_{l1} \cdot U^\star$.

Theorem 5.2 was proved independently by several authors (see Masani [Ma2, 4.3, 4.4], Helson and Lowdenslager [HL2, Theorem 15] and Rozanov [Roz1, Theorem 5]). The Beurling-Lax-Halmos Theorem (see Beurling [Be], Lax [La], Halmos [Hal] and also Masani [Ma2, Theorem 3.8.]) which describes the structure of shift invariant left (resp. right) submodules of $H_n^2(\mathbb{D})$ lies at the heart of the proof.

Theorem 5.3. *The identities*

$$\mathfrak{E}_{n,B,l}(\mathbb{D}) = \mathfrak{E}_{n,B,r}(\mathbb{D}) = \mathfrak{E}_n(\mathbb{D}) \cap H_n^2(\mathbb{D})$$

are valid.

Proof. First we show that

$$\mathfrak{E}_{n,B,r}(\mathbb{D}) = \mathfrak{E}_n(\mathbb{D}) \cap H_n^2(\mathbb{D}).$$

Our proof is based mainly on Theorem 3.2.
First assume that $E \in \mathfrak{E}_n(\mathbb{D}) \cap H_n^2(\mathbb{D})$. Then part (i) of Theorem 3.2 guarantees the existence of a sequence $(F_k)_{k\in\mathbb{N}}$ from $H_n^\infty(\mathbb{D})$ with the properties (α), (β) and (γ) formulated there. In view of property (γ), there exists a Borel subset B_0 of $\mathbb{T}$ with $m(B_0) = 0$ such that for all $k \in \mathbb{N}$ and all $t \in \mathbb{T} \setminus B_0$ the inequality

$$\|\underline{E}(t) \cdot \underline{F_k}(t) - I_n\| \leq \|\underline{E}(t) \cdot \underline{F_k}(t)\| + \|I_n\| \leq 2 \tag{5.1}$$

holds. In view of (α) and (5.1), an application of Lebesgue's dominated convergence theorem yields

$$\lim_{k\to\infty} \int_{\mathbb{T}} \|\underline{E}(t) \cdot \underline{F_k}(t) - I_n\|^2 \, m(dt) = 0.$$

Thus, $E \in \mathfrak{E}_{n,B}(\mathbb{D})$. Hence, the inclusion

$$\mathfrak{E}_n(\mathbb{D}) \cap H_n^2(\mathbb{D}) \subseteq \mathfrak{E}_{n,B}(\mathbb{D}) \tag{5.2}$$

holds true. Now assume that $E \in \mathfrak{E}_{n,B}(\mathbb{D})$. Then Definition 5.3 implies that

$$E \in H_n^2(\mathbb{D}). \tag{5.3}$$

We will show that E satisfies the conditions (α) and (β) in Theorem 3.2. In view of Definition 5.2, there exists a sequence $(F_k)_{k\in\mathbb{N}}$ from $H_n^\infty(\mathbb{D})$ for which

$$\lim_{k\to\infty}\int_{\mathbb{T}} \|\underline{E}(t)\cdot\underline{F_k}(t)-I_n\|^2\, m(dt)=0. \tag{5.4}$$

Obviously, for $k\in\mathbb{N}$ and $t\in\mathbb{T}$ the inequality

$$0\le \ln^+\|\underline{E}(t)\cdot\underline{F_k}(t)\|\le\|\underline{E}(t)\cdot\underline{F_k}(t)-I_n\| \tag{5.5}$$

is valid. From (5.4) and (5.5), it then follows that

$$\lim_{k\to\infty}\int_{\mathbb{T}} \ln^+\|\underline{E}(t)\cdot\underline{F_k}(t)\|\, m(dt)=0.$$

Hence, the family $(\ln^+\|\underline{E}\cdot\underline{F_k}\|)_{k\in\mathbb{N}}$ is uniformly m-integrable. In view of Remark 5.7 we see that $\det[E(z)]\neq 0$ for all $z\in\mathbb{D}$. Since $E\in H_n^2(\mathbb{D})\subseteq\mathfrak{N}_n(\mathbb{D})$ we now obtain $E^{-1}\in\mathfrak{N}_n(\mathbb{D})$. Hence, $\ln\|\underline{E}^{-1}\|=\ln\|\underline{E^{-1}}\|$ is m-integrable. Clearly, for $k\in\mathbb{N}$ and $t\in\mathbb{T}$ the inequality

$$\ln^+\|\underline{F_k}(t)\|\le\ln^+\|\underline{E}(t)\cdot\underline{F_k}(t)\|+\ln^+\|[\underline{E}(t)]^{-1}\| \tag{5.6}$$

holds true. Since the family $(\ln^+\|\underline{E}\cdot\underline{F_k}\|)_{k\in\mathbb{N}}$ is uniformly m-integrable and since $\ln\|\underline{E}^{-1}\|$ is m-integrable it follows from (5.6) that the family $(\ln^+\|\underline{F_k}\|)_{k\in\mathbb{N}}$ is uniformly m-integrable. Taking into account (5.4), the Theorem of F. Riesz-Fischer provides the existence of a subsequence $(F_{l_k})_{k\in\mathbb{N}}$ of $(F_k)_{k\in\mathbb{N}}$ such that

$$\lim_{k\to\infty}\underline{E}(t)\cdot\underline{F_{l_k}}(t)=I_n$$

for m-almost all $t\in\mathbb{T}$. Since the family $(\ln^+\|\underline{F_{l_k}}\|)_{k\in\mathbb{N}}$ is also uniformly m-integrable the conditions (α) and (β) in Theorem 3.2 are satisfied for the sequence $(F_{l_k})_{k\in\mathbb{N}}$. Thus, part (ii) of Theorem 3.2 implies that

$$E\in\mathfrak{E}_n(\mathbb{D}). \tag{5.7}$$

From (5.3) and (5.7) we obtain $\mathfrak{E}_{n,B}(\mathbb{D})\subseteq\mathfrak{E}_n(\mathbb{D})\cap H_n^2(\mathbb{D})$.
An application of (5.2) shows that

$$\mathfrak{E}_{n,B,r}(\mathbb{D})=\mathfrak{E}_n(\mathbb{D})\cap H_n^2(\mathbb{D}). \tag{5.8}$$

From (5.8) and Remarks 3.2 and 5.5 we then get

$$\mathfrak{E}_{n,B,l}(\mathbb{D})=\mathfrak{E}_n(\mathbb{D})\cap H_n^2(\mathbb{D}).$$

Thus, the theorem is proved. □

Theorem 5.4 (Inner-outer factorization in the Smirnov class $\mathfrak{N}_n^+(\mathbb{D})$).
Let $F \in \mathfrak{N}_n^+(\mathbb{D})$ be such that $\det F \not\equiv 0$. Then:

(i) *There exist functions $\Theta_r \in \mathfrak{I}_n(\mathbb{D})$ and $E_r \in \mathfrak{E}_n(\mathbb{D})$ such that*

$$F = \Theta_r \cdot E_r.$$

(ii) *Suppose that the functions $\Theta_{r1}, \Theta_{r2} \in \mathfrak{I}_n(\mathbb{D})$ and $E_{r1}, E_{r2} \in \mathfrak{E}_n(\mathbb{D})$ satisfy*

$$\Theta_{r1} \cdot E_{r1} = \Theta_{r2} \cdot E_{r2} = F.$$

Then there exists a unitary matrix $V \in \mathfrak{M}_n$ such that $\Theta_{r2} = \Theta_{r1} \cdot V$ and $E_{r2} = V^\star \cdot E_{r1}$.

(iii) *There exist functions $\Theta_l \in \mathfrak{I}_n(\mathbb{D})$ and $E_l \in \mathfrak{E}_n(\mathbb{D})$ such that*

$$F = E_l \cdot \Theta_l.$$

(iv) *Suppose that the functions $\Theta_{l1}, \Theta_{l2} \in \mathfrak{I}_n(\mathbb{D})$ and $E_{l1}, E_{l2} \in \mathfrak{E}_n(\mathbb{D})$ satisfy*

$$E_{l1} \cdot \Theta_{r1} = E_{l2} \cdot \Theta_{r2} = F.$$

Then there exists a unitary matrix $U \in \mathfrak{M}_n$ such that $\Theta_{l2} = U \cdot \Theta_{l1}$ and $E_{l2} = E_{l1} \cdot U^\star$.

Proof. We derive these results from Theorem 5.2.
(i) In view of Lemma 2.4 there exist functions $d \in \mathfrak{E}(\mathbb{D})$ and $\Phi \in \mathfrak{S}_{n\times n}(\mathbb{D})$ such that

$$F = \frac{1}{d} \cdot \Phi. \tag{5.9}$$

Since $\det F \not\equiv 0$, it follows from (5.9) that $\det \Phi \not\equiv 0$. Thus as $\mathfrak{S}_{n\times n}(\mathbb{D}) \subseteq H_n^2(\mathbb{D})$, Theorem 5.2 ensures the existence of functions $\Theta_r \in \mathfrak{I}_n(\mathbb{D})$ and $E_{r,B} \in \mathfrak{E}_{n,B}(\mathbb{D})$ such that

$$\Phi = \Theta_r \cdot E_{r,B}. \tag{5.10}$$

We set

$$E := d \cdot E_{r,B}. \tag{5.11}$$

According to Theorem 5.3 it follows that $E_{r,B} \in \mathfrak{E}_n(\mathbb{D})$. Since $d \in \mathfrak{E}(\mathbb{D})$ we get $E \in \mathfrak{E}_n(\mathbb{D})$ from (5.11). Thus (i) is proved.

(ii) The factorizations $F = \Theta_{r1} \cdot E_{r1} = \Theta_{r2} \cdot E_{r2}$ yield the factorizations

$$\Theta_{r1} \cdot E_{r1,B} = \Theta_{r2} \cdot E_{r2,B} = \Phi, \tag{5.12}$$

upon setting $E_{r1,B} := d \cdot E_{r1}$, $E_{r2,B} := d \cdot E_{r2}$ and invoking (5.9).
From $\Phi \in \mathfrak{S}_{n\times n}(\mathbb{D})$, (5.12) and its definition it is clear that

$$E_{r1,B}, E_{r2,B} \in \mathfrak{E}_n(\mathbb{D}) \cap \mathfrak{S}_{n\times n}(\mathbb{D}).$$

Thus, from Theorem 3.2 we get $E_{r1,B}, E_{r2,B} \in \mathfrak{E}_{n,B}(\mathbb{D})$. Now part (ii) of Theorem 5.2 provides the existence of a unitary matrix satisfying $\Theta_{r2} = \Theta_{r1} \cdot V$ and $E_{r2,B} = V^\star \cdot E_{r1,B}$. Hence,

$$E_{r2} = \frac{1}{d} \cdot E_{r2,B} = \frac{1}{d} \cdot V^\star \cdot E_{r1,B} = V^\star \cdot E_{r1}.$$

Thus, (ii) is proved.
Assertions (iii) and (iv) can be established analogously. □

Corollary 5.1. *Let $F \in \mathfrak{N}_n^+(\mathbb{D})$ be such that* $\det F \not\equiv 0$. *Then there exist functions $B_1 \in \mathfrak{J}_{n,B,l}(\mathbb{D}), S_1 \in \mathfrak{J}_{n,s}(\mathbb{D})$ and $E_1 \in \mathfrak{E}_n(\mathbb{D})$ (resp. $B_2 \in \mathfrak{J}_{n,B,r}(\mathbb{D}), S_2 \in \mathfrak{J}_{n,s}(\mathbb{D})$ and $E_2 \in \mathfrak{E}_n(\mathbb{D})$) such that*

$$F = B_1 \cdot S_1 \cdot E_1 \qquad (\text{resp. } F = E_2 \cdot S_2 \cdot B_2).$$

Proof. The assertion follows immediately by combining Theorem 4.1 and Theorem 5.4. □

It should be mentioned that using deep results and methods of V. Potapov [Pot] an alternate approach to Theorem 5.4 and Corollary 5.1 was presented by J.P. Ginzburg [Gi1]. His result contains also a multiplicative integral representation for the outer factor and the singular inner component.

The following theorem provides a useful characterization of the case that the inner component in the inner-outer factorization of a given function from $\mathfrak{N}_n^+(\mathbb{D})$ is a Blaschke-Potapov product.

Theorem 5.5. *Let $F \in \mathfrak{N}_n^+(\mathbb{D})$ be such that* $\det F \not\equiv 0$. *Suppose that the functions $\Theta_r, \Theta_l \in \mathfrak{J}_n(\mathbb{D})$ and $E_r, E_l \in \mathfrak{E}_n(\mathbb{D})$ satisfy $\Theta_r \cdot E_r = E_l \cdot \Theta_l = F$.*
Then the following statements are equivalent:

(i) $\Theta_r \in \mathfrak{J}_{n,B,r}(\mathbb{D})$.

(ii) $\Theta_l \in \mathfrak{J}_{n,B,l}(\mathbb{D})$.

(iii) $\lim\limits_{s\to 1-0} \int\limits_{\mathbb{T}} \ln |\det [F(st)]| \, m(dt) = \int\limits_{\mathbb{T}} \ln |\underline{F}(t)| \, m(dt)$.

Proof. In view of the fact that $E_r, E_l \in \mathfrak{E}_n(\mathbb{D})$, the functions $\det E_r$ and $\det E_l$ are outer. Moreover, since $\Theta_r, \Theta_l \in \mathfrak{J}_n(\mathbb{D})$, part (b) of Theorem 4.2 implies that the functions $\det \Theta_r, \det \Theta_l$ are inner. From part (d) of Theorem 4.2 it follows that (i) (resp. (ii)) holds if and only if $\det \Theta_r$ (resp. $\det \Theta_l$) is a Blaschke product. According to Lemma 4.4 this is equivalent to

$$\lim_{s\to 1-0} \int\limits_{\mathbb{T}} \ln |\det [\Theta_r(st)]| \, m(dt) = 0 \tag{5.13}$$

$$\left(\text{resp.} \quad \lim_{s\to 1-0} \int\limits_{\mathbb{T}} \ln |\det [\Theta_l(st)]| \, m(dt) = 0 \right). \tag{5.14}$$

From the multiplicative decomposition $F = \Theta_r \cdot E_r$ (resp. $F = E_l \cdot \Theta_l$) it follows immediately that (5.13) (resp. (5.14)) is equivalent to (iii).
Thus, the statements (i)–(iii) are equivalent. □

Remark 5.8. It is instructive to compare statement (iii) in Theorem 5.5 with the inequality (1.15) which is fulfilled for an arbitrary function F from $\mathfrak{N}_n^+(\mathbb{D})$.

6 An analogue of Frostman's theorem for matrix functions of the Smirnov class

Let f be a nonconstant function from the Smirnov class $\mathfrak{N}^+(\mathbb{D})$. For $\lambda \in \mathbb{C}$ the function

$$f_\lambda := f - \lambda \tag{6.1}$$

clearly belongs to $\mathfrak{N}^+(\mathbb{D})$ too. Thus, there exists an inner function θ_λ and an outer function e_λ such that

$$f_\lambda = \theta_\lambda \cdot e_\lambda. \tag{6.2}$$

It will turn out that in some sense "the typical situation" corresponds to the case that the inner function θ_λ in (6.2) is a Blaschke product. The set of all $\lambda \in \mathbb{C}$ for which θ_λ is not a Blaschke product is very thin. (A remarkable result of this type goes back to Frostman [Fr].) The corresponding notion of thinness can be formulated in terms of potential theory. For this reason, now we recall some notions of potential theory.

Suppose that ν is a nonnegative Borel measure with compact support. For all $\xi \in \mathbb{C}$ the integral

$$U^{(\nu)}(\xi) := \int_{\mathbb{C}} \ln|\xi - \lambda| \; \nu(d\lambda) \tag{6.3}$$

is then well-defined and takes its values in $[-\infty, \infty)$. The function $U^{(\nu)} : \mathbb{C} \to [-\infty, \infty)$ is called the *logarithmic potential of* ν. A Borel measure ν on $\mathbb{C}$ is said to be nontrivial if it is not the zero measure. If K is a Borel subset of $\mathbb{C}$, the Borel measure ν is said to be concentrated on K if $\nu(\mathbb{C} \setminus K) = 0$. By definition, a Borel subset K on $\mathbb{C}$ is called thin if for each nontrivial Borel measure ν which is concentrated on K the associated logarithmic potential $U^{(\nu)}$ is not bounded from below, or in other words if

$$\inf_{\xi \in \mathbb{C}} U^{(\nu)}(\xi) = -\infty.$$

If K is not thin, then there exists a nontrivial Borel measure ν which is concentrated on K and satisfies

$$\inf_{\xi \in \mathbb{C}} U^{(\nu)}(\xi) > -\infty. \tag{6.4}$$

The notion of *logarithmic capacity* is introduced in potential theory. More precisely, this means that with each Borel subset K of $\mathbb{C}$ there is associated a nonnegative number cap$K$ which is called *the logarithmic capacity of* K. It turns out that a Borel subset K of $\mathbb{C}$ is thin if and only if cap$K = 0$. In other words, if cap$K > 0$, then there exists a nontrivial Borel measure ν which is concentrated on K and satisfies condition (6.4). If cap$K > 0$, then amongst all the nontrivial Borel measures ν which are concentrated on K and satisfy (6.4) there is a distinguished probability measure ν_K, the so-called *equilibrium measure of* K. This measure ν_K is a solution of several natural extremal problems. (If cap$K = 0$ the equilibrium measure is not defined.)

The logarithmic potential is not always continuous on $\mathbb{C}$ but only upper semicontinuous on $\mathbb{C}$. More precisely, for all $\xi \in \mathbb{C}$,

$$\varlimsup_{\xi' \to \xi} U^{(\nu)}(\xi') \leq U^{(\nu)}(\xi).$$

Although it is bounded below, the logarithmic potential of the equilibrium measure need not be continuous on $\mathbb{C}$. If the set K is "bad" there are so-called irregular points. Nevertheless it can be proved (see de la Vallée Poussin [LVP2], [LVP3]) that if cap$K > 0$, then there exists a nontrivial nonnegative measure which is concentrated on K and for which the associated logarithmic potential is continuous on $\mathbb{C}$ (as already mentioned, the equilibrium measure ν_K can not generally be used for this purpose). We will not enter into such detailed and rather delicate potential-theoretical considerations. To avoid them we give the following definition.

Definition 6.1. A bounded Borel subset K of $\mathbb{C}$ *is said* to have positive logarithmic capacity *if there exists a nontrivial Borel measure* ν *which is concentrated on* K *and for which the associated logarithmic potential is continuous on* $\mathbb{C}$.

Clearly, if $K_1 \subseteq K_2$ and K_1 is a set of positive logarithmic capacity, then K_2 is also a set of positive logarithmic capacity.

Lemma on the capacity of an interval. *Every interval of the complex plane is a set of positive logarithmic capacity.*

Proof. Without loss of generality we can assume that the considered interval is a subinterval (α, β) of the real axis where $-\infty < \alpha < \beta < \infty$. Now we take for ν the restriction of one dimensional Lebesgue measure to this interval (α, β). The function $U^{(\nu)} : \mathbb{C} \to [-\infty, \infty)$ which is defined by the rule

$$\xi \to \int\limits_{(\alpha,\beta)} \ln|\xi - \lambda| \; \nu(d\lambda)$$

is continuous in $\mathbb{C}$. This can be checked in several ways, e.g. one can compute explicitly and then obtain the continuity of $U^{(\nu)}$ by direct estimates. □

W. Rudin [Ru1] (see also section 3.6 of the monograph [Ru2]) proved the following fact which generalizes Frostman's original result:
Let $f \in \mathfrak{N}^+(\mathbb{D})$ with $f \not\equiv 0$ and let K be some bounded Borel subset of $\mathbb{C}$ with positive logarithmic capacity. Then there exist a $\lambda \in K$ such that the inner factor in the multiplicative decomposition (6.2) is a Blaschke product. (Indeed, W. Rudin obtained a more general result which is formulated for the Smirnov class $\mathfrak{N}^+(\mathbb{D}^p)$ in the polydisc $\mathbb{D}^p$. This class is a natural analogue of $\mathfrak{N}^+(\mathbb{D})$ and coincides with it in the case $p = 1$.) It should be mentioned that S.A. Vinogradov [Vin] independently obtained such a generalization of Frostman's theorem too.

Remark 6.1. Let $F \in \mathfrak{N}_n^+(\mathbb{D})$ and define $F_\lambda := F - \lambda \cdot I_n$ for $\lambda \in \mathbb{C}$. Then the set $M_F := \{\lambda \in \mathbb{C} \ : \ \det(F_\lambda) \equiv 0\}$ is finite.

Now we formulate our main result.

Theorem 6.1. *Let $F \in \mathfrak{N}_n^+(\mathbb{D})$. Assume that for $\lambda \in \mathbb{C} \setminus M_F$ the functions $\Theta_{\lambda,r} \in \mathfrak{J}_n(\mathbb{D})$ and $E_{\lambda,r} \in \mathfrak{E}_n(\mathbb{D})$ are factors in the multiplicative decomposition*

$$F_\lambda = \Theta_{\lambda,r} \cdot E_{\lambda,r}.$$

Suppose that K is a bounded Borel subset of $\mathbb{C}$ with positive logarithmic capacity. Then there exists a point $\lambda \in K \cap (\mathbb{C} \setminus M_F)$ for which $\Theta_{\lambda,r}$ is a Blaschke-Potapov product.

Corollary 6.1. *The set of all $\lambda \in \mathbb{C} \setminus M_F$ for which $\Theta_{\lambda,r}$ is a Blaschke-Potapov product is dense in $\mathbb{C}$.*

Proof. Combine Theorem 6.1 and the Lemma on the capacity of an interval. □

In order to to follow the strategy of W. Rudin's proof we shell need to introduce a number of classes of scalar functions of several variables.

Definition 6.2. *A function $\sigma : \mathbb{C}^n \to \mathbb{R}$ is called* symmetric *if for all permutations $\begin{pmatrix} 1 & \dots & n \\ i_1 & \dots & i_n \end{pmatrix}$ and all $x = (x_1, \dots, x_n)^\top \in \mathbb{C}^n$ the relation*

$$\sigma((x_{i_1}, \dots, x_{i_n})^\top) = \sigma((x_1, \dots, x_n)^\top)$$

is valid.

In view of Definition 6.2 the following object is well-defined.

Definition 6.3. *Let $\sigma : \mathbb{C}^n \to \mathbb{R}$ be a symmetric function. Then the function $\varphi_\sigma : \mathfrak{M}_n \to \mathbb{R}$ which is defined by the rule*

$$A \to \sigma((l_1(A), \dots, l_n(A))^\top),$$

where $(l_j(A))_{j=1}^n$ are the roots of the characteristic polynomial of A (taking into account their algebraic multiplicities), is called the function of matrix argument which is generated by the symmetric function σ.

Lemma 6.1. *Suppose that $\sigma : \mathbb{C}^n \to \mathbb{R}$ is a continuous symmetric function. Then φ_σ is a continuous function.*

Proof. The lemma is an immediate consequence of Theorem 5.1 from Chapter II in Kato's monograph [Ka]. (See there especially formula (5.3) and the text following it.) □

If the symmetric function $\sigma : \mathbb{C}^n \to \mathbb{R}$ is a polynomial or a rational function in n variables $x_1, \dots, x_n$, then it can be expressed as a polynomial or a rational function of the elementary symmetric functions. In this case the function φ_σ is a polynomial or a rational function of the elements of the matrix variable.

We introduce now a *potential of the matrix argument.* Roughly speaking, we insert a matrix argument in formula (6.3) instead of the complex variable.

Remark 6.2. Suppose that $A \in \mathfrak{M}_n$. Then the function $h_A : \mathbb{C} \to \mathbb{R}$ which is defined by $\lambda \to |\det[A - \lambda I_n]|$ is continuous. Hence, the function $\ln h_A$ is continuous and locally bounded above. If ν is a finite Borel measure on $\mathbb{C}$ with compact support, then the function $\Phi^{(\nu)} : \mathfrak{M}_n \to [-\infty, \infty)$ with

$$\Phi^{(\nu)}(A) := \int_{\mathbb{C}} \ln|\det(A - \lambda I_n)|\, \nu(d\lambda) \tag{6.5}$$

is well-defined.

Definition 6.4. *Suppose that ν is a finite Borel measure on $\mathbb{C}$ with compact support. Then the function $\Phi^{(\nu)} : \mathfrak{M}_n \to [-\infty, \infty)$ which is defined by (6.5) is called the potential of the matrix argument associated with ν.*

Assume that ν is a finite Borel measure on $\mathbb{C}$ with compact support. Let $U^{(\nu)}$ denote the logarithmic potential of ν. Let $A \in \mathfrak{M}_n$ and let $(l_k(A))_{k=1}^n$ be roots of the characteristic polynomial of A. For $\lambda \in \mathbb{C}$ we then have

$$\ln|\det(A - \lambda I_n)| = \sum_{k=1}^{n} \ln|l_k(A) - \lambda|. \tag{6.6}$$

Hence, upon taking (6.3) into account we get

$$\Phi^{(\nu)}(A) = \sum_{k=1}^{n} U^{(\nu)}(l_k(A)). \tag{6.7}$$

We define $\sigma^{(\nu)} : \mathbb{C}_n \to [-\infty, \infty)$ via

$$\begin{pmatrix} x_1 \\ \vdots \\ x_n \end{pmatrix} \to \sum_{k=1}^{n} U^{(\nu)}(x_k). \tag{6.8}$$

Obviously, the function $\sigma^{(\nu)}$ is symmetric. From Definition 6.3 and formulas (6.7) and (6.8) we infer that

$$\Phi^{(\nu)}(A) = \varphi_{\sigma^{(\nu)}}(A). \tag{6.9}$$

Lemma 6.2. *Suppose that ν is a finite Borel measure on $\mathbb{C}$ with compact support such that the associated logarithmic potential $U^{(\nu)}$ is continuous on $\mathbb{C}$. Then the function $\Phi^{(\nu)} : \mathfrak{M}_n \to [-\infty, \infty)$ which is defined by (6.5) is continuous on $\mathfrak{M}_n$.*

Proof. Indeed, from (6.8) it follows that $\sigma^{(\nu)}$ is a continuous function on $\mathbb{C}^n$. Then in view of (6.9) and Lemma 6.1 the assertion follows. □

Definition 6.5. *Let ν be a finite Borel measure on $\mathbb{C}$ with compact support. Then the functions $\Phi_+^{(\nu)} : \mathfrak{M}_n \to [0, \infty)$ and $\Phi_-^{(\nu)} : \mathfrak{M}_n \to [0, \infty)$ are defined via the formulas*

$$\Phi_+^{(\nu)}(A) := \int_{\mathbb{C}} \ln^+ |\det(A - \lambda I_n)| \, \nu(d\lambda) \tag{6.10}$$

and

$$\Phi_-^{(\nu)}(A) := \int_{\mathbb{C}} \ln^- |\det(A - \lambda I_n)| \, \nu(d\lambda), \tag{6.11}$$

respectively.

Lemma 6.3. *Suppose that ν is a finite Borel measure on $\mathbb{C}$ with compact support. Then the function $\Phi_+^{(\nu)}$ defined by (6.10) is continuous on $\mathfrak{M}_n$.*

Proof. The function $f : \mathfrak{M}_n \times \mathbb{C} \to [0, \infty)$ which is defined by

$$f(A, \lambda) := |\det(A - \lambda I_n)|$$

is continuous on $\mathfrak{M}_n \times \mathbb{C}$. Since the function $\ln^+ := \max\{\ln, 0\}$ is continuous on $[0, \infty)$ the composition mapping $\ln^+ f$ is continuous on $\mathfrak{M}_n \times \mathbb{C}$. From this we infer that the function $\Phi_+^{(\nu)}$ is continuous on $\mathfrak{M}_n$. □

Lemma 6.4. *Suppose that ν is a finite Borel measure on $\mathbb{C}$ with compact support such that the associated logarithmic potential $U^{(\nu)}$ is continuous on $\mathbb{C}$. Then the function $\Phi_-^{(\nu)}$ which is defined by (6.11) is continuous on $\mathfrak{M}_n$; it is also bounded:*

$$\sup_{A \in \mathfrak{M}_n} \Phi_-^{(\nu)}(A) < +\infty. \tag{6.12}$$

Proof. From Definitions 6.4 and 6.5 we get the identity

$$\Phi^{(\nu)} = \Phi_+^{(\nu)} - \Phi_-^{(\nu)}. \tag{6.13}$$

In view of Lemma 6.2 the function $\Phi^{(\nu)}$ is continuous whereas Lemma 6.3 provides the continuity of $\Phi_+^{(\nu)}$. Thus, (6.13) shows the continuity of $\Phi_-^{(\nu)}$. We define the functions $U_+^{(\nu)} : \mathbb{C} \to [0, \infty)$ and $U_-^{(\nu)} : \mathbb{C} \to [0, \infty)$ by

$$U_+^{(\nu)}(\xi) := \int_{\mathbb{C}} \ln^+ |\xi - \lambda| \, \nu(d\lambda) \tag{6.14}$$

and

$$U_-^{(\nu)}(\xi) := \int_{\mathbb{C}} \ln^- |\xi - \lambda| \, \nu(d\lambda). \tag{6.15}$$

Combining (6.3), (6.14) and (6.15) we see that

$$U^{(\nu)} = U_+^{(\nu)} - U_-^{(\nu)}. \tag{6.16}$$

Since the function $U^{(\nu)}$ is continuous by assumption and since the function $U_+^{(\nu)}$ is always continuous (by Lemma 6.3 with $n = 1$) the continuity of $U_-^{(\nu)}$ follows from (6.16). If $(r_k)_{k=1}^n$ is a sequence from $[0, \infty)$, then clearly

$$\ln^- \left(\prod_{k=1}^n r_k \right) \leq \sum_{k=1}^n \ln^- r_k. \tag{6.17}$$

Let $A \in \mathfrak{M}_n$ and let $(l_k(A))_{k=1}^n$ be the roots of the characteristic polynomial of A. In view of (6.6) we get

$$\ln^- |\det (A - \lambda I_n)| = \ln^- \left(\prod_{k=1}^n |l_k(A) - \lambda| \right). \tag{6.18}$$

From (6.17), (6.18) and (6.15) we infer that

$$\Phi_-^{(\nu)}(A) \leq \sum_{k=1}^n U_-^{(\nu)}(l_k(A)).$$

Hence,

$$\sup_{A \in \mathfrak{M}_n} \Phi_-^{(\nu)}(A) \leq n \cdot \sup_{\xi \in \mathbb{C}} U_-^{(\nu)}(\xi). \tag{6.19}$$

Now it remains to show that our assumptions ensure that

$$\sup_{\xi \in \mathbb{C}} U_-^{(\nu)}(\xi) < \infty \tag{6.20}$$

is fulfilled. If $\xi \in \mathbb{C}$ satisfies

$$|\xi| \geq 1 + \sup_{\lambda \in \operatorname{supp} \nu} |\lambda|, \tag{6.21}$$

then using (6.15) we see that

$$U_-^{(\nu)}(\xi) = 0. \tag{6.22}$$

Now the continuity of $U_-^{(\nu)}$, (6.21), (6.22) and a classical theorem due to Weierstrass yield (6.20). The lemma is proved. □

Remark 6.3. If $a, b \in [0, \infty)$, then

$$\ln^+(a+b) \le \ln^+ a + \ln^+ b + \ln 2.$$

Remark 6.4. Let $A \in \mathfrak{M}_n$. Then $|\det A| \le \|A\|^n$.

Remark 6.5. Let $A \in \mathfrak{M}_n$ and $\lambda \in \mathbb{C}$. Then

$$\ln^+ |\det [A - \lambda I_n]| \le n \cdot [\ln^+ \|A\| + \ln^+ |\lambda| + \ln 2].$$

Indeed, using remarks 6.4 and 6.3 we obtain

$$\begin{aligned} \ln^+ |\det [A - \lambda I_n]| &\le \ln^+ [\|A - \lambda I_n\|^n] = n \cdot \ln^+ [\|A - \lambda I_n\|] \\ &\le n \cdot \ln^+ [\|A\| + \|\lambda I_n\|] = n \cdot \ln^+ [\|A\| + |\lambda|] \\ &\le n \cdot [\ln^+ \|A\| + \ln^+ |\lambda| + \ln 2]. \end{aligned}$$

Proof of Theorem 6.1. Let $\lambda \in \mathbb{C}$. For $r \in [0,1)$ we define

$$v_r(\lambda) := \int_{\mathbb{T}} \ln |\det [F(rt) - \lambda I_n]| \; m(dt). \tag{6.23}$$

Assume that $r_1, r_2 \in [0,1)$ satisfy $r_1 \le r_2$. Since the function $\det [F - \lambda I_n]$ is holomorphic we get $v_{r_1}(\lambda) \le v_{r_2}(\lambda)$. Thus, the limit

$$v_{1-0}(\lambda) := \lim_{r \to 1-0} v_r(\lambda) \tag{6.24}$$

exists. Define

$$v(\lambda) := \int_{\mathbb{T}} \ln |\det [\underline{F}(t) - \lambda I_n]| \; m(dt). \tag{6.25}$$

If we apply inequality (1.15) to the function $F - \lambda I_n$, then using (6.23)–(6.25) we obtain

$$v_{1-0}(\lambda) \le v(\lambda). \tag{6.26}$$

According to Theorem 5.5, equality holds in (6.26) for those and only those $\lambda \in \mathbb{C} \setminus M_F$ for which the inner factor $\Theta_{\lambda,r}$ is a Blaschke-Potapov product. Consequently, Theorem 5.5 reduces the question which is discussed in Theorem 6.1 to the study of the structure of the set of all $\lambda \in \mathbb{C} \setminus M_F$ for which the inequality in (6.26) is strict. More formally, we will show that if K is a bounded Borel subset of positive logarithmic capacity, then there exists a point $\lambda \in K \cap (\mathbb{C} \setminus M_F)$ such that equality holds true in (6.26). Furthermore, we will show that if K is such a set and if ν is a finite Borel measure on $\mathbb{C}$ which is concentrated on K, i.e.,

$$\nu(\mathbb{C} \setminus K) = 0,$$

and if the associated logarithmic potential $U^{(\nu)}$ (see (6.3)) is continuous in $\mathbb{C}$, then the identity

$$\int_{\mathbb{C}} [v(\lambda) - v_{1-0}(\lambda)]\, \nu(d\lambda) = 0 \tag{6.27}$$

is valid. Clearly, from (6.26) and (6.27) it will follow that $v(\lambda) = v_{1-0}(\lambda)$ for almost all λ with respect to ν. In particular, there exists a $\lambda \in K \cap (\mathbb{C} \setminus M_F)$ for which $v(\lambda) = v_{1-0}(\lambda)$ is satisfied. Now we are going to prove (6.27). According to (6.23) for $r \in [0,1)$ and $\lambda \in \mathbb{C}$ we have

$$\int_{\mathbb{T}} \ln^+ |\det[F(rt) - \lambda I_n]|\, m(dt) - \int_{\mathbb{T}} \ln^- |\det[F(rt) - \lambda I_n]|\, m(dt) = v_r(\lambda). \tag{6.28}$$

In view of Remark 6.5, the inequality

$$\ln^+ |\det[F(rt) - \lambda I_n]| \leq n \cdot [\ln^+ \|F(rt)\| + \ln^+ |\lambda| + \ln 2] \tag{6.29}$$

holds for $r \in [0,1), \lambda \in \mathbb{C}$ and $t \in \mathbb{T}$. For $\lambda \in \mathbb{C}$ and $r \in [0,1)$ the function $G_{\lambda,r} : \mathbb{T} \to [0,\infty)$ is defined by

$$G_{\lambda,r}(t) := \det[F(rt) - \lambda I_n]. \tag{6.30}$$

Suppose that $\lambda \in \mathbb{C}$ is fixed. Then from (6.29) and (6.30) we infer that the family $(\ln^+ |G_{\lambda,r}|)_{r\in[0,1)}$ is uniformly m-integrable. Clearly, for almost all $t \in \mathbb{T}$ with respect to m we have

$$\lim_{r\to 1-0} \ln^+ |\det[F(rt) - \lambda I_n]| = \ln^+ |\det[\underline{F}(t) - \lambda I_n]|.$$

Thus, using Vitali's convergence theorem again, we get

$$\lim_{r\to 1-0} \int_{\mathbb{T}} \ln^+ |\det[F(rt) - \lambda I_n]|\, m(dt) = \int_{\mathbb{T}} \ln^+ |\det[\underline{F}(t) - \lambda I_n]|\, m(dt). \tag{6.31}$$

Taking into account (6.31) we obtain the formula

$$\int_{\mathbb{T}} \ln^+ |\det[\underline{F}(t) - \lambda I_n]|\, m(dt) - \lim_{r\to 1-0} \int_{\mathbb{T}} \ln^- |\det[F(rt) - \lambda I_n]|\, m(dt)$$
$$= v_{1-0}(\lambda) \tag{6.32}$$

by letting $r \to 1-0$ in (6.28), where the limit of the second term on the left hand side of (6.32) necessarily exists. From (6.25) and (6.32) it follows that

$$\begin{aligned} v(\lambda) - v_{1-0}(\lambda) &= \lim_{r\to 1-0} \int_{\mathbb{T}} \ln^- |\det[F(rt) - \lambda I_n]|\, m(dt) \\ &\quad - \int_{\mathbb{T}} \ln^- |\det[\underline{F}(t) - \lambda I_n]|\, m(dt). \end{aligned} \tag{6.33}$$

In general, the family $(\ln^- |G_{\lambda,r}|)_{r\in[0,1)}$ is not uniformly m-integrable. For this reason, the right hand side in (6.33) is not necessarily zero. (However, according to Fatou's theorem this difference is nonnegative.) Nevertheless, it will turn out that after applying the following averaging procedure the right hand side of (6.33) vanishes. Suppose that ν is a finite nonnegative measure with compact support for which the associated logarithmic potential $U^{(\nu)}$ is continuous. We will prove that

$$\begin{aligned}\int_{\mathbb{C}} \Bigg(& \lim_{r\to 1-0} \int_{\mathbb{T}} \ln^- |\det [F(rt) - \lambda I_n]| \, m(dt) \\ & - \int_{\mathbb{T}} \ln^- |\det [\underline{F}(t) - \lambda I_n]| \, m(dt) \Bigg) \nu(d\lambda) = 0.\end{aligned} \tag{6.34}$$

Using Fubini's theorem and (6.11) we get

$$\begin{aligned}&\int_{\mathbb{C}} \left(\int_{\mathbb{T}} \ln^- |\det [\underline{F}(t) - \lambda I_n]| \, m(dt) \right) \nu(d\lambda) \\ &= \int_{\mathbb{T}} \left(\int_{\mathbb{C}} \ln^- |\det [\underline{F}(t) - \lambda I_n]| \, \nu(d\lambda) \right) m(dt) \\ &= \int_{\mathbb{T}} \Phi_-^{(\nu)}(\underline{F}(t)) \, m(dt).\end{aligned} \tag{6.35}$$

In view of (6.12) it follows that

$$\int_{\mathbb{C}} \left(\int_{\mathbb{T}} \ln^- |\det [\underline{F}(t) - \lambda I_n]| \, m(dt) \right) \nu(d\lambda) < \infty. \tag{6.36}$$

Now we integrate identity (6.33) with respect to ν and use (6.36) to rewrite the integral of the difference as the difference of integrals. Then we rewrite the second term using (6.35) and apply Fatou's theorem to the first one. Finally, we use Fubini's theorem and (6.11) to rewrite the first term. This leads us to the following estimate

$$\begin{aligned}&\int_{\mathbb{C}} [v(\lambda) - v_{1-0}(\lambda)] \, \nu(d\lambda) \\ &= \int_{\mathbb{C}} \Bigg[\lim_{r\to 1-0} \int_{\mathbb{T}} \ln^- |\det [F(rt) - \lambda I_n]| \, m(dt) \\ &\quad - \int_{\mathbb{T}} \ln^- |\det [\underline{F}(t) - \lambda I_n]| \, m(dt) \Bigg] \nu(d\lambda)\end{aligned}$$

$$
\begin{aligned}
&= \int_{\mathbb{C}} \Big[\lim_{r\to 1-0} \int_{\mathbb{T}} \ln^- |\det [F(rt) - \lambda I_n]| \; m(dt) \Big] \; \nu(d\lambda) \\
&\quad - \int_{\mathbb{C}} \Big[\int_{\mathbb{T}} \ln^- |\det [\underline{F}(t) - \lambda I_n]| \; m(dt) \Big] \; \nu(d\lambda) \\
&= \int_{\mathbb{C}} \Big[\lim_{r\to 1-0} \int_{\mathbb{T}} \ln^- |\det [F(rt) - \lambda I_n]| \; m(dt) \Big] \; \nu(d\lambda) - \int_{\mathbb{T}} \Phi_-^{(\nu)}(\underline{F}(t)) \; m(dt) \\
&\leq \underline{\lim}_{r\to 1-0} \int_{\mathbb{C}} \Big[\int_{\mathbb{T}} \ln^- |\det [F(rt) - \lambda I_n]| \; m(dt) \Big] \; \nu(d\lambda) - \int_{\mathbb{T}} \Phi_-^{(\nu)}(\underline{F}(t)) \; m(dt) \\
&= \underline{\lim}_{r\to 1-0} \int_{\mathbb{T}} \Big[\int_{\mathbb{C}} \ln^- |\det [F(rt) - \lambda I_n]| \; \nu(d\lambda) \Big] \; m(dt) - \int_{\mathbb{T}} \Phi_-^{(\nu)}(\underline{F}(t)) \; m(dt) \\
&= \underline{\lim}_{r\to 1-0} \int_{\mathbb{T}} \Phi_-^{(\nu)}(F(rt)) \; m(dt) - \int_{\mathbb{T}} \Phi_-^{(\nu)}(\underline{F}(t)) \; m(dt). \qquad (6.37)
\end{aligned}
$$

According to Lemma 6.4 and our choice of ν, the function $\Phi_-^{(\nu)}$ is continuous. Thus, for almost all $t \in \mathbb{T}$ with respect to m we get

$$\lim_{r\to 1-0} \Phi_-^{(\nu)}(F(rt)) = \Phi_-^{(\nu)}(\underline{F}(t)). \qquad (6.38)$$

Since the function $\Phi_-^{(\nu)}$ is also bounded (see Lemma 6.4), Lebesgue's theorem on dominated convergence and (6.38) guarantee that

$$\lim_{r\to 1-0} \int_{\mathbb{T}} \Phi_-^{(\nu)}(F(rt)) \; m(dt) = \int_{\mathbb{T}} \Phi_-^{(\nu)}(\underline{F}(t)) \; m(dt). \qquad (6.39)$$

Thus, combining (6.37) and (6.39) we obtain (6.27).
As explained above, this completes the proof. □

Comments on Theorem 6.1. These comments are intended to clarify the function-theoretic content of Theorem 6.1. Let $t \in \mathbb{T}$. Then the function $G : \mathbb{C} \to [-\infty, \infty)$ defined by

$$G(\lambda) := \ln |\det [\underline{F}(t) - \lambda I_n]|$$

is subharmonic. Let $r \in [0, 1)$ and let the function $G_r : \mathbb{C} \to [-\infty, \infty)$ be defined by $G_r(\lambda) := G_{\lambda,r}(t)$, where $G_{\lambda,r}$ is given in (6.30). Then G_r is subharmonic too. From standard theorems on integrating parametric families of subharmonic functions (see. e.g. Ronkin [Ron, Ch.I, §5] or Lelong and Gruman [LG, Appendix I, Proposition I.14]) it follows that the function v defined in (6.25) is subharmonic and that for each $r \in [0, 1)$ the function v_r defined in (6.23) is also subharmonic. Since the family $(v_r)_{r\in[0,1)}$ increases monotonically with r, the function v_{1-0} defined in

(6.24) is the upper envelope of this family. The function v_{1-0} is not necessarily subharmonic but its regularization $v_{1-0}^* : \mathbb{C} \to [-\infty, \infty)$ which is defined by

$$v_{1-0}^*(\lambda) := \varlimsup_{\lambda' \to \lambda} v_{1-0}(\lambda') \tag{6.40}$$

turns out to be subharmonic (see Ronkin [Ron, Ch.I, §5], Lelong and Gruman [LG, Appendix I, Proposition I.25] and Cartan [Car]). Clearly, for $\lambda \in \mathbb{C}$ the inequality

$$v_{1-0}(\lambda) \leq v_{1-0}^*(\lambda) \tag{6.41}$$

holds. According to an ingenious theorem of H. Cartan (see e.g. Ronkin [Ron, Ch.I, §5] or Cartan [Car]), the upper envelope of a family of subharmonic functions coincides with its regularization everywhere except for a set of logarithmic capacity zero. (For the exact formulation of Cartan's theorem and a proof we refer to Ronkin [Ron, Ch.I, §5]). In particular,

$$\text{cap}\ (\{\lambda : v_{1-0}^*(\lambda) > v_{1-0}(\lambda)\}) = 0. \tag{6.42}$$

Since the function v is upper semicontinuous (i.e., for $\lambda \in \mathbb{C}$, the inequality $v(\lambda) \geq \varlimsup_{\lambda' \to \lambda} v(\lambda')$ holds) the inequalities

$$v_{1-0}(\lambda) \leq v_{1-0}^*(\lambda) \leq v(\lambda) \tag{6.43}$$

follow for $\lambda \in \mathbb{C}$ from (6.26). Thus if we establish the equality $v_{1-0}(\lambda) = v(\lambda)$ for all λ belonging to some dense subset of $\mathbb{C}$, then in view of (6.43) we obtain

$$v_{1-0}^* \equiv v. \tag{6.44}$$

The identity $v_{1-0}(\lambda) = v(\lambda)$ for all λ belonging to some dense subset of $\mathbb{C}$ clearly follows from the identity

$$\int_I [v(\lambda) - v_{1-0}(\lambda)]\ \mu(d\lambda) = 0,$$

where I is an arbitrary one dimensional interval of $\mathbb{C}$ and μ is one dimensional Lebesgue measure. The use of H. Cartan's theorem on upper envelopes of families of subharmonic functions for proving the smallness (in the sense of capacity) of exceptional sets has many traditions in the theory of functions of one and several complex variables. The application of the recently created complex potential theory, in particular the analogue of H. Cartan's theorem for the upper envelope of a family of plurisubharmonic functions (see Bedford and Taylor [BT, Section 7] and Sadullaev's survey paper [Sad]) enables one to derive results on families of matrix-valued functions of a more general type, namely on families which depend holomorphically on p variables where $p \in \mathbb{N}$.

Finally, we turn our attention to the left version of our main result.

Theorem 6.2. *Let $F \in \mathfrak{N}_n^+(\mathbb{D})$. Assume that for $\lambda \in \mathbb{C} \setminus M_F$ the functions $\Theta_{\lambda,l} \in \mathfrak{J}_n(\mathbb{D})$ and $E_{\lambda,l} \in \mathfrak{E}_n(\mathbb{D})$ are factors in the multiplicative decomposition*

$$F_\lambda = E_{\lambda,l} \cdot \Theta_{\lambda,l}.$$

Suppose that K is a bounded Borel subset of $\mathbb{C}$ with positive logarithmic capacity. Then there exists a $\lambda \in K \cap (\mathbb{C} \setminus M_F)$ for which $\Theta_{\lambda,l}$ is a Blaschke-Potapov product.

Proof. Use Theorem 6.1, Remark 3.2 and Remark 4.2. □

Corollary 6.2. *The set of all $\lambda \in \mathbb{C} \setminus M_F$ for which $\Theta_{\lambda,l}$ is a Blaschke-Potapov product is dense in $\mathbb{C}$.*

For further matricial generalizations of the classical theorems of Frostman [Fr], Heins [Hei] and Rudin [Ru1], we refer the reader to the papers [Gi6] and [GiTa1]–[GiTa3].

References

[Ar1] Arov, D.Z.: *Darlington realization of matrix-valued functions* (in Russian), Izv. Akad. Nauk SSSR, Ser. Mat. **37** (1973), 1299–1331, Engl. transl. in: Math. USSR Izvestija **7** (1973), 1295–1326, MR 50#10287.

[Ar2] Arov, D.Z.: *Functions of class* Π (in Russian), Zap. Nauc. Sem. LOMI **135** (1984), 5–30, Engl. transl. in: J. Soviet. Math. **31** (1985), 2645–2659, MR 85h:47041.

[Ar3] Arov, D.Z.: *On regular and singular J-inner matrix-functions and related extrapolation problems* (in Russian), Funkcional. Anal. i. Prilozhen. **22** (1988), no. 1, 57–59, Engl. transl. in: Functional Analysis and its Applic. **22** (1988), 46–48, MR 89d:47082.

[Ar4] Arov, D.Z.: *γ-generating matrices, J-inner matrix-functions and related extrapolation problems I* (in Russian), Teor. Funkcii, Funkcional. Anal. i. Prilozhen. **51** (1989), 61–67, Engl. transl. in: J. Soviet Math. **52** (1990), 3487–3491, MR 92i:30034a.

[Ar5] Arov, D.Z.: *γ-generating matrices, J-inner matrix-functions and related extrapolation problems II* (in Russian), Teor. Funkcii, Funkcional. Anal. i. Prilozhen. **21** (1989), 103–109, Engl. transl. in: J. Soviet Math. **52** (1990), 3421–3425, MR 92i:30034b.

[Ar6] Arov, D.Z.: *γ-generating matrices, J-inner matrix-functions and related extrapolation problems III* (in Russian), Teor. Funkcii, Funkcional. Anal. i. Prilozhen. **53** (1990), 57–65, Engl. transl. in: J. Soviet Math. **52** (1992), 532–537, MR 92i:30034c.

[Ar7] Arov, D.Z.: *Regular J-inner matrix-function and related continuation problems*, in: Linear Operators in Function spaces (Eds.: Helson, H.; Sz.-Nagy, B.; Vasilescu, F.-H.), Operator Theory: Advances and Appl., vol. **43**, Basel: Birkhäuser Verlag 1990, 63–87, MR 93b:47028.

[Ar8] Arov, D.Z.: *The generalized bitangent Carathéodory-Nevanlinna-Pick problem and (j, J_0)-inner matrix-valued functions* (in Russian), Izv. Rossiiskoi Akad. Nauk **57** (1993), 3–32, Engl. transl. in: Russian Acad. Sci. Izv. Math. **42** (1994), 1–26, MR 94j:47023.

[Ar9] Arov, D.Z.: *Computation of the resolvent matrix for the generalized bitangential Schur- and Carathéodory-Nevanlinna-Pick interpolation problems in the strictly completely indeterminate case*, Integral Equations and Operator Theory **22** (1995), 253–272

[AFK1] Arov, D.Z.; Fritzsche, B.; Kirstein, B.: *On some completion problems for various subclasses of j_{pq}-inner functions*, Zeitschrift für Analysis und ihre Anwend. **11** (1992), 489–508, MR 95d:30073.

[AFK2] Arov, D.Z.; Fritzsche, B.; Kirstein, B.: *Completion problems for j_{pq}-inner functions, I*, Integral Equations and Operator Theory **16** (1993), 155–185, MR 93k:47027.

[AFK3] Arov, D.Z.; Fritzsche, B.; Kirstein, B.: *Completion problems for j_{pq}-inner functions, II*, Integral Equations and Operator Theory **16** (1993), 453–495, MR 94h:47024.

[AFK4] Arov, D.Z.; Fritzsche, B.; Kirstein, B.: *On block completion problems for various subclasses of j_{pq}-inner functions*, in: Challenge of a Generalized Systems Theory (Eds.: Dewilde, P.; Kaashoek, A.; Verhaegen, M.A.), North Holland, Amsterdam 1993, pp. 179–194, MR 95h:30061.

[AFK5] Arov, D.Z.; Fritzsche, B.; Kirstein, B.: *On block completion problems for $j_{qq} - J_q$-inner functions .I. The case of a given block column*, Integral Equations and Operator Theory **18** (1994), 1–29, MR 95b:47016a.

[AFK6] Arov, D.Z.; Fritzsche, B.; Kirstein, B.: *On block completion problems for $j_{qq} - J_q$-inner functions .II. The case of a given $q \times q$ block*, Integral Equations and Operator Theory **18** (1994), 245–260, MR 95b:47016b.

[AFK7] Arov, D.Z.; Fritzsche, B.; Kirstein, B.: *On some aspects of V.E. Katsnelson's investigations on interrelations between left and right Blaschke-Potapov products*, in: Operator Theory and Boundary Eigenvalue problems (Eds.: Gohberg, I.; Langer H.), Operator Theory: Advances and Appl., vol. **80**, Basel· Boston· Berlin: Birkhäuser Verlag 1995, pp. 21–41.

[BT] Bedford, E.; Taylor, B.A.: *A new capacity for plurisubharmonic functions*, Acta Math. **149** (1982), 1–40, MR 84d:32024.

[Be] Beurling, A.: *On two problems concerning linear transformation in Hilbert space*, Acta Math. **81** (1949), 239–255, MR 10, p. 381.

Reprinted in: *Collected Works of Arne Beurling, Volume* 1 *Complex Analysis*, Basel· Boston· Berlin: Birkhäuser Verlag 1989, pp. 147–163.

[Cam] Camera, G.A.: *Nonlinear superposition on spaces of analytic functions*, in: Harmonic Analysis and Operator Theory (Eds.: Marcantognini, S.A.M.; Mendoza, G.A.; Morán, M.D.; Octavio, A; Urbina, W.O.), Contemporary Mathematics, vol. **189** (1995), pp. 103–116, MR 95f:47093.

[Car] Cartan, H.: *Théorie du potentiel newtonian: énergie, capacité, suites de potentiels*, Bull. Soc. Math. France **73** (1945), 74–106, MR 7, p. 447.

[CoLo] Collingwood, E.F., Lohwater, A.J.: The Theory of Cluster Sets, Cambridge: Univ. Press 1966, Russ. transl.: Moscow: Mir 1971, MR 38#325.

[De] Devinatz, A.: *The factorization of operator valued functions*, Ann. Math. **73** (1961), 458–495, MR#A3997.

[Doo1] Doob, J.L.: *The boundary values of analytic functions*, I, Trans. Amer. Math. Soc. **34** (1932), 153–170.

[Doo2] Doob, J.L.: *The boundary values of analytic functions*, II, Trans. Amer. Math. Soc. **35** (1933), 418–451.

[DFK] Dubovoj, V.K.; Fritzsche, B.; Kirstein, B.: *Matricial Version of the Classical Schur Problem*, Teubner Texte zur Mathematik 129, Stuttgart: B.G. Teubner 1992, MR 93e:47021.

[Dur] Duren, P.L.: *Theory of H^p Spaces*, New York, London: Academic Press 1970, MR 42#3552.

[Fr] Frostman, O.: *Potentiel d' équilibre et capacité des ensembles avec quelques applications à la théorie des fonctions*, Medd. Lunds Univ. Math. Semin. **3** (1935), 1-118, MF 61, p. 1262 (MF – Jahrbuch über die Fortschritte der Mathematik).

[G] Garnett, J.B.: *Bounded analytic functions*, New York, London: Academic Press 1981, MR 83g:30037.

[Gi1] Ginzburg, Yu.P.: *The factorization of analytic matrix functions*, Dokl. Akad. Nauk SSSR **159** (1964), 489–492; Engl. transl. in: Soviet Math. Dokl. **5** (1961), 1510–1514, MR 30#3228.

[Gi2] Ginzburg, Yu.P.: *Multiplicative representations of bounded analytic operator functions* (in Russian), Dokl. Akad. Nauk SSSR **170** (1966), 23–26; Engl. transl. in: Soviet Math. Dokl. **7** (1966), 1125–1128, MR 34#611.

[Gi3] Ginzburg, Yu.P.: *Multiplicative representations of operator functions of bounded form*, Uspehi Mat. Nauk **22** (1967), no. 1, 163–165, MR 34#6511.

[Gi4] Ginzburg, Yu.P.: *Multiplicative representations and minorants of bounded analytic operator functions* (in Russian), Funkcional. Anal. i Prilozhen. **1** (1967), no. 3, 9-23, MR 36#4366 Engl. transl. in: Funkcional. Anal. and Appl. **1**:3 (1967), 180-192.

[Gi5] Ginzburg, Yu.P.: *Multiplicative representations of J-nonexpansive operator functions*, Mat. Issled. **2** (1967), no. 2, 52–83; no. 3, 20–51; Engl. transl. in: Amer. Math. Soc. Transl. (Series 2) **96** (1970), 189–254, MR 38#1551.

[Gi6] Ginzburg, Yu.P.: *The almost invariant spectral properties of contractions and the multiplicative properties of analytic operator-functions* (in Russian), Funkcional. Anal. i Prilozhen. **5** (1971), no. 3, 32–41, MR 44#834. Engl. transl. in: Funkcional. Anal. and Appl. **5**:3 (1971), 197–205.

[Gi7] Ginzburg, Yu.P.: *On the reconstruction of a multiplicative integral from its modulus* (in Russian), Teor. Funkcii, Funkcional. Anal. i Prilozhen. **41** (1984), MR 85k:47063.

[GiSh] Ginzburg, Yu.P.; Shevchuk, L.V.: *On the Potapov theory of multiplicative representations*, in: Matrix and Operator Valued Functions (Eds.: Gohberg, I.; Sakhnovich, L.A.), Operator Theory: Advances and Appl., vol. **72**, Basel·Boston·Berlin: Birkhäuser Verlag 1994, pp. 28–47, MR 95k:47020.

[GiTa1] Ginzburg, Yu.P.; Taljusch, N.A.: *A matricial analogue of a theorem of Heins and the typical special structure of contractions* (in Russian), Funkcional. Anal. i Prilozhen. **7** (1973), no. 1, 66–67, MR 47#4034 Engl. transl. in: Funkcional. Anal. and Appl. **7**:1 (1973), 56–57.

[GiTa2] Ginzburg, Yu.P.; Taljusch, N.A.: *On polynomial bundles of analytic matrix functions and families of contractive extensions of isometric operators* (in Russian), Izvestiya Vuzov. Matematika **26** no. 4, (1982), 19–27, MR 84d:47009. Engl. transl. in: Soviet Mathematics (Izvestiya VUZ.Matematika) **26**:4, 21–32.

[GiTa3] Ginzburg, Yu.P.; Taljusch, N.A.: *Exceptional sets of analytic matrix functions, contractive and dissipative operators* (in Russian), Izv. Vuzov 1984, no. 8, 9–14, Engl. transl. in: Soviet Mathematics (Izvestiya VUZ.Matematika) **28**:8, 10–16. MR 87d:47067.

[Hal] Halmos, P.: *Shifts on Hilbert spaces*, J. reine und angew. Math. **208** (1961), 102–112, MR 27#2868.

[Hei] Heins, M.: *On the Lindelöf Principle*, Ann. Math. **61** (1953), 440–473, MR 16, p. 1011.

[Hel1] Helson, H.: *Lectures on Invariant Subspaces*, New York, London: Acad. Press 1964, MR 30#1409.

[Hel2] Helson, H.: *Large analytic functions*, in: Linear Operators in Function Spaces (Eds.: Helson, H.; Sz.-Nagy, B.; Vasilescu, F.-H.), Operator Theory: Advances and Appl., vol. **43**, Basel · Boston · Berlin: Birkhäuser Verlag 1990, pp. 209–216, MR 92c:30038.

[Hel3] Helson, H.: *Large analytic functions*, II, in: Analysis and Partial Differential Equations: A Collection of Papers Dedicated to Mischa Cotlar (Ed.: Sadosky, C.), New York, Basel: Marcel Dekker 1990, pp. 217–220, MR 92c:30039.

[Hel4] Helson, H.: *Large analytic functions*, III, Colloq. Math., LX/LXI (1990), 221–223, MR 92c:30040.

[HL1] Helson, H.; Lowdenslager, D.: *Prediction theory and Fourier series in several variables*, I, Acta Math. **99** (1958), 165–202, MR 20#4155.

[HL2] Helson, H.; Lowdenslager, D.: *Prediction theory and Fourier series in several variables*, II, Acta Math. **106** (1961), 175–213, MR 31#562.

[HLP] Hardy, G.H.; Littlewood, J.E.; Pólya, G.: *Inequalities*, Cambridge: Cambridge Univ. Press 1934, MF 60, p. 169–170.

[Hoff] Hoffman, K.: *Banach spaces of analytic functions*, Englewood Cliffs, N.J.: Prentice Hall 1962, MR 24#A2844.

[Ka] Kato, T.: *Perturbation theory for linear operators*, Berlin, Heidelberg, New York: Springer Verlag 1966, MR 34#3324.

[Kats1] Katsnelson, V.E.: *A left Blaschke-Potapov product is not necessarily a right Blaschke-Potapov product* (in Russian), Dokl. Akad. Nauk Ukrainian SSR, Series A 10 (1989), 15–17, MR 90k:47030.

[Kats2] Katsnelson, V.E.: *Left and right Blaschke-Potapov products and Arov-singular matrix-valued functions*, Integral Equations and Operator Theory **13** (1990), 836–848, MR 91f:47021.

[Kats3] Katsnelson, V.E.: *Weighted spaces of pseudocontinuable functions and approximations by rational functions with prescribed poles*, Zeitschrift für Analysis und ihre Anwend. **12** (1993), 27–67, MR 94m:30072.

[Kats4] Katsnelson, V.E.: *Description of a class of functions which admit an approximation with preassigned poles I*, in: Matrix and Operator Valued Functions (Eds.: Gohberg, I.; Sakhnovich, L.A.), Operator Theory: Advances and Appl., vol. **72**, Basel-Boston-Berlin: Birkhäuser Verlag 1994, pp. 87–132, MR 96e:30095.

[Kh] Kheifets, A.J.: *On regularization of γ-generating pairs*, J. Functional Anal. **130** (1995), 310–333, MR 96b:47015.

[Kol] Kolmogorov, A.N.: *Stationary sequences in Hilbert space* (in Russian), Bull. Math. Univ. Moscow **2** (1941), 1- 40, MR#5101. Engl. transl. in: *Selected Works of A.N. Kolmogorov*. Vol. II: *Probability Theory and Mathematical Statistics.* (edited by A.N. Shiryaev), Dordrecht · Boston · London: Kluver Academic Publishers 1992, 228–271.

[Koo] Koosis, P.: *Introduction to H^p Spaces*, Cambridge: Cambridge University Press 1980, MR 81c:30062.

[Kr] Krein, M.G.: *On a generalization of some investigations of G. Szegö, V.I. Smirnov and A.N. Kolmogorov* (in Russian), Dokl. Akad. Nauk SSSR **46** (1945), 95–98, French transl. in: C.R. Acad. Sa. URSS (N.S.) **46** (1945), 91–94, MR#7156.

[LVP1] de la Vallée Poussin, C.: *Sur l'intégrale de Lebesgue*, Trans. Amer. Math. Soc. **16** (1915), 435–501, MF 45, p. 441–442

[LVP2] de la Vallée Poussin, C.: *Points irréguliers. Détermination des masses par les potentiels*, Bull. de la class d. Sci. Acad. Belgique **24** (1938) 368–384, 672–689, MF 64, p. 478, p. 1162.

[LVP3] de la Vallée Poussin, C.: *Le potentiel logarithmique, balayage et représentation conforme*, Paris: Gauthier-Villars 1949, Zbl. 37346.

[La] Lax, P.: *Translation invariant subspaces*, Acta Math. **101** (1959), 163–178, MR 21#4359.

[LG] Lelong, P.; Gruman, L.: *Entire functions of several complex variables*, Berlin·Heidelberg·New York: Springer Verlag 1986, MR 90i:32002.

[MM] Marcus, M.; Minc, H.: *A survey of matrix theory and matrix inequalities*, Boston: Allyn and Bacon 1964, MR 29#112.

[Ma1] Masani, P.R.: *Cramér's theorem on monotone matrix-valued functions and the Wold decomposition*, Probability and Statistics (U. Grenander, ed.) 175–189, New York: Wiley 1959, MR 23#A2236.

[Ma2] Masani, P.R.: *Shift invariant spaces and prediction theory*, Acta Math. **107** (1962), 275–290, MR 25#4344.

[Ma3] Masani, P.R.: *Wiener's contributions to generalized harmonic analysis, prediction theory and filter theory*, Bull. Amer. Math. Soc. **72** (1966), 73–125, MR 32#4773.

[Ma4] Masani, P.R.: *Recent trends in multivariate prediction theory*, in: Multivariate Analysis (Ed.: Krishnaiah, P.R.), New York, London: Academic Press 1966, pp. 351–382, MR 35#5079.

[McC] McCarthy, J.E.: *Topologies on the Smirnov class*, J. Functional Analysis **104** (1992), 229–241, MR 93a:30041.

[Mo1] Mochizuki, N.: *Algebras of holomorphic functions between* H^p *and* $\mathfrak{N}_*$, Proc. Amer. Math. Soc. **105** (1989), 889–902, MR 90a:46137.

[Mo2] Mochizuki, N.: *Nevanlinna and Smirnov classes on the upper half plane*, Hokkaido Math. J. **20** (1991), 609–620, MR 93b:30031.

[Na] Nagumo, M.: *Über die gleichmäßige Summierbarkeit und ihre Anwendung auf ein Variationsproblem*, Japan J. Math. **6** (1929), 173–182, MF 55, p. 156.

[Nic1] Nicolau, A.: *The coefficients of Nevanlinna's parametrization are not in* H^p, Proc. Amer. Math. Soc. **106** (1989), 115–117, MR 89k:30035.

[Nic2] Nicolau, A.: *A characterization of the leading coefficient of Nevanlinna's parametrization*, Illinois J. Math. **37** (1993), 284–301, MR 94c:30050.

[Nik1] Nikolskii, N.K.: *Selected problems on weighted approximation and spectral analysis* (in Russian), Trudy Math. Inst. Steklov **120** (1974), MR 57#7133. Engl. transl. in: Proceedings of the Steklov Inst. of Math. Number 120 (1973), Providence, R.I.: AMS 1976.

[Nik2] Nikolskii, N.K.: *Treatise on the Shift Operator*, Berlin, Heidelberg, New York: Springer Verlag 1986, MR 87i:47042.

[No] Noshiro, K.: *Cluster sets*, Berlin·Göttingen·Heidelberg: Springer Verlag 1960, Russ. transl.: Moscow: Izd. inostrann. literatury 1963, MR 24#A3295.

[Pot] Potapov, V.P.: *The multiplicative structure of J-contractive matrix functions* (in Russian), Engl. Transl. in: Amer. Math. Soc. Transl. (Series 2), **15** (1960), 131–243, MR 17, p. 958–959.

[Pri] Privalov, I.I.: *Boundary properties of analytic functions* (in Russian), MR 13, p. 926, German transl.: *Randeigenschaften analytischer Funktionen*, Berlin: Deutscher Verlag der Wissenschaften 1956, MR 18, p. 727.

[Rob] Roberts, J.W.: *The component of the origin in the Nevanlinna class*, Illinois J. Math. **19** (1975), 553–559, MR 52#3554 / MR 18#727.

[RoSt1] Roberts, J.W.; Stoll, M.: *Prime and principal ideals in the algebra* $\mathfrak{N}^+$, Arch. Math. **27** (1976), 387–393, Correction, ibid. **30** (1978), 672, MR 54#10625.

[RoSt2] Roberts, J.W.; Stoll, M.: *Composition operators on* F^+, Studia Math. **57** (1976), 217–228, MR 55#8773.

[Ron] Ronkin, L.I.: *Introduction to the theory of entire functions of several variables* (in Russian), Moscow: Nauka 1972, MR 47#8896, English transl. in: Translation of Math. Monographs, vol. **44**, AMS, Providence, R.I. 1974, MR 49#10901.

[RoRo1] Rosenblum, M.; Rovnyak, J.: *Hardy Classes and Operator Theory*, Oxford Mathematical Monographs, Oxford: Clarendon Press 1985, MR 87e:47001.

[RoRo2] Rosenblum, M.; Rovnyak, J.: *Topics in Hardy Classes and Univalent Functions*, Basel·Boston·Berlin: Birkhäuser Verlag 1994.

[Roz1] Rozanov, J.A.: *Spectral properties of multivariate stationary processes and boundary properties of analytic matrices*, Theory Probability Appl. (USSR), Engl. Transl. **5** (1960), 399–414, MR 24#A2432.

[Roz2] Rozanov, J.A.: *Stationary random processes*, Moscow: Fizmatgiz 1963 (in Russian), Engl. translation: San Francisco: Holdon-Day 1967, MR 35#4985.

[Ru1] Rudin, W.: *A generalization of a theorem of Frostman*, Math. Scand. **21** (1967), 136–143, MR 38#3463.

[Ru2] Rudin, W.: *Function Theory in Polydisc*, New York, Amsterdam: Benjamin 1969, MR 41#501.

[Ru3] Rudin, W.: *Real and Complex Analysis*, London: Mc Graw-Hill, London 1970, MR 32#230.

[Sad] Sadullaev, A.: *Plurisubharmonic Functions*, In: Several Complex Variables II: Encyclopaedia of Mathematical Sciences (Eds.: Khenkin, G.M.; Vitushkin, A.G.), Berlin, Heidelberg, New York: Springer Verlag 1994, pp. 59–106, MR 95e:32001.

[Sar1] Sarason, D.: *Exposed points in* H^1, in: The Gohberg Anniversary Collection, Volume II: Topics in Analysis and Operator Theory (Eds.: Dym, H.; Goldberg, S.; Kaashoek, M.A.; Lancaster, P.), Operator Theory: Advances and Apl., vol. **41**, Basel·Boston· Berlin: Birkhäuser Verlag 1989, pp. 485–496, MR 91h:46043.

[Sar2] Sarason, D.: *Sub-Hardy Hilbert Spaces in the Unit Disk*, University of Arkansas Lecture Notes in the Math. Sciences, vol. **10**, New York: Wiley 1995

[Sh] Shapiro, H.S.: *Weakly invertible elements in certain function spaces and generators in* L^1, Michigan Math. J., **11** (1964), 161–165, MR 29#3620.

[ShSh] Shapiro, J.H.; Shields, A.L.: *Unusual topological properties of the Nevanlinna class*, Amer. J. Math. **97** (1976), 915–936, MR 52#11053.

[Sm] Smirnov, V.I.: *Sur les formules de Cauchy et de Green et quelques problèms qui s'y rattachent* (in French), Izv. Akad. Nauk SSSR, Ser. Mat. **3** (1932), 338–372, Russian transl. in: Smirnov, V.I.: Selected Papers – Complex Analysis, Mathematical Theory of Diffraction (in Russian), Leningrad: Leningrad University Press 1988, pp. 82–111, MR 91g:01040.

[St1] Stoll, M.: *A characterization of* $F^+ \cap \mathfrak{N}$, Proc. Amer. Math. Soc **57** (1976), 97–98, MR 53#3315

[St2] Stoll, M.: *Mean growth and Taylor coefficients of some topological algebras of analytic functions*, Ann. Polon. Math. **35** (1977), 139–158, MR 57#3858.

[Sz1] Szegö, G.: *Beiträge zur Theorie der Toeplitzschen Formen I*, Math. Zeitschr. **6** (1920), 167–202, Reprinted in: Gabor Szegö – Collected Papers (Ed.: Askey, R.), vol. **1**, Boston, Basel, Stuttgart: Birkhäuser Verlag 1982, pp. 237–272, MF 47, p. 391

[Sz2] Szegö, G.: *Beiträge zur Theorie der Toeplitzschen Formen II*, Math. Zeitschr. **9** (1921), 167–190, Reprinted in: Gabor Szegö – Collected Papers (Ed.: Askey, R.), Volume 1, Boston, Basel, Stuttgart: Birkhäuser Verlag 1982, pp. 279–302,, MF 48, p. 376–378

[Sz3] Szegö, G.: *Über die Randwerte einer analytischen Funktion*, Math. Ann. **84** (1921), 232–244, Reprinted in: Gabor Szegö – Collected Papers (Ed.: Askey, R.), Volume 1, Boston·Basel·Stuttgart: Birkhäuser Verlag 1982, pp. 404–416, MF 48, p. 332

[SZNF] Sz.-Nagy B.; Foias, C.: *Analyse Harmonique des Opérateurs de l'Espace de Hilbert*, Budapest: Académiai Kiadó 1967, MR 37#778, Engl. transl.: *Harmonic Analysis of Operators on Hilbert Space*, Budapest: Académiai Kiadó and Amsterdam, London: North-Holland Publishing Company 1970, MR 43#947.

[Vin] Vinogradov, S.A.: *Properties of multipliers of Cauchy-Stieltjes integrals and some factorization problems for analytic functions* (in Russian), in: Math. Programming and Related Questions (Proc. Seventh Winter School, Drogobych, 1974: Theory of Functions and Functional Analysis), Moscow: Central Econ.-Math. Inst. Akad. Nauk SSSR 1976, pp. 5–39, Engl. transl. in: Amer. Soc. Transl. **115** (1980), 1–32, MR 58#28518.

[Vit] Vitali, G.: *Sull'integrazione per serie*, Rend. Circ. Matem. Palermo **23** (1907), 137–155, MF 38, p. 338.

[WM1] Wiener, N.; Masani, P.: *The prediction theory of multivariate stochastic processes I. The regularity condition*, Acta Math. **98** (1957), 111–150, Norbert Wiener – Collected Works (Ed.: Masani, P.R.), Volume III, Cambridge, MA; London: MIT Press 1981, pp. 164–203, MR 20#4323.

[WM2] Wiener, N.; Masani, P.: *The prediction theory of multivariate stochastic processes II. The linear predictor*, Acta Math. **99** (1958), 93–137, Norbert Wiener – Collected Works (Ed.: Masani, P.R.), Volume III, Cambridge, MA; London: MIT Press 1981, 204–248, MR 20#4325.

[Y1] Yanagihara, N.: *On a class of functions and their integrals*, Proc. London Math. Soc. (Third Series) **25** (1972), 550–576, MR 46#3801.

[Y2] Yanagihara, N.: *The second dual space for the space* $\mathfrak{N}^+$, Proc. Japan Academy **49** (1973), 33–36, MR 49#9600.

[Y3] Yanagihara, N.: *Multipliers and linear functionals for the class* $\mathfrak{N}^+$, Trans. Amer. Math. Soc. **180** (1973), 449–461, MR 49#3147.

[Y4] Yanagihara, N.: *The containing Frechét space for the class* $\mathfrak{N}^+$, Duke Math. J. **40** (1973), 93–103, MR 49#9599.

[Y5] Yanagihara, N.: *Bounded subsets of some spaces of holomorphic functions*, Scientific Papers of the College of General Education, University of Tokyo **23** (1973), no. 1, 19–28, MR 48#2403.

[Y6] Yanagihara, N.: *The class* $\mathfrak{N}^+$ *of holomorphic functions and its containing Frechét space* $\mathfrak{F}^+$, Boll. Un. Math. Ital. (Bologna) **8** (1973). 230–245, MR 48#11520.

[Y7] Yanagihara, N.: *Mean growth and Taylor coefficients of some classes of functions*, Ann. Polon. Math. **30** (1974), 37–48, MR 49#3148

[Y8] Yanagihara, N.: *Interpolation theorems for the class* $\mathfrak{N}^+$, Illinois J. Math. **18** (1974), 427–435, MR 50#10271

[Y9] Yanagihara, N.: *Generators and maximal ideals in some algebras of holomorphic functions*, Tohoku Math. J. **27** (1975), 31–47, MR 52#3555.

[Y10] Yanagihara, N.: *Variational methods for functions of bounded characteristic*, J. Math. Anal. Appl. **49** (1975), 561–574, MR 50#13493.

[YK] Yanagihara, N.; Kawase, S.: *On the characteristic of some classes of functions and their integrals*, Proc. London Math. Soc. (Third Series) **25** (1972), 577–585, MR 47#3681.

[YN] Yanagihara, N.; Nakamura, Y.: *Composition operators on the class* $\mathfrak{N}^+$, TRU Mathematics **14-2** (1978), 9–16, MR 80f:30024.

Victor E. Katsnelson
Department of Theoretical Mathematics
The Weizmann Institute of Science
Rehovot, 76100
Israel
E-mail:
katze@wisdom.weizmann.ac.il

Bernd Kirstein
Mathematisches Institut
Universität Leipzig
Leipzig, 04109
Bundesrepublik Deutschland
E-mail:
heide@mathematik.uni-leipzig.de

AMS Subject Classification: 30D50, 46E10.

Operator Theory
Advances and Applications, Vol. 95
© 1997 Birkhäuser Verlag Basel/Switzerland

Integral representation of functions of class $\mathcal{K}_a$

I.V. Kovalishina

Abstract. An extension problem of M.G. Krein, in which a continuous positive definite kernel of a certain form is specified on a square, is analyzed by a general method which V.P. Potapov developed for solving interpolation problems. The key step is to identify all solutions of the original extension problem with the solutions of an associated Fundamental Matrix Inequality. [Abstract added by editors.]

1 Formulation of the problem

A function $S(x)$ $(0 \leq x < 2a;\ a \leq +\infty)$ belongs to the class $\mathcal{K}_a$ if (1) it is continuous, (2) $S(0) = 0$ and (3)

$$\mathcal{K}(x, y) = \frac{1}{2}\{S(x+y) - S(|x-y|)\}\ (0 \leq x, y < a) \tag{1}$$

is a Hermitian positive kernel. We shall use the "exact" and "integral" definitions of the Hermitian positive kernel $\mathcal{K}(x, y)$:

1) A kernel $\mathcal{K}(x, y)$, defined on the quadrangle $(0 \leq x, y \leq l,\ \forall l < a,\ a \leq \infty)$, is said to be Hermitian positive if, for any choice of points

$$0 \leq x_1, x_2, ..., x_n \leq l\ ,$$

and any complex numbers $\xi_1, \xi_2, ..., \xi_n$

$$\sum_{k,j=1}^{n} \mathcal{K}(x_j, x_k)\xi_j\bar{\xi}_k \geq 0. \tag{2}$$

2) A kernel $\mathcal{K}(x, y)$, defined on the quadrangle $(0 \leq x, y \leq l,\ \forall l < a,\ a \leq \infty\ )$, is said to be Hermitian positive if

$$(K\varphi, \varphi) = \int_0^l \int_0^l \mathcal{K}(x, y)\varphi(y)\bar{\varphi}(x)dxdy \geq 0\ (\forall \varphi \in L^2[0, l]), \tag{3}$$

where

$$(K\varphi)(x) = \int_0^l \mathcal{K}(x, y)\varphi(y)dy \text{ and } (f, g) = \int_0^l f\bar{g}dx.$$

These definitions are equivalent.

M.G. Krein [1] proved that every function $S(x) \in \mathcal{K}_a$ admits an integral representation

$$S(x) = \int_{-\infty}^{+\infty} \frac{1-\cos(\sqrt{t}x)}{t} d\sigma(t), \quad (\sigma(t) \in \uparrow). \tag{4}$$

Conversely, if a function $S(x)$ $(0 \leq x < 2a)$ satisfies the condition (4) then it belongs to the class $\mathcal{K}_a$.

We formulate the following problem:

1) To prove the existence of at least one $\sigma(t)$ such that the function $S(x)$, which is defined by the relation (4), belongs to the class $\mathcal{K}_a$.

2) To describe all these $\sigma(t)$.

3) To establish criterion for the non-uniqueness of the integral representation.

Let us consider the so-called "associated" function related to the integral (4)

$$W(z) = \int_{-\infty}^{+\infty} \frac{d\sigma(t)}{t-z}. \tag{5}$$

This integral defines a pair of functions

$$W(z) = \begin{cases} W_1(z) & \mathrm{Im}z > 0, \\ W_2(z) & \mathrm{Im}z < 0, \end{cases}$$

which are holomorphic on the upper and the lower half-planes, respectively, and satisfy the general condition

$$\frac{W(z) - W^*(z)}{z - \bar{z}} = \int_{-\infty}^{+\infty} \frac{d\sigma(t)}{|t-z|^2}, \quad \mathrm{Im}z \neq 0\ , \tag{6}$$

and obey the symmetry relation

$$W_2(z) = W_1^*(\bar{z}). \tag{7}$$

Sometimes $W_1(z)$ and $W_2(z)$ become analytic continuations of each other and form one analytic function. Even though in general $W_1(z)$ and $W_2(z)$ are different analytic functions, connected only by the symmetry relation (7), we shall always regard the pair $W(z)$ as a single object.

The collection of functions (holomorphic pairs), characterized by the properties

$$\frac{W(z) - W^*(z)}{z - \bar{z}} \geq 0, \quad W(z) = W^*(\bar{z}), \quad \mathrm{Im}z \neq 0 \tag{8}$$

forms the class of Nevanlinna functions.

The introduction of the notion of "associated" functions makes it possible for us to solve the problem of integral representation of functions of class $\mathcal{K}_a$ by using the methods of the theory of functions of a complex variable.

We shall use the Potapov method [2,3] of building a general solution of classical discrete and continuous interpolation problems which is based on the theory of J-expansive analytic matrix-functions.

With this aim we shall build the Fundamental Matrix Inequality (FMI) for the associated function $W(z)$ and prove that every function $W(z)$ which satisfies the FMI admits a representation of the form

$$W(z) = \int_{-\infty}^{+\infty} \frac{d\sigma(t)}{t-z} \quad \left(\sigma(t) \in\uparrow, \ \int_{-\infty}^{+\infty} \frac{d\sigma(t)}{1+|t|} < +\infty\right)$$

and the corresponding integral

$$S(x) = \int_{-\infty}^{+\infty} \frac{1-\cos(\sqrt{t}x)}{t} d\sigma(t)$$

defines a function of the class $\mathcal{K}_a$.

We shall clarify the existence of a solution of the FMI, consider the case of non-unique representation and discuss the analytical properties of the resolvent matrix and criterion for non-uniqueness.

2 The fundamental matrix inequality (FMI)

Let $S(x)$ be a function of class $\mathcal{K}_a$; let

$$s(x) = \int_{-\infty}^{+\infty} \frac{1-\cos(\sqrt{t}x)}{t} d\sigma(t)$$

be any one of the integral representation of $S(x)$ and let

$$W(z) = \int_{-\infty}^{+\infty} \frac{d\sigma(t)}{t-z}$$

be its associated function.Then we have

Theorem 1. *The associated function $W(z)$ of an integral representation $s(x)$ of a function $S(x) \in \mathcal{K}_a$ satisfies the following matrix inequality (Fundamental Matrix Inequality)*

$$\left[\begin{array}{c|c} (K\varphi,\varphi) & (\frac{\sin(\sqrt{z}x)}{\sqrt{z}}W(z) - \int_0^x \cos(\sqrt{z}(x-t))s(t)dt, \varphi) \\ \hline * & \frac{W(z)-W^*(z)}{z-\bar{z}} \end{array}\right] \geq 0 \qquad (FMI)$$

for every $\varphi \in L^2[0,l]$ and every z with $\mathrm{Im} z \neq 0$.

Proof. The left hand side takes the form

$$H = \int_{-\infty}^{+\infty} \left[\begin{array}{c|c} \left| \left(\frac{\sin(\sqrt{t}x)}{\sqrt{t}}, \varphi \right) \right|^2 & \left(\frac{\sin(\sqrt{t}x)}{\sqrt{t}}, \varphi \right) \frac{1}{t-z} \\ \hline * & \frac{1}{|t-z|^2} \end{array} \right] d\sigma(t) \,,$$

whose non-negativity is obvious. □

We now consider the form

$$[1, \bar{\alpha}] H \begin{bmatrix} 1 \\ \alpha \end{bmatrix} \geq 0 \tag{9}$$

with

$$\varphi(x) = \psi(x) - \beta \cos(\sqrt{z}(l-x)) \quad \text{and} \quad \alpha = \beta \cos(\sqrt{z} l).$$

This yields the Transformed Fundamental Matrix Inequality (TFMI). The inequalities FMI and TFMI are equivalent.

Theorem 2. *If the conditions of Theorem 1 are fulfilled, then the associated function $W(z)$ satisfies the Transformed Fundamental Matrix Inequality (TFMI)*

$$\left[\begin{array}{c|c} (K\psi, \psi) & \begin{array}{c} \frac{1}{2}\left(\frac{\sin((l+x)\sqrt{z})}{\sqrt{z}} W(z) - \int_0^{l+x} \cos(\sqrt{z}(l+x-t)) s(t) dt, \psi \right) \\ -\frac{1}{2}\left(\frac{\sin((l-x)\sqrt{z})}{\sqrt{z}} W(z) - \int_0^{l-x} \cos(\sqrt{z}(l-x-t)) s(t) dt, \psi \right) \end{array} \\ \hline * & \begin{array}{c} \frac{1}{z-\bar{z}} [\cos^2(l\sqrt{z}) W(z) - \cos^2(l\sqrt{\bar{z}}) W^*(z)] \\ + \frac{1}{2(z-\bar{z})} \int_0^{2l} [\sqrt{z} \sin((2l-t)\sqrt{z}) - \sqrt{\bar{z}} \sin((2l-t)\sqrt{\bar{z}})] s(t) dt \end{array} \end{array} \right] \geq 0.$$

for every $\varphi \in L^2[0,l]$ and every z with $\mathrm{Im} z \neq 0$.

Theorem 1 is naturally supplemented by the following converse assertion:

Theorem 3. *If a function $W(z)$ satisfies the FMI, then:*

1. *It admits the following integral representation*

$$W(z) = \int_{-\infty}^{+\infty} \frac{d\sigma(t)}{t-z} \quad \left(\sigma(t) \in \uparrow, \ \int_{-\infty}^{+\infty} \frac{d\sigma(t)}{1+|t|} < +\infty \right).$$

2. *The integral*

$$\int_{-\infty}^{+\infty} \frac{1-\cos(\sqrt{t}x)}{t} d\sigma(t)$$

exists in the interval ($0 < x < 2a$) and the equality

$$s(x) = \int_{-\infty}^{+\infty} \frac{1-\cos(\sqrt{t}x)}{t} d\sigma(t)$$

is fulfilled.

The proof of the first part of Theorem 3 is based on fact that if a function $W(z)$ satisfies the FMI, then the asymptotic relation

$$\frac{\sin(\sqrt{z}x)}{\sqrt{z}}W(z) - \int_0^x \cos(\sqrt{z}(x-t))s(t)dt \longrightarrow 0 \tag{10}$$

is fulfilled as $z = iy (z = -iy) \to \infty$, $y > 0$ and all x $(0 \le x < 2a)$.

The proof of the second part of Theorem 3 uses the asymptotic relation (10), the Hamburger-Nevanlinna-Katsnelson Theorem [4,5] and the Phragmén-Lindelöf Principle.

Theorems 1 and 3 justify the contention that all information about the problem of integral representation of functions of class $\mathcal{K}_a$ is contained in the FMI. Because of this, the FMI is called the Fundamental Matrix Inequality of the problem.

3 The unification of the FMI and the Sakhnovich identity

Let us introduce the following notations:

$$b_z(x) = \frac{\sin(\sqrt{z}x)}{\sqrt{z}}, \tag{11}$$

$$c_z(x) = \int_0^x \cos(\sqrt{z}(x-t))s(t)dt, \tag{12}$$

$$(Kf)(x) = \int_0^l \frac{1}{2}\{s(x+t) - s(|x-t|)\}f(t)dt \quad (\forall l < a)\ . \tag{13}$$

Then the FMI acquires the form

$$\left[\begin{array}{c|c} (K\varphi,\varphi) & (b_z(x)W(z) - c_z(x),\varphi) \\ \hline * & \dfrac{W(z) - W^*(z)}{z - \bar{z}} \end{array}\right] \ge 0 \tag{FMI}$$

for $\varphi \in L^2[0,l]$ $(\forall l < a)$ and all z with $\mathrm{Im} z \ne 0$.

This form of the FMI does not differ from the Fundamental Matrix Inequality of well-known classical discrete interpolation problems such as the Nevanlinna-Pick problem, the Carathéodory problem, the power moment problem [2] and continuous problems such as the Krein problem on the integral representation of Hermitian positive function [3]. Consequently, we can use Potapov's method [2,3] to construct the solution of the FMI in our problem.

The Fundamental Matrix Inequalities of discrete interpolation problems [2] were solved with help of identities for the information blocks. In continuous problems an analogous role is played by the Sakhnovich identity [3].

$$(\mathcal{A}_z Kg,\varphi) - (K\mathcal{A}_{\bar{z}}^* g,\varphi) = (g, b_{\bar{z}}(x))(c_z(x),\varphi) - (g, c_{\bar{z}}(x))(b_z(x),\varphi) \tag{14}$$

for all $g, \varphi \in L^2[0,l]$ $(\forall l < a)$. Here $\mathcal{A}_z g$ is the bounded integral operator

$$(\mathcal{A}_z g)(x) = \int_0^x \frac{\sin(\sqrt{z}(x-t))}{\sqrt{z}} g(t)dt, \quad (\mathcal{A}_{\bar{z}}^* g)(x) = -\int_x^l \frac{\sin(\sqrt{\bar{z}}(x-t))}{\sqrt{\bar{z}}} g(t)dt \tag{15}$$

and the following relations of Hilbert are immediate

$$\mathcal{A}_z b_\zeta = -\frac{b_z - b_\zeta}{z-\zeta}, \quad \mathcal{A}_z c_\zeta = -\frac{c_z - c_\zeta}{z-\zeta}. \tag{16}$$

The first question in the analysis of the FMI is: does there exist at least one solution $W(z)$?

To answer this question we consider two cases: $(K\varphi_0, \varphi_0) = 0$ for some $\varphi_0 \not\equiv 0$ and $(K\varphi, \varphi) > 0$ for every $\varphi \not\equiv 0$

In the first case the solution of the FMI is checked by immediate calculation.

In the second case we approximate our problem by a discrete analog. The discrete analog inherits the structure of the original continuous kernel K and is a truncated power moment problem $\mathcal{H}_n$ with a strictly positive information block $\mathbf{S}_n$ of its Fundamental Matrix Inequality (FMI $\mathcal{H}_n$). It is known that the (FMI $\mathcal{H}_n$) has a solution $W_n(z)$. This solution $W_n(z)$ is also a solution of the truncated power moment problem $\mathcal{H}_n$.

Since the family of functions $W_n(z)$ is uniformly bounded on every compact set in the upper half-plane, it is possible to choose a subsequence $W_{n_\nu}(z)$ from the sequence $W_n(z)$ which converges to a holomorphic function $W(z)$ in the upper half-plane. Choosing n_ν so that $W_{n_\nu}(z) \to W(z)$, we take the limit in the inequality (FMI $\mathcal{H}_n$). Then the limiting function $W(z)$ will satisfy the original FMI.

Theorem 4. *The Fundamental Matrix Inequality for the problem of integral representation of functions of class $\mathcal{K}_a$ always has a solution.*

4 The case of more than one representation

As has already been remarked at the end the preceding section, we may restrict our attention to the case of a strictly positive information block:

$$(K\varphi, \varphi) > 0 \quad \text{for all } \varphi \neq 0, \quad \varphi \in L^2[0,l].$$

It is also known [2] that if the information block $\mathbf{S}_n$ is strictly positive then the truncated power moment problem has infinitely many solutions. An important role in the construction of the general solution is played by by the matrix-function

$$\mathcal{H}(\xi,\eta) = \frac{\xi-\eta}{i}\begin{bmatrix} c_{\tilde{\xi}}^* \\ -b_{\tilde{\xi}}^* \end{bmatrix} \mathbf{S}_n^{-1}[c_\eta, -b_\eta] = \frac{\xi-\eta}{i}\begin{bmatrix} C_{\tilde{\xi}}^* \\ -B_{\tilde{\xi}}^* \end{bmatrix} \mathbf{S}_n[C_\eta, -B_\eta],$$

where

$$\mathbf{S}_n B_\eta = b_\eta, \quad \mathbf{S}_n C_\eta = c_\eta \text{ and } \tilde{\xi} = \bar{\xi}.$$

The attempt to modifiy this construction to our problem encounters difficulties connected with the continuous nature of the problem. In contrast to the finite dimensional case such a construction can not always be carried out. The operator

$$(Kf)(x) = \int_0^l \mathcal{K}(x,t)f(t)dt$$

is completely continuous and the inverse K^{-1} is not defined on the whole space $L^2[0,l]$, but only on a domain $\mathcal{V} = \mathcal{R}(K)$ which is dense in $L^2[0,l]$. Moreover, the equations

$$KB_z = \frac{\sin(\sqrt{z}x)}{\sqrt{z}}, \quad KC_z = \int_0^x \cos(\sqrt{z}(x-t))s(t)dt \tag{17}$$

generally speaking do not have solutions in $L^2[0,l]$.

On the other hand, however, the strict positivity of the operator K leads to a natural extension the essence of which is in the introduction of a new metric

$$\langle f, g\rangle = (Kf, g)$$

in the space $L^2[0,l]$ and its completion (which consists of all Cauchy sequences in this new metric).

Therefore we shall obtain a new Hilbert space $\mathbf{H}_-$ and a uniquely defined extension $\tilde{K}$ of the operator K. The solutions of the equations (17) should be searched for in this space $\mathbf{H}_-$. But even this extension does not guarantee the solvability of the equations (17).

We shall show that if a function of class $\mathcal{K}_a$ admits more then one integral representation, then the equations (17) are solvable in $\mathbf{H}_-$ and the construction of a matrix $\mathcal{H}(\xi,\eta)$ is realized; if $S(x)$ admits only one representation, then the equations (17) have no solutions. (The connection between Rigged Spaces and classical continuous problems are discussed in the paper [3].)

The main conclusions are now summarized in the next theorem.

Theorem 5. *If the FMI has more than one solution, then it has infinitely many solutions, the set of which is equal to the set of fractional-linear transformations*

$$W(z) = [a(z)p(z) + b(z)q(z)][c(z)p(z) + d(z)q(z)]^{-1}$$

of all holomorphic non-singular Nevanlinna pairs col$[p(z), q(z)]$ *in* $\operatorname{Im} z > 0$.

The coefficient matrix of the fractional-linear transformation (i.e., the resolvent matrix)

$$A(z, z_0) = \begin{bmatrix} a(z) & b(z) \\ c(z) & d(z) \end{bmatrix}$$

is determined by the following relations:

$$A(z, z_0) = T(z, z_0)M_0, \quad \text{Im } z_0 \neq 0,$$

$$J = \begin{bmatrix} 0 & i \\ -i & 0 \end{bmatrix}, \quad T(z, z_0) = I + \mathcal{H}(z, z_0)J,$$

$$\mathcal{H}(z, z_0) = \frac{z - z_0}{i} \begin{bmatrix} (KC_{z_0}, C_{\bar{z}}) & (KB_{z_0}, C_{\bar{z}}) \\ (KC_{z_0}, B_{\bar{z}}) & (KB_{z_0}, B_{\bar{z}}) \end{bmatrix},$$

$$\begin{aligned} J - M_0^{*-1} J M_0^{-1} &= -J\mathcal{H}(\bar{z}_0, z_0)J \\ &= \frac{z_0 - \bar{z}_0}{i} \begin{bmatrix} (B_{z_0}, b_{z_0}(x)) & -(C_{z_0}, b_{z_0}(x)) \\ -(B_{z_0}, c_{z_0}(x)) & (C_{z_0}, c_{z_0}(x)) \end{bmatrix}. \end{aligned}$$

Therefore $A(z, z_0)$ is determined by the solutions of the two equations

$$KB_{z_0} = \frac{\sin(\sqrt{z_0}x)}{\sqrt{z_0}}, \quad KC_{z_0} = \int_0^x \cos(\sqrt{z_0}(x - t))s(t)dt.$$

The resolvent matrix possesses a number of remarkable properties:

1) $A(z, z_0)$ is a holomorphic matrix-function (or more precisely a pair of holomorphic matrix-functions) in the domains $\text{Im } z > 0$ and $\text{Im } z < 0$.

2) $A(z, z_0)$ is J-expansive in the upper half-plane and J-contractive in the lower half-plane, i.e.,

$$A^*(z, z_0)JA(z, z_0) - J \geq 0 \text{ if } \text{Im } z > 0 \text{ and } A^*(z, z_0)JA(z, z_0) - J \leq 0 \text{ if } \text{Im } z < 0.$$

3) The values of the matrix $A(z, z_0)$ in the half-planes $\text{Im } z > 0$, $\text{Im } z < 0$ are related through reflection-symmetry:

$$JA^*(\bar{z}, z_0)J = A^{-1}(z, z_0).$$

Under certain additional conditions (e.g., if Theorem 6 is fulfilled), the matrix-function $A(z, z_0)$ is holomorphic and J-unitary on the real axis. But then, the resolvent matrix $A(z, z_0)$ is an entire matrix-function, satisfying the J-properties, and we can choose $z_0 = 0$ as a "base" point for which $M_0 = I$; this greatly simplifies the structure of $A(z, z_0)$.

We next present a criterion for the non-uniqueness of the integral representation:

Theorem 6. *Suppose that $S(x) \in \mathcal{K}_a$ is a continuous function on the interval $([0, 2l]$ $(\forall\, l < a))$ and the operator*

$$(Kf)(x) = \int_0^l \mathcal{K}(x, t)f(t)dt$$

is strictly positive in the space $L^2[0,l]$. Then in order for $S(t)$ to admit more than one integral representation over the whole real axis it is necessary that the inequality

$$\sum_{j=1}^{\infty} \lambda_j \left| \left(\frac{\sin(\sqrt{z}x)}{\sqrt{z}}, \varphi_j(x) \right) \right|^2 < +\infty$$

is fulfilled for all z, and it is sufficient that the inequality is fulfilled for at least one non-real ζ_0. Here λ_j and $\varphi_j(x)$ are the characteristic values and the corresponding eigenfunctions of the integral operator K.

References

[1] Krein M.G., About one general method of decompositions of the positive-definite kernels on elementary products, *Dokl. Akad. Nauk SSSR* **53** (1950) 1125–1128.

[2] Kovalishina I.V., Analytic theory of a class of interpolation problems, *Izv. Akad. Nauk SSSR*, **47** (1983) 455–497.

[3] Kovalishina I.V. and Potapov V.P., Integral representation of Hermitian positive functions, deposited in VINITI.-1981.-1–120.

[4] Katsnelson V.E., Continual analogues of the Hamburger-Nevanlinna theorem and fundamental matrix inequalities for the classical problems 1., *Theor. Func. Anal. Pril.* **36** (1981) 31–48.

[5] Akhiezer N.I., *The Classical Problem of Moments and Some Related Topics in Analysis*, Fizmatgiz, Moskow, 1961. (Translation: Hafner, New York, 1965).

Department of Mathematics
Kharkov State Academy of Railway Transport
Feuerbach Square 7
Kharkov 310050, Ukraine

AMS Subject Classification: 42A82, 45H05, 47A57.

Operator Theory
Advances and Applications, Vol. 95
© 1997 Birkhäuser Verlag Basel/Switzerland

On the theory of entire matrix-functions of exponential type[1]

M.G. Krein

Abstract. It is shown that if an entire matrix function $U(\lambda)$ is J inner, then it is of exponential type and $(1+\lambda^2)^{-1}\ell n^+\|U(\lambda)\|$ is summable. These results are applied to the monodromy matrix of a canonical system. In particular, the rate of growth is calculated. [Abstract added by the editors.]

0 Introduction

Let n be a natural number. With each square matrix $A = \|a_{i,j}\|_{i,j=1}^n$ of order n we associate the linear transformation

$$\eta_j = \sum_{k=1}^{n} a_{j,k}\xi_k \qquad (j = 1, 2, \dots, n)$$

acting in the n-dimensional unitary space E_n. We denote this by

$$y = Ax,$$

for short. By definition[2],

$$\|A\| = \max_{x\in E_n} \frac{|Ax|}{|x|} \qquad \left(|x| = \sqrt{\sum_{j=1}^{n} |\xi_j|^2} \right).$$

Clearly,

$$\frac{1}{n}\sqrt{\sum_{j,k=1}^{n} |a_{j,k}|^2} \le \|A\| \le \sqrt{\sum_{j,k=1}^{n} |a_{j,k}|^2}. \tag{0.1}$$

1) (Translated from: Ukrainian Mathematical Journal, Vol.3, No.2 (1951), 164–137, MR **14**, p. 981.)

2) Because the linear space $\mathfrak{M}_n$ of all $n \times n$ matrices is finite dimensional, all the norms in $\mathfrak{M}_n$ are equivalent. The concrete choice of the norm is not important for us. In particular, the expression in the right hand side of (0.1) can be chosen as a norm of A.

We consider the matrix-function $F(z) = \|f_{j,k}(z)\|_{j,k}^n$ of the complex argument z. Such a function is said to be *entire* (*meromorphic*) if all its entries are entire (meromorphic) functions.

We prove the following statement.

Theorem 1. *Let $F(z) = \|f_{j,k}(z)\|_{j,k}^n$ be an entire $n \times n$ matrix-function which admits the representation*

$$F^{-1}(z) = \frac{C_{-1}}{z} + C_0 + C_1 z + \ldots + C_{p-1} z^{p-1} + z^p \sum_{k=1}^{\infty} \frac{A_k}{z - \alpha_k}, \tag{0.2}$$

where $p \geq 0$ is an integer, all the α_k $(k = 1, 2, \ldots)$ are real, $C_{-1}, C_0, \ldots C_{p-1}, A_k$ $(k = 1, 2, \ldots)$ are $n \times n$ matrices and

$$\sum_{k=1}^{\infty} \frac{\|A_k\|}{|\alpha_k|} < \infty.$$

Then the entire function F is a function of exponential type, i.e.,

$$\overline{\lim_{|z|\to\infty}} \frac{\log \|F(z)\|}{|z|} < \infty. \tag{0.3}$$

Moreover,

$$\int_{-\infty}^{\infty} \frac{\log^+ \|F(x)\|}{1 + x^2} dx < \infty. \tag{0.4}$$

For the scalar case ($n = 1$) this theorem was proved by us in [Kr1].

It has a number of applications in the theory of Hermitian operators [Kr2], [Kr3] and in particular, in the theory of regular and singular Sturm -Liouville operators [Kr4].

We shall show that Theorem 1 for matrix-functions allows us to establish a number of important properties of monodromy matrices for some classes of differential systems. We were led to these results while trying to generalize A.M. Lyapunov's investigations of differential equations with periodic coefficients ([Kr5], [Kr6]).

1 Proof of the main theorem

Definition. *Let us denote by (P) the class of all meromorphic functions $f(z)$ possessing the following property: The function $\log |f(z)|$ admits a positive harmonic majorant in each of the two halfplanes $\Im z > 0$ and $\Im z < 0$.*

The class (P) is a linear ring, i.e., if $f_1, f_2 \in (P)$, then $c_1 f_1 + c_2 f_2 \in (P)$ and $f_1 f_2 \in (P)$.

If $f \in (P)$ and f does not vanish outside the real axis then $f^{-1} \in (P)$ as well.

If f admits the representation

$$f(z) = \frac{c_{-1}}{z} + c_0 + c_1 z + \ldots + c_{p-1} z^{p-1} + z^p \sum_{k=1}^{\infty} \frac{a_k}{z - \alpha_k},$$

where $p \geq 0$, α_k $(k = 1, 2, \ldots)$ are real, and

$$\sum_{k=1}^{\infty} \left| \frac{a_k}{\alpha_k} \right| < \infty$$

then $f \in (P)$.

All these statements are contained in a general statement which was established in [Kr1]. In the same paper the following important fact was proved.

An entire function $f(z)$ belongs to the class (P) if and only if it satisfies the following conditions:

1) $$\lim_{|z|\to\infty} \frac{\log |f(z)|}{|z|} < \infty, \tag{1.1}$$

2) $$\int_{-\infty}^{\infty} \frac{\log^+ |f(x)|}{1 + x^2} dx < \infty. \tag{1.2}$$

After these remarks Theorem 1 can be proved without difficulties.

Proof of Theorem 1. Let

$$F^{-1}(z) = \|g_{j,k}\|_{j,k=1}^n.$$

The representations for $g_{j,k}$ $(j, k = 1, 2, \ldots, n)$ which follow from (0.2) show that $g_{j,k} \in (P)$. Hence, the determinant $|[g_{j,k}]_{j,k=1}^n| \in (P)$. Since $|[g_{j,k}(z)]_{j,k=1}^n| = \Delta^{-1}(z)$, where $\Delta(z) = |[f_{j,k}]_{j,k=1}^n|$, it follows that $\Delta^{-1} \in (P)$, and hence that $\Delta(z) \in (P)$ as well.
Since $f_{j,k}(z)$ $(j, k = 1, 2, \ldots, n)$ is proportional to the product of $\Delta(z)$ with a complementary minor of the corresponding entry $g_{k,j}(z)$ $(j, k = 1, 2, \ldots, n)$, we obtain that $f_{j,k}(z) \in (P)$ $(j, k = 1, 2, \ldots, n)$.

Thus, all the functions $f_{j,k}(z)$ $(j, k = 1, 2, \ldots, n)$ satisfy the conditions (1.1) and (1.2). Hence, the function

$$\|F(z)\| = \sqrt{\sum_{j,k=1}^{n} |f_{j,k}(z)|^2}$$

satisfies these conditions too. The theorem is proved. □

Remark 1. *Let us note in passing that* $\Delta(z) \in (P)$, *i.e.,*

$$\lim_{|z|\to\infty} \frac{\log|\Delta(z)|}{|z|} < \infty \quad \text{and} \quad \int_{-\infty}^{\infty} \frac{\log^+|\Delta(x)|}{1+x^2}dx < \infty. \tag{1.3}$$

Remark 2. *For the scalar case* $(n = 1)$ *Theorem 1 was generalized by us in the following direction:*

The condition "α_k $(k = 1, 2, \ldots)$ *are real" was replaced by the more general condition*[3]*:*

$$\sum_{k=1}^{\infty} \left| \frac{\Im \alpha_k}{\alpha_k^2} \right| < \infty.$$

Theorem 1 (for matrix-functions) admits such a generalization, too.

2 Applications of the theorem for the investigation of a monodromy matrix

Let $\mathfrak{J}$ be a nonsingular constant Hermitian matrix, let $H(t) = \|h_{j,k}\|_{j,k=1}^n$ be a Hermitian matrix-function whose entries are defined and integrable on an interval $(0, \omega)$ and let us consider the differential system

$$\frac{dx}{dt} = i\lambda\mathfrak{J}H(t)x. \tag{2.1}$$

Let $U(t;\lambda) = \|u_{j,k}(t;\lambda)\|_{j,k=1}^n$ be the solution of the initial value problem of the matricial differential system

$$\frac{dU}{dt} = i\lambda\mathfrak{J}H(t)U, \qquad U(0) = I.$$

The matrix-function $U(\omega;\lambda)$ *of the variable* λ *is said to be* **the monodromy matrix** *of the differential system (2.1).*

The monodromy matrix is an entire function of the variable λ.

Every solution $x = x(t)$ of the system (2.1) can be obtained by the formula

$$x(t) = U(t;\lambda)x_0, \qquad x_0 = x(0). \tag{2.2}$$

We are interested in the case in which the matrix $H(t)$ is *Hermitian-nonnegative*, i.e.,

$$(H(t)x, x) = \sum_{j,k=1}^{n} h_{j,k}(t)\xi_k\overline{\xi_j} \geq 0 \qquad (0 \leq t \leq \omega). \tag{!}$$

3) We made a mistake in the proof of such a generalization in [Kr1]. However, this mistake can be corrected.

In this case we derive from (2.1) that

$$\frac{d}{dt}(\mathfrak{J}^{-1}x, x) = i(\lambda - \overline{\lambda})(H(t)x, x) \begin{matrix} \leq 0 & \Im\lambda \geq 0 \\ \geq 0 & \Im\lambda \leq 0 \end{matrix}.$$

Thus,

$$\frac{1}{\Im\lambda}(\mathfrak{J}^{-1}x_\omega, x_\omega) \leq \frac{1}{\Im\lambda}(\mathfrak{J}^{-1}x_0, x_0), \qquad (x_\omega = x(\omega;\lambda),\ \Im\lambda \neq 0).$$

In view of (2.2), $x_\omega = U(\omega;\lambda)x_0$. Denoting $U(\omega;\lambda)$ by U_λ, we obtain

$$\frac{1}{\Im\lambda}(\mathfrak{J}^{-1}U_\lambda x_0, U_\lambda x_0) \leq \frac{1}{\Im\lambda}(\mathfrak{J}^{-1}x_0, x_0), \qquad (\Im\lambda \neq 0) \tag{2.3}$$

and

$$(\mathfrak{J}^{-1}Ux_0, Ux_0) = (\mathfrak{J}^{-1}x_0, x_0), \qquad (\Im\lambda = 0). \tag{2.4}$$

We note that

$$U(\omega;\lambda) = I + i\lambda\mathfrak{J}H_0 + C_2\lambda^2 + \dots, \tag{2.5}$$

where

$$H_0 = \int_0^\omega H(t)\, dt. \tag{2.5'}$$

Now let $\mathcal{E}$ be an arbitrary $\mathfrak{J}^{-1}$-unitary matrix, i.e.,

$$(\mathfrak{J}^{-1}\mathcal{E}x, \mathcal{E}x) = (\mathfrak{J}^{-1}x, x), \qquad (x \in E_n)$$

and assume that[4]

$$|U(\omega;\lambda) + \mathcal{E}| \not\equiv 0. \tag{2.6}$$

Let us introduce the *generalized Caley transform*

$$\Phi(\lambda) = i\{U(\omega;\lambda) - \mathcal{E}\}\{U(\omega;\lambda) + \mathcal{E}\}^{-1}.$$

Setting

$$y_0 = (U(\omega;\lambda) + \mathcal{E})x_0\ ,$$

we have

$$\Phi(\lambda)y_0 = i(U(\omega;\lambda) - \mathcal{E})x_0.$$

Hence,

$$U(\omega;\lambda)x_0 = \frac{1}{2}(I - i\Phi(\lambda))y_0 \ \text{ and } \ x_0 = \frac{\mathcal{E}^{-1}}{2}(I + i\Phi(\lambda))y_0.$$

---

4) By $|A|$ we denote the determinant of the matrix A. In view of (2.5) the condition (2.6) will be fulfilled if $\mathcal{E} = I$ (or more generally if $|I + \mathcal{E}| \neq 0$).

Inserting these expressions for x_0 and Ux_0 into (2.3) we find that

$$\frac{\Im(\mathfrak{J}^{-1}\Phi(\lambda)y_0, y_0)}{\Im\lambda} \leq 0 \qquad (\Im\lambda \neq 0). \tag{2.7}$$

If x_0 runs over all E_n, then y_0 runs over all E_n as well, unless λ is a root of the equation

$$|U(\omega;\lambda) + \mathcal{E}| = 0.$$

From this we conclude that (2.7) is fulfilled for every $y_0 \in E_n$ and for every nonreal λ.

From (2.7) it follows that[5] (see for example [ChMe], chapter IV]) that

$$(\mathfrak{J}^{-1}\Phi(\lambda)y_0, y_0) = \gamma_0 - \gamma_1\lambda - \lambda\sum_{k=1}^{\infty}\frac{\rho_k}{\alpha_k(\alpha_k-\lambda)} + \frac{\rho_0}{\lambda}, \tag{2.8}$$

where the α_k $(k = 1, 2, \ldots)$ are real poles ($\neq 0$) of the matrix-function $\Phi(\lambda)$, γ_0 is real and γ_1 and ρ_k $(k = 0, 1, 2, \ldots)$ are nonnegative values. Moreover,

$$\sum_{k=1}^{\infty}\frac{\rho_k}{\alpha_k^2} < \infty. \tag{2.9}$$

It is easy to see that γ_0, γ_1 and ρ_k $(k = 0, 1, 2, \ldots)$ are Hermitian forms with respect to the variable y_0, i.e., there exist $n \times n$ matrices Γ_0, Γ_1 and R_k $(k = 0, 1, 2, \ldots)$ such that

$$\gamma_0 = (\Gamma_0 y_0, y_0), \quad \gamma_1 = (\Gamma_1 y_0, y_0) \quad \text{and} \quad \rho_k = (R_k y_0, y_0) \quad (k = 0, 1, 2, \ldots).$$

Thus, the relation (2.8) means that

$$\mathfrak{J}^{-1}\Phi(\lambda) = \Gamma_0 - \Gamma_1\lambda - \lambda\sum_{k=1}^{\infty}\frac{R_k}{\alpha_k(\alpha_k-\lambda)} + \frac{R_0}{\lambda}. \tag{2.10}$$

Moreover, in view of (2.9),[6]

$$\sum_{k=1}^{\infty}\frac{\|R_k\|}{\alpha_k^2} \leq \infty. \tag{2.11}$$

Since

$$\Phi(\lambda) = iI - 2\mathcal{E}\{U(\omega;\lambda) + \mathcal{E}\}^{-1},$$

it follows from (2.10) that

$$\{U(\omega;\lambda) + \mathcal{E}\}^{-1} = C_0 + C_1\lambda + \lambda\sum_{k=1}^{\infty}\frac{B_k}{\alpha_k(\alpha_k-\lambda)} + \frac{B_0}{\lambda},$$

---

5) For simplicity we assume that the point $\lambda = 0$ is not a pole of the function $\Phi(\lambda)$.

6) Since $(R_k y, y) \geq 0$, $\|R_k\| \leq \mathrm{tr} R_k$ $(k = 0, 1, 2, \ldots)$.

where

$$C_0 = \frac{1}{2}\mathcal{E}^{-1}(iI - \mathfrak{I}\Gamma_0) \quad \text{and} \quad C_1 = \frac{1}{2}\mathcal{E}^{-1}\mathfrak{I}\Gamma_1, \quad B_k = \frac{1}{2}\mathcal{E}^{-1}\mathfrak{I}R_k \quad (k = 0, 1, 2, \ldots).$$

It is clear that[7]

$$\sum_{k=1}^{\infty} \frac{\|B_k\|}{\alpha_k^2} \le \infty.$$

Thus, Theorem 1 is applicable to the matrix-function $F(\lambda) = U(\omega; \lambda) + \mathcal{E}$. According to this theorem the function $F(\lambda)$ satisfies the conditions (0.3) and (0.4). Hence, the function $U(\omega; \lambda)$ satisfies these conditions as well. Thus, we have proved:

Theorem 2. Let $H(t)$ be a summable Hermitian matrix-function on $(0, \omega)$ which satisfies the condition (!) and let $\mathfrak{I}$ be a nonsingular Hermitian matrix. Then the monodromy matrix $U(\omega; \lambda)$ of the system (2.1) satisfies the following two conditions:

1)
$$\varlimsup_{|\lambda| \to \infty} \frac{\log \|U(\omega; \lambda)\|}{|\lambda|} < \infty.$$

2)
$$\int_{-\infty}^{\infty} \frac{\log^+ \|U(\omega; \lambda)\|}{1 + \lambda^2} d\lambda < \infty.$$

Let us remark that in order to establish the properties 1) and 2) of the entire matrix-function $U(\omega; \lambda)$ we used only property (2.3). We also proved in passing that the determinant
$\Delta(\lambda) = |U(\omega; \lambda) + \mathcal{E}|$ has only real zeros and satisfies the conditions (1.3).

Let us note that the zeros of the function $\Delta(\lambda)$ coincide with characteristic values of the boundary value problem[8]

$$\begin{cases} \frac{dx}{dt} = i\lambda\mathfrak{I}H(t)x \\ x(\omega) + \mathcal{E}x(0) = 0 \end{cases} \tag{2.12}$$

Moreover, the multiplicity of λ_0 as a zero of the function $\Delta(\lambda)$ coincide with the number of linearly independent solution of the system (2.12) for $\lambda = \lambda_0$. This can be derived from the simplicity of the poles of the function $\{U(\omega; \lambda) + \mathcal{E}\}^{-1}$.

If the function $\Delta(\lambda)$ is not a function of minimal type, i.e., if

$$h(\phi) = \varlimsup_{r \to \infty} \frac{\log |\Delta(re^{i\phi})|}{r} \not\equiv 0 \quad (0 \le \phi \le 2\pi),$$

then the zeros of the function $\Delta(\lambda)$ form a bilateral sequence

$$\ldots \le \lambda_{-2} \le \lambda_{-1} < \lambda_0 \le \lambda_1 \le \lambda_2 \le \ldots \quad (\lambda_{-1} < 0 < \lambda_0)$$

---

7)   Since $\|AB\| \le \|A\| \cdot \|B\|$.
8)   Using this it is possible to prove directly that the zeros of $\Delta(\lambda)$ are real.

(which tends to infinity in both directions). Moreover, in view of the condition

$$\int_{-\infty}^{\infty} \frac{\log^+ |\Delta(\lambda)|}{1+\lambda^2} d\lambda < \infty$$

(see [Kr4]), the following asymptotic formulas are valid:

$$\lim_{n\to\infty} \frac{\lambda_n}{n} = \lim_{n\to\infty} \frac{|\lambda_{-n}|}{n} = 2\pi \left[h\left(\frac{\pi}{2}\right) + h\left(-\frac{\pi}{2}\right)\right]^{-1} .$$

In particular, it is possible to choose the unit matrix I as the matrix $\mathcal{E}$. Then all the above stated conclusions are applicable to the "semiperiodic" boundary problem

$$\begin{cases} \frac{dx}{dt} = i\lambda \mathfrak{J} H(t)x, \\ x(\omega) = -x(0) \end{cases} . \tag{2.13}$$

If the matrix H_0 in (2.5′) is nonsingular, i.e.,

$$\int_0^{\omega} (H(t)x, x)\, dt > 0 \qquad (x \neq 0) \tag{!!}$$

is satisfied in addition to the condition (!), then it is easy to see ([Kr6] that strict inequality holds in (2.3) (if $x_0 \neq 0$). In this case, the condition (2.6) is fulfilled for every $\mathfrak{J}^{-1}$-unitary matrix $\mathcal{E}$.

In particular, under condition (!!), all of the above stated conclusions are applicable to the "periodic" boundary problem

$$\begin{cases} \frac{dx}{dt} = i\lambda \mathfrak{J} H(t)x, \\ x(\omega) = x(0) \end{cases} . \tag{2.14}$$

corresponding to $\mathcal{E} = -I$.

3 On a special case of the system (2.1)

The boundary problems (2.13) and (2.14) are of special interest in the case $n = 2$ for

$$\mathfrak{J} = \begin{pmatrix} 0 & -i \\ i & 0 \end{pmatrix} \quad \text{and} \quad H(t) = \begin{pmatrix} a(t) & b(t) \\ b(t) & c(t) \end{pmatrix},$$

where $a(t), b(t)$ and $c(t)$ $(0 \le t \le \omega)$ are real valued integrable functions. In this case the system (2.1) takes the form

$$\begin{cases} \frac{d\xi_1}{dt} = \lambda(b\xi_1 + c\xi_2), \\ \frac{d\xi_2}{dt} = -\lambda(a\xi_1 + b\xi_2) \end{cases} . \tag{3.1}$$

The condition (!) means that

$$a(t) \geq 0, \quad c(t) \geq 0, \quad a(t)c(t) - b^2(t) \geq 0 \quad (0 \leq t \leq \omega). \tag{3.2}$$

The extra-condition (!!) (if (3.2) is satisfied) is equivalent to the condition

$$\int_0^\omega a(t)\, dt \int_0^\omega c(t)\, dt - \left(\int_0^\omega b(t)\, dt \right)^2 > 0. \tag{3.3}$$

In this case, the characteristic values of the boundary value problems (2.13) and (2.14) interlace each other. These characteristic values serve also to mark the ends of the stability and unstability zones of the system (3.1) when the coefficient matrix is extended periodically with period ω from the interval $(0, \omega)$ onto the whole real axis (see [Kr4]).

Not without some difficulties we succeeded in proving that the equality

$$\overline{\lim_{|\lambda| \to \infty}} \frac{\log \|U(\omega; \lambda)\|}{|\lambda|} = \int_0^\omega \sqrt{ac - b^2}\, dx \tag{3.4}$$

holds true for the system (3.1) when it satisfies the conditions (3.2) and (3.3).

In the considered case, the equality (3.4) is equivalent to the statement that the exponential type of the determinant $\Delta(\lambda) = |U(\omega; \lambda) + \mathcal{E}|$ is equal to the right hand side of (3.4) (i.e., that in (3.4) the value $\|U(\omega; \lambda)\|$ can be replaced by $|\Delta(\lambda)|$).

If the boundary value problem (2.12) (with an arbitrary $\mathfrak{J}^{-1}$-unitary $\mathcal{E} = \begin{pmatrix} \alpha & \beta \\ \gamma & \delta \end{pmatrix}$) has infinitely many positive characteristic numbers,[9] then (3.4) implies that

$$\lim_{n \to \infty} \frac{\lambda_n}{n} = \pi \left(\int_0^\omega \sqrt{ac - b^2}\, dx \right)^{-1}.$$

An analogue is also true for the nonnegative characteristic numbers of the boundary value problem (2.12). In the considered case $(\mathfrak{J}x, x) = \frac{1}{i}(\xi_1 \overline{\xi_2} - \xi_2 \overline{\xi_1})$ and $\mathfrak{J}^{-1} = \mathfrak{J}$.

A matrix $\mathcal{E} = \begin{pmatrix} \alpha & \beta \\ \gamma & \delta \end{pmatrix}$ is $\mathfrak{J}$-unitary if and only if $|\alpha\delta - \beta\gamma| = 1$.

Let us denote

$$U(\omega; \lambda) = \begin{pmatrix} \varphi_1(\lambda) & \varphi_2(\lambda) \\ \psi_1(\lambda) & \psi_2(\lambda) \end{pmatrix},$$

where φ_i and ψ_i $(i = 1, 2)$ are entire functions of λ.

---

9) It is possible to show that the system (2.12) has finitely many characteristic numbers if and only if the functions a, b, c are piecewise constant and satisfy the condition $ac - b^2 = 0$ at all their continuity points.

Since $U(\omega; 0) = I$,

$$\varphi_1(0) = 1 \quad , \quad \varphi_2(0) = 0$$
$$\psi_1(0) = 0 \quad , \quad \psi_2(0) = 1.$$

In view of (2.4), $|U(\omega; \lambda)| = 1$ and hence

$$\text{(I)} \qquad \varphi_1(\lambda)\psi_2(\lambda) - \varphi_2(\lambda)\psi_1(\lambda) \equiv 1.$$

If we denote

$$\begin{cases} \eta_1 = \varphi_1(\lambda)\xi_1 + \varphi_2(\lambda)\xi_2, \\ \eta_2 = \psi_1(\lambda)\xi_1 + \psi_2(\lambda)\xi_2, \end{cases} \tag{3.5}$$

then the inequality (2.3) for $U(\omega; \lambda)$ means that

$$\frac{\eta_1\overline{\eta_2} - \eta_2\overline{\eta_1}}{\lambda - \overline{\lambda}} < \frac{\xi_1\overline{\xi_2} - \xi_2\overline{\xi_1}}{\lambda - \overline{\lambda}}.$$

(Equality in (2.3) is impossible because of (3.3).) Choosing arbitrary real values for ξ_1, ξ_2 we derive the following statement from (3.5):

(II) *For each real α, the function*

$$\omega(\lambda) = -\frac{\cos\alpha\varphi_1(\lambda) + \sin\alpha\varphi_2(\lambda)}{\cos\alpha\psi_1(\lambda) + \sin\alpha\psi_2(\lambda)}$$

maps the upper half plane $\Im\lambda > 0$ into itself.

This means that the functions $\cos\alpha\varphi_1(\lambda) + \sin\alpha\varphi_2(\lambda)$ and $\cos\alpha\psi_1(\lambda) + \sin\alpha\psi_2(\lambda)$ form a real pair in the sense of N.G. Chebotarev (see [ChMe]). Hence, their zeros are simple, real and interlace. However, then the pairs φ_1, φ_2 and ψ_1, ψ_2 are real pairs as well.

The properties (I) and (II) of the functions $\varphi_1, \varphi_2, \psi_1, \psi_2$ permit us to prove Theorem 2 for the considered case (n=2) based on Theorem 1 for the scalar case only. We have already followed this strategy in [Kr4]. Now let us explain it in more detail.

From (II) for $\alpha = 0, \frac{\pi}{2}$, it follows that

$$-\frac{\varphi_1(\lambda)}{\psi_1(\lambda)} = c_0' + c_1'\lambda - \frac{\rho_0'}{\lambda} - \lambda\sum_{j=1}^{\infty}\frac{\rho_j'}{\alpha_j'(\alpha_j' - \lambda)} \quad \left(c_1' \geq 0, \rho_j' > 0, \sum_{j=1}^{\infty}\frac{|\rho_j'|}{\alpha_j'^2} < \infty\right),$$

$$-\frac{\varphi_2(\lambda)}{\psi_2(\lambda)} = c_0'' + c_1''\lambda - \lambda\sum_{j=1}^{\infty}\frac{\rho_j''}{\alpha_j''(\alpha_j'' - \lambda)} \quad \left(c_1'' \geq 0, \rho_j'' > 0, \sum_{j=1}^{\infty}\frac{|\rho_j''|}{\alpha_j''^2} < \infty\right).$$

Subtracting the lower inequality from the upper and taking into account (I) we obtain

$$\frac{1}{\psi_1(\lambda)\psi_2(\lambda)} = c_0 + c_1\lambda + \frac{\rho_0}{\lambda} + \lambda\sum_{j=1}^{\infty}\frac{\rho_j}{\alpha_j(\alpha_j - \lambda)} \quad \left(\sum_{j=1}^{\infty}\frac{|\rho_j|}{\alpha_j^2} < \infty\right).$$

According to Theorem 1 (for (n=1)) we obtain $\psi_1\psi_2 \in (P)$. Since ψ_1, ψ_2 form a real pair, the function ψ_1/ψ_2 maps the whole upper (lower) halfplane into one of these halfplanes. However, then $\psi_1/\psi_2 \in (P)$. Hence, $\psi_1^2 = (\psi_1\psi_2) \cdot \psi_1/\psi_2 \in (P)$ and $\psi_1 \in (P)$. In an analogous way $\psi_2 \in (P)$, too. Thus $\varphi_i = \psi_i\,(\varphi_i/\psi_i) \in (P)$ as well. Hence, all these functions satisfy conditions (1) and (1). Hence, the matrix $U(\omega;\lambda)$ satisfies the conditions 1) and 2) in the formulation of Theorem 2.

Finally, we stress that in view of the properties (I) and (II) the theory of monodromy matrices for the system (3.1) is very closely related to our theory (see [Kr2], [Kr3]) of entire operators with defect indices (1) and in particular to the problem of the continuation of positive definite functions ([Kr7].

The deep relationships between these subjects which at first glance appear to be absolutely different, will be considered elsewhere.

References

[Kr1] Krein, M.G., *On the theory of entire functions of exponential type* (in Russian), Izvestiya Akad. Nauk SSSR, Ser. Matem., 11 (1947), 309–326. MR **9**, p.179.

[Kr2] Krein, M.G, *On a remarkable class of Hermitian operators* (in Russian), Doklady Akad. Nauk SSSR **44**:5 (1944), 191–195. MR **6**, p.269.

[Kr3] Krein, M.G, *Fundamental aspects of the representation theory of Hermitian operators with deficiency index* (m, m) (in Russian), Ukrainskii Matem. Zhurnal **1**:2 (1949), 3–66. English transl. in: Amer. Math. Soc. Transl. (ser.**2**) **34** (1963), 69–108. MR **14**, p.56.

[Kr4] Krein, M.G, *On the Sturm-Liouville boundary value problem on the interval* $(0.\infty)$ *and on a class of integral equations* (in Russian), Doklady Akad. Nauk SSSR **74**:6 (1950), 1125–1128. MR **12**, p.339.

[Kr5] Krein, M.G, *Generalizations of certain investigations of A.M.Lyapunov on linear differential equations with periodic coefficients* (in Russian), Doklady Akad. Nauk SSSR **73**:3 (1950), 445 -448. MR **12**, p.100.

[Kr6] Krein, M.G, *On the application of an algebraic proposition in the theory of monodromy matrices* (in Russian), Uspekhi Matem. Nauk **6**:1(41) (1951), 171–177. MR **14**, p.277.

[Kr7] Krein, M.G, *On the continuation problem for Hermitian-positive continuous functions* (in Russian), Doklady Akad. Nauk SSSR **26**:1 (1940), 17–21. MR **2**, p.361.

[ChMe] Chebotarev N.G. and N.N. Meĭman, *The Routh-Hurwitz Problem for Polynomials and Entire Functions* (in Russian), Trudy Matem. Instituta Steklova **26** (1949), 331 pp. MR **11**, p.509.

AMS Subject Classification: 30D15, 34B99, 34E99, 34L99, 47A55.

Operator Theory
Advances and Applications, Vol. 95
© 1997 Birkhäuser Verlag Basel/Switzerland

Analogs of the Nehari and Sarason theorems for character-automorphic functions and some related questions*

Stas Kupin and Peter Yuditskii

Abstract. An analog of the Nehari theorem for character-automorphic functions with respect to the action of Fuchsian groups of Widom-Carleson type is proved. A solvability criteria for the Nevanlinna-Pick problem in a Denjoy domain with homogeneous boundary is given as an example.

0 Introduction

In this paper we generalize some results of M. Abrahamse. In 1979 M. Abrahamse gave a solvability criteria for the Nevanlinna-Pick problem in finitely connected domains [1] in terms of a positivity condition on an infinite family of quadratic forms. This family is enumerated by the characters of the fundamental group of the domain. Around that time J. Ball proved a commutant lifting theorem for finitely connected domains under similar conditions [2]. Thus, he generalized the M. Abrahamse's criteria to this more general setting. Later D. Marshall showed (see [6]) that the classical Nevanlinna-Pick problem with automorphic data with respect to a Fuchsian group could have no automorphic solutions (even in the finitely connected case). This confirmed that the criteria due to Abrahamse was not redundant. We would like to mention here the studies on functional models of pairs of commuting nonselfadjoint operators [8] and on the extension of the Lax-Phillips scheme for spaces of automorphic functions [10] which are closely related to the subject. The most intriguing question of the subject is to obtain a description of the solution set of the problem. Some progress was achieved recently by J. Ball and V. Vinnikov [3] on this problem.

We generalize Abrahamse's theorem in two directions. First, we consider the Nevanlinna-Pick problem as a special case of the corresponding Nehari problem (from this point of view our approach is close to that of J. Ball's article [2]). Second, and what is more essential, we consider an important class of infinitely connected domains. By virtue of the uniformization theorem it is possible to consider the

*) This work was supported by the ISF Grant no U2Z000

Nevanlinna-Pick problem in a multiply connected domain or on an open Riemann surface as the Nevanlinna-Pick problem on the unit disk $\mathbb{D}$ of the complex plane in the class of character-automorphic functions with respect to the action of a certain Fuchsian group Γ.

Following the papers [4, 5, 11, 12, 13, 14] we consider the so called Fuchsian groups of Widom-Carleson type. Necessary facts concerning groups of this kind and spaces of character-automorphic functions are given in Section 1. Here we just briefly mention their main properties.

Let Γ^* be the group of unitary characters of a group Γ. Associate with an arbitrary character $\alpha \in \Gamma^*$ certain subspaces of the classical space L^2 on the unit circle $\mathbb{T}$

$$L^2(\alpha) = \{f \in L^2 | f \circ \gamma = \alpha(\gamma) f, \forall \gamma \in \Gamma\}, \; H^2(\alpha) = L^2(\alpha) \bigcap H^2.$$

If Γ is a group of Widom-Carleson type the space $H^2(\alpha)$ is not trivial for any character $\alpha \in \Gamma^*$, $H^2(\alpha) \neq \{\text{const}\}$. Moreover, there is an explicit description of the orthogonal complement $H^2_\perp(\alpha) = L^2(\alpha) \ominus H^2(\alpha)$ (see Lemma 1.3.2).

By a character-automorphic Nehari problem we mean the following question. Let $f_\perp \in H^2_\perp(\beta)$ for fixed $\beta \in \Gamma^*$. When does there exist a function $f_+ \in H^2(\beta)$ such that $f = f_+ + f_\perp$ belongs to the unit ball of $L^\infty(\beta)$ (i.e. $||f||_\infty \leq 1$).

In Section 2 we give a solvability criterion for this problem (Theorem 2.1.1). The proof is based on a description of the space $H^2_\perp(\alpha)$, $\forall \alpha \in \Gamma^*$; for the rest it is quite similar to one of the proofs of the classical Nehari theorem (see, for example, [9]).

We then prove a character-automorphic counterpart of the Sarason theorem (Theorem 2.2.1). As a special case of the latter, we obtain a solvability criterion for the Nevanlinna-Pick problem.

An important property of the groups of Widom-Carleson type is the continuous dependence of the reproducing kernel k^α_ζ of the space $H^2(\alpha)$ on α, for ζ fixed [5,13]. This property allows us to carry over the proof of the uniqueness theorem [1] for the Nevanlinna-Pick problem with finite data to the considered case (Proposition 3.5.1).

Finally, as an example we consider the Nevanlinna-Pick problem in a domain of the kind $\overline{\mathbb{C}} \backslash E$, where E is a real homogeneous compact in the sense of Carleson (see the definition in 4.1). It seems interesting to compare the criterion of solvability obtained here to the well-known conditions of solvability for the interpolation problems in generalized Stieltjes classes $S(E_m)$ (see [7, ch. 8, §9, sect. 5]).

We would like to thank A.Ya. Kheifets and M.L. Sodin for helpful discussions.

1 Groups of Widom-Carleson type and Hardy spaces of character-automorphic functions

1.1. The results, presented in this section, are not new and are well known to mathematicians dealing with the same topics. We are only going to give a summary of the facts concerning the Hardy spaces of character-automorphic functions which are necessary for the following considerations. The contents of this section are based, in the main, on the article of Ch. Pommerenke [11] and very close to the first part of [12]. Detailed presentation of the theory of Hardy spaces at infinitely-connected Riemann surfaces one can find in the monograph of M. Hasumi [5]. We shall be considering Fuchsian groups (discrete groups of linear-fractional automorphisms of the unit disk) without parabolic and elliptic elements. For this reason the definitions and theorems from the works [5,11,12] are not cited in the full generality, but in the form we need.

A Fuchsian group Γ is called a group of convergent type if the Blaschke product constructed with respect to the orbit $\{\gamma(0)\}_{\gamma\in\Gamma}$ of the origin under the action of Γ converges,

$$b(\zeta) = \prod_{\gamma\in\Gamma} \gamma(\zeta)\frac{|\gamma(0)|}{\gamma(0)}.$$

The function $b(\zeta)$ is called the Green function of the group Γ relative to the origin. Note that the products constructed with respect to the orbit of any point ζ_0 in the unit disk $\mathbb{D}$ also converge. The corresponding Green function is denoted by $b_{\zeta_0}(\zeta)$.

Denote the group of the unimodular characters of the group Γ by Γ^*. We associate with an arbitrary character $\alpha \in \Gamma^*$ spaces of character-automorphic functions. These spaces are subspaces of the classical L^p spaces, $1 \le p \le \infty$,

$$L^p(\alpha) = \{f \in L^p | f \circ \gamma = \alpha(\gamma) f, \forall \gamma \in \Gamma\}, \; H^p(\alpha) = L^p(\alpha) \bigcap H^p.$$

A group Γ is called a group of Widom type if for any $\alpha \in \Gamma^*$ the space $H^\infty(\alpha)$ is not trivial

$$H^\infty(\alpha) \neq \{\text{const}\} \; \forall \alpha \in \Gamma^*.$$

The following theorem gives a criterion for a group Γ to be a group of Widom type.

Theorem. (Pommerenke [11]). *A Fuchsian group Γ without elliptic and parabolic elements is a group of Widom type if and only if the derivative of the Green function b' is a function of bounded characteristic (can be presented as a quotient of two functions from H^∞). Moreover*

$$b' = \frac{\Delta}{\varphi}$$

where Δ is a Blaschke product, constructed with respect to the zeros of the function b', and φ is a bounded outer function.

The functions $b_{\zeta_0}(\zeta)$ and $\Delta(\zeta)$ are character-automorphic. Denote by μ_{ζ_0} ($\mu_0 = \mu$) and ν their characters

$$b_{\zeta_0} \circ \gamma = \mu_{\zeta_0}(\gamma) b_{\zeta_0}, \; \Delta \circ \gamma = \nu(\gamma)\Delta.$$

Note that there exists a Lebesgue measurable set E on the unit circle $\mathbb{T}$, fundamental with respect to the action of Γ on $\mathbb{T}$. More precisely, we can choose this set so that it does not contain only two Γ-equivalent points and

$$\int_{\bigcup_{\gamma\in\Gamma}\gamma(E)} dm = \int_{\mathbb{T}} dm$$

where dm is the standard Lebesgue measure on $\mathbb{T}$ (see [11, 12]).

1.2. For the standard duality between the spaces L^p and L^q $(1/p+1/q=1)$ we shall use the notation

$$\langle f,g\rangle = \int_{\mathbb{T}} \overline{g} f dm \tag{1.2.1}$$

with $f\in L^p$ and $g\in L^q$.

In what follows an important role is played by the following operator (see [11, 12]), originally defined on functions $f\in L^\infty$:

$$P^\alpha f = \frac{\sum_{\gamma\in\Gamma}\overline{\alpha(\gamma)}|\gamma'|f\circ\gamma}{\sum_{\gamma\in\Gamma}|\gamma'|} \in L^\infty(\alpha). \tag{1.2.2}$$

Note that

$$|b'(t)| = \sum_{\gamma\in\Gamma}|\gamma'(t)|$$

and the expression (1.2.2) can be rewritten as follows

$$P^\alpha f = \frac{b}{b'}\cdot\sum_{\gamma\in\Gamma}\overline{\alpha(\gamma)}\cdot\frac{\gamma'}{\gamma}\, f\circ\gamma. \tag{1.2.3}$$

The main properties of the operators P^α are:
1) $\langle P^\alpha f,g\rangle = \langle f,g\rangle\ \ \forall f\in L^\infty, g\in L^1(\alpha)$
2) $||P^\alpha f||_p \le ||f||_p$
The properties 1) and 2) imply the following:

Proposition ([12]).
1) *The space of antilinear functionals on* $L^p(\alpha)$, $1\le p<\infty$, *coincides with* $L^q(\alpha)$, $1/p+1/q=1$.
2) *The space* $L^p(\alpha)$ *is the closure of the image of the operator* P^α *in the* L^p *- metric*

$$\mathrm{clos}_{L^p}\{P^\alpha(L^\infty)\} = L^p(\alpha),$$

in particular,

$$\mathrm{clos}_{L^p}\{L^\infty(\alpha)\} = L^p(\alpha).$$

1.3. We turn now to the Hardy spaces of character-automorphic functions.

From the representation (1.2.3) of the operators P^α it follows that P^α maps the subspace $\Delta H^\infty\subset H^\infty$ to $H^\infty(\alpha)$:

$$P^\alpha(\Delta h)\in H^\infty(\alpha)\ \ \forall h\in H^\infty. \tag{1.3.1}$$

It is not difficult to see that

$$(P^{\alpha}\Delta h)(0) = \Delta(0)h(0) , \tag{1.3.2}$$

which implies that for every $\alpha \in \Gamma^*$ in the space $H^2(\alpha)$ there exists a function not vanishing at the origin.

Similarly one can show that in general, for arbitrary $\alpha \in \Gamma^*$ and ζ_0, there always exists a function $h \in H^{\infty}(\alpha)$ such that $h(\zeta_0) \neq 0$.

From (1.3.1) it is easy to deduce the following lemma.

Lemma 1.3.1. *The annihilator of $H^p(\alpha), 1 \leq p < \infty$, is contained in the subspace*

$$\Delta\overline{H_0^q(\nu\overline{\alpha})} = \{\Delta\overline{g} \mid g \in H^q(\nu\overline{\alpha}), g(0) = 0\} \subseteq L^q(\infty).$$

Proof. Let $f \in H_{\perp}^p(\alpha)$. Then for any function $h \in H^{\infty}$ we have

$$0 = \langle P^{\alpha}\Delta h, f\rangle = \langle \Delta h, f\rangle = \langle h, \overline{\Delta} f\rangle.$$

Hence, $\Delta\overline{f} \in H_0^q \cap L^q(\nu\overline{\alpha}) = H_0^q(\nu\overline{\alpha})$. □

Note that the spaces $H_{\perp}^p(\alpha)$ and $\Delta\overline{H_0^q(\nu\overline{\alpha})}$ do not coincide in general for groups of Widom type. It is possible to construct an example of a group of Widom type for which the space $H^2 \cap \Delta\overline{H_0^2}$ contains a nontrivial character-automorphic function [5].

However, if we require in addition that the zeros of the derivative of the Green function $b'(\zeta)$ form an interpolating sequence for the space H^{∞} (satisfy the Carleson condition), this disagreement vanishes.

Definition. A group Γ of Widom type is called a group of Widom-Carleson type if the zeros of $\Delta(\zeta)$ satisfy the Carleson condition.

The following proposition holds:

Proposition ([13]). *Let Γ be a group of Widom-Carleson type. Then for any function $f \in H_0^1(\nu)$*

$$\langle f, \Delta\rangle = 0. \tag{1.3.3}$$

Concerning the role of (1.3.3) see [5]. In particular, this relation yields

Lemma 1.3.2. *Let Γ be a group of Widom-Carleson type. Then*

$$H_{\perp}^p(\alpha) = \Delta\overline{H_0^q(\nu\overline{\alpha})},\ 1 \leq p < \infty .$$

Proof. It remains to prove the inclusion $\Delta\overline{H_0^q(\nu\overline{\alpha})} \subset H_{\perp}^p(\alpha)$. Let $f \in H^p(\alpha)$ and $g \in H_0^q(\nu\overline{\alpha})$. Then

$$\langle f, \Delta\overline{g}\rangle = \langle fg, \Delta\rangle = 0 ,$$

since $fg \in H_0^1(\nu)$. □

Remark. Since the proof of Lemma 1.3.1 actually proves that the annihilator of the (non-closed) subspace $P^\alpha(\Delta H^\infty) \subset H^p(\alpha)$ is $\Delta \overline{H_0^q(\nu\overline{\alpha})}$, it follows from Lemma 1.3.2 that

$$\mathrm{clos}_{L^p}\{P^\alpha(\Delta H^\infty)\} = H^p(\alpha)$$

and hence in particular $clos_{L^p} H^\infty(\alpha) = H^p(\alpha), 1 \le p < \infty$. This density property of $H^\infty(\alpha)$ will be used in the sequel.

2 Character-automorphic analogs of the Nehari and Sarason theorems

2.1. We recall that one can consider the classical Nehari theorem as an answer to the question: when can a function $f_- \in H^2_-$ be represented as the projection of a function f from the unit ball of the space L^∞ to $H^2_-, f_- = P_- f,\ ||f||_\infty \le 1$. We are going to consider a similar question with respect to spaces of character-automorphic functions.

By Lemma 1.3.2 for groups of Widom-Carleson type, the space $L^2(\alpha)$ admits the following orthogonal decomposition

$$L^2(\alpha) \doteq H^2(\alpha) \oplus H^2_\perp(\alpha) = H^2(\alpha) \oplus \Delta \overline{H_0^2(\nu\overline{\alpha})}. \tag{2.1.1}$$

Denote by $P_+(\alpha)$ and $P_\perp(\alpha)$ the orthoprojectors onto the subspaces $H^2(\alpha)$ and $H^2_\perp(\alpha)$ respectively.

Let a function $f_\perp \in H^2_\perp(\beta)$. We associate with an arbitrary character $\alpha \in \Gamma^*$ an operator from $H^2(\alpha)$ to $H^2_\perp(\alpha\beta)$; we define this operator on a dense subset of $H^2(\alpha)$ by the formula

$$F(\alpha)x = P_\perp(\alpha\beta)(f_\perp x),\ x \in H^\infty(\alpha).$$

Theorem 2.1.1. *The function $f_\perp \in H^2_\perp(\beta)$ is the projection of a function $f \in L^\infty(\beta), ||f||_\infty \le 1$, onto $H^2_\perp(\beta)$ if and only if*

$$\sup_{\alpha\in\Gamma^*} ||F(\alpha)|| \le 1. \tag{2.1.2}$$

The proof is quite similar to the proof of the classical Nehari theorem (see, for example, [9]). The only difference is that in the factorization of a function from the unit ball of an H^1-character-automorphic space into a product of two functions from unit balls of H^2-character-automorphic spaces, we do not know in advance the characters of the factors. We only know the product of these characters. This forces us to consider all possible characters in the condition (2.1.2).

Proof of Theorem 2.1.1. If $f_\perp = P_\perp(\beta)f,\ f \in L^\infty(\beta)$, then

$$F(\alpha)x = P_\perp(\alpha\beta)(f_\perp x) = P_\perp(\alpha\beta)(fx),\ x \in H^\infty(\alpha).$$

Since $||f||_\infty \leq 1$, it follows that

$$||F(\alpha)x||_2 = ||P_\perp(\alpha\beta)fx||_2 \leq ||fx||_2 \leq ||x||_2.$$

Conversely, consider the functional on $\overline{\Delta H_0^1(\nu\overline{\beta})} \subset L^1(\beta)$ defined on a dense subset of $\overline{\Delta H_0^1(\nu\overline{\beta})}$ by the formula

$$\langle f_\perp, \Delta\overline{x}\rangle,\ x \in H_0^\infty(\overline{\beta}\nu).$$

We recall that $\{\overline{\Delta H_0^1(\nu\overline{\beta})}\}_\perp = H^\infty(\beta)$.

We shall estimate the norm of this functional. Let the L^1 norm of x be less than or equal to one, $||x||_1 \leq 1$. Represent x as a product of character-automorphic functions from the unit ball of H^2. It is possible to do this, for example, if we put

$$x_1 = \sqrt{x_e},\ x_2 = x_i \cdot \sqrt{x_e},$$

where $x = x_i \cdot x_e$ is the inner-outer factorization of x. The function x_1 is obviously character-automorphic (see, for example, [11]). Denote by α the character of x_1, then $\overline{\alpha\beta}\nu$ is the character corresponding to x_2,

$$x_1 \in H^\infty(\alpha), ||x_1||_2 \leq 1;\quad x_2 \in H^\infty(\nu\overline{\alpha\beta}), ||x_2||_2 \leq 1.$$

For the norm of our functional we obtain

$$|\langle f_\perp, \Delta\overline{x_1x_2}\rangle| = |\langle f_\perp x_1, \Delta\overline{x_2}\rangle| = |\langle F(\alpha)x_1, \Delta\overline{x_2}\rangle| \leq ||F(\alpha)|| \leq 1.$$

Extend this functional to $L^1(\beta)$ by the Hahn-Banach theorem. It follows that there exists $f \in L^\infty(\beta),\ ||f||_\infty \leq 1$, satisfying

$$\langle f_\perp, \Delta\overline{x}\rangle = \langle f, \Delta\overline{x}\rangle\ \forall x \in H_0^\infty(\overline{\beta}\nu).$$

Hence $f - f_\perp \in H^2(\beta)$, that is $f_\perp = P_\perp(\beta)f,\ ||f||_\infty \leq 1$. □

Remark. Note the easy fact that if $\sup ||F(\alpha)|| \leq \rho$, then there exists a function $f \in L^\infty(\beta),\ ||f||_\infty \leq \rho$, with the property $P_\perp(\beta)f = f_\perp$.

2.2. An important special case of the Nehari theorem is the Sarason theorem (see [9]). We introduce some notation to formulate the character-automorphic counterpart of this theorem.

Let Θ be an inner character-automorphic function, $\Theta \circ \gamma = \delta(\gamma)\Theta, \forall\gamma \in \Gamma$. Associate with an arbitrary character $\alpha \in \Gamma^*$ a subspace $K_\Theta(\alpha) \subset H^2(\alpha)$

$$K_\Theta(\alpha) = H^2(\alpha) \ominus \Theta H^2(\overline{\delta}\alpha).$$

The orthoprojector onto $K_\Theta(\alpha)$ has the following form:

$$P_\Theta(\alpha) = \Theta P_\perp(\overline{\delta}\alpha)\overline{\Theta}. \tag{2.2.1}$$

Let $w_\Theta \in K_\Theta(\beta)$. Consider the question of when w_Θ can be complemented by a function from $\Theta H^2(\overline{\delta}\beta)$ to a contracting holomorphic function

$$w = w_\Theta + \Theta g, \; g \in H^2(\overline{\delta}\beta), \tag{2.2.2}$$

so that $||w||_\infty \leq 1$.

Obviously, to solve this problem it suffices to use Theorem 2.1.1 applied to the function $f_\perp = \overline{\Theta} w_\Theta \in H^2_\perp(\overline{\delta}\beta)$.

Theorem 2.2.1. *A function $w_\Theta \in K_\Theta(\beta)$ is a projection of a function $w \in H^\infty(\beta)$, $||w||_\infty \leq 1$, to $K_\Theta(\beta)$ if and only if*

$$\sup_{\alpha \in \Gamma^*} ||W(\alpha)|| \leq 1, \tag{2.2.3}$$

where the operator $W(\alpha) : K_\Theta(\alpha) \to K_\Theta(\alpha\beta)$ is given by the equality

$$W(\alpha) P_\Theta(\alpha) x = P_\Theta(\alpha\beta) w_\Theta x, \; x \in H^\infty(\alpha). \tag{2.2.4}$$

Proof. We first verify the correctness of the definition (2.2.4). If $P_\Theta(\alpha)x_1 = P_\Theta(\alpha)x_2$, then $x_2 - x_1 = \Theta h$, where h belongs to the space $H^2(\overline{\delta}\alpha)$ and hence to $H^\infty(\overline{\delta}\alpha)$. Therefore

$$P_\Theta(\alpha\beta) w_\Theta x_1 = P_\Theta(\alpha\beta) w_\Theta (x_1 + \Theta h) = P_\Theta(\alpha\beta) w_\Theta x_2.$$

By (2.2.3),

$$\begin{aligned} &||P_\perp(\overline{\delta}\alpha\beta)(\overline{\Theta} w_\Theta) x|| = ||\Theta P_\perp(\overline{\delta}\alpha\beta)\overline{\Theta} w_\Theta x|| \\ &= ||P_\Theta(\alpha\beta) w_\Theta x|| \leq ||P_\Theta(\alpha) x|| \leq ||x||, \; \forall x \in H^\infty(\alpha). \end{aligned} \tag{2.2.5}$$

From Theorem 2.1.1 and (2.2.5) it follows that there exists a function $g \in H^2(\overline{\delta}\beta)$ such that

$$||\overline{\Theta} w_\Theta + g||_\infty = ||w_\Theta + \Theta g||_\infty \leq 1.$$

The proof of the converse implication is easy. If $w = w_\Theta + \Theta g \in H^\infty(\beta)$, $||w||_\infty \leq 1$, then

$$\begin{aligned} &||W(\alpha) P_\Theta(\alpha) x|| = ||P_\Theta(\alpha\beta) w_\Theta x|| \\ &= ||P_\Theta(\alpha\beta) w x|| = ||P_\Theta(\alpha\beta) w P_\Theta(\alpha) x|| \leq ||P_\Theta(\alpha) x||. \end{aligned}$$ □

Remark. If $w \in H^\infty(\beta)$ is a function of the form (2.2.2), then the operators $W(\alpha)$ of (2.2.4) are correctly defined on all of $K_\Theta(\alpha)$ by the formula

$$W(\alpha) x = P_\Theta(\alpha\beta) w x, \forall x \in K_\Theta(\alpha).$$

It is easy to compute the adjoint operator. For every $x \in K_\Theta(\alpha)$ and $y \in K_\Theta(\alpha\beta)$ we have

$$\langle W(\alpha)x, y\rangle = \langle P_\Theta(\alpha\beta) w x, y\rangle = \langle wx, y\rangle = \langle x, \overline{w} y\rangle = \langle x, P_+(\alpha)\overline{w} y\rangle.$$

We show that $P_+(\alpha)\overline{w}y$ already lies in $K_\Theta(\alpha)$. Indeed,

$$\langle P_+(\alpha)\overline{w}y, \Theta g\rangle = \langle y, \Theta wg\rangle = 0, \ \forall g \in H^2(\overline{\delta}\alpha),$$

since $y \in K_\Theta(\alpha\beta)$ and $wg \in H^2(\overline{\delta}\alpha\beta)$. Thus

$$W^*(\alpha)y = P_+(\alpha)\overline{w}y, \forall y \in K_\Theta(\alpha\beta).$$

3 A solvability criterion for the character-automorphic Nevanlinna-Pick problem and the uniqueness of solution

3.1. Let $\zeta_1, \ldots, \zeta_n$ be a collection of points in the unit disk $\mathbb{D}$, which are pairwise nonequivalent with respect to the action of Γ. Associate a value w_j to each point ζ_j. Let β be some fixed character. The character-automorphic Nevanlinna-Pick problem consists in the description of functions $w \in H^\infty(\beta)$ satisfying the interpolation conditions

$$w(\zeta_j) = w_j, \ 1 \le j \le n$$

and the norm restriction $||w||_\infty \le 1$.

We shall give a solvability condition for the Nevanlinna-Pick problem and establish a criterion for the uniqueness of solution.

3.2. We will need some properties of reproducing kernels for the spaces $H^2(\alpha)$. The linear functional $x \mapsto x(\zeta), \zeta \in \mathbb{D}$, is bounded and is therefore represented by some vector from $H^2(\alpha)$. This vector is called the reproducing kernel, and we denote it by k_ζ^α or $k^\alpha(t, \zeta)$,

$$\langle x, k_\zeta^\alpha\rangle = x(\zeta), \ \forall x \in H^2(\alpha).$$

Since for every $\zeta \in \mathbb{D}$ there are functions in $H^2(\alpha)$ which do not vanish at this point (see 1.3), these functionals are non-trivial, that is $k^\alpha(\zeta, \zeta) = ||k_\zeta^\alpha||^2 > 0$.

The investigation of the uniqueness of solution for the Nevanlinna-Pick problem is based on the Hayashi theorem [5] stating continuous dependence of the kernel $k^\alpha(t, \zeta)$ on the character $\alpha \in \Gamma^*$ for fixed t, ζ. Note that the map from Γ^* to L^2 associating to any character α the function k_ζ^α (with ζ fixed) is continuous in the strong sense [13].

3.3. Associate with the points $\zeta_1, \ldots, \zeta_n$ a character-automorphic Blaschke product

$$B = b_{\zeta_1} \ldots b_{\zeta_n}, \ B \circ \gamma = \delta(\gamma)B \ (\delta = \mu_{\zeta_1} \ldots \mu_{\zeta_n}).$$

Lemma 3.3.1. *The space $K_B(\alpha) = H^2(\alpha) \ominus BH^2(\overline{\delta}\alpha)$ consists of the linear combinations of the reproducing kernels $\{k^\alpha_{\zeta_j}\}$.*

The system $\{k^\alpha_{\zeta_j}\}$ is linearly independent, and is dual (biorthogonal) to the basis $\{l^\alpha_{\zeta_j}\}$ consisting of vectors of the following form:

$$l^\alpha_{\zeta_j} = c_j \frac{B}{b_{\zeta_j}} k^{\alpha_j}_{\zeta_j},$$

where $\alpha_j = \overline{\delta}\mu_j\alpha$ and $c_j = \left[\left(\frac{B}{b_{\zeta_j}}\right)(\zeta_j)k^{\alpha_j}(\zeta_j,\zeta_j)\right]^{-1}$.

Proof. The vector $k^\alpha_{\zeta_j}$ lies in $K_B(\alpha)$, since for an arbitrary function $g \in H^2(\overline{\delta}\alpha)$ we have

$$\langle Bg, k^\alpha_{\zeta_j}\rangle = (Bg)(\zeta_j) = 0.$$

On the other hand, every vector x from $K_B(\alpha)$ which is orthogonal to the vectors $\{k^\alpha_{\zeta_j}\}$, is necessarily divisible by B since

$$0 = \langle x, k^\alpha_{\zeta_j}\rangle = x(\zeta_j).$$

But the spaces $K_B(\alpha)$ and $BH^2(\overline{\delta}\alpha)$ are orthogonal, hence $x = 0$.

The vectors $l^\alpha_{\zeta_j}$ lie in $K_B(\alpha)$ since one has

$$\langle Bg, l^\alpha_{\zeta_j}\rangle = \overline{c_j}\langle b_{\zeta_j}g, k^{\alpha_j}_{\zeta_j}\rangle = \overline{c_j}(b_{\zeta_j}g)(\zeta_j) = 0,\ \forall g \in H^2(\overline{\delta}\alpha).$$

The systems $\{k^\alpha_{\zeta_j}\}$ and $\{l^\alpha_{\zeta_j}\}$ are biorthogonal since

$$\langle l^\alpha_{\zeta_j}, k^\alpha_{\zeta_i}\rangle = l^\alpha_{\zeta_j}(\zeta_i) = \begin{cases} 1\,, & i = j \\ 0\,, & i \neq j, \end{cases}$$

which concludes the proof of the Lemma. □

Corollary. *For an arbitrary vector $x \in H^2(\alpha)$,*

$$P_B(\alpha)x = \sum_j x(\zeta_j) l^\alpha_{\zeta_j}.$$

Proof. Write the decomposition of $P_B(\alpha)x$ with respect to a biorthogonal pair of bases

$$P_B(\alpha)x = \sum_j \langle P_B(\alpha)x, k^\alpha_{\zeta_j}\rangle l^\alpha_{\zeta_j} = \sum_j \langle x, k^\alpha_{\zeta_j}\rangle l^\alpha_{\zeta_j} = \sum_j x(\zeta_j) l^\alpha_{\zeta_j}\,.$$ □

3.4. We obtain the solvability criterion for the Nevanlinna-Pick problem from the character-automorphic counterpart of the Sarason theorem (Theorem 2.2.1) by setting $\Theta = B$.

Theorem 3.4.1. *Let $\{\zeta_j, w_j\}_{j=1}^n$ be the interpolation data. The Nevanlinna-Pick problem has a solution with the character $\beta \in \Gamma^*$ if and only if the matrices*

$$||k^{\alpha\beta}(\zeta_j, \zeta_i) - w_j\overline{w_i}k^{\alpha}(\zeta_j, \zeta_i)||_{i,j}$$

are nonnegative for all $\alpha \in \Gamma^*$.

Proof. Let $w \in H^\infty(\beta)$ be a solution of the problem. Then the projection w onto the space $K_B(\beta)$ has the following form

$$w_B = P_B(\beta)w = \sum_j w(\zeta_j)l^\beta_{\zeta_j} = \sum_j w_j l^\beta_{\zeta_j}.$$

Conversely, if a function $w_B = \sum_j w_j l_j^\beta \in K_B(\beta)$ satisfies the assumptions of Theorem 2.2.1, then there exists a function $w \in H^\infty(\beta)$ with $||w||_\infty \leq 1$ such that

$$w = w_B + Bg, g \in H^2(\overline{\delta}\beta)$$

and $w(\zeta_j) = w_B(\zeta_j) = w_j$.

The operators $W(\alpha)$ appearing in Theorem 2.2.1 have the following form

$$W(\alpha)\{P_B(\alpha)x\} = W(\alpha)\{\sum_j x(\zeta_j)l^\alpha_{\zeta_j}\} = P_B(\alpha\beta)w_Bx = \sum_j w_jx(\zeta_j)l^{\alpha\beta}_{\zeta_j}.$$

Therefore the condition that $W(\alpha)$ is a contraction is equivalent to the non-negativity of the quadratic form

$$\sum_{i,j}\{\langle l^\alpha_{\zeta_i}, l^\alpha_{\zeta_j}\rangle - \overline{w_j}w_i\langle l^{\alpha\beta}_{\zeta_i}, l^{\alpha\beta}_{\zeta_j}\rangle\}x(\zeta_i)\overline{x(\zeta_j)} \geq 0. \tag{3.4.1}$$

In terms of the biorthogonal system this condition corresponds to the non-negativity of the matrix

$$||\langle k^{\alpha\beta}_{\zeta_j}, k^{\alpha\beta}_{\zeta_i}\rangle - w_j\overline{w_i}\langle k^\alpha_{\zeta_j}, k^\alpha_{\zeta_i}\rangle||_{i,j} \geq 0. \tag{3.4.2}$$

□

3.5. Let us consider now the uniqueness of solution of the character-automorphic Nevanlinna-Pick problem with finite interpolation data.

It is more convenient to express our reasoning in terms of the operators $W(\alpha)$ corresponding to the interpolation data.

Lemma 3.5.1. *In the case of the Nevanlinna-Pick problem with finite interpolation data the value*

$$\sup_{\alpha\in\Gamma^*} ||W(\alpha)||$$

is attained at some point α_0.

Proof. One needs to prove in fact that the function

$$\Phi(\alpha,\xi)=\frac{\sum k^{\alpha}(\zeta_j,\zeta_i)\overline{w_i}\xi_i w_j\overline{\xi_j}}{\sum k^{\alpha\beta}(\zeta_j,\zeta_i)\xi_i\overline{\xi_j}}$$

reaches its supremum when $\alpha\in\Gamma^*,\xi=[\xi_j]_{j=1}^n\in S^n=\{\xi\in\mathbb{C}^n:||\xi||=1\}$.

Observe that the function

$$\sum k^{\alpha\beta}(\zeta_j,\zeta_i)\xi_i\overline{\xi_j}$$

is continuous on $\Gamma^*\times S^n$ and its domain of definition is compact (Γ^* is equipped with the topology dual to discrete). Hence this function reaches its infimum. Since the matrix of the quadratic form $||k^{\alpha\beta}(\zeta_j,\zeta_i)||$ is in fact the Gram matrix of a system of linearly independent vectors $\{k^{\alpha\beta}_{\zeta_j}\}$, it is non-degenerate $\forall\alpha\in\Gamma^*$. Therefore

$$\inf_{\xi\in S^n,\alpha\in\Gamma^*}\sum k^{\alpha\beta}(\zeta_j,\zeta_i)\xi_i\overline{\xi_j}>0\ .$$

This implies that the function $\Phi(\alpha,\xi)$ is continuous and reaches its supremum. □

Lemma 3.5.2. *Let $\rho=\sup_{\alpha\in\Gamma^*}||W(\alpha)||$. The following alternative takes place:*
1) *If $\rho<1$, then the Nevanlinna-Pick problem is not uniquely solvable.*
2) *If $\rho=1$, then the problem is uniquely solvable and the corresponding function $w\in H^\infty(\beta)$ is unimodular, i.e., $|w(t)|=1$ almost everywhere on $\mathbb{T}$.*

Proof. Let $\rho<1$. There always exists a solution w_0 such that its L^∞ norm is less than ρ (see the remark from Section 2.1), so every function

$$w=w_0+gB,\ g\in H^\infty(\overline{\delta}\beta),\ ||g||_\infty\le 1-\rho$$

is also a solution of our problem.

Consider the case $\rho=1$. Let w be some solution of the problem, let α_0 be the character at which the supremum of $||W(\alpha)||$ is attained and let x_0 be a vector for which

$$||W(\alpha_0)x_0||=||x_0||,x_0\neq 0.$$

We recall that the space $K_B(\alpha_0)$ is finite-dimensional and use the easily verified identity

$$\langle(I-W^*(\alpha)W(\alpha))x,\rangle=\langle(1-\overline{w}w)x,x\rangle+\langle(w-W(\alpha))x,(w-W(\alpha))x\rangle$$

$$\forall x\in K_\Theta(\alpha),\ \forall\alpha\in\Gamma^*.$$

Put $x=x_0$, $\alpha=\alpha_0$ in the identity. The left-hand side of the equality vanishes while the right-hand side is the sum of two nonnegative terms. Therefore each term also vanishes. The relation

$$\langle(1-\overline{w}w)x_0,x_0\rangle=0,\ x_0\neq 0,$$

implies that $|w(t)| = 1$ almost everywhere on $\mathbb{T}$. From the other relation

$$\langle (w - W(\alpha_0))x_0, (w - W(\alpha_0))x_0 \rangle = 0,$$

it follows that

$$wx_0 = W(\alpha_0)x_0 \ ,$$

that is

$$w = \frac{W(\alpha_0)x_0}{x_0}.$$

The right-hand side of the last equality does not depend on the choice of w, hence the solution is unique. □

We formulate the result in terms of the quadratic forms (3.4.2).

Proposition 3.5.1. *The character-automorphic Nevanlinna-Pick problem with finite interpolation data $\{\zeta_j, w_j\}_{j=1}^n$ has a unique solution with character β if and only if at least one of matrices*

$$||k^{\alpha\beta}(\zeta_j, \zeta_i) - w_j\overline{w_i}k^{\alpha}(\zeta_j, \zeta_i)||_{i,j}, \alpha \in \Gamma^*,$$

is degenerate while all the other matrices are non-negative.

4 A solvability criterion for the Nevanlinna-Pick problem in the plane with a homogeneous compact removed

4.1. Definition. A compact

$$E = [b_0, a_0] \backslash \bigcup_j (a_j, b_j) \subset \mathbb{R}$$

is called homogeneous if there exists such a constant $M = M(E)$ such that $\forall x \in E, \forall \eta < \mathrm{diam}(E)$

$$|(x - \eta, x + \eta) \bigcap E| \geq M\eta,$$

where $|F|$ is the Lebesgue measure of the set $F \subset \mathbb{R}$.

By the Koëbe-Poincaré uniformization theorem, the domain $\overline{\mathbb{C}} \backslash E$ is conformally equivalent to the quotient of the unit disk $\mathbb{D}$ by the action of a Fuchsian group Γ, $\overline{\mathbb{C}} \backslash E \simeq \mathbb{D}/\Gamma$. In other words, there exists a function $z : \mathbb{D} \to \overline{\mathbb{C}} \backslash E$ meromorphic in $\mathbb{D}$ with the following properties :

1) z is automorphic with respect to $\Gamma, z \circ \gamma = z \ \forall \gamma \in \Gamma$.
2) z maps $\mathbb{D}$ onto the domain $\overline{\mathbb{C}} \backslash E$

$$\forall z_0 \in \overline{\mathbb{C}} \backslash E \ \exists \zeta_0 \in \mathbb{D} : z(\zeta_0) = z_0$$

in such a way that any two preimages of z_0 are Γ-equivalent

$$z(\zeta_1) = z(\zeta_2) \Longrightarrow \exists \gamma \in \Gamma : \zeta_1 = \gamma(\zeta_2).$$

Normalize this mapping by the conditions $z(0) = \infty, (\zeta z)(0) > 0$.

As Jones and Marshall have shown [4], in the case we consider (when E is a homogeneous compact) the Fuchsian group Γ is a group of Widom-Carleson type. Hence the results of Section 3 are applicable to the investigation of the Nevanlinna-Pick problem in the domains of this kind.
The problem may be formulated as follows. *Let a collection of data* $\{z_j, w_j\}_{j=1}^n$, $z_j \in \overline{\mathbb{C}}\backslash E$ *be given. We are interested in the following question: Do there exist holomorphic single-valued functions in* $\overline{\mathbb{C}}\backslash E$ *not exceeding one in absolute value, i.e.,* $|w(z)| \le 1\ \forall z \in \overline{\mathbb{C}}\backslash E$, *which satisfy the interpolation conditions* $w(z_j) = w_j$.

We will give a solvability criterion for this problem.

4.2. In what follows we shall assume that Γ is a group of Widom-Carleson type associated with a homogeneous compact E as above.

As before, b stands for the Green function of the group Γ relative to the origin. Observe that $-ln|b| = G \circ z$, where G is the Green function of the domain $\overline{\mathbb{C}}\backslash E$ relative to infinity. Since

$$\sup_{\zeta \in \mathbb{D}} |zb| = \sup_{z \in \overline{\mathbb{C}}\backslash E} \left\{ |z| e^{-G(z)} \right\} < \infty$$

the function bz belongs to H^∞. In particular, the function z has boundary values on $\mathbb{T}$ such that $z \in L^\infty$, moreover $z(t) = \overline{z(t)}, t \in \mathbb{T}$.

The reproducing kernels for the spaces $H^2(\alpha), \alpha \in \Gamma^*$ are closely connected to the resolvent of the operator of multiplication by z in $L^2(\alpha)$ (for details see [13]).

Lemma 4.2.1. *The operator* $P_+(\alpha)z \cdot |H^2(\alpha)$ *is given by the formula*

$$P_+(\alpha)zx = zx - C(\alpha)\frac{k^{\alpha\mu}}{b}x(0), \tag{4.2.1}$$

where $k^{\alpha\mu} = k_0^{\alpha\mu}$ *is the reproducing kernel relative to the origin,*

$$C(\alpha) = \frac{(zb)(0)}{k^{\alpha\mu}(0)} > 0.$$

Proof. Since $bz \in H^\infty(\mu)$ it follows that $bzx \in H^2(\alpha\mu)$ for any $x \in H^2(\alpha)$. We use the decomposition $H^2(\alpha\mu) = K_b(\alpha\mu) \oplus bH^2(\alpha)$ to obtain

$$bzx = (zb)(0)\frac{k^{\alpha\mu}}{k^{\alpha\mu}(0)}x(0) + bh,\ \ h \in H^2(\alpha)$$

Since $k^{\alpha\mu}/b \in H^2_\perp(\alpha)$ we have

$$P_+(\alpha)zx = h = zx - \frac{(zb)(0)}{k^{\alpha\mu}(0)}\frac{k^{\alpha\mu}}{b}x(0). \qquad \square$$

Lemma 4.2.2. *The reproducing kernel for the space $H^2(\alpha)$ has the form*

$$k_\zeta^\alpha = C(\alpha)\left[\frac{k^{\alpha\mu}}{b}\overline{k^\alpha(\zeta)} - \overline{\frac{k^{\alpha\mu}}{b}(\zeta)}k^\alpha\right]/[z - \overline{z(\zeta)}]. \tag{4.2.2}$$

Proof. Use the equality

$$\langle (z - z(\zeta))x, k_\zeta^\alpha\rangle = \{P_+(\alpha)(z - z(\zeta))x\}(\zeta). \tag{4.2.3}$$

By virtue of (4.2.1) we have

$$\{P_+(\alpha)(z - z(\zeta))x\}(\zeta) = z(\zeta)x(\zeta) - C(\alpha)\frac{k^{\alpha\mu}}{b}(\zeta)x(\zeta) - z(\zeta)x(\zeta) =$$

$$= -C(\alpha)\frac{k^{\alpha\mu}}{b}(\zeta)\langle x, k^\alpha\rangle = \langle x, -C(\alpha)\overline{\frac{k^{\alpha\mu}}{b}(\zeta)}k^\alpha\rangle. \tag{4.2.4}$$

On the other hand, since the function z is real on $\mathbb{T}$,

$$\langle (z - z(\zeta))x, k_\zeta^\alpha\rangle = \langle x, (z - \overline{z(\zeta)})k_\zeta^\alpha\rangle = \langle x, P_+(\alpha)(z - \overline{z(\zeta)})k_\zeta^\alpha\rangle. \tag{4.2.5}$$

Comparing (4.2.4) to (4.2.5) , we get

$$P_+(\alpha)(z - \overline{z(\zeta)})k_\zeta^\alpha = -C(\alpha)\overline{\frac{k^{\alpha\mu}}{b}(\zeta)}k^\alpha,$$

or

$$(z - \overline{z(\zeta)})k_\zeta^\alpha - C(\alpha)\frac{k^{\alpha\mu}}{b}k_\zeta^\alpha(0) = -C(\alpha)\overline{\frac{k^{\alpha\mu}}{b}(\zeta)}k^\alpha.$$

Since $k_\zeta^\alpha(0) = \overline{k_0^\alpha(\zeta)}$ one has

$$(z - \overline{z(\zeta)})k_\zeta^\alpha = C(\alpha)\{\frac{k^{\alpha\mu}}{b}\overline{k^\alpha(\zeta)} - \overline{\frac{k^{\alpha\mu}}{b}(\zeta)}k^\alpha\}.$$

The lemma is proved. □

4.3. Let $F(z)$ be a meromorphic function in $\overline{\mathbb{C}}\backslash E$. Then $f(\zeta) = F(z(\zeta))$ is a Γ-automorphic meromorphic function in $\mathbb{D}$, $f \circ \gamma = f\ \forall \gamma \in \Gamma$. Conversely, if f is a Γ-automorphic function in $\mathbb{D}$, it has the form $f = F \circ z$. From this remark it follows that every solution of the Nevanlinna-Pick problem with the data $\{z_j, w_j\}_{j=1}^n$ in the domain $\overline{\mathbb{C}}\backslash E$ corresponds uniquely to a Γ-automorphic solution of the Nevanlinna-Pick problem with the data $\{\zeta_j, w_j\}_{j=1}^n$ where ζ_j is some preimage of z_j, $z(\zeta_j) = z_j$. Hence the following proposition holds.

Proposition 4.3.1. *Let $\{z_j, w_j\}_{j=1}^n$ be the data of the Nevanlinna-Pick problem in the domain $\overline{\mathbb{C}}\backslash E$. This problem is solvable if and only if the matrices*

$$||(1 - w_j\overline{w_i})\left[\frac{k^{\alpha\mu}}{b}(\zeta_j)\overline{k^{\alpha}(\zeta_i)} - \overline{\frac{k^{\alpha\mu}}{b}(\zeta_i)}k^{\alpha}(\zeta_j)\right]/[z(\zeta_j) - \overline{z(\zeta_i)}]||_{i,j} \tag{4.3.1}$$

are non-negative for each $\alpha \in \Gamma^*$.

It seems interesting to give a solvability criterion for this problem in terms of the domain $\overline{\mathbb{C}}\backslash E$. Now we shall make a first step in this direction.

Consider the Γ-automorphic meromorphic functions $k^{\alpha\mu}/(bk^{\alpha})$ and define the functions $r^{\alpha}(z)$ by the equality

$$r^{\alpha} \circ z = \frac{k^{\alpha\mu}}{bk^{\alpha}}.$$

From the inequality

$$k^{\alpha}(\zeta, \zeta) = C(\alpha)\left[\frac{k^{\alpha\mu}}{b}(\zeta)\overline{k^{\alpha}(\zeta)} - \overline{\frac{k^{\alpha\mu}}{b}(\zeta)}k^{\alpha}(\zeta)\right]/[z(\zeta) - \overline{z(\zeta)}]$$

$$= C(\alpha)|k^{\alpha}(\zeta)|^2[r^{\alpha}(z(\zeta)) - \overline{r^{\alpha}(z(\zeta))}]/[z(\zeta) - \overline{z(\zeta)}] > 0,$$

it follows that the imaginary part of $r^{\alpha}(z)$ is positive in the upper half-plane and negative in the lower half-plane. This means that $r^{\alpha}(z)$ is a so called Nevanlinna function. In particular, the poles of this function can lie only on the real axis.

If $k^{\alpha}(\zeta_j) \neq 0$ (or, equivalently, if z_j is not a pole of $r^{\alpha}(z)$), one can rewrite the conditions (4.3.1) in the form

$$||(1 - w_j\overline{w_i})\frac{r^{\alpha}(z_j) - \overline{r^{\alpha}(z_i)}}{z_j - \overline{z_i}}||_{i,j} \geq 0. \tag{4.3.2}$$

Further progress is connected with an explicit description of the functions $\{r^{\alpha}(z)\}_{\alpha\in\Gamma^*}$.

4.4. For the given set E, we shall define an infinite-dimensional (in the general case) torus $D(E)$. Associate with every complementary interval $[a_j, b_j]$ its double covering

$$I_j = \{(x_j, \varepsilon_j)|x_j \in [a_j, b_j], \varepsilon_j = \pm 1\},$$

where, by definition, $(a_j, 1) \equiv (a_j, -1), (b_j, 1) \equiv (b_j, -1)$.

The set I_j is endowed with the natural topology, in which I_j is homeomorphic to a circle. The torus $D(E)$ is a direct product of $I_j, D(E) = \prod_{j\geq 1} I_j$, equiped with the topology of the direct product of circles.

We associate a Nevanlinna function with every point

$$D = \{(x_j, \varepsilon_j)\}_{j\geq 1} \in D(E)$$

(for details see [13]). First we construct a function depending on $\{x_j\}_{j\geq 1}$ only:

$$R^D(z) = \sqrt{(z-a_0)(z-b_0)} \prod_{j\geq 1} \frac{\sqrt{(z-a_j)(z-b_j)}}{z-x_j}. \tag{4.4.1}$$

Then for a point x_j lying inside the interval (a_j, b_j) we either remove the pole of the function $R^D(z)$ or we double the residue value according to the sign of ε_j. To be precise, we introduce the following function

$$\tilde{r}^D(z) = z + R^D(z) + \sum_{j\geq 1} \frac{\sigma_j \varepsilon_j}{z-x_j}, \quad D \in D(E), \tag{4.4.2}$$

where $\sigma_j = Res_{x_j} R^D(z)$.

Proposition 4.4.1. ([13]). *There exists a homeomorphism from $D(E)$ to Γ^* for which the function $r^{\alpha(D)}, D \in D(E)$, has the form*

$$r^{\alpha(D)}(z) = C_1(D)\tilde{r}^D(z) + C_2(D),$$

where $C_1(D) > 0, C_2(D) \in \mathbb{R}$.

Now we are able to complete our considerations with the following

Theorem 4.4.1. *Let $\{z_j, w_j\}_{j=1}^n$ be the data of the Nevanlinna-Pick problem in the domain $\overline{\mathbb{C}}\backslash E$. Then this problem is solvable if and only if the matrices*

$$||(1-w_j\overline{w_i})\frac{\tilde{r}^D(z_j) - \overline{\tilde{r}^D(z_i)}}{z_j - \overline{z_i}}||_{i,j} \geq 0 \tag{4.4.3}$$

are non-negative for each $D \in D(E)$ for which no point z_j is a pole of $\tilde{r}^D(z)$.

Proof. The set of the points D considered is obviously dense in $D(E)$. By Proposition 4.4.1, the inequalities (4.3.2) hold. They are equivalent to (4.3.1) for a set of characters which is dense in Γ^*. The continuous dependence on the character implies that the matrices (4.3.1) are non-negative for arbitrary $\alpha \in \Gamma^*$. Hence the Nevanlinna-Pick problem is solvable.

The proof of the other direction is trivial. □

Remark. Using an obvious passage to the limit, we could define a form of the kind (4.4.3) for any $D \in D(E)$. On the other hand, the non-negativity of the forms (4.4.3) for any dense subset in $D(E)$ is sufficient for the solvability of the problem.

References

[1] Abrahamse M., The Pick interpolation theorem for finitely connected domains, *Michigan Math. J.* **26** (1979) 195–203.

[2] Ball J., A lifting theorem for operator models of finite rank on multiply connected domains, *J. Operator Theory* **1** (1979), 3–25.

[3] Ball J. and Vinnikov V., Zero-pole interpolation for matrix meromorphic functions on an algebraic curve with determinantal representation, *Acta Math. Applicandae* (to appear).

[4] Jones P. and Marshall D., Critical points of Green's function, harmonic measure, and the corona problem, *Arkiv för Matematik* **23** (1985) 281–314.

[5] Hasumi M., *Hardy Classes on Infinitely Connected Riemann Surfaces*, Lecture Notes in Math. **1027**, Springer Verlag, Berlin and New York, 1983.

[6] Heins M., On an example of Donald Marshall concerning automorphic Pick-Nevanlinna interpolation problems *Complex Variables Theory Appl.* **7** (1986) 71–78.

[7] Krein M.G. and Nudelman A.A., *Markov Moment Problem and Extremal Problems*, Nauka, Moscow, 1973 (Russian).

[8] Livsic M., Kravitsky N., Markus A., Vinnikov V., *Commuting Nonselfadjoint Operators and their Applications to System Theory*, Kluwer, 1995

[9] Nikolskii N.K., *Treatise on the Shift Operator*, Springer Verlag. Berlin and New York, 1986.

[10] Pavlov B. and Fedorov S., The group of shifts and harmonic analysis on a Riemann surface of genus one, *Algebra i Analiz* **1** (1989) 132–168, (Russian).

[11] Pommerenke Ch., On the Green's function of Fuchsian groups, *Ann. Acad. Sci. Fenn.* **2** (1976) 409–427.

[12] Samokhin M., Some classical problems in the theory of analytic functions in domains of Parreau-Widom type, *Math. USSR Sbornik* **73** (1992) 273–288.

[13] Sodin M.L. and Yuditskii P.M., Infinite-dimensional Jacobi inversion problem, almost-periodic Jacobi matrices with homogeneous spectrum, and Hardy classes of character-automorphic functions, *Journ. of Geometr. Analysis* (to appear).

[14] Widom H., H^p sections of vector bundles over Riemann surfaces, *Ann. of Math.* **94** (1971) 304–323.

Mathematical Division
Institute for Low Temperature
Physics and Engineering
47 Lenin's Ave.
Kharkov 310164, Ukraine
E-mail: kupin@ilt.kharkov.ua

Mathematical Division
Institute for Low Temperature
Physics and Engineering
47 Lenin's Ave.
Kharkov 310164, Ukraine
E-mail: yuditskii@ilt.kharkov.ua

AMS Subject Classification: Primary: 30E05, 30F35; Secondary: 30F20, 30D50.

Operator Theory
Advances and Applications, Vol. 95
© 1997 Birkhäuser Verlag Basel/Switzerland

The Blaschke-Potapov factorization theorem and the theory of nonselfadjoint operators

M.S. Livšic

Abstract. The origins and main properties of the characteristic function of a bounded linear operator with finite dimensional imaginary part is briefly sketched. The motivation for the Potapov theory on the multiplicative decomposition of J-contractive matrix-valued functions is discussed. [Abstract added by editors.]

0 Preface

I am grateful to the organizers of this conference for inviting me to give a lecture at Leipzig University on my scientific collaboration with V.P. Potapov in Odessa during the years 1945–1957.

Dear friends and colleagues!

In 1934 V.P. Potapov at the age of 20 abandoned the piano class at the Odessa Conservatory and entered the Mathematics Department of Odessa University. I had entered the same department a year earlier and we soon became close friends. Our teachers were M.G. Krein and B.Ja. Levin. However, our scientific collaboration began only after World War II, when we came back to our native town of Odessa. Later we were joined by M.S. Brodskii, and then there were three of us, who formed a small group involved in the development of a new field on the border between Operator Theory and Complex Analysis. We often discussed problems which arose in our investigations with M.G. Krein and his advice and remarks were very useful. At that time, many mathematicians tried to generalize the Weierstrass theory of elementary divisors to the infinite dimensional case. There were two main schools of thought. One school believed that the infinite dimensional analog of the Weierstrass theory would come from commutative Banach algebras, but the other believed more in close ties between Operator Theory and the theory of analytic functions.

1 Characteristic functions

Consider the Volterra operator

$$g(x) = \int_a^x K(x,s)f(s)ds \ ;$$

in their famous book on Functional Analysis, Riesz and Sz.-Nagy wrote: "The existence of such a variety of linear transformations, having the same spectrum, concentrated in a single point, brings out the difficulties of characterization of linear transformations of general type by means of their spectra."

Of course, the Riesz decomposition of the resolvent with respect to disjoint spectral components does not work in the case of Volterra operators. In 1943–44, generalizing the extension theory of J. von Neumann, I introduced a certain analytic matrix-valued function which characterized the nonselfadjoint extensions of a symmetric operator, and named it the characteristic function of a nonselfadjoint operator. In particular a nonselfadjoint operator may be recovered from its characteristic function up to unitary equivalence. Later the notion of characteristic function was introduced also for bounded operators, and I will consider here only this case. Perhaps the easiest way to explain what a characteristic function is, is to consider the Moëbius transformation

$$w = \frac{\bar{a} - \lambda}{a - \lambda} \ ,$$

which maps the upper (lower) halfplane onto the unit disk. Let us try to generalize this linear fractional transformation to the case when $a = A$ is an operator in a Hilbert space. Consider the fraction

$$(A - \lambda I)^{-1}(A^* - \lambda I) = I - i(A - \lambda I)^{-1}\frac{A - A^*}{i} \ .$$

Assume that the non-Hermitian subspace

$$G = (A - A^*)H$$

is finite dimensional and define the operator-valued function

$$W(\lambda) = I_G - iP_G(A - \lambda I)^{-1}\frac{A - A^*}{i}|_G \ ,$$

which acts in the non-Hermitian subspace G. *The function $W(\lambda)$ is said to be the characteristic function of the operator A.*

Let H_1 be a subspace of H and let $A_1 = P_1 A|_{H_1}$. Then $\dim G_1 \leq \dim G$, and we can construct the operator-valued function $W_1(\lambda) = P_1[W(\lambda)]$, which acts in G_1, but it is difficult to compare W and W_1 since they act in different subspaces. To overcome this difficulty we will consider a more general object: Let E be a finite dimensional space (not necessarily a subspace of H) equipped with a scalar product, and let $\Phi : H \to E$ be a linear mapping of H into E. Let $J = J^*$ be a given selfadjoint operator in E. A collection

$$X = (A; H, \Phi, E; J)$$

is said to be a *colligation* if the condition

$$\frac{1}{i}(A - A^*) = \Phi^* J \Phi$$

holds:

$$\begin{array}{ccccc} & H & \xrightarrow{\Phi} & E & \\ \frac{1}{i}(A-A^*) & \downarrow & & \downarrow & J \\ & H & \xleftarrow{\Phi^*} & E & \end{array}$$

The operator-valued function

$$W(\lambda) = I_E - i\Phi(A - \lambda I)^{-1}\Phi^* J$$

is said to be the characteristic function of the colligation X. If, for instance, $E = G$ and $\Phi = P_G$, then we obtain the characteristic function of A which was mentioned above. In this case,

$$J = \frac{1}{i}(A - A^*)|_G \; .$$

There is another possibility: let us define

$$\Phi = |\frac{1}{i}(A - A^*)|^{1/2} \; .$$

Then

$$\frac{1}{i}(A - A^*) = \Phi^* J \Phi \; ,$$

where

$$J = \operatorname{sign}[\frac{1}{i}(A - A^*)]_G \; .$$

In this case $J^2 = I$ and the characteristic function is

$$W(\lambda) = I_G - i\left|\frac{A - A^*}{i}\right|^{1/2}(A - \lambda I)^{-1}\left|\frac{A - A^*}{i}\right|_G^{1/2} J \; .$$

As an example, let us consider the *Leibnitz-Newton* integral:

$$(Af)(x) = i\int_a^x f(t)dt \quad (f \in L_2(a, b)) \; .$$

It is evident that

$$(A^* f)(x) = -i\int_x^b f(t)dt \; ,$$

and, therefore,

$$\frac{1}{i}(A - A^*)f = \int_a^b f(t)dt = (f, g_0)g_0 ,$$

where $g_0(x) = 1$ $(a \le x \le b)$. Hence $\dim(A - A^*)H = 1$ in this case and, calculating the characteristic function, we obtain

$$W(\lambda) = \exp\left[i\frac{b-a}{\lambda}\right] .$$

It is evident that

$$\begin{aligned} |W(\lambda)| &> 1 \quad \text{if} \quad \mathrm{Im}\lambda > 0 , \\ |W(\lambda)| &= 1 \quad \text{if} \quad \mathrm{Im}\lambda = 0 \quad (\lambda \neq 0) , \\ |W(\lambda)| &< 1 \quad \text{if} \quad \mathrm{Im}\lambda < 0 . \end{aligned}$$

In the general case one can prove the following relations:

$$\frac{W^*(u)JW(\lambda) - J}{i(\overline{u} - \lambda)} = R^*(u)R(\lambda) ,$$

where

$$R(\lambda) = (A - \lambda I)^{-1}\Phi^* J .$$

Therefore, the characteristic function of a colligation has the following properties with respect to the *metric which is defined by the imaginary part of A*:

$$W^*(\lambda)JW(\lambda) - J \begin{cases} \ge 0 & \text{if } \mathrm{Im}\lambda > 0 , \\ = 0 & \text{if } \mathrm{Im}\lambda = 0 , \end{cases}$$

where λ is a regular point of the resolvent.

It can be proved also that, as an analytic function of λ, the characteristic function has the following properties:

1. All the nonreal singularities of $W(\lambda)$ are poles.
2. $W(\lambda)$ is regular in a neighborhood of infinity and $W(\infty) = 1$.

The metric conditions together with the analytic conditions are also sufficient for a given matrix-function to be the characteristic function of a colligation [3].

It can be proved [3] that the spectrum of A coincides with the set of singular points of its characteristic function. Another remarkable property of the characteristic function of a colligation is its *multiplicative behaviour* with respect to invariant subspaces [2, 3].

If $H = H_1 \oplus H_2 \oplus \cdots \oplus H_n$, where H_1, $H_1 \oplus H_2, \ldots, H_1 \oplus H_2 \oplus \cdots \oplus H_{n-1}$, are invariant subspaces of A, then

$$W(\lambda) = W_1(\lambda)W_2(\lambda)\cdots W_n(\lambda) ,$$

where $W_k(\lambda) = P_{H_k}[W]$. This factorization theorem opens a way to obtain *triangular models* of A even in the case of a single point spectrum. In the case

$\frac{1}{i}(A - A^*) \geq 0$ and $\dim(A - A^*) = 1$, the multiplicative representation of an expansive scalar function has the form

$$W(\lambda) = \prod \frac{\lambda - \overline{\lambda}_k}{\lambda - \lambda_k} \cdot \exp\, i \int_a^b \frac{dt}{\lambda - \alpha(t)} \quad (\mathrm{Im}\lambda_k > 0)\,.$$

Here $\alpha(t)$ is a bounded nondecreasing function, which can be obtained from the Riesz-Herglotz theorem. Using this result we obtained the triangular model for the case $\dim(A - A^*)H = 1$. It is interesting to note that if $\dim(A - A^*) = 1$ and the spectrum of A is concentrated at the single point $\lambda = 0$, then the Riesz-Herglotz formula implies that

$$W(\lambda) = \exp(i\frac{\ell}{\lambda}) \quad (\ell \geq 0)\,,$$

and the operator A must be unitarily equivalent to the Leibnitz-Newton integral with $\ell = b - a$. To be precise, there exists a possibility to add a selfadjoint "appendix" A_0 and to consider an operator of the form

$$\begin{pmatrix} A & 0 \\ 0 & A_0 \end{pmatrix},$$

which has the same characteristic function. Therefore, the *Leibnitz-Newton integral is the unique operator* (up to unitary equivalence and a trivial appendix $A_0 = 0$) *with one-dimensional imaginary part and* $\lambda = 0$.

2 Potapov's factorization theorem

In the case $\dim(A - A^*) > 1$, we need *Potapov's Factorization Theorem* for J-expansive matrix-functions, which in our case has the form [1, 3]

$$W(\lambda) = \overset{\curvearrowright}{\Pi} \left(I - i\frac{\varphi_k \varphi_k^*}{\lambda_k - \lambda} J\right) \overset{\curvearrowright}{\int_0^\ell} \exp i \frac{dE(t)}{t - \alpha(t)} J\,,$$

where the φ_k are column vectors and $E(t)$ is a non-decreasing matrix-function such that $\mathrm{tr}\, E(t) = t$. By applying this multiplicative representation to the characteristic function, we obtained the triangular models [3] of a nonselfadjoint operator with a finite dimensional imaginary part. This triangular model can be considered as a natural generalization of the theorem of Schur which states that every finite square matrix is unitarily equivalent to an upper triangular matrix. The correspondence between the elementary factors (discrete or continuous) and the invariant subspaces of an operator has proved to be very fruitful.

A second application of Potapov's theorem was to the chain synthesis of electrical multipoles. He obtained (together with Efimov) the complete classification of all possible elementary lossless multipoles. It turned out that the transfer matrix-function of a lossless multipole is a J-expansive matrix-function with

$$J = \begin{pmatrix} 0 & I_p \\ I_p & 0 \end{pmatrix},$$

which satisfies some additional "realization" conditions.

A third application, and this was the last one in his life, was the application of his theory to interpolation problems, both discrete and continuous.

To obtain the Factorization Theorem in the matrix case, V. Potapov had to overcome many difficulties. The first difficulty stemmed from the noncommutativity of the factors in the matrix case [1]. Additional difficulties arose in the case of an indefinite metric because in this case the maximum principle could not be used. New problems arose in the case of a multiplicative integral. In this case, V. Potapov had to approximate a given J-expansive matrix-function having singularities on the boundary, with the help of rational J-expansive matrix-functions. In the imagination of his friends at that time, V. Potapov was like a skilled carpenter who was splitting off elementary factors from a big and expansive "log".

Finally, he proved his Factorization Theorem [1], which has been widely recognized as one of the most beautiful and important theorems in the Theory of Analytic Matrix-Functions and its applications.

References

[1] V.P. Potapov, *Multiplicative structure of J-expansive matrix-functions*, Trud. Moscow Mat. Obs. **4** (1955) 125–326. English transl: Amer. Math. Soc. Transl. (2) **15** (1960), 131–243.

[2] M.S. Livšic and V.P. Potapov, *A factorization theorem for characteristic matrix-functions*, Doklady Akad. Nauk. USSR **72**, 4 (1950), 625–628.

[3] M.S. Brodskii and M.S. Livšic, *Spectral analysis of nonselfadjoint operators and intermediate systems*, Uspekhi Mat. Nauk **13:** 1 (1958), 1–85. English transl: Amer. Math. Soc. Transl. (2) **13** (1960), 265–346.

Department of Mathematics
Ben-Gurion University of the Negev
Beer-Sheva 84105, Israel

AMS Subject Classification: Primary: 47A45, 47A48; Secondary: 30D50, 30D55.

Operator Theory
Advances and Applications, Vol. 95
© 1997 Birkhäuser Verlag Basel/Switzerland

Weyl matrix circles as a tool for uniqueness in the theory of multiplicative representation of J-inner matrix functions[1]

Irina V. Mikhailova

Abstract. It is shown that the class of 2×2 J-inner matrix functions of the form

$$\mathfrak{A}(z) = e^{izA_1} \cdot e^{izA_2} \cdot \dots \cdot e^{-izA_n} ,$$

where $A_kJ \geq 0$ for $k = 1, \dots, n$, have an essentially unique multiplicative decomposition. This is a very special case of a deep theorem of de Branges. However, the method of proof is new and elementary. [Abstract added by editors.]

1. One of the main achievements of Potapov's J-theory is the theorem on the multiplicative structure of a J-inner matrix-function. Here J is an $m \times m$ matrix satisfying the conditions

$$J = J^* \quad \text{and} \quad J^2 = I, \tag{1}$$

where I is the identity matrix. Let us recall that a matrix $\mathfrak{A}$ is said to be J-*expansive* if $\mathfrak{A}^* J \mathfrak{A} - J \geq 0$ (or, equivalently, $\mathfrak{A} J \mathfrak{A}^* \geq 0$). A matrix $\mathfrak{A}$ is said to be J-*unitary* if $\mathfrak{A}^* J \mathfrak{A} - J = 0$ (or, what is the same, $\mathfrak{A} J \mathfrak{A}^* - J = 0$).

Definition 1. *An entire $m \times m$ matrix valued function $\mathfrak{A}(z)$ is said to be J-**inner** if it is J-expansive in the upper half-plane and J-unitary on the real axis:*

$$\begin{aligned} \mathfrak{A}^*(z) J \mathfrak{A}(z) - J &\geq 0 \qquad (\Im z > 0) \\ \mathfrak{A}^*(z) J \mathfrak{A}(z) - J &= 0 \qquad (\Im z = 0) . \end{aligned} \tag{2}$$

---

1) Translated from: Analiz v beskonečnomernyh prostranstvah i teorija operatorov. ("Naukova Dumka" publishing house, Kiev, 1983, V.A.Marchenko – ed.), 101–117.

Theorem (V.P.Potapov, [1]). *An entire J-inner matrix-function is representable by a multiplicative integral*

$$\mathfrak{A}(z) = \int_0^{\overset{l}{\curvearrowright}} e^{-izH(t)\,dt}\,\mathfrak{A}(0), \tag{3}$$

where $l \in (0, \infty)$ and $H(t)$ is a summable matrix-function on $(0, l)$ satisfying the condition $H(t) \geq 0,\ t \in (0, l)$.

The function $H(t)$ which appears in (3) is said to be *an exponent of the multiplicative integral.*

Is the exponent $H(t)$ of the multiplicative integral (3) uniquely determined by the matrix function $\mathfrak{A}(z)$? This is the so-called *uniqueness problem* for the multiplicative integral (3) and it turns out to be very complicated. This problem was fully solved only for 2×2 matrix-functions by de Branges (see [2] and the references therein). de Branges showed that under a natural normalizing condition the exponent $H(t)$ in (3) is determined uniquely. The proof of the uniqueness theorem given by de Branges arises from a long chain of analytic constructions and is based on deep results from the theory of entire functions. In the present paper a new approach to the uniqueness problem of the multiplicative representation of entire (and not only entire) J-expansive matrix-functions is presented. This approach is based on the construction of the so-called *Weyl circles related to these matrix-functions.* Here we apply the Weyl circles method to give a simple proof of the de Branges uniqueness theorem in a special case. Namely, we consider J-inner matrix-functions of the special form:

$$\mathfrak{A}(z) = e^{-izA_1} \cdot e^{-izA_2} \cdot \ldots \cdot e^{-izA_n}, \tag{4}$$

where $A_j \quad (j = 1, 2, \cdots, n)$ are constant J-positive $(AJ \geq 0)$ matrices. The problem of the representation of a matrix-function $\mathfrak{A}(z)$ in the form (4) is an essential element of the J-theory. In effect, we solve here the problem of splitting off entire discrete factors from entire matrix functions. For entire matrix functions this problem plays the same role that is played in the rational case by the problem of splitting off Blaschke-Potapov factors. In what follows matrix-functions of the form e^{-izA} are called *elementary factors.* For convenience, we fix the concrete matrix

$$J = \begin{bmatrix} 0 & i \\ -i & 0 \end{bmatrix} \tag{5}$$

This does not lead to loss of generality because an arbitrary 2×2 matrix J, $J \neq \pm I$ with the properties (1) is equivalent to the matrix (5).

2. Let us recall the notion of *the Weyl circle* $\mathcal{W}(\mathfrak{A})$ *related to* an arbitrary J-*expansive matrix* $\mathfrak{A}$, $(\mathfrak{A}J\mathfrak{A}^* - J \geq 0)$.

A matrix $\mathfrak{A}$ with entries

$$\mathfrak{A} = \begin{bmatrix} a & b \\ c & d \end{bmatrix} \tag{6}$$

generates a fractional-linear transformation

$$w = (a\omega + b)(c\omega + d)^{-1} \tag{7}$$

whose argument is ω. If a fractional-linear transformation is the superposition of two fractional-linear transformations, then its matrix is the product of the coefficient matrices of these fractional-linear transformations. The fractional-linear transformation (7) maps the upper half-plane onto a circle. If its coefficient matrix $\mathfrak{A}$ is J-expansive, then this circle lies in the upper half-plane.

Definition. ***The Weyl circle*** $\mathcal{W}(\mathfrak{A})$ ***of a*** J***-contractive invertible matrix*** $\mathfrak{A}$ *(6) is the image of the fractional-linear transformation (7) with the coefficient matrix* $\mathfrak{A}$ *when the parameter* ω *rans over the upper half plane* $\Im z \geq 0$.

Since $2\Im w = \begin{bmatrix} \overline{w} & 1 \end{bmatrix} J \begin{bmatrix} w \\ 1 \end{bmatrix}$, it is not difficult to see, that the Weyl circle of the matrix $\mathfrak{A}$ consist precisely of these points w which satisfy the condition

$$\begin{bmatrix} \overline{w} & 1 \end{bmatrix} (\mathfrak{A}^{-1})^* J \mathfrak{A}^{-1} \begin{bmatrix} w \\ 1 \end{bmatrix} \geq 0. \tag{8}$$

The general proof of this fact can be found in [3]. Using the definition (8), we shall show that the Weyl circle of a J-contractive matrix $\mathfrak{A}$ lies in the upper half-plane. We shall also find the center and the radius of this circle. From the relation $\mathfrak{A}^* J \mathfrak{A} - J \geq 0$ it follows that $(\mathfrak{A}^{-1})^* (\mathfrak{A}^* J \mathfrak{A} - J) \mathfrak{A}^{-1} \geq 0$, i.e., $J - \mathfrak{A}^{-1} J (\mathfrak{A}^{-1})^* \geq 0$. From here it follows that

$$\begin{bmatrix} \overline{w} & 1 \end{bmatrix} J \begin{bmatrix} w \\ 1 \end{bmatrix} - \begin{bmatrix} \overline{w} & 1 \end{bmatrix} (\mathfrak{A}^{-1})^* J \mathfrak{A}^{-1} \begin{bmatrix} w \\ 1 \end{bmatrix} \geq 0,$$

i.e.,

$$\frac{w - \overline{w}}{i} \geq \begin{bmatrix} \overline{w} & 1 \end{bmatrix} (\mathfrak{A}^{-1})^* J \mathfrak{A}^{-1} \begin{bmatrix} w \\ 1 \end{bmatrix}.$$

Thus, if w satisfies (8), then $\Im w > 0$. Hence, the Weyl circle lies in the upper half plane.

Denoting the entries of a *Weyl matrix* by

$$(\mathfrak{A}^{-1})^* J \mathfrak{A}^{-1} = \begin{bmatrix} -R & S \\ \overline{S} & -T \end{bmatrix},$$

we obtain

$$\begin{bmatrix} \overline{w} & 1 \end{bmatrix} \begin{bmatrix} -R & S \\ \overline{S} & -T \end{bmatrix} \begin{bmatrix} w \\ 1 \end{bmatrix} \geq 0, \quad \text{or} \quad -R\overline{w}w + \overline{S}w + S\overline{w} - T \geq 0.$$

From the J-contractivity of the matrix $\mathfrak{A}^{-1}$, it follows that

$$\begin{bmatrix} R & i-S \\ -i-\overline{S} & T \end{bmatrix} \geq 0,$$

and hence that $R \geq 0$ and $T \geq 0$. Moreover, if $R = 0$, then $S = i$. Thus, if $R > 0$, then the inequality (8) is equivalent to the inequality

$$\left|w - SR^{-1}\right|^2 \leq \left(S\overline{S} - TR\right) \Big/ R^2 .$$

From this it follows that if $R > 0$, then the inequality (8) defines a circle with center

$$C = SR^{-1} \tag{9}$$

and radius

$$\rho = \sqrt{S\overline{S} - TR} \Big/ R = \sqrt{-\det \mathfrak{A}^* J \mathfrak{A}} \Big/ R = 1/R \det \mathfrak{A}. \tag{10}$$

If $R = 0$, then the Weyl circle degenerates to the half-plane $\dfrac{w - \overline{w}}{i} \geq T$.

3. Let us introduce the notion of *divisibility* for nondegenerate J-expansive matrices: *a J-expansive matrix $\mathfrak{A}$ is said to be* ***a left divisor*** *of a J-expansive matrix $\mathfrak{C}$ if their quotient $\mathfrak{A}^{-1}\mathfrak{C}$ is a J-expansive matrix as well.*

The main role in our considerations is the fact that the divisibility of two J-expansive matrices is equivalent to the nesting of their Weyl circles. The precise formulation of this principle is formulated in the following two lemmas:

Lemma 1. *If $\mathfrak{A}$ and $\mathfrak{B}$ are J-expansive matrices, then*

$$\mathcal{W}(\mathfrak{A}) \supseteq \mathcal{W}(\mathfrak{A}\mathfrak{B}) . \tag{11}$$

Proof. From the inequality $\mathfrak{B}^* J \mathfrak{B} - J \geq 0$ we obtain successively

$$J - (\mathfrak{B}^{-1})^* J \mathfrak{B}^{-1} \geq 0,$$

$$\begin{bmatrix} \overline{w} & 1 \end{bmatrix} \left(\mathfrak{A}^{-1}\right)^* \left(J - (\mathfrak{B}^{-1})^* J \mathfrak{B}^{-1}\right) \mathfrak{A}^{-1} \begin{bmatrix} w \\ 1 \end{bmatrix} \geq 0,$$

$$\begin{bmatrix} \overline{w} & 1 \end{bmatrix} \left(J - (\mathfrak{A}^{-1})^* J \mathfrak{A}^{-1}\right) \begin{bmatrix} w \\ 1 \end{bmatrix} \geq \begin{bmatrix} \overline{w} & 1 \end{bmatrix} \left(J - ((\mathfrak{A}\mathfrak{B})^{-1})^* J (\mathfrak{A}\mathfrak{B})^{-1}\right) \begin{bmatrix} w \\ 1 \end{bmatrix} .$$

The embedding (11) follows from here. The lemma is proved. □

If in addition we assume that the matrices $\mathfrak{A}$ and $\mathfrak{B}$ are symplectic:

$$\mathfrak{A} J \mathfrak{A}^{\tau} = J \quad \text{and} \quad \mathfrak{C} J \mathfrak{C}^{\tau} = J \tag{12}$$

(τ indicates the transpose), then Lemma 1 admits a converse.

Lemma 2. *Let $\mathfrak{A}$ and $\mathfrak{C}$ be J-expansive symplectic matrices, and let their Weyl circles be nested:*

$$\mathcal{W}(\mathfrak{A}) \supseteq \mathcal{W}(\mathfrak{C}). \tag{13}$$

Then the matrix $\mathfrak{B} = \mathfrak{A}^{-1}\mathfrak{C}$ is J-expansive.

Proof. From (13) it follows that that the fractional-linear transformation with coefficient matrix

$$\mathfrak{B} = \mathfrak{A}^{-1}\mathfrak{C} = \begin{bmatrix} a & b \\ c & d \end{bmatrix}$$

maps the upper half-plane into itself, i.e., $\mathfrak{B}$ is a nonsingular symplectic *plus-matrix*. Then it follows from Theorem 11 of the paper [4] that the matrix $\mathfrak{B}$ is J-expansive. The lemma is proved. □

Now let $\mathfrak{A}$ be an arbitrary entire J-inner real matrix-function. (As usual, an entire matrix-function is said to be *real* if its entries take real values on the real axis.) Then $\det \mathfrak{A}(z) \equiv \pm 1$, and the equality

$$\mathfrak{A}(z) J \mathfrak{A}(z)^{\tau} = J \tag{14}$$

holds. This identity can be verified first on the real axis, and then can be extended to the whole complex plane. Thus, a real entire J-inner matrix-function is symplectic. We denote the class of all real entire J-inner matrix-functions by $\mathfrak{M}$.

Let us associate with each fixed point $z = z_0$ from the upper half-plane ***the Weyl circle***

$$\mathcal{W}_{z_0}(\mathfrak{A}) \stackrel{\text{def}}{=} \mathcal{W}(\mathfrak{A}(z_0))$$

Thus, if the point z varies in the upper half-plane, we have a family $\{\mathcal{W}_z(\mathfrak{A})\}$ of Weyl circles related to the matrix-function $\mathfrak{A}$, floating in the upper half-plane together with the point z.

Definition. *Let $\mathfrak{A}(z)$ and $\mathfrak{B}(z)$ be entire real J-inner matrix-functions (i.e., $\mathfrak{A}(z)$, $\mathfrak{B}(z) \in \mathfrak{M}$). Then $\mathfrak{A}(z)$* ***is said to divide*** *$\mathfrak{B}(z)$* ***from the left*** *if the quotient $(\mathfrak{A}(z))^{-1}\mathfrak{B}(z)$ belongs to the class $\mathfrak{M}$ as well.*

The following theorem is a consequence of Lemmas 1 and 2.

Theorem 1. *Let $\mathfrak{A}(z)$ and $\mathfrak{B}(z) \in \mathfrak{M}$. Then, in order for a matrix-function $\mathfrak{A}(z)$ to be a left divisor of a matrix-function $\mathfrak{B}(z)$ it is necessary and sufficient that at each point $z : \Im z \geq 0$ the embedding*

$$\mathcal{W}_z(\mathfrak{A}) \supseteq \mathcal{W}_z(\mathfrak{B})$$

holds.

4. Going directly to the study of the elementary factor e^{-izH}, we will clarify the structure of a J-positive exponent H $(HJ \geq 0)$. Among the set of all J-positive matrices, the ***J-projectors*** stand out:

$$P : PJ \geq 0, P^2 = P; \quad Q : QJ \geq 0, Q^2 = -Q; \quad E : EJ \geq 0, E^2 = 0,$$

as well as matrices satisfying the conditions $HJ \geq 0$, $\operatorname{sp} H = 0$.

The J-projectors admit the parametrization:

$$\begin{bmatrix} \alpha \\ \beta \end{bmatrix} [\,\overline{\alpha}\,,\overline{\beta}\,]\,J; \quad Q = \begin{bmatrix} \overline{\alpha} \\ \overline{\beta} \end{bmatrix} [\,\alpha\,,\beta\,]\,J; \quad E = \begin{bmatrix} \xi \\ \eta \end{bmatrix} [\,\overline{\xi}\,,\overline{\eta}\,]\,J, \tag{15}$$

where the numbers α, β satisfies the conditions $\dfrac{\alpha\overline{\beta} - \beta\overline{\alpha}}{i} = 1$ and $\dfrac{\xi\overline{\eta} - \eta\overline{\xi}}{i} = 0$.

It turns out, that projectors of such a form are components of an arbitrary 2×2 J-positive matrix. It is not difficult to verify that the following statement is true:

Lemma 3. *If $\Phi J \geq 0$, then either $\Phi^2 = 0$, or the matrix Φ has the form*

$$\Phi = kQ + lP, \tag{16}$$

where $k, l \in [0, \infty)$ and

$$PJ \geq 0,\ QJ \geq 0,\ P^2 = P,\ Q^2 = -Q,\ PQ = QP = 0,\ P - Q = I.$$

From (15) one further representation of a J-positive matrix Φ follows:

$$\Phi = \frac{1}{2}\mathrm{sp}\Phi \cdot I + H; \qquad H = \kappa(P + Q); \qquad HJ \geq 0,\ \mathrm{sp}H = 0. \tag{17}$$

From here we derive

Lemma 4. *An arbitrary elementary factor $e^{-iz\Phi}$, $\Phi J \geq 0$ is representable in the form*

$$e^{-iz\Phi} = e^{-iz\,\mathrm{sp}\Phi/2} \cdot e^{-izH},$$

where $HJ \geq 0$, $\mathrm{sp}H = 0$.

Let us remark that from $\mathrm{sp}H = 0$ it follows that $\det e^{-izH} \equiv 1$ and that the elementary factor e^{-izH} is symplectic. Lemma 4 implies that an arbitrary finite product of elementary J-inner matrix-functions can be reduced to symplectic form by multiplication by some scalar exponential factor.

Remark 1. *If $\Phi = kQ + lP$; $k \geq 0$, $P - Q = I$ and if ρ is chosen such that $-l \leq \rho \leq k$, then the product $e^{-iz\rho}e^{-iz\Phi}$ remains a J-inner matrix-function.*

Indeed, because $\Phi + \rho I = (k - \rho)Q + (l + \rho)P$, then $(\Phi + \rho I)J = (k - \rho)QJ + (l + \rho)PJ \geq 0$.

Remark 2. *If elementary factors in the multiplicative representation (4) of the matrix-function $\mathfrak{A}(z)$ are not symplectic, the formulation of the uniqueness problem of this representation may not be meaningful.*

Example. Let $\mathfrak{A}(z) = e^{-iz\Phi_1}e^{-iz\Phi_2}$, where $\Phi_1 = k_1Q_1 + l_1P_1, \Phi_2 = k_2Q_2 + l_2P_2$, $\Phi_1\Phi_2 \neq \Phi_2\Phi_1$ and the numbers k and l are strictly positive. Let us choose ρ such that

$$-l_1 < \rho < k_1 \quad \text{and} \quad -l_2 < \rho < k_2.$$

Then

$$e^{-iz\Phi_1}\,e^{-iz\Phi_2} = e^{-iz\tilde{\Phi}_1}\,e^{-iz\tilde{\Phi}_2},$$

where $\tilde{\Phi}_1 = \Phi_1 + \rho$ and $\tilde{\Phi}_2 = \Phi_2 + \rho$. *The uniqueness is violated.*

Thus the uniqueness problem is meaningful only under some additional normalizing condition. *The symplecticity condition is an example of such a normalizing condition.*

Let us derive a representation for a matrix $\mathfrak{A} : HJ \geq 0, \mathrm{sp}H = 0$ similar to the representation (15).

Theorem 2. *A J-positive matrix H, $HJ \geq 0$ such that $spH = 0$ admits the representation*

$$H = \begin{bmatrix} \alpha\overline{\alpha} & \dfrac{\alpha\overline{\beta} + \beta\overline{\alpha}}{2} \\ \dfrac{\beta\overline{\alpha} + \alpha\overline{\beta}}{2} & \beta\overline{\beta} \end{bmatrix} J, \qquad \frac{\alpha\overline{\beta} - \beta\overline{\alpha}}{2i} = \kappa \geq 0. \tag{18}$$

Proof. The proof of the theorem follows from the representation (15) and Lemma 3. According to this lemma, either the matrix H is a J-projector of the third kind: $H = E$, $E^2 = 0$, or it decomposes into the sum of two mutually orthogonal orthoprojectors (17): $H = \kappa(P + Q)$.

5. Let us calculate the entries of the Weyl matrix of the elementary factor $b(z) = e^{-izH}$, where $HJ \geq 0$ and $\mathrm{sp}H = 0$:

$$\mathcal{W}(z) = (b^{-1})^*(z)Jb^{-1}(z) = \begin{bmatrix} -R(z) & S(z) \\ \overline{S}(z) & -T(z) \end{bmatrix}.$$

First we find the expression for $b^{-1}(z)$. Since $H^2 = \kappa^2 I$ from (17), it follows that

$$\begin{aligned} e^{\alpha H} &= I + \alpha H + \frac{(\alpha H)^2}{2!} + \frac{(\alpha H)^3}{3!} + \cdots \\ &= I\left(1 + \frac{\alpha^2\kappa^2}{2!} + \cdots\right) + H\left(\alpha + \frac{\alpha^3\kappa^2}{3!} + \cdots\right) \\ &= \mathrm{ch}(\alpha\kappa)I + \frac{\mathrm{sh}(\alpha\kappa)}{\kappa}H. \end{aligned}$$

Thus,

$$b^{-1}(z) = e^{+izH} = \mathrm{ch}(\kappa\, z)\, I + \frac{\mathrm{sh}(\kappa\, z)}{\kappa} H, \tag{19}$$

where by definition $(\sin\, \kappa z)/\kappa = z$ for $\kappa = 0$ and $e^{-izE} = I + izE$ (since $E^2 = 0$). From (19) we obtain

$$\mathcal{W}(z) = (b^{-1})^*(z)Jb^{-1}(z) = \mathrm{ch}\left(\frac{z-\overline{z}}{i}\kappa\right)J + \frac{1}{\kappa}\mathrm{sh}\left(\frac{z-\overline{z}}{i}\kappa\right)JH.$$

Setting $\dfrac{z-\overline{z}}{i} = y$ and taking (18) into account, we finally obtain

$$\mathcal{W}(z) = \begin{bmatrix} -\dfrac{\mathrm{sh}\ 2\kappa y}{\kappa}\beta\overline{\beta} & \dfrac{\mathrm{sh}\ 2\kappa y}{\kappa}\dfrac{\alpha\overline{\beta}+\overline{\alpha}\beta}{2} + i\ \mathrm{ch}\ 2\kappa y \\ \dfrac{\mathrm{sh}\ 2\kappa y}{\kappa}\dfrac{\alpha\overline{\beta}+\overline{\alpha}\beta}{2} - i\ \mathrm{ch}\ 2\kappa y & -\dfrac{\mathrm{sh}\ 2\kappa y}{\kappa}\alpha\overline{\alpha} \end{bmatrix}. \tag{20}$$

Using (20), we find the center and the radius of the Weyl circle $\mathcal{W}_z(b)$ by the formulas (9), (10):

$$\begin{aligned} C_z &= S(z)R(z)^{-1} = \frac{\alpha\overline{\beta}+\beta\overline{\alpha}}{2\beta\overline{\beta}} + i\frac{\kappa}{\beta\overline{\beta}\,\mathrm{th}(2\kappa y)} \\ &= \frac{\alpha}{\beta} + i\frac{\kappa}{\beta\overline{\beta}}(\mathrm{cth}(2\kappa y) - 1); \\ \rho_z &= \frac{1}{R(z)\det b(z)} = \frac{1}{R(z)} = \frac{1}{\beta\overline{\beta}}\frac{\kappa}{\mathrm{sh}(2\kappa y)}. \end{aligned}$$

If $\beta = 0$, then the Weyl circle degenerates to the halfplane

$$\Im\omega \geq \alpha\overline{\alpha}\Im z.$$

Thus, we obtain the following picture of the behavior of the Weyl circle $\mathcal{W}_z(b)$ of an elementary factor according to the properties of the matrix H.

If $H^2 \neq 0$, then the Weyl circle lies strictly inside the upper half plane and, as $\Im z \to +\infty$, these circles shrink to the point $\frac{\alpha}{\beta}$, where $\Im\frac{\alpha}{\beta} = \frac{\kappa}{\beta\overline{\beta}} > 0$.

If $H^2 = 0$ and $\beta \neq 0$, then all the Weyl circles are tangent to the real axis at the point $\dfrac{\alpha}{\beta} = \dfrac{\overline{\alpha}}{\overline{\beta}}$ and, as $\Im z \to +\infty$, these circles shrink to this point.

If $\beta = 0$ (in this case $H^2 = 0$), the Weyl circles are half planes $\Im\omega \geq \alpha\overline{\alpha}\Im z$ and, as $\Im z \to \infty$, these circles shrink to the point $z = \infty$.

The asymptotic behavior of the Weyl circles of an elementary factor can be summarized as follows:

Lemma 5. *Suppose that the matrix H, $HJ \geq 0$ admits the representation (18). Then the Weyl circles of the elementary factor e^{-izH} shrink to the point α/β ($\alpha/\beta = \infty$ if $\beta = 0$)* as $\Im z \to +\infty$.

The separation of the asymptotics of Weyl circles of two different factors $b_1 = e^{izH_1}$, $b_2 = e^{izH_2}$ ($H_1 \neq kH_2$ where k is a scalar) has the following consequence:

if $y = \Im z$ is large enough, then the Weyl circles $\mathcal{W}_z(b_1)$ and $\mathcal{W}_z(b_2)$ do not intersect. Because the inclusion

$$\mathcal{W}_z(b_1) \supseteq \mathcal{W}_z(b_1\mathfrak{A}), \qquad \mathcal{W}_z(b_2) \supseteq \mathcal{W}_z(b_2\mathfrak{B})$$

holds for every pair of J expansive matrix functions $\mathfrak{A}, \mathfrak{B}$, the Weyl circles of the matrix functions $\mathfrak{A}_1(z) = e^{izH_1}\mathfrak{A}(z)$ and $\mathfrak{B}_1(z) = e^{izH_2}\mathfrak{B}(z)$ do not intersect either. Hence, $\mathfrak{A}_1(z) \neq \mathfrak{B}_1(z)$. Thus, we have proved:

Theorem 3. *Let $\mathfrak{A}(z), \mathfrak{B}(z) \in \mathfrak{M}$ be matrix functions. Then from the equality*

$$e^{-izH_1}\mathfrak{A}(z) = e^{-izH_2}\mathfrak{B}(z) \qquad (H_iJ \geq 0, \mathrm{spH_i} = 0, \mathrm{H_i} \neq 0, \mathrm{i} = 1, 2)$$

it follows that $H_1 = kH_2$ where $k > 0$ is a scalar.

From this we obtain the following uniqueness result:

Theorem 4. *Suppose that the matrix function $\mathfrak{A}(z)$ admits two multiplicative representations*

$$\begin{array}{ll} \mathfrak{A}(z) = e^{-izA_1}e^{-izA_2} \cdot \ldots \cdot e^{-izA_n} & (A_jJ \geq 0,\ \mathrm{spA_j} = 0,\ \mathrm{j} = 1, 2, \ldots, \mathrm{n}) \\ \mathfrak{B}(z) = e^{-izB_1}e^{-izB_2} \cdot \ldots \cdot e^{-izB_m} & (B_jJ \geq 0,\ \mathrm{spB_j} = 0,\ \mathrm{j} = 1, 2, \ldots, \mathrm{m}) \end{array}, \tag{21}$$

where $A_j \neq kA_{j+1}$ $(j = 1, \ldots, n-1), A_n \neq 0$; $B_j \neq kB_{j+1}$ $(j = 1, \ldots, m-1), B_m \neq 0$ (k is a scalar). Then $n = m$ and $B_j = A_j$ $(j = 1, 2, \ldots, n)$.

Proof. According to Theorem 2, the exponents A_1 and B_1 are colinear $A_1 = lB_1$. If $l \neq 1$, (for example if $l > 1$), then from the representation

$$e^{izB_1}\mathfrak{A}(z) = e^{-iz(l-1)B_1}e^{-izA_2} \cdot \ldots \cdot e^{-izA_n} = e^{-izB_2} \cdot \ldots \cdot e^{-izB_m}$$

it follows that the matrices B_1 and B_2 are colinear. This contradicts the assumptions of the theorem. Hence, $A_1 = B_1$. Now considering the matrix function $e^{izA_1}\mathfrak{A}(z)$ we infer that $A_2 = B_2$. Continuing this way step by step we conclude that $A_j = B_j$ and $m = n$.
The theorem is proved. □

Theorem 5. *Suppose that the matrix function $\mathfrak{A}(z) \in \mathfrak{M}$ admits the multiplicative representation*

$$\mathfrak{A}(z) = e^{-izA_1}e^{-izA_2} \cdots e^{-izA_n}\mathfrak{B}(z)\ , \tag{22}$$

where $\mathfrak{B}(z) \in \mathfrak{M}$ and the exponents A_j $(j = 1, 2, \ldots, n)$ satisfy the assumptions of Theorem 4. Then the Weyl circle $\mathcal{W}_z(\mathfrak{A})$ of the matrix $\mathfrak{A}(z)$ is embedded in the Weyl circle $\mathcal{W}_z(e^{-izA_1})$ and, as $\Im z \to +\infty$, it shrinks to a point ξ_1 $(\Im\xi_1 \geq 0)$. Moreover, the matrix A_1 can be recovered from the point ξ_1 up to a scalar factor $k > 0$:

(1) *If $\xi_1 = \infty$, then*

$$A_1 = kE_0 = k\begin{bmatrix} 1 & 0 \\ 0 & 0 \end{bmatrix} J \qquad (E_0^2 = 0). \tag{23}$$

(2) *If* $\xi_1 = \overline{\xi_1} < \infty$, *then*

$$A_1 = k\tilde{E} = k\frac{1}{\xi_1^2}\begin{bmatrix} \xi_1^2 & \xi_1 \\ \xi_1 & 1 \end{bmatrix} J \qquad (\tilde{E}^2 = 0). \tag{24}$$

(3) *If* $\Im\xi_1 > 0$, *then*

$$A_1 = k\tilde{H} = k\frac{1}{\Im\xi_1}\begin{bmatrix} \xi_1\overline{\xi_1} & \Re\xi_1 \\ \Re\xi_1 & 1 \end{bmatrix} J \qquad (\tilde{H}^2 = I). \tag{25}$$

6. Suppose that the matrix-function $\mathfrak{A}(z)$ has the form (22). If V is a J-unitary matrix, i.e., if $V^*JV = J$, then the matrix $\tilde{\mathfrak{A}}(z) = V^{-1}\mathfrak{A}(z)V$ has the form

$$\tilde{\mathfrak{A}}(z) = e^{-iz\tilde{A}_1} \cdot e^{-iz\tilde{A}_2} \ \cdots \ e^{-iz\tilde{A}_n} \cdot \tilde{\mathfrak{B}}(z) \tag{26}$$

(where $\tilde{A}_j = V^{-1}A_jV$ for $j = 1, 2, \ldots, n$ and $\tilde{\mathfrak{B}}(z) = V^{-1}\mathfrak{B}(z)V \in \mathfrak{M}$). Let us choose the matrix V in such a way that the first factor $e^{-iz\tilde{A}_1}$ has one of the two standard forms:

$$\begin{array}{rcll} V^{-1}e^{-izA_1}V & = & e^{-iz\kappa E_0}, & \text{if } \ \mathrm{A}_1^2 = 0, \\ V^{-1}e^{-izA_1}V & = & e^{-iz\kappa J}, & \text{if } \ \mathrm{A}_1^2 \neq 0. \end{array} \tag{27}$$

Such a matrix V could be constructed from the limit point ξ of the circle $\mathcal{W}_z(\mathfrak{A})$: if $\xi_1 = \overline{\xi_1} \neq 0$, then

$$V = \begin{bmatrix} 1 & 0 \\ \frac{1}{\xi_1} & 1 \end{bmatrix};$$

if $\xi = 0$, then

$$V = \begin{bmatrix} 0 & -1 \\ 1 & 0 \end{bmatrix};$$

if $\Im\xi_1 > 0$, then

$$V = \frac{1}{\sqrt{\Im\xi_1}}\begin{bmatrix} \Im\xi_1 & \Re\xi_1 \\ 0 & 1 \end{bmatrix}.$$

Let us study the properties of the standard factors $e^{-iz\kappa E_0}$ and $e^{-iz\kappa J}$. For this purpose we introduce the notion of a **functional Weyl circle**.

Let

$$\mathfrak{A}(z) = \begin{bmatrix} a(z) & b(z) \\ c(z) & d(z) \end{bmatrix} \tag{28}$$

belong to $\mathfrak{M}$ and consider the fractional linear transformation with coefficient matrix $\mathfrak{A}(z)$:

$$w(z) = (a(z)\omega(z) + b(z))\,(c(z)\omega(z) + d(z))^{-1}\,, \tag{29}$$

where $\omega(z)$ is a function from the Nevanlinna class $\mathcal{N}$. (Let us recall that an analytic function ω in the upper half plane *belongs to the Nevanlinna class* if

$\Im\omega(z) \geq 0$ for $\Im z > 0$.) In view of the J-contractivity of the matrix-function $\mathfrak{A}(z)$, the function in (29) belongs to the Nevanlinna class.

The set of functions $w(z) \in \mathcal{N}$ which are representable in the form (29) with $\omega \in \mathcal{N}$ is said to be ***the functional Weyl circle of the matrix function*** $\mathfrak{A}(z)$.

Let us denote the functional Weyl circle of the matrix-function $\mathfrak{A}(z)$ by $\mathcal{W}\{\mathfrak{A}(z)\}$. Clearly, the functional Weyl circle $\mathcal{W}\{\mathfrak{A}(z)\}$ consists exactly of those functions $w \in \mathcal{N}$ whose values at every point z $(\Im z > 0)$, belong to the appropriate Weyl circle $\mathcal{W}_z(\mathfrak{A})$. It is not difficult to verify that a functional Weyl circle $\mathcal{W}\{\mathfrak{A}(z)\}$ is determined by the inequality

$$[\overline{w(z)}, 1](\mathfrak{A}^{-1})^*(z)J\mathfrak{A}^{-1}(z)\begin{bmatrix} w(z) \\ 1 \end{bmatrix} \geq 0 \quad (\Im z > 0)\,. \tag{30}$$

The criterion for the divisibility of matrix-functions can be translated into the language of functional Weyl circles:

Theorem 6. *Let $\mathfrak{A}(z)$ and $\mathfrak{B}(z) \in \mathfrak{M}$. Then the matrix-function $\mathfrak{B}(z)$ is a left divisor of the matrix-function $\mathfrak{A}(z)$ if and only if*

$$\mathcal{W}\{\mathfrak{A}(z)\} \subseteq \mathcal{W}\{\mathfrak{B}(z)\}.$$

Let us note that if two functions $w_1(z), w_2(z)$ from the Nevanlinna class belong to the same functional Weyl circle (for some $\mathfrak{A}(z) \in \mathfrak{M}$) then the inequality

$$|w_1(z) - w_2(z)| \leq 2\rho_z \tag{31}$$

holds, where ρ_z is the radius of the Weyl circle determined by (10) (if $R(z) \neq 0$). Calculating the entry $R(z)$ of the Weyl matrix $(\mathfrak{A}^{-1})^*(z)J\mathfrak{A}^{-1}(z)$, we obtain

$$\rho_z = \frac{1}{R(z)} = \frac{i}{d(z)\overline{c(z)} - c(z)\overline{d(z)}} = \frac{1}{|c(z)|^2\, 2\Im(\frac{d(z)}{c(z)})}. \tag{32}$$

Now we formulate four lemmas without proofs. (We omit the proofs for reasons of space). These proofs are based on the Schwarz-Pick inequality (5) for a matrix $\mathfrak{A}(z) \in \mathfrak{M}$. (They will be presented in another paper.)

Lemma 6. *Let $\mathfrak{A}(z) \in \mathfrak{M}$ and suppose that the entry $R(z)$ of the Weyl matrix*

$$(\mathfrak{A}^{-1})^*(z)J\mathfrak{A}^{-1}(z) = \begin{bmatrix} -R(z) & S(z) \\ S^*(z) & -T(z) \end{bmatrix} \tag{33}$$

vanishes at least at one point $z = z_0$, $\Im z_0 > 0$ $(R(z_0) = 0)$. Then R vanishes identically $(R \equiv 0)$ and

$$\mathfrak{A}(z) = e^{-iz\kappa E_0} \cdot \mathfrak{A}(0)$$

for some $\kappa \in [0, \infty)$.

Lemma 7. *Let $\mathfrak{A}(z) \in \mathfrak{M}$ and suppose that the entry R from (33) is bounded on the imaginary half axis:*

$$R(iy) \leq C < \infty.$$

Then $R(z) \equiv 0$ and $\mathfrak{A}(z) = e^{-iz\kappa E_0} \cdot \mathfrak{A}(0)$ for some $\kappa \in [0, \infty)$.

Lemma 8. *Let $\mathfrak{A}(z), \mathfrak{B}(z) \in \mathfrak{M}$ and suppose that $\mathfrak{A}(z)$ is either a left or right divisor of $\mathfrak{B}(z)$ in the class $\mathfrak{M}$. Then the estimate*

$$\|\mathfrak{A}(z)\| \leq C \cdot |z| \cdot \|\mathfrak{B}(z)\|$$

holds in the half plane $\Im z \geq 1$, where $C < \infty$ is a constant.

Lemma 9. *Let $\mathfrak{A}(z) \in \mathfrak{M}$ be a function of exponential type L. Then the following estimate holds true:*

$$R(iy) \geq C(\epsilon) \cdot e^{2y(L-\epsilon)} \qquad (y > 1,\ C(\epsilon) < \infty\ \forall \epsilon > 0).$$

7. Now we will study the properties of the functional Weyl circle of an elementary factor $b_0(z) = e^{-iz\kappa E_0}$.

Theorem 7. *The functional Weyl circle of the elementary factor $b_0(z) = e^{-iz\kappa E_0}$ consists exactly of those functions*

$$w(z) = pz + q + \int_{-\infty}^{\infty} \left\{ \frac{1}{t-z} - \frac{t}{1+t^2} \right\} d\sigma(t) \quad \left(\int_{-\infty}^{\infty} \frac{d\sigma(t)}{1+t^2} < \infty \right) \tag{34}$$

from the Nevanlinna class for which the inequality

$$p \geq \kappa \tag{35}$$

holds.

Proof. In view of the identity

$$b_0(z) = e^{-iz\kappa E_0} = I - iz\kappa E_0 = \begin{bmatrix} 1 & \kappa z \\ 0 & 1 \end{bmatrix},$$

the functional Weyl circle $\mathcal{W}\{b_0(z)\}$ consists of those functions $w(z)$ which are representable in the form

$$w(z) = \omega(z) + \kappa z \qquad (w \in \mathcal{N}). \tag{36}$$

Clearly, the representation (36) is equivalent to (34) and (35).
The theorem is proved. □

We remark that for a Nevanlinna function $w(z)$, the conditions (34) and (35) are equivalent to the condition

$$\lim_{y\to\infty} \frac{w(iy)}{iy} \geq \kappa. \tag{37}$$

Lemma 10. *Let $\mathfrak{A}(z) \in \mathfrak{M}$, and let $w_0(z)$ be a function from the functional Weyl circle $\mathcal{W}\{\mathfrak{A}(z)\}$ which satisfies the condition*

$$\lim_{y\to\infty} \frac{w_0(iy)}{iy} = \kappa. \tag{38}$$

Then either $\mathfrak{A}(z) = e^{-izlE_0}$ for some $l \leq \kappa$, or the condition (38) holds for every function $w(z) \in \mathcal{W}\{\mathfrak{A}(z)\}$.

Proof. If $\mathfrak{A}(z) = e^{-izlE_0}$, then, in view of Theorem 7, (38) holds only if $l \leq \kappa$. Let $\mathfrak{A}(z) \neq e^{-izlE_0}$. Then from (31) and (32) we obtain the inequality

$$|w(z) - w_0(z)| \leq 2\rho_z = \frac{2}{R(z)}$$

which holds for every function $\omega(z)$ from $\mathcal{W}\{\mathfrak{A}(z)\}$. In view of Lemma 7, there exist a sequence $(y_j)_{j\in\mathrm{N}}$ with $\lim\limits_{j\to\infty} y_j = +\infty$ such that $\lim\limits_{j\to\infty} \rho_{iy_j} = 0$. Hence, the function $\omega(z)$ satisfies the inequality (38) together with the function $\omega_0(z)$.
The lemma is proved. □

Thus, if a matrix-function $\mathfrak{A}(z)$ satisfies the assumptions of Lemma 10, then its functional Weyl circle is embedded in the Weyl circle of the matrix-function $b_0(z) = e^{-izlE_0}$. Translating this fact into the language of divisibility in the class $\mathfrak{M}$ we obtain the following statement:

Theorem 8. *Let $\mathfrak{A}(z) \in \mathfrak{M}$. Assume that there exists a parameter $\omega(z) \in \mathcal{N}$ such that the image of the fractional linear transformation (29) satisfies the condition (38). Then either $\mathfrak{A}(z) = e^{-izlE_0}$ for some $l \leq \kappa$ or the left quotient $b_0^{-1}\mathfrak{A}$ belongs to the class $\mathfrak{M}$.*

Theorem 7 and formulas (23), (24) yield a criterion for splitting off a factor of the form e^{-izE_0} $(E_0 J \geq 0, E_0^2 = 0)$ from the matrix-function $\mathfrak{A}(z)$.

Theorem 9. (i) *A matrix-function $\mathfrak{A}(z) \in \mathfrak{M}$ admits a representation of the form*

$$\mathfrak{A}(z) = e^{-iz\kappa E_0} \cdot \mathfrak{B}(z) \quad \textit{with} \quad \kappa = \overline{\kappa} > 0, E_0 = \begin{bmatrix} 0 & i \\ 0 & 0 \end{bmatrix} \quad \textit{and} \quad \mathfrak{B}(z) \in \mathfrak{M},$$

if and only if

$$\lim_{y\to\infty} \frac{a(iy)}{iy\ c(iy)} \geq \kappa.$$

(ii) *A matrix-function $\mathfrak{A}(z) \in \mathfrak{M}$ admits a representation of the form*

$$\mathfrak{A}(z) = e^{-izE} \cdot \mathfrak{B}(z) \ \ with \ \ E = \begin{bmatrix} \alpha\overline{\alpha} & \alpha\overline{\beta} \\ \overline{\alpha}\beta & \beta\overline{\beta} \end{bmatrix} J, \ \alpha\overline{\beta} = \overline{\alpha}\beta, \ \beta \neq 0 \ \ and \ \ \mathfrak{B}(z) \in \mathfrak{M},$$

if and only if the following two conditions are both satisfied:

(a)
$$\lim_{y\to\infty} \frac{a(iy)}{c(iy)} = \xi_1 = \overline{\xi_1} \qquad (\xi_1 = \frac{\alpha}{\beta})$$

(b) *The matrix-function*

$$\tilde{\mathfrak{A}}(z) = V^{-1}\mathfrak{A}(z)V = \begin{bmatrix} \tilde{a}(z) & \tilde{b}(z) \\ \tilde{c}(z) & \tilde{d}(z) \end{bmatrix}, \tag{39}$$

which is defined by formula (26), *satisfies the condition*

$$\lim_{y\to\infty} \frac{\tilde{a}(iy)}{iy\ \tilde{c}(iy)} \geq \kappa \qquad (\kappa = \alpha\overline{\alpha}). \tag{40}$$

8. Previously we studied the characteristic properties of functions $w(z)$ from the Weyl circle of the elementary factor $b_0 = e^{-iz\kappa E_0}$. Next we investigate the properties of functions $\omega(z)$ from the Weyl circle of the factor $B_0(z) = e^{-izlJ}$.

First we introduce the notion of a function associated with a function $\omega(z)$.

Let $w(z) \in \mathcal{N}$, and let the number p in the representation (34) be equal to zero, i.e.,

$$w(z) = q + \int_{-\infty}^{\infty} \left\{ \frac{1}{t-z} - \frac{t}{1+t^2} \right\} d\sigma(t) \quad \left(\int_{-\infty}^{\infty} \frac{d\sigma(t)}{1+t^2} < \infty, \ q = \overline{q} \right). \tag{41}$$

The function $g : \mathbb{R} \to \mathbb{C}$ which is constructed from the measure σ and the number q by the formula

$$g(x) = g_\omega(x) = -iqx + \int_{-\infty}^{\infty} \left\{ 1 - e^{-itx} - \frac{itx}{1+t^2} \right\} \frac{d\sigma(t)}{t^2}, \tag{42}$$

is said to be a function associated with $w(z)$. Functions which are representable in the form (42) are said to belong to the class $\mathcal{G}$. Clearly, g is a continuous function and $g(-x) = \overline{g(x)}$.

We remark that a function $w(z) \in \mathcal{N}$ is of the form (41) if and only if

$$\lim_{\Im z\to\infty} \frac{w(z)}{z} = 0. \tag{43}$$

Lemma 11. *Let $\mathfrak{A}(z) \in \mathfrak{M}$ be a matrix-function of exponential type L. Suppose that for some function $w_0(z) \in \mathcal{W}\{\mathfrak{A}(z)\}$ the condition (43) holds true. Then for every choice of $w_1(z)$ and $w_2(z)$ in $\mathcal{W}\{\mathfrak{A}(z)\}$ the condition*

$$w_1(iy) - w_2(iy) = O\left(e^{-2y(L-\delta)}\right) \qquad (y \to \infty,\ \forall \delta > 0) \tag{44}$$

is fulfilled and the corresponding associated functions g_1 and g_2 agree on the interval $(-2L, 2L)$:

$$g_1(x) = g_2(x) \qquad (\forall x \in (-2L, 2L))\ . \tag{45}$$

Proof. The estimate (44) follows from (31), (32) and Lemma 9. The equivalence of (44) and (45) follows from the fact that for an arbitrary function $w(z)$ of the form (41) and its associated function $g_w(z)$ in (42) and for an arbitrary function $s(x)$ the equality

$$g_w(x) = s(x) \qquad (\forall x \in (-2L, 2L)) \tag{46}$$

holds true if and only if for each $\delta > 0$ the function $\omega(z)$ and $s(z)$ are related by the condition

$$-i\omega(iy)\left(e^{-2yL} - 1\right)y + \int_0^{2L} ye^{uy}s(u)du = O\left(e^{-2y(L-\delta)}\right) \quad (y \to \infty). \tag{47}$$

(A proof can be found in [7]). □

Now we are ready to characterize the Weyl circle of the factor

$$B_0(z) = e^{-izlJ} = \begin{bmatrix} \cos lz & \sin lz \\ -\sin lz & \cos lz \end{bmatrix} = \begin{bmatrix} a(z) & b(z) \\ c(z) & d(z) \end{bmatrix}. \tag{48}$$

Theorem 10. *The Weyl circle of the matrix-function $B_0(z)$ in (48) consists exactly of those matrix-functions $w(z) \in \mathcal{N}$ which satisfy the asymptotic condition*

$$w(iy) - i = O\left(e^{-2y(l-\delta)}\right) \qquad (y \to \infty,\ \forall \delta > 0) \tag{49}$$

or the equivalent condition

$$g_w(x) = |x| \qquad (\forall x \in (-2l, 2l). \tag{50}$$

Proof. Upon setting $\omega(z) \equiv i$ it follows from the formula

$$w(z) = (a(z)i + b(z))(c(z)i + d(z))^{-1} = (i \cos lz + \sin lz)(-i \sin lz + \cos lz)^{-1} = i,$$

and the definition of functional Weyl circles that the function $w(z) \equiv i$ belongs to $\mathcal{W}\{B_0(z)\}$. Now the asymptotic conditions follow from Lemma 11. Since the function $w(z) = i$ belongs to the Nevanlinna class it admits a representation of the form (41), in fact we may choose $d\sigma(t) = dt$ and $q = 0$. Calculating, we obtain $g_i(x) = |x|$ (where g_i is the function associated with the function $w(z) = i$). Again using Lemma 11 we obtain (50).

Now we will show that the Weyl circle $\mathcal{W}\{B_0(z)\}$ contains all those functions $\omega(z) \in \mathcal{N}$ which satisfy the condition (50). For that purpose we refer to a remarkable relation from the general theory of interpolation problems. Namely, we refer to the Fundamental Matrix Inequality (FMI) for the class $\mathcal{G}$ (see [8], p. 38 of the Russian original or pp. 55–56 of the English translation). Suppose that a function $\tilde{w}(z)$ satisfies the condition (49). According to a theorem in [8] (p. 40) this function satisfies the FMI($\mathcal{G}_{2L}$) together with the associated function $s(x) = g(x) = |x|$ $(-l \leq x \leq 2l)$. By choosing the parameter $\varphi(x)$ in the FMI($\mathcal{G}$) in a special way, it follows that the function $\tilde{w}(z)$ satisfies the inequality

$$\begin{bmatrix} \dfrac{e^{4yl} - 1}{4y} & \dfrac{(\tilde{w}(z) - i)(e^{4yl} - 1)}{4y} \\[2ex] \dfrac{(\overline{\tilde{w}(z)} + i)(e^{4yl} - 1)}{4y} & \dfrac{\tilde{w}(z) - \overline{\tilde{\omega}(z)}}{2iy} \end{bmatrix} \geq 0 \quad (\forall z : \Im z > 0).$$

After a transformation the last inequality can be reexpressed in the form

$$|\tilde{w}(z) - i\mathrm{cth}(2yl)| \leq \frac{1}{\mathrm{sh}(2yl)}. \tag{51}$$

Now we remark that the inequality (51) is equivalent to the Weyl inequality (30) with $\mathfrak{A}(z) = B_0(z)$. Since the function $\tilde{w}(z)$ satisfies (49) it now follows that $\tilde{w}(z)$ belongs to the Weyl circle $\mathcal{W}\{B_0(z)\}$. The theorem is proved. □

Since the function $g(x) = |x|$ is associated with $\omega(z) = i$, we obtain the following results.

Lemma 12. *Let $\mathfrak{A}(z) \in \mathfrak{M}$ be a matrix-function of exponential type L $(L \geq l)$ and suppose that some function $w_0(z) \in \mathcal{W}\{\mathfrak{A}(z)\}$ satisfies the asymptotic relation*

$$w(iy) - i = O\left(e^{-2y(l-\delta)}\right) \qquad (y \to \infty, \forall \delta > 0). \tag{52}$$

Then every function $w(z) \in \mathcal{W}\{\mathfrak{A}(z)\}$ satisfies the condition (52) *and the corresponding associated function g_w satisfies the condition* (50).

Proof. The asymptotic relation (52) follows from the estimate (44) which is a consequence of Lemma 11. The equivalence of (52) and (50) can be proved in the same way that we proved the equivalence of (46) and (47) in Lemma 11. The lemma is proved. □

Thus, if a matrix- function $\mathfrak{A}(z)$ satisfies the assumptions of Lemma 12, then its functional Weyl circle is embedded in the Weyl circle of the matrix $B_0(z)$. Let us translate this fact into the language of divisibility in the class $\mathfrak{M}$.

Theorem 11. *Let $\mathfrak{A}(z) \in \mathfrak{M}$ be a matrix-function of exponential type L $(L \geq l)$. Assume that there exists a function $w_0(z) \in \mathcal{W}\{\mathfrak{A}(z)\}$ for which* (52) *is fulfilled* (*or equivalently that the associated function $g_{w_0}(x)$ satisfies* (50)). *Then the matrix-function $\mathfrak{A}(z)$ has the form*

$$\mathfrak{A}(z) = e^{-izlJ} \cdot \mathfrak{B}(z) \quad \textit{where} \quad \mathfrak{B}(z) \in \mathfrak{M}. \tag{53}$$

From Theorem 11 and formula (25) we obtain the following criterion for being able to split off a factor e^{-izH} $(HJ \geq 0, \mathrm{spH} = 0)$ from a matrix-function $\mathfrak{A}(z) \in \mathfrak{M}$.

Theorem 12. (i) *In order for a matrix-function $\mathfrak{A}(z) \in \mathfrak{M}$ of exponential type L $(L \geq l)$ to admit a representation of the form (53) it is necessary and sufficient that the quotient $w_0(z) = \dfrac{a(z)}{c(z)}$ of its entries in formula (2.8) satisfy the condition*

$$w_0(iy) - i = O\left(e^{-2y(l-\delta)}\right) \qquad (y \to \infty, \forall \delta > 0).$$

(ii) *In order for a matrix-function $\mathfrak{A}(z) \in \mathfrak{M}$ to admit a representation of the form*

$$\mathfrak{A}(z) = e^{-izH} \cdot \mathfrak{B}(z)\ , \qquad \mathfrak{B}(z) \in \mathfrak{M},$$

where

$$H = \begin{bmatrix} \alpha\overline{\alpha} & \dfrac{\overline{\alpha}\beta + \alpha\overline{\beta}}{2} \\ \dfrac{\alpha\overline{\beta} + \overline{\alpha}\beta}{2} & \beta\overline{\beta} \end{bmatrix} J, \quad \frac{\alpha\overline{\beta} - \overline{\alpha}\beta}{2i} = \kappa > 0$$

($\kappa = \sqrt{-\det H} \leq l$), *it is sufficient that both the following two necessary conditions are satisfied:*

(a) $$\lim_{y\to\infty} \frac{a(iy)}{c(iy)} = \xi_1 \quad \textit{where} \quad \Im\xi_1 > 0 \quad \textit{and} \quad \xi_1 = \frac{\alpha}{\beta}\ .$$

(b) *For each fixed $\delta > 0$ the entries of the matrix-function* (39) (*where V is chosen from* (27)) *satisfy the condition*

$$\frac{\tilde{a}(iy)}{\tilde{c}(iy)} - i = O\left(e^{-2y(1-\delta)}\right) \quad \text{as} \quad y \to \infty\ . \tag{54}$$

9. The main theorem now follows from the statements proved in Sections 7 and 8 and a general uniqueness theorem for matrix-functions which admit a representation of the form (21).

Main Theorem. *Assume that a matrix-function $\mathfrak{A}(z) \in \mathfrak{M}$ admits a multiplicative decomposition of the form* (21) *and another representation in the form of a multiplicative integral*

$$\mathfrak{A}(z) = \overset{\curvearrowright}{\int_0^T} e^{iz\Phi(\tau)d\tau}, \tag{55}$$

where $\Phi(\tau)J \geq 0, \mathrm{sp}\Phi(\tau) = 0$ and $\Phi(\tau) \in \mathcal{L}^1(0,T)$. Then there exists a partition

$$0 = \tau_0 < \tau_1 < \tau_2 < \ldots < \tau_n = T$$

of the interval $[0,T]$ such that $\Phi(\tau) = \varphi(\tau)A_j$ for $\tau \in (\tau_{j-1}, \tau_j)$, where $\varphi(\tau) \geq 0$ for $\tau \in (\tau_{j-1}, \tau_j)$ and $\int_{\tau j-1}^{\tau_j} \varphi(\tau)d\tau = 1$. Hence,

$$\overset{\curvearrowright}{\int_{\tau_{j-1}}^{\tau_j}} e^{-iz\Phi(\tau)d\tau} = e^{-izA_j} \quad (j = 1, 2, \ldots, n).$$

Proof. It remains to prove that there exists a number τ_1 with $0 < \tau_1 \leq T$ such that the equality

$$\overset{\curvearrowright}{\int_0^{\tau_1}} e^{-iz\Phi(\tau)d\tau} = e^{-izA_1} \tag{56}$$

holds. Suppose that the matrix A_1 from the decomposition (21) coincides with one of the two matrices: $A_1 = \kappa E_0$ or $A_1 = lJ$ ($\kappa, l \in (0, \infty)$). We will show that in either of these two cases (56) takes place. Since for each τ with $0 < \tau < T$ the matrix-function

$$\mathfrak{A}(\tau, z) := \overset{\curvearrowright}{\int_0^{\tau}} e^{-iz\Phi(u)du}$$

is a divisor of the matrix-function $\mathfrak{A}(z)$ it follows that

$$\mathcal{W}\{\mathfrak{A}(z)\} \subseteq \mathcal{W}\{\mathfrak{A}(z, \tau)\}. \tag{57}$$

On the other hand from the representation (21) it follows that

$$\mathcal{W}\{\mathfrak{A}(z)\} \subseteq \mathcal{W}\{e^{-izA_1}\}. \tag{58}$$

First let us consider the case $A_1 = \kappa E_0$. From the multiplicative decomposition (21) it follows (for $n = 1$ directly and for $n \geq 2$ by applying (24) and (25) to the factor e^{-izA_2}) that the functional Weyl circle $\mathcal{W}\{\mathfrak{A}(z)\}$ contains a function $\omega(z)$ which satisfies the condition (38). In view of (57), the functional Weyl circle $\mathcal{W}\{\mathfrak{A}(z,\tau)\}$ contains also this function $\omega(z)$. According to Lemma 8 we obtain either

$$\mathfrak{A}(z,\tau) = e^{-izlE_0} \qquad (l \leq \kappa)$$

or

$$\mathfrak{A}(z,\tau) = e^{-i\kappa E_0} \cdot \mathfrak{B}(z) \qquad (\mathfrak{B}(z) \in \mathfrak{M}).$$

Suppose that the two points τ' and τ'' satisfy $\tau' < \tau''$. Then the matrix-function $\mathfrak{A}(z,\tau')$ is a divisor of the matrix-function $\mathfrak{A}(z,\tau'')$. Since the matrix-function $e^{-iz\kappa E_0} \cdot \mathfrak{B}(z)$ can not be a divisor of the matrix-function e^{-izlE_0} with $l < \kappa$ it follows that the interval $(0,T)$ must split into two subintervals $(0,\tau_1)$ and (τ_1,T) such that

$$\mathfrak{A}(z,\tau) = e^{-izl(\tau)E_0}$$

where $l(\tau) < \kappa$ for $\tau \in (0,\tau_1)$ and

$$\mathfrak{A}(z,\tau) = e^{-iz\kappa E_0} \cdot \mathfrak{B}(z,\tau),$$

with $\mathfrak{B}(z,\tau) \in \mathfrak{M}$ for $\tau \in [\tau_1,T)$. Since the matrix-function $\mathfrak{A}(z,\tau)$ is continuous with respect to τ and satisfies the condition $\mathfrak{A}(z,0) = I$, the number τ_1 is strictly positive and we obtain the formula

$$\mathfrak{A}(z,\tau_1) = e^{-iz\kappa E_0} = e^{-izA_1}.$$

Hence,

$$\mathfrak{A}(z,\tau) = e^{-iz\lambda(\tau)A_1},$$

where $\lambda(\tau) \leq 1$ for $\tau \in (0,\tau_1)$.
Now let us consider the case $A_1 = \kappa J$. Let $\varphi(\tau)$ denote the exponential type of the matrix-function $\mathfrak{A}(z,\tau)$. The function $\varphi(\tau)$ is in general nonstrictly increasing. In view of the inequality

$$\|\mathfrak{A}(z,\tau'')\mathfrak{A}(z,\tau')^{-1}\| \leq \exp\left(|z| \int_{\tau'}^{\tau''} \|\Phi(\tau)\| d\tau\right),$$

the function $\varphi(\tau)$ is continuous. Hence, there exists a number τ_1 such that $\varphi(\tau_1) = \kappa$ and $\varphi(\tau) < \kappa$ for $\tau < \tau_1$. Let $\tau < \tau_1$. Then the representation

$$\mathfrak{A}(z,\tau) = e^{-iz\varphi(\tau)J} M(z,\tau),$$

follows from (57), (58) and Theorem 11, where $M(z,\tau) \in \mathfrak{M}$ is a matrix-function of exponential type zero. The estimates

$$\|\mathfrak{A}(z)\| \leq C|z|^n \qquad (z = x + i, -\infty < x < \infty)$$

and

$$\|M(z)\| \leq C|z|^{n+1} \qquad (z = x + i, -\infty < x < \infty)$$

follow from (21) and Lemma 8, respectively. Taking into account the Phragmén-Lindelöf principle we conclude that $M(z)$ is a polynomial. Hence, $M(z)$ admits a representation of the form

$$M(z) = e^{-izE_1} \cdot e^{-izE_2} \cdot \ldots \cdot e^{-izE_m},$$

where $E_j J \geq 0$ and $E_j^2 = 0$. Hence,

$$\mathfrak{A}(z) = e^{-iz\varphi(\tau)J} \cdot e^{-izE_1} \cdot e^{-izE_2} \cdot \ldots \cdot e^{-izE_m} \cdot \mathfrak{B}(z),$$

where $\mathfrak{B}(z) \in \mathfrak{M}$ and $\varphi(\tau) < \kappa$. Combining the last representation with (21) we see that

$$e^{-izE_1} \cdot e^{-izE_2} \cdot \ldots \cdot e^{-izE_m} \cdot \mathfrak{B}(z) = e^{-iz(\kappa - \varphi(\tau))J} \cdot e^{-izA_2} \cdot \ldots \cdot e^{-izA_n},$$

where $(\kappa - \varphi(\tau)) > 0$. From Theorem 3 it follows that $m = 0$. Hence, $M(z) = I$ and

$$\mathfrak{A}(z, \tau) = e^{-iz\varphi(\tau)J} \qquad (\tau < \tau_1).$$

Thus, $\mathfrak{A}(z, \tau_1) = e^{-izA_1}$. The theorem is proved. □

In conclusion, we remark that Theorems 9, 12 and formulas (23) - (25) contain the algorithm for recovering the factor e^{-izA_1} from the representation (21). Then it is possible to recover the factor e^{-izA_2} from the matrix-function $\mathfrak{A}_1(z) = e^{izA_1} \cdot \mathfrak{A}(z)$ and so on. Thus, it is possible to recover the multiplicative decomposition (21) (i.e., all of the factors e^{-izA_j} $(j = 1, 2, \ldots, n)$) constructively from the matrix-function $\mathfrak{A}(z)$.

References

[1] Potapov, V.P., *The multiplicative structure of J-contractive matrix-functions*, (in Russian), Trudy Mosk. Matem. Obshch. **4** (1955), 125–236. English transl.: Amer. Math. Soc. Transl. (ser. **2**) **15** (1960), 131–243.

[2] de Branges, L., *Hilbert Spaces of Entire Functions*, Prentice Hall, Englewood Cliffs, New York, 1968.

[3] Kovalishina, I.V. and V.P. Potapov, *An indefinite metric in the Nevanlinna-Pick problem* (in Russian), Akad Nauk Armyan. SSR Doklady **59**:1 (1974), 17–22. English. transl.: Amer. Math. Soc. Transl. (ser. **2**) **138** (1988), 37–54.

[4] Potapov, V.P., *Linear-fractional transformations of matrices* (in Russian). In: Issledovanija po teorii operatorov i ih priloženijam, "Naukova Dumka" publishing house, Kiev 1979 (Marchenko, V.A.– ed.). English translation in: Amer. Math. Soc. Transl. (ser.**2**) **138** (1988) (*Seven papers translated from the Russian*), 21–36.

[5] Efimov, A.V. and V.P. Potapov, *J-expansive matrix-functions and their role in the theory of electrical circuits* (in Russian), Uspekhi Matem. Nauk **28**:1 (1973), 65–130. English transl.: Russian Math. Surveys **28**:1 (1973), 69–140.

[6] Krein, M.G., *On the logarithm of an infinitely decomposable Hermitian-positive function* (in Russian), Doklady Akad. Nauk SSSR **45** (1944), 99–102.

[7] Katsnelson, V.E., *Continuous analogues of the Hamburger-Nevanlinna theorem and fundamental matrix inequalities of classical problems. II.* (in Russian), Teorija finkciĭ, funkcional'niĭ analiz i ih priloženija **37** (1982) ("Kharkov university" publishing house, Marchenko, V.A. – ed), 31–48. English translation in: Amer. Math. Soc. Transl. (ser.**2**) **136** (*Fourteen Papers Translated from the Russian*), 67–83.

[8] Katsnelson, V.E., *Continuous analogues of the Hamburger-Nevanlinna theorem and fundamental matrix inequalities of classical problems. I.* (in Russian), Teorija finkciĭ, funkcional'niĭ analiz i ih priloženija **36** (1981) ("Kharkov university" publishing house, Marchenko, V.A. – ed), 31–48. English translation in: Amer. Math. Soc. Transl. (ser.**2**) **136** (*Fourteen Papers Translated from the Russian*), 49–65.

Irina Mikhailova
Meierstraße 1 A
Steinbergkirche
24972
Bundesrepublik Deutschland

AMS Subject Classification: 30D50, 46E10.

Operator Theory
Advances and Applications, Vol. 95
© 1997 Birkhäuser Verlag Basel/Switzerland

On a criterion of positive definiteness[1]

I.V. Mikhailova and V.P. Potapov

Abstract. Let $s(x)$ be a continuous function on $(-a, a)$ such that (i) $s(x) = \overline{s(-x)}$ for $-a < x < a$; (ii) $s(x)$ is twice continuously differentiable on $[0, a)$; (iii) $s'(+0) + \overline{s'(+0)} = -\lambda < 0$. M.G. Krein has shown that if $s(0) > 0$, then there exists a neighborhood of 0 such that $s(x)$ is positive definite in this neighborhood and admits more than one positive definite continuation from this neighborhood onto the whole real axis. The aim of the paper is the computation of the largest such neighborhood. [Abstract added by editors.]

0 Introduction

A function $s(x)$ defined on an interval $[-\xi, \xi]$ (where $0 < \xi < \infty$) is said to be *positive definite on this interval* if the kernel $s(x-y)_{0\leq x,y\leq \xi}$ is positive definite, i.e., if for every choice of $x_j \in [0, \xi]$, $(j = 1, 2, \dots, n$, where $n \in \mathbb{N}$ is an arbitrary natural number) the hermitian form based on the matrix $\|s(x_j - x_k)\|_{1\leq j,k\leq n}$ is nonnegative:

$$\sum_{1\leq j,k\leq n} s(x_j - x_k) f_j \overline{f_k} \geq 0 \qquad \forall\ (f_1, \dots f_n) \in \mathbb{C}^n .$$

If the function $s(x)$ is continuous on $[-\xi, \xi]$, the last condition is equivalent to the condition

$$\int_0^\xi \int_0^\xi s(x-y) f(x) \overline{f(y)}\, dx\, dy \geq 0 \qquad \forall\ f(x) \in C[0, \xi] .$$

Let us consider a function $s(x)$ which is continuous in $(-a, a)$ and has the following properties [2]:

(i) $s(x) = \overline{s(-x)}, \qquad -a < x < a;$

(ii) $s(x)$ *is twice continuously differentiable in* $[0, a)$;

(iii) $s'(+0) + \overline{s'(+0)} = -\lambda < 0.$

---

1) Translated from: Teorija funkciĭ funkcional'niĭ analiz i ih priloženija (The Kharkov University Publishing House, V.A. Marchenko, ed.) **36** (1981) 65–89. **MR** 84a:47051

2) The condition (ii) can be essentially weakened.

A positive definite (p.d.) function $s(x)$ defined on a closed interval $[-\xi,\xi]$ is said to *admit a continuation to the whole real axis* if there exists a function defined and p.d. on the whole real axis such that it coincides with the original given function $s(x)$ on the interval $[-\xi,\xi]$.

A positive definite (p.d.) function $s(x)$ defined on the interval $[-\xi,\xi]$ is said to be *nonuniquely continuable* if there exist at least two different p.d. continuations of the function $s(x)$ to the real axis. (In fact, if a function $s(x)$ is nonuniquely continuable, then there exist infinitely many continuations.)

A positive definite (p.d.) function $s(x)$ defined on an open interval $(-l,l)$ *is said to be nonuniquely continuable from* $(-l,l)$ if the function $s(x)$ is nonuniquely continuable from each closed subinterval $[-\xi,\xi]\subset(-l,l)$.[3]

It was shown by M.G. Krein [1] that if a function $s(x)$, which is defined and continuous in the interval $(-a,a)$, satisfies $s(0)>0$ and conditions (i)–(iii), then this function is positive definite in a rather small neighborhood of zero and is nonuniquely continuable from this neighborhood. *The main goal of this paper is to determine the maximal such neighborhood* $(-l,l)$.

E. Gorin has pointed out that, in addition to being interesting in itself, this problem is interesting also because it is related to far-reaching generalizations of Bernstein-type inequalities.

The problem of the continuation of a positive definite function from a finite interval to the real axis and of the description of all its p.d. continuations was first formulated and solved by M.G. Krein [2]. He also obtained a refined criterion for a given p.d. function $s(x)$ on a finite interval to be nonuniquely continuable.

Criterion of M.G. Krein. *Let $s(x)$ be a continuous p.d. function on $[-\xi,\xi]$, $\xi<\infty$. Let $\boldsymbol{S}_\xi$ be the integral operator acting in $L^2(0,\xi)$:*

$$(\boldsymbol{S}_\xi f)(x)\overset{\text{def}}{=}\int_0^\xi s(x-u)f(u)du \qquad (0\le x\le\xi)\,.$$

Let us assume that an eigenfunction system $\{\varphi_j\}$ of the operator $\boldsymbol{S}_\xi$ is complete in $L^2([0,\xi])$, and let $0<\lambda_1\le\lambda_2\le\ldots$ denote the corresponding set of eigenvalues: $\lambda_j\boldsymbol{S}_\xi\varphi_j=\varphi_j$. Let $\alpha_j(z)$ be Fourier coefficients of the function e^{-izx}:

$$\alpha_j(z)=\langle e^{-izx},\varphi_j(x)\rangle=\int_0^\xi e^{-izx}\overline{\varphi_j(x)}\,dx.$$

Then for the function s to admit a nonunique p.d. continuation from the interval $[-\xi,\xi]$ to the whole real axis it is necessary that the series

$$\sum_{j=1}^{\infty}|\alpha_j(z)|^2\lambda_j \tag{1}$$

3) It could happen that a function is nonuniquely continuable from an open interval $(-l,l)$, but is uniquely continuable from the closed interval $[-l,l]$.

be convergent for all complex z and it is sufficient that the series to be convergent for at least one nonreal z.

As before, $s(x)$ is a function satisfying the conditions (i)–(iii). Following M.G. Krein, we associate with the function s the operator $\left(\lambda \boldsymbol{I}_\xi - \boldsymbol{K}_\xi\right)$ acting in the space $C\big([0,\xi]\big)$ of continuous functions:

$$\left(\lambda \boldsymbol{I}_\xi - \boldsymbol{K}_\xi\right) f(x) \stackrel{\text{def}}{=} \lambda f(x) - \int_0^\xi s''(x-u) f(u) du \,, \qquad f \in C([0,\xi]) \,.$$

This operator depends on the upper limit ξ of the integral. Because

$$\langle \boldsymbol{S}_\xi \psi, \psi \rangle = \big\langle (\lambda \boldsymbol{I}_\xi - \boldsymbol{K}_\xi) f, f \big\rangle$$

for functions ψ of the form $\psi(x) = f'(x), f(0) = f(\xi) = 0$, it follows that

$$\big\langle (\lambda \boldsymbol{I} - \boldsymbol{K}_\xi)\, \varphi, \varphi \big\rangle \geq 0 \qquad \forall \quad \varphi \in C\big([0,l]\big) \,. \tag{2}$$

In [1] it was proved that if equality holds in (2) for some $\varphi \not\equiv 0$, then there exists only one p.d. continuation of the function s from the interval $[-\xi,\xi]$ onto the whole real axis. Thus, for a function s which is p.d. on an interval $[-l,l]$ to be continuable nonuniquely onto the whole real axis (and hence to admit infinitely many p.d. continuations) it is necessary that the strict inequality

$$\big\langle (\lambda \boldsymbol{I}_\xi - \boldsymbol{K}_\xi) \varphi, \varphi \big\rangle > 0 \qquad \forall \quad \varphi \in C[0,\xi] \,, \; \varphi \not\equiv 0 \,, \qquad \forall \xi \in (0,l) \tag{A}$$

hold. Because of (i)–(iii), the inequality (A) is certainly true for ξ from some neighborhood of zero. However, the condition (A) alone is not sufficient for the function s to be nonuniquely continuable onto the whole real axis. One more necessary condition, the condition (B), will be formulated below. The conditions (A) and (B) together will be sufficient for a given function s to be both positive definite on $[0,l]$ and nonuniquely continuable.

1 The triad

In what follows $s(x)$ is a function on the interval $(-l,l)$ which satisfies the conditions (i)–(iii) there and the strict inequality (A). For simplicity, we assume that $\lambda = 1$. Then the integral equation of the second kind

$$f(x) - \int_0^\xi s''(x-t) f(t)\, dt = -s''(t) \qquad (0 \leq x \leq \xi) \tag{3}$$

is uniquely solvable for each $\xi \in (0,l)$. Let $\chi_\xi(x)$ denote its solution. It is clear that for each fixed ξ, the function $\chi_\xi(x)$ is continuous with respect to $x \in [0,\xi]$.

Let us introduce two values depending on ξ :

$$\begin{aligned} k(\xi) &\stackrel{\text{def}}{=} s'(+0) + \int_0^\xi \overline{s'(t)}\chi_\xi(t)\,dt = s'(x) - \int_0^\xi s'(x-t)\chi_\xi(t)\,dt, \\ b(\xi) &\stackrel{\text{def}}{=} s(0) - \int_0^\xi \overline{s(t)}\chi_\xi(t)dt\ . \end{aligned} \tag{4}$$

The triple $(\chi_\xi(x), k(\xi), b(\xi))$ is called a *triad*. It is clear that (χ, k, b) can be determined from the equation

$$\int_0^\xi s(x-t)\chi_\xi(t)dt = s(x) - k(\xi)x - b(\xi) \qquad (0 < x < \xi)\ . \tag{5}$$

Equation (5) is called *the triad identity*. In the following we need not only the identity for $\chi_\xi(x)$:

$$\chi_\xi(x) - \int_0^\xi s''(x-t)\chi_\xi(t)dt = -s''(x)\ , \tag{6}$$

but also the 'adjoint' identity

$$\overline{\chi_\xi(\xi-x)} - \int_0^\xi s''(x-t)\overline{\chi_\xi(\xi-t)}\,dt = -s''(x-\xi)\ . \tag{6'}$$

To obtain (6′), we change variables $x \to \xi - x$, $t \to \xi - t$ and pass to the complex conjugate. Analogously, from (5) we obtain the equality

$$\int_0^\xi s(x-t)\overline{\chi_\xi(\xi-t)}dt = s(x-\xi) - \overline{k(\xi)}(\xi - x) - \overline{b(\xi)} \qquad (0 < x < \xi)\ . \tag{5'}$$

Let us investigate the behavior of the triad elements as $\xi \to +0$. From (6) it follows that

$$\int_0^\xi |\chi_\xi(x)|\,dx \le \int_0^\xi \left| \int_0^\xi s''(x-t)\chi_\xi(t)\,dt \right| dx + \int_0^\xi |s''(x)|\,dx\ .$$

Because

$$\int_0^\xi \left| \int_0^\xi s''(x-t)\chi_\xi(t)\,dt \right| dx \le \int_0^\xi |\chi_\xi(t)| \int_0^\xi |s''(x-t)|\,dx\,dt$$

and

$$\int_0^\xi |\chi_\xi(t)| \int_{-t}^{\xi-t} |s''(u)|\,du\,dt \le \int_0^\xi |\chi_\xi(t)| \int_{-\xi}^{\xi} |s''(u)|\,du\,dt\ ,$$

it follows that

$$\int_0^\xi |\chi_\xi(x)|\,dx \le \int_0^\xi |s''(x)|\,dx \frac{1}{1 - \int_{-\xi}^\xi |s''(x)|\,dx} \longrightarrow 0 \qquad (\xi \to +0)\ .$$

Hence

$$\lim_{\xi\to+0}\int_0^\xi |\chi_\xi(x)|\,dx = 0\ . \tag{7}$$

From (7) it follows that

$$\lim_{\xi\to+0} k(\xi) = s'(+0) \quad \text{and} \quad \lim_{\xi\to+0} b(\xi) = s(0)\ . \tag{8}$$

2 The value $\Delta(\xi)$

The value

$$\Delta(\xi) = \xi k(\xi)\overline{k(\xi)} + k(\xi)\overline{b(\xi)} + \overline{k(\xi)}b(\xi) \tag{Δ}$$

will be the main object of our further considerations.

Theorem 1. *The function $\Delta(\xi)$ is continuously differentiable and*

$$\frac{d}{d\xi}\Delta(\xi) = k(\xi)\overline{k(\xi)} \qquad (0 \le \xi < l)\ .$$

Proof. First of all we show that for $\xi \le \xi_0 < l$ the norms of the operators $(\boldsymbol{I}_\xi - \boldsymbol{K}_\xi)^{-1}$ are uniformly bounded in the appropriate spaces $C[0,\xi] : \left\|(\boldsymbol{I}_\xi - \boldsymbol{K}_\xi)^{-1}\right\| \le M_{\xi_0}$. Changing variables $x = \xi\tilde{x}\ ,\quad t = \xi\tilde{t}$ we obtain the equality

$$\begin{aligned}(\boldsymbol{I}_\xi - \boldsymbol{K}_\xi)f(x) &= f(x) - \int_0^\xi s''(x-t)f(t)\,dt \\ &= f(\xi\tilde{x}) - \int_0^1 \xi s''(\xi(\tilde{x}-\tilde{t}))\,d\tilde{t} = (\boldsymbol{I}_1 - \widetilde{\boldsymbol{K}}_\xi)f(\xi\tilde{x})\ .\end{aligned}$$

The operators $(\boldsymbol{I}_1 - \widetilde{\boldsymbol{K}}_\xi)$ now act in the fixed space $C[0,1]$. The invertibility of the operator $(\boldsymbol{I}_\xi - \boldsymbol{K}_\xi)$ guarantees the existence of the inverse operator $(\boldsymbol{I}_1 - \widetilde{\boldsymbol{K}}_\xi)^{-1}$ for each ξ. Since the kernel $\xi s''(\xi(\tilde{x}-\tilde{t})$ of the operator $\widetilde{\boldsymbol{K}}_\xi$ is continuous with respect to ξ, the inverse operator $(\boldsymbol{I}_1 - \widetilde{\boldsymbol{K}}_\xi)^{-1}$ depends on ξ continuously as well. Hence,

$$\left\|(\boldsymbol{I}_\xi - \boldsymbol{K}_\xi)^{-1}\right\|_{C_{[0,\xi]}} = \left\|(\boldsymbol{I}_1 - \widetilde{\boldsymbol{K}}_\xi)^{-1}\right\|_{C_{[0,1]}} \le M_{\xi_0}.\ \ 0 < \xi \le \xi_0\ .$$

From here it follows that the functions $\chi_\xi(x)$ are uniformly bounded:

$$\|\chi_\xi(x)\| = \left\|(\boldsymbol{I}_\xi - \boldsymbol{K}_\xi)^{-1}s''\right\| \le M_{\xi_0}\max_{x\in[0,\xi_0]}\|s''(x)\| \qquad (\forall\xi, x : 0 < \xi \le \xi_0 < l)\ .$$

Next, we consider the difference $\{\xi_{\xi+h}(x) - \xi_{\xi-h}(x)\}$ for h_1 and h_2 small enough. One immediate consequence of (6) is the identity

$$\begin{aligned}\{\chi_{\xi+h_1}(x) - \chi_{\xi-h_2}(x)\} &- \int_0^{\xi-h_2} s''(x-t)\{\chi_{\xi+h_1}(t) - \chi_{\xi-h_2}(t)\}\,dt \\ &= \int_{\xi-h_2}^{\xi+h_1} s''(x-t)\chi_{\xi+h_1}(t)dt\ , \qquad (x \in [0, \xi-h_2])\ .\end{aligned} \tag{9}$$

The right hand side of (9) tends to zero as $h_1, h_2 \to 0$. Since the operator family $\|(\boldsymbol{I}_\xi - \boldsymbol{K}_\xi)^{-1}\|$ is uniformly bounded,

$$\lim_{h_1,h_2\to 0} \left|\chi_{\xi+h_1}(x) - \chi_{\xi-h_2}(x)\right| = 0 \qquad (0 \le x < \xi)\ .$$

(For $x = \xi$ the one-sided limit exists: $\lim_{h_1\to 0} \left|\chi_{\xi+h_1}(\xi) - \chi_\xi(\xi)\right| = 0$). Hence, the function χ_ξ is a continuous function of the two variables x, ξ in the closed triangle $0 \le x \le \xi \le \xi_0$. Finally, dividing (9) by $h_1 + h_2$, we obtain the equality

$$\frac{\chi_{\xi+h_1}(x) - \chi_{\xi-h_2}(x)}{h_1 + h_2} = (\boldsymbol{I}_\xi - \boldsymbol{K}_{\xi-h_2})^{-1} \left\{ \frac{1}{h_1 + h_2} \int_{\xi-h_2}^{\xi+h_1} s''(x-t)\chi_{\xi+h_1}(t)\, dt \right\} . \tag{10}$$

From (6′) it follows that $(\boldsymbol{I}_\xi - \boldsymbol{K}_\xi)^{-1}\{s''(x-\xi)\} = -\overline{\chi_\xi(\xi - x)}$. Since

$$\lim_{h_1,h_2\to 0} \frac{1}{h_1+h_2} \int_{\chi_{\xi-h_2}}^{\chi_{\xi+h_1}} s''(x-t)\chi_{\xi+h_1}(t)\, dt = s''(x-\xi)\chi_\xi(\xi)\ ,$$

it is clear that the right hand side of (10) tends to a limit which is equal to $-\chi_\xi(\xi)\overline{\chi_\xi(\xi - x)}$. From this it follows that the limit $\frac{\partial}{\partial\xi}\chi_\xi(x)$ of the left hand side exists as well, and

$$\frac{\partial}{\partial\xi}\chi_\xi(x) = -\chi_\xi(\xi)\overline{\chi_\xi(\xi - x)} \qquad (0 \le x \le \xi)\ .$$

Setting

$$\theta(\xi) \stackrel{\text{def}}{=} \chi_\xi(\xi),$$

we obtain the formula[4]

$$\frac{\partial}{\partial\xi}\chi_\xi(x) = -\theta(\xi)\overline{\chi_\xi(\xi - x)}\ . \tag{11}$$

Now we are able to calculate the derivatives $\frac{d}{d\xi}k(\xi)$, $\frac{d}{d\xi}b(\xi)$:

$$\frac{d}{d\xi}k(\xi) = \frac{d}{d\xi}\int_0^\xi \overline{s'(t)}\chi_\xi(t)\, dt = \theta(\xi)\overline{s'(\xi)} - \theta(\xi)\int_0^\xi \overline{s'(t)}\,\overline{\chi_\xi(\xi - t)}\, dt = \theta(\xi)\overline{k(\xi)},$$

$$\begin{aligned} \frac{d}{d\xi}b(\xi) &= -\frac{d}{d\xi}\int_0^\xi \overline{s(t)}\chi_\xi(t)\, dt \\ &= -\theta(\xi)\overline{s(\xi)} + \theta(\xi)\int_0^\xi \overline{s(t)}\,\overline{\chi_\xi(\xi - t)}\, dt = -\theta(\xi)\Big(\overline{\overline{k(\xi)}\xi + b(\xi)}\Big). \end{aligned}$$

4) Formula (11) was first obtained by M.G. Krein [3].

Thus,

$$\frac{d}{d\xi}k(\xi) = \theta(\xi)\overline{k(\xi)} \quad \text{and} \quad \frac{d}{d\xi}b(\xi) = -\theta(\xi)\Big(\overline{k(\xi)}\xi + \overline{b(\xi)}\Big). \tag{12}$$

Using the formulas (12), we obtain

$$\begin{aligned}\frac{d}{d\xi}\Delta(\xi) &= \frac{d}{d\xi}\Big\{k(\xi)\overline{k(\xi)}\xi + k(\xi)\overline{b(\xi)} + \overline{k(\xi)}b(\xi)\Big\}\\ &= \theta(\xi)\overline{k(\xi)}\,\overline{k(\xi)}\xi + k(\xi)\theta(\xi)k(\xi)\xi + k(\xi)\overline{k(\xi)} + \theta(\xi)\overline{k(\xi)}b(\xi)\\ &\quad - k(\xi)\overline{\theta(\xi)}\Big(k(\xi)\xi + b(\xi)\Big) + \overline{\theta(\xi)}k(\xi)b(\xi) - \overline{k(\xi)}\theta(\xi)\Big(\overline{k(\xi)}\xi + \overline{b(\xi)}\Big)\\ &= k(\xi)\overline{k(\xi)}\,.\end{aligned}$$

The theorem is proved. □

In particular, it follows from (12) that $k(\xi) \neq 0$ for $\xi \in [0,l)$. Consequently, $\Delta(\xi)$ is a strictly increasing function on $[0,l)$. The following theorem reveals the role of the value of $\Delta(\xi)$:

Theorem 2. *Let $s(x)$ be a function which is positive definite on $(-l,l)$, satisfies the conditions (i)–(iii) and is nonuniquely continuable from $(-l,l)$. Then $\Delta(\xi)$ is negative for $\xi \in (-l,l)$.*

First we prove the following lemma.

Lemma 1. *Let g be a continuous positive definite function on an interval $[-\xi,\xi]$ which is nonuniquely continuable from this interval. Then the equation*

$$\int_0^{\xi} g(x-t)f(t)\,dt = \alpha g(x) + \beta g(x-\xi) \qquad (|\alpha| + |\beta| > 0) \tag{13}$$

has no summable solution f.

Proof. By Mercer's Theorem, the representation

$$g(x-t) = \sum_{j=1}^{\infty} \frac{\varphi_j(x)\,\overline{\varphi_j(t)}}{\lambda_j} \qquad (x,t \in [0,\xi]) \tag{14}$$

holds, where the series (which has continuous terms) converges absolutely and uniformly. Let a function $f \in L^1$ satisfy (13) and let $f_j \overset{\text{def}}{=} \langle f,\varphi_j\rangle = \int_0^{\xi} f(t)\overline{\varphi_j(t)}\,dt$. Then

$$\begin{aligned}\int_0^{\xi} g(x-t)f(t)\,dt &= \int_0^{\xi}\sum_{j=1}^{\infty}\frac{\varphi_j(x)\,\overline{\varphi_j(t)}}{\lambda_j}f(t)\,dt\\ &= \sum_{j=1}^{\infty}\frac{\varphi_j(x)}{\lambda_j}\int_0^{\xi}\overline{\varphi_j(t)}f(t)\,dt = \sum_{j=1}^{\infty}\frac{f_j}{\lambda_j}\varphi_j(x).\end{aligned}$$

On the other hand, from (14)

$$\alpha g(x)+\beta g(x-\xi)=\sum_{j=1}^{\infty}\left\{\alpha\frac{\overline{\varphi_j(0)}}{\lambda_j}+\beta\frac{\overline{\varphi_j(\xi)}}{\lambda_j}\right\}\varphi_j(x),$$

and hence, $f_j=\alpha\overline{\varphi_j(0)}+\beta\overline{\varphi_j(\xi)}$. Since the function $s(x)$ is nonuniquely continuable, the Fourier coefficients $\alpha_j(z)\stackrel{\text{def}}{=}\langle e^{-izx},\varphi_j(x)\rangle$ satisfy the condition (1): $\sum_j|\alpha_j(z)|^2<\infty$, by M.G. Krein's criterion. This condition guarantees the absolute convergence of the Fourier series to the function e^{-izx}:

$$e^{-izx}=\sum_0^{\infty}\alpha_j(z)\varphi_j(x), \tag{15}$$

because

$$\sum_{j=m}^{n}|\alpha_j(z)\varphi_j(x)|\le\left\{\sum_{j=m}^{n}|\alpha_j(z)|^2\lambda_j\cdot\sum_{j=m}^{n}\frac{|\varphi_j(x)|2}{\lambda_j}\right\}^{1/2}$$

$$\le\left\{\sum_{j=m}^{\infty}|\alpha_j(z)|^2\lambda_j\cdot\sum_{j=m}^{\infty}\frac{|\varphi_j(x)|2}{\lambda_j}\right\}^{1/2}$$

$$=\left\{g(0)\sum_{j=m}^{\infty}|\alpha_j(z)|^2\lambda_j\right\}^{\frac{1}{2}}\to 0\ \text{ as }\ m\to\infty\ .$$

Let us consider the scalar product $\langle e^{-izx},f(x)\rangle$. As the series (15) converges uniformly and absolutely,

$$\langle e^{-izx},f(x)\rangle=\int_0^{\xi}\sum_{j=1}^{\infty}\alpha_j(z)\varphi_j(x)\overline{f(x)}\,dx=\sum_{j=1}^{\infty}\alpha_j(z)\int_0^{\xi}\varphi_j(x)\overline{f(x)}\,dx$$

$$=\sum_{j=1}^{\infty}\alpha_j(z)\overline{f_j}=\sum_{j=1}^{\infty}\alpha_j(z)\left[\overline{\alpha}\varphi_j(0)+\overline{\beta}\varphi_j(\xi)\right]$$

$$=\overline{\alpha}\sum_{j=1}^{\infty}\alpha_j(z)\varphi_j(0)+\overline{\beta}\sum_{j=1}^{\infty}\alpha_j(z)\varphi_j(\xi).$$

Taking into account (15), we obtain $\langle e^{-izx},f(x)\rangle=\overline{\alpha}1+\overline{\beta}e^{-iz\xi}$. Inserting $z=\omega$ $(\Im\omega=0)$ and letting ω tend to ∞, we conclude that $\alpha=0$, $\beta=0$. The lemma is proved. □

Proof of Theorem 2. Multiplying (5) by $\left(\overline{k(\xi)}\xi + \overline{b(\xi)}\right)$, (5′) by $-b(\xi)$ and combining, we obtain the identity

$$\int_0^\xi s(x-t)\left\{ \left(\overline{k(\xi)}\xi + \overline{b(\xi)}\right) \chi_\xi(t) - b(\xi)\overline{\chi_\xi(\xi - t)}\right\} dt$$
$$= \left(\overline{k(\xi)}\xi + \overline{b(\xi)}\right) s(x) - b(\xi)s(x-\xi) - \Delta(\xi)x.$$

Let us suppose that $\Delta(\xi) = 0$. Then by Lemma 1, $\overline{k(\xi)}\xi + \overline{b(\xi)} = 0$ and $b(\xi) = 0$, i.e., $k(\xi) = 0$ and $b(\xi) = 0$. Now the identity (5) takes the form

$$\int_0^\xi s(x-t)\chi_\xi(t)\, dt = s(x) \qquad (0 < x < \xi) .$$

By Lemma 1, the last equality is impossible.

This contradiction shows that $\Delta(\xi) \neq 0$ for each $\xi \in (0, l)$. At the point $\xi = +0$, the behavior of Δ can be determined from the formulas (8):

$$\Delta(+0) = k(+0)\overline{b(+0)} + \overline{k(+0)}b(+0) = s'(+0)\overline{s(+0)} + \overline{s'(+0)}s(+0) = -\lambda s(0) < 0.$$

The inequality $\Delta(\xi) < 0$ now follows from the continuity of the function $\Delta(\xi)$ in the interval $(-l, l)$. □

Thus, to this point we have shown that a function $s(x)$ which is positive definite on the interval $(-l, l)$ and is nonuniquely continuable from this interval and satisfies the conditions (i)–(iii) *must* satisfy the following two conditions:

$$\langle (\lambda I - K_\xi)\varphi, \varphi \rangle > 0, \quad \forall \quad \varphi \in C[0, l],\ \varphi \not\equiv 0, \qquad \forall \quad \xi \in (0, l) \tag{A}$$

$$\Delta(l - 0) \leq 0 . \tag{B}$$

3 The fundamental antilinear forms $B_\xi(f)$, $L_\xi(f)$, $C_\xi(f)$

Our goal is to show that the conditions (A) and (B) together are sufficient for a function $s(x)$ to be both positive definite on an interval $(-l, l)$ and nonuniquely continuable from this interval.

To show this we introduce three antilinear forms whose arguments are arbitrary function $f(x)$ which are continuously differentiable on $[0, l)$. These forms depend on a parameter ξ and are defined for all $\xi \in (0, l)$ except those points ξ at which $\Delta(\xi) = 0$ (if such points exist) as follows:

$$B_\xi(f) = \frac{1}{\Delta(\xi)} \left\{ \overline{k(\xi)}\,\overline{f(0)} + k(\xi)\overline{f(\xi)} - \overline{k(\xi)}\langle \chi_\xi(u), f(u)\rangle_\xi \right.$$
$$\left. - k(\xi)\langle \overline{\chi_\xi(\xi - u)}, f(u)\rangle_\xi \right\},$$

$$\begin{aligned} L_\xi(f) = \frac{i}{\Delta(\xi)} \Big\{ \Big(\overline{k(\xi)}\xi + \overline{b(\xi)}\Big)\, \overline{f(0)} - b(\xi)\overline{f(\xi)} - \\ - \Big(\overline{k(\xi)}\xi + \overline{b(\xi)}\Big) \langle \chi_\xi(u), f(u)\rangle_\xi + b(\xi) \langle \overline{\chi_\xi(\xi-u)}, f(u)\rangle_\xi \Big\}, \end{aligned} \tag{16}$$

$$C_\xi(f) = -\Delta(\xi)\Big\{L(f) + B(M_0)B(f) - B\Big(A_0 f(x)\Big)\Big\},$$

where A_0 is the integration operator

$$A_0 f(x) \stackrel{\text{def}}{=} i\int_0^x f(u)\,du$$

and

$$M_0(x) \stackrel{\text{def}}{=} A_0 s(x) = i\int_0^x s(u)\,du\ .$$

As usually, $\langle\ ,\ \rangle_\xi$ is the standard scalar product[5] in $L^2([0,\xi])$:

$$\langle g(u), f(u)\rangle_\xi \stackrel{\text{def}}{=} \int_0^\xi g(u)\overline{f(u)}\,du$$

In what follows we omit ξ in the notations $k(\xi)$, $b(\xi), \Delta(\xi)$, $\chi(x) = \chi_\xi(x)$, $\langle\ ,\ \rangle_\xi$ (whenever it will not lead to confusion).

Theorem 3. *Assume that the conditions* (i)–(iii) *and* (A) *are satisfied. Then the following formulas hold:*

$$\begin{aligned} &\frac{d}{d\xi}B_\xi(f) = -\frac{k(\xi)}{\Delta(\xi)}\Big\{\overline{k(\xi)}B_\xi(f) - \overline{f'(\xi)} + \langle\overline{\chi(\xi-u)}, f'(u)\rangle_\xi\Big\} \\ &\frac{d}{d\xi}C_\xi(f) = \Big\{\overline{k(\xi)}B_\xi(f) - \overline{f'(\xi)} + \langle\overline{\chi(\xi-u)}, f'(u)\rangle_\xi\Big\}\Big\{k(\xi)B_\xi(M_0) - ib(\xi)\Big\} \\ &\Delta(\xi)B_\xi(1) = -1 \\ &\Delta(\xi)B_\xi(M_0) = \int_0^\xi \frac{k(v)\overline{b(v)} - \overline{k(v)}b(v)}{2i}\,dv\ - \frac{i\Delta\xi}{2}\ . \end{aligned} \tag{17}$$

Proof. The formulas (17) are proved by means of simple calculations which are based on formulas (11) and (12). First we show that

$$\frac{d}{d\xi}\Big\{\Delta B(f)\Big\} = k\Big\{\overline{f'(\xi)} - \langle\overline{\chi(\xi-u)}, f'(u)\rangle\Big\}. \tag{18}$$

---

5) The function f, which appears in the scalar product $\Big\langle g(u), f(u)\Big\rangle_\xi$ is defined from the very beginning on the interval $[0,l)$. In this scalar product we consider the restriction of the function f to the interval $[0,\xi]$.

We have

$$\frac{d}{d\xi}\Big\{\Delta B(f)\Big\}=\frac{d}{d\xi}\Big\{\overline{k}\,\overline{f(0)}+k\overline{f(\xi)}-\overline{k}\langle\chi,f\rangle-k\langle\overline{\chi(\xi-u)},f(u)\rangle\Big\}.$$

Let us calculate the derivative of each summand separately:

$$\frac{d}{d\xi}\Big\{\overline{k}\,\overline{f(0)}\Big\}=\overline{\theta}k\overline{f(0)};\qquad \frac{d}{d\xi}\Big\{k\,\overline{f(\xi)}\Big\}=\theta\overline{k}\,\overline{f(\xi)}+k\overline{f'(\xi)}.$$

Then, calculating

$$\begin{aligned}\frac{d}{d\xi}\langle\chi(u),f(u)\rangle&=\frac{d}{d\xi}\int_0^\xi\chi_\xi(u)\overline{f(u)}\,du\\&=\theta\,\overline{f(\xi)}+\int_0^\xi\frac{\partial}{\partial\xi}\Big\{\chi_\xi(u)\overline{f(u)}\Big\}\,du\\&=\theta\,\overline{f(\xi)}-\theta\langle\overline{\chi(\xi-u)},f(u)\rangle,\end{aligned}\tag{19}$$

we obtain

$$-\frac{d}{d\xi}\Big\{\overline{k}\langle\chi(u),f(u)\rangle\Big\}=-\overline{\theta}k\langle\chi(u),f(u)\rangle-\overline{k}\theta\overline{f(\xi)}+\theta\overline{k}\langle\overline{\chi(\xi-u)},f(u)\rangle\;.$$

Finally, because

$$\begin{aligned}\frac{d}{d\xi}\langle\overline{\chi_\xi(\xi-u)},f(u)\rangle&=\frac{d}{d\xi}\int_0^\xi\overline{\chi_\xi(\xi-u)}\,\overline{f(u)}du=\frac{d}{d\xi}\int_0^\xi\overline{\chi_\xi(u)}\;\overline{f(\xi-u)}du\\&=\overline{\theta}\,\overline{f(0)}+\int_0^\xi\Big\{\frac{\partial}{\partial\xi}\Big[\overline{(\chi_\xi(u))}\Big]\overline{f(\xi-u)}+\overline{\chi_\xi(u)f'(\xi-u)}\Big\}du\\&=\overline{\theta}\,\overline{f(0)}-\overline{\theta}\langle\chi_\xi(u),f(u)\rangle+\langle\overline{\chi_\xi(\xi-u)},f'(u)\rangle\;,\end{aligned}\tag{20}$$

it follows that

$$\begin{aligned}-\frac{d}{d\xi}\Big\{k\langle\overline{\chi(\xi-u)},f(u)\rangle\Big\}=&-\theta\overline{k}\langle\overline{\chi(\xi-u)},f(u)\rangle-\overline{k}\theta\;\overline{f(0)}\\&+k\overline{\theta}\langle\chi(u),f(u)\rangle-k\langle\overline{\chi(\xi-u)},f'(u)\rangle\;.\end{aligned}$$

Adding, we obtain (18). Using the expression for $\frac{d}{dx}\{\Delta B(f)\}$, we arrive at the first of the formulas (17):

$$\begin{aligned}\frac{d}{d\xi}\{B(f)\}=\frac{d}{d\xi}\Big\{\frac{1}{\Delta}\Delta B(f)\Big\}&=-\frac{k\overline{k}}{\Delta^2}B(f)+\frac{k}{\Delta}\Big\{\overline{f'(\xi)}-\langle\overline{\chi(\xi-u)},f'(u)\rangle\Big\}\\&=\Big\{-\overline{k}B(f)+\overline{f'(\xi)}-\langle\overline{\chi(\xi-u)}\;,\;f'(u)\rangle\Big\}\frac{k}{\Delta}\;.\end{aligned}$$

Setting $f(x)\equiv 1$ into (18), we get $\frac{d}{d\xi}\{\Delta B(1)\}=0$.

Taking into account the behavior of the elements at the point $\xi = +0$,

$$\overline{\Delta B(1)} \longrightarrow \overline{k(+0)} + k(+0) = -\lambda = -1 \quad \text{as} \quad \xi \longrightarrow +0\,,$$

we see that $\Delta B(1) = -1$. Inserting $f(x) = M_0(x) = (A_0 s)(x)$ into (18), we obtain

$$\begin{aligned}\frac{d}{d\xi}\{\Delta B(M_0)\} &= k\Big\{-\overline{is(\xi)} - \langle\overline{\chi(\xi-u)}, is(u)\rangle\Big\}\\ &= ik\Big\{\int_0^\xi \overline{s(u)\chi(\xi-u)}du - \overline{s(\xi)}\Big\}\\ &= ik\Big\{\int_0^\xi \overline{s(\xi-u)\chi(u)}du - \overline{s(\xi)}\Big\}\,.\end{aligned}$$

By identity (5), the last expression is equal to:

$$\frac{d}{d\xi}\{\Delta B(M_0)\} = ik(-\overline{k}\xi - \overline{b}) = -i(k\overline{k}\xi + k\overline{b})\,. \tag{21}$$

Collecting the real and imaginary parts, we obtain

$$\frac{d}{d\xi}\{\Delta B(M_0)\} = \frac{k\overline{b} - \overline{k}b}{2i} - \frac{i}{2}(k\overline{k}\xi + \Delta) = \frac{k\overline{b} - \overline{k}b}{2i} - i\frac{\Delta'\xi + \Delta}{2} = \frac{k\overline{b} - \overline{k}b}{2i} - \frac{i}{2}(\Delta\xi)'\,.$$

Since $B(M_0) \longrightarrow 0$ as $\xi \longrightarrow +0$, we get

$$\Delta B(M_0) = \int_0^\xi \frac{k(u)\overline{b(u)} - \overline{k(u)}b(u)}{2i}du - \frac{i}{2}\Delta\xi\,.$$

We turn now to the calculation of the derivative $\frac{d}{d\xi}C(\xi)$. First we find $\frac{d}{d\xi}\{\Delta L(f)\}$:

$$\frac{d}{d\xi}\{\Delta L(f)\} = i\frac{d}{d\xi}\Big\{(\overline{k}\xi + \overline{b})\overline{f(0)} - \overline{bf(\xi)} - (\overline{k}\xi + \overline{b})\langle\chi, f\rangle + b\langle\overline{\chi(\xi-u)}, f(u)\rangle\Big\}\,.$$

Since

$$\frac{d}{d\xi}(\overline{k}\xi + \overline{b}) = \overline{\theta}k\xi + \overline{k} - \overline{\theta}(k\xi + b) = \overline{k} - \overline{\theta}b \quad \text{and} \quad \frac{d}{d\xi}b = -\theta(\overline{k}\xi + \overline{b})\,,$$

we find, with the help of (19) and (20), that

$$\begin{aligned}\frac{d}{d\xi}\{\Delta L(f)\} = i\Big\{&(\overline{k} - \overline{\theta}b)\overline{f(0)} + \theta(\overline{k}\xi + \overline{b})\overline{f(\xi)} - b\overline{f'(\xi)} - (\overline{k} - \overline{\theta}b)\langle\chi, f\rangle\\ &- (\overline{k}\xi + \overline{b})\Big[\theta\,\overline{f(\xi)} - \theta\langle\overline{\chi(\xi-u)}, f(u)\rangle\Big] - \theta(\overline{k}\xi + \overline{b})\langle\overline{\chi(\xi-u)}, f(u)\rangle\\ &+ b\Big[\overline{\theta f(0)} - \overline{\theta}\langle\chi, f\rangle + \langle\overline{\chi(\xi-u)}, f'(u)\rangle\Big]\Big\}\\ = ik\Big\{&\overline{f(0)} - \langle\chi, f\rangle\Big\} - ib\Big\{\overline{f'(\xi)} - \langle\overline{\chi(\xi-u)}, f'(u)\rangle\Big\}\,.\end{aligned}$$

Taking into account the last inequality and formulas (18), (21), we calculate:

$$\begin{aligned}
\frac{d}{d\xi}C(f) &= \frac{d}{d\xi}\Big\{ -\Delta L(f) - \Delta B(M_0)B(f) - i\Delta B\Big(\int_0^x f(u)du\Big)\Big\} \\
&= -\frac{d}{d\xi}\{\Delta L(f)\} - \frac{d}{d\xi}\{\Delta B(M_0)\}B(f) - \Delta B(M_0)\frac{d}{d\xi}B(f) \\
&\quad - i\frac{d}{d\xi}\Big\{\Delta B\Big(\int_0^x f(u)du\Big)\Big\} \\
&= -i\bar{k}\Big\{\overline{f(0)} - \langle \chi, f\rangle\Big\} + ib\Big\{f'(\xi) - \langle \overline{\chi(\xi-u)}, f'(u)\rangle\Big\} + i(k\bar{k}\xi + k\bar{b})B(f) \\
&\quad + B(M_0)k\Big\{\bar{k}B(f) - \overline{f'(\xi)} + \langle \overline{\chi(\xi-u)}, f'(u)\rangle\Big\} \\
&\quad - ik\Big\{\overline{f(\xi)} - \langle \overline{\chi(\xi-u)}, f(u)\rangle\Big\} \\
&= -i\Delta B(f) + \Big\{\overline{f'(\xi)} - \langle \overline{\chi(\xi-u)}, f'(u)\rangle\Big\}\{ib - kB(M_0)\} \\
&\quad + b(f)\{B(M_0)k\bar{k} + ik\bar{k}\xi + ik\bar{b}\} \\
&= \Big\{\bar{k}B(f) - \overline{f'(\xi)} + \langle \overline{\chi(\xi-u)}, f'(u)\rangle\Big\}\{kB(M_0) - ib\}\,.
\end{aligned}$$

This completes the proof of the theorem. □

4 The matrix function $W(z,\xi)$ and the related canonical system

As a basis for further considerations we need the values of the forms $B(f)$ and $C(f)$ for the functions

$$f(x) = e^{-izx} \quad \text{and} \quad f(x) = M_{\bar{z}}(x) = i\int_0^x e^{-i\bar{z}(x-u)}s(u)du\,.$$

Let us introduce the entire functions of the variable z depending on the parameter ξ:

$$a_{11}(z,\xi) = 1 - zB(M_{\bar{z}})\,,\; a_{12}(z,\xi) = zC(M_{\bar{z}})\,,$$
$$a_{21}(z,\xi) = -zB(e^{-i\bar{z}x})\,, \qquad a_{22}(z,\xi) = 1 + zC(e^{-i\bar{z}x})\,,$$

and the matrix-valued function

$$W(z,\xi) = \begin{bmatrix} a_{11}(z,\xi) & a_{12}(z,\xi) \\ a_{21}(z,\xi) & a_{22}(z,\xi) \end{bmatrix}\,.$$

Let us consider also the rank one matrix

$$Q(\xi) = J\begin{bmatrix} \bar{\alpha} \\ \bar{\beta} \end{bmatrix}[\alpha \quad \beta]\,,$$

where

$$J = \begin{bmatrix} 0 & i \\ -i & 0 \end{bmatrix}, \quad \alpha = \frac{k}{\Delta} \quad \text{and} \quad \beta = kB(M_0) - ib = \frac{k}{\Delta}\int_0^\xi \frac{k\bar{b} - \bar{k}b}{2i} dv - \frac{ik\xi}{2} - ib .$$

It is not difficult to verify that Q is a J-projector, i.e.,

$$JQ \geq 0 \quad \text{and} \quad Q^2 = -Q .$$

The first claim is clear. The second is a consequence of the equality

$$\operatorname{Im} B(M_0) = -\frac{\xi}{2} .$$

Indeed,

$$\begin{aligned} \bar{\alpha}\beta - \alpha\bar{\beta} &= \frac{\bar{k}}{\Delta}\{kB(M_0) - ib\} - \frac{k}{\Delta}\left\{\bar{k}\,\overline{B(M_0)} + i\bar{b}\right\} \\ &= \frac{k\bar{k}}{\Delta} 2i \operatorname{Im} B(M_0) - i\frac{k\bar{b}}{\Delta} - i\frac{\bar{k}b}{\Delta} \\ &= -i . \end{aligned} \tag{22}$$

Hence,

$$\begin{aligned} Q^2 &= J\begin{bmatrix} \bar{\alpha} \\ \bar{\beta} \end{bmatrix} \left\{ [\alpha \quad \beta] J \begin{bmatrix} \bar{\alpha} \\ \bar{\beta} \end{bmatrix} \right\} [\alpha \quad \beta] \\ &= J\begin{bmatrix} \bar{\alpha} \\ \bar{\beta} \end{bmatrix} \{-i(\bar{\alpha}\beta - \alpha\bar{\beta})\} [\alpha \quad \beta] \\ &= J\begin{bmatrix} \bar{\alpha} \\ \bar{\beta} \end{bmatrix} \{-1\} [\alpha \quad \beta] \\ &= -Q . \end{aligned}$$

The following theorem plays a central role in our considerations.

Theorem 4. *Assume that the conditions (i)–(iii) and (A) are fulfilled. Then the matrix function $W(z, \xi)$ satisfies the differential equation*

$$\frac{d}{d\xi} W(z, \xi) = -izW(z, \xi)Q(\xi)$$

and the initial condition

$$W(z, 0) = \begin{bmatrix} 1 & 0 \\ -z/s(0) & 1 \end{bmatrix} .$$

Proof. Let us verify that $W(z,\xi)$ satisfies the differential equation

$$\frac{d}{d\xi}W(z,\xi) = -izW(z,\xi)Q(\xi)\,. \tag{23}$$

Indeed,

$$\frac{d}{d\xi}W(z,\xi) = \frac{d}{d\xi}\begin{bmatrix} a_{11}(z,\xi) & a_{12}(z,\xi) \\ a_{21}(z,\xi) & a_{22}(z,\xi) \end{bmatrix}$$
$$= z\frac{d}{d\xi}\begin{bmatrix} -B(M_{\bar z}) & C(M_{\bar z}) \\ -B(e^{-i\bar z x}) & C(e^{-i\bar z x}) \end{bmatrix}.$$

The identities (17) allow us to represent the result in the form

$$\frac{d}{d\xi}W(z,\xi)=z\begin{bmatrix} \bar kB(M_{\bar z}) - M_{\bar z}'(\xi) + \langle\overline{\chi(\xi-u)}, M_{\bar z}'(u)\rangle \\ \bar kB(e^{-i\bar z x}) - ize^{iz\xi} + iz\langle\overline{\chi(\xi-u)}, e^{-i\bar z u}\rangle \end{bmatrix}\begin{bmatrix} \frac{k}{\Delta} & kB(M_0) - ib \end{bmatrix}.$$

Since

$$Q = J\begin{bmatrix} \bar k/\Delta \\ \overline{kB(M_0) - ib} \end{bmatrix}\begin{bmatrix} \frac{k}{\Delta} & kB(M_0) - ib \end{bmatrix},$$

it remains only to check the identity

$$\begin{bmatrix} \bar kB(M_{\bar z}) - \overline{M_{\bar z}'(\xi)} + \langle\overline{\chi(\xi-u)}, M_{\bar z}'(u)\rangle \\ \bar kB(e^{-i\bar z x}) - ize^{iz\xi} + iz\langle\overline{\chi(\xi-u)}, e^{-i\bar z u}\rangle \end{bmatrix} = -iW(z,\xi)J\begin{bmatrix} \bar k/\Delta \\ \overline{kB(M_0) - ib} \end{bmatrix}. \tag{24}$$

Let us transform the right hand side of (24). Taking into account the fact that

$$\overline{B(M_0)} = B(M_0) + i\xi\,,$$

we obtain

$$-iJ\begin{bmatrix} \bar k/\Delta \\ \overline{kB(M_0) - ib} \end{bmatrix} = \begin{bmatrix} 0 & 1 \\ -1 & 0 \end{bmatrix}\begin{bmatrix} \bar k/\Delta \\ \bar kB(M_0) + i\bar k\xi + i\bar b \end{bmatrix}$$
$$= \begin{bmatrix} \bar kB(M_0) + i\bar k\xi + i\bar b \\ -\bar k/\Delta \end{bmatrix}.$$

It follows immediately that

$$-iW(z,\xi)J\left[\frac{\bar{k}/\Delta}{kB(M_0)-ib}\right]$$

$$=\begin{bmatrix}1-zB(M_{\bar{z}}) & zC(M_{\bar{z}})\\ -zB(e^{-i\bar{z}x}) & 1+zC(e^{-i\bar{z}x})\end{bmatrix}\begin{bmatrix}\bar{k}B(M_0)+i\bar{k}\xi+i\bar{b}\\ -\bar{k}/\Delta\end{bmatrix}$$

$$=\begin{bmatrix} i\bar{k}\xi+i\bar{b}-izB(M_{\bar{z}})(\bar{k}\xi+\bar{b})+\bar{k}B(M_0)-z\bar{k}B(M_0)B(M_{\bar{z}})-z\frac{\bar{k}}{\Delta}C(M_{\bar{z}})\\ -izB(e^{-i\bar{z}x})(\bar{k}\xi+\bar{b})-z\bar{k}B(M_0)B(e^{-i\bar{z}x})-\frac{\bar{k}}{\Delta}-z\frac{\bar{k}}{\Delta}C(e^{-i\bar{z}x})\end{bmatrix}.$$

Looking at the formula for $C(f)$,

$$C(f)=-\Delta\{L(f)+B(M_0)B(f)-B(A_0f)\},$$

we note that

$$z\bar{k}B(M_0)B(M_{\bar{z}})+z\frac{k}{\Delta}C(M_{\bar{z}})=-z\bar{k}L(M_{\bar{z}})+z\bar{k}B(A_0M_{\bar{z}})$$

and

$$z\bar{k}B(M_0)B(e^{-i\bar{z}x})+z\frac{\bar{k}}{\Delta}C(e^{-i\bar{z}x})=-z\bar{k}L(e^{-iz\bar{x}})+z\bar{k}B(A_0e^{-i\bar{z}x}).$$

Using the relations

$$A_0e^{-i\bar{z}x}=i\int_0^x e^{-i\bar{z}v}dv=\frac{1-e^{-i\bar{z}x}}{\bar{z}}$$

and

$$A_0M_{\bar{z}}(x)=-\int_0^x\int_0^u e^{-i\bar{z}(u-v)}s(v)dv\,du=\frac{M_0(x)-M_{\bar{z}}(x)}{\bar{z}},$$

we obtain the equality

$$-iW(z,\xi)J\left[\frac{\bar{k}/\Delta}{kB(M_0)-ib}\right]$$

$$=\begin{bmatrix} i\bar{k}\xi+i\bar{b}-izB(M_{\bar{z}})(\bar{k}\xi+\bar{b})+\bar{k}B(M_0)+z\bar{k}L(M_{\bar{z}})-\bar{k}B(M_0-M_{\bar{z}})\\ -izB(e^{-i\bar{z}x})(\bar{k}\xi+\bar{b})-\frac{\bar{k}}{\Delta}+z\bar{k}L(e^{-i\bar{z}x})-\bar{k}B(1-e^{-i\bar{z}x})\end{bmatrix}.$$

Upon taking into account the fact that $B(1) = -1/\Delta$ and collecting similar terms, we see that the right hand side of (24) is equal to

$$
\begin{aligned}
&-iW(z,\xi)J\begin{bmatrix} \bar{k}/\Delta \\ kB(M_0)-ib \end{bmatrix} \\
&= \begin{bmatrix} i\bar{k}\xi + i\bar{b} - izB(M_{\bar{z}})(\bar{k}\xi+\bar{b}) + z\bar{k}L(M_{\bar{z}}) + \bar{k}B(M_{\bar{z}}) \\ -izB(e^{-i\bar{z}x})(\bar{k}\xi+\bar{b}) + z\bar{k}L(e^{-i\bar{z}x}) + \bar{k}B(e^{-i\bar{z}x}) \end{bmatrix}.
\end{aligned}
\tag{25}
$$

To transform the left hand side of (24), we calculate the derivative

$$-\overline{M'_{\bar{z}}(x)} = \frac{d}{dx} i \int_0^x e^{iz(x-v)}\overline{s(v)}dv = i\,\overline{s(x)} - iz\,\overline{M_{\bar{z}}(x)}\,.$$

In view of the triad identity (5),

$$
\begin{aligned}
&-\overline{M'_{\bar{z}}(\xi)} + \langle \overline{\chi(\xi-u)}, M'_{\bar{z}}(u)\rangle \\
&= i\,\overline{s(x)} - i\int_0^\xi \overline{\chi(\xi-x)}\,\overline{s(x)}dx - iz\overline{M_{\bar{z}}(\xi)} + iz\langle\overline{\chi(\xi-u)}, M_{\bar{z}}(u)\rangle \\
&= i\bar{k}\xi + i\bar{b} - iz\overline{M_{\bar{z}}(\xi)} + iz\langle\overline{x(\xi-u)}, M_{\bar{z}}(u)\rangle\,.
\end{aligned}
$$

From the last relation and from (25) it follows that (24) can be reduced to the equality

$$\begin{bmatrix} -i\overline{M_{\bar{z}}(\xi)} + i\langle\overline{x(\xi-u)}, M_{\bar{z}}(u)\rangle \\ -ie^{iz\xi} + i\langle\overline{x(\xi-u)}, e^{-i\bar{z}u}\rangle \end{bmatrix} = \begin{bmatrix} -i(\bar{k}\xi+\bar{b})B(M_{\bar{z}}) + \bar{k}L(M_{\bar{z}}) \\ -i(\bar{k}\xi+\bar{b})B(e^{-i\bar{z}x}) + \bar{k}L(e^{-i\bar{z}x}) \end{bmatrix}.$$

This equality is easily verified by using the obvious identity

$$\bar{k}L(f) - i(\bar{k}\xi+\bar{b})B(f) = -i\overline{f(\xi)} + i\langle\overline{\chi(\xi-u)}, f(u)\rangle\,,$$

which follows from (16). Thus, equation (23) is established.

The initial condition for $W(z,\xi)$ can be derived from formulas (7) and (8), which determine the behavior of the triad at the point $\xi = +0$. We have

$$
\begin{aligned}
\lim_{\xi\to+0} B(f) &= \lim_{\xi\to+0}\left\{\frac{\bar{k}}{\Delta}\overline{f(0)} + \frac{k}{\Delta}\overline{f(\xi)} - \frac{k}{\Delta}\langle\chi, f\rangle - \frac{k}{\Delta}\langle\overline{\chi(\xi-u)}, f(u)\rangle\right\} \\
&= \left\{\frac{\overline{k(+0)}}{\overline{\Delta(+0)}}\overline{f(0)} + \frac{k(+0)}{\Delta(+0)}\overline{f(0)}\right\} \\
&= \frac{-\lambda}{-\lambda s(0)}\overline{f(0)} = \frac{\overline{f(0)}}{s(0)}\,,
\end{aligned}
$$

from which we get

$$\lim_{\xi\to+0} B(e^{-i\bar{z}x}) = \frac{1}{s(0)} \quad \text{and} \quad \lim_{\xi\to+0} B(M_{\bar{z}}) = \frac{1}{s(0)}\overline{M_{\bar{z}}(0)} = 0\,.$$

Furthermore, since

$$\lim_{\xi\to+0} L(f)$$
$$= \lim_{\xi\to+0} i\left\{\frac{\bar{k}\xi+\bar{b}}{\Delta}\overline{f(0)} - \frac{b}{\Delta}\overline{f(\xi)} - \frac{\bar{k}\xi+\bar{b}}{\Delta}\langle\chi, f\rangle + \frac{b}{\Delta}\overline{\langle\chi(\xi-u), f(u)\rangle}\right\} = 0\,,$$

it follows that

$$\lim_{\xi\to+0} C(f) = -\Delta(+0)\left\{\lim_{\xi\to+0} L(f) + \lim_{\xi\to+0}\{B(M_0)B(f)\} - \lim_{\xi\to+0} B(A_0 f)\right\} = 0\,.$$

Thus,

$$\lim_{\xi\to+0} W(x,\xi) = \lim_{\xi\to+0}\begin{bmatrix} 1 - zB(M_{\bar{z}}) & zC(M_{\bar{z}}) \\ -zB(e^{-i\bar{z}x}) & 1 + zC(e^{-i\bar{z}x}) \end{bmatrix} = \begin{bmatrix} 1 & 0 \\ -\dfrac{z}{s(0)} & 1 \end{bmatrix}.$$

This completes the proof of the theorem. □

The result proved in Theorem 4 can be formulated as follows:

Theorem 5. *Under the conditions (i)–(iii), (A) and (B), the matrix function $W(z,\xi)$ admits the following multiplicative representation:*

$$W(z,\xi) = e^{-izE} \overset{\curvearrowleft}{\int_0^{\xi}} \exp\{-izQ(t)dt\}\,,$$

where the matrix E,

$$E \overset{\text{def}}{=} \begin{bmatrix} 0 & 0 \\ -\dfrac{i}{s(0)} & 0 \end{bmatrix},$$

satisfies the conditions

$$EJ \geq 0 \quad \text{and} \quad E^2 = 0\,,$$

and the matrix valued function

$$Q : \overset{\text{def}}{=} J \begin{bmatrix} \dfrac{k\bar{k}}{\Delta^2} & \dfrac{k\bar{k}}{\Delta}\displaystyle\int_0^{\xi}\frac{k\bar{b}-\bar{k}b}{2i}dv - \frac{k\bar{b}-\bar{k}b}{2i\Delta} - \frac{i}{2} \\ * & \left|\dfrac{k}{\Delta}\displaystyle\int_0^{\xi}\frac{k\bar{b}-\bar{k}b}{2i}dv - \frac{ik\xi}{2} - ib\right|^2 \end{bmatrix}$$

satisfies the conditions

$$JQ \geq 0 \quad \text{and} \quad Q^2 = -Q\,.$$

The condition (B) is necessary and sufficient for the existence of the multiplicative integral.

Corollary 1. *The matrix function* $W(\xi, z)$ *possesses the following* J*-properties* ([4]–[7]):

$$W(z,\xi)JW^*(z,\xi) - J \begin{cases} >0\,, & \operatorname{Im} z > 0\,, \\ =0\,, & \operatorname{Im} z = 0\,, \\ <0\,, & \operatorname{Im} z < 0\,. \end{cases}$$

From the J-properties of the matrix function $W(z,\xi)$ it follows:

Corollary 2. *The fractional linear transformation*

$$w(z) = \frac{a_{11}(z,\xi)\omega(z) + a_{12}(z,\xi)}{a_{21}(z,\xi)\omega(z) + a_{22}(z,\xi)}\,, \tag{26}$$

which is based on the coefficient matrix $W(z,\xi)$, *maps the class*[6] *of all* R *functions into itself:*

$$\frac{w(z) - \overline{w(z)}}{2i} \geq 0 \;\; for \;\; \operatorname{Im} z > 0 \;\; if \;\; \frac{\omega(z) - \overline{\omega(z)}}{2i} \geq 0 \;\; for \;\; \operatorname{Im} z > 0\,.$$

5 The relationship between the matrix function $W(z,\xi)$ and the original problem

To discover the relationship between the constructed matrix-function $W(z,\xi)$ and the original function $s(x)$, we derive the following asymptotic relation:

Theorem 6. *Let* $\xi \in (0,l)$ *and assume that for this interval the conditions* (i)–(iii), (A) *and* (B) *are satisfied, and let* $w(z)$ *be an* R*-function. Then the function*

$$w(z) \stackrel{\text{def}}{=} \frac{a_{11}(z,\xi)\omega(z) + a_{12}(z)}{a_{21}(z,\xi)\omega(z) + a_{22}(z)}$$

satisfies the condition

$$\lim_{y\to+\infty} e^{xy}\left\{w(y) - i\int_0^x e^{-yu}s(u)du\right\} = 0 \tag{27}$$

for every point $x \in (0,\xi)$.

---

6) Following M.G. Krein, a function which is analytic in the open upper half plane, with a nonnegative imaginary part there, is said to be an R-function.

Proof. We calculate the limit as $y \to +\infty$ of the expression

$$e^{yx}\left\{w(iy) - i\int_0^x e^{-yu}s(u)du\right\}$$

$$= e^{yx}\left\{\frac{a_{11}(iy,\xi)\omega(iy) + a_{12}(iy,\xi)}{a_{21}(iy,\xi)\omega(iy) + a_{22}(iy,\xi)} - i\int_0^x e^{-yu}s(u)du\right\}$$

$$= e^{yx}\frac{a_{11}(iy,\xi) - i\Big(\int_0^x e^{-yu}s(u)du\Big)\,a_{21}(iy,\xi)}{a_{21}(iy,\xi)\omega(iy) + a_{22}(iy,\xi)}\omega(iy)$$

$$+ e^{yx}\frac{a_{12}(iy,\xi) - i\Big(\int_0^x e^{-yu}s(u)du\Big)\,a_{22}(iy,\xi)}{a_{21}(iy,\xi)\omega(iy) + a_{22}(iy,\xi)}$$

in two steps.

In Step 1 we show that the relations

$$\frac{e^{yx}}{y}\left\{a_{11}(iy,\xi) - i\Big(\int_0^x e^{-yu}s(u)du\Big)\,a_{21}(iy,\xi)\right\} \to 0 \quad\text{as}\quad y \to +\infty$$

and (28)

$$\frac{e^{yx}}{y}\left\{a_{12}(iy,\xi) - i\Big(\int_0^x e^{-yu}s(u)du\Big)\,a_{22}(iy,\xi)\right\} \to 0 \quad\text{as}\quad y \to +\infty$$

are satisfied. In Step 2 we estimate the values of the terms

$$\frac{1}{y}\left|\frac{a_{21}(iy,\xi)\omega(iy) + a_{22}(iy,\xi)}{\omega(iy)}\right|$$

and

$$\frac{1}{y}|a_{21}(iy,\xi)\omega(iy) + a_{22}(iy,\xi)|$$

from below for y large enough for an arbitrary R-function $\omega(z)$.

Step 1. Let us consider the first of the expressions (28):

$$\frac{e^{yx}}{y}\left\{a_{11}(iy,\xi) - i\Big(\int_0^x e^{-yu}s(u)du\Big)\,a_{21}(iy,\xi)\right\}$$

$$= \frac{e^{yx}}{y}\left\{[1 - iyB(M_{iy}(u))] - i\int_0^x e^{-yu}s(u)du[-iyB(e^{-yu})]\right\}$$

$$= \frac{e^{yx}}{y} - ie^{yx}\left\{B(M_{iy}(u)) - i\int_0^x e^{-yu}s(u)du\,B(e^{-yu})\right\}.$$

Writing

$$
\begin{aligned}
&i\left\{B(M_{iy}(u)) - i\int_0^x e^{-yt}s(t)dt\,B(e^{-yu})\right\}\\
&= i\frac{k}{\Delta}\overline{M_{-iy}(\xi)} - \frac{i\bar{k}}{\Delta}\langle\chi, M_{-iy}\rangle - i\frac{k}{\Delta}\langle\overline{\chi(\xi-u)}, M_{-iy}(u)\rangle\\
&\quad + \frac{\bar{k}}{\Delta}\int_0^x e^{-yt}s(t)dt + \frac{k}{\Delta}e^{-y\xi}\int_0^x e^{-yt}s(t)dt\\
&\quad - \frac{\bar{k}}{\Delta}\langle\chi(u), e^{-yu}\rangle\int_0^x e^{-yt}s(t)dt - \frac{k}{\Delta}\langle\overline{\chi(\xi-u)}, e^{-yu}\rangle\int_0^x e^{-yt}s(t)dt\\
&= \frac{\bar{k}}{\Delta}A_1 + \frac{k}{\Delta}A_2\,,
\end{aligned}
$$

we will show that the estimates

$$
\begin{aligned}
A_1 &= -i\langle\chi, M_{-iy}\rangle + \int_0^x e^{-yt}s(t)dt - \langle\chi(u), e^{-yu}\rangle\int_0^x e^{-yt}s(t)dt\\
&= \frac{1}{y}\left\{b + \frac{k}{y}\right\} + e^{-yx}o(1)
\end{aligned}
$$

and

$$
\begin{aligned}
A_2 &= i\,\overline{M_{-iy}(\xi)} - i\langle\overline{\chi(\xi-u)}, M_{-iy}(u)\rangle + e^{-y\xi}\int_0^x e^{-yt}s(t)dt\\
&\quad - \langle\overline{\chi(\xi-u)}, e^{-yu}\rangle\int_0^x e^{-yt}s(t)dt\\
&= \frac{1}{y}\left\{\bar{k}\xi + \bar{b} - \frac{\bar{k}}{y}\right\} + e^{-yx}o(1) \qquad (29)
\end{aligned}
$$

hold, as $y \uparrow \infty$.

Let us establish the first one:

$$
\begin{aligned}
A_1 &= -\int_0^\xi \chi(u)\int_0^u e^{-(u-t)y}\overline{s(t)}dt\,du + \int_0^x e^{-yt}s(t)dt\\
&\quad - \int_0^\xi \chi(u)e^{-yu}du\int_0^x e^{-yt}s(t)dt\\
&= \int_0^x e^{-yt}s(t)dt - \int_0^\xi \chi(u)\int_{-u}^0 e^{-(u+\tau)y}s(\tau)d\tau\,du\\
&\quad - \int_0^\xi \chi(u)\int_0^x e^{-(u+t)y}s(t)dt\,du
\end{aligned}
$$

$$= \int_0^x e^{-yt}s(t)dt - \int_0^\xi \chi(u) \int_{-u}^{x-u} e^{-(u+\tau)y} s(\tau)d\tau\, du$$
$$- \int_0^\xi \chi(u) \int_{x-u}^{x} e^{-(u+\tau)y} s(\tau)d\tau\, du .$$

Since

$$\int_0^\xi h(u)e^{-yu}du \to 0 \text{ as } y \to +\infty \tag{30}$$

for every summable function $h(u)$, we see that

$$\int_0^\xi \chi(u) \int_{x-u}^{x} e^{-(u+\tau)y} s(\tau)d\tau\, du = e^{-xy} \int_0^\xi \chi(u) \int_0^u e^{-ty} s(t-u+x)dt\, du$$
$$= e^{-xy} o(1) ,$$

and hence that

$$A_1 = \int_0^x e^{-yt}s(t)dt - \int_0^\xi \chi(u) \int_{-u}^{x-u} e^{-(u+\tau)y} s(\tau)d\tau\, du + e^{-xy}o(1)$$
$$= \int_0^x e^{-yt}s(t)dt - \int_0^\xi \chi(u) \int_0^x e^{-ty} s(t-u)dt\, du + e^{-xy}o(1)$$
$$= \int_0^x e^{-ty} \left\{ s(t) - \int_0^\xi \chi(u)s(t-u)du \right\} dt + e^{-xy}o(1) .$$

Using the triad identity (5), we obtain

$$A_1 = \int_0^x e^{-ty}\{k(\xi)t + b(\xi)\}dt + e^{-xy}o(1)$$
$$= \frac{1}{y}\left\{ b(\xi) + \frac{k(\xi)}{y} \right\} + e^{-xy}o(1) .$$

To estimate A_2, we first observe that in view of (30),

$$e^{-y\xi} \int_0^x e^{-yt}s(t)dt = o(1)e^{-xy} .$$

Hence

$$A_2 = \int_0^\xi e^{-y(\xi-t)}\overline{s(t)}dt - \int_0^\xi \overline{\chi(\xi-u)} \int_0^u e^{-(u-t)y}\, \overline{s(t)}dt\, du$$
$$- \int_0^\xi \overline{\chi(\xi-u)}e^{-yu}du \int_0^x e^{-yt}s(t)dt + e^{-yx}o(1)$$

$$
\begin{aligned}
&= \int_0^\xi e^{-yt}\overline{s(\xi-t)}dt - \int_0^\xi \overline{\chi(\xi-u)} \int_{-u}^0 e^{-(u+\tau)y}s(\tau)d\tau\, du \\
&\quad - \int_0^\xi \overline{\chi(\xi-u)} \int_0^x e^{-(u+t)y}s(t)dt\, du + e^{-yx}o(1) \\
&= \int_0^\xi e^{-yt}\overline{s(\xi-t)}dt - \int_0^\xi \overline{\chi(\xi-u)} \int_{-u}^{x-u} e^{-(u+\tau)y}s(\tau)d\tau\, du \\
&\quad - \int_0^\xi \overline{\chi(\xi-u)} \int_{x-u}^x e^{-(u+\tau)y}s(\tau)d\tau\, du + e^{-yx}o(1) \\
&= \int_0^x e^{-yt}\overline{s(\xi-t)}dt + \int_x^\xi e^{-yt}\overline{s(\xi-t)}dt \\
&\quad - \int_0^\xi \overline{\chi(\xi-u)} \int_0^x s(t-u)e^{ty}dt\, du \\
&\quad - e^{-xy}\int_0^\xi \overline{\chi(\xi-u)} \int_0^u e^{-ty}s(t-u+x)dt\, du + e^{-yx}o(1) \\
&= \int_0^x e^{-yt}s(t-\xi)dt + e^{-yx}\int_0^{\xi-x} e^{-y\tau}s\left(\overline{\xi-x-\tau}\right)d\tau \\
&\quad - \int_0^x e^{-ty}\int_0^\xi s(t-u)\overline{\chi(\xi-u)}du\, dt + e^{-yx}o(1) \\
&= \int_0^x e^{-yt}\left\{s(t-\xi) - \int_0^\xi s(t-u)\overline{\chi(\xi-u)}du\right\}dt + e^{-yx}o(1) \\
&= \int_0^x e^{-yt}\left\{\overline{k(\xi)}(\xi-t) + \overline{b(\xi)}\right\}dt + e^{-yx}o(1) \\
&= \frac{1}{y}\left\{\overline{k(\xi)}\xi + \overline{b(\xi)} - \frac{\overline{k(\xi)}}{y}\right\} + e^{-yx}o(1)\,.
\end{aligned}
$$

Thus

$$
\begin{aligned}
&i\left\{B(M_{-iy}) - i\int_0^x e^{-yt}s(t)dt\, B(e^{-yu})\right\} \\
&= \frac{\bar{k}}{\Delta}A_1 + \frac{k}{\Delta}A_2 \\
&= \frac{\bar{k}}{\Delta}\left\{\frac{b}{y} + \frac{k}{y^2}\right\} + \frac{k}{\Delta}\left\{\frac{\bar{k}\xi}{y} + \frac{\bar{b}}{y} - \frac{\bar{k}}{y^2}\right\} + e^{-yx}o(1) \\
&= \frac{1}{y} + e^{-xy}o(1)\,,
\end{aligned}
\tag{31}
$$

from which the first of the relations (28) follows.

Let us turn to the second of the relations (28). Using the estimates (29), we obtain

$$\left\{L(M_{-iy}) - i\int_0^x e^{-yt}s(t)dt\, L(e^{-yu})\right\}$$

$$= \frac{\bar{k}\xi+\bar{b}}{\Delta}A_1 - \frac{b}{\Delta}A_2$$

$$= \frac{\bar{k}\xi+\bar{b}}{\Delta}\left\{\frac{b}{y}+\frac{k}{y^2}\right\} - \frac{b}{\Delta}\left\{\frac{\bar{k}\xi}{y}+\frac{\bar{b}}{y}-\frac{\bar{k}}{y^2}\right\} + e^{-yx}o(1)$$

$$= \frac{1}{y^2} + e^{-yx}o(1)\,.$$

Invoking the last inequality and (31), we calculate:

$$C(M_{-iy}) - i\int_0^x e^{-yt}s(t)dt\, C(e^{-yu})$$

$$= -\Delta\left\{\left[L(M_{-iy}) - i\int_0^x e^{-yt}s(t)dt\, L(e^{-yu})\right]\right.$$

$$+B(M_0)\left[B(M_{-iy}) - i\int_0^x e^{-yt}s(t)dt\, B(e^{-yu})\right]$$

$$\left.-\left[B\left(\frac{M_0-M_{-iy}}{-iy}\right) - i\int_0^x e^{-yt}s(t)dt\, B\left(\frac{1-e^{-yu}}{-iy}\right)\right]\right\}$$

$$= -\Delta\left\{\frac{1}{y^2} - iB(M_0)\frac{1}{y} + e^{-yx}o(1) - \frac{1}{iy}\left[B(M_0) - i\int_0^x e^{-yt}s(t)dt\, B(1)\right]\right.$$

$$\left.+\frac{1}{iy}\left[B(M_{-iy}) - i\int_0^x e^{-yt}s(t)dt\, B(e^{-yu})\right]\right\}$$

$$= -\Delta\left\{\frac{1}{y^2} - \frac{1}{\Delta y}\int_0^x e^{-yt}s(t)dt - \frac{1}{y^2} + e^{-yx}o(1)\right\}$$

$$= \frac{1}{y}\int_0^x e^{-xt}s(t)dt + e^{-yx}o(1)\,.$$

Thus

$$\frac{e^{yx}}{y}\left\{a_{12}(iy,\xi) - i\int_0^x e^{-yt}s(t)dt\, a_{22}(iy,\xi)\right\}$$

$$= \frac{e^{yx}}{y}\left\{iyC(M_{-iy}) - i\int_0^x e^{-yi}s(t)dt[1+iyC(e^{-yu}]\right\}$$

$$= ie^{yx}\left\{C(M_{-iy}) - i\int_0^x e^{-yt}s(t)dt\, C(e^{-yu}) - \frac{1}{y}\int_0^x e^{-yt}s(t)dt\right\}$$

$$\to 0 \quad (y\to+\infty)\,.$$

The relations (28) are established.

Step 2. The aim is to find an estimate from below for the term

$$|a_{21}(z,\xi)\omega(z)+a_{22}(z,\xi)| = |a_{21}(z,\xi)|\cdot\left|\omega(z)+\frac{a_{22}(z,\xi)}{a_{21}(z,\xi)}\right|$$

$$\geq |a_{21}(z,\xi)|\,\mathrm{Im}\left\{\omega(z)+\frac{a_{22}(z,\xi)}{a_{21}(z,\xi)}\right\}.$$

From the J-properties of the matrix $W(z,\xi)$ it follows that $a_{22}(z,\xi)/a_{21}(z,\xi)$ is an R-function, and hence that

$$\mathrm{Im}\left\{\omega(z)+\frac{a_{22}(z,\xi)}{a_{21}(z,\xi)}\right\} = \mathrm{Im}\,\omega(z)+\mathrm{Im}\frac{a_{22}(z,\xi)}{a_{21}(z,\xi)}$$

$$\geq \mathrm{Im}\frac{a_{22}(z,\xi)}{a_{21}(z,\xi)}$$

$$= \frac{a_{22}(z,\xi)\overline{a_{21}(z,\xi)}-\overline{a_{22}(z,\xi)}a_{21}(z,\xi)}{2i|a_{21}(z,\xi)|^2}.$$

Thus,

$$|a_{21}(z,\xi)\omega(z)+a_{22}(z,\xi)| \geq \frac{a_{22}(z,\xi)\overline{a_{21}(z,\xi)}-\overline{a_{22}(z,\xi)}\,a_{21}(z,\xi)}{2i|a_{21}(z,\xi)|}.$$

Analogously,

$$\left|\frac{a_{21}(z,\xi)\omega(z)+a_{22}(z,\xi)}{\omega(z)}\right| = |a_{22}(z,\xi)|\cdot\left|\frac{a_{21}(z,\xi)}{a_{22}(z,\xi)}+\frac{1}{\omega(z)}\right|$$

$$\geq |a_{22}(z,\xi)|\cdot\left|\mathrm{Im}\left\{\frac{a_{21}(z,\xi)}{a_{22}(z,\xi)}+\frac{1}{\omega(z)}\right\}\right|$$

$$= -|a_{22}(z,\xi)|\cdot\mathrm{Im}\left\{\frac{a_{21}(z,\xi)}{a_{22}(z,\xi)}+\frac{1}{\omega(z)}\right\}$$

$$\geq -|a_{22}(z,\xi)|\cdot\mathrm{Im}\frac{a_{21}(z,\xi)}{a_{22}(z,\xi)}$$

$$= \frac{a_{22}(z,\xi)a_{21}(z,\xi)-\overline{a_{22}(z,\xi)}a_{21}(z,\xi)}{2i|a_{22}(z,\xi)|}.$$

We now set $z=iy$, and evaluate the limits

$$\lim_{y\to+\infty}\frac{1}{y}a_{21}(iy,\xi) = \lim_{y\to+\infty}\frac{1}{y}\{-iyB(e^{-yx})\}$$

$$= -i\lim_{y\to+\infty}\left\{\frac{\bar{k}}{\Delta}+\frac{k}{\Delta}e^{-y\xi}-\frac{\bar{k}}{\Delta}(x(u),e^{-yu})-\frac{k}{\Delta}\left(\overline{x(\xi-u)},e^{-yu}\right)\right\}$$

$$= -i\frac{\bar{k}}{\Delta} \quad = -i\overline{\alpha}\,,$$

$$\begin{aligned}\lim_{y\to+\infty}\frac{1}{y}a_{22}(iy,\xi) &= \lim_{y\to+\infty}\frac{1}{y}\{1+iyC(e^{-yx})\}\\ &= -i\lim_{y\to+\infty}\Delta\{L(e^{-yx})+B(M_0)B(e^{-yx})-B(A_0e^{-yx})\}\\ &= -i\Delta\left\{i\,\frac{\bar{k}\xi+\bar{b}}{\Delta}+\frac{\bar{k}}{\Delta}B(M_0)\right\}\\ &= -i\{\bar{k}B(M_0)+i\bar{k}\xi+i\bar{b}\}\\ &= -i\left\{\overline{kB(M_0)-ib}\right\}\\ &= -i\overline{\beta}\,,\end{aligned}$$

and thus conclude with the help of (22) that

$$\lim_{y\to+\infty}\frac{1}{y^2}\left\{a_{22}(iy,\xi)\overline{a_{21}(iy,\xi)}-\overline{a_{22}(iy,\xi)}a_{21}(iy,\xi)\right\}=\overline{\beta}\alpha-\beta\overline{\alpha}=i\,.$$

Finally, we obtain

$$\lim_{y\to+\infty}\frac{1}{y}\left\{\frac{a_{22}(iy,\xi)\overline{a_{21}(iy,\xi)}-\overline{a_{22}(iy,\xi)}\;a_{21}(iy,\xi)}{2i|a_{21}(iy,\xi)|}\right\}=\frac{1}{2|\alpha|}$$

and

$$\lim_{y\to+\infty}\frac{1}{y}\left\{\frac{a_{22}(iy,\xi)\overline{a_{21}(iy,\xi)}-\overline{a_{22}(iy,\xi)}\;a_{21}(iy,\xi)}{2i|a_{22}(iy,\xi)|}\right\}=\frac{1}{2|\beta|}\,.$$

Hence, for y large enough, the inequalities

$$\frac{1}{y}|a_{21}(iy,\xi)\omega(iy)+a_{22}(iy,\xi)|\geq\frac{1}{4|\alpha|}$$

and (32)

$$\frac{1}{y}\left|\frac{a_{21}(iy,\xi)\omega(iy)+a_{22}(iy,\xi)}{\omega(iy)}\right|\geq\frac{1}{4|\beta|}$$

hold. The asymptotic relation (27) follows from (28) and (32). The theorem is proved.

□

The asymptotic relation established in Theorem 6 allows us to sum up our investigations. Namely, both the positive definiteness of the function $s(x)$ on $(-l,l)$ and the nonuniqueness of its p.d. continuation from $(-l,l)$ follow from this asymptotic relation.

Theorem 7. *Let $w(z)$ be an R-function, and let $s(x)$ be a function which is continuous on $(0,\xi)$, and for which the asymptotic relation*

$$\lim_{\substack{z=iy\\ y\to+\infty}} e^{-izx}\left\{w(z)-i\int_0^x e^{izv}s(v)dv\right\}=0\,,\quad x\in(0,\xi) \tag{33}$$

is satisfied. Then the following three conclusions hold:

1. *The function $s(x)$ is determined uniquely.*
2. *The function w is representable in the form*

$$w(z)=\int_{-\infty}^{\infty}\frac{d\sigma(t)}{t-z}\,, \tag{34}$$

where $d\sigma(t)\geq 0$ and $\displaystyle\int_{-\infty}^{\infty} d\sigma(t)<\infty$.

3. *The function $s(x)$ admits the integral representation*

$$s(x)=\int_{-\infty}^{\infty} e^{-ixt}d\sigma(t)\,,\quad for\quad x\in(0,\xi)\,, \tag{35}$$

in terms of this measure $d\sigma$.

Proof. 1. Let $s_1(x)$ and $s_2(x)$ be two functions for which the asymptotic relation (33) holds. Then

$$\lim_{\substack{z=iy\\ y\to+\infty}} e^{-izx}\int_0^x e^{izv}\{s_1(v)-s_2(v)\}dv=0\,,\quad x\in(0,\xi)\,. \tag{36}$$

The function

$$\phi(z)=\int_0^x e^{-iz(x-v)}\{s_1(v)-s_2(v)\}dv$$

is an entire function of exponential type, which is bounded on the real axis and on the negative imaginary half-axis. From (36) it follows that ϕ tends to zero on the positive imaginary half-axis. Therefore, by the Phragmén-Lindelöf principle, $\phi\equiv 0$. Hence, $s_1(x)=s_2(x)$ for $0<x<\xi$.

2. Rewriting the asymptotic relation (33) in the form

$$\lim_{y\to+\infty}\frac{e^{yx}}{y}\left\{yw(iy)-i\int_0^x ye^{-yu}s(u)du\right\}=0\,,$$

we conclude that

$$yw(iy)-i\int_0^x ye^{-yu}s(u)du\to 0\quad\text{as}\quad y\to+\infty\,.$$

Therefore, since $\int_0^x ye^{-yu}s(u)du$ is bounded, it follows that

$$|yw(iy)|\leq\text{const}\quad\text{for}\quad y\geq 1\,.$$

The representability of $w(z)$ in the form (34) follows from this estimate.
3. From conclusion 2 it follows that the Khinchin-Bochner integral

$$s_0(x) = \int_{-\infty}^{\infty} e^{-ixt} d\sigma(t)$$

makes sense. Let us consider

$$\begin{aligned}
F(iy) &= e^{yx} \left\{ w(iy) - i \int_0^x e^{-yv} s_0(v) dv \right\} \\
&= e^{yx} \left\{ \int_{-\infty}^{\infty} \frac{d\sigma(t)}{t - iy} - i \int_0^x e^{-yv} \int_{-\infty}^{\infty} e^{-ivt} d\sigma(t) dv \right\} \\
&= e^{yx} \int_{-\infty}^{\infty} \left\{ \frac{1}{t - iy} - i \int_0^x e^{-i(t-iy)v} dv \right\} d\sigma(t) \\
&= e^{yx} \int_{-\infty}^{\infty} \left\{ \frac{1}{t - iy} - \frac{1}{t - iy} \left(1 - e^{-i(t-iy)x}\right) \right\} d\sigma(t) \\
&= \int_{-\infty}^{\infty} \frac{e^{itx}}{t - iy} d\sigma(t) \,.
\end{aligned}$$

Next, from the estimate

$$|F(iy)| \le \int_{-\infty}^{\infty} \frac{d\sigma(t)}{\sqrt{t^2 + y^2}} \le \int_{-\infty}^{\infty} \frac{d\sigma(t)}{y} \to 0 \quad \text{as} \quad y \to +\infty \,,$$

we conclude that $s_0(x)$ satisfies the asymptotic relation (33). Finally, from the uniqueness, it follows that $s(x) = s_0(x)$ for $0 < x < \xi$, i.e.,

$$s(x) = \int_{-\infty}^{\infty} e^{-ixt} d\sigma(t) \quad \text{for} \quad x \in (0, \xi) \,.$$

The theorem is proved. □

6 The formulation of the main theorem

The following result is a consequence of Theorems 2, 5, 6 and 7.

Main theorem. *Let $s(x)$ be a function defined on $(-l, l)$ and satisfying the properties (i)–(iii) there. For the function s to be positive definite on this interval and to be nonuniquely continuable from there, it is necessary and sufficient that the function $s(x)$ satisfies both conditions:*
(A) $\langle (\lambda \boldsymbol{I}_\xi - \boldsymbol{K}_\xi)\varphi, \varphi \rangle > 0 \ \forall \varphi \in C[0, \xi] \,, \quad \varphi \neq 0 \,; \quad \forall \xi \in (0, l);$
(B) $\Delta(l - 0) \le 0.$

7 The discussion of the main theorem

Our Main Theorem gives the answer to the question:

What is the maximal value l_0 such that a function $s(x)$ satisfying (i)–(iii) is nonuniquely continuable from $(-l_0, l_0)$?

This value l_0 can be characterized in the following way:

1) This is the least value l for which the strict inequality (A) is violated, and $\Delta(l_0 - 0) \leq 0$,

or

2) The condition (A) is satisfied for all $\xi \in [0, l_0)$ (including the value $\xi = l_0$), but $\Delta(l_0) = 0$.

In both of these cases the continuation of the function $s(x)$ from the interval $[-l_0, l_0]$ is unique.

The meaning of the value $\Delta(\xi)$ can be clarified from the following formula. Let us consider a generalized function B:

$$B = B_\xi(x) = \frac{\bar{k}}{\Delta}\delta_0(x) + \frac{k}{\Delta}\delta_\xi(x) - \frac{\bar{k}}{\Delta}\chi_\xi(x) - \frac{k}{\Delta}\overline{\chi_\xi(\xi - x)}\,, \tag{37}$$

which is a linear combination of two Dirac δ-functions $\delta_\xi(x) = \delta(x - \xi)$, $\delta_0(x) = \delta(x)$ and a summable function. Let us calculate the scalar product $\langle \boldsymbol{S}_\xi B, B \rangle$. Calculating formally the action of the integral operator $\boldsymbol{S}_\xi$ (with a continuous kernel) on B_ξ, we obtain

$$\begin{aligned}
\boldsymbol{S}_\xi B_\xi(x) &= \int_0^\xi s(x-u)B_\xi(u)du = \frac{\bar{k}}{\Delta}s(x) + \frac{k}{\Delta}s(x-\xi) \\
&\quad - \frac{\bar{k}}{\Delta}\int_0^\xi s(x-u)\chi(u)du - \frac{k}{\Delta}\int_0^\xi s(x-u)\overline{\chi(\xi-u)}du \\
&= \frac{\bar{k}}{\Delta}\left\{s(x) - \int_0^\xi s(x-u)\chi(u)du\right\} \\
&\quad + \frac{k}{\Delta}\left\{s(x-\xi) - \int_0^\xi s(x-u)\overline{\chi(\xi-u)}du\right\} \\
&= \frac{\bar{k}}{\Delta}\{kx + b\} + \frac{k}{\Delta}\{\bar{k}(\xi - x) + \bar{b}\} = 1\,.
\end{aligned} \tag{38}$$

Let us multiply both sides of formula (4) for k: $k = s'(x) - \int\limits_0^\xi s'(x-t)\chi(t)dt$, by $\overline{\chi(x)}$ and integrate:

$$k\int_0^\xi \overline{\chi(x)}\,dx = \int_0^\xi s'(x)\overline{\chi(x)}dx - \int_0^\xi \overline{\chi(x)}\int_0^\xi s'(x-t)\chi(t)dt\,dx\,.$$

Adding the last inequality to its conjugate, we obtain

$$k\int_0^{\xi}\overline{\chi(x)}dx+\bar{k}\int_0^{\xi}\chi(x)dx=\int_0^{\xi}s'(x)\overline{\chi(x)}dx+\int_0^{\xi}\overline{s'(x)}\chi(x)dx\ .$$

Because

$$\int_0^{\xi}\overline{s'(x)}\chi(x)dx=k-s'(+0)\ \text{ and }\int_0^{\xi}s'(x)\overline{\chi(x)}dx=\bar{k}-\overline{s'(+0)}\ ,$$

it follows that

$$k+\bar{k}-k\int_0^{\xi}\overline{\chi(x)}dx-\bar{k}\int_0^{\xi}\chi(x)dx=s'(+0)+\overline{s'(+0)}\ .$$

Thus

$$\begin{aligned}\langle \boldsymbol{S}_{\xi}B,B\rangle=\langle 1,B\rangle&=\frac{k}{\Delta}+\frac{\bar{k}}{\Delta}-\frac{k}{\Delta}\int_0^{\xi}\overline{\chi(x)}dx-\frac{\bar{k}}{\Delta}\int_0^{\xi}\chi(x)dx\\&=\frac{s'(+0)+\overline{s'(+0)}}{\Delta}=-\frac{\lambda}{\Delta}\ .\end{aligned}$$

In particular, if $s(x)$ is a positive definite function, then from the last formula it follows that Δ can not be positive.[7] Since t he function $\Delta(\xi)$ is strictly increasing, we conclude that in Case 2) the function $s(x)$ is no longer positive definite on each interval larger than $(-l_0,l_0)$.

We turn to the generalized function B for a reason. The forms $B(f)$ and $L(f)$ introduced in (16) are scalar products of the generalized functions B and

$$L=L_{\xi}(x)=i\frac{\bar{k}\xi+\bar{b}}{\Delta}\delta_0(x)-i\frac{b}{\Delta}\delta_{\xi}(x)-i\frac{\bar{k}\xi+\bar{b}}{\Delta}\chi_{\xi}(x)+i\frac{b}{\Delta}\overline{\chi_{\xi}(\xi-x)} \tag{39}$$

with the function $f(x)$.
According to (38), the generalized function B satisfies the equality $\boldsymbol{S}_{\xi}B_{\xi}(x)=1$. Calculating the corresponding expression for L, we obtain

$$\begin{aligned}\boldsymbol{S}_{\xi}L_{\xi}(x)=&i\frac{\bar{k}\xi+\bar{b}}{\Delta}\left\{s(x)-\int_0^{\xi}s(x-t)\chi_{\xi}(t)dt\right\}-i\frac{b}{\Delta}\left\{s(x-\xi)\right.\\&\left.-\int_0^{\xi}s(x-t)\overline{\chi_1(\xi-t)}dt\right\}=ix\ .\end{aligned}$$

With the third antilinear form $C(f)$ from (10), one should associate the generalized function

$$C=C_{\xi}(x)=-\Delta\{L_{\xi}(x)+B(M_0)B_{\xi}(x)A_0^*B_{\xi}(x)\}\ , \tag{40}$$

7) From the relation $\langle SB,B\rangle=-\lambda/\Delta$, it follows that Theorem 2 could be proved without using M.G. Krein's criterion (1) for nonunique continuations of a p.d. function from an interval to the whole real axis.

where the adjoint of the operator A_0, $A_0^* f(x) = -i \int_x^\xi f(u)du$ acts on the δ-function $\delta_{x_0}(x)$ as follows:

$$A_0^* \delta_{x_0}(x) = \begin{cases} -i\,, & x \le x_0\,, \\ 0\,, & x > x_0\,. \end{cases}$$

With such a definition the equality $\langle A_0 h, f\rangle = \langle h, A_0^* f\rangle$ holds. It is possible to check that $SC_\xi(x) = M_0(x)$.

The notion of generalized functions B and C as solutions of the equations $\boldsymbol{S}_\xi B = 1$ and $\boldsymbol{S}_\xi C = M_0$, respectively, can be formalized in terms of a rigged Hilbert space related to the p.d. operator $\boldsymbol{S}_\xi$: $H_- \supseteq H = L^2 = [0,\xi] \supseteq H_+$. Because $s(x)$ is nonuniquely continuable from $[-\xi,\xi]$, it follows that $e^{-i\bar{z}x}$ and $M_{\bar{z}}(x)$ belong to H_+ for each fixed x. The generalized functions B, L and C, defined by (37), (39) and (40), can be considered as elements of the generalized space H_-. The values $B(e^{-i\bar{z}x})$ and $C(e^{-i\bar{z}x})$ of the antilinear forms can be considered as scalar products of the elements B and C from H_- with the functions $e^{-i\bar{z}x}, M_{\bar{z}}(x) \in H_+$.

A more general investigation of these problems by I.V. Kovalishina and V.P. Potapov[8], allows us to conclude that the fractional-linear transformation (26) with coefficient matrix $W(z,\xi)$, not only gives functions $w(z)$ which are associated with $s(x)$ by relations (34) and (35), but describes *all such functions.*

Finally, we remark that the condition of the continuity of the second derivative in (ii) can be weakened. We may assume only that the first derivative of s is absolutely continuous. Under condition (A) for such functions $s(x)$, the triad (χ, k, b) and the value $\Delta(\xi) = k\bar{k}\xi + \bar{k}b + u\bar{b}$ can still be introduced. However, now the function $\chi_\xi(x)$ is only summable (with respect to x). In spite of this, the validity of Theorem 3 can be verified by approximation arguments. All the other proofs remain true.

8 An example

In closing, we give examples illustrating the situations 1) and 2). Let us consider a family $\{s_\beta = x^2 - \beta x + 1\}_{\beta>0}$ of quadratic functions. In the figure, maximal sections of parabolas are presented such that (after the even extension) the corresponding functions are positive definite on the appropriate intervals and nonuniquely continuable from these intervals. These sections are intercepted on these parabolas by the straight line $y = 1$ and by the curve

$$y = \frac{1}{2}x^2 - \sqrt{-\frac{1}{12}x^4 + x^2 + 1}\,;$$

see the figure below.

8) See the supplementary reference [8], which appeared after this paper was published, and [9] for another approach and additional references.

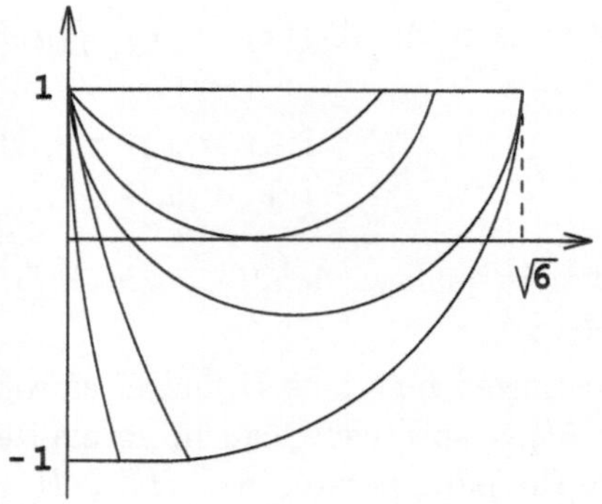

1. If $\beta \in (0, \sqrt{6})$, then Case 1 takes place:
Let $l_0 = \beta$. This l_0 satisfies the equality $s_\beta(l_o) = 1 (= s(0))$. The parabola s_β is continuable nonuniquely from the open interval $[0, l_0)$ and becomes continuable uniquely from the closed interval $[0, l_0]$, because the operator

$$(\lambda I - K_{l_0}^{(\beta)})f = 2\beta f(x) - 2\int_0^{l_0} f(u)du$$

is no longer positive definite. However, $\Delta(l_0) = \Delta(l_0 - 0) < 0$.

2. If $\beta > \sqrt{6}$, then Case 2 takes place.
The parabolas become continuable uniquely at the points l_0 of its intersections with the curve

$$y = \frac{1}{12}x^2 - \sqrt{-\frac{1}{12}x^4 + x^2 + 1}\ .$$

At these points $\Delta(l_0)$ vanishes.

The case $\beta = \sqrt{6}$ is a subject of special interest: at the point $l_0 = \sqrt{6}$ both conditions are violated: (A) and (B): $\Delta(l_0 - 0) = 0$.

References

[1] Krein, M.G., *On the logarithm of an infinitely decomposable Hermitian-positive function* (in Russian), Doklady Akad. Nauk SSSR **45**:3 (1944), 91–94. **MR 6**, p. 269.

[2] Krein, M.G., *On the continuation problem for Hermitian-positive continuous functions* (in Russian), Doklady Akad. Nauk SSSR **26**:1 (1940), 17–21. **MR 2**, p. 361.

[3] Krein, M.G., *On integral equations that generate second-order differential equations* (in Russian), Doklady Akad. Nauk SSSR **97**:1 (1954), 21–24. **MR 16**, p. 372.

[4] Potapov, V.P., *The multiplicative structure of J-contractive matrix-functions*, (in Russian), Trudy Mosk. Matem. Obshch. **4** (1955), 125–236. **MR 17**,

p. 958–959. English transl.: Amer. Math. Soc. Transl. (ser. **2**) **15** (1960), 131–243. **MR 22**#5733.

[5] Efimov, A.V. and V.P. Potapov, *J-expansive matrix-functions and their role in the theory of electrical circuits* (in Russian), Uspekhi Matem. Nauk **28**:1 (1973), 65–130. English transl.: Russian Math. Surveys **28**:1 (1973), 69–140. **MR 52** #15090.

[6] Kovalishina, I.V. and V.P. Potapov, *An indefinite matrix in the Nevanlinna-Pick problem* (in Russian), Akad. Nauk Armyan. SSR Doklady **59**:1 (1974), 17–22. English. transl.: Amer. Math. Soc. Transl. (ser. **2**) **138** (1988), 37–54. **MR 53** #13577.

[7] Kovalishina, I.V., *J-expansive matrix-functions in the Carathéodory problem* (in Russian), Akad Nauk Armyan. SSR Doklady **59**:3 (1974), 129–135. **MR 51** #5939.

Supplementary references added in translation

[8] Kovalishina, I.V. and V.P. Potapov, *The triad method in the theory of extension of Hermitian positive functions*, (in Russian), Izv. Akad. Nauk Armyan. SSR Ser. Mat. **24** (1989), No. 3, 269–292. English. transl.: Soviet J. Contemporary Math. Anal. **24** (1989), No. 3, 61–84.

[9] Krein, M.G. and H.K. Langer, *On some continuation problems which are closely related to the theory of operators in the space* Π_κ. *IV*. J. Operator Theory **13** (1985), 299–417.

Irina Mikhailova
Meierstrasse 1A
Steinbergkirche
24972 Bundesrepublik Deutschland

AMS Subject Classification: 42A82, 34L50.

Operator Theory
Advances and Applications, Vol. 95
© 1997 Birkhäuser Verlag Basel/Switzerland

Matrix boundary value problems with eigenvalue dependent boundary conditions (the linear case)

E. Russakovskii

Abstract. A selfadjoint extension of a matrix boundary value problem in which the eigenvalue parameter enters the boundary conditions linearly is constructed by adjoining a finite dimensional space with an indefinite scalar product. A related generalized Lagrange identity involving the V-Bezoutian of the linear matrix polynomials from the boundary conditions is derived.

0 Introduction

Let $L^2_d(0,1)$ be the space of d-dimensional vector valued functions such that each component of the column vector belongs to $L^2(0,1)$.

We consider in this space a regular matrix boundary value problem

$$\left.\begin{aligned} & l[u(x)] \equiv -(\mathcal{P}(x)u'(x))' + \mathcal{Q}(x)u(x) = \lambda u(x), \quad x\in(0,1), \\ & F(\lambda)\cdot\begin{pmatrix} u(x)|_{x=0} \\ u(x)|_{x=1}\end{pmatrix} - G(\lambda)\cdot\begin{pmatrix} \mathcal{P}(x)u'(x)|_{x=0} \\ -\mathcal{P}(x)u'(x)|_{x=1}\end{pmatrix} = 0, \end{aligned}\right\} \tag{0.1}$$

where the $d\times d$ matrix valued functions $\mathcal{P}^{-1}(x)$ and $\mathcal{Q}(x)$ are selfadjoint and summable on the closed interval [0,1] and the linear $2d\times 2d$ matrix polynomials $F(\lambda) = F_1\lambda + F_0$ and $G(\lambda) = G_1\lambda + G_0$ satisfy the following conditions:

$$(\forall\,\lambda\in\mathbf{R})\qquad F(\lambda)\cdot[G(\lambda)]^* - G(\lambda)\cdot[F(\lambda)]^* = 0, \tag{0.2}$$

$$(\forall\,\lambda\in\mathbf{C}\cup\infty)\ F(\lambda)\cdot[F(\lambda)]^* + G(\lambda)\cdot[G(\lambda)]^* > 0 \tag{0.3}$$

(we adopt the convention that if $\lambda = \infty$ in (0.3), then the value of the matrix polynomial is replaced by its leading coefficient).

Some special cases of the problem (0.1)–(0.3) were considered by the author in [1, 2, 3].

A related problem (with $F(\lambda)\equiv F$, $G(\lambda)\equiv G$) was studied by F.S. Rofe-Beketov [4]. In that case the conditions (0.2)–(0.3) turn into the selfadjointness conditions for the problem (0.1) (see [4]):

$$F \cdot G^* - G \cdot F^* = 0,$$
$$F \cdot F^* + G \cdot G^* > 0.$$

Similar boundary value problems can be obtained when using the Fourier method of separation of variables on mixed problems of partial differential equations with boundary conditions containing the first time derivative. Different aspects of such problems were studied by A.V. Strauss, L. Collatz, J. Walter, A. Dijksma, H. Langer, H.S.V. de Snoo, B. Textorius, C.T. Fulton, D.S. Cohen, B.E. Goodwin, F.W. Schäfke, A. Schneider, H.D. Nießen, A.A. Shkalikov, G.V. Radzievskii, R. Mennicken, M. Möller, H. Röh, and some others (see [5]–[27] and also the detailed references in [1, 2, 3]). In all the special cases considered by the author (the Sturm-Liouville problem, the generalized periodic problem and the so-called crossed boundary problem [1, 2, 3]), it turned out that the problem (0.1)–(0.3) yielded a selfadjoint (in some indefinite scalar product) problem in a finite-dimensional extension $\mathcal{L}^2_d(0,1)$ of the space $L^2_d(0,1)$. This extension could be regarded as some Pontryagin space Π_κ; the appropriate indefinite scalar product in it (in which the problem (0.1)–(0.3) is selfadjoint) could be chosen (in a unique way) with the help of V-Bezoutians of matrix polynomials entering the boundary conditions.

In this paper we obtain similar results for the case of general linear polynomials $F(\lambda), G(\lambda)$ satisfying the conditions (0.2)–(0.3). At the same time we obtain a generalized Lagrange identity related to the problem (0.1)–(0.3). The nonlinear case of the problem (0.1)–(0.3) will be considered in another paper. We note only that if max (deg $F(\lambda)$, deg $G(\lambda)$) > 1, then the generalized Lagrange identity has a much more complicated structure.

1 Notation and remarks

Let V be a $4d \times 4d$ matrix.

The notions of a V-neutral pair of matrix polynomials, the resultant, and the V-Bezoutian were considered by the author in [1, 28, 2, 3] (see also [29]; the latter paper is now almost inaccessible for western readers). These notions were also studied by B.D.O. Anderson, E.J. Jury, I.C. Gohberg, G. Heinig, L. Lerer, T. Kailath, R.R. Bitmead, S.-Y. Kung, M.A. Kaashoek, L. Rodman, M. Tismenetsky, P. Fuhrmann, T. Shalom, and some others (see [30]–[42] and also the detailed references in [3]).

In the case under consideration, the linear matrix polynomials $F(\lambda) = F_1\lambda + F_0$ and $G(\lambda) = G_1\lambda + G_0$ form a V-neutral pair (where $V = \begin{pmatrix} 0 & I \\ -I & 0 \end{pmatrix}$), their resultant

$$\mathcal{R} = \begin{pmatrix} F_1 & G_1 \\ F_0 & G_0 \end{pmatrix}$$

is an invertible $4d \times 4d$ matrix, their V-Bezoutian

$$\mathcal{B} = F_1 \cdot G_0^* - G_1 \cdot F_0^* = G_0 \cdot F_1^* - F_0 \cdot G_1^*$$

is a Hermitian invertible $2d \times 2d$ matrix.

The formula

$$\mathcal{R} \cdot V \cdot \mathcal{R}^* = \begin{pmatrix} 0 & \mathcal{B} \\ -\mathcal{B} & 0 \end{pmatrix} \tag{1.1}$$

which connects the resultant and the V-Bezoutian of a V-neutral pair of linear matrix polynomials (see [29, 1, 28, 2, 3]) will be used in Section 3.

We consider a Pontryagin space $\mathcal{L}_d^2(0,1)$ of aggregates $\mathcal{U} = \text{col}\ (u(x); \hat{u}\ )$ where $u(x) \in L_d^2(0,1)$ and $\hat{u}$ is a $2d\times1$ matrix. The indefinite scalar product in $\mathcal{L}_d^2(0,1)$ is given by the following formula:

$$\langle \mathcal{U}, \mathcal{V} \rangle = \int_0^1 v^*(x) u(x)\ dx + \hat{v}^{\ *} \cdot \mathcal{A} \cdot \hat{u}\ , \tag{1.2}$$

where $\mathcal{A}$ is some Hermitian invertible $2d \times 2d$ matrix. Evidently, the Pontryagin space $\mathcal{L}_d^2(0,1)$ is a $2d$-dimensional extension of the source space $L_d^2(0,1)$.

2 The operator $\mathcal{T}$ in the space $\mathcal{L}_d^2(0,1)$

We define a linear operator $\mathcal{T}$ in the space $\mathcal{L}_d^2(0,1)$ with domain

$$\mathcal{D}_\mathcal{T} = \Big\{ \mathcal{U} = \text{col}\ (u(x); \hat{u}\ ) \in \mathcal{L}_d^2(0,1) :$$

a) $u(x)$, $\mathcal{P}(x)u'(x)$ are absolutely continuous on the closed interval $[0,1]$, $l[u(x)] \in L_d^2(0,1)$;

b) $\hat{u} = F_1 \cdot \begin{pmatrix} u(x)|_{x=0} \\ u(x)|_{x=1} \end{pmatrix} - G_1 \cdot \begin{pmatrix} \mathcal{P}(x)u'(x)|_{x=0} \\ -\mathcal{P}(x)u'(x)|_{x=1} \end{pmatrix} \Big\}$

by the rule

$$\mathcal{T}(\mathcal{U}) = \text{col}\ \left(l[u(x)];\ -F_0 \cdot \begin{pmatrix} u(x)|_{x=0} \\ u(x)|_{x=1} \end{pmatrix} + G_0 \cdot \begin{pmatrix} \mathcal{P}(x)u'(x)|_{x=0} \\ -\mathcal{P}(x)u'(x)|_{x=1} \end{pmatrix} \right)$$

for $\mathcal{U} \in \mathcal{D}_\mathcal{T}$.

Evidently, the problem (0.1)–(0.3) is equivalent to the eigenvalue problem for the operator $\mathcal{T}$ acting in the space $\mathcal{L}_d^2(0,1)$ (cf. with [1, 28, 3]).

3 The generalized Lagrange identity for the operator $\mathcal{T}$

Let $\mathcal{U} = \mathrm{col}\,(u(x); \hat{u}\,),\ \mathcal{V} = \mathrm{col}\,(v(x); \hat{v}\,) \in \mathcal{D}_{\mathcal{T}}$,

$$\theta_{\mathcal{U}} = (u(x)|_{x=0},\ u(x)|_{x=1},\ \mathcal{P}(x)u'(x)|_{x=0}, -\mathcal{P}(x)u'(x)|_{x=1})\,,$$

$$\theta_{\mathcal{V}} = (\,v(x)|_{x=0},\ v(x)|_{x=1}, \mathcal{P}(x)v'(x)|_{x=0},\ -\mathcal{P}(x)v'(x)|_{x=1})\,.$$

A relation of the form

$$\langle \mathcal{T}(\mathcal{U}),\ \mathcal{V}\rangle\ -\ \langle \mathcal{U},\ \mathcal{T}(\mathcal{V})\rangle\ =\ \mathcal{S}\,(\theta_{\mathcal{U}},\ \theta_{\mathcal{V}})$$

(where $\mathcal{S}(\theta_{\mathcal{U}},\ \theta_{\mathcal{V}})\ =\ \theta_{\mathcal{V}}^* \cdot \mathcal{S}_{\mathcal{T},\,\mathcal{A}} \cdot \theta_{\mathcal{U}}$ is a bilinear form in $\theta_{\mathcal{U}},\ \theta_{\mathcal{V}}$) is called a generalized Lagrange identity for the operator $\mathcal{T}$ with respect to the indefinite scalar product (1.2).

Evaluating the expressions $\langle \mathcal{T}(\mathcal{U}),\ \mathcal{V}\rangle$ and $\langle \mathcal{U},\ \mathcal{T}(\mathcal{V})\rangle$, we obtain

$$\langle \mathcal{T}(\mathcal{U}), \mathcal{V}\rangle = \int_0^1 \{[\mathcal{P}(x)v'(x)]^* \cdot \mathcal{P}^{-1^*}(x) \cdot [\mathcal{P}(x)u'(x)] + v^*(x)\mathcal{Q}(x)u(x)\}\,dx$$

$$-([\,v(x)|_{x=0}]^*,\ [\,v(x)|_{x=1}]^*) \cdot F_1^* \mathcal{A} F_0 \cdot \begin{pmatrix} u(x)|_{x=0} \\ u(x)|_{x=1} \end{pmatrix}$$

$$+([\,v(x)|_{x=0}]^*,\ [\,v(x)|_{x=1}]^*) \cdot (F_1^* \mathcal{A} G_0 + I) \cdot \begin{pmatrix} \mathcal{P}(x)u'(x)|_{x=0} \\ -\mathcal{P}(x)u'(x)|_{x=1} \end{pmatrix}$$

$$+([\,\mathcal{P}(x)v'(x)|_{x=0}]^*,\ [\,-\mathcal{P}(x)v'(x)|_{x=1}]^*) \cdot G_1^* \mathcal{A} F_0 \cdot \begin{pmatrix} u(x)|_{x=0} \\ u(x)|_{x=1} \end{pmatrix}$$

$$-([\,\mathcal{P}(x)v'(x)|_{x=0}]^*,\ [\,-\mathcal{P}(x)v'(x)|_{x=1}]^*) \cdot G_1^* \mathcal{A} G_0 \cdot \begin{pmatrix} \mathcal{P}(x)u'(x)|_{x=0} \\ -\mathcal{P}(x)u'(x)|_{x=1} \end{pmatrix}$$

and a similar expression for $\langle \mathcal{U},\ \mathcal{T}(\mathcal{V})\rangle$.

Thus the remainder $\langle \mathcal{T}(\mathcal{U}),\ \mathcal{V}\rangle\ -\ \langle \mathcal{U},\ \mathcal{T}(\mathcal{V})\rangle$ can be written in the form

$$\langle \mathcal{T}(\mathcal{U}), \mathcal{V}\rangle - \langle \mathcal{U}, \mathcal{T}(\mathcal{V})\rangle = \theta_{\mathcal{V}}^* \cdot \left[\mathcal{R}^* \cdot \begin{pmatrix} 0 & \mathcal{A} \\ -\mathcal{A}^* & 0 \end{pmatrix} \cdot \mathcal{R} + V\right] \cdot \theta_{\mathcal{U}} = \theta_{\mathcal{V}}^* \cdot \mathcal{S}_{\mathcal{T},\mathcal{A}} \cdot \theta_{\mathcal{U}}\,.$$

Using the relation (1.1), we obtain:

$$\mathcal{S}_{\mathcal{T},\,\mathcal{A}} = V^{-1} \cdot \mathcal{R}^{-1} \cdot \begin{pmatrix} 0 & \mathcal{B} \\ -\mathcal{B} & 0 \end{pmatrix} \cdot \left[\begin{pmatrix} 0 & \mathcal{A} \\ -\mathcal{A}^* & 0 \end{pmatrix}\right.$$

$$\left. - \begin{pmatrix} 0 & -\mathcal{B}^{-1} \\ \mathcal{B}^{-1} & 0 \end{pmatrix}\right] \cdot \begin{pmatrix} 0 & \mathcal{B} \\ -\mathcal{B} & 0 \end{pmatrix} \cdot \mathcal{R}^{-1^*} \cdot V^{-1}.$$

So the generalized Lagrange identity takes the form

$$\langle \mathcal{T}(\mathcal{U}), \mathcal{V}\rangle - \langle \mathcal{U}, \mathcal{T}(\mathcal{V})\rangle = \theta_{\mathcal{V}}^{*} \cdot \Big\{ V^{-1} \cdot \mathcal{R}^{-1} \cdot \begin{pmatrix} 0 & \mathcal{B} \\ -\mathcal{B} & 0 \end{pmatrix}$$
$$\times \left[\begin{pmatrix} 0 & \mathcal{A} \\ -\mathcal{A}^{*} & 0 \end{pmatrix} - \begin{pmatrix} 0 & -\mathcal{B}^{-1} \\ \mathcal{B}^{-1} & 0 \end{pmatrix} \right] \cdot \begin{pmatrix} 0 & \mathcal{B} \\ -\mathcal{B} & 0 \end{pmatrix} \cdot \mathcal{R}^{-1*} \cdot V^{-1} \Big\} \cdot \theta_{\mathcal{U}}. \tag{3.1}$$

4 The appropriate indefinite scalar product in the Pontryagin space $\mathcal{L}_d^2(0,1)$

Let us find an appropriate indefinite scalar product (1.2) in which the operator $\mathcal{T}$ is symmetric. Using formula (3.1), we obtain a necessary and sufficient condition for symmetry:

$$\mathcal{S}_{\mathcal{T},\mathcal{A}} = 0 \iff \begin{pmatrix} 0 & \mathcal{A} \\ -\mathcal{A}^{*} & 0 \end{pmatrix} = \begin{pmatrix} 0 & -\mathcal{B}^{-1} \\ \mathcal{B}^{-1} & 0 \end{pmatrix} \iff \mathcal{A} = -\mathcal{B}^{-1} \tag{4.1}$$

(evidently, the relation (4.1) defines a Hermitian invertible $2d \times 2d$ matrix).

Further we assume that the scalar product in the space $\mathcal{L}_d^2(0,1)$ is chosen in accordance with the condition (4.1):

$$\langle \mathcal{U}, \mathcal{V}\rangle = \int_0^1 v^*(x)u(x)\,dx + \hat{v}^{\,*} \cdot (-\mathcal{B}^{-1}) \cdot \hat{u}\,. \tag{4.2}$$

In this case the rank of indefiniteness of the scalar product (see (4.2)) is equal to $\kappa = n_-(-\mathcal{B})$ ($n_-(C)$ denotes the number of negative eigenvalues of the matrix C); $0 \le \kappa \le 2d$.

5 Properties of the operator $\mathcal{T}$

Let λ_0 be an eigenvalue of the operator $\mathcal{T}$, $\mathbf{c}_1, \mathbf{c}_2, ..., \mathbf{c}_s$ be eigenchains (consisting of an eigenvector and adjoint vectors) corresponding to the eigenvalue λ_0 of the operator $\mathcal{T}$ and let l_k be the length of the chain $\mathbf{c}_k$. Let $\rho(\lambda_0) = \sum_{k=1}^{s}[l_k/2]$ if λ_0 is real, and let $\rho(\lambda_0) = \sum_{k=1}^{s} l_k$ in the opposite case (see [43]).

Theorem 5.1. *The operator $\mathcal{T}$ is selfadjoint in the indefinite scalar product (4.2). Its resolvent is a compact operator in the space $\mathcal{L}_d^2(0,1)$. The operator $\mathcal{T}$ has a discrete spectrum; only a finite number of its eigenvalues can be nonreal or multiple. For any system $\{\lambda_m\}$ of eigenvalues of the operator $\mathcal{T}$ with nonnegative imaginary parts*

$$\textstyle\sum_m \rho(\lambda_m) \le \kappa.$$

The system of all eigenvectors and adjoint vectors of the operator $\mathcal{T}$ is complete in the space $\mathcal{L}^2_d(0,1)$ and forms a Riesz basis in $\mathcal{L}^2_d(0,1)$.

We give an outline of the proof of this theorem.

The selfadjointness of the operator $\mathcal{T}$ can be checked by direct calculation of the adjoint operator $\mathcal{T}^c$ (with respect to the scalar product (4.2)).

Further, it is not difficult to obtain an explicit formula for the resolvent R_λ of the operator $\mathcal{T}$. It is easy to see that this resolvent is a meromorphic operator valued function in λ; its poles coincide with the eigenvalues of the operator $\mathcal{T}$; if λ is not an eigenvalue of the operator $\mathcal{T}$, then the value of R_λ is a compact operator in the space $\mathcal{L}^2_d(0,1)$. Thus the operator $\mathcal{T}$ has a discrete spectrum.

All the other properties of the spectrum of the operator $\mathcal{T}$ which are stated in Theorem 5.1 follow from a well known theorem of L.S. Pontryagin (Theorem 3 in [43]).

Finally, let $\tilde{\lambda}$ be a real regular point of the operator $\mathcal{T}$. Then the operator $(\mathcal{T}-\tilde{\lambda}I)^{-1}$ exists and is a compact operator in the Pontryagin space $\mathcal{L}^2_d(0,1)$. It is easy to see that this operator is also selfadjoint in $\mathcal{L}^2_d(0,1)$; its eigenvalues $\{\mu_n\}$ are connected with the eigenvalues $\{\lambda_n\}$ of the operator $\mathcal{T}$ by the relation $\mu_n = 1/(\lambda_n - \tilde{\lambda})$; μ=0 is not an eigenvalue of the operator $(\mathcal{T}-\tilde{\lambda}I)^{-1}$. The root subspace of the operator $\mathcal{T}$ corresponding to its eigenvalue λ_n coincides with that of the operator $(\mathcal{T}-\tilde{\lambda}I)^{-1}$ corresponding to its eigenvalue μ_n. In accordance with a well known theorem of T.Ya. Azizov and I.S. Iohvidov [44], the system of all eigenvectors and adjoint vectors of the operator $(\mathcal{T}-\tilde{\lambda}I)^{-1}$ (and, therefore, also that of the operator $\mathcal{T}$) is complete in the space $\mathcal{L}^2_d(0,1)$ and forms a Riesz basis in it.

Corollary. *The system of all the eigenfunctions and adjoint functions of the problem (0.1)–(0.3) is complete in the space $\mathcal{L}^2_d(0,1)$ but is not minimal; it is redundant by 2d functions.*

6 Comparison with earlier results

Let us compare the obtained formula (4.1) with similar ones in three special cases (considered in the paper [2]).

a) The Sturm-Liouville problem (SLP) (see [2], Sect.5).

$$\text{Let } F(\lambda) = \begin{pmatrix} A(\lambda) & 0 \\ 0 & D(\lambda) \end{pmatrix} \text{ and } G(\lambda) = \begin{pmatrix} -B(\lambda) & 0 \\ 0 & C(\lambda) \end{pmatrix}.$$

Then (0.2) and (0.3) hold if and only if

$$(\forall\, \lambda \in \mathbf{R}) \quad \begin{aligned} (\,A(\lambda),\ B(\lambda)\,) \cdot V_{SLP} \cdot (\,A(\lambda),\ B(\lambda)\,)^* &= 0\,, \\ (\,C(\lambda),\ D(\lambda)\,) \cdot V_{SLP} \cdot (\,C(\lambda),\ D(\lambda)\,)^* &= 0 \end{aligned}$$

and

$$(\forall\,\lambda\in\mathbf{C}\cup\infty)\qquad \begin{array}{c} A(\lambda)\cdot[A(\lambda)]^* + B(\lambda)\cdot[B(\lambda)]^* > 0, \\ C(\lambda)\cdot[C(\lambda)]^* + D(\lambda)\cdot[D(\lambda)]^* > 0, \end{array} \tag{6.1}$$

where

$$V_{SLP} = \begin{pmatrix} 0 & I \\ -I & 0 \end{pmatrix}.$$

Let $\mathcal{B}_0$ be the V_{SLP}-Bezoutian of $A(\lambda)$ and $B(\lambda)$ and let $\mathcal{B}_1$ be the V_{SLP}-Bezoutian of $C(\lambda)$ and $D(\lambda)$. Then

$$\mathcal{B} = \begin{pmatrix} -\mathcal{B}_0 & 0 \\ 0 & -\mathcal{B}_1 \end{pmatrix} \text{ and } \mathcal{A} = -\mathcal{B}^{-1} = \begin{pmatrix} \mathcal{B}_0^{-1} & 0 \\ 0 & \mathcal{B}_1^{-1} \end{pmatrix}.$$

b) The generalized periodic problem (GPP) (see [2], Sect.6).

$$\text{Let } F(\lambda) = \begin{pmatrix} A(\lambda) & B(\lambda) \\ 0 & 0 \end{pmatrix} \text{ and } G(\lambda) = \begin{pmatrix} 0 & 0 \\ -D(\lambda) & C(\lambda) \end{pmatrix}.$$

Then (0.2) and (0.3) hold if and only if

$$(\forall\,\lambda\in\mathbf{R})\qquad (\,A(\lambda),\,B(\lambda)\,)\cdot V_{GPP}\cdot(\,C(\lambda),\,D(\lambda)\,)^* = 0,$$

where

$$V_{GPP} = \begin{pmatrix} 0 & I \\ -I & 0 \end{pmatrix}$$

and the condition (6.1) is also satisfied. Let $\mathcal{B}_0$ be the V_{GPP}-Bezoutian of the quadruple $A(\lambda)$, $B(\lambda)$, $C(\lambda)$, $D(\lambda)$ (see [2], Sect.2). Then $\mathcal{B} = \begin{pmatrix} 0 & -\mathcal{B}_0 \\ -\mathcal{B}_0^* & 0 \end{pmatrix}$ and $\mathcal{A} = -\mathcal{B}^{-1} = \begin{pmatrix} 0 & \mathcal{B}_0^{-1^*} \\ \mathcal{B}_0^{-1} & 0 \end{pmatrix}$.

c) The crossed boundary problem (CBP) (see [2], Sect.7).

$$\text{Let } F(\lambda) = \begin{pmatrix} A(\lambda) & 0 \\ 0 & C(\lambda) \end{pmatrix} \text{ and } G(\lambda) = \begin{pmatrix} 0 & B(\lambda) \\ -D(\lambda) & 0 \end{pmatrix}.$$

Then (0.1) and (0.2) hold if and only if

$$(\forall\,\lambda\in\mathbf{R})\qquad (\,A(\lambda),\,B(\lambda)\,)\cdot V_{CBP}\cdot(\,C(\lambda),\,D(\lambda)\,)^* = 0,$$

where

$$V_{CBP} = \begin{pmatrix} 0 & I \\ I & 0 \end{pmatrix}$$

and the condition (6.1) is also satisfied. Let $\mathcal{B}_0$ be the V_{CBP}-Bezoutian of the quadruple $A(\lambda)$, $B(\lambda)$, $C(\lambda)$, $D(\lambda)$.

Then $\mathcal{B} = \begin{pmatrix} 0 & -\mathcal{B}_0 \\ -\mathcal{B}_0^* & 0 \end{pmatrix}$ and $\mathcal{A} = -\mathcal{B}^{-1} = \begin{pmatrix} 0 & \mathcal{B}_0^{-1^*} \\ \mathcal{B}_0^{-1} & 0 \end{pmatrix}$.

References

[1] Russakovskii, E.M.: Operator approaches to a boundary value problem with the eigenvalue parameter in the boundary conditions, Kharkov State University Ph.D.Thesis, 1990 (Russian).

[2] Russakovskii, E.M.: The theory of V-Bezoutians and its applications, Linear Algebra Appl. 212/213 (1994), 437–460.

[3] Russakovskii, E.M.: A matrix Sturm-Liouville problem with the eigenvalue parameter in the boundary conditions. Algebraic and operator aspects, Trudy MMO 57 (1995), to appear (Russian).

[4] Rofe-Beketov, F.S.: Selfadjoint extensions of differential operators in a space of vector functions, Soviet Math. Dokl. 10 (1969), 188–192.

[5] Strauss, A.V.: On spectral functions of a differential operator of an even order, Dokl. Akad. Nauk SSSR 115 (1957), 767–770 (Russian).

[6] Strauss, A.V.: On spectral functions of the differentiation operator, Usp. Mat. Nauk 13(84) (1958), 185–191 (Russian).

[7] Strauss, A.V.: On some extension families of a symmetric operator, Dokl. Akad. Nauk SSSR 139 (1961), 316–319 (Russian).

[8] Collatz, L.: Eigenwertaufgaben mit Technischen Anwendungen, Akademische Verlagsgesellschaft Geest & Portig, Leipzig 1963.

[9] Walter, J.: Regular eigenvalue problems with eigenvalue parameter in the boundary conditions, Math. Zeitschr. 133 (1973), 301–312.

[10] Dijksma, A.: Eigenfunction expansions for a class of J-selfadjoint ordinary differential operators with boundary conditions containing the eigenvalue parameter, Proc. Roy. Soc. Edinburgh 86A (1980), 1–27.

[11] Dijksma, A., Langer, H., and de Snoo, H.S.V.: Selfadjoint extensions of symmetric subspaces: an abstract approach to boundary problems with spectral parameter in the boundary conditions, Int. Equat. Oper. Theory 7 (1984), 459–515.

[12] Dijksma, A., Langer, H., and de Snoo, H.S.V.: Symmetric Sturm-Liouville operators with eigenvalue depending boundary conditions, Canadian Math. Soc. Conference Proc. 8 (1987), 87–116.

[13] Langer, H. and Textorius, B.: L-resolvent matrices of symmetric linear relations with equal defect numbers; applications to canonical differential relations, Int. Equat. Oper. Theory 5 (1982), 208–243.

[14] Fulton, C.T.: Two-point boundary value problems with eigenvalue parameter contained in the boundary conditions, Proc. Roy. Soc. Edinburgh 77A (1977), 293–308.

[15] Fulton, C.T.: Singular eigenvalue problems with eigenvalue parameter contained in the boundary conditions, Proc. Roy. Soc. Edinburgh 87A (1980), 1–34.

[16] Cohen, D.S.: On integral transform associated with boundary conditions containing an eigenvalue parameter, SIAM J. Appl. Math. 14 (1966), 1164–1175.

[17] Goodwin, B.E.: On the realization of the eigenvalues of integral equations whose kernels are entire or meromorphic in the eigenvalue parameter, SIAM J. Appl. Math. 14 (1966), 65–85.

[18] Schneider, A.: A note on eigenvalue problems with eigenvalue parameter in the boundary conditions, Math. Zeitschr. 136 (1974), 163–167.

[19] Schneider, A.: On spectral theory for the linear selfadjoint equation $Fy = \lambda Gy$, in "Ordinary and Partial Differential Equations, Proc., Dundee, Scotland 1980", Lecture Notes in Mathematics 846 (1981), 306–332.

[20] Schäfke, F.W. and Schneider, A.: S-hermitesche Rand-Eigenwertprobleme. I, Math. Ann. 162 (1966), 9–26.

[21] Schäfke, F.W. and Schneider, A.: S-hermitesche Rand-Eigenwertprobleme. II, Math. Ann. 165 (1966), 236–260.

[22] Schäfke, F.W. and Schneider, A.: S-hermitesche Rand-Eigenwertprobleme. III, Math. Ann. 177 (1968), 67–94.

[23] Niessen, H.-D.: Singuläre S-hermitesche Rand-Eigenwertprobleme, Manuscripta Math. 3 (1970), 35–68.

[24] Shkalikov, A.A.: Boundary value problems for ordinary differential equations with a parameter in the boundary conditions, Soviet Math. 33 (1986), 1311–1342.

[25] Radzievskii, G.V.: On a method of proving completeness of the root vectors of operator-valued functions, Soviet Math. Dokl. 15 (1974), 138–142.

[26] Mennicken, R. and Möller, M.: Boundary eigenvalue problems, Notas de Algebra y Analisis No.14, Universidad Nacional del Sur, Instituto de Matematica, Bahia Blanca (Argentina) 1986.

[27] Röh, H.: Self-adjoint subspace extensions satisfying λ-linear boundary conditions, Proc. Roy. Soc. Edinburgh 90A (1981), 107–124.

[28] Russakovskii, E.M.: A matrix Sturm-Liouville problem with the eigenvalue parameter in the boundary conditions, Funct. Anal. Appl. 27 (1993), 73–74.

[29] Russakovskii, E.M.: On Bezoutian and resultant theory of matrix polynomials, Deposited paper No. 5321, VINITI, Moscow 1981 (Russian).

[30] Anderson, B.D.O. and Jury, E.J.: Generalized Bezoutians and Sylvester matrices in multivariable linear control, IEEE Trans. Automat. Control 21 (1976), 551–556.

[31] Gohberg, I.C. and Heinig, G.: The resultant matrix and its generalizations, I. The resultant operator of matrix polynomials, Acta Sci. Math. Szeged 37 (1975), 41–61 (Russian).

[32] Gohberg, I.C. and Lerer, L.E.: Resultants of matrix polynomials, Bull. Amer. Math. Soc. 82 (1976), No.4, 565–567.

[33] Bitmead, R.R., Kung, S.-Y., Anderson, B.D.O., and Kailath, T.: Greatest common divisors via generalized Sylvester and Bezout matrices, IEEE Trans. Automat. Control 23 (1978), 1043–1047.

[34] Gohberg, I.C., Kaashoek, M.A., Lerer, L., and Rodman, L.: Common multiples and common divisors of matrix polynomials, I. Spectral method, Indiana Univ. Math. J. 30 (1981), 321–355.

[35] Gohberg, I.C., Kaashoek, M.A., Lerer, L., and Rodman, L.: Common multiples and common divisors of matrix polynomials, II. Vandermonde and resultant, Linear and Multilinear Algebra 12 (1982), 159–203.

[36] Lerer, L. and Tismenetsky, M.: The Bezoutian and the eigenvalue-separation problem for matrix polynomials, Int. Equat. Oper. Theory 5 (1982), 386–445.

[37] Lerer, L., Rodman, L., and Tismenetsky, M.: Bezoutian and Schur-Cohn problem for operator polynomials, J. Math. Anal. Appl. 103 (1984), 83–102.

[38] Fuhrmann, P.A.: Block Hankel inversion – the polynomial approach, Oper. Theory: Adv. Appl. 19 (1986), 207–230.

[39] Lerer, L. and Tismenetsky, M.: Generalized Bezoutian and the inversion problem for block matrices, I. General scheme, Int. Equat. Oper. Theory 9 (1986), 790–819.

[40] Gohberg, I.C. and Lerer, L.: Matrix generalizations of M.G.Krein theorems on orthogonal polynomials, Oper. Theory: Adv. Appl. 34 (1988), 137–202.

[41] Lerer, L. and Tismenetsky, M.: Generalized Bezoutian and matrix equations, Linear Algebra Appl. 99 (1988), 123–160.

[42] Gohberg, I.C. and Shalom, T.: On Bezoutians of nonsquare matrix polynomials and inversion of matrices with nonsquare blocks, Linear Algebra Appl. 137/138 (1990), 249–323.

[43] Pontryagin, L.S.: Hermitian operators in a space with indefinite scalar product, Izv. Akad. Nauk SSSR. Ser. mat. 8 (1944), N6, 243–280 (Russian).

[44] Azizov, T.Ya. and Iohvidov, I.S.: A criterium of completeness and basicity of root vectors of a compact J-selfadjoint operator in a Pontryagin space Π_κ, Matem. Issled. (Kishinev) 6 (1971), N1, 158–161 (Russian).

E.M. Russakovskii
Kharkov State Automobile Highway Technical University
Department of Mathematics
25 Petrovskii Street
310078 Kharkov
Ukraine
E-mail: arus@ilt.kharkov.ua

AMS Subject Classification: 15A24, 47A75, 47B50

Operator Theory
Advances and Applications, Vol. 95
© 1997 Birkhäuser Verlag Basel/Switzerland

Weyl-Titchmarsh functions of the canonical periodical system of differential equations

L.A. Sakhnovich

Abstract. The analytical properties of the Weyl-Titchmarsh function of a periodic canonical differential system of order $2m \times 2m$ are studied.

0 Introduction

Let $H(x)$ be a matrix-function of order $n \times n$ $(n = 2m)$, where the elements of $H(x)$ belong to $L_1(0,1)$,

$$H(x) \geq 0, \quad \int_0^1 H(x)dx > 0 \quad \text{and} \quad H(x+1) = H(x). \tag{0.1}$$

Let us also introduce the matrix

$$J = \begin{bmatrix} 0 & E_m \\ E_m & 0 \end{bmatrix} . \tag{0.2}$$

The corresponding canonical system of differential equations has the form

$$\frac{dY}{dx} = iz\, JH(x)Y, \qquad -\infty < x < \infty , \tag{0.3}$$

where $Y(x,z)$ is a vector-column of order $n \times 1$.

Conditions (0.1), (0.2) mean that the system (0.3) is of positive type [1] and periodic. We let $W(x,z)$ denote the matrix-solution of the system (0.3) which satisfies the normalization condition

$$W(0,z) = E_n . \tag{0.4}$$

In the articles [2], [3] we introduced an analogue of the Weyl-Titchmarsh function $v(z)$ for systems of the form (0.3) and clarified the connection between $v(z)$ and the spectral characteristics of the system (0.3). In the same articles we gave a procedure for solving the inverse problem under additional conditions, i.e., a procedure for reconstructing $H(x)$ from the known matrix-function $v(z)$. In the present article we investigate the Weyl-Titchmarsh matrix-function $v(z)$ in the periodic case. Here the character of the singularities of $v(z)$ (i.e., the branch points) are defined by the Jordan structure of the monodromy matrix

$$W(z) = W(l, z) \tag{0.5}$$

of the system (0.3).

Together with the system (0.3) we consider in this article the system

$$\frac{dY}{dx} = iz\, JH(x + \tau)Y, \qquad -\infty < x < \infty\,, \tag{0.6}$$

with a shift in the argument and find the connection between the Weyl-Titchmarsh function $v(\tau, z)$ of the system (0.6) and $v(z) = v(0, z)$.

In particular, we shall prove that the branch points of $v(z)$ are invariant with respect to the shift in the argument of the function $H(x)$.

1 On multipliers

1. This paragraph is an auxiliary one. Here the notions and results of the theory of multipliers which are contained in the famous article by G. Ja Ljubarski and M.G. Krein [5] are stated.

The eigenvalues $\rho_k(z)$ of the monodromy matrix $W(z)$ are called the multipliers of system (0.3). We denote the corresponding eigenvectors by $h_k(z)$ and the order of the corresponding Jordan cell by $q_k(z)$.

Theorem 1.1 [5]. *In the neighbourhood of the point $z_0 = \bar{z}_0$ the multipliers $\rho_k(z)$ of system (0.1)–(0.3) permit the expansion*

$$\rho_k(z) = \rho_{k,0} + \sum_{s=1}^{\infty} c_{k,s}(z - z_0)^{s/q_k}\,, \tag{1.1}$$

where $q_k = q_k(z)$ and

$$c_{k1} \neq 1\,. \tag{1.2}$$

It follows from expansion (1.1) that the corresponding eigenvectors $h_k(z)$ can be represented in the form

$$h_k(z) = \sum_{s=1}^{\infty} h_{k,s}(z - z_0)^{s/q_k}\,. \tag{1.3}$$

Let us also expand $W(z)$ in a series

$$W(z) = W_0 + \sum_{s=1}^{\infty} W_s(z - z_0)^s\,, \qquad W_0 = W(z_0)\,. \tag{1.4}$$

Corollary 1.1 [5]. *There exist numbers $l_1 < 0$ and $l_2 > 0$ such that the relations*

$$|\rho_k(z)| = 1, \quad q_k(z) = 1, \quad 1 \leq k \leq n, \quad z \in (l_1, l_2) \tag{1.5}$$

are valid in the interval (l_1, l_2).

From the equalities (1.1)–(1.4) and the relation

$$W(z)h_k(z) = \rho_k(z)h_k(z) \tag{1.6}$$

we deduce that

$$W_0 h_{k,s} = \rho_{k,0} + c_{k,s}h_{k,0} + c_{k,s-1}h_{k,1} + \cdots + c_{k,1}h_{k,s-1} \ , \tag{1.7}$$

where $1 \leq s \leq q_k - 1$.

It follows from (1.7) that $h_{k,0}$ is an eigenvector corresponding to the multiplier $\rho_{k,0} = \rho_k(z_0)$ and $h_{k,s}(1 \leq s \leq q_k - 1)$ are the corresponding adjunct vectors. From the results of article [5] we obtain the following assertion.

Theorem 1.2. *Let relations (0.1)–(0.3) be fulfilled and suppose that*

$$|\rho_{k,0}| = 1 \ . \tag{1.8}$$

Then the equalities

$$(Jh_{k,p}, h_{k,s}) = 0 \ , \qquad p + s < q_k - 1 \ , \tag{1.9}$$

and

$$(Jh_{k,p}, h_{k,s}) = e^{i\alpha_k s} c_k \ , \qquad p + s = q_k - 1 \ , \tag{1.10}$$

are valid, where

$$c_k = (JW_1 h_{k,0}, h_{k,0})/c_{k,1} \neq 0 \quad \text{and} \quad e^{i\alpha_k} = -\frac{\rho_{k,0}\bar{c}_{k,1}}{\bar{\rho}_{k,0}c_{k,1}} \ . \tag{1.11}$$

It follows from (1.10) that

$$(Jh_{k,s}, h_{k,p}) = e^{i\alpha_k p} c_k \ , \qquad p + s = q_k - 1 \ . \tag{1.12}$$

As

$$(Jh_{k,s}, h_{k,p}) = \overline{(Jh_{k,p}, h_{k,s})}$$

then from (1.10), (1.12) we deduce the equality

$$e^{i\alpha_k(q_k - 1)} = \bar{c}_k / c_k \ . \tag{1.13}$$

2. Now let us consider the case when

$$z_0 = \bar{z}_0, \quad \rho_{k,0} = \rho_{r,0}, \quad k \neq r, \quad |\rho_{k,0}| = 1 \ . \tag{1.14}$$

Theorem 1.3. *Let the relations (0.1)–(0.3) and (1.14) be fulfilled. Then the equalities*

$$(Jh_{k,p}, h_{r,s}) = 0, \qquad p + s \leq q_k - 1\ , \tag{1.15}$$

are true.

Remark 1.1. Relations (1.9), (1.10), (1.15) were deduced in [5] under the condition that $s = 0$. The assertions of Theorems 1.2 and 1.3 follow easily from this fact, if we take into consideration the equalities (1.7) and the relation

$$W_0^* J W_0 = J\ . \tag{1.16}$$

At least for the case

$$z_0 = \bar{z}_0\ , \qquad \rho_k \bar{\rho}_0 \neq 1 \tag{1.17}$$

the following assertion [5] is known:

Theorem 1.4. *Let the relations (0.1)—(0.3) and (1.17) be fulfilled. Then the equalities*

$$(Jh_{k,p}, h_{r,s}) = 0, \quad 0 \leq p \leq q_k - 1, \quad 0 \leq s \leq q_r - 1 \tag{1.18}$$

are true.

2 Weyl-Titchmarsh functions

1. Let r be a natural number. We introduce the matrix-functions

$$a_r(z) = J[W^r(z) + W^{-r}(z)]J\ , \tag{2.1}$$

$$b_r(z) = -J[W^r(z) - W^{-r}(z)]J \tag{2.2}$$

and write down the linear fractional transformation

$$v_r(z) = iJ[a_r(z)JP(z) + b_r(z)Q(z)]\ [b_r(z)JP(z) + a_r(z)Q(z)]^{-1}\ , \tag{2.3}$$

where $P(z)$, $Q(z)$ are a pair of meromorphic matrix-function such that

$$\det[P^*(z)P(z) + Q^*(z)Q(z)] \not\equiv 0, \qquad \text{Im } z > 0\ , \tag{2.4}$$

$$P^*(z)Q(z) + Q^*(z)P(z) \geq 0\ , \qquad \text{Im } z > 0\ . \tag{2.5}$$

The functions $v_r(z)$ form a Weyl circle; the limit

$$v(z) = \lim_{r\to\infty} v_r(z), \qquad \text{Im } z > 0, \tag{2.6}$$

exists and does not depend on the choice of $P(z)$ and $Q(z)$. We shall call the matrix-function $v(z)$ the Weyl-Titchmarsh function of system (0.3). The connection between $v(z)$ and the spectral characteristics of system (0.3) was clarified in [2], [3].

It is known that the function $W(z)$ has the following J-properties [6]:

$$W(z)JW^*(z) < J\ , \qquad \text{Im } z > 0\ , \tag{2.7}$$

$$W(z)JW^*(z) > J\ , \qquad \text{Im } z < 0\ , \tag{2.8}$$

$$W(z)JW^*(z) = J\ , \qquad \text{Im } z = 0\ . \tag{2.9}$$

It follows from (2.7) that $W(z)$ admits the representation

$$W(z) = U(z)\begin{bmatrix} D_1(z) & 0 \\ 0 & D_2(z) \end{bmatrix} U^{-1}(z), \qquad \text{Im } z > 0\ , \tag{2.10}$$

where $D_1(z)$ and $D_2(z)$ are matrices of order $m \times m$ and

$$\| D_1(z) \| < 1\ , \qquad \| D_2^{-1}(z) \| < 1\ , \qquad \text{Im } z > 0\ . \tag{2.11}$$

Theorem 2.1. *The following equality*

$$v(z) = iU(z)jU^{-1}(z)J\ , \qquad \text{Im } z > 0\ , \tag{2.12}$$

where

$$j = \begin{bmatrix} E_m & 0 \\ 0 & -E_m \end{bmatrix}, \tag{2.13}$$

is valid.

Proof. Let us rewrite formula (2.3) in the form

$$v_r(z) = iJb_r(z)[q_r(z)JP(z) + Q(z)]\ [q_r^{-1}(z)JP(z) + Q(z)]^{-1}a_r(z)^{-1}\ , \tag{2.14}$$

where

$$q_r(z) = b_r^{-1}(z)a_r(z)\ . \tag{2.15}$$

From (2.1), (2.2) and (2.10) we obtain that

$$q_r(z) = -JU(z)\begin{bmatrix} A_1(r,z) & 0 \\ 0 & A_2(r,z) \end{bmatrix} U^{-1}(z)J\ , \tag{2.16}$$

where

$$A_k(r,z) = [D_k^r(z) - D_k^{-r}(z)]^{-1}[D_k^r(z) + D_k^{-r}(z)]\ . \tag{2.17}$$

Let $r \to \infty$. Taking (2.11) and (2.17) into account, we deduce that

$$q(z) = \lim_{r\to\infty} q_r(z) = JU(z)jU^{-1}(z)J\ , \tag{2.18}$$

i.e.,

$$q(z) = q^{-1}(z)\ . \tag{2.19}$$

The validity of (2.12) now follows from (2.14), (2.18) and (2.19). □

2. Let us prove that the matrix $v(z)$ admits an analytic continuation.

Theorem 2.2. *The matrix-function $v(z)$ admits an analytic continuation on the whole complex plane with the cuts along the rays of the real axis $(-\infty, l_1]$, $[l_2, +\infty)$, where $l_1 < 0 < l_2$.*

Proof. Let us denote the multipliers of system (0.1)–(0.3) by $\rho_k(z)$ and the corresponding eigenvectors by $h_k(z)$. It is known [5] that there exist numbers $l_1 < 0 < l_2$ such that the matrix $W(z)$ does not have Jordan cells and $|\rho_k(z)| = 1$ $(1 \le k \le 2m)$ on the interval (l_1, l_2). This means that when $z \in (l_1, l_2)$, there are m multipliers of the first kind $\rho_k(z)$ $(1 \le k \le m)$ and m multipliers of the second kind $\rho_k(z)$ $(m+1 \le k \le 2m)$, i.e.,

$$\begin{aligned} &(Jh_k(z),\ h_k(z)) > 0, \qquad 1 \le k \le m\ ; \\ &(Jh_k(z), h_k(z)) < 0, \qquad m+1 \le k \le 2m\ . \end{aligned} \tag{2.20}$$

Let

$$\begin{aligned} D_1(z) &= \operatorname{diag}\{\rho_1(z), \cdots, \rho_m(z)\}\ , \\ D_2(z) &= \operatorname{diag}\{\rho_{m+1}(z), \cdots, \rho_{2m}(z)\}\ . \end{aligned}$$

Now we write down the matrix $v(z)$ in the form (2.10). The matrix-functions $U(z)$, $U^{-1}(z)$, $D_1(z)$, $D_2(z)$ will be analytic in a certain domain G which contains the interval (l_1, l_2). Let $z \in G$ and $\operatorname{Im}\ z < 0$ then the representation (2.10) remains true. However, relations (2.11) are replaced by the new bounds [5]

$$\| D_1^{-1}(z) \| < 1, \quad \| D_2(z) \| < 1, \quad \operatorname{Im}\ z < 0, \quad z \in G. \tag{2.21}$$

Since $W(z)$ does not have eigenvalues with modulus equal to 1 when $\operatorname{Im}\ z < 0$, the matrix-functions $U(z)$, $U^{-1}(z)$ satisfying relations (2.10), (2.21) can be chosen analytically at any boundary part of the low half-plane. This proves the theorem. □

3. Let the columns of the matrix $W(z)$ be formed by the vectors $h_k(z)$. We introduce the notation

$$U^*(z)JU(z) = S(z), \qquad z = \bar{z}. \tag{2.22}$$

We consider the case when on a certain interval of the real axis

$$a < z < b$$

all the multipliers are off the circle $|z| = 1$, i.e.,

$$|\rho_k(z)| < 1, \quad 1 \le k \le m; \qquad |\rho_k(z)| > 1, \quad m+1 < k < 2m. \tag{2.23}$$

Let us denote by L_1 and L_2 the invariant subspaces of $W(z)$ corresponding to the eigenvalues $|\rho_k(z)| < 1$ and $|\rho_k(z)| > 1$.

It is known that

$$\xi_k^* J \xi_k = 0, \qquad \xi_k \in L_k, \qquad (k = 0, 1). \tag{2.24}$$

By (2.24) the matrix $S(z)$ has the form

$$S(z) = \begin{bmatrix} 0 & s_1(z) \\ s_1^*(z) & 0 \end{bmatrix}.$$

This means that the equality

$$jS^{-1}(z) = \begin{bmatrix} 0 & s_1^{*-1}(z) \\ -s_1^{-1}(z) & 0 \end{bmatrix}$$

is valid. It follows from (2.22) that the considered case (2.12) can be rewritten in the form

$$v(z) = U(z) \begin{bmatrix} 0 & is_1^{*-1}(z) \\ -is_1^{-1}(z) & 0 \end{bmatrix} U^*(z) \tag{2.25}$$

i.e.,

$$v(z) = v^*(z), \qquad a < z < b. \tag{2.26}$$

From (2.26) and the results of article [3] we deduce the following theorem.

Theorem 2.3. *The interval (a, b) belongs to a gap in the spectrum of system (0.1)–(0.3) if and only if all the multipliers of $W(z)$ are off the circle $|\zeta| = 1$ when $z \in (a, b)$.*

Similarly to (2.25) it can be proved that under the condition

$$|\rho_k(z)| = 1, \qquad 1 \le k \le 2m, \qquad z \in (a, b), \tag{2.27}$$

the representation

$$v(z) = iU(z)jS^{-1}(z)U^*(z), \qquad jS^{-1}(z) > 0, \tag{2.28}$$

is true.

Thus, in case (2.27) we have

$$[v(z) - v^*(z)]/i > 0. \tag{2.29}$$

3 Singular points of the Weyl matrix function

In this paragraph we shall study the character of the singular points of the Weyl-Titchmarsh matrix-function $v(z)$ of the system (0.1)–(0.3).

1. Let the relations

$$z_0 = \bar{z}_0, \qquad |\rho_{k,0}| = 1, \qquad q_k = q_k(z_0) > 1 \tag{3.1}$$

be fulfilled in (1.1). We denote by $\eta_{k,r}(z)$ different branches of the multivalued function

$$\eta_{k,r}(z) = \sum_{s=0}^{\infty} h_{ks} e^{i2\pi rs/q_k}(z-z_0)^{s/q_k}, \ 0 \le r \le q_k - 1. \tag{3.2}$$

We shall assume that

$$z > z_0 \quad \text{and} \quad \arg(z-z_0)^{s/q_k} = 0. \tag{3.3}$$

By (1.9), (1.10) the asymptotic equality

$$\begin{aligned} &(J\eta_{ku}(z),\ \eta_{k,v}(z)) \\ &= (z-z_0)^{(q_k-1)/q_k} \sum_{p+s=q_k-1} e^{2\pi i(up-vs)/q_k} e^{i\alpha_k s} c_k + O(|z-z_0|)\ , \end{aligned} \tag{3.4}$$

where

$$c_k = (JW_1 h_{k,0}, h_{k,0})/c_{k,1} \neq 0\ , \tag{3.5}$$

is valid.

2. To begin with, we shall suppose that

$$e^{i\alpha_k q_k} \neq 1\ . \tag{3.6}$$

In this case

$$\sum_{p+s=q_k-1} e^{i2\pi(up-vs)/q_k} e^{i\alpha_k s} = S_{uv}\ , \tag{3.7}$$

where

$$S_{uv} = (1-e^{i\alpha_k q_k})/(e^{i2\pi u/q_k} - e^{-i2\pi v/q_k} e^{i\alpha_k})\ . \tag{3.8}$$

From (3.4), (3.7) we deduce that

$$(J\eta_{ku}(z),\ \eta_{kv}(z)) = (z-z_0)^{(q_k-1)/q_k} c_k s_{uv} + O(z-z_0). \tag{3.9}$$

Let us introduce the matrix

$$S = \{s_{vu}\}_{v,u=0}^{q_k-1} = \beta_k \left\{ \frac{1}{a_v + b_u} \right\}_{v,u=0}^{q_k-1} \tag{3.10}$$

where

$$a_v = -e^{-i2\pi v/q_k} e^{i\alpha_k},\ b_u = e^{i2\pi u/q_k},\ \beta_k = (1-e^{i\alpha_k q_k})\ . \tag{3.11}$$

Lemma 3.1. *Let condition (3.6) be fulfilled and let*

$$T = \{t_{pr}\}_{p,r=0}^{q_k-1} = S^{-1}\ . \tag{3.12}$$

Then

$$t_{p,r} = \beta_k e^{i\alpha_k(q_k-1)}/q_k^2 (e^{i2\pi r/q_k} - e^{-i2\pi p/q_k} e^{i\alpha_k})\ . \tag{3.13}$$

Proof. Now we write down the Cauchy formula (see [7]):

$$\Delta = \det\{1/(a_v + b_u)\}_{v,u=0}^{q-1} = \prod_{q-1\geq j>p\geq 0} (a_j - a_p)(b_j - b_p) / \prod_{u,v=0}^{q-1} (a_v + b_u) \,. \tag{3.14}$$

Let $A_{p,r}$ denote the complementary minor of the p, r entry in the Cauchy matrix. It follows from (3.14) that

$$A_{p,r}/\Delta = \frac{\prod_{v=0}^{q_k-1}(a_v + b - r)\prod_{u=0}^{q_k-1}(a_p + b_u)}{(a_p + b_r)\prod_{j\neq p}(a_j - a_p)\prod_{j\neq r}(b_j - b_r)} \,. \tag{3.15}$$

Taking into account (3.11) we have

$$\prod_{j\neq p}(a_j - a_p) = e^{i2\pi p/q_k} e^{i\alpha_k(q_k-1)} \prod_{s=1}^{q_k-1} (1 - e^{i2\pi s/q_k}) \tag{3.16}$$

and

$$\prod_{j\neq p}(b_j - b_r) = e^{i2\pi r/q_k} \prod_{s=1}^{q_k-1} (e^{i2\pi s/q_k} - 1) \,. \tag{3.17}$$

Analogously we deduce the relations

$$\prod_{v=0}^{q_k-1} (a_v + b_r) = \prod_{s=0}^{q_k-1} (1 - e^{i\alpha\pi s/q_k}) \tag{3.18}$$

and

$$\prod_{u=0}^{q_k-1} (a_p + b_u) = e^{i\alpha_k q_k} \prod_{s=0}^{q_k-1} (e^{i2\alpha\pi s/q_k} e^{i\alpha_k} - 1)) \,. \tag{3.19}$$

From the equality

$$\prod_{s=0}^{q-1} (z - e^{i2\pi s/q}) = z^q - 1$$

we obtain the relations

$$\prod_{s=1}^{q_k-1} (1 - e^{i2\pi s/q}) = q_k \tag{3.20}$$

and

$$\prod_{s=0}^{q_k-1} (e^{i\alpha_k} - e^{i2\pi s/q_k}) = e^{i\alpha_k q} - 1 \,. \tag{3.21}$$

In view of (3.20), formulas (3.16) and (3.17) take the forms

$$\prod_{j\neq p}(a_j - a_p) = e^{i2\pi p/q_k} e^{i\alpha_k(q_k-1)} q_k \tag{3.22}$$

and

$$\prod_{j\neq p}(b_j - b_r) = e^{i2\pi r/q_k} q_k(-1)^{q_k-1} , \tag{3.23}$$

respectively. According to (3.21), formulas (3.18), (3.19) can be written down in the following form

$$\prod_{v=0}^{q_k-1} (a_v + b_r) = \beta_k , \quad \prod_{u=0}^{q_k-1} (a_p + b_u) = \beta_k(-1)^{q_k-1} . \tag{3.24}$$

Here we use the notations of (3.11). Substituting (3.22)–(3.24) in (3.15) we obtain the equality

$$A_{p,r}/\Delta = \beta_k^2 e^{-i\alpha_k(q_k-1)}/[q^2(e^{i2\pi p/q_k} - e^{i2\pi r/q_k}e^{i\alpha_k})] . \tag{3.25}$$

Relation (3.13) follows from (3.25). □

Lemma 3.2. *The equality*

$$J_r = \sum_{p=0}^{q-1}(e^{i2\pi r/q} - e^{i2\pi p/q}e^{i\alpha})^{-1} = e^{-i2\pi r/q}q/(1 - e^{i\alpha q}) \tag{3.26}$$

is valid.

Proof. Let us write down J_r in the form

$$J_r = Q_r(z)/\prod_{p=0}^{q-1}(e^{i2\pi r/q} - e^{i2\pi p/q}e^{i\alpha}) , \tag{3.27}$$

where

$$Q_r(z) = \sum_{p=0}^{q-1}\prod_{s\neq p}(e^{i2\pi p/q} - e^{i2\pi p/q}\cdot z), \quad z = e^{i\alpha} . \tag{3.28}$$

Thus $Q_r(z)$ is a polynomial of degree not higher than $(q-1)$. By (3.28) the equalities

$$Q_r(e^{i2\pi(k+r)/q}) = e^{i2\pi r/q}\prod_{s\neq k}(1 - e^{i2\pi s/q}) = e^{i2\pi r/q}q , \tag{3.29}$$

$$(0 \leq k \leq q-1)$$

are true. From (3.29) we deduce that for all z the relation

$$Q_r(z) = e^{-i2\pi r/q}q \tag{3.30}$$

is valid. According to (3.21) we have

$$\prod_{p=0}^{q-1}(e^{i2\pi r/q} - e^{i2\pi p/q}e^{i\alpha}) = 1 - e^{i\alpha q} . \tag{3.31}$$

Thus equality (3.26) follows directly from (3.27), (3.30) and (3.31). □

3. Now let us consider the case when condition (3.6) is not fulfilled, i.e.,

$$e^{i\alpha_k} = e^{i2\pi r/q_k}, \qquad 0 \le r \le q_k - 1 . \tag{3.32}$$

In this case (3.9) holds again and

$$s_{uv}^{(k)} = \begin{cases} 0 & u+v \not\equiv r \pmod{q_k} \\ q_k e^{i2\pi u/q_k} & u+v \equiv r \pmod{q_k} \end{cases} . \tag{3.33}$$

Let us introduce the matrix

$$S_k = \left\{ s_{uv}^{(k)} \right\}_{v,u=0}^{q_k-1} .$$

It is easy to see that the elements t_{vu} of the matrix $S^{-1} = T$ are such that

$$t_{vu} = \begin{cases} 0 & u+v \not\equiv r \pmod{q_k} \\ e^{i2\pi u/q_k}/q_k & u+v \equiv r \pmod{q_k} . \end{cases} \tag{3.34}$$

We denote by $\vec{e}$ the vector

$$\vec{e} = \text{col}[1, 1, \cdots, 1] .$$

The equality

$$T\vec{e} = e^{i2\pi r/q_k} \text{col}[1, e^{-i2\pi/q_k}, \cdots, e^{i2\pi(q_k-1)/q_k}] \tag{3.35}$$

is obtained from (3.34).

4. We shall numerate the eigenvalues ρ_{k0} of $W(z_0)$ in such a way that

$$|\rho_{k,0}| = 1, \qquad 1 \le k \le N; \qquad |\rho_{k,0}| \ne 1, \qquad k > N. \tag{3.36}$$

We denote the dimensions of the root subspaces of $W(z_0)$ corresponding to $\rho_{k,0}$ by $q_k = q_k(z_0)$. As $h_k(z)$ (see (1.3)) is a multivalued vector-function, the number q_k of linearly independent vectors correspond to every number "k" when $z \ne z_0$. We denote by $L_1(z)$ the subspace with a basis constructed from the vectors $h_k(z)$ for which

$$|\rho_{k,0}| < 1 . \tag{3.37}$$

Let us also introduce the subspace $L_2(z)$ with a basis constructed from the vectors $h_k(z)$ for which

$$|\rho_{k,0}| > 1 . \tag{3.38}$$

Further let $\dim\ L_p(z) = Q_p \ \ (p = 1, 2)$. The bases chosen by us degenerate when $q_k > 1$. However, according to the Riesz spectral theorem, there exist analytical bases $\tilde{h}_k(z) \ \ (q < k \le Q_1 + q)$ and $\tilde{h}_k(z) \ \ (Q_1 + q < k \le 2m)$ of the subspaces $L_1(z)$ and $L_2(z)$ which do not degenerate. We shall introduce the matrix $\tilde{U}(z)$ of order $2m \times 2m$ the first q columns which are constructed from the vectors $h_k(z)$ while the others are constructed from $\tilde{h}_k(z)$.

The matrix

$$\tilde{S}(z) = \tilde{U}^*(z) J \tilde{U}(z) \tag{3.39}$$

will play an essential role.

From (2.10), (2.12) and (3.39) we deduce the relation

$$v(z) = i\tilde{U}(z)\tilde{j}\tilde{S}^{-1}(z)\tilde{U}^*(z), \qquad \text{Im } z > 0\ , \tag{3.40}$$

where

$$\tilde{j} = \text{diag}\,\{\varepsilon_1, \varepsilon_2, \cdots, \varepsilon_{2m}\}\ , \qquad \varepsilon_r = \pm 1\ . \tag{3.41}$$

Here $\varepsilon_r = 1$ if the r-th column of $\tilde{U}(z)$ is an eigenvector of the first kind and $r \le q$ whereas $\varepsilon_r = -1$ if the r-th column of $\tilde{U}(z)$ is an eigenvector of the second kind and $r \le q$. The equalities

$$\varepsilon_r = 1 \quad (q < r \le q + Q_1), \qquad \varepsilon_r = -1 \quad (q + Q_1 < r \le 2m)$$

hold when $r > q$.

We remark that equality (3.40) is valid only in a certain neighbourhood of z_0.

Let us write down the matrices $\tilde{S}(z)$, $v(z)$ and $\tilde{j}$ in the block form

$$\tilde{S}(z) = \{s_{pr}(z)\}_{p,r=1}^{N+1}\ , \qquad v(z) = \{v_{p,r}(z)\}_{p,r=1}^{N+1}$$

and

$$\tilde{j} = \text{diag}\,\{j_1, j_2, \cdots, j_{N+1}\}\ ,$$

respectively. The matrices $S_{rr}(z), v_{rr}(z)$ and j_r are of order q_r order $r \le N$ and the matrices $S_{N+1,N+1}(z), v_{N+1,N+1}(z)$ and j_{N+1} are of order $Q_1 + Q_2 = Q$.

Lemma 3.3. *The relations*

$$s_{p,r}(z) = 0(|z - z_0|)\ , \qquad p \neq r\ , \tag{3.42}$$

$$s_{kk}(z) = (z - z_0)^{(q_k - 1)/q_k} c_k S_k + 0(|z - z_0|), \qquad k \le N\ , \tag{3.43}$$

$$S_{N+1,N+1}(z) = S_{N+1} + 0(|z - z_0|) \tag{3.44}$$

are valid and

$$\det\ S_k \neq 0, \qquad 1 \le k \le N + 1. \tag{3.45}$$

Proof. The relations (3.42) follow from Theorems 1.3 and 1.4. By (3.9) and (3.10) equalities (3.43) are valid. According to (3.12) and (3.34) the inequalities (3.45) hold when $1 \leq k \leq N$. As the vectors $\tilde{h}_k(z)$ are linear independent and the analytical asymptotic equation (3.44) is true and

$$S_{N+1} = \left\{(J\tilde{h}_{q+r}(z_0),\ h_{q+p}(z_0))\right\}_{p,r=1}^{Q} , \tag{3.46}$$

$$\det\ S_{N+1} \neq 0 . \tag{3.47}$$

□

Let

$$T(z) = \{t_{p,r}(z)\}_{p,r=1}^{N+1} = \tilde{S}^{-1}(z) .$$

Lemma 3.4. *When $z > z_0$ the relations*

$$t_{p,p}(z) = (z - z_0)^{-(q_p-1)/q_p}\ S_p^{-1}\ \frac{1}{c_p}\ [E + 0(1)], \qquad p \leq N , \tag{3.48}$$

$$t_{N+1,N+1}(z) = S_{N+1}^{-1}[E + 0(1)] , \tag{3.49}$$

$$t_{p,r} = (z - z_0)^{-m_{p,r}}[E + 0(1)] , \qquad p \neq r , \tag{3.50}$$

are valid, where

$$m_{p,r} = \begin{cases} \min\{q_p, q_r\} , & p \leq N,\ r \leq N \\ 0 & p = N + 1 \\ 0 & r = N + 1 . \end{cases}$$

Proof. By Lemma 3.3 the matrix $\tilde{S}(z)$ can be written down in the form

$$\tilde{S}(z) = S_0(z) + 0(|z - z_0|) , \tag{3.51}$$

where $S_0(z)$ is a block-diagonal matrix with blocks

$$(z - z_0)^{(q_k-1)/q_k} c_k S_k \qquad (k \leq N)$$

and with the block S_{N+1} when $k = N + 1$.

It follows from (3.51) that

$$T(z) = S_0^{-1}(z)[E + 0(1)] = [E + 0(1)]S_0^{-1}(z). \tag{3.52}$$

From (3.52) we obtain the assertion of the lemma. □

5. The matrix $\tilde{U}(z_0)$ can be written down in the block form

$$\tilde{U}(z_0) = [U_1(z_0), U_2(z_0), \cdots, U_n(z_0), U_{N+1}(z_0)] \ ,$$

where the block $U_k(z_0)$ has the form

$$U_k(z_0) = \underbrace{[h_{k,0}, h_{k,0}, \cdots, h_{k,0}]}_{q_k} \ , \qquad k \leq N \ ,$$

and the block $U_{N+1}(z)$ has the form

$$U_{N+1}(z_0) = [\tilde{h}_{q+1,0}, \cdots, \tilde{h}_{2m,0}] \ .$$

We shall also assume that the q_k are enumerated in decreasing order and that q_1 is repeated N_1 times. Now we formulate the main theorem of this subsection.

Theorem 3.1. *The asymptotic equality*

$$v(z) = i\left[\sum_{k=1}^{N_1} U_{1k}(z_0) j_k S_k^{-1} U_{1k}^*(z_0)/C_k\right](z - z_0)^{-(q_1-1)/q_1}$$

$$+(z - z_0)^{-(q_1-1)/q_1} \cdot O(1) \tag{3.54}$$

is valid in the neighbourhood of z_0 $(z > z_0)$ *and*

$$\sum_{k=1}^{N_1} U_{ik}(z_0) j_k S_k^{-1} U_{ik}^*(z_0)/C_k \neq 0 \ . \tag{3.55}$$

Proof. Relation (3.54) follows directly from Lemma 3.4 and equality (3.40).

Let us denote

$$h_{k,0} = \mathrm{col}[h_1^{(k)}, h_2^{(k)}, \cdots, h_{2m}^{(k)}] \ , \qquad 1 \leq k \leq N \ ,$$

$$j_k = \mathrm{diag}\,\{\varepsilon_{1,k}, \varepsilon_{2,k}, \cdots, \varepsilon_{q_1,k}\} \ , \qquad \varepsilon_{p,r} = \pm 1 \ .$$

From equalities (1.13), (3.9) and (3.26) we deduce that

$$U_{1,k}(z_0) j_k S_k^{-1} U_{1,k}^*/C_k = \left\{h_p^{(k)} \cdot \bar{h}_s^{(k)}\right\}_{p,s=1}^{2m} \sum_{r=1}^{q_1} e^{-i2\pi(r-1)/q_k}, \varepsilon_{rk}/\bar{C}_k \ . \tag{3.56}$$

From formula (1.1) and the connection of the ε_{rk} with the kind of the corresponding eigenvalue $\rho_k(z)$ we obtain the inequality

$$\arg \rho_{k,0} - \arg C_{k1} - \frac{\pi}{2} \leq \arg\, e^{i(r-1)2\pi/q_k} \varepsilon_{rk}/\bar{C}_k \leq \arg \rho_{k,0} - \arg C_{k,1} + \frac{\pi}{2} \ . \tag{3.57}$$

This means that

$$\sum_{r=1}^{q_1} e^{-2i\pi(r-1)/q_k} \varepsilon_{rk} \neq 0 \ . \tag{3.58}$$

Since the vectors $h_{k,0}$ are linear independent, equation (3.55) follows from (3.56) and (3.58).

This proves the theorem. □

4 Transformation of the Weyl-Titchmarsh function

1. Let us consider the system of the form

$$\frac{dY}{dx} = iJH(x+\tau)Y, \qquad \tau=\bar{\tau}\ , \tag{4.1}$$

where $H(x)$ is a continuous matrix-function and

$$H(x) \geq 0\ . \tag{4.2}$$

In this section we shall find a connection between the Weyl-Titchmarsh function $v(\tau,z)$ of the system (4.1) and $v(z)=v(0,z)$. An analogous problem for the Sturm-Liouville equation is solved in [4]. This result proved to be useful for constructing finite-band solutions [4].

2. Let us introduce the notations

$$H_{\pm}(x,\tau) = H(x \pm \tau), \qquad\qquad x \geq 0\ . \tag{4.3}$$

We shall define the matrix functions $W_{\pm}(x,z)$ by the relations

$$\frac{dW_{\pm}(x,\tau)}{dx} = \pm izJH_{\pm}(x,\tau)W_{\pm}(x,\tau,z), \qquad W_{\pm}(0,\tau,z)=E_{2m}\ . \tag{4.4}$$

By (4.4) we have

$$W_{+}(x,\tau,z) = \int_0^{\overset{x}{\curvearrowleft}} \exp[izJH(x+\tau)dx]. \tag{4.5}$$

We introduce the matrix function

$$A(\tau,z) = \int_0^{\overset{\tau}{\curvearrowleft}} \exp[izJH(x)dx]. \tag{4.6}$$

From formulas (4.5) and (4.6) we have the relation

$$W_{+}(x,\tau,z) = W_1(x+\tau,z)A^{-1}(\tau,z), \qquad x+\tau>0\ , \tag{4.7}$$

where

$$W_1(x,z) = W_{+}(x,0,z). \tag{4.8}$$

Similarly we deduce the equality

$$W_{-}(x,\tau,z) = W_2(x-\tau,z)A^{-1}(\tau,z), \qquad x-\tau>0\ , \tag{4.9}$$

where

$$W_2(x,z) = W_{-}(x,0,z). \tag{4.10}$$

The Weyl-Titchmarsh functions $v(\tau,z)$ of the system (4.1) are defined by the inequalities [3]:

$$\int_0^{\infty} [J \pm iv^*(\tau,z)]W_{\pm}^*(x,t,z)H_{\pm}(x,\tau)W_{\pm}(x,\tau,z)[J \mp iv(\tau,z)]dx < \infty, \quad \text{Im } z>0\ . \tag{4.11}$$

Theorem 4.1. *Assume that the system (4.1) has a unique Weyl-Titchmarsh function when $\tau = 0$. Then the system (4.1) has a unique Weyl-Titchmarsh function for every choice of $\tau = \bar{\tau}$ and the equality*

$$v(\tau, z) = A(\tau, z)[v(z)J]A^{-1}(\tau, z)J \tag{4.12}$$

is valid.

Proof. Using formulas (4.3) and (4.7) we make a shift $x + \tau = u$ in (4.11):

$$\int_{\tau}^{\infty} [J+iv^*(\tau,z)]A^{*-1}(\tau,z)W_+^*(u,z)H_+(u)W_+(u,z)A^{-1}(\tau,z)[J-iv(\tau,z)]du<\infty \,.$$

As a result of the shift $x - \tau = u$ we have

$$\int_{-\tau}^{\infty} [J-iv^*(\tau,z)]A^{*-1}(\tau,z)W_-^*(u,z)H_-(u)W_-(u,z)A^{-1}(\tau,z)[J+iv(\tau,z)]du<\infty \,.$$

Since the Weyl-Titchmarsh function is unique when $\tau = 0$ it follows from the last two formulas that

$$v(z) = A^{-1}(\tau, z)v(\tau, z)JA(\tau, z)J \,.$$

Hence we get the assertion of the theorem. □

Corollary 4.1. *The singular points $z_k \neq \infty$ of the matrix functions $v(z)$ and $v(\tau, z)$ coincide.*

The validity of the corollary follows directly from formula (4.12) if we take into account the fact that $A(\tau, z)$ and $A^{-1}(\tau, z)$ are entire matrix functions of z.

References

[1] I. Gohberg, M.G. Krein, *Theory and applications of Volterra operators in Hilbert space*, Nauka, Moscow, 1967. English translation: Amer. Math. Soc., Providence, RI, 1970.

[2] L.A. Sakhnovich, *Factorization problems and operator identities*, Russian Math. Surv. **41** (1986). 1–64.

[3] L.A. Sakhnovich, *Method of operator identities and analysis problems*, Algebra Analiz. **5** (1993), 30–80. English translation: St. Petersburg Math. Journ. **5** 1 (1994), 1–69.

[4] B.M. Levitan, *Obratnye Zadachi Shturma-Liuvillia*, Nauka, Moscow, 1984. English translation: *Inverse Sturm-Liouville Problems*, VNU Science Press, Utrecht, The Netherlands, 1987.

[5] G. Ja Ljubarski, M.G. Krein, *Analytic properties of multipliers of periodic canonical differential systems of positive type*, Izv. Akad. Nauk SSSR, Ser.

Mat. **26** (1962), 549–572. English translation: Amer. Math. Soc. Transl. **89** (1970), 1–28.

[6] V.P. Potapov, *The multiplicative structure of j-contractive matrix-functions*, Amer. Math. Society Translations **15** (1960), 131–243.

[7] G. Polya, G. Szegö, *Problems and Theorems in Analysis*, **Vol 2**, Springer, Berlin, 1970.

L.A. Sakhnovich
pr. Dobrovolskogo 154, ap. 199
Odessa 270111,
Ukraine

AMS Subject Classification: 47A57, 34L05.

Operator Theory
Advances and Applications, Vol. 95
© 1997 Birkhäuser Verlag Basel/Switzerland

On boundary values of functions regular in a disk*

Vladimir I. Smirnov

Abstract. In this paper the parametric representation of a holomorphic function $f(z)$ from the Hardy class H_δ $(0 < \delta < \infty)$ as a product of the form $f(z) = b(z)\sigma(z)D(z)$, where (in the modern terminology) $b(z)$ is a Blaschke product, $\sigma(z)$ is a singular inner function and $D(z)$ is an outer function is established. A new proof of a theorem by F. & R. Nevanlinna on the representation of a holomorphic function of bounded characteristic as the quotient of two contractive holomorphic functions is also presented. [Abstract added by editors.]

1. Let $f(z)$ be a function which is regular in the unit disk $\boldsymbol{K}$ $(|z| < 1)$. In this paper some problems on the relationship between the properties of the function $f(z)$ inside the unit disk and the properties of its boundary values on the unit circle $\boldsymbol{C}$ $(|z| = 1)$ are considered. Our starting point is a class of functions which was considered by F. Riesz in his paper [R1]. Let us present first some known properties of functions from this class which will be needed below.

A function $f(z) = u(re^{i\varphi}) + iv(re^{i\varphi})$ which is regular in the disk $\boldsymbol{K}$ *belongs to the class* H_δ, $\delta > 0$, if the integrals

$$\int_0^{2\pi} |f(re^{i\varphi})|^\delta d\varphi,$$

which are increasing with r, are bounded as $r \uparrow 1$.

The boundary values $f(e^{i\varphi})$ of a function $f(z)$ belonging to H_δ exist almost everywhere, and

$$\lim_{r\uparrow 1} \int_M |f(re^{i\varphi})|^\delta d\varphi = \int_M |f(e^{i\varphi})|^\delta d\varphi, \tag{1}$$

*) Translated from French. French original: *Sur les valeurs limités des fonctions régulières à l'intérieur d'un cercle.* Journal de la Société Phys.-Math. de Léningrade. Tom **2**, fasc.2 (1928), pp.22–37.

where M is an arbitrary measurable subset of the interval $[0, 2\pi]$. Moreover,

$$\lim_{r \uparrow 1} \int_M |f(re^{i\varphi}) - f(e^{i\varphi})|^\delta d\varphi = 0 . \tag{2}$$

Let us assume that $f(z)$ has infinitely many zeros $\alpha_1, \alpha_2, \ldots, \alpha_n, \ldots$. Then the infinite product $\prod_{n=1}^{\infty} |\alpha_n|$ converges and the function

$$b(z) = \prod_{n=1}^{\infty} |\alpha_n| \, \frac{1 - z/\alpha_n}{1 - \bar{\alpha}_n z} . \tag{3}$$

($\bar{\alpha}_n$ is the number conjugate to α_n), which is termed *a Blaschke function*, has the properties: $|b(z)| < 1$ in points of the disk $\boldsymbol{K}$ and b has boundary values which equal one in modulus. The original function $f(z)$ can be represented in the form

$$f(z) = b(z)g(z) \tag{4}$$

where g does not vanish in $\boldsymbol{K}$ and belongs to H_δ as well. The procedure leading to formula (4) is said to be the *splitting off a Blaschke factor from* $f(z)$. If the function $f(z)$ has only finitely many zeros, then the product (3) will be finite as well.

If a function $f(z)$ belongs to the class H_1, then it follows from the formula (2) that the function is representable by *its* Cauchy integral

$$f(z) = \frac{1}{2\pi i} \int_C \frac{f(e^{i\varphi})}{e^{i\varphi} - z} \, d(e^{i\varphi}) \tag{5}$$

as well as by *its* Poisson integral

$$f(re^{i\theta}) = \frac{1}{2\pi} \int_0^{2\pi} f(e^{i\varphi}) \, \frac{1 - r^2}{1 - 2r\cos(\varphi - \theta) + r^2} \, d\varphi . \tag{6}$$

2. Let us state two theorems which are of great importance for function theory.

Theorem 1. *Let a function $f(z)$ belong to the class H_δ $(\delta > 0)$ and let the function $|f(e^{i\varphi})|^\lambda$ be summable for some $\lambda > \delta$. Then the function $f(z)$ belongs to the class H_λ.*

Proof. Splitting off a Blaschke function, let us write down the formula (4). Clearly, the function $|g(e^{i\varphi})|^\lambda$ is summable, and we need only to prove that $g(z) \in H_\lambda$. Thus, from the very beginning we may assume that the function f does not vanish in the disk $\boldsymbol{K}$. Replacing now $f(z)$ with $[f(z)]^{1/\delta}$, we see that we may restrict ourselves to the case $\delta = 1$, $\lambda > 1$. Since $f(z)$ belongs to H_1, formula (6) holds and the harmonic functions $u(re^{i\theta})$ and $v(re^{i\theta})$ are representable by Poisson integrals. Applying the well-known Hölder inequality to these integrals, we obtain

$$\int_0^{2\pi} |u(re^{i\varphi})|^\lambda d\varphi \le \int_0^{2\pi} |u(e^{i\varphi})|^\lambda d\varphi \ , \quad \int_0^{2\pi} |v(re^{i\varphi})|^\lambda d\varphi \le \int_0^{2\pi} |v(e^{i\varphi})|^\lambda d\varphi \, .$$

From here it follows that $f(z)$ belongs to the class H_λ. □

Theorem 2. *If a function $f(z)$ is represented by a Cauchy integral*

$$f(z) = \frac{1}{2\pi i} \int_C \frac{\psi(e^{i\varphi})}{e^{i\varphi} - z}\, d\left(e^{i\varphi}\right) \tag{7}$$

with a summable function $\psi(e^{i\varphi})$ or by a Cauchy-Stieltjes integral

$$f(z) = \frac{1}{2\pi i} \int_0^{2\pi} \frac{dF(\varphi)}{e^{i\varphi} - z} \tag{8}$$

with a function F of bounded variation, then the function $f(z)$ belongs to every class H_δ with $\delta < 1$.

Proof. Let us consider functions $f(z)$ which are representable by formula (7). It is clear that we may restrict ourselves to the case where $\psi(e^{i\varphi})$ is a real valued nonnegative function which is separated from zero: $\psi(e^{i\varphi}) \ge m > 0$, where m is some number. The corresponding function $u(re^{i\varphi})$ is representable by the Poisson integral, thus $u(re^{i\varphi}) \ge m > 0$. It follows from here that $f(z)$ does not vanish in the disk $\boldsymbol{K}$ and that the integrals

$$\int_0^{2\pi} |u(re^{i\varphi})| d\varphi \tag{9}$$

are bounded as $r \uparrow 1$. Actually, in the considered case these integrals do not depend on r and are equal $2\pi u(0)$. Thus, we have only to prove that the integrals

$$\int_0^{2\pi} |v(re^{i\varphi})|^\delta d\varphi \tag{10}$$

are bounded for every $0 < \delta < 1$. Let us consider the modulus $R(e^{i\varphi})$ and the argument $\Omega(e^{i\varphi})$ of the function $f(z)$. We may consider only the case $-\frac{\pi}{2} < \Omega(re^{i\varphi}) < \frac{\pi}{2}$. Let us turn to the function

$$[f(z)]^\delta = u_\delta(re^{i\varphi}) + iv_\delta(re^{i\varphi}) = R^\delta \cos\delta\Omega \ + iR^\delta \sin\delta\Omega$$

which is regular in the disk $\boldsymbol{K}$. For fixed δ $(0 < \delta < 1)$ and an appropriate constant k the inequality $\cos \delta\Omega \geq k > 0$ holds. From this inequality it follows that

$$|v(re^{i\varphi})|^{\delta} \leq R^{\delta} \leq (1/k)\, u_{\delta}(re^{i\varphi}).$$

Therefore, the integrals (10) do not exceed the value $(2\pi/k)\, u_{\delta}(0)$.

To obtain formula (8), it suffices to separate the real and imaginary parts of the function F and to represent them as the differences of increasing functions. The function $u(re^{i\varphi})$ will be positive again, the integrals (9) remain bounded and the proof given above remains valid. □

3. Let us note some immediate consequences of the theorems established above.

1. *If a function $f(z)$ is representable by a Cauchy integral (7) with some summable function $\psi(e^{i\varphi})$ and if the boundary values of the function $f(z)$ are summable, then the function f is representable by its Cauchy integral (formula (5)).*

From here it follows, among other things, that if a function $f(z)$ is representable by *its* Cauchy integral, then it belongs to H_1 and is representable as *its* Poisson integral [1].

2. *If a series which is conjugate to some Fourier-Lebesgue series is not a Fourier-Lebesgue series itself, then its generalized sum is a function which is not summable in the Lebesgue sense.*

Integrating by parts in formula (8) and taking into account the existence of the boundary values $f(e^{i\varphi})$, we arrive at the following known result[2]: *the series obtained by means of term by term differentiation of the Fourier series of a function of bounded variation as well as of its conjugate series, are Poisson summable.*

Let us make a remark related to Theorem 2. This theorem claims that from formulas (7) and (8) it follows that the function $f(z)$ belongs to all the classes H_{δ}, $\delta < 1$. However, the converse statement is not true. The following example demonstrates this:

$$f(z) = \frac{\log(1-z)}{1-z} = -\sum_{1 \leq n \leq \infty} \left(1 + \frac{1}{2} + \cdots + \frac{1}{n}\right) z^n.$$

This function belongs to all the classes H_{δ}, $\delta < 1$, but it is not representable by formulas (7) and (8) because its McLaurin coefficients are not bounded.

4. Let us make several additional remarks on functions $f(z)$ which are representable by a Cauchy integral of the form (7). Let us write (7) down in the form

$$f(z) = \int_0^{2\pi} \frac{\pi_1(\varphi) + i\pi_2(\varphi)}{e^{i\varphi} - z}\, d(e^{i\varphi}), \tag{11}$$

---

1) This statement was obtained by G.M. Fichtenholz [F] by means of a uniqueness theorem for analytic functions.

2) A.I. Plessner, [P]

where the functions $\pi_1(\varphi)$ and $\pi_2(\varphi)$ are real valued and summable. Let us introduce a new notion which we will need below. Let $\lambda(\varphi)$ be a real valued summable function on the interval $[0, 2\pi]$. If the trigonometric series which is conjugate to the Fourier series of this function is a Fourier series as well, then we will denote the latter by the symbol $\overline{\lambda}(\varphi)$ and will say that *the function* $\lambda(\varphi)$ *has the conjugate function* $\overline{\lambda}(\varphi)$. Let us note two obvious facts. In order for the Cauchy integral

$$\int_0^{2\pi} \frac{\lambda(\varphi) - i\mu(\varphi)}{e^{i\varphi} - z}\, d(e^{i\varphi})$$

to represent a function which is identically zero inside the unit disk, it is necessary and sufficient that $\mu(\varphi)$ be a function which is conjugate to the function $\lambda(\varphi)$ and that the Fourier series of these functions do not contain constant terms. Moreover, if formula (11) holds, then the function $f(z)$ is representable as *its* Cauchy integral (formula (5)) if and only if the functions $\pi_1(\varphi)$ and $\pi_2(\varphi)$ possess conjugate functions.

Let us turn again to functions $f(z)$ which are representable by formula (11). Introducing the absolutely continuous function

$$\pi(\varphi) = \int_0^{\varphi} [\pi_1(\tau) + i\,\pi_2(\tau)]\, d(e^{i\tau})$$

and putting $\pi(2\pi) = \alpha\, 2\pi i$, we can rewrite formula (11) in the form

$$f(z) = \frac{1}{2\pi i}\int_0^{2\pi} \frac{\pi(\varphi) - i\alpha\varphi}{(e^{i\varphi} - z)^2}\, d(e^{i\varphi}) = \frac{1}{2\pi i}\int_0^{2\pi} \frac{\omega_1(\varphi) + i\omega_2(\varphi)}{(e^{i\varphi} - z)^2}\, d(e^{i\varphi})\,,$$

where $\omega_1(\varphi)$ and $\omega_2(\varphi)$ are absolutely continuous periodic functions. Let us consider a primitive of the function $f(z)$:

$$F(z) = \frac{1}{2\pi i}\int_0^{2\pi} \frac{\pi(\varphi) - i\alpha\varphi}{e^{i\varphi} - z}\, d(e^{i\varphi}).$$

This function is representable as *its* Cauchy integral because there exist functions conjugate[3] to the functions $\omega_1(\varphi)$ and $\omega_2(\varphi)$. However, in general, the function $\omega_1(\varphi) + i\omega_2(\varphi)$ differs from the boundary values $F(e^{i\varphi})$. Nevertheless, the formula

$$0 = \frac{1}{2\pi i}\int_0^{2\pi} \frac{\omega_1(\varphi) + i\omega_2(\varphi) - F(e^{i\varphi})}{e^{i\varphi} - z}\, d(e^{i\varphi})$$

---

3) It is easy to prove that $F(z)$ belongs to all classes H_δ, $\delta > 0$.

holds in any case. That is, *there exists a real valued summable function* $\sigma(\varphi)$ *with a conjugate function* $\overline{\sigma}(\varphi)$ *such that the function*

$$F(e^{i\varphi}) + \sigma(\varphi) - i\overline{\sigma}(\varphi)$$

is absolutely continuous and periodic. It is easy to see that *this condition is not only necessary but also sufficient for the function* $f(z)$ *to be representable by a Cauchy integral.*

Observe that *the absolute continuity of the boundary values themselves is necessary for the function* $f(z)$ *to be representable by its Cauchy integral (formula (5)).*[4] Let us consider[5] an example of a function $f(z)$ which is representable by (11) but not by (5):

$$f(z) = \sum_{n=2}^{\infty} \frac{z^n}{\log n}$$

In this case the boundary values $F(e^{i\varphi})$ is an unbounded function which can be represented as a series

$$F(e^{i\varphi}) = \sum_{n=3}^{\infty} \frac{\cos n\varphi + i \sin n\varphi}{n \log(n-1)}$$

which converges uniformly outside each neighborhood of the point $\varphi = 0$. Moreover,

$$\pi_1(\varphi) \sim \sum_{n=2}^{\infty} \frac{\cos n\varphi}{\log n} \quad \text{and} \quad \pi_2(\varphi) = 0\ .$$

From here it follows that $\alpha = 0$ and

$$\begin{aligned}\pi(\varphi) = {} & \frac{1}{2}\sum_{n=2}^{\infty} \frac{1}{(n^2-1)\log n} - \frac{\cos\varphi}{\log 2} - \frac{\cos 2\varphi}{2\log 3} \\ & + \sum_{n=3}^{\infty} \frac{\cos n\varphi}{n}\left[\frac{1}{\log(n-1)} - \frac{1}{\log(n+1)}\right] \\ & + i\left\{\frac{\sin\varphi}{\log 2} + \frac{\sin 2\varphi}{2\log 3} + \sum_{n=3}^{\infty}\frac{\sin n\varphi}{n}\left[\frac{1}{\log(n-1)} + \frac{1}{\log(n+1)}\right]\right\}.\end{aligned}$$

5. Let us establish a formula giving a parametric representation of arbitrary regular functions from some class H_δ, $\delta > 0$. We first make one preliminary remark. Using Harnak's theorem on increasing sequence of harmonic functions and the Gauss mean value theorem (on harmonic functions), it is easy to show that a function $f(z)$ belongs to the class H_δ if and only if there exists a harmonic function in $\boldsymbol{K}$

4) F. & M. Riesz, [R2].
5) G.M. Fichtenholz, [F].

which majorizes the function $|f(z)|^\delta$ at every point. From here it follows immediately that a conformal mapping of the disk $\boldsymbol{K}$ onto itself transforms functions from the class H_δ to functions from the same class.

Let us turn to the problem of analytic representations of positive functions. It is known that if a function $f(z)$ belongs to the class H_δ, then the functions $|f(e^{i\varphi})|^\delta$ and $\log|f(e^{i\varphi})|$ are summable. Conversely, if $p(\varphi) \geq 0$ and if the functions $|p(\varphi)|^\delta$ and $\log p(\varphi)$ are summable then there exist infinitely many functions in the class H_δ whose boundary values coincide with $p(\varphi)$ in absolute value almost everywhere. Among these functions there exists one which is bigger in absolute value than the others inside the disk $\boldsymbol{K}$. (We do not consider that multiplying a function by a unimodular constant leads to an essentially new function.) In the paper [S] G. Szegö has found the form of this function using the theory of Toeplitz forms. We will do this without using Toeplitz forms and shall give a general formula for all of the above mentioned functions from the class H_δ.

Let us consider first the special case $\delta = 1$ and $p(\varphi) = 1$. If a function $f(z)$ belongs to H_1 and its boundary values are equal to one in absolute value, then $f(0) = \frac{1}{2\pi}\int_0^{2\pi} f(e^{i\varphi})\,d\varphi$ and $|f(0)| \leq 1$. Because any arbitrary prescribed point from the unit disk can be mapped into the point zero by an appropriate conformal mapping of the unit disk onto itself and because after such a substitution the function $f(z)$ remains in the class H_1, we conclude that $|f(z)| \leq 1$ everywhere in $\boldsymbol{K}$. Moreover, it is easy to see that the equality here is impossible unless the function $f(e^{i\varphi})$ is equal to a constant almost everywhere. Of course, in the last case the function $f(z)$ is representable by its Cauchy integral which is equal to the constant.

Assume now that $\delta = 1$ and $p(\varphi) \geq m > 0$. Let us prove that the required maximal function has the form

$$D(z) = \exp\frac{1}{2\pi}\int_0^{2\pi} \log p(x)\,\frac{e^{ix}+z}{e^{ix}-z}\,dx. \tag{12}$$

It is clear that $D(z)$ belongs to H_1. Observing that the function $1/p(\varphi)$ is summable and using the Jensen inequality, we conclude that

$$\int_0^{2\pi} |D^{-1}(re^{i\varphi})|d\varphi \leq \frac{1}{2\pi}\int_0^{2\pi}\frac{1}{p(x)}\,dx,$$

i.e., $D^{-1}(z)$ belongs to H_1 as well. Furthermore, if a function $f(z)$ has zeros in the disk $\boldsymbol{K}$ then, splitting off the Blaschke function $b(z)$ according to formula (4), we obtain the function $g(z)$ whose boundary values equal $|f(e^{i\varphi})|$ in absolute value but which is bigger than $|f(z)|$ in absolute value in the disk $\boldsymbol{K}$. Thus, when seeking the maximal function $f(z)$, one can assume that it does not vanish in the disk $\boldsymbol{K}$. The function $\sqrt{f(z)D^{-1}(z)}$ belongs to H_1 too, and its boundary values are equal to one in absolute value almost everywhere. It follows from the previous discussion that $|f(z)| \leq |D(z)|$; moreover the equality can hold only in the case $f(z) = cD(z)$, $|c| = 1$. Hence, $D(z)$ is the (unique) maximal function. Retaining the condition $\delta = 1$, consider now the general case in which $p(\varphi)$ is an arbitrary nonnegative summable function such that $\log p(\varphi)$ is summable also.

Let us introduce the auxiliary functions $p_m(\varphi)$ which are equal to $p(\varphi)$ if $p(\varphi) \geq m$ and to m if $p(\varphi) < m$. Set

$$D_m(z) = \exp \frac{1}{2\pi} \int_0^{2\pi} \log\, p_m(x)\, \frac{e^{ix} + z}{e^{ix} - z}\, dx.$$

The functions $D_m(z)$ and $D_m(z)^{-1}$ belong to H_1. The function $\sqrt{f(z) D_m^{-1}(z)}$ also belongs to H_1, and its boundary values do not exceed one in absolute value. As in the case $p(\varphi) \equiv 1$, we conclude that

$$|f(z)| \cdot |D_m^{-1}(z)| \leq 1,$$

i.e.,

$$|f(z)| \leq \exp \frac{1}{2\pi} \int_0^{2\pi} \log\, p_m(x)\, \frac{1 - r^2}{1 - 2r\cos(x - \theta) + r^2}\, dx,$$

where $z = re^{i\theta}$. Passing to the limit as $m \to \infty$, we obtain the inequality

$$|f(z)| \leq |D(z)|$$

for the function (12). It remains only for us to investigate when equality holds in this inequality. However the quotient $f(z)/D(z)$ is bounded and belongs to H_1. Thus, as was already shown, equality can hold only in the case $f(z) = cD(z)$, $|c| = 1$.

The general form of functions with prescribed absolute values $p(\varphi)$ of their boundary values can be represented as the product of three functions

$$f(z) = D(z) b(z) \sigma(z)\,, \tag{13}$$

where $D(z)$ is the maximal function which has been already defined, $b(z)$ is a Blaschke function and σ is a function which satisfies the inequality $0 < |\sigma(z)| \leq 1$ in the disk $\boldsymbol{K}$ whose boundary values are equal to one in absolute value. Let us consider the regular function $\log \sigma(z)$. Its real part is harmonic and negative in the disk $\boldsymbol{K}$. Such a function is representable as a Poisson-Stieltjes integral[6]

$$\log|\sigma(z)| = \frac{1}{2\pi} \int_0^{2\pi} \frac{1 - r^2}{1 - 2\cos(x - \theta) + r^2}\, d\psi(x)$$

6) See Fichtenholz [F].

with some decreasing function $\psi(x)$ whose derivative vanishes almost everywhere. The function $\sigma(z)$ itself has the form

$$\sigma(z) = \exp \frac{1}{2\pi} \int_0^{2\pi} \frac{e^{ix} + z}{e^{ix} - z}\, d\psi(x) \,. \tag{14}$$

Conversely, any function $\psi(x)$ with the above mentioned properties generates a function $\sigma(z)$ with the required properties by formula (14).

This allows one to conclude that *every function from the class H_1 has the form (13); the independent parameters corresponding to such a function are:*

1) *the function $p(x)$ which appears in formula* (12);
2) *the zeros $\alpha_1, \ldots, \alpha_n, \ldots,$ from which the function $b(z)$ is constructed*;
3) *the function $\psi(x)$ which appears in formula* (14).

Let us formulate once more the conditions *which the parameters must satisfy:*

1) *the function $p(x)$ and $\log p(x)$ are summable*;
2) *the set of zeros α_n is finite or infinite, in the last case the product $\prod_{k=1}^{\infty} |\alpha_k|$ converges*;
3) *the function $\psi(x)$ decreases and its derivative vanishes almost everywhere.*

The parametric representation of functions from the class H_δ is the same, the only distinction is that the requirement of the summability of the function $p(x)$ have to be replaced by the requirement of the summability of $[p(x)]^\delta$.

6. The above established parametric representation of functions from the class H_δ is closely related to the problem of an analogous representation of functions regular in the unit disk such that the integrals

$$\int_0^{2\pi} \log^+ |f(re^{i\varphi})|\, d\varphi \tag{15}$$

are bounded as $r \to 1$; here, by definition, $\log^+ a = \log a$, if $a \geq 1$ and $\log^+ a = 0$ if $0 < a < 1$. The set of all such functions is said to be *the class* **(A)** . It is known that if a function $f(z)$ from the class (**A**) has infinitely many zeros $\alpha_1, \alpha_2, \ldots, \ldots,$ then the infinite product $\prod_{n=1}^{\infty} |\alpha_n|$ converges[7] and so it is possible to split off the Blaschke function and write down the formula (4).

It is easy to check that the boundedness of the integrals (15) is equivalent to an analogous condition for the integrals

$$\int_0^{2\pi} |\log |f(re^{i\varphi})||\, d\varphi.$$

7) A. Ostrovski, [O.]

From here it follows immediately that if $f(z)$ belongs to the class (**A**), then the function $g(z)$ from formula (4) also satisfy such a property, that is the integrals $\int_0^{2\pi} |\log|g(re^{i\varphi})|\,|\,d\varphi$ are bounded. Furthermore, it is known[8] that the boundedness of the integral $\int_0^{2\pi} |u(re^{i\varphi})|\,d\varphi$ is a necessary and sufficient condition for a harmonic function $u(re^{i\varphi})$ to be representable as a Poisson-Stieltjes integral

$$u(re^{i\varphi}) = \frac{1}{2\pi}\int_0^{2\pi} \frac{1-r^2}{1-2r\cos(\varphi - x)+r^2}\,d\omega(x)$$

in terms of a function $\omega(x)$ of bounded variation. Hence, $g(z)$ has the form

$$g(z) = c\exp\frac{1}{2\pi}\int_0^{2\pi}\frac{e^{ix}+z}{e^{ix}-z}\,d\omega(x)\ , \tag{16}$$

where $\omega(x)$ is a function of bounded variation. Conversely, if $g(z)$ is defined by formula (16), then $g(z)$ and $f(z)$ belong to the class (**A**). Thus *the parametric representation of function from the class (***A***) is given by the formula*

$$f(z) = b(z)\exp\frac{1}{2\pi}\int_0^{2\pi}\frac{e^{ix}+z}{e^{ix}-z}\,d\omega(x)\ , \tag{17}$$

where $b(z)$ is a Blaschke function and $\omega(x)$ is an arbitrary function of bounded variation. Let us decompose $\omega(x)$ into the difference of two decreasing functions

$$\omega(x) = \omega_1(x) - \omega_2(x)\ . \tag{18}$$

Then formula (17) turns into the representation

$$f(z) = f_1(z)/f_2(z)\ , \tag{19}$$

where the functions

$$f_1(z) = b(z)\exp\frac{1}{2\pi}\int_0^{2\pi}\frac{e^{ix}+z}{e^{ix}-z}\,d\omega_1(x)$$

and

$$f_2(z) = \exp\frac{1}{2\pi}\int_0^{2\pi}\frac{e^{ix}+z}{e^{ix}-z}\,d\omega_2(x)$$

do not exceed one in absolute value and $f_2(z)$ does not vanish in the disk $\boldsymbol{K}$. Conversely, if $f(z)$ is representable as the quotient (19) of functions $f_1(z)$ and

8) G.M. Fichtenholz, [F].

$f_2(z)$ with the above mentioned properties, then it is not difficult to see that $f(z)$ belongs to the class (**A**).

Hence, *the representability of a function $f(z)$ in the form (19) with the above mentioned properties is a necessary and sufficient condition for the function f to be in the class* (**A**). This is a well-known theorem of F. & R. Nevanlinna [N].[9]

Finally, let us remark that a function $\omega(x)$ of bounded variation can be represented in the form (18) in various ways. If we choose the total variation of the function ω on the interval $(0,x)$ as the function $\omega_2(x)$, then the function $f_2(x)$ in formula (19) will be maximal in absolute value in the disk $\boldsymbol{K}$.[10]

7. Using the results of the previous section, it is easy to prove the following theorem which was stated by A. Zygmund in [Zyg]. *Let $\psi(x)$ be a positive function such that $\psi(x)/x \to +\infty$ as $x \to +\infty$. If for some function $f(z)$ which is holomorphic in K, its boundary values $f(e^{i\theta})$ satisfy the condition*

$$\int_0^{2\pi} |f(e^{i\varphi})|^\delta \, d\varphi < \infty$$

for some positive δ and if moreover the integrals

$$\int_0^{2\pi} \psi(\log^+ |f(re^{i\varphi})|) \, d\varphi$$

are bounded as $r \to 1$, then the integrals

$$\int_0^{2\pi} |f(re^{i\varphi})|^\delta \, d\varphi < \infty$$

are bounded as $r \to 1$ as well.

Proof. From the assumptions of the theorem it follows immediately that $f(z)$ belongs to the class (**A**). Assume first that this function does not vanish in the disk $\boldsymbol{K}$. Then

$$f(z) = \exp \frac{1}{2\pi} \int_0^{2\pi} \frac{e^{ix} + z}{e^{ix} - z} \, d\omega(x) \ , \tag{20}$$

where $\omega(x)$ is a real valued function of bounded variation. Setting

$$u(re^{i\varphi}) = \log |f(z)| = \frac{1}{2\pi} \int_0^{2\pi} \frac{1 - r^2}{1 - 2r\cos(\varphi - x) + r^2} \, d\omega(x) \ ,$$

---

9) This paper is missing in our libraries, so I unfortunately had no way to read it. The method of the proof used by these authors is unknown to me.

10) However, then $\omega_2(x)$ will be increasing, not decreasing. [Editorial note.]

we obtain, as is well-known,

$$\omega(x) = \lim_{r\to 1} \int_0^x u(re^{i\varphi})\, d\varphi.$$

Let $u^+(re^{i\varphi}) = u(re^{i\varphi})$ if $u(re^{i\varphi}) \geq 0$ and $u^+(re^{i\varphi}) = 0$ otherwise; the function $u^-(re^{i\varphi})$ is defined analogously. Let us consider the function

$$\omega_1(x) = \lim_{r\to 1} \int_0^x u^+(re^{i\varphi})\, d\varphi .$$

According to a well-known theorem of de la Vallée-Poussin,[11] the nondecreasing function $\omega_1(x)$ is absolutely continuous. Let us consider now the nonincreasing function

$$\omega_2(x) = \omega(x) - \omega_1(x) = \lim_{r\to 1} \int_0^x u^-(re^{i\varphi})\, d\varphi$$

and decompose it into the sum of its absolutely continuous and singular components. Thus the function $\omega(x)$ is representable as a sum of three functions. Replacing $\omega(x)$ by this sum in (20), we obtain

$$f(z) = \exp \frac{1}{2\pi} \int_0^{2\pi} q(x) \frac{e^{ix}+z}{e^{ix}-z} dx \; \exp \frac{1}{2\pi} \int_0^{2\pi} \frac{e^{ix}+z}{e^{ix}-z}\, d\omega_3(x) \tag{21}$$

where $\omega_3(x)$ is a nonincreasing function whose derivative vanishes almost everywhere and the function $e^{\delta q(x)}$ is summable (according to the assumptions of the theorem). From formula (21) it follows that $f(z)$ belongs to the class H_δ. □

Assume now that the function $f(z)$ has zeros in $\boldsymbol{K}$. We split off the Blaschke function $b(z)$ so that

$$f(z) = b(z)g(z)$$

Using the same theorem of de la Vallée-Poussin, we obtain

$$\lim_{r\to 1} \int_0^{2\pi} \log^+ |f(re^{i\varphi})|\, d\varphi = \int_0^{2\pi} \log^+ |g(e^{i\varphi})| d\varphi . \tag{22}$$

Furthermore, from the inequality $|b(z)| < 1$ ($|z| < 1$) we obtain that

$$\lim_{r\to 1} \int_0^{2\pi} \log^+ |g(re^{i\varphi})|\, d\varphi \geq \int_0^{2\pi} \log^+ |g(e^{i\varphi})| d\varphi . \tag{23}$$

11) [VP, pp. 447 and p. 451 – *editorial note*]

Let $b_n(z)$ denote the partial product of the first n factors of the product (3) which represents the function $b(z)$ and set

$$g_n(z) = f(z)/b_n(z).$$

It is clear that $|b_n(z)| < 1$ for $|z| < 1$ and that $|b_n(re^{i\varphi})| \to 1$ uniformly with respect to φ as $r \to 1$. Since the integrals

$$\int_0^{2\pi} \log^+ |f(re^{i\varphi}| \, d\varphi$$

increase as r increases, we obtain

$$\lim_{r\to 1} \int_0^{2\pi} \log^+ |g_n(re^{i\varphi})| \, d\varphi = \lim_{r\to 1} \int_0^{2\pi} \log^+ |f(re^{i\varphi})| \, d\varphi.$$

Furthermore,

$$\lim_{r\to 1} \int_0^{2\pi} \log^+ |g_n(re^{i\varphi})| \, d\varphi = \int_0^{2\pi} \log^+ |g(e^{i\varphi})| \, d\varphi$$

according to formula (22). From here it follows that

$$\int_0^{2\pi} \log^+ |g_n(re^{i\varphi})| \, d\varphi \le \int_0^{2\pi} \log^+ |g(e^{i\varphi})| \, d\varphi$$

for $r < 1$. Passing to the limit , we obtain

$$\int_0^{2\pi} \log^+ |g(re^{i\varphi})| \, d\varphi \le \int_0^{2\pi} \log^+ |g(e^{i\varphi})| \, d\varphi.$$

This inequality and inequality (23) show that

$$\lim_{r\to 1} \int_0^{2\pi} \log^+ |g(re^{i\varphi})| \, d\varphi = \int_0^{2\pi} \log^+ |g(e^{i\varphi})| \, d\varphi \, .$$

Now, using the de la Vallée-Poussin theorem again, we see that the function $g(z)$ belongs to the class H_δ, as does the function $f(z) = b(z)g(z)$. □

References

[F] Fichtenholz, G.M., *Sur l'intégrale de Poisson et quelques questions qui s'y rattachent.* Fundamenta Mathematicae **13** (1929), 1–33.

[N] Nevanlinna, F.& R., *Über die Eigenschaften analytischer Funktionen in der Umgebung einer singulären Stelle oder Linie.* Acta Societatis Scientiarum Fennicae. Tom **50**. No. 5 (1922), Helsingfors. 46 p.

[A] Ostrovski, A., *Über die Bedeutung der Jensenschen Formel für einige Fragen der komplexen Funktionen Theorie.* Acta Litterarum ac Scientiarum Regiae Universitatis Hungaricae Francisco-Josephinae. Tomus **1**, (1922–1923), 80–87.
Reprinted in: Ostrovski, A., *Collected Mathematical Papers.* Vol. 5 (*Complex Function Theory*). Birkhäuser Verlag. 1984, pp. 52–59.

[P] Plessner, A.I., *Zur Theorie der konjugierten trigonometrischen Reihen.* Mitteilungen des mathematischen Seminars Giessen. Heft X. Giessen. 1923. 36 pp.

[R1] Riesz, F., *Sur les valeurs du module des fonctions harmoniques et des fonctions analytiques.* Acta Litterarum ac Scientiarum Regiae Universitatis Hungaricae Francisco-Josephinae. Tomus **1**, (1922–1923), 27–32.
Reprinted in: Riesz, F., *Oeuvres complètes.* Vol. 1. Acad. Kiado. Budapest. 1960, pp. 661–666.

[R2] Riesz, F. & M., *Über die Randwerte einer analytischen Funktion.* Comptes Rendus du 4. Congr. des Math. Scand., Stockholm 1916 (1920), pp.27–44.
Reprinted in: Riesz, F., *Oeuvres complètes.* Vol. 1. Acad. Kiado. Budapest. 1960, pp. 537–554.
Reprinted also in: Riesz, M.: *Collected papers.* Vol. 1. Springer-Verlag. 1988, pp. 195–212.

[S] Szegö, G., *Über die Randwerte einer analytischen Funktion.* Mathematische Annalen **84** (1921), 232–244.
Reprinted in: Segö, G., *Collected.papers.* Vol. 1. Birkhäuser. 1982, pp. 400–416.

[V-P] de la Vallée-Poussin, C., *Sur l'intégrale de Lebesgue.* Trans. Amer. Math. Soc. **16** (1915), 435–501.

[Z] Zygmund, A., *Sur les fonctions conjuguées.* Fundamenta Mathematicae **13** (1929), 284–303.

The editors have added a number of collected works to the original list of references.

AMS Subject Classification: 30D50, 30D55.

Titles previously published in the series

OPERATOR THEORY: ADVANCES AND APPLICATIONS

BIRKHÄUSER VERLAG

Edited by
I. Gohberg,
School of Mathematical Sciences, Tel-Aviv University, Ramat Aviv, Israel

This series is devoted to the publication of current research in operator theory, with particular emphasis on applications to classical analysis and the theory of integral equations, as well as to numerical analysis, mathematical physics and mathematical methods in electrical engineering.

45. **J.A. Ball, I. Gohberg, L. Rodman:** Interpolation of Rational Matrix Functions, 1990, (3-7643-2476-7)
46. **P. Exner, H. Neidhardt** (Eds.): Order, Disorder and Chaos in Quantum Systems, 1990, (3-7643-2492-9)
47. **I. Gohberg** (Ed.): Extension and Interpolation of Linear Operators and Matrix Functions, 1990, (3-7643-2530-5)
48. **L. de Branges, I. Gohberg, J. Rovnyak** (Eds.): Topics in Operator Theory. Ernst D. Hellinger Memorial Volume, 1990, (3-7643-2532-1)
49. **I. Gohberg, S. Goldberg, M.A. Kaashoek:** Classes of Linear Operators, Volume I, 1990, (3-7643-2531-3)
50. **H. Bart, I. Gohberg, M.A. Kaashoek** (Eds.): Topics in Matrix and Operator Theory, 1991, (3-7643-2570-4)
51. **W. Greenberg, J. Polewczak** (Eds.): Modern Mathematical Methods in Transport Theory, 1991, (3-7643-2571-2)
52. **S. Prössdorf, B. Silbermann:** Numerical Analysis for Integral and Related Operator Equations, 1991, (3-7643-2620-4)
53. **I. Gohberg, N. Krupnik:** One-Dimensional Linear Singular Integral Equations, Volume I, Introduction, 1992, (3-7643-2584-4)
54. **I. Gohberg, N. Krupnik:** One-Dimensional Linear Singular Integral Equations, Volume II, General Theory and Applications, 1992, (3-7643-2796-0)
55. **R.R. Akhmerov, M.I. Kamenskii, A.S. Potapov, A.E. Rodkina, B.N. Sadovskii:** Measures of Noncompactness and Condensing Operators, 1992, (3-7643-2716-2)
56. **I. Gohberg** (Ed.): Time-Variant Systems and Interpolation, 1992, (3-7643-2738-3)
57. **M. Demuth, B. Gramsch, B.W. Schulze** (Eds.): Operator Calculus and Spectral Theory, 1992, (3-7643-2792-8)

58. **I. Gohberg** (Ed.): Continuous and Discrete Fourier Transforms, Extension Problems and Wiener-Hopf Equations, 1992, (3-7643-2809-6)
59. **T. Ando, I. Gohberg** (Eds.): Operator Theory and Complex Analysis, 1992, (3-7643-2824-X)
60. **P.A. Kuchment:** Floquet Theory for Partial Differential Equations, 1993, (3-7643-2901-7)
61. **A. Gheondea, D. Timotin, F.-H. Vasilescu** (Eds.): Operator Extensions, Interpolation of Functions and Related Topics, 1993, (3-7643-2902-5)
62. **T. Furuta, I. Gohberg, T. Nakazi** (Eds.): Contributions to Operator Theory and its Applications. The Tsuyoshi Ando Anniversary Volume, 1993, (3-7643-2928-9)
63. **I. Gohberg, S. Goldberg, M.A. Kaashoek:** Classes of Linear Operators, Volume 2, 1993, (3-7643-2944-0)
64. **I. Gohberg** (Ed.): New Aspects in Interpolation and Completion Theories, 1993, (3-7643-2948-3)
65. **M.M. Djrbashian:** Harmonic Analysis and Boundary Value Problems in the Complex Domain, 1993, (3-7643-2855-X)
66. **V. Khatskevich, D. Shoiykhet:** Differentiable Operators and Nonlinear Equations, 1993, (3-7643-2929-7)
67. **N.V. Govorov †:** Riemann's Boundary Problem with Infinite Index, 1994, (3-7643-2999-8)
68. **A. Halanay, V. Ionescu:** Time-Varying Discrete Linear Systems Input-Output Operators. Riccati Equations. Disturbance Attenuation, 1994, (3-7643-5012-1)
69. **A. Ashyralyev, P.E. Sobolevskii:** Well-Posedness of Parabolic Difference Equations, 1994, (3-7643-5024-5)
70. **M. Demuth, P. Exner, G. Neidhardt, V. Zagrebnov** (Eds): Mathematical Results in Quantum Mechanics. International Conference in Blossin (Germany), May 17–21, 1993, 1994, (3-7643-5025-3)
71. **E.L. Basor, I. Gohberg** (Eds): Toeplitz Operators and Related Topics. The Harold Widom Anniversary Volume. Workshop on Toeplitz and Wiener-Hopf Operators, Santa Cruz, California, September 20–22, 1992, 1994 (3-7643-5068-7)
72. **I. Gohberg, L.A. Sakhnovich** (Eds): Matrix and Operator Valued Functions. The Vladimir Petrovich Potapov Memorial Volume, (3-7643-5091-1)
73. **A. Feintuch, I. Gohberg** (Eds): Nonselfadjoint Operators and Related Topics. Workshop on Operator Theory and Its Applications, Beersheva, February 24–28, 1994, (3-7643-5097-0)
74. **R. Hagen, S. Roch, B. Silbermann:** Spectral Theory of Approximation Methods for Convolution Equations, 1994, (3-7643-5112-8)
75. **C.B. Huijsmans, M.A. Kaashoek, B. de Pagter**: Operator Theory in Function Spaces and Banach Lattices. The A.C. Zaanen Anniversary Volume, 1994 (ISBN 3-7643-5146-2)
76. **A.M. Krasnosellskii**: Asymptotics of Nonlinearities and Operator Equations, 1995, (ISBN 3-7643-5175-6)

77. **J. Lindenstrauss, V.D. Milman** (Eds): Geometric Aspects of Functional Analysis Israel Seminar GAFA 1992–94, 1995, (ISBN 3-7643-5207-8)
78. **M. Demuth, B.-W. Schulze** (Eds): Partial Differential Operators and Mathematical Physics: International Conference in Holzhau (Germany), July 3–9, 1994, 1995, (ISBN 3-7643-5208-6)
79. **I. Gohberg, M.A. Kaashoek, F. van Schagen**: Partially Specified Matrices and Operators: Classification, Completion, Applications, 1995, (ISBN 3-7643-5259-0)
80. **I. Gohberg, H. Langer** (Eds): Operator Theory and Boundary Eigenvalue Problems. International Workshop in Vienna, July 27–30, 1993, 1995, (ISBN 3-7643-5275-2)
81. **H. Upmeier**: Toeplitz Operators and Index Theory in Several Complex Variables, 1996, (ISBN 3-7643-5282-5)
82. **T. Constantinescu**: Schur Parameters, Factorization and Dilation Problems, 1996, (ISBN 3-7643-5285-X)
83. **A.B. Antonevich**: Linear Functional Equations. Operator Approach, 1995, (ISBN 3-7643-2931-9)
84. **L.A. Sakhnovich**: Integral Equations with Difference Kernels on Finite Intervals, 1996, (ISBN 3-7643-5267-1)

85/ **Y.M. Berezansky, G.F. Us, Z.G. Sheftel**: Functional Analysis, Vol. I + Vol. II, 1996,
86. Vol. I (ISBN 3-7643-5344-9), Vol. II (3-7643-5345-7)

87. **I. Gohberg, P. Lancaster, P.N. Shivakumar** (Eds): Recent Developments in Operator Theory and Its Applications. International Conference in Winnipeg, October 2–6, 1994, 1996, (ISBN 3-7643-5414-5)
88. **J. van Neerven** (Ed.): The Asymptotic Behaviour of Semigroups of Linear Operators, 1996, (ISBN 3-7643-5455-0)
89. **Y. Egorov, V. Kondratiev**: On Spectral Theory of Elliptic Operators, 1996, (ISBN 3-7643-5390-2)
90. **A. Böttcher, I. Gohberg** (Eds): Singular Integral Operators and Related Topics. Joint German-Israeli Workshop, Tel Aviv, March 1–10, 1995, 1996, (ISBN 3-7643-5466-6)
91. **A.L. Skubachevskii**: Elliptic Functional Differential Equations and Applications, 1997, (ISBN 3-7643-5404-6)
92. **A.Ya. Shklyar**: Complete Second Order Linear Differential Equations in Hilbert Spaces, 1997, (ISBN 3-7643-5377-5)
93. **Y. Egorov, B.-W. Schulze**: Pseudo-Differential Operators, Singularities, Applications, 1997, (ISBN 3-7643-5484-4)
94. **M.I. Kadets, V.M. Kadets**: Series in Banach Spaces. Conditional and Unconditional Convergence, 1997, (ISBN 3-7643-5401-1)

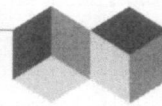

Mathematics with Birkhäuser

OT 72
Operator Theory: Advances and Applications

I. Gohberg / L.A. Sakhnovich, Raymond and Beverly Sackler Faculty of Exact Sciences, School of Mathematical Sciences, Tel Aviv University, Israel (Eds)

Matrix and Operator Valued Functions

The Vladimir Petrovich Potapov Memorial Volume

1994. 240 pages. Hardcover
ISBN 3-7643-5091-1

This book is dedicated to the memory of an outstanding mathematician and personality, Vladimir Petrovich Potapov, who made important contributions and exerted considerable influence in the areas of operator theory, complex analysis and their points of juncture.
The book commences with insightful biographical material, and then presents a collection of papers on different aspects of operator theory and complex analysis covering those recent achievements of the Odessa-Kharkov school in which Potapov was very active.

The papers deal with interrelated problems and methods. The main topics are the multiplicative structure of contractive matrix and operator functions, operators in spaces with indefinite scalar products, inverse problems for systems of differential equations, interpolation and approximation problems for operator and matrix functions.

The book will appeal to a wide group of mathematicians and engineers, and much of the material can be used for advanced courses and seminars.

For orders originating from all over the world except USA and Canada:
Birkhäuser Verlag AG
P.O Box 133
CH-4010 Basel/Switzerland
Fax: +41/61/205 07 92
e-mail: farnik@birkhauser.ch

For orders originating in the USA and Canada:
Birkhäuser
333 Meadowland Parkway
USA-Secaucus, NJ 07094-2491
Fax: +1 201 348 4033
e-mail: orders@birkhauser.com

Birkhäuser

Birkhäuser Verlag AG
Basel · Boston · Berlin

VISIT OUR HOMEPAGE **http://www.birkhauser.ch**